Breaking Tolerance to Anti-Cancer Cell-Mediated Immunotherapy

NK CELLS IN CANCER IMMUNOTHERAPY: SUCCESSES AND CHALLENGES

VOLUME 4

Breaking Tolerance to Anti-Cancer Cell-Mediated Immunotherapy Series

Series Editor: Benjamin Bonavida, PhD

Volume 1: Autophagy in Immune Response: Impact on Cancer Immunotherapy
Edited by Salem Chouaib

Volume 2: Immunotherapy in Resistant Cancer: From the Lab Bench Work to Its Clinical Perspectives
Edited by Jorge Morales-Montor and Mariana Segovia-Mendoza

Volume 3: Immunotherapeutic Strategies for the Treatment of Glioma
Edited by Christopher Jackson and Michael Lim

Volume 4: NK Cells in Cancer Immunotherapy: Successes and Challenges
Edited by Anahid Jewett and Yuman Fong

Upcoming Volumes:

Breaking Tolerance to Unresponsiveness to Immunotherapy by Natural Killer Cells
Edited by Sandro Matosevic

T Cell Metabolism and Cancer Immunotherapy
Edited by Jianxun Song

Gamma Delta T Cells in Cancer Immunotherapy
Edited by Ilan Bank

Principles of Immunotherapy in Breast and Gastrointestinal Cancers: Activity, Mechanisms of Resistance and New Sensitizing Strategies
Edited by Michele Ghidini

Breaking Tolerance to Pancreatic Unresponsiveness to Immunotherapy
Edited by Hideki Kasuya and Itzel Bustos Villabolos

Gamma Delta Cell Cancer Immunotherapy: Evidence-based Perspectives for Clinical Translation
Edited by Marta Barisa

Breaking Tolerance to Anti-Cancer Cell-Mediated Immunotherapy

NK CELLS IN CANCER IMMUNOTHERAPY: SUCCESSES AND CHALLENGES

VOLUME 4

Edited by

ANAHID JEWETT, Ph.D., M.P.H.

Division of Oral Biology and Medicine, The Jane and Jerry Weintraub Center for Reconstructive Biotechnology, University of California Los Angeles (UCLA), Los Angeles, CA, United States
The Jonsson Comprehensive Cancer Center, UCLA School of Dentistry and Medicine, Los Angeles, CA, United States

YUMAN FONG, M.D.

Department of Surgery, City of Hope Comprehensive Cancer Center, Duarte, CA, United States

ELSEVIER

ACADEMIC PRESS

An imprint of Elsevier

Academic Press is an imprint of Elsevier
125 London Wall, London EC2Y 5AS, United Kingdom
525 B Street, Suite 1650, San Diego, CA 92101, United States
50 Hampshire Street, 5th Floor, Cambridge, MA 02139, United States
The Boulevard, Langford Lane, Kidlington, Oxford OX5 1GB, United Kingdom

ISBN: 978-0-12-822620-9

For information on all Academic Press publications
visit our website at https://www.elsevier.com/books-and-journals

Publisher: Stacy Masucci
Acquisitions Editor: Linda Versteeg-Buschman
Editorial Project Manager: Samantha Allard
Production Project Manager: Punithavathy Govindaradjane
Cover Designer: Vicky Pearson

Typeset by STRAIVE, India

Cover Image Insert

Hypothetical role of NK cells in differentiation of tumor cells, increased susceptibility of tumor cells to chemotherapy or radiation and lower risk of metastasis

Three potential treatment outcomes can be envisioned depending on the functional capacity of NK cells in patients. Since targets of NK cells are cancer stem cells/poorly differentiated tumors, when competent NK cells, having increased ability to kill and secrete IFN-g and TNF-a (A), encounter cancer stem cells/poorly differentiated tumors they will kill a large number of these tumors and differentiate the remaining tumor cells by the secretion of IFN-g and TNF-a, resulting in the generation of well-differentiated tumors. Such NK-differentiated tumors are then highly susceptible to chemotherapy or radiation and therefore are eliminated in cancer patients by the use of chemotherapy or radiation, thereby exhibiting lower risk of metastasis. Patients who may maintain high NK cytotoxicity but have low or decreased ability to secrete IFN-g and TNF-a (B) will be able to kill tumors but will have deficiency in differentiating tumors resulting in a moderately differentiated tumors. These tumors may have moderate susceptibility to chemotherapy or radiation and will pose a moderate risk for metastasis. Patients who have very low levels of cytotoxicity and secretion of IFN-g and TNF-a (C) will not be able to kill or differentiate tumors and therefore such tumors will be resistant to chemotherapy or radiation and pose a strong risk for metastasis.

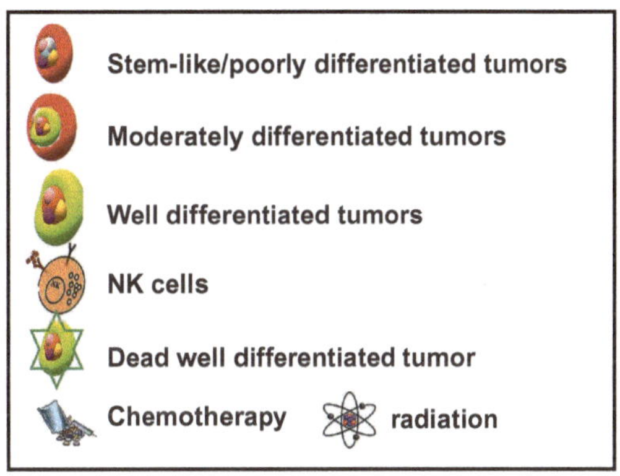

Aims and Scope of Series "Breaking Tolerance to Anti-Cancer Cell-Mediated Immunotherapy"

The role of the immune system in the eradication of cancers has been investigated for several decades with controversial findings. The controversy was the result of a poor understanding of the underlying mechanisms that govern responsiveness and unresponsiveness. Hence, significant advances have been made with respect to the regulation of the host immune response against cancer and several immunotherapeutics have been recently introduced and used clinically. These include both antibody- and cell-mediated immunity targeting the cancer cells. Such immunotherapies led to significant clinical responses in various cancer types that were unresponsive to conventional therapies. However, only a subset of cancer patients responds to such immunotherapeutics and also there is a responding subset that develops resistance to further treatment. Various studies have examined potential underlying mechanisms involved in resistance and identified a variety of gene products that play pivotal roles in maintaining the resistant phenotype of the cancer cells to cell-mediated immunotherapy.

The main objective of the proposed series "Breaking Tolerance to Anti-Cancer Cell-Mediated Immunotherapy" is the development of individual volumes that are focused on the application of particular sensitizing agents that, when used in combination with cell-mediated immunotherapy, result in the reversal of resistance.

A variety of different classes of immunosensitizing agents has been reported. Each individual volume will focus on one class of immunosensitizing agents and their effects on the reversal of cell-mediated immune resistance in different cancers. Emphasis will be on biochemical, molecular, and genetic mechanisms by which the sensitizing agents mediate their effects individually and/or in combination with immunotherapy. Each editor will compile nonoverlapping review chapters on the therapeutic role of specific sensitizing agents used in combination with conventional immunotherapy and the reversal of resistance. There will also be an emphasis on discrimination of responses obtained in various cancer types.

The scope of the series is to provide updated information to scientists and clinicians that is valuable in their quest to gather information, carry out new investigations, and develop novel immunosensitizing agents that are both more potent and might also be active in contrast to existing ones that were not active.

Benjamin Bonavida, PhD (Series Editor)

About the Series Editor

Dr. Benjamin Bonavida, PhD (Series Editor), is currently distinguished research professor at the University of California, Los Angeles (UCLA). He is affiliated with the Department of Microbiology, Immunology and Molecular Genetics, UCLA David Geffen School of Medicine. His research career, thus far, has focused on investigations in the fields of basic immunochemistry and cancer immunobiology. His research investigations have ranged from the biochemical, molecular, and genetic mechanisms of cell-mediated killing and tumor cell resistance to chemo-immuno cytotoxic drugs. The reversal of tumor cell resistance was investigated by the use of various selected sensitizing agents based on molecular mechanisms of resistance. In these investigations, there was the newly characterized dysregulated NF-κB/Snail/YY1/RKIP/PTEN loop in many cancers that was reported to regulate cell survival, proliferation, invasion, metastasis, and resistance. Emphasis was focused on the roles of the tumor suppressor Raf kinase inhibitor protein (RKIP), the tumor promoter Yin Yang 1 (YY1) and the role of nitric oxide as a chemo-immuno-sensitizing factor. Many of the aforementioned studies are centered on the clinical challenging features of cancer patients' failure to respond to both conventional and targeted therapies.

The editor has been active in the organization of regular sequential international miniconferences that are highly focused on the roles of YY1, RKIP, and nitric oxide in cancer and their potential therapeutic applications. Several books edited or coedited by the editor have been published. In addition, the editor has been the series editor of books (over 23) published by Springer on *Resistance to Anti-Cancer Targeted Therapeutics*. In addition, the editor is presently the series editor of three series published by Elsevier/Academic Press on *Cancer Sensitizing Agents for Chemotherapy*, *Breaking Tolerance to Anti-Cancer Cell-Mediated Immunotherapy*, and *Breaking Tolerance to Antibody-Mediated Immunotherapy*. Lastly, the editor is the editor-in-chief of the journal *Critical Reviews in Oncogenesis*. The editor has published over 500 research publications and reviews in various scientific journals of high impact.

Acknowledgments: The editor wishes to acknowledge the excellent editorial assistance of Ms. Inesa Navasardyan, who has worked diligently for the completion of this volume, by editing and formatting the various contributions to this volume. Ms. Navasardyan has recently graduated from UCLA (June 2020) and has also contributed a review chapter in this volume on triple-negative breast cancer.

The editor acknowledges the Department of Microbiology, Immunology and Molecular Genetics and the UCLA David Geffen School of Medicine for their continuous support. The editor also acknowledges the assistance of Mr. Rafael Teixeira, Acquisitions Editor for Elsevier/Academic Press, and the excellent assistance of Ms. Samantha Allard, Editorial Project Manager for Elsevier/Academic Press, for their continuous cooperation throughout the development of this book.

Aims and Scope of Volume

This year, nearly 20 million new cases of cancer will occur and 10 million people will die worldwide. Although the treatment strategies have steadily been advancing in cancer, we still lack comprehensive and durable treatment modalities to successfully prevent, treat, or cure cancer. In the past 5 years, we have witnessed approvals of four cellular therapies for clinical use, such as chimeric antigen receptor T cell (CAR-T), natural killer cell (NK), tumor-infiltrating lymphocyte (TIL), and NK-CAR therapies. We are also expecting additional cellular therapies in the coming years, and cellular immunotherapy is no longer a laboratory science but an important clinical discipline.

Despite the advancements in cancer therapy, we still have challenges to overcome since many of the current immunotherapeutic strategies are only effective in certain cancers, do not provide long-lasting remissions, and are effective in only a fraction of the cancer patients. Not only are the costs of such treatments largely prohibitive, they are also only available to certain subsets of cancer patients, and in some individuals they inflict significant collateral damage that compromises the patients' quality of life. What we have learned from many years of investigations focusing on the in vivo immune system in cancer is a that successful treatment may require the supplementation of different immune effectors such as NK cells, CD8+ T cells, TILs, NKT, CAR-T, or CAR-NK cells to work in synergy to correct or overcome the defective nature of patients' immune cells.

The aim of this book is to summarize and highlight the state of the field of cellular immunotherapy in cancer. The different chapters in the book have been authored by authoritative members in this field who are engaged in preclinical investigations, clinical trials, and key scientific studies. This book is divided into three sections. Section A summarizes the basics of cellular immunotherapies, delineating the basic immunobiology of NK cells and T cells in targeting cancer. In Section B of the book, the authors present processes for trial optimization including the challenges and promises of efficient production, in vivo tracking of cellular therapies, and novel diagnostics and biologic readouts. Section C is dedicated to reviewing trials and combinational strategies in cellular immunotherapy. Many of the chapters are focused on introducing cellular therapies with standard therapeutic modalities in order to chart a way forward for more effective, less toxic, and less costly cell therapies.

This book is intended to present the successes and the challenges of cellular immunotherapies for current and future translational researchers at all levels. It is meant to be a guide for graduate students and postdoctoral fellows entering the field. It is intended to summarize the current state-of-the art for the investigators and trialists active in the development and research of cellular therapies for cancer. Most of all, we intend to present this book as one of the first manuals for the emerging field of clinical cellular therapies for cancer. We hope that our audience of scientists, oncologists, radiologists, and colleagues in industry will benefit greatly from this book in not only understanding the state of technology but also effectively using the information provided in this book to advance their preclinical and clinical outcomes.

Dr. Anahid Jewett
Dr. Yuman Fong

About the Volume Editors

Dr. Anahid Jewett is Professor and Director of the Tumor Immunology Laboratory in the Division of Oral Biology and Medicine and Weintraub Center for Reconstructive Biotechnology at the UCLA School of Dentistry and Medicine. She has membership in the Jonsson Comprehensive Cancer Center (JCCC) and is a member of the UCLA Tumor Immunology subgroup. Her extensive educational and training background in the fields of pathology, microbiology, and immunology as well as infectious disease epidemiology positioned her well to resolve the future scientific challenges of infectious diseases as well as diseases based on inflammation, autoimmunity, and cancer in which she has a significant track record of productive publications. She is well known nationally and internationally for her contributions to the fields of NK cell biology, tumor immunology, and cancer immunotherapy. She has received a large number of honors and awards and holds memberships in many professional organizations and societies. She has chaired important committees at UCLA and elsewhere and has been instrumental in shaping the graduate studies for the health professionals at UCLA. She chairs and lectures in several graduate-level courses, and her laboratory is sought out by many foreign and domestic scholars who spend several years to receive training in her laboratory. She serves on the editorial boards of many prestigious journals and has been a reviewer on the board of the National Institute of Health and many other national and international cancer institutes, foundations, and charities. She holds many patents, has given more than 300 invited lectures and presentations nationally and internationally, and has published more than 150 high-impact journal articles, reviews, commentaries, and book chapters in the field of cancer. She has research collaborations with investigators from China, Israel, Slovenia, Mexico, Poland, Germany, Thailand, Japan, Portugal, South Korea, Austria, and Sweden to name a few. She has organized a number of conferences on cancer immunity nationally and internationally. She has trained more than 200 graduate students and health professionals in her laboratory, many of whom are leaders in their respective institutions. She has served on review panels for grants from many countries including England, France, the Netherlands, Qatar, Poland, and Israel to name a few. She has received grants from NIH as well as other foundations and institutes. She serves on the scientific boards of many companies and is a consultant to a number of investment and venture capital firms. She has served as editor-in-chief, associate editor, or guest editor for a number of journals, and has edited several journal sections for a number of distinct publishing companies. Her public outreach and education includes many interviews with celebrities, organizations, and societies in the field of cancer and infectious diseases.

One of Dr. Jewett's major contributions to science and NK cell biology was the identification, characterization, and the establishment of the concept of split anergy in NK cells. Equally important was Dr. Jewett's discovery demonstrating that NK cells were important for the elimination, selection, and differentiation of cancer stem cells as well as healthy stem cells. Most recently, she has identified, characterized, and patented a novel technology to expand large numbers of super-charged NK cells, which has been licensed and is on track to be used in clinical trials of cancer patients. In addition, she has developed disease-specific formulations of probiotic bacteria to restore the damaged microbiome and prevent and treat inflammation, autoimmunity, and cancer. Her novel approach to cell therapy uses a number of key therapeutic strategies used in combination to eliminate all different clones of cancer in an attempt to cure rather than short-term treatment of cancer.

Dr. Yuman Fong is the Sangiacomo Chair and Chairman of the Department of Surgery at the City of Hope Medical Center.

Dr. Fong's clinical expertise is in oncology and in the field of liver and pancreatic surgery. He helped usher in robotic techniques for HPB surgery. He is editor of the *SAGES Atlas of Robotic Surgery*. For his clinical work, he was awarded the Layton F. Rikkers Master Clinician Award from the SSAT. He has assisted in the design and deployment of many novel surgical tools. For his contributions to surgical research, he has been awarded the Stanley Dudrick Award from the American Society of Parenteral and Enteral Nutrition, The Shipley Award from the Southern Surgical Association, and the Flance-Karl Award from the American Surgical Association. His work in medical engineering has led to his election to the American Institute of Medical and Biologic Engineering.

His research has long been in the field of immunology. He was part of the team that discovered the biologic effects of tumor necrosis factor that helped establish the field of innate immunity. His current laboratory focus is on the design of gene and immune cell therapies for cancer. His leadership at the national level has included serving as the Chair of the Recombinant DNA Advisory Committee (RAC) of the National Institutes of Health. He is Founding Editor-in-Chief of *Molecular Therapy Oncolytics* (Cell Press, official journal of the ASGCT). He has coauthored over 1000 papers (h-index = 143. Citations 85,000) and 22 textbooks. He has also been elected to the American Society of Clinical Investigation and to the National Academy of Medicine.

In his spare time, he enjoys music, sandboarding, and ballroom dancing.

Preface—Cellular immunotherapies: Evolution from laboratory studies to effective human therapies

Cellular immunology is central to all facets of human health and pathology including cancer, infection, autoimmune disease, and recovery after trauma. Our understanding of the basic mechanisms of cellular immunity has been growing exponentially over the last half century. In oncology, until recently, studies of cellular immunity and immunotherapy have been restricted to preclinical studies and clinical trials. The last 5 years have witnessed FDA approvals of four cellular therapies for clinical use: tisagenlecleucel (Kymriah; Novartis), axicabtagene ciloleucel (Yescarta; Kite Pharma, Gilead Sciences Inc), lisocabtagene maraleucel (Breyanzi; Juno Therapeutics), and brexucabtagene autoleucel. These are all chimeric antigen receptor T cell (CAR-T) therapies and have been approved to treat a number of hematologic malignancies, including diffuse large B-cell lymphoma, follicular lymphoma, mantle cell lymphoma, multiple myeloma, and B-cell acute lymphoblastic leukemia (ALL). These new human medicines validate the field of cellular immune-therapeutics as a clinical field and hold forth the promise that many other cellular therapies [NK, NK-CAR, and tumor-infiltrating lymphocytes (TILs)] may soon become human therapies for both hematologic and solid tumors.

Cancer is in part a disease of the immune system. An effective and balanced functioning immune system should be able to recognize and remove the cells that have phenotypic changes due to gene deletions or mutations. However, when the function of immune cells, in particular NK cells and CD8+ T cells, becomes compromised, an opportunity arises for the unwanted mutated cells to survive and give rise to cancer. Although advances have been made in the field of cancer immunotherapy, we still have significant challenges to overcome since few of the current immunotherapeutic strategies provide long-lasting remissions, and all are effective in only specific cancers and a fraction of cancer patients. The high levels of adverse effects and high cost of these treatments at present are shown to result in collateral damage and significant financial constraints, respectively. The main objective of this book is to summarize the latest advancements and novel strategies for NK and T cell immunotherapies and to chart a way forward for more effective, less toxic, and less costly cell therapies.

The heterogeneous nature of cancer requires targeting by several different strategies since one strategy may be inadequate to remove all the existing clones of cancer. In this book, we introduce such combinatorial strategies. Since the discovery of NK cells, NK cell immunobiologists have been studying these cells tirelessly. The identification of cancer stem-like cells (CSCs) or poorly differentiated tumors as targets of NK cells has finally shown the indispensable role these effectors play in successful cancer therapy. Since NK cells are responsible for elimination and differentiation of the aggressive CSCs/poorly differentiated tumors, it is logical that when the function of these cells is compromised, the patient could greatly be predisposed to tumor growth, expansion, invasion, and metastasis. In addition, by the aid of IFN-γ-mediated differentiation of tumor cells and increased expression of MHC class I, NK cells facilitate the targeting of differentiated tumors by CD8+ T cells. Moreover, NK cells are important in mediating antibody-mediated cellular cytotoxicity and lysis of oncolytic virus-infected tumor cells among other important functions.

This book consists of contributions from established leaders in the fields of NK and T cell biology and immunotherapy. The reviews address the latest advances in cancer immunotherapy using NK cells and T cells and highlight the challenges and difficulties in breaking the treatment-resistance of tumor cells surrounded by a suppressive tumor stroma that contain inhibitory immune factors. This book also reviews the current clinical applications of classical NK and T cell immunotherapies as well as the adoption of universal therapeutic NK and T cell preparations.

The book is separated into three sections. The first section summarizes the basics of cellular immunotherapies, delineating the basic immunobiology of NK cells, and differing roles of NK and T cells in targeting cancer. Also discussed is the intimate, synergistic interaction of NK cells and T cells and the role of NK cells in the expansion and functional activation of CD8+ T cells. Both CAR-NK and CAR-T therapeutics are now in many clinical trials, and their biology and significance in cancer treatments are also discussed in this section. Finally, the use of TILs in cancer treatment has been advocated by many investigators and is discussed in this section.

In Section I, Drs. Senjor, Jewett, and their colleagues discuss the basic biology of NK cells and their function in cancer as well as the impact of the tumor microenvironment (TME) on NK cell function. In addition, they introduce their novel engineering methodology for generating super-charged NK cells for use in tumor immunotherapy. Drs. Navin and Parihar describe different suppressive immune effectors within the TME and methods to overcome the suppressive nature of the TME. Dr. Matosevic discusses the use of NK cells in solid tumor treatments listing the advantages and the disadvantages of allogeneic NK cells in cancer therapy. Dr. Wu reviews the significance of NKG2D receptor signaling in NK activation and provides evidence for the use of this axis in successful immunotherapeutic strategies. In the chapter titled "Chimeric antigen receptor-modified cells for the treatment of solid tumors: First steps in a thousand-mile March," Drs. Rafei, Basar, Rezvani, and Daher describe the successes and challenges of CAR-T therapy for hematological and solid tumors. Drs. Mahuron and Fong introduce TILs and discuss their significance in cancer therapy and the challenges encountered in the use of these cells in cancer therapy. Drs. Murad, Park, and Priceman discuss the status of current FDA-approved CAR-T cell products and their clinical impact, highlight unique hurdles for CAR-T cells for solid tumors, and provide perspectives on the future of CAR-engineered cell therapies.

In Section II, the authors present processes for trials optimization including novel diagnostics, assay readouts, influence of route of administration, and production optimization for cell therapy. Drs. Wang and Rivière delineate "Optimization of production of cell therapies," including preclinical proof-of-concept studies, technology transfer, and process development. Drs. Volpe, Blasberg, Serganova, and Ponomarev discuss

"Imaging the immune cell in immunotherapy," providing a comprehensive summary of state-of-the-art imaging modalities for tracking of cancer immunotherapy. Drs. Feldman and Badie discuss the effective cell delivery systems for CNS cancers and list the challenges that cell therapists face when designing strategies to effectively target aggressive CNS tumors. Central to their discussion is the concept of regional delivery of cellular therapies that enhance their efficacy while offering the possibility of cellular therapy without the need for lymphodepletion. Finally, in the chapter titled "Diagnostic methods to assess the numbers, phenotype and function of primary and engineered NK cells: Methods to predict prognosis and treatment outcome," Drs. Ko, Kaur, Chen, Breznik, Senjor, Chovatiya, Wong, Turnsek, Kos, and Jewett discuss novel diagnostic tests to assess the survival, expansion, and function of both primary NK cells and those that have been super-charged.

Section III is dedicated to reviewing trials and combinational strategies in cellular immunotherapy. Historically, most papers, books, and grant funding have been dedicated to only specific cell subsets. This has resulted in a wealth of knowledge for single cell type. However, it has fallen short in providing effective cancer therapeutics. We hope, going forward, academic focus and funding will be leveraged to understand how key immune effectors work in synergy. The key is to understand fully the multiple interactions between different immune cell subsets for the successful eradication of the tumors. The authors in this section present a number of important concepts and methodologies regarding patient trials and combinational strategies in cellular immunotherapy and indicate that successful cell therapy may depend on the selection of other treatment strategies synergistic with immune cell therapy. Drs. Sadeghi, Chen, Jewett, and Kaur discuss the differential targeting of undifferentiated vs well-differentiated tumors by the NK cells and combinatorial use of NK cells therapy with chemotherapy and radiation for maximum treatment efficacy. Drs. Chaurasiya, Woo, Choong, Deshpande, and Fong discuss the synergistic effect of oncolytic viral therapy with different cell therapies, listing the successes and challenges encountered in such therapies. Drs. Singh and Gumrukcu discuss different stages of cancer with its relevance in tailored therapeutic strategies and successful patient therapies using NK and T cells. Drs. Breznik, Novak, Majc, Habič, and Jewett discuss the existing trials and list the successes and challenges of NK cell therapy in glioblastoma and offer combinatorial treatments for the effective eradication of these aggressive tumors. Drs. Zhu, Bellis, Saini, Fong, and Adusumilli review the field of CAR-T therapy and provide future perspectives on the use of these cells in cancer therapy. Drs. Zeng, Li, Lee, and Yang discuss their novel NK T cell therapeutic strategies and provide the list of successes and challenges of using such cells in cancer immunotherapy. Finally, Drs. Maharaj, Polineni, Abreu, and Gouvea discuss "Molecular remission using low-dose immunotherapy with minimal toxicities for poor prognosis hematological and solid tumor cancers."

The goals of this book are to review (1) the theory and practice of cancer cellular therapies; (2) the challenges and promises of efficient production, in vivo tracking of cellular therapies, and related biologic readouts; and (3) to summarize current trials data and the state of knowledge of combination therapies. This book is intended to summarize the field for current and future translational researchers at all levels. It is meant to be a guide for graduate students and postdoctoral fellows entering the field. It is meant to summarize the current state-of-the art for the investigators and trialists active in development and research in cellular therapies for cancer. Most of all, we are hoping this is one of the first clinical manuals for the emerging field of clinical cellular therapies for cancer. We hope that our audience of scientists, oncologists, radiologists, and colleagues in industry find this useful.

A work like this is only possible because of the contributions of many individuals. The authorship of this work includes scientific leaders in cell biology, cell engineers, trialists, cell production experts, and experts in molecular imaging. We thank them for their contributions and efforts to collaborate in the creation of this comprehensive and special work.

We also thank our teachers, graduate students, postdoctoral fellows, and colleagues who have shared their knowledge and experience with us. We thank our patients who inspire us to be superior clinicians and constantly strive to improve the field. We thank our Assistant Editor at Elsevier, Samantha Allard, for her efforts in keeping us on track and on time. Finally, we thank our families, especially our spouses Keith and Nicole, for the patience and support they have given us daily for our scientific work and then to complete this scientific book.

Anahid Jewett
Yuman Fong

Contents

9. Lymphodepletion and cellular immunotherapy

Yuan Yuan, Supriya Deshpande, Jamie Rand, and Yuman Fong

10. Imaging the immune cell in immunotherapy

Alessia Volpe, Ronald Blasberg, Inna Serganova, and Vladimir Ponomarev

11. Radiologic assessment of tumor response to immunotherapy and its complications

Leonid Roshkovan and Sharyn I. Katz

12. Novel cell delivery systems: Intracranial and intrathecal

Lisa Feldman and Behnam Badie

13. Diagnostic methods to assess the numbers, phenotype, and function of primary and engineered NK cells: Methods to predict prognosis and treatment outcome

Meng-Wei Ko, Kawaljit Kaur, Po-Chun Chen, Barbara Breznik, Emanuela Senjor, Nishant Chovatiya, Paul Wong, Tamara Lah Turnsek, Janko Kos, and Anahid Jewett

III

Patient trials and combinational strategies in cellular immunotherapy; Successful cell therapy may depend on the selection of treatment strategies synergistic with cell therapy

14. Combination of NK cell immunotherapy with chemotherapy and radiation enhances NK cell therapy and provides improved prognosis in cancer patients and in humanized BLT mouse model system

Saeed Sadeghi, Po-Chun Chen, Anahid Jewett, and Kawaljit Kaur

15. Combining oncolytic viruses with immune cell therapy as treatments for cancer: OV, CAR T-cell, and NK combinations

Shyambabu Chaurasiya, Yanghee Woo, Kevin Choong, Supriya Deshpande, and Yuman Fong

16. Natural killer cells in the treatment of glioblastoma: Diverse antitumor functions and potential clinical applications

Barbara Breznik, Metka Novak, Bernarda Majc, Anamarija Habič, and Anahid Jewett

17. Immunotherapy using CAR T: What we have learned from trials and where we are heading

Amy Zhu, Rebecca Bellis, Jasmeen Saini, Christina Fong, and Prasad S. Adusumilli

18. NKT cell: Success and promises in transplantation and immunotherapy

Samuel Zeng, Zhe Li, Derek Lee, and Lili Yang

19. Tumor-infiltrating lymphocytes: Prognostic considerations and current trials as adoptive cell therapy

Timothy D. Folsom, Branden S. Moriarity, Timothy K. Starr, Emil Lou, and Beau R. Webber

20. Molecular remission using personalized low-dose immunotherapy with minimal toxicities for poor prognosis hematological and solid tumor cancers

Dipnarine Maharaj, Vineet Polineni, Maria M. Abreu, and Jacqùeline Gouvea

Contributors

Maria M. Abreu Institute for NeuroImmune Medicine, Nova Southeastern University, Ft. Lauderdale, FL, United States

Prasad S. Adusumilli Thoracic Service, Department of Surgery; Center for Cell Engineering, Memorial Sloan Kettering Cancer Center, New York, NY, United States

Behnam Badie Division of Neurosurgery, City of Hope National Medical Center, Duarte, CA, United States

Rafet Basar Department of Stem Cell Transplantation and Cellular Therapy, The University of Texas MD Anderson Cancer Center, Houston, TX, United States

Rebecca Bellis Thoracic Service, Department of Surgery, Memorial Sloan Kettering Cancer Center, New York, NY, United States

Ronald Blasberg Department of Radiology; Department of Neurology; Molecular Pharmacology and Chemistry Program; Memorial Sloan Kettering Cancer Center, New York, NY, United States

Barbara Breznik Division of Oral Biology and Medicine, The Jane and Jerry Weintraub Center for Reconstructive Biotechnology, University of California Los Angeles (UCLA), Los Angeles, CA, United States; Department of Genetic Toxicology and Cancer Biology, National Institute of Biology, Ljubljana, Slovenia

Shyambabu Chaurasiya Department of Surgery, City of Hope Comprehensive Cancer Center, Duarte, CA, United States

Po-Chun Chen Division of Oral Biology and Medicine, The Jane and Jerry Weintraub Center for Reconstructive Biotechnology, University of California Los Angeles (UCLA), Los Angeles, CA, United States

Kevin Choong Department of Surgery, City of Hope Comprehensive Cancer Center, Duarte, CA, United States

Nishant Chovatiya Division of Oral Biology and Medicine, The Jane and Jerry Weintraub Center for Reconstructive Biotechnology, University of California Los Angeles (UCLA), Los Angeles, CA, United States

May Daher Department of Stem Cell Transplantation and Cellular Therapy, The University of Texas MD Anderson Cancer Center, Houston, TX, United States

Supriya Deshpande Department of Surgery, City of Hope Comprehensive Cancer Center, Duarte, CA, United States

Lisa Feldman Division of Neurosurgery, City of Hope National Medical Center, Duarte, CA, United States

Timothy D. Folsom Department of Pediatrics; Masonic Cancer Center; Center for Genome Engineering; Stem Cell Institute, University of Minnesota, Minneapolis, MN, United States

Christina Fong Department of Pediatrics, Memorial Sloan Kettering Cancer Center, New York, NY, United States

Yuman Fong Department of Surgery, City of Hope Comprehensive Cancer Center, Duarte, CA, United States

Jacqueline Gouvea The Maharaj Institute of Immune Regenerative Medicine, Boynton Beach, FL, United States

Anamarija Habič Department of Genetic Toxicology and Cancer Biology, National Institute of Biology; Jozef Stefan International Postgraduate School, Ljubljana, Slovenia

Anahid Jewett Division of Oral Biology and Medicine, The Jane and Jerry Weintraub Center for Reconstructive Biotechnology, University of California Los Angeles (UCLA); The Jonsson Comprehensive Cancer Center, UCLA School of Dentistry and Medicine, Los Angeles, CA, United States

Sharyn I. Katz Division of Thoracic Radiology, Department of Radiology, University of Pennsylvania Perelman School of Medicine, Philadelphia, PA, United States

Kawaljit Kaur Division of Oral Biology and Medicine, The Jane and Jerry Weintraub Center for Reconstructive Biotechnology, University of California Los Angeles (UCLA), Los Angeles, CA, United States

Meng-Wei Ko Division of Oral Biology and Medicine, The Jane and Jerry Weintraub Center for Reconstructive Biotechnology, University of California Los Angeles (UCLA), Los Angeles, CA, United States

Janko Kos Department of Biotechnology, Jožef Stefan Institute; Faculty of Pharmacy, University of Ljubljana, Ljubljana, Slovenia

Derek Lee Department of Microbiology, Immunology, and Molecular Genetics, University of California Los Angeles (UCLA), Los Angeles, CA, United States

Zhe Li Department of Microbiology, Immunology, and Molecular Genetics, University of California Los Angeles (UCLA), Los Angeles, CA, United States

Emil Lou Masonic Cancer Center, University of Minnesota; Department of Medicine, Division of Hematology, Oncology, and Transplantation, Minneapolis, MN, United States

Dipnarine Maharaj The Maharaj Institute of Immune Regenerative Medicine, Boynton Beach, FL, United States

Kelly Mahuron Department of Surgery, City of Hope Comprehensive Cancer Center, Duarte, CA, United States

Bernarda Majc Department of Genetic Toxicology and Cancer Biology, National Institute of Biology; Jozef Stefan International Postgraduate School, Ljubljana, Slovenia

Sandro Matosevic Department of Industrial and Physical Pharmacy, Purdue University, West Lafayette, IN, United States

Branden S. Moriarity Department of Pediatrics; Masonic Cancer Center; Center for Genome Engineering; Stem Cell Institute, University of Minnesota, Minneapolis, MN, United States

John P. Murad Department of Hematology and Hematopoietic Cell Transplantation, City of Hope, Duarte, CA, United States

Ishwar Navin Department of Immunology and Microbiology, Baylor College of Medicine; Center for Cell and Gene Therapy, Texas Children's Hospital, Houston Methodist Hospital, and Baylor College of Medicine, Houston, TX, United States

Metka Novak Department of Genetic Toxicology and Cancer Biology, National Institute of Biology, Ljubljana, Slovenia

Robin Parihar Department of Immunology and Microbiology, Baylor College of Medicine; Center for Cell and Gene Therapy, Texas Children's Hospital, Houston Methodist Hospital, and Baylor College of Medicine; Section of Hematology/Oncology, Department of Pediatrics, Texas Children's Hospital, Houston, TX, United States

Anthony K. Park Department of Hematology and Hematopoietic Cell Transplantation, City of Hope, Duarte, CA, United States

Vineet Polineni The Maharaj Institute of Immune Regenerative Medicine, Boynton Beach, FL, United States

Vladimir Ponomarev Department of Radiology; Molecular Pharmacology and Chemistry Program, Memorial Sloan Kettering Cancer Center, New York, NY, United States

Saul J. Priceman Department of Hematology and Hematopoietic Cell Transplantation, City of Hope, Duarte, CA, United States

Hind Rafei Department of Stem Cell Transplantation and Cellular Therapy, The University of Texas MD Anderson Cancer Center, Houston, TX, United States

Jamie Rand Department of Surgery, City of Hope Comprehensive Cancer Center, Duarte, CA, United States

Katayoun Rezvani Department of Stem Cell Transplantation and Cellular Therapy, The University of Texas MD Anderson Cancer Center, Houston, TX, United States

Isabelle Rivière Cell Therapy and Cell Engineering Facility; Center for Cell Engineering; Molecular Pharmacology Program, Memorial Sloan-Kettering Cancer Center, New York, NY, United States

Leonid Roshkovan Division of Abdominal Imaging, Department of Radiology, University of Pennsylvania Perelman School of Medicine, Philadelphia, PA, United States

Saeed Sadeghi Division of Hematology-Oncology, Department of Medicine, UCLA David Geffen School of Medicine, Los Angeles, CA, United States

Jasmeen Saini Thoracic Service, Department of Surgery, Memorial Sloan Kettering Cancer Center, New York, NY, United States

Emanuela Senjor Division of Oral Biology and Medicine, The Jane and Jerry Weintraub Center for Reconstructive Biotechnology, University of California Los Angeles (UCLA), Los Angeles, CA, United States; Department of Biotechnology, Jožef Stefan Institute; Faculty of Pharmacy, University of Ljubljana, Ljubljana, Slovenia

Inna Serganova Department of Neurology, Memorial Sloan Kettering Cancer Center; Department of Medicine, Division of Hematology and Medical Oncology, Weill Cornell Medicine; Human Oncology and Pathogenesis Program, Memorial Sloan Kettering Cancer Center, New York, NY, United States

Timothy K. Starr Masonic Cancer Center, University of Minnesota; Department of Obstetrics and Gynecology and Women's Health, Minneapolis, MN, United States

Tamara Lah Turnsek Department of Genetic Toxicology and Cancer Biology, National Institute of Biology, Ljubljana, Slovenia

Alessia Volpe Department of Radiology, Memorial Sloan Kettering Cancer Center, New York, NY, United States

Xiuyan Wang Cell Therapy and Cell Engineering Facility; Center for Cell Engineering; Molecular Pharmacology Program, Memorial Sloan-Kettering Cancer Center, New York, NY, United States

Beau R. Webber Department of Pediatrics; Masonic Cancer Center; Center for Genome Engineering; Stem Cell Institute, University of Minnesota, Minneapolis, MN, United States

Paul Wong Division of Oral Biology and Medicine, The Jane and Jerry Weintraub Center for Reconstructive Biotechnology, University of California Los Angeles (UCLA), Los Angeles, CA, United States

Yanghee Woo Department of Surgery, City of Hope Comprehensive Cancer Center, Duarte, CA, United States

Jennifer Wu Feinberg School of Medicine, Northwestern University, Chicago, IL, United States

Lili Yang Department of Microbiology, Immunology, and Molecular Genetics; Eli and Edythe Broad Center of Regenerative Medicine and Stem Cell Research; Jonsson Comprehensive Cancer Center, David Geffen School of Medicine; Molecular Biology Institute, University of California Los Angeles (UCLA), Los Angeles, CA, United States

Yuan Yuan Department of Medical Oncology, Cedars-Sinai Medical Center, Los Angeles, CA, United States

Samuel Zeng Department of Microbiology, Immunology, and Molecular Genetics, University of California Los Angeles (UCLA), Los Angeles, CA, United States

Amy Zhu Thoracic Service, Department of Surgery, Memorial Sloan Kettering Cancer Center, New York, NY, United States

SECTION I

Basics of cellular immunotherapy: Differing roles of NK and T cells in targeting cancer and their intimate synergistic interaction underscores the requirement for the use of both cell types in successful cancer treatment

CHAPTER

1

Multifaceted nature of natural killer cells: Potential mode of interaction and shaping of stem cells

Emanuela Senjor[a,b,c], *Meng-Wei Ko*[a], *Kawaljit Kaur*[a], *Po-Chun Chen*[a], *Barbara Breznik*[a,d], *Nishant Chovatiya*[a], *Janko Kos*[b,c], *and Anahid Jewett*[a,e]

[a]Division of Oral Biology and Medicine, The Jane and Jerry Weintraub Center for Reconstructive Biotechnology, University of California Los Angeles (UCLA), Los Angeles, CA, United States [b]Department of Biotechnology, Jožef Stefan Institute, Ljubljana, Slovenia [c]Faculty of Pharmacy, University of Ljubljana, Ljubljana, Slovenia [d]Department of Genetic Toxicology and Cancer Biology, National Institute of Biology, Ljubljana, Slovenia [e]The Jonsson Comprehensive Cancer Center, UCLA School of Dentistry and Medicine, Los Angeles, CA, United States

Abstract

Natural killer (NK) cells are the key immune effectors with the ability to lyse stem-like/poorly differentiated tumor cells and mediate differentiation through their cytotoxic activity and via secretion of IFN-γ and TNF-α, respectively. In this book chapter, we describe the basic biology of NK cells and their function in cancer as well as the impact of the tumor microenvironment on NK cell function. Furthermore, we present a novel strategy for the expansion of allogenic NK cells, which results in the production of large numbers of highly functional supercharged NK cells. Moreover, we discuss the recent advances in the application of NK cells in cancer immunotherapy. The combination of standard anticancer treatments with recently established immunotherapeutic approaches with NK cells should provide the optimal conditions for the eradication of tumors.

Abbreviations

ADCC	antibody-dependent cellular cytotoxicity
BiKE	bispecific killer cell engager
CAR	chimeric antigen receptors
CCL	chemokine (C-C motif) ligand

https://doi.org/10.1016/B978-0-12-822620-9.00020-3

CD	cluster of differentiation
COX2	cyclooxygenase 2
CSCs	cancer stem-like cells
CTLA-4	cytotoxic T-lymphocyte-associated protein 4
DC	dendritic cell
EGFR	epidermal growth factor receptor
GFP	green fluorescent protein
GM-CSF	granulocyte-macrophage colony-stimulating factor
Hu-BLT	humanized-bone marrow/liver/thymus
IDO	indoleamine 2,3-dioxygenase
IFN	interferon
IL	interleukin
ILC	innate lymphoid cells
iPSCs	induced pluripotent stem cells
KIR	killer cell immunoglobulin-like receptors
LTi cells	lymphoid-tissue inducer cells
MDSC	myeloid derived suppressor cell
MHC	major histocompatibility complex
NF-κB	nuclear factor kappa B
NK cell	natural killer cell
NSG mouse	NOD SCID gamma mouse
OCs	osteoclasts
OV	oncolytic virus
PAMP	pathogen-associated molecular pattern
PD-1	programmed cell death protein 1
PGE	prostaglandin E
TGF	transforming growth factor
TIGIT	T cell immunoglobulin and ITIM domain
TIM-3	T cell immunoglobulin and mucin-domain containing-3
TLR	Toll-like receptor
TME	tumor microenvironment
TNF	tumor necrosis factor
TRAIL	TNF-related apoptosis-inducing ligand
TriKE	tri-specific killer cell engager

Conflict of interest

No potential conflicts of interest were disclosed.

Natural killer cells: Overview and background

Natural killer (NK) cells differ morphologically and phenotypically from T and B lymphocytes and were defined traditionally as cells capable of spontaneously killing tumor cells [1]. NK cells are defined as large granular lymphocytes [2], which comprise 5%–20% of the peripheral blood lymphocytes [3]. NK cells are generated in the bone marrow and are crucial in lysing virally infected cells as well as malignantly transformed targets [4]. They are defined by the surface expression of CD56 [5] and CD16 [6] and lack the expression of CD3 surface receptors [7]. CD16 receptors, also known as FcγRIII receptors, expressed on the surface of NK cells constitute one of the main receptors on NK cells important in mediating both direct cytotoxicity and antibody-dependent cellular cytotoxicity (ADCC) [8,9]. In addition to circulating NK cells, many are also found in different tissues, such as lymphoid organs, healthy skin, gut, liver, lung, and uterus to name a few [10]. The missing-self hypothesis proposed in 1986 speculated that the presence of MHC class-I on the surface of the NK-interacting cells is responsible for the inhibitory signals delivered to the NK cells [11]. Since then many inhibitory receptors, such as NKG2A [12], p58 [13], KIR2D [14], and activating receptors, such as NKp46 [15], NKp30 [16], NKp44 [17], and NKG2D [18], have been defined. The net sum of activating and inhibitory signals that NK cells receive determines the levels of NK cell activation. The threshold for activation of NK cells is lowered in a suitable cytokine microenvironment provided by IL-2, IL-12, IL-15, IL-18, and type I interferons [19].

NK cells exhibit two main or primary effector functions: cytotoxicity and release of cytokines (Fig. 1). NK cells mediate cytotoxicity via two different mechanisms: granule-mediated cytotoxicity and cytotoxicity mediated by the death receptors. Release of cytotoxic granules induces both the necrotic and the apoptotic cell death via the activation of caspases. The granules contain perforin, a calcium-dependent pore forming protein, and serine peptidases known as granzymes [20]. Both perforin and granzymes are synthesized in their precursor forms and are required to be cleaved by cathepsins to their active forms. Processing of perforin is catalyzed by cathepsin L [21], whereas cathepsins C and H activate the granzymes A and B [22–24]. The activity of these cathepsins is further regulated by cystatin F [25]. NK cells express ligands for death receptors, such as Fas ligand, TNF-α, and TRAIL (Apo2L). Upon interaction with the death ligands on NK cells, the target cells receive signals to undergo cell death [26]. Both pathways can also be activated via ADCC induced by the binding of Fc portion of the antibody on target cells to CD16 receptors on NK cells [27].

Upon activation, NK cells secrete an array of cytokines, chemokines, and growth factors, among which IFN-γ is one of the key cytokines [28]. IFN-γ enables the activation of dendritic cells (DCs) and monocytes. This mechanism is important for the activation and expansion of adaptive immune responses [29,30]. In addition, IFN-γ produced by the NK cells stimulates the expression of MHC class-I and -II molecules on the cell surface of their interacting cells [31], mediating inhibition of tumor proliferation [32], and upregulation of adhesion molecules, which increases the sensitivity of tumors to CD8+ T cell cytotoxic effectors [33] (Fig. 1).

NK cells were found to have specific targets, as they were only able to kill certain tumor cells, but not those highly susceptible to T cell killing [34]. It was later discovered that they preferentially targeted cancer stem-like cells (CSCs)/poorly-differentiated tumor cells that expressed lower levels of MHC class-I molecules on their surface [35–38] (Fig. 1).

NK cells form part of the recently discovered family of innate lymphoid cells (ILCs). Latest agreement on nomenclature divides ILCs into NK cells, ILCs type 1, 2, and 3 (also referred to as helper ILCs), and lymphoid-tissue inducer cells (LTis) [39]. ILCs do not have antigen-specific receptors, but they do mirror the function of T lymphocytes. ILC1s respond to intracellular pathogens, similar to CD4+ Th1 cells. ILC2s, similar to CD4+ Th2 cells, react to extracellular parasites and allergens, while ILC3s react to extracellular bacteria and fungi [40]. By that analogy, NK cells are the innate counterparts of CD8+ T cells. In contrast to T lymphocytes, the time of action of ILCs is at an early phase in the immune response [41].

Although all ILCs are developed in the bone marrow, ILC1, 2, and 3 are localized in mucosa-associated tissues, whereas NK cells are localized in the blood stream and secondary lymphoid organs. As helper ILCs are mainly tissue resident, they are in an ideal location to fine-tune antitumor immunity [42]. While NK cells undergo constant replenishment from the bone marrow, other ILCs remain in the tissues throughout life and are replenished by local proliferation of tissue resident progenitor cells [43]. Among the ILCs, only NK cells have cytotoxic properties

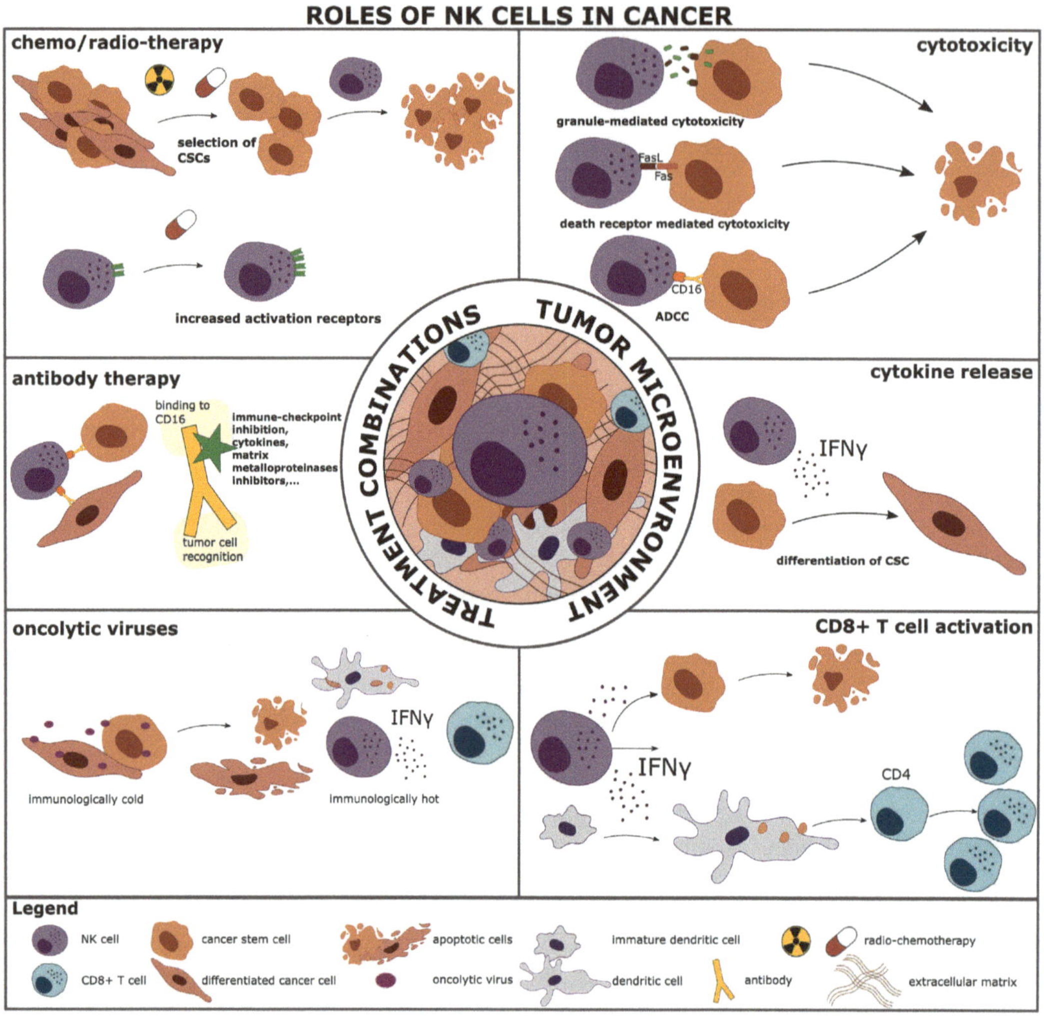

FIG. 1 Roles of NK cells in cancer.

NK cells have different roles in cancer. In the tumor microenvironment, they are able to eliminate CSCs by three different mechanisms: granule-dependent cytotoxicity, death receptor-mediated cytotoxicity, and antibody-dependent cellular cytotoxicity. In addition to eliminating CSCs by killing, they eliminate CSCs by differentiation caused by released cytokines, especially IFN-γ. Finally, NK cells help in the activation of CD8+ T cells. After CSCs are killed, tumor antigens are released. Secreted IFN-γ is important for the differentiation and activation of dendritic cells. Dendritic cells present tumor antigens to CD_4+ T cells, which leads to activation of CD8+ T cells. NK cells can act synergistically with various cancer treatment strategies. Chemotherapy and radiotherapy are usually more efficient in eliminating differentiated cancer cells. Treatment with NK cells would be beneficial in eliminating the remaining CSCs. NK cells also mediate their cytotoxicity through ADCC. By designing antibodies that bind to the CD16 receptor on NK cells and tumor-specific antigens, we can bring together effector cells and target cells. In addition, therapeutic antibodies can be further engineered to express supportive cytokines and matrix metalloproteinase inhibitors or inhibit immune checkpoints. Oncolytic viruses also represent a good treatment option as they switch the tumor environment from immunologically cold to immunologically hot, allowing immune cells to function.

against virally and malignantly transformed cells. Some reports indicate that ILC1s are weakly cytotoxic, but mainly they represent the first line of defense against some viruses and bacteria [44,45]. ILC2s mediate the response to extracellular helminth parasites, while ILC3s are important for maintaining tolerance to symbiotic microbiota [42]. Both ILC2 and 3 can present antigens via MHC class-II, but ILC3 do not possess necessary costimulatory factors [46,47]. LTis are important in the fetal developmental stage as they induce the formation of secondary lymphoid organs and Peyer's patches [48]. Subsets of ILCs are identified by differentially expressed surface receptors and by the secreted cytokines. Both NK cells and ILC1s secrete IFN-γ and TNF-α, while ILC2s secrete type 2 cytokines (IL-4, IL-5, IL-9, IL-13, and amphiregulin), and ILC3s and LTis secrete IL-17, IL-22, and GM-CSF [42].

The role of ILCs in tumor defense differs depending on the tumor type and the microenvironment [49]. Depending on the signals from the microenvironment, they can convert from one subset to another [50–52]. NK cells in their fully functional state are able to eliminate the most resistant CSCs/poorly differentiated tumors. Under the influence of TGF-β, NK cells can be converted to intermediate ILC1s, which do not have antitumor properties. Helper ILCs were shown to promote immunosuppressive microenvironment in cancer patients. Different factors, such as increased CTLA-4 and GM-CSF, low IFN-γ levels, CCL5, and TGF-β, contribute to the pro-tumoral microenvironment [51,53]. ILC2s, which produce suppressive IL-5 and IL-13 cytokines [54], are found to be increased in gastric [55], prostate [56], and breast cancers [57]. In addition, ILC2s produce amphiregulin that promotes EGFR expressing tumor cells leading to tumor invasion and metastasis [58]. ILC3s promote tumorigenesis of colorectal cancers by secretion of IL-22 and IL-17 [59]. In breast cancer, increased numbers of ILC3s correlated with the increased metastasis to lymph nodes [60]. In contrast, other reports suggested a beneficial role for ILC3s in cancer. In B16 melanoma mouse model, ILC3s were able to recruit CD8+ T cells and NK cells to the tumor site [61]. Future studies should be directed to defining the exact similarities and differences between ILCs and those of other traditionally defined immune effectors.

Two faces of NK cells: Concept of split anergy in NK cells and its potential role in tumor differentiation

NK cells enter various stages of development after their generation in the bone marrow. Subsequent to generation of NK cells, the licensing takes place, which selects for NK cells expressing MHC class-I receptors and are therefore thought to be tolerant to self [62]. Two subsets of NK cells have been identified based on surface expression of CD56 and CD16. The minor subset makes up approximately 10% of human NK cells and is defined as $CD56^{bright}CD16^{dim}$. This subset is speculated either to be a less mature form of NK cells or to develop from the $CD56^{dim}CD16^{bright}$ subtype. This subtype is able to secrete cytokines, but possesses no or low cytotoxic ability. Fully functional cytotoxic and cytokine releasing $CD56^{dim}$ and $CD16^{bright}$ NK cells represent the major subset of NK cells in humans (about 90%), also defined as stage I NK cells [27]. After interaction with target CSCs or activation with IL-2 and CD16 monoclonal antibodies, NK cells enter in stage II and become split anergized cells, exhibiting the phenotype of $CD16^{low}CD56^{bright}CD69^{bright}$. The cytotoxic ability of split-anergized NK cells is low, but they secrete higher levels of cytokines such as IFN-γ and TNF-α [63–65]. In the absence of rescue signals, NK cells enter stage III and lose the ability to secrete IFN-γ and TNF-α. Instead, these NK cells will likely secrete cytokines such as IL-6 and IL-10, similar to those seen by NK92 cell line [66]. Finally, NK cells enter stage IV and become unresponsive, exhausted, and eventually undergo senescence and apoptosis (Fig. 2) [67]. Even though split anergized NK cells lose their cytotoxicity, they are still important in the tumor microenvironment (TME). Owing to their secretion of IFN-γ, split anergized NK cells contribute to the differentiation of CSC/poorly differentiated tumors by increasing MHC class-I, CD54, and B7H1 (PD-L1) and decreasing CD44 surface expression levels on tumor cells (Fig. 1). Differentiated tumors are known to be susceptible to T cell-mediated killing [68–70]. In addition, differentiated tumor cells have a lower ability to proliferate and metastasize [63,71,72]. Overall, these results indicated that NK cells are important in shaping the TME.

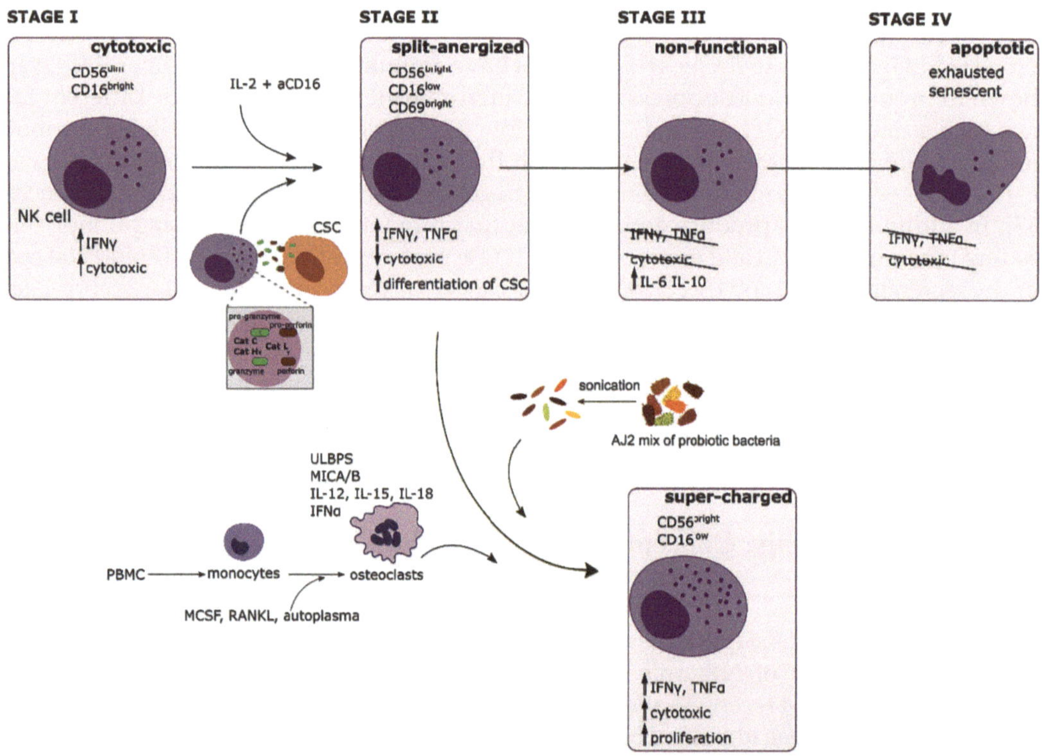

FIG. 2 Stages of NK cell phenotype and function.

Peripheral blood NK cells mediate cytotoxicity against CSCs and secrete cytokines, especially IFN-γ (stage I). Cytotoxic granules containing granzyme B and perforin facilitate one of the possible cytotoxic pathways. After interaction with CSCs or in vitro activation with IL-2 and anti-CD16 monoclonal antibodies, NK cells become split-anergized and enter the stage II. Split-anergized NK cells are no longer cytotoxic, but they have higher cytokine secretion, which allows differentiation of CSC to more differentiated phenotypes. Without rescue signals, the cells enter stage III and become nonfunctional and harmful in the TME as they secrete IL-6 and IL-10. Eventually, NK cells become apoptotic and enter stage IV. We can expand supercharged NK cells from split anergized cells. By coculturing split anergized NK cells with monocyte-derived osteoclasts expressing NKG2D ligands (ULBPS and MICA/B) and cytokines IL-12, IL-15, IL-18, and IFN-α, with the addition of sonicated probiotic bacteria sAJ2, we can produce highly cytotoxic supercharged NK cells with a higher ability to secrete cytokines and a higher proliferation potential.

The rationale for the functional activation of NK cells in many gene knockout mice: Potential common mechanism of activation

NK cells are shown to be activated in many knock-out or knock-down mice (Fig. 1). We have previously reported a comprehensive list of genes that upon deletion are implicated in the triggering of augmented NK cell function in in vitro culture conditions and in mice models [73]. For example, blockage of NF-κB in oral tumor cells not only augmented the proliferation and activation of NK cells but also of CD8+ T cells [74]. Furthermore, the secretion of immunosuppressive cytokine, IL-6, was decreased, which benefited the activation of NK cells. Moreover, the knockdown of NF-κB downregulated the expression of MHC class-I on Hep2 tumor cells, rendering these tumors more susceptible to NK cell-mediated killing [75]. Similarly, knock down of COX-2 in nontransformed healthy myeloid cells and mouse embryonic fibroblasts augmented NK cell function and expansion [76]. Expression of CD44, a cell surface glycoprotein that facilitates binding of NK cells to tumor cells, correlates with stem-like/poorly differentiated phenotype of tumor cells. Knock-down of CD44 in breast cancer cell line MDA-MB231 and melanoma cell line A375 resulted in the downmodulation of CD54, PD-L1, and MHC class-I and surprisingly increased the susceptibility to NK cell-mediated killing [70] (Fig. 1). Based on the results of our work on gene knockout mice and tumors and those reported by many other laboratories we hypothesized that knocking key genes in mice or in tumors may decrease cellular differentiation, correlating with an overall decrease in MHC class-I, CD54, and PD-L1 and subsequently activating the NK cells to compensate for the decreased differentiation of these cells. Our results demonstrated that the relationship between tumor cell differentiation and their susceptibility to NK cell-mediated effects may not be linear, indicating that the lower the differentiation of the tumor cells, the higher the activation of NK cells will likely be. Indeed, the opposite may also hold true, since the higher the differentiation of the tumors, the lower the capacity of these tumors to activate NK cells [73].

Dysfunctional NK cells in cancer patients: Defects in NK cells to lyse and differentiate CSCs

Several studies have compared the function of NK cells between healthy individuals and patients with cancer. The consensus is that NK cells from cancer patients are less reactive against target cells. The numbers of NK cells are lower in certain cancer patients' PBMCs, and the cytotoxic function declines with the progression of cancer [77–82]. In terms of the four stages of NK cells described earlier, the numbers of stage III and stage IV NK cells increase substantially in cancer patients [83]. Patients' NK cells secrete lower amounts of IFN-γ, therefore they are not able to differentiate CSCs effectively. In addition, in vitro studies have shown that IFN-γ secreted by cancer patients' NK cells is also dysfunctional. When using the same amounts of IFN-γ secreted by the NK cells from healthy individuals and those of the cancer patients, the ability to differentiate tumors was substantially lower by patients' NK cells secreted IFN-γ when compared to healthy individuals [84].

Suppression of antitumor immune function and change in NK cell phenotype in tumor microenvironment: Could compromised NK function lead to dysregulated immune function in tumor microenvironment?

The numbers of tumor infiltrating immune cells are found to be low at the later stages of cancer when effective immune surveillance might have been lost to tumor escape, or the tumors have lost their ability to differentiate in the microenvironment and are then populated by the stem-like/poorly-differentiated tumors [85]. Similar to cancer patients' peripheral blood NK cells, tumor-infiltrating NK cells were also shown to have reduced effector function characterized by increased expression of inhibitory receptors, decreased expression of activating receptors, and downregulation of various effector molecules such as IFN-γ, CD107a, granzyme B, FasL, TRAIL, and perforin, as shown in breast cancer [86], nonsmall cell lung carcinoma [87], prostate cancer [88], ovarian cancer [89], gastric cancer [90], lung cancer [91], and hepatocellular carcinoma [92] among others. Several factors including TGF-β and IL-10 induced in the TME were found to be partly responsible for the changes mentioned earlier [86,92]. Additional proof that TME influences the NK cell function came from the studies which demonstrated restoration of activating receptors on NK cells in breast cancer patients in remission [86]. Due to their interactions with the suppressive factors within the TME, NK cells can enter a state of unresponsiveness and exhaustion. Phenotypically, this is marked with a decrease in activating receptors and increase in the inhibitory receptors on NK cell surface, skewing the balance toward inhibitory response of NK cells [93]. However, in our previous studies, we have shown that both NK activating and inhibitory ligands on tumor cells were decreased, suggesting a more complex picture for the functional inactivation of NK cells within TME [84].

NK cells also interact with and regulate other immune effectors. DCs promote NK cell proliferation and activation by releasing IL-12 during their interaction with NK cells resulting in increased NK-mediated cytotoxicity and the release of IFN-γ. Secretion of IFN-γ promotes differentiation of naïve CD4+ T cells to Th1 cells. NK cells promote the function of DCs by releasing antigens from killed target cells, resulting in increased CD8+ T cell responses (Fig. 1). By killing immature DCs, NK cells participate in DC-editing ensuring the survival of optimal DCs expressing adequate costimulatory molecules that will prime successful T cell responses [94,95]. NK cells expand and are activated by osteoclasts (OCs) substantially, leading to supercharged NK cells (please see below) [84]. Finally, NK cells are able to expand CD8+ T cells by elimination of activated CD4+ T cells (please see later) [96]. Therefore, by eliminating naive, immature, and activated immune cells, NK cells may select for diverse differentiated/activated immune cells within the TME to promote lysis, differentiation, and curtailment of tumor growth (Fig. 1). Therefore, since the NK cells' function in cancer is suppressed, this may also have significant impact on the activation and regulation of other immune cells.

Tumor-associated stromal cells may shape the function of NK cells

Diverse tumor-associated stromal cells have an important function in determining the fate of the NK cells. Tumor-associated macrophages, myeloid-derived suppressor cells (MDSCs), neutrophils, tolerogenic DCs, and tumor-associated fibroblasts promote immune suppression by secreting TGF-β and IL-10. These cytokines inhibit the expression of activation receptors NKp30 and NKG2D on NK cell surface [97–101] and reduce secretion of IFN-γ [102,103]. As mentioned earlier, TGF-β also mediates transformation of NK cells into ILC1 cells that lack cytotoxic properties [51,53]. Additional immunosuppressive metabolites that reduce the proliferation and function of NK cells are PGE2 secreted by the mesenchymal stem cells [104] and IDO and adenosine [105]. IDO causes the accumulation of L-kynurenine that prevents the upregulation of NKp46 and NKG2D after IL-2 stimulation [106]. Melanoma cell lines [107] and melanoma-associated fibroblasts [108] cause decrease in killing ability of NK cells via the induction of IDO and PGE2 in vitro. MDSCs also inhibit NK cell function in direct cell contact via the interaction with NKp30 receptor on NK cells [109]. Regulatory T cells (Tregs) downmodulate NKG2D expression on NK cells via membrane bound TGF-β [110]. In addition, tumor sites are usually hypoxic that affects NK cell function adversely [111].

Function of NK cells in tumor-bearing humanized-BLT mice mirrors those of the cancer patients

Preclinical in vivo studies are necessary to test anticancer treatment strategies in a complex TME. Mice represent a good model as they exhibit high tumor growth and can be genetically modified. However, in order for the model to reliably predict the outcomes in humans, the model should have similar characteristics in terms of TME, tumor heterogeneity, and state of immune system as humans do [112]. Our studies have shown that humanized-bone marrow/liver/thymus (hu-BLT) mice are the appropriate model to study NK–tumor interactions. To prepare this model, human CD34+ progenitor cells are injected into NSG mice. Human fetal liver and thymus fragments are implanted under the renal capsule, which provides the appropriate microenvironment for immune cell development [113,114]. Hu-BLT mice were found to exhibit greater than 90% human CD45+ immune cells in their peripheral blood. The percentages of CD4+, CD8+ T cells, and B cells are identical, but the frequency of NK cells in hu-BLT mice is approximately half of those in human peripheral blood [115]. NK cells in tumor bearing hu-BLT mice have decreased numbers and cytotoxic ability and secrete less IFN-γ, similarly to those seen in NK cells from cancer patients [84,113,115,116]. Similar to humans, stem-like tumors grow rapidly, metastasize, and resist chemotherapy, whereas differentiated tumors do not rapidly proliferate, nor metastasize and are highly susceptible to chemotherapeutic drugs in hu-BLT mice [84,115]. Using supercharged NK cells as immunotherapy, we are able to inhibit the growth and metastasis of tumors in hu-BLT mice [115,116].

Novel strategy to expand supercharged NK cells for immunotherapy using osteoclasts as feeder cells: Different efficacy of supercharged NK cell expansion and function using allogeneic vs autologous NK cells from healthy or cancer patients

We have developed a novel ex vivo expansion strategy that produces large numbers of highly functional NK cells for adoptive cell therapy of cancer. In our platform, we use OCs as feeder cells. OCs, derived from monocytes, are potent activators of NK cells, as they secrete IL-15, IL-12, IL-18, and IFN-α, and they express lower levels of MHC classes-I and -II, CD14, CD11b, CD54 and higher levels of NKG2D ligands, ULBPs, and MICA/B (Fig. 2). These characteristics make OCs more effective feeder cells than monocytes or DCs in NK activation [117]. Best results are obtained when healthy NK cells are used for the expansion with autologous OCs. On the contrary, the combination of autologous NK cells and OCs derived from cancer patients has severe defects in NK expansion and functions. Similar conclusions were drawn using cells purified from tumor-bearing hu-BLT mice [118]. When cancer patients' NK cells are used for expansion in our platform, the success is limited due to faster proliferation of the small fraction (0.2%–1%) of contaminating T cells [84]. Human NK cells express TLR-3 and TLR-9 and can thus become activated by the recognition of PAMPs. This represents an alternative mode of activation without the necessity for the recognition of target cells [119]. We used sonicated probiotic bacteria AJ2, a mix of 7–8 gram-positive bacterial strains for the expansion and activation of NK cells in combination with IL-2 and anti-CD16 monoclonal antibodies [72]. Activated NK cells are then cultured with OCs and IL-2, which produces large numbers of highly cytotoxic supercharged NK cells that secrete high amounts of IFN-γ (Fig. 2) [84].

Supercharged NK cells differ from primary NK cells in phenotype and function

Primary NK cells change their morphology in the process of expansion to supercharged NK cells. Primary NK cells change from round to elongated polymorphic shape and become bigger in size. Supercharged NK cells grow attached to the plate or float in the culture medium in clumps. As described earlier, supercharged NK cells expand faster than primary cells. This is also reflected in the increased expression of Ki-67 proliferation marker in supercharged NK cells compared to primary NK cells. Additionally, several activating and inhibitory receptors and effector molecules are elevated in supercharged NK cells [84]. Phenotypical changes are reflected in the functional differences between primary and supercharged NK cells. Supercharged NK cells have increased cytotoxic ability and secrete higher amounts of IFN-γ and other cytokines. We have also shown that supercharged NK cells are more polyfunctional, as one supercharged NK cell is able to secrete more diverse cytokines compared to single primary NK cell. Supercharged NK cells' supernatants induce higher levels of differentiation and death of tumor cells (unpublished observations). Unlike primary NK cells, super-charged NK cells are split anergized much less as they exhibit less inactivation in cytotoxicity after co-culture with tumor cells (Fig. 3) [120].

Supercharged NK cells preferentially and rapidly expand CD8+ T cells

It is known that NK cells regulate adaptive immunity. Although we know a lot about the interaction of NK cells with T and B cells, the details of such interactions during activation of NK cells are not well understood. Supercharged NK cells are able to expand for longer period of time (30–60 days), before the contaminating fraction of T cells starts to proliferate, resulting

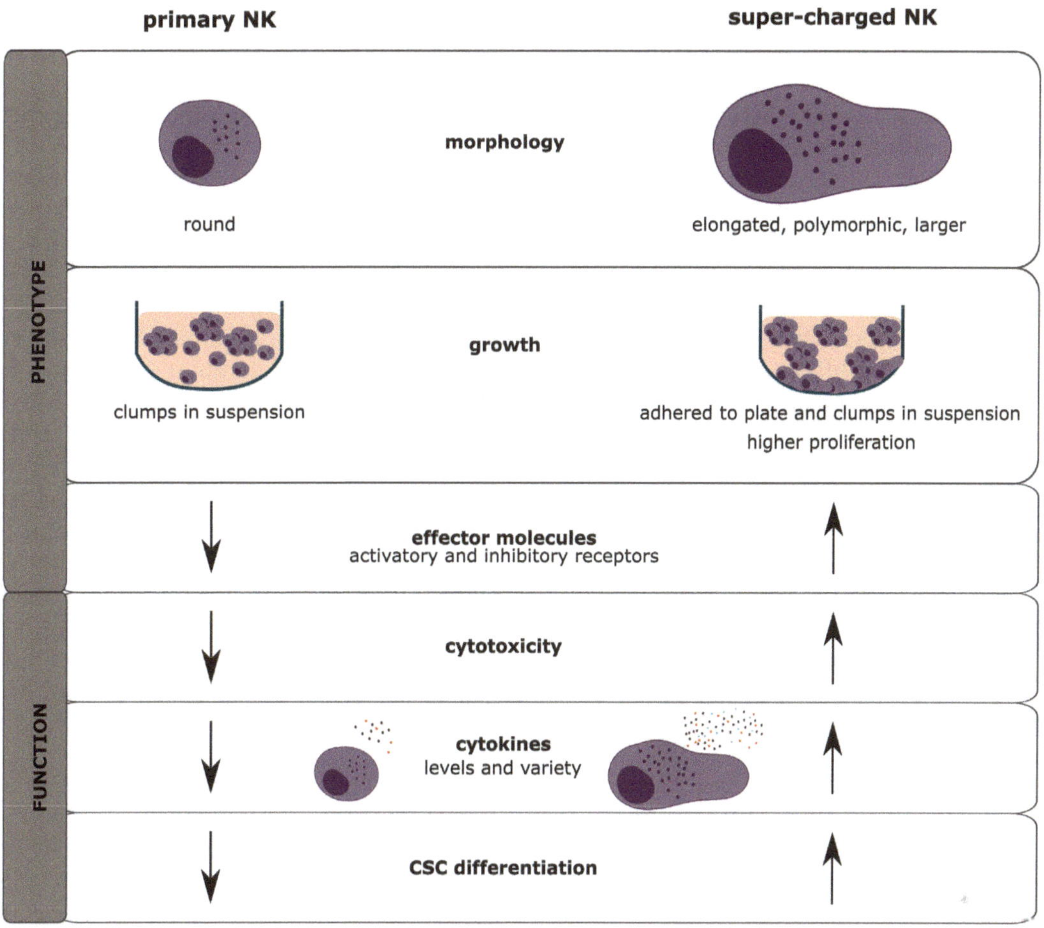

FIG. 3 Primary and supercharged cells differ in phenotype and function.
Primary NK cells are round, small, and grow as suspension cells that form clumps. Supercharged NK cells are larger than primary NK cells, have a more elongated and polymorphic phenotype. They grow adhered to the plate. When they proliferate, they form clumps that detach from the plate and float in suspension. Supercharged NK cells exhibit higher expression of effector molecules compared to primary NK cells. Supercharged NK cells exhibit higher cytotoxicity and cytokine secretion profile and are therefore more successful in differentiating CSCs.

in the NK cell decline. Interestingly, the CD8+ T cells are the subset of T lymphocytes that is preferentially expanded in the cultures of super-charged NK cells (Fig. 1). Using target visualization assay, we have demonstrated that super-charged NK cells are able to lyse activated CD4+ T cells but not CD8+ T cells. We found that DCs promote faster expansion of T cells compared to OCs [73,96]. Oral and pancreatic tumor-bearing hu-BLT mice have increased CD8+ T cell numbers in peripheral blood, bone-marrow, and spleen after being injected with supercharged NK cells [96,116] (Fig. 1).

Functional differences of NK cells in different NK expansion platforms against CSCs/poorly differentiated tumor cells: Comparison with supercharged NK cells

Different sources of NK cells for the preparation of the adoptive transfer therapy have been used. NK cells are collected from peripheral blood, cord blood, human embryonic stem cells or induced pluripotent stem cells (iPSC), or derived from patient tumor cells. After purification process, several different strategies are used for the NK expansion [105]. Although there are attempts to restore function of cancer patients' NK cells using different cytokines and feeder cells, the allogeneic transfer of NK cells from healthy individuals is preferable [105], since the autologous transfer of NK cells to treat cancer does not give promising results [121].

Several factors have hampered the development of NK cell transfer therapies. NK cell numbers in peripheral blood are low and are difficult to obtain without contamination of T cells. In addition, the expansion process is difficult. It requires the support of many cytokines and/or feeder cells and can be overwhelmed by the fast proliferation of contaminating T cells. Allogeneic T cells can also cause acute graft-vs-host disease [122]. Following the success of engineered T cells with chimeric antigen receptors (CAR) [123], the CAR strategy is being applied to NK cells. NK cells possess several advantages over the use of T cells. Major concerns in the usage of CAR-T cells are the possibility of autoimmunity or malignant transformation of CAR-T cells due to the long-term survival, and even more importantly the deleterious effect of cytokine-release syndrome, which could jeopardize the patients' life. However, CAR-NK cells are shorter-lived effectors than CAR-T cells, which can potentially be removed earlier than CAR-T cells. Also, their cytokine profile, consisting of higher secretion of IFN-γ, is safer. Another advantage is that, even if tumor cells downregulate the CAR-targeted antigen, NK cells are still able to kill tumor cells [105].

We believe that the standardization among all different NK cell platforms is necessary in order to choose the most effective products for immunotherapy. We compared NK function and efficacy of expansion of supercharged NK cells to NK cells generated from cord blood and iPSC-derived NK cells. Early studies suggested that reactivity of cord blood NK cells was comparable to that of adults [124]. However, additional studies have shown that cord blood NK cells lack in all aspects of functionalities degranulation, cytokine production, and ADCC compared to adult NK cells [125,126]. Similarly, we observed extremely low cytotoxicity toward CSCs and low IFN-γ secretion of cord blood-derived NK cells. Similar results to those of cord blood derived NK cells were also obtained for iPSC-derived NK cells [73].

NK92 malignantly transformed cell line [127] enables the opportunity to avoid the difficulties in obtaining NK cells in sufficient numbers. As NK92 cells are easy to transfect, many CAR-NK92 cells are currently in development [128]. Nevertheless, we found that NK92 cells are functionally inferior to peripheral blood-derived NK cells. In addition, NK92 cells are not only unable to produce IFN-γ, but they also produce IL-6 and IL-10 that may limit immune activation [66]. Although we were not able to compare the efficacy of supercharged NK cells to all existing platforms, we were able to show that supercharged NK cells are superior to those of cord blood, iPSC-derived, NK-92 and K562, and cytokine expanded NK cells. It is needless to say that only the most appropriate and effective strategies for NK expansion will be successful in overcoming the suppressive elements within TME to control the growth and expansion of CSCs/poorly differentiated tumors [73].

Immunotherapy is essential in combination with chemotherapy: Chemotherapy targets differentiated tumors more than cancer stem-like cells

Conventional treatment of cancer consists of surgical intervention, radiotherapy, and different chemotherapies. It is now known that well-established radio- and chemotherapies deliver their effects also in part via the immune system. Due to cellular stress of tumor cells caused by conventional therapies, for example, 5-fluorouracil [129] and erlotinib [130], the ligands of NK cell activation receptors are upregulated [131]. Some agents contribute to promoting NK cell function. Paclitaxel stimulates NK cells to infiltrate breast carcinoma [132], pemetrexed stimulates IFN-γ secretion in pancreatic cancer patients [133], and imatinib promotes expansion of peripheral NK cells and CTLs and elevates IFN-γ secretion [134,135]. In addition, several chemotherapeutic agents were shown to deplete immunosuppressive Tregs and MDSCs [136]. Sulindac mediates its anticancer effects via the downregulation of NF-κB and JNK, resulting in the decrease of tumor-promoting IL-6 and IL-8 and the activation of NK cells [70]. Targeting of NF-κB activates NK and CD8+ T cells to mediate anticancer effects [74,137]. Nevertheless, chemotherapeutic agents target differentiated cells, while CSCs are resistant to chemotherapy [138]. Using NK cells to kill CSCs/poorly differentiated tumors represents a synergistic approach to ensure complete removal of cancer in conjunction with chemotherapy (Fig. 1). Since NK cells are important for the removal of CSCs/poorly differentiated tumor cells and are able to prime the adaptive immune response via IFN-γ, it is important to optimize therapeutic strategies to preserve or enhance the function of NK cells in cancer patients [139].

Combination of NK cell and antibody therapy: NK cells can target CSCs and their differentiated counterparts through direct lysis and/or ADCC, respectively

Primary cytokine-activated NK cells predominantly eliminate CSCs through direct lysis, but they are not able to target differentiated cells directly as they possess inhibitory ligands. Nevertheless, NK cells can eliminate differentiated tumor cells through ADCC. Several anti-tumor antibodies are used to treat patients. The antigen-specific Fab fragment of the monoclonal antibodies binds to the antigen on tumor cell, and the Fc fragment binds to the immune effector cells. CD16 receptor on NK cells is able to bind the Fc fragments resulting in the immediate activation of NK cells. This effect can be improved by enhancing the affinity to CD16 [140], fusing the Fc fragments with IL-2 [141], or combining the monoclonal antibody treatment and IL-2 infusion [142]. The major threat to the success of this anticancer strategy is the shedding of CD16 caused by matrix metalloproteinases, such as ADAM-17 [143,144]. Further modifications lead to bi- or tri-specific killer-cell engagers (BiKEs, TriKEs) [145]. They are developed by fusing Fv domains recognizing tumor antigens and CD16 [146]. The third domain either binds another tumor antigen that is fused to metalloproteinase inhibitor [147] or incorporates NK supporting cytokine such as IL-15 [148]. ADCC can also be achieved by the use of aptamers—single-stranded oligonucleotides engineered to bind tumor antigens and CD16 receptor [149]. Combinations of NK infusions with monoclonal antibodies are already tested in clinical settings with encouraging results [150–152] (Fig. 1).

NK cells and the oncolytic viruses

Oncolytic viruses (OVs) are prepared to selectively infect, replicate in, and lyse cancer cells, based on their interaction with tumor-specific receptors [153]. The induction of immune response is necessary for the success of OV immunotherapy. The direct lysis of cancer cells, induced by replication dependent cell death, releases antigens in the TME, resulting in the stimulation of NK cells and antigen-presenting cells [154]. Released inflammatory cytokines, such as IFN-γ and TNF-α, induce the antiviral/antitumor response and recruit other immune cells to the TME and enable the transition of TME from "cold" to immunologically "hot" environment. Activated immune effectors are able to additionally eliminate virally infected cancer cells, causing immunogenic cell death [155]. OVs can be further genetically modified to enhance immune responses [156], by expressing NK supporting cytokines such as IL-12, IL-15, and IL-18 [157,158]. Even though NK cells might limit the OV replication by targeting virally infected cells [159], the combination of OV therapy with adoptive NK cell transfer can be beneficial in cancer treatment [160–162] (Fig. 1). It is important to note that not all OVs have the same effects. In our studies, infection of differentiated pancreatic cells PL-12 with various preparations of OVs resulted in different effects on decrease in MHC class-I expression and subsequent activation of NK cells. The 33-GFP OV was the most effective in decreasing MHC class-I, PD-L1, and CD54, compared to 189 OV and Mokola virus. The 33-GFP OV was also more effective in inducing IFN-γ secretion by purified NK cells [118].

Combination therapy with NK cells and immune checkpoint inhibitors

Cancer immunotherapy using checkpoint inhibitors to reactivate T cells has enabled a substantial progress in the treatment of certain cancers [163]. Similar to T cells, NK cells express immune checkpoint receptors that mark for the exhaustion of NK cells. Several immune checkpoint receptors have been proposed to be expressed on NK cell surface including PD-1 [164], TIGIT [165], TIM-3 [166], NKG2A [167], and CD96 [168]. Interactions of immune checkpoint receptors with their ligands suppress cytotoxic activity and IFN-γ production, which is important in regulating NK cell responses in acute inflammation but leads to exhausted state of NK cells in tumors [169–171]. We have shown that loss of NK cytotoxicity and IFN-γ secretion can be restored in tumor bearing mice after injection of supercharged NK cells in combination with anti-PD-1 antibody [116] (Fig. 1). Since many immune checkpoint inhibitors are antibodies, they have additional beneficial effects, such as enabling the activation of NK cell mediated ADCC [172,173].

Conclusions

Despite recent advances, success in cancer therapy still remains elusive. Conventional therapy and novel immunological treatments are effective against differentiated tumor cells but do not eliminate CSCs. As NK cells are able to target CSCs/poorly differentiated tumors, NK cells have the indispensable role in successful cancer therapy. In addition to their direct cytotoxic effect, they are able to target cancer cells through ADCC and secreted IFN-γ, which sets in motion the differentiation of cancer cells. As a result of differentiation, cancer cells become susceptible to CTL-mediated lysis and conventional chemotherapy. Therefore, the design of effective NK expansion strategies that will generate sufficient numbers of highly active NK cells is needed. One of these approaches is the expansion of supercharged NK cells, which shows superior functional properties compared to primary NK cells. The combination of allogenic supercharged NK cells with other therapies, such as chemotherapy and radiotherapy, as well as the use of antibodies, checkpoint inhibitors, OVs and microbiome restoration and modulating agents could be effective in optimal cancer treatment (Fig. 1).

Acknowledgments

The authors are grateful to numerous undergraduate and graduate students, post-doctoral fellows, visiting scholars, and faculty members for their excellent contributions to the work presented in this chapter. In addition, we thank our funding agencies and donors for supporting the work.

References

[1] Kiessling R, Klein E, Pross H, Wigzell H. "Natural" killer cells in the mouse. II. Cytotoxic cells with specificity for mouse Moloney leukemia cells. Characteristics of the killer cell. Eur J Immunol 1975;5:117–21.

[2] Timonen T, Saksela E, Ranki A, Häyry P. Fractionation, morphological and functional characterization of effector cells responsible for human natural killer activity against cell-line targets. Cell Immunol 1979;48:133–48.

[3] Trinchieri G. Biology of natural killer cells. Adv Immunol 1989;47:187–376. Academic Press.

[4] Haller O, Kiessling R, Orn A, Wigzell H. Generation of natural killer cells: an autonomous function of the bone marrow. J Exp Med 1977;145:1411–6.

[5] Lanier LL, Testi R, Bindl J, Phillips JH. Identity of Leu-19 (CD56) leukocyte differentiation antigen and neural cell adhesion molecule. J Exp Med 1989;169:2233–8.

[6] Perussia B, Trinchieri G, Jackson A, Warner NL, Faust J, Rumpold H, et al. The Fc receptor for IgG on human natural killer cells: phenotypic, functional, and comparative studies with monoclonal antibodies. J Immunol 1984;133.

[7] Hercend T, Schmidt RE. Characteristics and uses of natural killer cells. Immunol Today 1988;9:291–3. Elsevier Current Trends.

[8] Wisløff F, Frøland SS, Michaelsen TE. Antibody-sependent cytotoxicity mediated by human Fc-receptor-bearing cells lacking markers for B- and T-lymphocytes. Int Arch Allergy Immunol 1974;47:139–54.

[9] Van Boxel JA, Stobo JD, Paul WE, Green I. Antibody-dependent lymphoid cell-mediated cytotoxicity: no requirement for thymus-derived lymphocytes. Science (80–) 1972;175:194–6. American Association for the Advancement of Science.

[10] Carrega P, Ferlazzo G. Natural killer cell distribution and trafficking in human tissues. Front Immunol 2012;3:1–6.

[11] Kärre K, Ljunggren HG, Piontek G, Kiessling R. Selective rejection of H-2-deficient lymphoma variants suggests alternative immune defence strategy. Nature 1986;319:675–8. Nature Publishing Group.

[12] Mingari MC, Vitale C, Cantoni C, Bellomo R, Ponte M, Schiavetti F, et al. Interleukin-15-induced maturation of human natural killer cells from early thymic precursors: selective expression of CD94/NKG2-A as the only HLA class I-specific inhibitory receptor. Eur J Immunol 1997;27:1374–80. Wiley-VCH Verlag.

[13] Moretta A, Vitale M, Bottino C, Orengo AM, Morelli L, Augugliaro R, et al. P58 molecules as putative receptors for major histocompatibility complex (MHC) class I molecules in human natural killer (NK) cells. Anti-p58 antibodies reconstitute lysis of MHC class I-protected cells in NK clones displaying different specificities. J Exp Med 1993;178:597–604. Rockefeller University Press.

[14] Fan QR, Mosyak L, Winter CC, Wagtmannt N, Long EO, Wiley DC. Structure of the inhibitory receptor for human natural killer cells resembles haematopoietic receptors. Nature 1997;389:96–100. Nature Publishing Group.

[15] Sivori S, Vitale M, Morelli L, Sanseverino L, Augugliaro R, Bottino C, et al. p46, a novel natural killer cell-specific surface molecule that mediates cell activation. J Exp Med 1997;186:1129–36. Rockefeller University Press.
[16] Pende D, Parolini S, Pessino A, Sivori S, Augugliaro R, Morelli L, et al. Identification and molecular characterization of NKp30, a novel triggering receptor involved in natural cytotoxicity mediated by human natural killer cells. J Exp Med 1999;190:1505–16.
[17] Vitale M, Bottino C, Sivori S, Sanseverino L, Castriconi R, Marcenaro E, et al. NKp44, a novel triggering surface molecule specifically expressed by activated natural killer cells, is involved in non-major histocompatibility complex-restricted tumor cell lysis. J Exp Med 1998;187:2065–72.
[18] Houchins JP, Yabe T, McSherry C, Bach FH. DNA sequence analysis of NKG2, a family of related cDNA clones encoding type II integral membrane proteins on human natural killer cells. J Exp Med 1991;173:1017–20.
[19] Papamichail M, Perez SA, Gritzapis AD, Baxevanis CN. Natural killer lymphocytes: biology, development, and function. Cancer Immunol Immunother 2004;176–86. Springer.
[20] Blink EJ, Trapani JA, Jans DA. Perforin-dependent nuclear targeting of granzymes: a central role in the nuclear events of granule-exocytosis-mediated apoptosis? Immunol Cell Biol 1999;77:206–15.
[21] Konjar Š, Sutton VR, Hoves S, Repnik U, Yagita H, Reinheckel T, et al. Human and mouse perforin are processed in part through cleavage by the lysosomal cysteine proteinase cathepsin L. Immunology 2010;131:257–67.
[22] D'Angelo ME, Bird PI, Peters C, Reinheckel T, Trapaniand JA, Sutton VR. Cathepsin H is an additional convertase of pro-granzyme B. J Biol Chem 2010;285:20514–9. American Society for Biochemistry and Molecular Biology.
[23] Thiele DL, Lipsky PE. The action of leucyl-leucine methyl ester on cytotoxic lymphocytes requires uptake by a novel dipeptide-specific facilitated transport system and dipeptidyl peptidase I-mediated conversion to membranolytic products. Enzym Protein 1990;172.
[24] Pham CTN, Ley TJ. Dipeptidyl peptidase I is required for the processing and activation of granzymes A and B in vivo. Proc Natl Acad Sci U S A 1999;96:8627–32. National Academy of Sciences.
[25] Perišić Nanut M, Sabotič J, Švajger U, Jewett A, Kos J. Cystatin F affects natural killer cell cytotoxicity. Front Immunol 2017;8:1459. Frontiers.
[26] Ashkenazi A. Targeting death and decoy receptors of the tumour-necrosis factor superfamily. Nat Rev Cancer 2002;420–30. European Association for Cardio-Thoracic Surgery.
[27] Cooper MA, Fehniger TA, Caligiuri MA. The biology of human natural killer-cell subsets. Trends Immunol 2001;22:633–40.
[28] Wu J, Lanier LL. Natural killer cells and cancer. Adv Cancer Res 2003;90:127–56. Academic Press Inc.
[29] Biron CA, Nguyen KB, Pien GC, Cousens LP, Salazar-Mather TP. Natural killer cells in antiviral defense: function and regulation by innate cytokines. Annu Rev Immunol 1999;189–220.
[30] Fernandez NC, Lozier A, Flament C, Ricciardi-Castagnoli P, Bellet D, Suter M, et al. Dendritic cells directly trigger NK cell functions: cross-talk relevant in innate anti-tumor immune responses in vivo. Nat Med 1999;5:405–11. Nature Publishing Group.
[31] Kaplan DH, Shankaran V, Dighe AS, Stockert E, Aguet M, Old LJ, et al. Demonstration of an interferon γ-dependent tumor surveillance system in immunocompetent mice. Proc Natl Acad Sci U S A 1998;95:7556–61. National Academy of Sciences.
[32] Chin YE, Kitagawa M, Su WCS, You ZH, Iwamoto Y, Fu XY. Cell growth arrest and induction of cyclin-dependent kinase inhibitor p21WAF1/CIP1 mediated by STAT1. Science (80–) 1996;272:719–22. American Association for the Advancement of Science.
[33] Boehm U, Klamp T, Groot M, Howard JC. Cellular responses to interferon-γ. Annu Rev Immunol 1997;749–95.
[34] Kiessling R, Petranyi G, Kärre K, Jondal M, Tracey D, Wigzell H. Killer cells: a functional comparison between natural, immune T-cell and antibody-dependent in vitro systems. J Exp Med 1976;143:772–80.
[35] Grossenbacher SK, Canter RJ, Murphy WJ. Natural killer cell immunotherapy to target stem-like tumor cells. J Immunother Cancer 2016;19. BioMed Central Ltd.
[36] Jewett A, Tseng H-C, Arasteh A, Saadat S, Christensen RE, Cacalano NA. Natural killer cells preferentially target cancer stem cells: role of monocytes in protection against NK cell mediated lysis of cancer stem cells. Curr Drug Deliv 2012;9:5–16.
[37] Tallerico R, Todaro M, Di Franco S, Maccalli C, Garofalo C, Sottile R, et al. Human NK cells selective targeting of colon cancer-initiating cells: a role for natural cytotoxicity receptors and MHC class I molecules. J Immunol 2013;190:2381–90.

[38] Castriconi R, Daga A, Dondero A, Zona G, Poliani PL, Melotti A, et al. NK cells recognize and kill human glioblastoma cells with stem cell-like properties. J Immunol 2009;182:3530–9. The American Association of Immunologists.
[39] Vivier E, Artis D, Colonna M, Diefenbach A, Di Santo JP, Eberl G, et al. Innate lymphoid cells: 10 years on. Cell 2018;1054–66. Cell Press.
[40] Cording S, Medvedovic J, Aychek T, Eberl G. Innate lymphoid cells in defense, immunopathology and immunotherapy. Nat Immunol 2016;17(7):755–7.
[41] Gasteiger G, Fan X, Dikiy S, Lee SY, Rudensky AY. Tissue residency of innate lymphoid cells in lymphoid and nonlymphoid organs. Science (80–) 2015;350:981–5. American Association for the Advancement of Science.
[42] Artis D, Spits H. The biology of innate lymphoid cells. Nature 2015;293–301. Nature Publishing Group.
[43] Klose CSN, Artis D. Innate lymphoid cells as regulators of immunity, inflammation and tissue homeostasis. Nat Immunol 2016;765–74. Nature Publishing Group.
[44] Klose CSN, Flach M, Möhle L, Rogell L, Hoyler T, Ebert K, et al. Differentiation of type 1 ILCs from a common progenitor to all helper-like innate lymphoid cell lineages. Cell 2014;157:340–56. Cell Press.
[45] Abt MC, Lewis BB, Caballero S, Xiong H, Carter RA, Susac B, et al. Innate immune defenses mediated by two ILC subsets are critical for protection against acute Clostridium difficile infection. Cell Host Microbe 2015;18:27–37. Cell Press.
[46] Hepworth MR, Monticelli LA, Fung TC, Ziegler CGK, Grunberg S, Sinha R, et al. Innate lymphoid cells regulate CD4 + T-cell responses to intestinal commensal bacteria. Nature 2013;498:113–7. Nature Publishing Group.
[47] Oliphant CJ, Hwang YY, Walker JA, Salimi M, Wong SH, Brewer JM, et al. MHCII-mediated dialog between group 2 innate lymphoid cells and CD4+ T cells potentiates type 2 immunity and promotes parasitic helminth expulsion. Immunity 2014;41:283–95. Cell Press.
[48] Mebius RE, Rennert P, Weissman IL. Developing lymph nodes collect CD4+CD3- LTβ+ cells that can differentiate to APC, NK cells, and follicular cells but not T or B cells. Immunity 1997;7:493–504. Cell Press.
[49] Bruchard M, Ghiringhelli F. Deciphering the roles of innate lymphoid cells in cancer. Front Immunol 2019;10:656. Frontiers Media S.A.
[50] Bernink JH, Peters CP, Munneke M, Te Velde AA, Meijer SL, Weijer K, et al. Human type 1 innate lymphoid cells accumulate in inflamed mucosal tissues. Nat Immunol 2013;14:221–9. Nature Publishing Group.
[51] Gao Y, Souza-Fonseca-Guimaraes F, Bald T, Ng SS, Young A, Ngiow SF, et al. Tumor immunoevasion by the conversion of effector NK cells into type 1 innate lymphoid cells. Nat Immunol 2017;18:1004–15. Nature Publishing Group.
[52] Bald T, Wagner M, Gao Y, Koyasu S, Smyth MJ. Hide and seek: plasticity of innate lymphoid cells in cancer. Semin Immunol 2019;101273. Academic Press.
[53] Cortez VS, Ulland TK, Cervantes-Barragan L, Bando JK, Robinette ML, Wang Q, et al. SMAD4 impedes the conversion of NK cells into ILC1-like cells by curtailing non-canonical TGF-β signaling. Nat Immunol 2017;18:995–1003. Nature Publishing Group.
[54] Suzuki M, Morita R, Hirata Y, Shichita T, Yoshimura A. Spred1, a suppressor of the Ras–ERK pathway, negatively regulates expansion and function of group 2 innate lymphoid cells. J Immunol 2015;195:1273–81. The American Association of Immunologists.
[55] Bie Q, Zhang P, Su Z, Zheng D, Ying X, Wu Y, et al. Polarization of ILC2s in peripheral blood might contribute to immunosuppressive microenvironment in patients with gastric cancer. J Immunol Res 2014;2014. Hindawi Publishing Corporation.
[56] Trabanelli S, Chevalier MF, Martinez-Usatorre A, Gomez-Cadena A, Salomé B, Lecciso M, et al. Tumour-derived PGD2 and NKp30-B7H6 engagement drives an immunosuppressive ILC2-MDSC axis. Nat Commun 2017;8:1–14. Nature Publishing Group.
[57] Salimi M, Wang R, Yao X, Li X, Wang X, Hu Y, et al. Activated innate lymphoid cell populations accumulate in human tumour tissues. BMC Cancer 2018;18:341. BioMed Central Ltd.
[58] Busser B, Sancey L, Brambilla E, Coll JL, Hurbin A. The multiple roles of amphiregulin in human cancer. Biochim Biophys Acta-Rev Cancer 2011;119–31. Elsevier B.V.
[59] Kirchberger S, Royston DJ, Boulard O, Thornton E, Franchini F, Szabady RL, et al. Innate lymphoid cells sustain colon cancer through production of interleukin-22 in a mouse model. J Exp Med 2013;210:917–31. The Rockefeller University Press.

[60] Irshad S, Flores-Borja F, Lawler K, Monypenny J, Evans R, Male V, et al. RORγt+ innate lymphoid cells promote lymph node metastasis of breast cancers. Cancer Res 2017;77:1083–96. American Association for Cancer Research Inc.
[61] Nussbaum K, Burkhard SH, Ohs I, Mair F, Klose CSN, Arnold SJ, et al. Tissue microenvironment dictates the fate and tumorsuppressive function of type 3 ILCs. J Exp Med 2017;214:2331–47. Rockefeller University Press.
[62] Elliott JM, Yokoyama WM. Unifying concepts of MHC-dependent natural killer cell education. Trends Immunol 2011;32:364–72. Elsevier Ltd.
[63] Tseng HC, Cacalano N, Jewett A. Split anergized natural killer cells halt inflammation by inducing stem cell differentiation, resistance to NK cell cytotoxicity and prevention of cytokine and chemokine secretion. Oncotarget 2015;6:8947–59.
[64] Jewett A, Cavalcanti M, Bonavida B. Pivotal role of endogenous TNF-alpha in the induction of functional inactivation and apoptosis in NK cells. J Immunol 1997;159.
[65] Jewett A, Teruel A, Romero M, Head C, Cacalano N. Rapid and potent induction of cell death and loss of NK cell cytotoxicity against oral tumors by F(ab′)2 fragment of anti-CD16 antibody. Cancer Immunol Immunother 2008;57:1053–66.
[66] Magister Š, Tseng H-C, Bui VT, Kos J, Jewett A. Regulation of split anergy in natural killer cells by inhibition of cathepsins C and H and cystatin F. Oncotarget 2015;6:22310–27.
[67] Kaur K, Nanut MP, Ko M-W, Safaie T, Kos J, Jewett A. Natural killer cells target and differentiate cancer stem-like cells/undifferentiated tumors: strategies to optimize their growth and expansion for effective cancer immunotherapy. Curr Opin Immunol 2018;51:170–80. Elsevier Current Trends.
[68] Tseng HC, Bui V, Man YG, Cacalano N, Jewett A. Induction of split anergy conditions natural killer cells to promote differentiation of stem cells through cell-cell contact and secreted factors. Front Immunol 2014;5. Frontiers Research Foundation.
[69] Tseng HC, Inagaki A, Bui VT, Cacalano N, Kasahara N, Man YG, et al. Differential targeting of stem cells and differentiated glioblastomas by NK cells. J Cancer 2015;6:866–76.
[70] Kozlowska A, Topchyan P, Kaur K, Tseng HC, Teruel A, Hiraga T, et al. Differentiation by NK cells is a prerequisite for effective targeting of cancer stem cells/poorly differentiated tumors by chemopreventive and chemotherapeutic drugs. J Cancer 2017;8:537–54.
[71] Jewett A, Man Y, Cacalano N, Kos J, Tseng H-C. Natural killer cells as effectors of selection and differentiation of stem cells: role in resolution of inflammation. J Immunotoxicol 2014;11:297–307.
[72] Bui VT, Tseng HC, Kozlowska A, Maung PO, Kaur K, Topchyan P, et al. Augmented IFN-γ and TNF-α induced by probiotic bacteria in NK cells mediate differentiation of stem-like tumors leading to inhibition of tumor growth and reduction in inflammatory cytokine release; regulation by IL-10. Front Immunol 2015;6:2. Frontiers Media S.A.
[73] Jewett A, Kos J, Kaur K, Turnsek TL, Breznik B, Senjor E, et al. Multiple defects of natural killer cells in cancer patients: anarchy, dysregulated systemic immunity, and immunosuppression in metastatic cancer. Crit Rev Immunol 2020;40:93–133.
[74] Jewett A, Nakamura H, Wang M, Teruel A, Paranjpe A, Romero M, et al. Dedifferentiation of epithelial tumors enhances cytotoxicity, survival and expansion of allogeneic CD8+ T cells and natural killer cells. J Carcinogen Mutagen 2012;, S1:007.
[75] Jewett A, Wang M-YY, Teruel A, Poupak Z, Bostanian Z, NHN-H P. Cytokine dependent inverse regulation of CD54 (ICAM1) and major histocompatibility complex class I antigens by nuclear factor κB in HEp2 tumor cell line: effect on the function of natural killer cells. Hum Immunol 2003;64:505–20. Elsevier Inc.
[76] Tseng HC, Arasteh A, Kaur K, Kozlowska A, Topchyan P, Jewett A. Differential cytotoxicity but augmented IFN-γ secretion by NK cells after interaction with monocytes from humans, and those from wild type and myeloid-specific COX-2 knockout mice. Front Immunol 2015;6:1–21.
[77] Takasugi M, Mickey MR, Terasaki PI. Reactivity of lymphocytes from normal persons on cultured tumor cells. Cancer Res 1973;33.
[78] Takasugi M, Ramseyer A, Takasugi J. Decline of natural nonselective cell-mediated cytotoxicity in patients with tumor progression. Cancer Res 1977;37.
[79] Balch CM, Tilden AB, Dougherty PA, Cloud GA. Depressed levels of granular lymphocytes with natural killer (NK) cell function in 247 cancer patients. Ann Surg 1983;198(2):192–9.
[80] Forbes JT, Greco FA, Oldham RK. Human natural cell-mediated cytotoxicity—II. Levels in neoplastic disease. Cancer Immunol Immunother 1981;11:147–53. Springer-Verlag.

[81] Imai K, Matsuyama S, Miyake S, Suga K, Nakachi K. Natural cytotoxic activity of peripheral-blood lymphocytes and cancer incidence: an 11-year follow-up study of a general population. Lancet 2000;356:1795–9.
[82] Peng YP, Zhu Y, Zhang JJ, Xu ZK, Qian ZY, Dai CC, et al. Comprehensive analysis of the percentage of surface receptors and cytotoxic granules positive natural killer cells in patients with pancreatic cancer, gastric cancer, and colorectal cancer. J Transl Med 2013;11:262. BioMed Central.
[83] Shurin MR, Umansky V, Malyguine A, Hurwitz AA, Apte RN, Whiteside T, et al. Cellular and molecular pathways in the tumor immunoenvironment: 3rd cancer immunotherapy and immunomonitoring (CITIM) meeting, 22–25 April 2013, Krakow, Poland. Cancer Immunol Immunother 2014;73–80. Springer.
[84] Kaur K, Cook J, Park SH, Topchyan P, Kozlowska A, Ohanian N, et al. Novel strategy to expand super-charged NK cells with significant potential to lyse and differentiate cancer stem cells: differences in NK expansion and function between healthy and cancer patients. Front Immunol 2017;8:297.
[85] Paul S, Kulkarni N, Shilpi N, Lal G. Intratumoral natural killer cells show reduced effector and cytolytic properties and control the differentiation of effector Th1 cells. Oncoimmunology 2016;5:e1235106. Taylor and Francis Inc.
[86] Mamessier E, Sylvain A, Thibult ML, Houvenaeghel G, Jacquemier J, Castellano R, et al. Human breast cancer cells enhance self tolerance by promoting evasion from NK cell antitumor immunity. J Clin Invest 2011;121:3609–22.
[87] Platonova S, Cherfils-Vicini J, Damotte D, Crozet L, Vieillard V, Validire P, et al. Profound coordinated alterations of intratumoral NK cell phenotype and function in lung carcinoma. Cancer Res 2011;71:5412–22. American Association for Cancer Research.
[88] Pasero C, Elle Gravis G, Guerin M, Granjeaud S, Thomassin-Piana J, Rocchi P, et al. Inherent and tumor-driven immune tolerance in the prostate microenvironment impairs natural killer cell antitumor activity. Cancer Res 2016;76(8):2153–65.
[89] Carlsten M, Norell H, Bryceson YT, Poschke I, Schedvins K, Ljunggren H-G, et al. Primary human tumor cells expressing CD155 impair tumor targeting by down-regulating DNAM-1 on NK cells. J Immunol 2009;183:4921–30. The American Association of Immunologists.
[90] Saito H, Osaki T, Ikeguchi M. Decreased NKG2D expression on NK cells correlates with impaired NK cell function in patients with gastric cancer. Gastric Cancer 2012;15:27–33. Springer.
[91] Hodge G, Barnawi J, Jurisevic C, Moffat D, Holmes M, Reynolds PN, et al. Lung cancer is associated with decreased expression of perforin, granzyme B and interferon (IFN)-γ by infiltrating lung tissue T cells, natural killer (NK) T-like and NK cells. Clin Exp Immunol 2014;178:79–85. Blackwell Publishing Ltd.
[92] Sun C, Xu J, Huang Q, Huang M, Wen H, Zhang C, et al. High NKG2A expression contributes to NK cell exhaustion and predicts a poor prognosis of patients with liver cancer. Oncoimmunology 2017;6.
[93] Bi J, Tian Z. NK cell exhaustion. Front Immunol 2017;760. Frontiers Media S.A.
[94] Pallmer K, Oxenius A. Recognition and regulation of T cells by NK cells. Front Immunol 2016;1. Frontiers Research Foundation.
[95] Moretta A, Marcenaro E, Sivori S, Della Chiesa M, Vitale M, Moretta L. Early liaisons between cells of the innate immune system in inflamed peripheral tissues. Trends Immunol 2005;668–75. Elsevier Ltd.
[96] Kaur K, Ko MW, Ohanian N, Cook J, Jewett A. Osteoclast-expanded super-charged NK-cells preferentially select and expand CD8+ T cells. Sci Rep 2020;10. Nature Research.
[97] Vitale M, Cantoni C, Pietra G, Mingari MC, Moretta L. Effect of tumor cells and tumor microenvironment on NK-cell function. Eur J Immunol 2014;1582–92. Wiley-VCH Verlag.
[98] Li H, Han Y, Guo Q, Zhang M, Cao X. Cancer-expanded myeloid-derived suppressor cells induce anergy of NK cells through membrane-bound TGF-β1. J Immunol 2009;182:240–9. The American Association of Immunologists.
[99] Trzonkowski P, Szmit E, Myśliwska J, Dobyszuk A, Myśliwski A. CD4+CD25+ T regulatory cells inhibit cytotoxic activity of T CD8+ and NK lymphocytes in the direct cell-to-cell interaction. Clin Immunol 2004;112:258–67.
[100] Cekic C, Day YJ, Sag D, Linden J. Myeloid expression of adenosine A2A receptor suppresses T and NK cell responses in the solid tumor microenvironment. Cancer Res 2014;74:7250–9.
[101] Costa D, Venè R, Benelli R, Romairone E, Scabini S, Catellani S, et al. Targeting the epidermal growth factor receptor can counteract the inhibition of natural killer cell function exerted by colorectal tumor-associated fibroblasts. Front Immunol 2018;9. Frontiers Media S.A.
[102] Otegbeye F, Ojo E, Moreton S, Mackowski N, Lee DA, de Lima M, et al. Inhibiting TGF-beta signaling preserves the function of highly activated, in vitro expanded natural killer cells in AML and colon cancer models. Gangopadhyay N, editor. PLoS One 2018;13:e0191358. Public Library of Science.

[103] Lazarova M, Steinle A. Impairment of NKG2D-mediated tumor immunity by TGF-β. Front Immunol 2019;2689. Frontiers Media S.A.
[104] Galland S, Vuille J, Martin P, Letovanec I, Caignard A, Fregni G, et al. Tumor-derived mesenchymal stem cells use distinct mechanisms to block the activity of natural killer cell subsets. Cell Rep 2017;20:2891–905. Elsevier B.V.
[105] Guillerey C, Huntington ND, Smyth MJ. Targeting natural killer cells in cancer immunotherapy. Nat Immunol 2016;1025–36. Nature Publishing Group.
[106] Della Chiesa M, Carlomagno S, Frumento G, Balsamo M, Cantoni C, Conte R, et al. The tryptophan catabolite L-kynurenine inhibits the surface expression of NKp46- and NKG2D-activating receptors and regulates NK-cell function. Blood 2006;108:4118–25. American Society of Hematology.
[107] Pietra G, Manzini C, Rivara S, Vitale M, Cantoni C, Petretto A, et al. Melanoma cells inhibit natural killer cell function by modulating the expression of activating receptors and cytolytic activity. Cancer Res 2012;72:1407–15. American Association for Cancer Research.
[108] Balsamo M, Scordamaglia F, Pietra G, Manzini C, Cantoni C, Boitano M, et al. Melanoma-associated fibroblasts modulate NK cell phenotype and antitumor cytotoxicity. Proc Natl Acad Sci U S A 2009;106:20847–52. National Academy of Sciences.
[109] Hoechst B, Voigtlaender T, Ormandy L, Gamrekelashvili J, Zhao F, Wedemeyer H, et al. Myeloid derived suppressor cells inhibit natural killer cells in patients with hepatocellular carcinoma via the NKp30 receptor. Hepatology 2009;50:799–807.
[110] Ghiringhelli F, Ménard C, Martin F, Zitvogel L. The role of regulatory T cells in the control of natural killer cells: relevance during tumor progression. Immunol Rev 2006;229–38.
[111] Balsamo M, Manzini C, Pietra G, Raggi F, Blengio F, Mingari MC, et al. Hypoxia downregulates the expression of activating receptors involved in NK-cell-mediated target cell killing without affecting ADCC. Eur J Immunol 2013;43:2756–64.
[112] Chulpanova DS, Kitaeva KV, Rutland CS, Rizvanov AA, Solovyeva VV. Mouse tumor models for advanced cancer immunotherapy. Int J Mol Sci 2020;1–15. MDPI AG.
[113] Kozlowska AK, Kaur K, Topchyan P, Jewett A. Novel strategies to target cancer stem cells by NK cells: studies in humanized mice. Front Biosci-Landmark 2017;22:370–84. Frontiers in Bioscience.
[114] Vatakis DN, Bristol GC, Kim SG, Levin B, Liu W, Radu CG, et al. Using the BLT humanized mouse as a stem cell based gene therapy tumor model. J Vis Exp 2012;, e4181.
[115] Kaur K, Kozlowska AK, Topchyan P, Ko MW, Ohanian N, Chiang J, et al. Probiotic-treated super-charged NK cells efficiently clear poorly differentiated pancreatic tumors in Hu-BLT mice. Cancers (Basel) 2019;12:63.
[116] Kaur K, Topchyan P, Kozlowska AK, Ohanian N, Chiang J, Maung PO, et al. Super-charged NK cells inhibit growth and progression of stem-like/poorly differentiated oral tumors in vivo in humanized BLT mice: effect on tumor differentiation and response to chemotherapeutic drugs. Oncoimmunology 2018;7:1–14. Taylor & Francis.
[117] Tseng HC, Kanayama K, Kaur K, Park SH, Park S, Kozlowska A, et al. Bisphosphonate-induced differential modulation of immune cell function in gingiva and bone marrow in vivo: role in osteoclastmediated NK cell activation. Oncotarget 2015;6:20002–25. Impact Journals LLC.
[118] Jewett A, Kos J, Kaur K, Safaei T, Sutanto C, Chen W, et al. Natural killer cells: diverse functions in tumor immunity and defects in pre-neoplastic and neoplastic stages of tumorigenesis. Mol Ther Oncolytics 2020;41–52. Cell Press.
[119] Sivori S, Falco M, Della Chiesa M, Carlomagno S, Vitale M, Moretta L, et al. CpG and double-stranded RNA trigger human NK cells by toll-like receptors: induction of cytokine release and cytotoxicity against tumors dendritic cells. Proc Natl Acad Sci U S A 2004;101:10116–21. National Academy of Sciences.
[120] Kaur K, Safaie T, Ko MW, Wang Y, Jewett A. ADCC against MICA/B is mediated against differentiated oral and pancreatic and not stem-like/poorly differentiated tumors by the NK cells: loss in cancer patients due to down-modulation of cd16 receptor. Cancers (Basel) 2021;13:1–19. MDPI AG.
[121] Parkhurst MR, Riley JP, Dudley ME, Rosenberg SA. Adoptive transfer of autologous natural killer cells leads to high levels of circulating natural killer cells but does not mediate tumor regression. Clin Cancer Res 2011;17:6287–97.
[122] Klingemann H. Challenges of cancer therapy with natural killer cells. Cytotherapy 2015;245–9. Elsevier Inc.
[123] Gill S, June CH. Going viral: chimeric antigen receptor T-cell therapy for hematological malignancies. Immunol Rev 2015;263:68–89. Blackwell Publishing Ltd.
[124] Herberman RB, Holden HT. Natural cell-mediated immunity. Adv Cancer Res 1978;27:305–77. Academic Press.

[125] Strauss-Albee DM, Liang EC, Ranganath T, Aziz N, Blish CA. The newborn human NK cell repertoire is phenotypically formed but functionally reduced. Cytom Part B Clin Cytom 2017;92:33–41. John Wiley and Sons Inc.
[126] Campbell AC, Waller C, Wood J. Lymphocyte subpopulations in the blood of newborn infants. Clin Exp Immunol 1974;18:469–82. Wiley-Blackwell.
[127] Gong JH, Maki G, Klingemann HG. Characterization of a human cell line (NK-92) with phenotypical and functional characteristics of activated natural killer cells. Leukemia 1994;8:652–8. Nature Publishing Group.
[128] Boissel L, Betancur M, Lu W, Wels WS, Marino T, Van Etten RA, et al. Comparison of mRNA and lentiviral based transfection of natural killer cells with chimeric antigen receptors recognizing lymphoid antigens. Leuk Lymphoma 2012;53:958–65.
[129] Khallouf H, Märten A, Serba S, Teichgräber V, Büchler MW, Jäger D, et al. 5-Fluorouracil and interferon-α immunochemotherapy enhances immunogenicity of murine pancreatic cancer through upregulation of NKG2D ligands and MHC class I. J Immunother 2012;35:245–53.
[130] Kim H, Kim SH, Kim MJ, Kim SJ, Park SJ, Chung JS, et al. EGFR inhibitors enhanced the susceptibility to NK cell-mediated lysis of lung cancer cells. J Immunother 2011;34:372–81.
[131] Hodgins JJ, Khan ST, Park MM, Auer RC, Ardolino M. Killers 2.0: NK cell therapies at the forefront of cancer control. J Clin Invest 2019;3499–510. American Society for Clinical Investigation.
[132] Ladoire S, Arnould L, Apetoh L, Coudert B, Martin F, Chauffert B, et al. Pathologic complete response to neoadjuvant chemotherapy of breast carcinoma is associated with the disappearance of tumor-infiltrating Foxp3 + regulatory T cells. Clin Cancer Res 2008;14:2413–20.
[133] Davis M, Conlon K, Bohac GC, Barcenas J, Leslie W, Watkins L, et al. Effect of pemetrexed on innate immune killer cells and adaptive immune T cells in subjects with adenocarcinoma of the pancreas. J Immunother 2012;35:629–40.
[134] Ménard C, Blay JY, Borg C, Michiels S, Ghiringhelli F, Robert C, et al. Natural killer cell IFN-γ levels predict long-term survival with imatinib mesylate therapy in gastrointestinal stromal tumor-bearing patients. Cancer Res 2009;69:3563–9.
[135] Riva G, Luppi M, Lagreca I, Barozzi P, Quadrelli C, Vallerini D, et al. Long-term molecular remission with persistence of BCR-ABL1-specific cytotoxic T cells following imatinib withdrawal in an elderly patient with Philadelphia-positive ALL. Br J Haematol 2014;299–302.
[136] Galluzzi L, Zitvogel L, Kroemer G. Immunological mechanisms underneath the efficacy of cancer therapy. Cancer Immunol Res 2016;4.
[137] Jewett A, Cacalano NA, Teruel A, Romero M, Rashedi M, Wang M, et al. Inhibition of nuclear factor kappa B (NFκB) activity in oral tumor cells prevents depletion of NK cells and increases their functional activation. Cancer Immunol Immunother 2006;55:1052–63. Springer.
[138] Alison MR, Lin WR, Lim SML, Nicholson LJ. Cancer stem cells: in the line of fire. Cancer Treat Rev 2012;589–98.
[139] Zingoni A, Fionda C, Borrelli C, Cippitelli M, Santoni A, Soriani A. Natural killer cell response to chemotherapy-stressed cancer cells: role in tumor immunosurveillance. Front Immunol 2017. Frontiers Media S.A.
[140] Gerdes CA, Nicolini VG, Herter S, Van Puijenbroek E, Lang S, Roemmele M, et al. GA201 (RG7160): a novel, humanized, glycoengineered anti-EGFR antibody with enhanced ADCC and superior in vivo efficacy compared with cetuximab. Clin Cancer Res 2013;19:1126–38. American Association for Cancer Research.
[141] Yang RK, Kalogriopoulos NA, Rakhmilevich AL, Ranheim EA, Seo S, Kim K, et al. Intratumoral treatment of smaller mouse neuroblastoma tumors with a recombinant protein consisting of IL-2 linked to the Hu14.18 antibody increases intratumoral CD8+ T and NK cells and improves survival. Cancer Immunol Immunother 2013;62:1303–13.
[142] Gluck WL, Hurst D, Yuen A, Levine AM, Dayton MA, Gockerman JP, et al. Phase I studies of interleukin (IL)-2 and rituximab in B-cell non-Hodgkin's lymphoma: IL-2 mediated natural killer cell expansion correlations with clinical response. Clin Cancer Res 2004;10:2253–64.
[143] Wu J, Mishra HK, Walcheck B. Role of ADAM17 as a regulatory checkpoint of CD16A in NK cells and as a potential target for cancer immunotherapy. J Leukoc Biol 2019;105:1297–303. John Wiley and Sons Inc.
[144] Romee R, Foley B, Lenvik T, Wang Y, Zhang B, Ankarlo D, et al. NK cell CD16 surface expression and function is regulated by a disintegrin and metalloprotease-17 (ADAM17). Blood 2013;121:3599–608. American Society of Hematology.
[145] Gleason MK, Verneris MR, Todhunter DA, Zhang B, McCullar V, Zhou SX, et al. Bispecific and trispecific killer cell engagers directly activate human NK cells through CD16 signaling and induce cytotoxicity and cytokine production. Mol Cancer Ther 2012;11:2674–84.

[146] Schmohl JU, Gleason MK, Dougherty PR, Miller JS, Vallera DA. Heterodimeric bispecific single chain variable fragments (scFv) killer engagers (BiKEs) enhance NK-cell activity against CD133+ colorectal cancer cells. Target Oncol 2016;11:353–61. Springer-Verlag France.
[147] Wiernik A, Foley B, Zhang B, Verneris MR, Warlick E, Gleason MK, et al. Targeting natural killer cells to acute myeloid leukemia in vitro with a CD16 × 33 bispecific killer cell engager and ADAM17 inhibition. Clin Cancer Res 2013;19:3844–55.
[148] Vallera DA, Felices M, McElmurry R, McCullar V, Zhou X, Schmohl JU, et al. IL15 trispecific killer engagers (TriKE) make natural killer cells specific to CD33+ targets while also inducing persistence, in vivo expansion, and enhanced function. Clin Cancer Res 2016;22:3440–50. American Association for Cancer Research Inc.
[149] Boltz A, Piater B, Toleikis L, Guenther R, Kolmar H, Hock B. Bi-specific aptamers mediating tumor cell lysis. J Biol Chem 2011;286:21896–905.
[150] Federico SM, McCarville MB, Shulkin BL, Sondel PM, Hank JA, Hutson P, et al. A pilot trial of humanized anti-GD2 monoclonal antibody (hu14.18K322A) with chemotherapy and natural killer cells in children with recurrent/refractory neuroblastoma. Clin Cancer Res 2017;23:6441–9. American Association for Cancer Research Inc.
[151] Modak S, Le Luduec JB, Cheung IY, Goldman DA, Ostrovnaya I, Doubrovina E, et al. Adoptive immunotherapy with haploidentical natural killer cells and anti-GD2 monoclonal antibody m3F8 for resistant neuroblastoma: results of a phase I study. Oncoimmunology 2018;7. Taylor and Francis Inc.
[152] Bachanova V, Sarhan D, DeFor TE, Cooley S, Panoskaltsis-Mortari A, Blazar BR, et al. Haploidentical natural killer cells induce remissions in non-Hodgkin lymphoma patients with low levels of immune-suppressor cells. Cancer Immunol Immunother 2018;67:483–94. Springer Science and Business Media Deutschland GmbH.
[153] Aitken AS, Roy DG, Bourgeois-Daigneault MC. Taking a stab at cancer: oncolytic virus-mediated anti-cancer vaccination strategies. Biomedicines 2017;5(1):3. MDPI AG.
[154] Shi T, Song X, Wang Y, Liu F, Wei J. Combining oncolytic viruses with cancer immunotherapy: establishing a new generation of cancer treatment. Front Immunol 2020;683. Frontiers Media S.A.
[155] de Graaf JF, de Vor L, Fouchier RAM, van den Hoogen BG. Armed oncolytic viruses: a kick-start for anti-tumor immunity. Cytokine Growth Factor Rev 2018;28–39. Elsevier Ltd.
[156] Kaufman HL, Kohlhapp FJ, Zloza A. Oncolytic viruses: a new class of immunotherapy drugs. Nat Rev Drug Discov 2015;642–62. Nature Publishing Group.
[157] Choi KJ, Zhang SN, Choi IK, Kim JS, Yun CO. Strengthening of antitumor immune memory and prevention of thymic atrophy mediated by adenovirus expressing IL-12 and GM-CSF. Gene Ther 2012;19:711–23. Nature Publishing Group.
[158] Stephenson KB, Barra NG, Davies E, Ashkar AA, Lichty BD. Expressing human interleukin-15 from oncolytic vesicular stomatitis virus improves survival in a murine metastatic colon adenocarcinoma model through the enhancement of anti-tumor immunity. Cancer Gene Ther 2012;19:238–46.
[159] Alvarez-Breckenridge CA, Yu J, Kaur B, Caligiuri MA, Chiocca EA. Deciphering the multifaceted relationship between oncolytic viruses and natural killer cells. Adv Virol 2012;2012, 702839.
[160] Kim Y, Yoo JY, Lee TJ, Liu J, Yu J, Caligiuri MA, et al. Complex role of NK cells in regulation of oncolytic virus–bortezomib therapy. Proc Natl Acad Sci U S A 2018;115:4927–32. National Academy of Sciences.
[161] Ogbomo H, Zemp FJ, Lun X, Zhang J, Stack D, Rahman MM, et al. Myxoma virus infection promotes NK lysis of malignant gliomas in vitro and in vivo. PLoS One 2013;8:66825. Public Library of Science.
[162] Bhat R, Dempe S, Dinsart C, Rommelaere J. Enhancement of NK cell antitumor responses using an oncolytic parvovirus. Int J Cancer 2011;128:908–19.
[163] Robert C. A decade of immune-checkpoint inhibitors in cancer therapy. Nat Commun 2020;11:3801. Nature Research.
[164] Hsu J, Hodgins JJ, Marathe M, Nicolai CJ, Bourgeois-Daigneault MC, Trevino TN, et al. Contribution of NK cells to immunotherapy mediated by PD-1/PD-L1 blockade. J Clin Invest 2018;128:4654–68. American Society for Clinical Investigation.
[165] Zhang Q, Bi J, Zheng X, Chen Y, Wang H, Wu W, et al. Blockade of the checkpoint receptor TIGIT prevents NK cell exhaustion and elicits potent anti-tumor immunity. Nat Immunol 2018;19:723–32. Nature Publishing Group.
[166] Da Silva IP, Gallois A, Jimenez-Baranda S, Khan S, Anderson AC, Kuchroo VK, et al. Reversal of NK-cell exhaustion in advanced melanoma by Tim-3 blockade. Cancer Immunol Res 2014;2:410–22. American Association for Cancer Research Inc.

[167] Braud VM, Allan DSJ, O'Callaghan CA, Soderstrom K, D'Andrea A, Ogg GS, et al. HLA-E binds to natural killer cell receptors CD94/NKG2A, B and C. Nature 1998;391:795–9.

[168] Chan CJ, Martinet L, Gilfillan S, Souza-Fonseca-Guimaraes F, Chow MT, Town L, et al. The receptors CD96 and CD226 oppose each other in the regulation of natural killer cell functions. Nat Immunol 2014;15:431–8. Nature Publishing Group.

[169] Kamiya T, Seow SV, Wong D, Robinson M, Campana D. Blocking expression of inhibitory receptor NKG2A overcomes tumor resistance to NK cells. J Clin Invest 2019;129:2094–106. American Society for Clinical Investigation.

[170] Bi J, Zhang Q, Liang D, Xiong L, Wei H, Sun R, et al. T-cell Ig and ITIM domain regulates natural killer cell activation in murine acute viral hepatitis. Hepatology 2014;59:1715–25. John Wiley and Sons Ltd.

[171] Sun H, Huang Q, Huang M, Wen H, Lin R, Zheng M, et al. Human CD96 correlates to natural killer cell exhaustion and predicts the prognosis of human hepatocellular carcinoma. Hepatology 2019;70:168–83. John Wiley and Sons Inc.

[172] André P, Denis C, Soulas C, Bourbon-Caillet C, Lopez J, Arnoux T, et al. Anti-NKG2A mAb is a checkpoint inhibitor that promotes anti-tumor immunity by unleashing Both T and NK cells. Cell 2018;175:1731–1743. e13. Cell Press.

[173] Juliá EP, Amante A, Pampena MB, Mordoh J, Levy EM. Avelumab, an IgG1 anti-PD-L1 immune checkpoint inhibitor, triggers NK cell-mediated cytotoxicity and cytokine production against triple negative breast cancer cells. Front Immunol 2018;9:2140. Frontiers Media S.A.

CHAPTER

2

Reversing the NK inhibitory tumor microenvironment by targeting suppressive immune effectors

Ishwar Navin[a,b] *and Robin Parihar*[a,b,c]

[a]Department of Immunology and Microbiology, Baylor College of Medicine, Houston, TX, United States [b]Center for Cell and Gene Therapy, Texas Children's Hospital, Houston Methodist Hospital, and Baylor College of Medicine, Houston, TX, United States [c]Section of Hematology/Oncology, Department of Pediatrics, Texas Children's Hospital, Houston, TX, United States

Abstract

The efficacy of natural killer (NK) cell-based therapies against solid tumors has been poor. Solid tumors generate a hostile tumor microenvironment (TME) comprising multiple cell types including inhibitory immune cells that play a key role in depressing the antitumor functions of therapeutic NK cells. Understanding how inhibitory immune effectors operate within the confines of the TME is critical to developing therapies to reverse this inhibition. In this chapter, we cover the suppressive mechanisms employed by TME-resident inhibitory immune cells and discuss methodologies to assess their composition and functionality within the TME. We also highlight novel therapeutic strategies that target inhibitory cells of the TME to improve the antitumor functions of endogenous or adoptively transferred NK cells. Multimodal approaches to overcome inhibitory immune cells within the TME will drive the development of personalized NK cell therapeutics with optimal activity, leading to improved clinical outcomes in patients with solid tumors.

Abbreviations

ADCC	antibody-dependent cellular cytotoxicity
APC	antigen-presenting cell
ARG-1/2	arginase-1/2
ASO	antisense oligonucleotide
BiKE	bispecific killer cell engager
BM	bone marrow
CAF	cancer-associated fibroblast
CAR	chimeric antigen receptor

https://doi.org/10.1016/B978-0-12-822620-9.00011-2

CB	cord blood
ccRCC	clear cell renal carcinoma
cDC	conventional dendritic cell
CRS	cytokine-release syndrome
CSF-1	colony-stimulating factor
DC	dendritic cell
ECAR	extracellular acidification rate
ECM	extracellular matrix
FFFPE	formalin-fixed-paraffin-embedded
FGF	fibroblast growth factor
FOXP3	forkhead box protein 3
GM-CSF	granulocyte-macrophage colony-stimulating factor
GSEA	gene set enrichment analysis
GVHD	graft-vs-host disease
HCC	hepatocellular carcinoma
HDAC	histone deacetylase inhibitor
HSCT	hematopoietic stem cell transplant
ICAM	intercellular adhesion molecule
IDO	indoleamine 2,3-dioxygenase
IF	immunofluorescence
IFNγ	interferon-gamma
IHC	immunohistochemistry
iNOS	inducible nitric oxide synthase
KIR	killer cell Ig-like receptors
KO	knockout
mAb	monoclonal antibody
mb	membrane bound
MDSC	myeloid-derived suppressor cell
MMP	matrix metalloproteinase
MR	magnetic resonance
NK	natural killer
NO	nitric oxide
NOTA	^{64}Cu-labeled 1,4,7-triazacyclononane-triacetic acid
OCR	oxygen consumption rate
OV	oncolytic virus
PD-1	programed cell death
PDAC	pancreatic ductal adenocarcinoma
pDC	plasmacytoid dendritic cell
PDGF	platelet-derived growth factor
PEG-FA-Lip	pegylated-folic acid liposome
PET	positron electron tomography
PGE	prostaglandin E
PMN	polymorphonuclear
PVR	poliovirus receptor
ROS	reactive oxygen species
S1P	sphingosine 1-phosphate
TAM	tumor-associated macrophage
TGFβ	transforming growth factor-beta
TIL	tumor-infiltrating lymphocyte
TLR	toll-like receptor
TME	tumor microenvironment
TNFα	tumor necrosis factor-alpha
Treg	regulator T cell

TriKE	trispecific killer cell engager
TSA	tyramide signal amplification
VCAM	vascular cell adhesion molecule
VEGF	vascular endothelial growth factor

Conflict of interest

No potential conflicts of interest were disclosed.

Introduction

Natural killer (NK) cells are innate effector lymphocytes in the control of viral infections and tumor growth. Unlike T cells, which recognize targets via T cell receptors in an MHC-dependent manner, NK cells use a range of germline encoded surface receptors, activating and inhibiting killer cell Ig-like receptors (KIRs) and interact with targets without the need for prior antigen sensitization. The combination of activating and inhibitory signals originating these activating and inhibitory receptors ultimately determines the target cell's fate. Upon activation, NK cells are capable of exhibiting cytotoxicity and secretion of proinflammatory cytokines (IFN-γ and TNF-α). In humans, two distinct peripheral NK cell subsets have been identified based on CD56 surface expression [1]. The $CD56^{dim}$ NK cells subset is cytotoxic and expresses high levels of the Fcγ receptor CD16, which initiates antibody-dependent cellular cytotoxicity (ADCC). $CD56^{bright}$ NK cells, which represents the majority of NK cells in peripheral blood, specializes in the secretion of cytokines and chemokines in response to IL-12, IL-15, and IL-18 [1,2]. NK cells have been implicated in tumor immunobiology. NK cells that circulate in peripheral blood, and those which reside in the spleen and bone marrow, are both capable of infiltrating other tissues, including tumor. In murine models, transplanted tumors were found to be more aggressive with increased metastasis following NK cell depletion [3]. Additionally, multiple studies have highlighted NK cells as a prognostic marker for tumor progression underscoring its significance in controlling tumors [4–8]. However, adoptive transfer of autologous and allogeneic activated NK cell-based therapies have not resulted in robust clinical outcomes in patients due to the generation of the tumor microenvironment (TME) by solid tumors [9,10]. NK cell activation, cytotoxicity, and persistence are impaired in the TME, which poses various biochemical and physical challenges and lead to NK cell dysfunction. Hence, understanding how the mechanisms by which the TME operates to inhibit the antitumor response of NK cells as well as designing strategies to help NK cells overcome TME-mediated suppression will lead to improved and effective NK cell-based therapies to treat patients with solid tumors. In this chapter, we will review the mechanisms by which the TME suppresses NK cell activity and discuss the role of inhibitory immune effectors in mediating this process. In addition, we will highlight strategies to enhance NK cell-based therapies that target TME-intrinsic and extrinsic inhibitory immune cells and discuss techniques to evaluate NK cell fitness, viability, and antitumor function in the TME. Finally, we will discuss future directions concerning TME-targeting NK cell-based therapies for improved and effective treatment of patients with solid tumors.

Dysregulation of NK cell immune responses by the tumor microenvironment

Solid tumors manipulate and reorganize the areas around them to generate a hostile and immunosuppressive surrounding called the tumor microenvironment (TME) (Fig. 1). The tumor is deeply shaped by the interactions that it makes with the TME which culminates in the establishment, progression, and metastasis of the tumor mass [11,12]. In addition to supporting tumorigenesis, the TME presents unique physical and biochemical challenges to tumor-infiltrating lymphocytes (TILs), including NK cells. It comprises secreted factors like inhibitory cytokines and metabolites which impact NK cell activity and persistence, hypoxic zones which affect NK cell metabolism, and extracellular matrix (ECM) proteins which remold the environment to promote tumor progression.

In addition to immune infiltrates, the TME comprises nonimmune components that affect immune function. ECM proteins in the TME, including matrix metalloproteinase (MMP)-7, cleaves Fas and Fas-L, preventing their interactions as well as cleaving Fcγ receptors on NK cells [13,14]. Endothelial cells influenced by tumor-derived vascular endothelial growth factor (VEGF) lead to the formation of chaotic networks of blood vessels which impacts blood flow [15]. Such vasculature is structurally unstable and leaky and results in limited nutrient delivery to the TME [16,17]. This creates a hypoxic environment and subsequent acidosis in

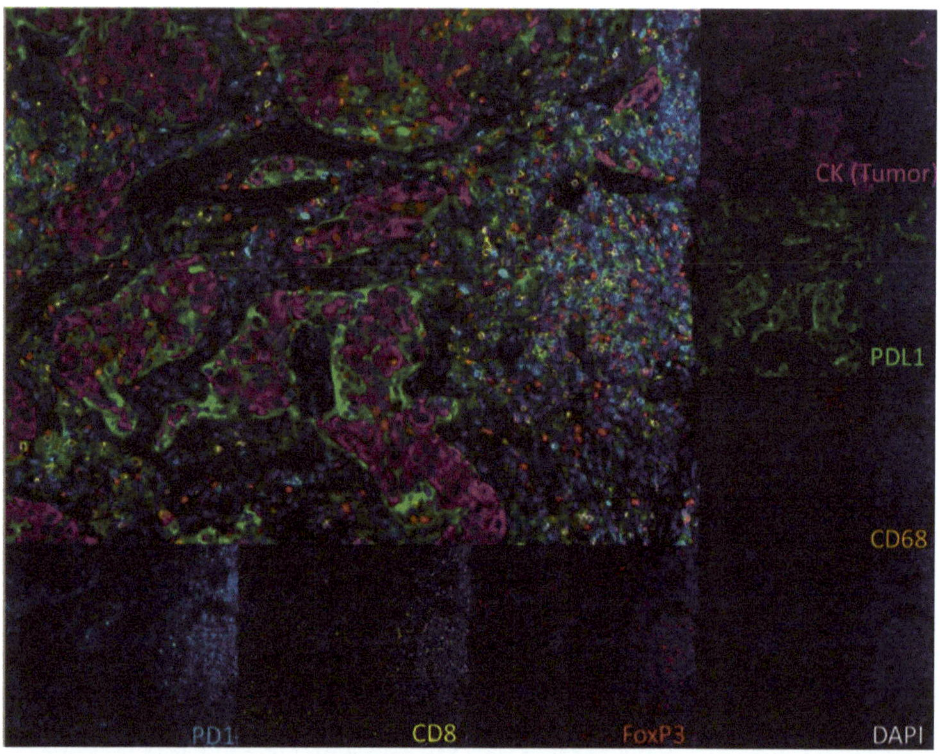

FIG. 1 NK cell suppression in the tumor microenvironment. Schematic depicting the mechanisms by which the antitumor activity of NK cells are inhibited in the solid tumor microenvironment.

the TME due to increased anaerobic glycolysis of the tumor cells [18]. Also, tumor-associated endothelial cells have clustering defects in adhesion molecules like intercellular-adhesion molecule (ICAM)-1 and vascular cell adhesion protein (VCAM)-1, which impairs immune cell extravasation [19]. Tumors also drive cancer-associated fibroblast (CAF) activation via secreted factors like transforming growth factor growth factor (TGF)-β, platelet-derived growth factor (PDGF), and fibroblast growth factor (FGF) [20]. While contributing to ECM remodeling, CAFs create physical barriers that inhibit immune effector cell infiltration. Melanoma-associated CAFs also been reported to inhibit NK cell cytotoxicity by blocking the upregulation of activating receptors like NKp30, NKp44, and DNAM-1 and the accumulation of intracellular cytotoxic granules via prostaglandin E (PGE)2 [21]. Alternatively, endometrial CAFs suppressed NK cell cytotoxicity in a PGE2 independent manner by downregulating poliovirus receptor (PVR), which normally binds DNAM-1 which displays the heterogeneity of CAF responses against NK cells [22]. While nonimmune cellular components are indeed an important contributor to tumorigenesis, the TME also contains inhibitory immune cells, which are critical mediators of immunosuppression and important prognostic tools for clinical outcomes in patients.

Suppressive immune effectors of the tumor microenvironment

Myeloid-derived suppressor cells

Myeloid-derived suppressor cells (MDSCs) are a heterogenous population of suppressive immature myeloid cells that develop in the bone marrow (BM) and are recruited to sites of infection or injury to control inflammation [23]. However, in the context of solid tumors, they are activated and recruited by the solid tumor as part of the suppressive tumor microenvironment. This pathological activation and migration of MDSCs toward the tumor stems from the continuous stimulation of the myeloid compartment by tumor-derived factors such as granulocyte-macrophage colony-stimulating factor (GM-CSF), IL-6, TGF-β, as well as chemokines like CCL2 and CXCL12 [24–29]. Additionally, evidence in murine models suggests that recruited MDSCs actively support tumor development, neovascularization, and metastasis through the production of vascular endothelial growth factor (VEGF) and basic fibroblast growth factor (bFGF) [30]. Apart from supporting tumor progression, MDSCs play a key role in the suppression of immune responses via multiple mechanisms [31,32]. MDSCs have also been reported to secrete inhibitory cytokines like TGF-β and IL-10 to suppress the immune responses of T cells and NK cells while also inducing the generation of regulatory T cells (Tregs) [33,34]. This capacity to induce the de novo generation of Tregs was found to be mediated by interferon-γ (IFN-γ) and IL-10. Zhao et al demonstrated that MDSCs isolated from patients with hepatocellular carcinoma (HCC) upon co-culture with autologous T cells induced the generation of CD4+CD25+Foxp3+ Tregs [35]. Along with Treg induction, MDSCs have also been reported to convert macrophages to an immunosuppressive M2-like phenotype to promote tumor growth [36].

In addition to inhibitory cytokines, MDSCs are known to secrete reactive oxygen species (ROS) and contribute to the elimination of TILs. Most ROS comprise superoxide anions, hydroxyl radicals, hydrogen peroxide, and singlet oxygen. Liu et al demonstrated that

ROS inhibition via catalase addition impaired the immunosuppressive nature of MDSCs [37]. Additionally murine MDSCs lacking the NADPH isoform NOX2 released lower levels of ROS and failed to inhibit CD8+ T cell proliferation and IFN-γ responses [38]. Interestingly, MDSCs themselves were partially protected from the deleterious effects of ROS through the expression of Nrf2, a mediator of cellular antioxidant responses. MDSCs also produce high levels of nitric oxide (NO) via the upregulation of inducible NO synthase (iNOS) leading to suppression of antitumor responses of lymphocytes [39]. They also decrease the availability of metabolites like L-arginine that are critical to survival of TILs. MDSCs upregulate arginases (ARG-1 and ARG-2), which contribute to the conversion of L-arginine to L-ornithine and urea, but also aids the production of extracellular matrix proteins and tumor growth [40]. Finally, MDSCs are reported to express ligands to immune checkpoint receptors on incoming lymphocytic populations. MDSCs that express the ligand to programmed cell death protein-1 (PD-1), PD-L1, are enriched in patients with nonsmall cell lung cancer and are mediators of inhibition against PD-1 expressing T and NK cells [41,42].

Tumor-associated macrophages

Tumor-associated macrophages (TAMs) are a key immune component of the TME and are thought to primarily originate from BM monocytes with minor contributions from splenic monocytes [43]. Increased TAM infiltration is correlated with poor clinical outcomes in multiple types of cancer including breast, lung, and ovarian cancer [6,44,45]. TAM subsets are categorized as tumoricidal (M1) or tumor-promoting (M2) macrophages based on the expression of specific markers and phenotypes [46]. The primary lineage regulator of a majority of macrophage populations, including TAMs, is colony-stimulating factor-1 (CSF-1) [47]. Interestingly, it also acts as a chemotactic factor for macrophages. High CSF-1 concentrations are associated with poor prognosis and whose expression is located at the periphery of tumors [45,48]. Direct evidence for the role of CSF-1 was provided using a murine model of breast cancer with CSF-1 expressing mammary epithelium and resulted in localized macrophage recruitment and acceleration of tumor progression [49,50]. VEGF-A also functions as an independent M2 chemoattractant as demonstrated by Lin et al using a gain-of-function VEGF-A over a loss-of-function CSF1 in the same murine breast cancer model [51]. VEGF-A was also shown to recruit macrophage progenitors to the tumor site that then could be differentiated to M2s under the influence of IL-4 using a xenograft model of skin cancer [52]. Similar roles for CCL18 and CCL9 as a potential M2 chemoattractant were also established with ablation of either protein leading to deficiencies in tumor growth and tumor invasion [53–55]. M2s, like MDSCs, directly contribute to the growth and progression of the solid tumor. CSF-1-regulated macrophages promote neovascularization by stimulating vascular endothelial cells to secrete VEGF [51]. The angiogenic capacity of M2s has since been established in multiple xenograft models [56–58]. M2s also produce several other secreted factors that promote tumor cell invasion including SPARC leading to improved tumor cell migration [59]. In addition, they secrete cathepsin proteases that remodel the extracellular matrix and release trapped growth factors, and TGF-β that drives epithelial-to-mesenchymal transition in tumor cells [48].

M2s also exhibit the capacity to actively suppress the immune response in conjunction with its role as a direct promoter of tumor growth. Macrophages express classical HLA molecules like HLA-C and HLA-E, and nonclassical versions like HLA-G as membrane bound or in soluble form which inhibits NK cell activation and distinct T cell subsets that express CD94 [60]. Also, macrophages expressing members of the LILRB inhibitory receptors family can bind HLA-G on neighboring macrophages which promotes IL-10 and TGF-β secretion. In addition to inhibitory HLA molecule expression, macrophages express the ligands to PD-1 [61]. Noman et al showed that M2s residing in hypoxic regions of the TME-upregulated PD-L1 expression following hypoxia inducible factor 1-α (HIF-1α) signaling [62]. Finally, M2s can also suppress TIL activity through the depletion of L-arginine in the TME. Arginase 1 (ARG1) expressing TAMs have been identified in both human and mouse cancers and is considered a hallmark feature of M2 macrophages [63–65].

Regulatory T cells

Tregs are an inhibitory effector subset of CD4+ T cells and are often characterized by the expression of the transcription factor forkhead box protein P3 (FOXP3). They are responsible for preventing hyperinflammation by suppressing immune effector responses at localized sites of infection [66]. They are also critical to the maintenance of self-tolerance thus preventing autoimmunity [67]. However, increased numbers of tumor Tregs are associated with poor patient survival in multiple solid tumors and play a major role in both supporting tumor growth by inhibiting antitumor immune responses by taking up positions within the TME [68–70]. Tregs account for 10%–50% of all tumor-associated CD4+ T cells and this accumulation within the TME occurs via multiple mechanisms. Tregs are chemoattracted to the tumor in response to tumor-derived secreted factors like CCL17, CCL5, CCL28, CXCL9, CXCL10, and CXCL11 [70]. Other mechanisms of Treg recruitment are still being uncovered. Priceman et al showed that sphingosine 1-phosphate (S1P) receptor 1 signaling was required for Treg accumulation in the TME, acting through the JAK/STAT-3 signaling pathway in preclinical models [71]. Tregs have also been found to proliferate in response to myeloid derived secreted factors like TGF-β and IL-10 within the TME [34,72,73]. Generation of suppressive Tregs from noninhibitory CD4+ T cells within the TME is found to be regulated by hypoxia driven factors like HIF-1α. Increased HIF-1α expression was correlated with increased expression of FOXP3 [74,75]. Interestingly, HIF-1α also indirectly promotes the conversion of T cells to Tregs by regulating the secretion of TGF-β in surrounding tumor tissues. The binding of the TGF-β to its receptor on Tregs leads to the induction of downstream signaling events via the SMAD3/SMAD4 complex and subsequent FOXP3 expression [76]. Once recruited into the tumor, they can be maintained in the TME due to the abundance of self-antigens expressed which are preferentially recognized by Tregs leading to a continuous state of activation. Altogether, the signals provided by various components of the TME prime Tregs to mediate inhibitory effects toward incoming TILs.

One of the most well-defined models of Treg-mediated immunosuppression describes the secretion of inhibitory soluble or membrane-bound mediators, which require cell-to-cell contact. Tregs use membrane-bound CTLA4 to bind B7 molecules, causing DCs to secrete the tryptophan catabolic agent indoleamine 2,3-dioxygenase (IDO) to starve T cells of necessary metabolites in the TME [77]. Tregs also express inhibitory receptors like TIGIT whose expression is correlated with poor prognosis in patients [78]. However, its functional role in mediating immunosuppression is unclear. The major suppressive soluble mediators responsible are IL-10, TGF-β, and IL-35, which are known inhibitors of T and NK cell function, thus promoting tumor progression [69,79]. Additionally, Treg-derived cytokines polarize dendritic cells (DCs) toward a more tolerogenic phenotype thereby preventing them from efficiently activating effector T cells [79]. Tregs also act as a sink for the T cell activating cytokine IL-2 by expressing high levels of high affinity IL-2 receptor CD25 leading to cytokine deprivation (IL-2 sink) mediated apoptosis of effector CD4+ T cell [80]. While Tregs are considered immunosuppressive, studies suggest they may possess cytolytic activity. This is evidenced by the fact that Tregs have shown the capacity to utilize granzyme B, TRAIL, and galectin-1 by inducing cell death in T cells and NK cells [81–85]. Importantly, Cao et al successfully demonstrated that adoptive transfer of granzyme B and perforin-deficient Tregs were handicapped in their ability to impede NK and T cell antitumor response in murine tumor models [85].

Dendritic cells

Dendritic cells are potent antigen-presenting cells (APCs) and are critical to the regulation of the adaptive immune response. DCs arise from and differentiate in the BM prior to migrating to the peripheral tissues [86]. DCs play an important immunogenic role in activating priming T cells against tumor. However, exposure to inhibitory cytokines as well interactions with the tumor and other inhibitory immune cells within the TME drive the development of a tolerogenic DC state [87–90]. This switch to a more regulatory role contributes to a stunted antitumor T cell response. DCs are a heterogenous population of cells that comprise distinct subsets that are either localized in or recruited to the TME. Two of these subsets that are harnessed by the tumor are: plasmacytoid DCs (pDCs) and conventional DCs (cDCs). The pDC subsets are producers of type 1 interferons and direct viral immunity [91]. However, antigen presentation by pDCs largely evokes a tolerogenic response that drives the activation of Tregs [92]. Additionally, pDCs have also been reported to engage inhibitory receptors on T cells like PD-1 leading to suppressed immune response [93].

Multiple studies have correlated pDC infiltration with poor prognosis in multiple solid tumors, due to their impaired capacity to produce type 1 interferons in the TME [94–96]. The cDC subsets are the most powerful inducers of T cell activation and can infiltrate tumors, although they constitute a minor subset of immune cells in the TME. The recruitment of cDCs into the tumor is dependent on the availability of the chemokine CCL13 [97]. Although cDC populations do shape immune responses from within the TME, they are also a target of TME-derived factors like VEGF and TGF-β which impact the cDC activation, cytokine expression, and T cell priming capacity [98]. Similarly, infiltrating TAMs and tumor-associated Tregs that secrete IL-10 and impair the capacity of cDCs to secrete the activating cytokine IL-12, which limits the response of CD8+ T cells and type 1 helper T cell responses [99,100]. Interestingly, tumor-infiltrating cDCs express PD-1, which on binding PD-L1, limits cDC T cell priming capacity [101]. While immature tolerogenic DCs actively impair T cell activation, their lack of MHC expression makes them susceptible to NK cells killing. However, tumor-associated dendritic cells may induce and activate suppressive Treg populations which in turn inhibit antitumor NK cell responses.

In summary, the TME, composed of immune cellular components and a pleiotropic chemical and biological mediator (cytokines, reactive oxygen species, etc.)-rich milieu, is a highly active niche that coordinates to affect numerous immune functions within patients with solid tumors. Inhibitory immune cells like MDSCs, M2-TAMs, Tregs, and DCs represent key immunologic barriers to effective antitumor immune responses in the setting of engineered and nonengineered NK cell therapies. Therefore, novel strategies designed to target these immune suppressors may lead to improved clinical outcomes for patients with solid tumors.

Techniques to evaluate the tumor microenvironment

Currently, the number of therapies that target the TME are increasing due to the recognition of the significant role it plays in promoting tumor progression and coordinating antitumor immune responses. Additionally, there is growing evidence to show that antitumor immune effector cell infiltration is associated with better clinical outcomes in patients with solid tumors [102,103]. For therapies that count on activation of endogenous intratumoral cells or adoptive cell therapies, it will be critical to be able to assess the composition, metabolic fitness, phenotype, and spatial localization of antitumor immune cells and inhibitory immune cells within the TME. Currently, many groups have designed and implemented a variety of tools to analyze resident immune and nonimmune cell populations in the TME (Table 1). Such analyses may inform the development of novel NK cell-based therapies to improve clinical outcomes of patients with solid tumors.

TABLE 1 Multiparametric assessment of tumor microenvironment.

Modality	Technique	Modification	Improvement	Information provided
Imaging	Immunofluorescence	Tyramide signal amplification	– Increase in detectable markers – No antibody cross-reactivity	– Spatial distribution of immune TME components
	Immunohistochemistry/ Immunocytochemistry	Metal-tagged antibodies CyTOF mass cytometry	– Increase in detectable markers detection – Improved cell resolution	
	Hybridization	Padlock probes RNAScope	– Amplified signal intensity – Increase in detectable RNA species	

Continued

TABLE 1 Multiparametric assessment of tumor microenvironment—cont'd

Modality	Technique	Modification	Improvement	Information provided
	Computed tomography	Nano-radiomics (Lipid nanoparticles)	– High resolution imaging – 3D spatial analysis	
	Positron electron tomography/magnetic resonance	^{64}Cu-labeled NOTA-monoclonal antibodies	– Noninvasive – Quantitative analysis	
Cytometric analysis	Time-of-flight mass cytometry	Metal-tagged antibodies/ metal nanoparticles	– Amplified signal intensity – Multidimensional analysis – Compatibility with other techniques	– Characterization of TME composition – Correlative biomarker identification
Gene expression analysis	Gene profiling	xCell platform	– Discriminates between closely related cell types	– Characterization of TME composition – Identification of mechanisms of therapeutic resistance
		TIminer software	– Contains multiple bioinformatics tools – Simple installation – Integrative immunogenomic analysis	
		MCP-counter	– Calculate abundance of immune and stromal TME components – Quantitative analysis	
		scRNA-seq (barcoded primers)	– Increased resolution of RNA sequencing – Assess spatial distribution of transcripts	
Metabolic fitness	Cellular metabolism	Seahorse extracellular flux analyzer	– Simultaneous measurement of glycolysis and oxidative phosphorylation	– Assess metabolic reprogramming of immune cells in the TME -Correlative functional

TABLE 1 Multiparametric assessment of tumor microenvironment—cont'd

Modality	Technique	Modification	Improvement	Information provided
			– No radioactive substrate requirement – Requires less starting material	analysis

Summary of techniques and tools for multiparametric analysis of inhibitory immune cells and tumor infiltrating NK cells in the tumor microenvironment.

Improvements in tissue analysis and imaging

One method to analyze the TME is to use immunohistochemistry (IHC) or immunofluorescence (IF) to directly quantify various cellular populations. Both these methods require the use of a primary antibody targeted against an antigen of interest. A secondary antibody conjugated to a catalytic agent (IHC) or a fluorophore (IF) is used to generate and amplify the intensity of the signal. These methodologies often provide information on the distribution of the target molecule in the sample. These stained samples can be observed using a microscope and scanned images can then be analyzed using various histopathology software to quantify the marker expression. While both approaches are flexible due to the wide variety of antigens that can be targeted by antibodies, only a small number of markers can be assessed simultaneously due to cross-reactivity between primary and secondary antibodies. This prolongs and complicates the process to gather descriptive data in a complex TME system. However, recent efforts have made it possible to assess increased numbers of markers on a single slide by using IF automation enabling simultaneous detection of up to seven colors [104,105]. One multiplexing method utilized by this approach is the tyramide signal amplification (TSA) system. In this system, a fluorophore conjugated to tyramide is catalyzed by horseradish peroxidase conjugated to a secondary antibody. This leads to the generation and release of short-lived labeled tyramide radicals that covalently bind to tyrosine residues proximal to the original epitope binding site. This allows both primary and secondary antibodies to be stripped from the tissue sample, thus preventing cross-reactivity during the next round of staining [106]. This multiplexed analysis allows for the detection of several markers of the same cell, and information on their distribution across the whole tissue [107].

Other studies have alternatively used metal-tagged antibodies and mass cytometry to detect tissue staining. These methods offer a detection of up to 32 markers on formalin-fixed-paraffin-embedded (FFPE) tumor sections [108,109]. Such systems that drastically expand the number of simultaneous detection of markers in tissue will positively impact TME analysis. In addition, strategies to assess and visualize mRNA in a multiplexed manner have been developed using sequence-specific "padlock" probes against amplified cDNA-converted transcripts. These padlocked sequences are then hybridized to fluorescent detection probes for detection in human cells and tissue [110,111]. An alternative multiplex strategy called RNAscope utilized a hybridization-mediated signal amplification followed by exposure to a fluorophore/HRP conjugated label probe to detect multiple RNA species in FFPE slides [112,113]. Additionally, a modified RNAscope platform coupled with antibody staining for mass cytometric imaging enabled simultaneous multiplexed detection of both mRNA and proteins in tissues from patients with breast cancer [114].

While such development in IHC- and IF-based techniques have garnered attention, analysis of large numbers of patients is prolonged and labor intensive. The development and integration of machine learning programs will improve automated analysis of histopathological images [115]. Sophisticated mathematical models like neural networks can be trained on tumor slides to identify tumor cells, assess its pathological state, and predict patient survival probability. Such machine learning models aim to analyze cellular populations within the TME, with a large focus on lymphocytes at a comparable level to pathologists. This would enable expedited and reproducible analysis of the TME in large numbers as well as quantification of tumor-infiltrating NK cells in patients. Finally, Devkota et al used a noninvasive nano-radiomics approach to detect changes in tumor morphology in response to MDSC-targeted NK cell therapy [116]. Liposome nanoparticles containing contrasting agents supported radiomic analysis of the TME and generated high-resolution enhanced CT imaging. The extended half-life and extravasation capacity of nanoparticles improved their ability to accumulate in perivascular regions of the TME and enabled a quantitative three-dimensional analysis of MDSC-influenced vascular architecture in the TME.

A salient feature of the TME is its capacity to recruit inhibitory immune cells to the tumor and is a key determinant in setting up an immunosuppressive region around the tumor [117]. Hoffmann et al developed an approach to track MDSC migration in primary and metastatic cancers by employing PET as a quantitative noninvasive imaging modality [118]. Polymorphonuclear (PMN-) and monocytic (M-) MDSCs were marked using a ^{64}Cu-labeled 1,4,7-triazacyclononane-triacetic acid (NOTA)-modified CD11b-specific monoclonal antibody in vitro. The binding of these antibodies to their target would result in internalization of the antibody–antigen complex leading to subsequent stable radiolabeling of the cells. Radiolabeled PMN- and M-MDSCs adoptively transferred into mice with primary and metastatic breast cancer and melanoma could then be detected using positron electron tomography (PET) and magnetic resonance (MR) imaging up to 48h postinfusion.

Cytometric analysis of TME composition and function

Cytometry is a technique that enables analysis of several markers on a population or single-cell basis and can be used to precisely assess their phenotype and functionality of cellular populations in the TME. Cells are labeled with fluorophore-conjugated antibodies against markers, which is then read by a laser. This allows for the acquisition and recording of thousands of events per second. Each population can be subsequently analyzed based on their expression of various markers. Cytometry has been routinely used to study the TME, and the TILs and inhibitory immune cells that reside within it. This composition analysis is crucial to the prescription of effective therapies in patients. However, traditional cytometry suffers from loss of resolution due to autofluorescence and spectral spillover. This limits traditional cytometry when higher numbers of parameters need to be analyzed. To overcome this limitation, new methods for antibody detection must be considered. Mass cytometry is one such detection approach and utilizes a time-of-flight mass spectrometer. In mass cytometry, cells are labeled with antibodies tagged with highly purified, stable isotypes and are then quantified using time-of-flight mass spectrometry [119]. It measures the abundance of the metal isotopes conjugated to the antibodies, and amplification of signal intensity is generated using a polymeric metal-chelating reagent or metal nanoparticles [120,121]. Mass cytometry greatly increased the number of parameters that can be analyzed thus enabling multidimensional analysis of samples. Lavin et al employed mass cytometry to describe the immune cell profile in early lung tumors. Importantly, they showed that Stage 1 lung adenocarcinoma lesions had already altered intratumoral T and NK cells as well as the enrichment of inhibitory TAMs in the TME [122]. Similarly, Chevrier et al conducted in-depth immune profiling of 73 patients with clear cell renal cell carcinoma (ccRCC) using mass cytometry [123]. They reported multiple phenotypes of TAMs and T cells and also identified an immune signature that was correlated to shorter progression-free survival in patients. Berrien-Elliott et al distinguished in vivo differentiated adoptively transferred NK cells from peripheral blood and BM NK cells acutely exposed to cytokine treatment in patients with acute myeloid leukemia [124]. Using a 37-marker NK cell-mass cytometry panel, key phenotypic distinctions between donor and patient NK cells were found 7 days post-NK cell infusion in all 11 patients. Gonzalez-Gugel et al used mass cytometry to assess exhaustion marker expression on peripheral blood NK cells and evaluate changes in their expression upon exposure to an immunosuppressive environment [125]. Mass cytometry has also been paired with other high-throughput techniques like single cell (sc) RNA-seq to bridge genomic and proteomic tissue analysis [114,122]. Additionally, mass cytometry has also been combined with IHC to image proteins modifications in FFPE breast cancer samples and human mammary epithelial cells [108]. Mass cytometry provides insight into tumor-driven phenotypic changes in immune effectors and the identification of potential biomarkers of tumor progression and is thus a powerful tool to aid in design and development of NK cell-based therapies.

Estimating TME composition using gene expression data

Cytometry and other imaging-based approaches are able to identify distinct immune cell populations in the TME. However, they become cumbersome when applied to a large series of tumor samples. Profiling of gene expression via high-throughput approaches provides greater resolution, speed, reduced costs, and novel tools to better understand a complex TME system [126]. In fact, several methods have now been designed to assess TME composition by mining transcriptomic data. Gene set enrichment analysis (GSEA) that identifies enriched or over-represented gene sets based on a predefined list of ranked genes. Bindea et al used GSEA analysis to not only identify intratumoral immune infiltrate populations but to also determine which populations were important to tumor progression or control over time [127]. Although GSEA is flexible and can be applied across multiple data types, it faces difficulties in discriminating populations of closely related cell types. To overcome this limitation, novel gene signature-based approach based on GSEA with improved discriminatory capacity was developed [128]. The xCell platform successfully identified gene signatures for

64 cell types, which included TME-resident immune and nonimmune cells derived from thousands of expression profiles. Such tools can effectively mine immunologic and genomic data relevant to various cancers. However, computational software available to analyze large amounts of information are limited, and analysis with existing computational models is laborious and time consuming. TIminer, a computational software, was developed to overcome these limitations. The easy-to-install software was packaged with various bioinformatic tools allowed for integrative immunogenic analyses derived from next-generation-sequencing data including HLA typing, neoantigen prediction, immune infiltrate characterization, and quantification of tumor immunogenicity [129]. As an alternative output of TIminer, Charoentong et al designed immunophenoscores based on the gene expression of immune checkpoints and antigen processing and presentation and the cellular composition of effector and inhibitory immune cells [130]. They correlated generated immunophenoscores to patient survival in multiple solid tumors. This score was also associated to clinical responses to CTLA-4 and PD-1 checkpoint blockade in metastatic melanoma patient cohorts. Finally, MCP-counter is a computational approach that analyzes transcriptomic marker expression in a cell population and provides an arbitrary score that corresponds the cell proportion in the sample [131,132]. The approach was used to identify TME resident immune and stromal cell populations in several malignancies, and the scores generated by MCP-counter were capable of predicting efficacy of immune checkpoint blockade and murine cancer models [132].

Recently, interest in the application of scRNA-seq to describe TME composition has gained traction [133]. Analyzing immune cells in the TME at a single-cell scale will provide a detailed determination of its phenotype and functionality. Additionally, scRNA-seq can be used to study rare populations of cells whose expression would normally be hard to differentiate from a bulk cellular population. Such analysis using scRNA-seq was used to characterize immune profiles in the TME of patients with metastatic melanoma and hepatocellular carcinoma [134,135]. This would alleviate the limitations of other expression-based techniques that use lymphocytes from peripheral blood of healthy blood donors to generate reference sequence libraries [136,137]. One of the major issues surrounding transcriptomic approaches is their inability to assess spatial information of identified cell types in the TME. However, Stahl et al developed an approach by using barcoded-primers associated with distinct regions of analyzed tissue prior to RNA-seq analysis to generate a two-dimensional cellular map of tumor tissues [138,139]. Altogether, such analysis that allows for the portrayal of a full TME landscape across multiple samples leads to an improved understanding of cellular heterogeneity in the TME and the development of therapies to overcome them.

Assessing metabolic fitness in the TME

Metabolism has increasingly gained recognition as an important aspect of immune cell differentiation, homeostasis, and function. Until recently, the role of metabolism in NK cells was not well studied. Cancer cells and immune cells display changes in metabolic programs in the TME, which can ultimately influence NK cell antitumor function [140,141]. Thus, it would be useful to develop tools to monitor the NK cell metabolic profile to better understand their effects on NK cell activity in the TME. Evaluations of cellular metabolism can be made using a Seahorse extracellular flux analyzer (EFA). The Seahorse EFA is capable of simultaneously evaluating glycolysis by analyzing extracellular acidification rate (ECAR) and mitochondrial oxidative phosphorylation based on oxygen consumption rate (OCR) in real time [142,143]. In contrast to traditional assays that directly measure substrate to assess glycolysis and oxidative phosphorylation, the Seahorse EFA does not require the use of radioactive substrates while using much fewer cells for analysis. The need for less starting material may prove useful in the analysis of rare populations of NK cells in the TME, as well as NK cells directly ex vivo without the need for extended culture period. Additionally, the seahorse assay has also been applied to measure colon cancer tumor cells in three-dimensional spheroid cultures [144]. Considering that three-dimensional approaches are now being used to model the TME, the Seahorse assay may be utilized to model NK cell metabolic reprogramming in this suppressive environment to correlate this to NK cell antitumor function [145].

The TME varies depending on the type of cancer and patient. Deciphering the composition of the TME is crucial to the success of cancer therapies. The development of novel tools and the utilization of existing techniques to analyze large numbers of samples, parameters, and readouts will aid in this process.

Advancements in therapeutic modalities to overcome TME suppression

Cancer therapy research has primarily focused on eliminating tumor cells. However, the effect of the TME has been identified as a key player in tumor progression and multidrug resistance [146]. A two-fold approach may be taken when designing and developing therapeutic approaches to target the TME and improve clinical outcomes in patients. One approach directly targets the TME by depleting it of inhibitory immune cells thereby reducing its inhibitory capacity. The other approach aims to equip immune cells with machinery to exploit the suppressive mechanisms of inhibitory immune cells to enhance its own therapeutic capacity. Understanding how effective these therapeutic modalities are is crucial to future development of strategies to overcome the TME and ultimately provide better clinical outcomes to patients.

Direct targeting of inhibitory immune cells of the TME

Small molecules: Small-molecule inhibitors have been extensively applied as a treatment option for various cancers over the past two decades [147]. More recently, synthetic compounds are now utilized against inhibitory immune cells of the TME to improve the efficacy of other immunotherapy-based approaches like checkpoint blockade therapy and adoptive cell therapy [148].

A current limitation of small drugs is their nonselective nature leading to off-target toxicities in patients. To overcome this challenge, Deng et al developed a PEGylated-folate-liposome (PEG-FA-Lip), which contains the immunogenic cell death inducer doxorubicin [149]. The PEG-FA-Lip efficiently targeted tumor cells and TAMs via folate receptor-mediated endocytosis resulting in entry of the cytotoxic doxorubicin and their eventual elimination. PEG-FA-Lip treatment against 4T1 tumor cells resulted in high levels of tumor apoptosis in vitro. Additionally, the cell death induced by PEG-FA-Lip led to increased release of other immunogenic molecules like HMGB1. Importantly, PEG-FA-Lip treatment of mice containing breast cancer tumor cells resulted in high levels of tumor cell death and HMGB1 release. This may lead to the reshaping of the suppressive TME to a more inflammatory setting to improve the efficacy of other immunotherapeutic approaches.

Groups have attempted to reshape the TME by interfering with the differentiation and migratory capacities of MDSCs to the TME. Ko et al demonstrated that treatment with the tyrosine kinase inhibitor sunitinib led to reduced MDSC burden in metastatic renal carcinoma patients [150]. This depletion of MDSCs was thought to be in part due to the Sunitinib-mediated blockade of VEGFR and c-kit whose signaling cascades play key roles in shaping MDSC development in the TME [150–152]. Interestingly, sunitinib treatment was also found to deplete Tregs in renal carcinoma patients although the mechanism of this depletion is still unknown. Stiff et al reported that the Bruton's tyrosine kinase inhibitor ibrutinib successfully inhibited generation of human MDSCs from healthy donor monocytes in vitro [153]. Additionally, ibrutinib was found to inhibit the NO production as well as the migratory capacity of human MDSCs toward the breast cancer cell line EMT6. The inhibitory antagonistic capacity of ibrutinib on MDSC migration toward the tumor was recapitulated in vivo using an EMT6 mammary carcinoma-bearing mouse model. A similar study conducted by Varikuti et al showed ibrutinib treatment led to reduced frequency of MDSCs in both the spleen and tumor sites and led to stunted tumor progression in an orthotopic mouse breast cancer model [154]. Interestingly, ibrutinib-treated MDSCs isolated from these tumors were reported to switch to a more active dendritic cell phenotype ex vivo. As mentioned earlier, MDSCs recruited by the tumor play an important role in tumor progression and suppress tumor-infiltrating NK cell antitumor activity. Chemokine receptors on MDSCs like CCR5 demonstrate ligand redundancy (CCL3, CCL4, and CCL5) which enables to them to effectively home to tumor sites [155]. However, Blattner et al reported that blockade of these ligands using a soluble CCR5-Ig fusion protein led to reduced MDSC homing and improved survival of melanoma-bearing mice. Similarly, blockade of CXCR2 on MDSCs using a short CXCR2 peptide "pepducin" led to impaired metastasis and tumor growth in a murine model of pancreatic cancer, thus demonstrating the viability of recruitment blockers as a therapeutic strategy [156]. Alfaro et al successfully interfered with MDSC chemoattraction using the small-molecule CXCR1/2 blocking agent reparixin [157]. Lymphocyte-deficient mice injected with colorectal adenocarcinoma tumor line to increase MDSC numbers were subjected to hydrodynamic gene transfer of IL-8 to the liver, which subsequently lead to increased IL-8-mediated MDSC recruitment to the liver specifically. However, upon reparixin treatment, the frequency of granulocytic MDSCs was reduced. Sun et al explored the efficacy of a CXCR1/CXCR2 small molecule inhibitor SX-682 to block MDSC recruitment [158]. SX-682 successfully inhibited CXCR2+ MDSC trafficking toward tumor sites in mice with oral cancer or lung cancer. Importantly, SX-682 treatment led to increased accumulation of tumor-infiltrating T cells in tumor sites. Additionally, SX-682 treatment also enhanced tumor infiltration of adoptively transferred OT-1 T cells in RAG1-deficient mice-bearing OVA expressing oral cancer or lung cancer. However, trafficking of CXCR1+ macrophages was unaffected possibly due to their co-expression of the chemokine receptor CSF-1R. While CSF-1R does not bind chemokines, it is heavily implicated in MDSC recruitment by the tumor [159–161]. Holmgaard et al showed that CSF-1R inhibition via a potent small-molecule inhibitors PLX647 and PLX5622 lead to inhibited MDSC infiltration in mice with melanoma tumor cells [160]. PLX647 was also used in conjunction with an IDO inhibitor in a murine melanoma tumor model. This led to decreased tumor growth and increased accumulation of tumor-infiltrating CD4+ and CD8+ effector T cells within the TME.

TLR9 agonists: Toll like receptors (TLRs) are vital to host pathogen-sensing capacity and can be divided into two families based on their localization. TLR-mediated activation of innate immune cells leads to the production of proinflammatory cytokines and chemokines, which potentiates the activity of the adaptive immune system. This has prompted multiple groups to attempt to harness the potent immunostimulatory property of TLRs for immunotherapy against cancer [162]. TLR agonists administered in conjunction with tumor vaccines to impair tumor growth enhanced cytotoxic T cell activity [163]. However, they failed to completely eradicate established tumors. TLR agonist therapies now aim to focus on targeting inhibitory immune cells of the tumor microenvironment like MDSCs and TAMs to improve the efficacy of cancer treatments.

TLR9 detects and is activated by CpG motifs normally present in high frequency on bacterial DNA. While CpG-based treatments targeting TLR9 have been clinically employed as adjuvants in tumor vaccines, their role in modulating components of the TME were not studied [163]. Recently, Liu et al demonstrated that CpG treatment of macrophages enabled metabolic changes that enabled them to overcome CD47-mediated inhibition by tumor cells leading to their eventual engulfment [164]. Treatment of macrophages with ODN1826, a CpG oligonucleotide was found to enhance their phagocytic capacity against pancreatic ductal adenocarcinoma (PDAC) tumor cells when compared to nontreated controls in vitro. CpG treatment also inhibited tumor growth in two different mouse models of PDAC and improved their survival. Importantly, the therapeutic efficacy of CpG oligonucleotides was found to be dependent on a subpopulation of TAMs that expressed CSF-1R. Depletion of this subpopulation via anti-CSF-1R treatment blocked the therapeutic response of CpG treatment. Interestingly, CpG treatment also metabolically rewired macrophages leading to enhanced fatty acid oxidation which was found to be crucial to their antitumor activity in vivo against PDAC tumor cells. A similar study by Zoglmeier et al also reported that the short-term stimulation of TLR9 using CpG blocked the suppressive capacity of MDSCs in tumor-bearing mice [165]. CpG treatment was found to rescue impaired T cell proliferation in murine models of colon adenocarcinoma and gastric carcinoma. Analysis of MDSCs harvested for tumor-bearing mice from both cancer models revealed a more proinflammatory profile when exposed to CpG treatment. This was characterized by the expression of DC maturation markers MHCII, CD11c, and CD80. Moreira et al demonstrated that CpG oligonucleotides can also be tethered to small molecule drugs to inhibit intracellular processes while simultaneously stimulating TLR9 on MDSCs [166]. A CpG motif was successfully conjugated to an antisense oligonucleotide (ASO) to inhibit STAT3, a key signaling molecule governing the suppressive activity of MDSCs [167]. CpG-STAT3ASO effectively knocked down STAT3 expression in murine myeloid cells in vitro. Uptake of the ASO was mediated by scavenger receptors. Interestingly, the CpG moiety improved this process by enhancing clathrin-mediated internalization. Additionally, CpG-STAT3ASO treated TLR9+ MDSCs from prostate cancer patients were unable to suppress CD3+ T cell proliferation in vitro. Additionally, the CPG-STAT3ASO conjugate stimulated CD8+ T cell granzyme B and IFN-γ production. Analysis of MDSCs posttreatment revealed a 90% reduction in STAT3 mRNA. However, neither STAT3ASO alone nor CpG-scramble controls mediated any improvement in CD8+ T cell activity indicating the necessity of both components in overcoming MDSC-mediated suppression. These results underscored the therapeutic potential of CpG oligonucleotides to prime immune responsiveness against cancer cells and improve the prospects of follow-up immunotherapeutic approaches to treat patients with cancer.

Monoclonal antibody therapy: Monoclonal antibody (mAb) therapy has resulted in durable clinical responses in patients with cancer. However, a majority of patients develop resistance to these therapies and eventually relapse. While mAb-based therapy has traditionally targeted immune checkpoint receptors on cytotoxic lymphocytes like T cells and NK cells, development has now focused on generating antibodies that also target inhibitory immune cells of the TME to improve the efficacy of cancer treatments [168,169].

MacDonald et al developed a novel neutralizing antibody against CSF-1R. The administration of the anti-CSF-1R antibody selectively depleted monocyte precursors of tissue resident macrophages and was capable of penetrating most tissues barring the lung and female reproductive organs in mice. In syngeneic murine models of mesothelioma and lung carcinoma, CSF-1R antibody blockade selectively depleted the TAM population in the TME while tumor growth was unaffected. Ries et al successfully developed a humanized anti-CSF-1R monoclonal antibody (RG7155) capable of inhibiting CSF-1R activation by impairing receptor dimerization [169]. CSF-1R blockade resulted in the apoptosis of CSF-1R+ CD163+ TAMs but not CD80+ pro-inflammatory macrophages in vitro. Administration of RG7155 in mice with colorectal adenocarcinoma led to increased accumulation of CD8+ T cells relative to CD4+ Tregs. In a phase I trial, administration of RG7155 resulted in clinical responses in patients with diffuse-type giant cell tumor with an objective response rate of 70% at 6 weeks posttreatment. The clinical activity was associated with a reduction in intratumoral CSF1R+CD163+ macrophages.

Aside from TAMs, Tregs are regarded as a major obstacle to successful clinical application of cancer immunotherapy. The FDA-approved anti-CD25 monoclonal antibody (daclizumab) is currently used to deplete inhibitory CD25+ Tregs in metastatic breast cancer patients in combination with a multipeptide cancer vaccine (NCT00573495). Initial results showed that daclizumab combination therapy was well tolerated with a robust and prolonged depletion of CD25+ FOXP3+ CD4+ Tregs after a single infusion.

Epigenetic regulators: Histone deacetylase inhibitors (iHDAC) are a new class of chemical agents developed as a therapeutic modality against hematological malignancies [170]. They modify the acetylation state of histones in chromatin resulting in altered gene expression leading to cell cycle arrest and apoptosis of targeted cells [170,171]. iHDACs contribute to antitumor immunity via direct tumoricidal effects [172]. Evidence also suggests that they augment the immunogenicity of tumors by upregulating the expression of MHC class 1 and class 2 and MHC class 1-like molecules MICA and MICB leading to enhanced T and NK cell immune responses [172,173]. Recently, studies investigating the effects of iHDACs on other inhibitory immune cells of the TME have also yielded promising preclinical results. Wang et al reported that iHDAC vorinostat (SAHA) contributed to antitumor responses in mice by through the depletion of MDSCs [174]. SAHA was successfully able to inhibit the expansion of GM-CSF-treated BM MDSCs in vitro. Additionally, MDSCs isolated from mice injected with 4T1 mammary tumor cells were also depleted to undetectable levels after SAHA treatment. This depletion was found to be due to an increase in ROS-mediated apoptosis of Gr-1+

MDSCs. The increase in apoptosis was reversed when SAHA-treated BM Gr-1+ cells were also exposed to *N*-acetylcysteine, an ROS inhibitor. Finally, upon treatment with SAHA, mice-bearing 4T1 mammary tumor cells displayed a decreased proportion of splenic and peripheral blood Gr-1+ MDSCs. Additionally, SAHA administration also diminished the total number of intratumoral Gr-1+ MDSCs in these mice and resulted in a significantly lowered tumor burden compared to nontreated controls. Orillion et al targeted immunosuppressive MDSCs using the selective class 1 iHDAC, entinostat in combination with anti-PD-1 immunotherapy [175]. Entinostat improved the antitumor effect of PD-1 blockade in two syngeneic mouse tumor models by lowering tumor burden and improving survival. Entinostat inhibited the suppressive activity of PMN-MDSCs. Analysis of MDSC responses to entinostat revealed significantly reduced arginase-1, iNOS, and COX-2 levels, which suggests potential mechanisms for their altered function. The authors also observed significant alterations in the TME cytokine/chemokine profile of patients subjected to entinostat alone and in combination with anti-PD-1 therapy. The TME profile of these cohorts contained reduced levels of MDSC-activating cytokines including G-CSF, GM-CSF, and IL-10 as well as upregulate on of proinflammatory cytokines IL-1 and IL-12 compared to the TME profile of control cohorts. Currently, clinical trials have been initiated evaluating the efficacy of entinostat-checkpoint blockade combination therapy in patients with breast cancer and ovarian carcinoma. However, initial results of clinical trials evaluating the efficacy of entinostat-pembrolizumab combination therapy showed only modest antitumor activity [176,177]. Hashimoto et al also demonstrated that suppressive activity of entinostat was limited to PMN-MDSCs with no effect on M-MDSCs or TAMs, which express high levels of class 2 HDACs [178]. However, treatment using the class 2 iHDAC ricolinostat in combination with entinostat resulted in lowered tumor burden in two murine tumor models. Dual treatment of ricolinostat and entinostat led to the depletion of TAMs and also significantly reduced the suppressive activity of both MDSC populations.

Viral vectors: Oncolytic viruses (OV) are vectors based on viral strains and are designed to selectively replicate in and eliminate tumor cells. Interestingly, the current generation of oncolytic viral vectors that have succeeded in randomized clinical trials expressed GM-CSF, which in the context of the TME promote MDSC generation [179–181]. However, it has been observed that the level of prevalence of MDSCs in the tumor in different murine tumor models correlated closely to the sensitivity of the model to OV therapy [182]. Thus, the inability of OVs to infect tumors with high levels of intratumoral MDSCs (resistant tumors) and induce a therapeutic effect necessitated the need to design viral vectors aimed at disrupting MDSC-promoting mechanisms.

Multiple reports have identified COX-2-mediated production of the prostaglandin PGE_2 was a key to maintaining MDSC persistence and their suppressive activity [183–188]. Hou et al developed an OV-based approach to reduce PGE_2 levels through the viral expression of a PGE_2 inhibitor, HPGD (hydroxyprostaglandin dehydrogenase 15-(NAD), 15-PGDH) [182]. It was found that viral expression of HPGD was nontoxic and selectively depleted intra tumoral granulocytic MDSCs in mice-bearing Renca tumor tumors. Interestingly, a reduction in intratumoral Tregs was also noted upon treatment. Comparative analysis using a variety of COX2 and PGE_2 inhibitors revealed that only OV-mediated expression of HPGD was able to block immunosuppression. HPGD expression also led to the remodeling of the TME chemokine profile with decreased levels of the MDSC recruiting chemokine CXCL12 and increased levels of the T cell recruiting chemokine CXCL11 [182,189]. The benefits of such TME remodeling in attracting adoptively transferred T cells were investigated in a bilateral mouse Renca tumor model. Adoptively transferred T cells migrated more frequently toward the tumor injected with HPGD+ OVs compared to a HPGD-OV control. This indicated that the expression of HPGD also influenced the homing capacity of TILs in addition to their PGE_2 inhibiting activity. A separate mouse Renca tumor model revealed that the injection of HPGD+ OVs also improved their survival compared to nontreated controls.

Reversing NK cell inhibition by the TME through genetic and nongenetic methods

Techniques that utilized the unique functionality of NK cells for anticancer therapy have included administration of monoclonal antibodies to initiate antibody-dependent cellular cytotoxicity (ADCC), exploiting MHC-KIR mismatches to improve graft vs leukemia effects after hematopoietic stem cell transplant (HSCT), and more recently, genetically modifying NK cells ex vivo to express chimeric antigen receptors (CARs) that target tumor-associated antigens. A recent trial using CAR-NK cells that target the CD19 antigen lead to impressive clinical responses in patients with acute leukemia. Additionally, no observed graft-vs-host disease (GVHD) or cytokine-release syndrome (CRS) were reported [190]. However, the effectiveness of NK cell-based treatment of solid tumors which utilize such approaches is limited due to the generation of the TME (Fig. 2). To overcome this challenge, novel NK cell-based approaches have been developed to combat TME-mediated suppression and improve the effectiveness of these therapies in treating patients.

Dominant negative receptors: The TME is surrounded by various inhibitory cytokines, including TGF-β which is involved in immunosuppression and supports tumor progression in patients [191]. Elevated TGF-β levels in patients with solid tumors correlated with poor patient prognosis and increased chances of tumor metastasis [192–194]. TGF-β produced by tumor cells, as well as inhibitory tumor immune cells, inhibits the antitumor response of tumor-infiltrating NK cells [31,195–197]. It suppresses NK cell cytotoxicity by limiting IFN-γ production and downregulating the NK cell activating receptors NKG2D and NKp30 [198–201]. To improve the efficacy of adoptively transferred NK cells in patients with higher TGF-β levels, NK cells have been genetically modified to express a "dominant negative" receptor against TGF-β to overcome TGF-β-mediated inhibition.

Dominant negative receptors do not contain intracellular signaling domains, and so are unable to transduce signals initiated from the cell surface. However, they still retain the capacity to bind/neutralize the external cytokine. Yvon et al successfully engineered a cytoplasmic domain deficient TGF-β receptor II (DNRII) to improve the antitumor function of cord-blood derived (CB) primary NK cells [202]. CB NK cells transduced with DNRII and expanded ex vivo using a modified feeder cell line expressing 4-1BBL and membrane bound (mb) IL-21. DNRII transduced NK cells presented an activating receptor phenotype similar to that of nontransduced controls. Importantly, exposure to TGF-β did not impact the surface expression of the activating receptors NKG2D and DNAM-1 on DNRII transduced NK cells. While the proliferation capacity of DNRII-NK cells was dampened compared to unmodified NK cell controls, their numbers were deemed sufficient for clinical use (NCT01729091). The ability of DNRII to block TGF-β induced signaling was assessed by measuring intracellular

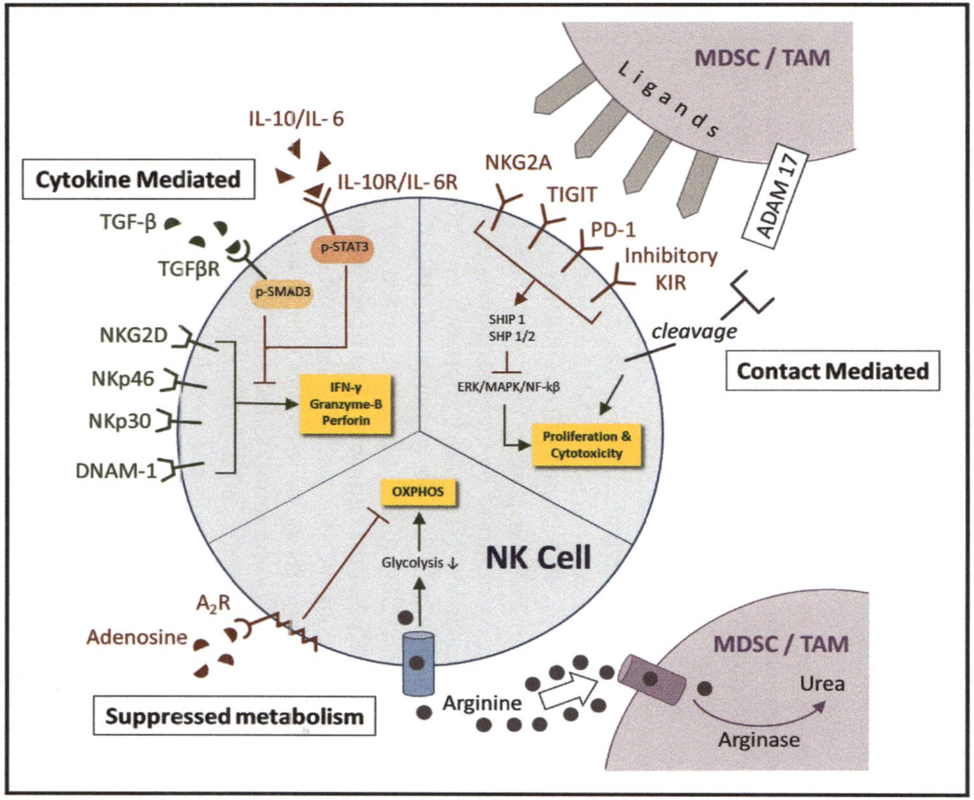

FIG. 2 NK cell suppression in the tumor microenvironment. Schematic depicting the mechanisms by which the antitumor activity of NK cells are inhibited in the solid tumor microenvironment.

levels of phosphorylated SMAD2 (p-SMAD2), an important mediator of TGF-β-mediated inhibition. Importantly, p-SMAD2 was not detected in DNRII-NK cells in contrast to that in unmodified NK cells that demonstrated a proof-of-concept. DNRII-NK cells exposed to TGF-β-enabled enhanced cytotoxicity against glioblastoma tumor cells and IFN-γ, perforin, and granzyme-B production compared to TGF-β-treated unmodified NK cells, thus achieving the desired biologic effect. DNRII was also used in a study by Yang et al as a strategy for the adoptive transfer of TGF-β-insensitive NK-92 cells, a clinically approved leukemic NK cell line [203]. DNRII-modified NK-92 cells were able to resist the inhibitory effects of TGF-β with a marked absence in p-SMAD2 levels. These genetically modified cells were able to exhibit enhanced cytotoxicity against lung carcinoma in vitro and in xenograft mouse models when compared to unmodified NK-92 cells.

Switch receptors: "Switch receptors" are chimeric receptors that contain an inhibitory extracellular domain fused to a signaling cytoplasmic domain of an activating receptor. This design allows an extracellular inhibitory stimulus to be converted to an internal downstream activating signal leading to cell activation. Switch receptors are an alternative option to dominant negative receptors to avoid cytokine-mediated inhibition in the TME. Burga et al compared the antitumor activity of ex vivo expanded CB NK cells genetically engineered to express versions of a TGF-β receptor II-based switch receptor containing the NK activating domain DAP12 (NKA) or a Notch minimal regulatory region containing a RELA-binding domain (SynNotch) [204]. The transduction of NK cells with either variant did not affect its activation receptor phenotype when compared to DNRII transduced NK cells and unmodified NK cells. However, upon TGF-β exposure, both SynNotch-and NKA-NK cells upregulated DNAM-1 and NKG2D, accompanied by increased levels of phosphorylated AKT compared to DNRII- and nontransduced NK cells. Importantly, p-SMAD2/3 was not detectable in NKA- or SynNotch-NK cells in the presence of TGF-β. Pretreatment of NKA- and SynNotch-NK cells with TGF-β displayed antitumor activity comparable to DNRII-NK cells and increased cytotoxicity against neuroblastoma tumor targets compared to nontransduced controls. Repeated infusions of NKA- or SynNotch- NK cells improved the survival of mice with neuroblastoma xenografts when compared to unmodified controls. However, mice injected with NKA-NK cells were found to have the best survival rates and significantly reduced tumor burdens when compared to SynNotch- and DNRII-NK cells. Additionally, unlike the other variants, NKA-NK cells successfully persisted in mice for over 2 months. This suggested that the activation of the DAP12 domain in the NKA switch receptor mediated powerful survival signals while also improving antitumor therapeutic efficacy.

Wang et al conducted a similar study using NK-92 cells genetically engineered to express a switch-receptor containing a fusion of the extracellular domain of TGF-β receptor II and the intracellular domain of NKG2D (TN) [205]. TN-transduced NK-92 cells exposed to TGF-β were resistant to TGF-β signaling and characterized by the lack of intracellular p-SMAD2 when compared to unmodified NK-92 controls. Upon phorbol myristate acetate (PMA) and ionomycin exposure, TGF-β-treated TN-NK-92 cells continued to secrete high levels of IFN-γ. Additionally, TGF-β-treated TN-NK-92-mediated improved cytotoxicity against hepatocellular carcinoma and prostate tumor targets unlike empty vector NK-92 cells. In addition to improved NK-92 cytotoxicity, TN-NK-92 cells displayed improved their homing capacity toward TGF-β secreting tumor cells in vitro. This suggested that the TN transgene may improve the trafficking capacity of NK-92 cells into TGF-β-rich tumor microenvironments.

Unfortunately, adoptive transfer of TN-NK-92 cells only slightly improved outcomes in nude mice containing hepatocellular xenografts suggesting that the future TGF-β switch receptor designs may preclude NKG2D endodomains.

Enhanced NK receptors and genetic deletions: Tumor and TME-resident inhibitory immune cells are able employ multiple mechanisms to impede NK cell antitumor responses. Contact-dependent inhibition is carried out through the expression of ligands toward inhibitory receptors on tumor-infiltrating NK cells. Additionally, the TME contains secreted factors such as soluble ligands that block activating receptors on NK cells, inhibitory cytokines, and matrix metalloproteinases. These mechanisms can induce downregulation of activating receptors and key adaptor molecules required for upregulating inhibitory receptors like PD-1, TIGIT, and TIM-3 resulting in dampened NK cell cytotoxicity and interferon/chemokine responses [196,206–213]. Thus, strategies to improve NK cell-based therapies need to counter these suppressive modalities of the TME to enhance their effectiveness.

Parihar et al genetically engineered primary NK cells to express a chimeric NKG2D receptor that comprised an extracellular NKG2D domain fused with a CD3ζ cytoplasmic domain (NKG2D.ζ) [214]. This design negates the requirement of the DAP10 signaling molecule, whose downregulation in the TME has been reported. NKG2D.ζ NK cells could be expanded using modified K562 feeder cells that expressed 4-BBL and mbIL-15. These expanded NKG2D.ζ NK cells exhibited enhanced cytotoxicity successfully eliminated NKG2D ligand expressing tumor targets. NKG2D.ζ NK cells were also capable of exhibiting this cytotoxicity when cultured with human MDSCs over short- and long-term timepoints unlike unmodified NK cells. This cytotoxicity was retained when NKG2D.ζ NK cells were cultured with MDSCs isolated from patients with neuroblastoma. Importantly, Parihar et al determined that exposure to TGF-β and soluble NKG2D ligands like MICA and MICB did not affect the expression of the NKG2D.ζ receptor indicating potential resistance to TME inhibition when compared to unmodified NK cells which suffered from downregulated NKG2D resulting in impaired cytotoxicity against NKG2D ligand expressing K562 targets. When injected into immunodeficient mice with xenografts containing NKG2D-L+ MDSCs and NKG2DL-neuroblastoma tumor, NKG2D.ζ NK cells successfully improved their survival while also lowering the tumor burden in contrast to unmodified NK cells. Interestingly, when infused in combination with GD2 antigen-directed CAR T-cells, NKG2D.ζ NK cells not only enabled GD2 CAR-T cell recruitment and infiltration into MDSC containing tumor xenografts but also improved their antitumor activity after NK cell-mediated elimination of MDSCs. A similar study employed a chimeric NKG2D receptor that included both DAP10 and CD3ζ cytoplasmic domains (NKG2D.DAP10.ζ) in primary expanded NK cells [215]. Interestingly, incorporation of the DAP10 signaling domain resulted in increased chimeric NKG2D receptor expression when compared to NKG2D.ζ transduced NK cells. NKG2D.ζ NK cells exhibited enhanced cytotoxicity consistently across a variety of solid tumor cells lines and was accompanied by elevated levels of in IFN-γ secretion in contrast to mock transduced NK cells. Importantly, ligand specific stimulation of the NKG2D.DAP10.ζ chimeric receptor resulted in the activation of the DAP10 downstream signaling effectors, NF-κβ and AKT. NKG2D.DAP10.ζ NK cells were also effective in mediating tumor cytotoxicity in NSG mice with osteosarcoma xenografts, which resulted in reduced tumor burden compared to mock transduced NK cells.

Xiao et al also designed a novel chimeric NKG2D comprising an NKG2D extracellular domain fused to a DAP12 intracellular domain (NKG2D.DAP12) in primary NK cells [216].

NKG2D.DAP12 NK cells displayed enhanced cytotoxicity in vitro against colon, ovarian, and pharyngeal cancer cell lines. Multiple infusions of NKG2D.DAP12 NK cells in mice with colorectal xenografts improved their survival while also impeding tumor growth compared to mock transduced NK cells. The clinical efficacy of NKG2D.DAP12 was assessed in a pilot clinical study with three patients with metastatic colorectal cancer using NK cells isolated from haploidentical or autologous sources. NKG2D.DAP12 was introduced in NK cells using mRNA electroporation instead of viral vectors to enable transient receptor expression while also easing the burden on clinical manufacturing. Importantly, repeated doses of NKG2D. DAP12 NK cells did not cause severe adverse events during the 4-week toxicity assessment parents. However, while the therapy did control the growth of malignant ascites, the best overall response was limited to only 1 out of 3 patients and was classified with stable disease.

An alternative strategy to boost native NK cell receptors and strengthen NK cell antitumor responses in the TME is to enhance NK cell-mediated ADCC against tumor cells. Clinical trials have attempted to enhance endogenous and adoptively transferred NK cell expansion and ADCC activity in patients by administering tumor-targeting antibodies in conjunction with cytokines and drugs. However, these approaches only resulted in modest therapeutic responses in patients [217–224]. These modest clinical outcomes may be explained due to the presence of TME-associated matrix metalloproteinases which cleave Fcγ receptors (FcγRIIIa or CD16) which inhibits the capacity of NK cells to mediate ADCC [13]. To overcome this challenge, Snyder et al reported the design of a novel recombinant Fcγ receptor comprising the extracellular section of CD64 (the highest affinity Fcγ) fused with the transmembrane and cytoplasmic signaling domains of CD16A (CD64/CD16A) [225]. This recombinant receptor was implemented in NK-92 and iPSC-derived NK cell (iNK) to improve their ADCC capacity. Importantly, the CD64/CD16A chimeric receptor was designed to lack the binding site for the matrix metalloproteinase ADAM17, thus conferring resistance to ADAM17 induced cleavage in the TME. NK-92 cells and iNK cells that expressed CD64/CD16A showcased enhanced recognition and conjugation capacity to HER2+ ovarian tumor cell upon treatment with the anti-HER2 monoclonal antibody (mAB) trastuzumab, resulting in enhanced IFN-γ secretion. Unlike the endogenous CD16 receptor, CD64/CD16A surface expression was not downregulated after engaging trastuzumab-coated targets. These experimental results were recapitulated in a similar study that also implemented an engineered noncleavable high-affinity CD16A receptor (hnCD16A) in iNK cells [226]. hnCD16A expressing iNK cells were capable of heightened degranulation and IFN-γ secretion in vitro against antibody-coated ovarian and squamous carcinoma cells compared to unmodified iNK cells. Additionally, hnCD16A was stable expressed and resistant to iNK activation induced downregulation. Administration of anti-HER2 mAb in combination with hnCD16A-iNK cells led to improved survival and lowered tumor burden in HER2+ tumor-bearing mice, which demonstrated that hnCD16A induced a stronger ADCC response than the native CD16 receptor.

While the studies listed above have focused on developing improved versions of NK activating receptors, interest in genetic engineering approaches to generate NK cells that lack inhibitory receptors is gaining traction. Pomeroy et al recently implemented CRISPR/Cas9 technology to genetically ablate the expression of the inhibitory receptor PD-1 in primary NK cells [227]. Using this method, the authors were able to achieve a PD-1 knockout (KO) efficiency of 82% in primary expanded NK cells. Furthermore, PD-1 KO NK cells could be secondarily expanded to clinically relevant numbers. Preclinical studies revealed that PD-1 KO in NK cells enhanced their cytotoxicity against prostate and ovarian carcinoma cell lines in vitro and that a single infusion of PD-1 KO NK cells in mice with ovarian xenografts led to lowered tumor burden and moderate improvements in survival compared to unmodified NK cells.

NK checkpoint inhibitors and engagers: NK cells can be genetically engineered to display enhanced cytotoxicity and immune modulation in the TME. However, genetic manipulation of NK cells poses technical challenges that make it less readily available as a therapeutic option [228–231]. There are other methods that enhance NK cell capabilities but bypass genetic-based approaches and proven ex vivo NK cell expansion techniques. These are NK cell-specific checkpoint inhibitors and chimeric proteins that coengage NK cell receptors with tumor targets (i.e., NK cell engagers). Monalizumab is a monoclonal antibody that neutralizes NKG2A, an inhibitory NK receptor. It has shown robust cytotoxicity against cancer cells expressing NKG2A, and against an antibody-coated squamous cell carcinoma of the head and neck (SCCHN) cell line [232]. In a Phase II clinical trial of monalizumab with cetuximab for previously treated squamous cell carcinoma of the head and neck, interim results showed a 31% objective response rate (NCT02643550). The trial suggests that endogenous NK cells and, by extension, adoptively transferred NK cells are inhibited by TME-expressed ligands for NKG2A. This TME inhibition can be countered by introducing a checkpoint blockade against NKG2A. Common side effects of this therapy were mild: fatigue (17%), fever (13%), and headache (10%). BiKEs (Bispecific Natural Killer Engagers) and TriKEs (Trispecific Natural Killer Engagers) are alternative antibody-based molecules that enhance antitumor activity of NK cells. BiKEs have two monoclonal antibody fragments connected by a linker. One fragment targets endogenous CD16 on NK cells, while the other fragment is specific to a tumor-associated antigen. TriKEs are similarly designed molecules that additionally include a stimulatory cytokine. These cytokines such as IL-15 or IL-21 improve NK expansion, viability, and function in the TME. BiKEs and TriKEs enhance NK cell cytotoxicity against tumor via CD16 activation, with the TriKE system additionally increasing NK cell survival. They also enhanced NK cell cytotoxicity against CD33+ MDSCs, a component of the TiME [233–235]. Also, TriKEs performed better than BiKEs in improving NK cell survival and proliferation capacity in tumor-bearing NSG mice [235]. Versions of these agents are now being tested in clinical trials (NCT03214666; NCT00560794).

The presence of immunosuppressive cells of the TME has an adverse effect on the antitumor response of tumor-infiltrating NK cells, which leads to tumor progression and poor patient survival [236–238]. Therapeutic approaches designed to destroy these immunoregulators may deprive the tumor and the TME of key soluble growth factors and suppressive cytokines leading to enhanced antitumor responses [236]. Alternatively, therapies that enable infiltrating NK cells to evade or exploit the TME to enhance their own recruitment to, persistence, and cytotoxicity can lead to more robust and sustained immune responses. In order to address the problems of tumor therapeutic resistance, it is important to recognize the diversity in TME composition amongst patients. Therefore, appropriate strategy selection should be made with a focus on combination therapies to treat patients with solid tumors more effectively.

Perspectives and conclusions

The data presented in this chapter highlight the suppressive mechanisms of TME-resident inhibitory immune cells, current therapies to reverse this suppression, and techniques that can be used to study the role of NK cells in the TME. Since multiple suppressive modalities are capable of being employed by immune and nonimmune components of the TME, the therapies that are designed to overcome them must be equally diverse, innovative, and resilient to improve the efficacy of NK cells in the TME. While such therapies have been developed and assessed in preclinical studies under optimal controlled conditions, they possess limitations that may impact their ability to function in the TME of patients. These limitations include bioavailability of administered agents, concentrations/availability of exogenous ligands that trigger NK cell activity, and ability of TME components to "sequester" immune-stimulatory agents. For example, the effectiveness of therapies that utilize chimeric receptors to target specific molecules is largely dependent on the availability of these molecules in the TME. Similarly, NK cells transduced with either dominant negative or switch receptors presented in this chapter both rely on their access to TGF-β. However, the TMEs of different types of tumors are unlikely to contain the same levels of TGF-β in their cytokine milieu. In addition, the effectiveness of chimeric NKG2D receptors depends on the expression of NKG2D ligands tumor and components of the TME which are known to shed NKG2D ligands as a mechanism of immune evasion [239]. In the context of monoclonal antibody-based therapies, their efficacy if often hindered by stromal cellular components on the TME periphery that deprive these therapeutic antibodies access to the TME [240]. These antibodies may also be sequestered at least transiently by Fc receptors of TME-resident MDSCs and M2 macrophages [155,241]. Such limitations also limit the efficacy of NK cell-based therapies that utilize tumor-targeting antibodies to mediate cytotoxicity such as high affinity CD16 receptors as well as BiKEs and TriKEs.

It is imperative that novel NK cell-based therapies address other key deficiencies of NK cells in the TME. MDSCs accumulate in perivascular regions of the TME and in these areas may serve to impede trafficking of NK cells into the TME [116]. Ng et al genetically modified CAR NK cells to express CXCR1 and improved their infiltration and antitumor activity in mice-bearing ovarian carcinoma xenografts. MDSCs are capable of regulating NK cell activation by producing high levels of reactive oxygen species to induce apoptosis [242–244]. They can also impact NK cell immunometabolism by secreting inhibitory metabolites like adenosine and by depriving NK cells of essential amino acids via arginases [245–249]. Recently, studies have aimed to improve NK cell activity by targeting inhibitory monocytes via chemical inhibitors, antibody-blockade, and enzyme inhibitors [244,250,251]. However, some of these therapeutic options also inhibit shared pathways, which result in the downregulation of NKG2D ligand expression in tumor cells. It would be important to develop alternative strategies that only target inhibitory immune cells specially or boost NK cell hypoxic resistance and metabolism to improve their cytotoxicity and persistence within the TME.

Recent advances in technology have made it possible to study NK cell populations in the TME providing high-throughput and large-scale analysis in imaging, phenotyping, and transcriptomics. However, metabolic assays have not been applied in the same vein. While high-throughput metabolic screening do exist, they have yet to be applied to study tumor-infiltrating NK cell populations in the TME [252]. While multiparametric testing has unearthed new data on the immune TME composition and their correlations with tumor progression and control, the functional state of many immune cells, including NK cells in this context have not been elucidated. It would be beneficial to describe these functional differences in different TME tissues to better understand the role of inhibitory immune cells in suppressing NK cell cytotoxicity in the TME. NK cells are a vital component of the antitumor response. Uncovering the mechanisms that govern NK cell suppression by inhibitory immune cells in the TME will aid the development of sophisticated therapeutic modalities to treat patients with solid tumors more effectively.

Clinical trials evaluating the efficacy of TME-altering approaches are currently underway (Table 2). While patient survival and assessment of tumor burden are the primary objectives for these trials, evaluating the composition of inhibitory immune cells within the TME is not always considered. Tissue assessment and imaging using multiplex IHC and flow cytometry could be performed on tumor biopsies from patients prior to treatment to assess the

TABLE 2 Therapeutic strategies targeting tumor microenvironment.

Modality	Agent	Target cell (target molecule)	Clinical trail	Reference
Small molecule	Doxorubicin containing Pegylated-Folate-Liposome (Lytic agent)	M2 Macrophage (Folate Receptor)	–	[149]
	SX-682 (Blocking recruitment)	MDSC (CXCR2)	NCT03161431	
	Reparixin (Blocking recruitment)	MDSC (CXCR2)	NCT02370238	
	AZD5069 (Blocking recruitment)	MDSC (CXCR2)	NCT02499328	
	Pexidartinib (Blocking recruitment)	MDSC (CSF-1R)	NCT01349036	
	Sunitinib (Tyrosine kinase)	MDSC	–	[253]
	Ibrutinib (Tyrosine kinase)	MDSC	–	[153,154]
TLR9 agonist	CpG Oligodeoxynucleotides	M2 Macrophages	–	[164]
		MDSC	NCT03445533	
			NCT03831295	
			NCT02680184	
	AZD9150 (STAT3 inhibitor)	MDSC	NCT03421353	
Monoclonal antibody therapy	Anti-CD25 antibody	Tregs	NCT00573495	
	Anti-CSF-1R antibody	Macrophage	NCT01494688	
Epigenetic regulator	Entinostat (HDAC inhibitor)	MDSC	NCT02708680	
			NCT02915523	
Viral therapy	Hydroxyprostaglandin dehydrogenase	MDSC (Prostaglandin E2)	–	[182]

Summary of implementation strategies and approaches to target inhibitory immune cells of the solid tumor microenvironment.

'immunosuppressive strength' of their TMEs. Additionally, such testing may also indicate which patients are likely responders and could be included in the eligibility criteria for patient recruitment. Multidimensional testing after treatment may provide additional insight on the secondary effects of therapies on TME composition, TME reorganization fitness. Importantly, monitoring of immune cell populations of patient TMEs posttreatment may indicate which patients are more likely to relapse after completion of treatment. Over the past decade, there is a growing number of clinical trials that evaluating therapies aimed at activating endogenous NK cells or infusing donor-derived NK cells to combat tumors in the TME. Multiparametric approaches could be utilized to monitor and assess various NK cell parameters like antitumor activity, exhaustion marker profiles, and infiltration capacity in the TME. Such testing would provide a more robust analysis and determination of effects of the selected therapeutic approach in patients.

The complexity of the TME poses a major hurdle to effective antitumor immune responses. Therefore, a fundamental understanding of the TME is required for the development and validation of novel TME-targeting strategies. The implementation of sophisticated technologies to assess patient responsiveness and guide the implementation of therapies will aid the development of personalized TME targeting therapeutic approaches and subsequently lead to improved clinical outcomes in patients with solid tumors.

References

[1] Poli A, et al. CD56bright natural killer (NK) cells: an important NK cell subset. Immunology 2009;126:458–65.
[2] Fauriat C, Long EO, Ljunggren H-G, Bryceson YT. Regulation of human NK-cell cytokine and chemokine production by target cell recognition. Blood 2010;115:2167–76.
[3] Melaiu O, Lucarini V, Cifaldi L, Fruci D. Influence of the tumor microenvironment on NK cell function in solid tumors. Front Immunol 2020;10:3038.
[4] Denkert C, et al. Tumour-infiltrating lymphocytes and prognosis in different subtypes of breast cancer: a pooled analysis of 3771 patients treated with neoadjuvant therapy. Lancet Oncol 2018;19:40–50.
[5] Stevens A, Klöter I, Roggendorf W. Inflammatory infiltrates and natural killer cell presence in human brain tumors. Cancer 1988;61:738–43.
[6] Tartter PI, Steinberg B, Barron DM, Martinelli G. The prognostic significance of natural killer cytotoxicity in patients with colorectal cancer. Arch Surg 1987;122:1264–8.
[7] Takanami I, Takeuchi K, Giga M. The prognostic value of natural killer cell infiltration in resected pulmonary adenocarcinoma. J Thorac Cardiovasc Surg 2001;121:1058–63.
[8] Fregni G, Perier A, Avril M-F, Caignard A. NK cells sense tumors, course of disease and treatments. Onco Targets Ther 2012;1:38–47.
[9] Murciano-Goroff YR, Warner AB, Wolchok JD. The future of cancer immunotherapy: microenvironment-targeting combinations. Cell Res 2020;30:507–19.
[10] Habif G, Crinier A, André P, Vivier E, Narni-Mancinelli E. Targeting natural killer cells in solid tumors. Cell Mol Immunol 2019;16:415–22.
[11] Quail D, Joyce J. Microenvironmental regulation of tumor progression and metastasis. Nat Med 2013;19:1423–37.
[12] Qi D, Wu E. Cancer prognosis: considering tumor and its microenvironment as a whole. EBioMedicine 2019;43:28–9.
[13] Rossi GR, Trindade ES, Souza-Fonseca-Guimaraes F. Tumor microenvironment-associated extracellular matrix components regulate NK cell function. Front Immunol 2020;11:73.
[14] Mitsiades N, Yu W, Poulaki V, Tsokos M, Stamenkovic I. Matrix metalloproteinase-7-mediated cleavage of Fas ligand protects tumor cells from chemotherapeutic drug cytotoxicity. Cancer Res 2001;61:577–81.
[15] Schaaf MB, Garg AD, Agostinis P. Defining the role of the tumor vasculature in antitumor immunity and immunotherapy. Cell Death Dis 2018;9:1–14.

[16] Padera TP, et al. Pathology: cancer cells compress intratumour vessels. Nature 2004;427:695.
[17] Welti J, Loges S, Dimmeler S, Carmeliet P. Recent molecular discoveries in angiogenesis and antiangiogenic therapies in cancer. J Clin Invest 2013;123:3190–200.
[18] Chouaib S, Noman MZ, Kosmatopoulos K, Curran MA. Hypoxic stress: obstacles and opportunities for innovative immunotherapy of cancer. Oncogene 2017;36:439–45.
[19] Bouzin C, Brouet A, De Vriese J, Dewever J, Feron O. Effects of vascular endothelial growth factor on the lymphocyte-endothelium interactions: identification of caveolin-1 and nitric oxide as control points of endothelial cell anergy. J Immunol 2007;178:1505–11.
[20] Barrett R, Puré E. Cancer-associated fibroblasts: key determinants of tumor immunity and immunotherapy. Curr Opin Immunol 2020;64:80–7.
[21] Balsamo M, et al. Melanoma-associated fibroblasts modulate NK cell phenotype and antitumor cytotoxicity. PNAS 2009;106:20847–52.
[22] Inoue T, et al. Cancer-associated fibroblast suppresses killing activity of natural killer cells through downregulation of poliovirus receptor (PVR/CD155), a ligand of activating NK receptor. Int J Oncol 2016;49:1297–304.
[23] Magri S, et al. Human MDSCs derived from the bone marrow maintain their functional ability but have a reduced frequency of induction in the elderly compared to pediatric donors. Immun Ageing 2020;17:27.
[24] Barreda DR, Hanington PC, Belosevic M. Regulation of myeloid development and function by colony stimulating factors. Dev Comp Immunol 2004;28:509–54.
[25] Hamilton JA. GM-CSF as a target in inflammatory/autoimmune disease: current evidence and future therapeutic potential. Expert Rev Clin Immunol 2015;11:457–65.
[26] Morales JK, Kmieciak M, Knutson KL, Bear HD, Manjili MH. GM-CSF is one of the main breast tumor-derived soluble factors involved in the differentiation of CD11b-Gr1- bone marrow progenitor cells into myeloid-derived suppressor cells. Breast Cancer Res Treat 2010;123:39–49.
[27] Umansky V, Blattner C, Gebhardt C, Utikal J. The role of myeloid-derived suppressor cells (MDSC) in cancer progression. Vaccines (Basel) 2016;4:36.
[28] Huang B, et al. CCL2/CCR2 pathway mediates recruitment of myeloid suppressor cells to cancers. Cancer Lett 2007;252:86–92.
[29] OuYang L-Y, et al. Tumor-induced myeloid-derived suppressor cells promote tumor progression through oxidative metabolism in human colorectal cancer. J Transl Med 2015;13:47.
[30] Yang L, et al. Expansion of myeloid immune suppressor gr+CD11b+ cells in tumor-bearing host directly promotes tumor angiogenesis. Cancer Cell 2004;6:409–21.
[31] Kumar V, Patel S, Tcyganov E, Gabrilovich DI. The nature of myeloid-derived suppressor cells in the tumor microenvironment. Trends Immunol 2016;37:208–20.
[32] Bruno A, Mortara L, Baci D, Noonan DM, Albini A. Myeloid derived suppressor cells interactions with natural killer cells and pro-angiogenic activities: roles in tumor progression. Front Immunol 2019;10:771.
[33] Li H, Han Y, Guo Q, Zhang M, Cao X. Cancer-expanded myeloid-derived suppressor cells induce anergy of NK cells through membrane-bound TGF-beta 1. J Immunol 2009;182:240–9.
[34] Huang B, et al. Gr-1+CD115+ immature myeloid suppressor cells mediate the development of tumor-induced T regulatory cells and T-cell anergy in tumor-bearing host. Cancer Res 2006;66:1123–31.
[35] Zhao F, Korangy F, Greten TF. Cellular immune suppressor mechanisms in patients with hepatocellular carcinoma. Dig Dis 2012;30:477–82.
[36] Beury DW, et al. Cross-talk among myeloid-derived suppressor cells, macrophages, and tumor cells impacts the inflammatory milieu of solid tumors. J Leukoc Biol 2014;96:1109–18.
[37] Liu Y, Wei J, Guo G, Zhou J. Norepinephrine-induced myeloid-derived suppressor cells block T-cell responses via generation of reactive oxygen species. Immunopharmacol Immunotoxicol 2015;37:359–65.
[38] Corzo CA, et al. Mechanism regulating reactive oxygen species in tumor-induced myeloid-derived suppressor cells. J Immunol 2009;182:5693–701.
[39] Gabrilovich DI, Nagaraj S. Myeloid-derived-suppressor cells as regulators of the immune system. Nat Rev Immunol 2009;9:162–74.
[40] Rodríguez PC, Ochoa AC. Arginine regulation by myeloid derived suppressor cells and tolerance in cancer: mechanisms and therapeutic perspectives. Immunol Rev 2008;222:180–91.
[41] Yamauchi Y, et al. Circulating and tumor myeloid-derived suppressor cells in resectable non-small cell lung cancer. Am J Respir Crit Care Med 2018;198:777–87.
[42] Pesce S, et al. PD/1-PD-ls checkpoint: insight on the potential role of NK cells. Front Immunol 2019;10, 1242.

[43] Shand FHW, et al. Tracking of intertissue migration reveals the origins of tumor-infiltrating monocytes. Proc Natl Acad Sci U S A 2014;111:7771–6.
[44] Qian B, et al. A distinct macrophage population mediates metastatic breast cancer cell extravasation, establishment and growth. PLoS One 2009;4, e6562.
[45] Qian B, Pollard JW. Macrophage diversity enhances tumor progression and metastasis. Cell 2010;141:39–51.
[46] Laviron M, Boissonnas A. Ontogeny of tumor-associated macrophages. Front Immunol 2019;10, 1799.
[47] Chitu V, Stanley ER. Colony-stimulating factor-1 in immunity and inflammation. Curr Opin Immunol 2006;18:39–48.
[48] Laoui D, et al. Tumor-associated macrophages in breast cancer: distinct subsets, distinct functions. Int J Dev Biol 2011;55:861–7.
[49] Wyckoff JB, et al. Direct visualization of macrophage-assisted tumor cell intravasation in mammary tumors. Cancer Res 2007;67:2649–56.
[50] Lin EY, Pollard JW. Tumor-associated macrophages press the angiogenic switch in breast cancer. Cancer Res 2007;67:5064–6.
[51] Lin EY, et al. Vascular endothelial growth factor restores delayed tumor progression in tumors depleted of macrophages. Mol Oncol 2007;1:288–302.
[52] Linde N, et al. Vascular endothelial growth factor-induced skin carcinogenesis depends on recruitment and alternative activation of macrophages. J Pathol 2012;227:17–28.
[53] Kitamura T, et al. SMAD4-deficient intestinal tumors recruit CCR1+ myeloid cells that promote invasion. Nat Genet 2007;39:467–75.
[54] Kitamura T, et al. Inactivation of chemokine (C-C motif) receptor 1 (CCR1) suppresses colon cancer liver metastasis by blocking accumulation of immature myeloid cells in a mouse model. Proc Natl Acad Sci U S A 2010;107:13063–8.
[55] Su S, et al. A positive feedback loop between mesenchymal-like cancer cells and macrophages is essential to breast cancer metastasis. Cancer Cell 2014;25:605–20.
[56] Du R, et al. HIF1alpha induces the recruitment of bone marrow-derived vascular modulatory cells to regulate tumor angiogenesis and invasion. Cancer Cell 2008;13:206–20.
[57] Coffelt SB, Hughes R, Lewis CE. Tumor-associated macrophages: effectors of angiogenesis and tumor progression. Biochim Biophys Acta 2009;1796:11–8.
[58] Nucera S, Biziato D, De Palma M. The interplay between macrophages and angiogenesis in development, tissue injury and regeneration. Int J Dev Biol 2011;55:495–503.
[59] Sangaletti S, et al. Macrophage-derived SPARC bridges tumor cell-extracellular matrix interactions toward metastasis. Cancer Res 2008;68:9050–9.
[60] Borrego F, Ulbrecht M, Weiss EH, Coligan JE, Brooks AG. Recognition of human histocompatibility leukocyte antigen (HLA)-E complexed with HLA class I signal sequence-derived peptides by CD94/NKG2 confers protection from natural killer cell-mediated lysis. J Exp Med 1998;187:813–8.
[61] Noy R, Pollard JW. Tumor-associated macrophages: from mechanisms to therapy. Immunity 2014;41:49–61.
[62] Noman MZ, et al. PD-L1 is a novel direct target of HIF-1α, and its blockade under hypoxia enhanced MDSC-mediated T cell activation. J Exp Med 2014;211:781–90.
[63] Sharda DR, et al. Regulation of macrophage arginase expression and tumor growth by the Ron receptor tyrosine kinase. J Immunol 2011;187:2181–92.
[64] Doedens AL, et al. Macrophage expression of HIF-1α suppresses T cell function and promotes tumor progression. Cancer Res 2010;70:7465–75.
[65] Sica A, Mantovani A. Macrophage plasticity and polarization: in vivo veritas. J Clin Invest 2012;122:787–95.
[66] Romano M, Fanelli G, Albany CJ, Giganti G, Lombardi G. Past, present, and future of regulatory T cell therapy in transplantation and autoimmunity. Front Immunol 2019;10.
[67] Zhang X, Olsen N, Zheng SG. The progress and prospect of regulatory T cells in autoimmune diseases. J Autoimmun 2020;111, 102461.
[68] Fridman WH, Zitvogel L, Sautès-Fridman C, Kroemer G. The immune contexture in cancer prognosis and treatment. Nat Rev Clin Oncol 2017;14:717–34.
[69] Ghiringhelli F, et al. CD4+CD25+ regulatory T cells inhibit natural killer cell functions in a transforming growth factor-β-dependent manner. J Exp Med 2005;202:1075–85.
[70] Chaudhary B, Elkord E. Regulatory T cells in the tumor microenvironment and cancer progression: role and therapeutic targeting. Vaccines (Basel) 2016;4:28.

[71] Priceman SJ, et al. S1PR1 is crucial for accumulation of regulatory T cells in tumors via STAT3. Cell Rep 2014;6:992–9.
[72] Ghiringhelli F, et al. Tumor cells convert immature myeloid dendritic cells into TGF-β-secreting cells inducing CD4+CD25+ regulatory T cell proliferation. J Exp Med 2005;202:919–29.
[73] Hoechst B, et al. A new population of myeloid-derived suppressor cells in hepatocellular carcinoma patients induces CD4(+)CD25(+)Foxp3(+) T cells. Gastroenterology 2008;135:234–43.
[74] Wu J, et al. Effect of HIF1α on Foxp3 expression in CD4+ CD25- T lymphocytes. Microbiol Immunol 2014;58:409–15.
[75] Ben-Shoshan J, Maysel-Auslender S, Mor A, Keren G, George J. Hypoxia controls CD4+CD25+ regulatory T-cell homeostasis via hypoxia-inducible factor-1alpha. Eur J Immunol 2008;38:2412–8.
[76] Deng B, et al. Intratumor hypoxia promotes immune tolerance by inducing regulatory T cells via TGF-β1 in gastric cancer. PLoS One 2013;8, e63777.
[77] Fallarino F, et al. Modulation of tryptophan catabolism by regulatory T cells. Nat Immunol 2003;4:1206–12.
[78] Fourcade J, et al. CD226 opposes TIGIT to disrupt Tregs in melanoma. JCI Insight 2018;3, e121157.
[79] Vignali DAA, Collison LW, Workman CJ. How regulatory T cells work. Nat Rev Immunol 2008;8:523–32.
[80] Pandiyan P, Zheng L, Ishihara S, Reed J, Lenardo MJ. CD4 + CD25 + Foxp3 + regulatory T cells induce cytokine deprivation-mediated apoptosis of effector CD4 + T cells. Nat Immunol 2007;8:1353–62.
[81] Grossman WJ, et al. Differential expression of granzymes a and B in human cytotoxic lymphocyte subsets and T regulatory cells. Blood 2004;104:2840–8.
[82] Gondek DC, Lu L-F, Quezada SA, Sakaguchi S, Noelle RJ. Cutting edge: contact-mediated suppression by CD4 +CD25+ regulatory cells involves a granzyme B-dependent, perforin-independent mechanism. J Immunol 2005;174:1783–6.
[83] Ren X, et al. Involvement of cellular death in TRAIL/DR5-dependent suppression induced by CD4(+)CD25(+) regulatory T cells. Cell Death Differ 2007;14:2076–84.
[84] Garín MI, et al. Galectin-1: a key effector of regulation mediated by CD4+CD25+ T cells. Blood 2007;109:2058–65.
[85] Cao X, et al. Granzyme B and perforin are important for regulatory T cell-mediated suppression of tumor clearance. Immunity 2007;27:635–46.
[86] Alvarez D, Vollmann EH, von Andrian UH. Mechanisms and consequences of dendritic cell migration. Immunity 2008;29:325.
[87] Lombardi VC, Khaiboullina SF, Rizvanov AA. Plasmacytoid dendritic cells, a role in neoplastic prevention and progression. Eur J Clin Investig 2015;45:1–8.
[88] Wiseman BS, Werb Z. Stromal effects on mammary gland development and breast cancer. Science 2002;296:1046–9.
[89] Curiel TJ, et al. Blockade of B7-H1 improves myeloid dendritic cell-mediated antitumor immunity. Nat Med 2003;9:562–7.
[90] Curiel TJ, et al. Dendritic cell subsets differentially regulate angiogenesis in human ovarian cancer. Cancer Res 2004;64:5535–8.
[91] Zitvogel L, Galluzzi L, Kepp O, Smyth MJ, Kroemer G. Type I interferons in anticancer immunity. Nat Rev Immunol 2015;15:405–14.
[92] Young LJ, et al. Differential MHC class II synthesis and ubiquitination confers distinct antigen-presenting properties on conventional and plasmacytoid dendritic cells. Nat Immunol 2008;9:1244–52.
[93] Matta BM, Castellaneta A, Thomson AW. Tolerogenic plasmacytoid DC. Eur J Immunol 2010;40:2667–76.
[94] Wylie B, Macri C, Mintern JD, Waithman J. Dendritic cells and cancer: from biology to therapeutic intervention. Cancers (Basel) 2019;11:521.
[95] Hartmann E, et al. Identification and functional analysis of tumor-infiltrating plasmacytoid dendritic cells in head and neck cancer. Cancer Res 2003;63:6478–87.
[96] Faith A, et al. Plasmacytoid dendritic cells from human lung cancer draining lymph nodes induce Tc1 responses. Am J Respir Cell Mol Biol 2007;36:360–7.
[97] Diao J, Zhao J, Winter E, Cattral MS. Recruitment and differentiation of conventional dendritic cell precursors in tumors. J Immunol 2010;184:1261–7.
[98] Hargadon KM, et al. Melanoma-derived factors alter the maturation and activation of differentiated tissue-resident dendritic cells. Immunol Cell Biol 2016;94:24–38.
[99] Heufler C, et al. Interleukin-12 is produced by dendritic cells and mediates T helper 1 development as well as interferon-gamma production by T helper 1 cells. Eur J Immunol 1996;26:659–68.

[100] Ruffell B, et al. Macrophage IL-10 blocks CD8+ T cell-dependent responses to chemotherapy by suppressing IL-12 expression in intratumoral dendritic cells. Cancer Cell 2014;26:623–37.
[101] Lim TS, et al. PD-1 expression on dendritic cells suppresses CD8+ T cell function and antitumor immunity. Onco Targets Ther 2015;5, e1085146.
[102] Mandal R, et al. The head and neck cancer immune landscape and its immunotherapeutic implications. JCI Insight 2016;1, e89829.
[103] Coca S, et al. The prognostic significance of intratumoral natural killer cells in patients with colorectal carcinoma. Cancer 1997;79:2320–8.
[104] Lim JCT, et al. An automated staining protocol for seven-colour immunofluorescence of human tissue sections for diagnostic and prognostic use. Pathology 2018;50:333–41.
[105] Blom S, et al. Systems pathology by multiplexed immunohistochemistry and whole-slide digital image analysis. Sci Rep 2017;7:15580.
[106] Bobrow MN, Shaughnessy KJ, Litt GJ. Catalyzed reporter deposition, a novel method of signal amplification: II. Application to membrane immunoassays. J Immunol Methods 1991;137:103–12.
[107] Stack EC, Wang C, Roman KA, Hoyt CC. Multiplexed immunohistochemistry, imaging, and quantitation: a review, with an assessment of Tyramide signal amplification, multispectral imaging and multiplex analysis. Methods 2014;70:46–58.
[108] Giesen C, et al. Highly multiplexed imaging of tumor tissues with subcellular resolution by mass cytometry. Nat Methods 2014;11:417–22.
[109] Angelo M, et al. Multiplexed ion beam imaging of human breast tumors. Nat Med 2014;20:436–42.
[110] Larsson C, Grundberg I, Söderberg O, Nilsson M. In situ detection and genotyping of individual mRNA molecules. Nat Methods 2010;7:395–7.
[111] Ke R, et al. In situ sequencing for RNA analysis in preserved tissue and cells. Nat Methods 2013;10:857–60.
[112] Wang F, et al. RNAscope: a novel in situ RNA analysis platform for formalin-fixed, paraffin-embedded tissues. J Mol Diagn 2012;14:22–9.
[113] Player AN, Shen L-P, Kenny D, Antao VP, Kolberg JA. Single-copy gene detection using branched DNA (bDNA) in situ hybridization. J Histochem Cytochem 2001;49:603–11.
[114] Schulz D, et al. Simultaneous multiplexed imaging of mRNA and proteins with subcellular resolution in breast cancer tissue samples by mass cytometry. Cell Syst 2018;6:25–36.e5.
[115] Litjens G, et al. A survey on deep learning in medical image analysis. Med Image Anal 2017;42:60–88.
[116] Devkota L, et al. Detection of response to tumor microenvironment-targeted cellular immunotherapy using nano-radiomics. Sci Adv 2020;6, eaba6156.
[117] Yang Y, Li C, Liu T, Dai X, Bazhin AV. Myeloid-derived suppressor cells in tumors: from mechanisms to antigen specificity and microenvironmental regulation. Front Immunol 2020;11:1371.
[118] Hoffmann SHL, et al. Visualization and quantification of in vivo homing kinetics of myeloid-derived suppressor cells in primary and metastatic cancer. Theranostics 2019;9:5869–85.
[119] Spitzer MH, Nolan GP. Mass cytometry: single cells, many features. Cell 2016;165:780–91.
[120] Lou X, et al. Polymer-based elemental tags for sensitive bioassays. Angew Chem Int Ed 2007;46:6111–4.
[121] Bandura DR, et al. Mass cytometry: technique for real time single cell multitarget immunoassay based on inductively coupled plasma time-of-flight mass spectrometry. Anal Chem 2009;81:6813–22.
[122] Lavin Y, et al. Innate immune landscape in early lung adenocarcinoma by paired single-cell analyses. Cell 2017;169:750–765.e17.
[123] Chevrier S, et al. An immune atlas of clear cell renal cell carcinoma. Cell 2017;169:736–749.e18.
[124] Berrien-Elliott MM, et al. Multidimensional analyses of donor memory-like NK cells reveal new associations with response after adoptive immunotherapy for leukemia. Cancer Discov 2020;10:1854–71.
[125] Gonzalez-Gugel E, et al. Abstract A191: novel approaches to the study of NK cell exhaustion in humans. Cancer Immunol Res 2019;7:A191.
[126] van Dijk EL, Auger H, Jaszczyszyn Y, Ten Thermes C. Years of next-generation sequencing technology. Trends Genet 2014;30:418–26.
[127] Bindea G, et al. Spatiotemporal dynamics of intratumoral immune cells reveal the immune landscape in human cancer. Immunity 2013;39:782–95.
[128] Aran D, Hu Z, Butte AJ. xCell: digitally portraying the tissue cellular heterogeneity landscape. Genome Biol 2017;18:220.

[129] Tappeiner E, et al. TIminer: NGS data mining pipeline for cancer immunology and immunotherapy. Bioinformatics 2017;33:3140–1.
[130] Charoentong P, et al. Pan-cancer immunogenomic analyses reveal genotype-immunophenotype relationships and predictors of response to checkpoint blockade. Cell Rep 2017;18:248–62.
[131] Becht E, et al. Estimating the population abundance of tissue-infiltrating immune and stromal cell populations using gene expression. Genome Biol 2016;17:218.
[132] Petitprez F, et al. Transcriptomic analysis of the tumor microenvironment to guide prognosis and immunotherapies. Cancer Immunol Immunother 2018;67:981–8.
[133] Qian J, et al. A pan-cancer blueprint of the heterogeneous tumor microenvironment revealed by single-cell profiling. Cell Res 2020;30:745–62.
[134] Zheng C, et al. Landscape of infiltrating T cells in liver cancer revealed by single-cell sequencing. Cell 2017;169:1342–1356.e16.
[135] Tirosh I, et al. Dissecting the multicellular ecosystem of metastatic melanoma by single-cell RNA-seq. Science 2016;352:189–96.
[136] Schelker M, et al. Estimation of immune cell content in tumour tissue using single-cell RNA-seq data. Nat Commun 2017;8:2032.
[137] Racle J, de Jonge K, Baumgaertner P, Speiser DE, Gfeller D. Simultaneous enumeration of cancer and immune cell types from bulk tumor gene expression data. elife 2017;6, e26476.
[138] Ståhl PL, et al. Visualization and analysis of gene expression in tissue sections by spatial transcriptomics. Science 2016;353:78–82.
[139] Salmén F, et al. Multidimensional transcriptomics provides detailed information about immune cell distribution and identity in HER2+ breast tumors. bioRxiv 2018;, 358937. https://doi.org/10.1101/358937.
[140] Li Y, Wan YY, Zhu B. Immune cell metabolism in tumor microenvironment. Adv Exp Med Biol 2017;1011:163–96.
[141] Zaiatz-Bittencourt V, Finlay DK, Gardiner CM. Canonical TGF-β signaling pathway represses human NK cell metabolism. J Immunol 2018;200:3934–41.
[142] van der Windt GJW, Chang C-H, Pearce EL. Measuring bioenergetics in T cells using a seahorse extracellular flux analyzer. Curr Protoc Immunol 2016;113:3.16B.1–3.16B.14.
[143] Zhang J, Zhang Q. Using seahorse machine to measure OCR and ECAR in cancer cells. In: Haznadar M, editor. Cancer metabolism: methods and protocols. Springer; 2019. p. 353–63. https://doi.org/10.1007/978-1-4939-9027-6_18.
[144] Russell S, Wojtkowiak J, Neilson A, Gillies RJ. Metabolic profiling of healthy and cancerous tissues in 2D and 3D. Sci Rep 2017;7, 15285.
[145] Tomás-Bort E, Kieler M, Sharma S, Candido JB, Loessner D. 3D approaches to model the tumor microenvironment of pancreatic cancer. Theranostics 2020;10:5074–89.
[146] Tsai M-J, Chang W-A, Huang M-S, Kuo P-L. Tumor microenvironment: a new treatment target for cancer. ISRN Biochem 2014;2014, 351959.
[147] Bedard PL, Hyman DM, Davids MS, Siu LL. Small molecules, big impact: 20 years of targeted therapy in oncology. Lancet 2020;395:1078–88.
[148] Law AMK, Valdes-Mora F, Gallego-Ortega D. Myeloid-derived suppressor cells as a therapeutic target for cancer. Cell 2020;9:561.
[149] Deng C, et al. Tumors and their microenvironment dual-targeting chemotherapy with local immune adjuvant therapy for effective antitumor immunity against breast cancer. Adv Sci (Weinh) 2019;6, 1801868.
[150] Ko JS, et al. Sunitinib mediates reversal of myeloid-derived suppressor cell accumulation in renal cell carcinoma patients. Clin Cancer Res 2009;15:2148–57.
[151] Vetsika E-K, Koukos A, Kotsakis A. Myeloid-derived suppressor cells: major figures that shape the immunosuppressive and angiogenic network in cancer. Cell 2019;8:1647.
[152] Kao J, et al. Targeting immune suppressing myeloid-derived suppressor cells in oncology. Crit Rev Oncol Hematol 2011;77:12–9.
[153] Stiff A, et al. Myeloid-derived suppressor cells express Bruton's tyrosine kinase and can be depleted in tumor bearing hosts by ibrutinib treatment. Cancer Res 2016;76:2125–36.
[154] Varikuti S, et al. Ibrutinib treatment inhibits breast cancer progression and metastasis by inducing conversion of myeloid-derived suppressor cells to dendritic cells. Br J Cancer 2020;122:1005–13.

[155] Weber R, et al. Myeloid-derived suppressor cells hinder the anti-cancer activity of immune checkpoint inhibitors. Front Immunol 2018;9, 1310.
[156] Steele CW, et al. CXCR2 inhibition profoundly suppresses metastases and augments immunotherapy in pancreatic ductal adenocarcinoma. Cancer Cell 2016;29:832–45.
[157] Alfaro C, et al. Tumor-produced interleukin-8 attracts human myeloid-derived suppressor cells and elicits extrusion of neutrophil extracellular traps (NETs). Clin Cancer Res 2016;22:3924–36.
[158] Sun, L. et al. Inhibiting myeloid-derived suppressor cell trafficking enhances T cell immunotherapy. JCI Insight 4, e126853.
[159] Priceman SJ, et al. Targeting distinct tumor-infiltrating myeloid cells by inhibiting CSF-1 receptor: combating tumor evasion of antiangiogenic therapy. Blood 2010;115:1461–71.
[160] Holmgaard RB, Zamarin D, Lesokhin A, Merghoub T, Wolchok JD. Targeting myeloid-derived suppressor cells with colony stimulating factor-1 receptor blockade can reverse immune resistance to immunotherapy in indoleamine 2,3-dioxygenase-expressing tumors. EBioMedicine 2016;6:50–8.
[161] Zhu Y, et al. CSF1/CSF1R blockade reprograms tumor-infiltrating macrophages and improves response to T cell checkpoint immunotherapy in pancreatic cancer models. Cancer Res 2014;74:5057–69.
[162] Adams S. Toll-like receptor agonists in cancer therapy. Immunotherapy 2009;1:949–64.
[163] Shirota H, Tross D, Klinman DM. CpG oligonucleotides as cancer vaccine adjuvants. Vaccines (Basel) 2015;3:390–407.
[164] Liu M, et al. Metabolic rewiring of macrophages by CpG potentiates clearance of cancer cells and overcomes tumor-expressed CD47-mediated 'don't eat me signal. Nat Immunol 2019;20:265–75.
[165] Zoglmeier C, et al. CpG blocks immunosuppression by myeloid-derived suppressor cells in tumor-bearing mice. Clin Cancer Res 2011;17:1765–75.
[166] Moreira D, et al. STAT3 inhibition combined with CpG immunostimulation activates antitumor immunity to eradicate genetically distinct castration-resistant prostate cancers. Clin Cancer Res 2018;24:5948–62.
[167] Su Y-L, Banerjee S, White SV, Kortylewski M. STAT3 in tumor-associated myeloid cells: multitasking to disrupt immunity. Int J Mol Sci 2018;19, 1803.
[168] MacDonald KPA, et al. An antibody against the colony-stimulating factor 1 receptor depletes the resident subset of monocytes and tissue- and tumor-associated macrophages but does not inhibit inflammation. Blood 2010;116:3955–63.
[169] Research, A. A. for C. A CSF1R-targeting antibody inhibits tumor-associated macrophages. Cancer Discov 2014;4:OF16.
[170] Khan ANH, Gregorie CJ, Tomasi TB. Histone deacetylase inhibitors induce TAP, LMP, Tapasin genes and MHC class I antigen presentation by melanoma cells. Cancer Immunol Immunother 2008;57:647–54.
[171] Marks PA, Richon VM, Breslow R, Rifkind RA. Histone deacetylase inhibitors as new cancer drugs. Curr Opin Oncol 2001;13:477–83.
[172] Bolden JE, Peart MJ, Johnstone RW. Anticancer activities of histone deacetylase inhibitors. Nat Rev Drug Discov 2006;5:769–84.
[173] Ni L, et al. The histone deacetylase inhibitor valproic acid inhibits NKG2D expression in natural killer cells through suppression of STAT3 and HDAC3. Sci Rep 2017;7, 45266.
[174] Wang H-F, et al. Histone deacetylase inhibitors deplete myeloid-derived suppressor cells induced by 4T1 mammary tumors in vivo and in vitro. Cancer Immunol Immunother 2017;66:355–66.
[175] Orillion A, et al. Entinostat neutralizes myeloid derived suppressor cells and enhances the antitumor effect of PD-1 inhibition in murine models of lung and renal cell carcinoma. Clin Cancer Res 2017;23:5187–201.
[176] Ordentlich P, et al. Abstract CT041: identification of gene signatures associated with response in a phase II trial of entinostat (ENT) plus pembrolizumab (PEMBRO) in non-small cell lung cancer (NSCLC) patients whose disease has progressed on or after anti-PD-(L)1 therapy. Cancer Res 2019;79:CT041.
[177] Sullivan RJ, et al. Abstract CT072: efficacy and safety of entinostat (ENT) and pembrolizumab (PEMBRO) in patients with melanoma previously treated with anti-PD1 therapy. Cancer Res 2019;79:CT072.
[178] Hashimoto A, Fukumoto T, Zhang R, Gabrilovich D. Selective targeting of different populations of myeloid-derived suppressor cells by histone deacetylase inhibitors. Cancer Immunol Immunother 2020;69:1929–36.
[179] Andtbacka RHI, et al. OPTiM: a randomized phase III trial of talimogene laherparepvec (T-VEC) versus subcutaneous (SC) granulocyte-macrophage colony-stimulating factor (GM-CSF) for the treatment (tx) of unresected stage IIIB/C and IV melanoma. J Clin Oncol 2013;31, LBA9008.

[180] Heo J, et al. Randomized dose-finding clinical trial of oncolytic immunotherapeutic vaccinia JX-594 in liver cancer. Nat Med 2013;19:329–36.
[181] Ma N, Liu Q, Hou L, Wang Y, Liu Z. MDSCs are involved in the protumorigenic potentials of GM-CSF in colitis-associated cancer. Int J Immunopathol Pharmacol 2017;30:152–62.
[182] Hou W, Sampath P, Rojas JJ, Thorne SH. Oncolytic virus-mediated targeting of PGE2 in the tumor alters the immune status and sensitizes established and resistant tumors to immunotherapy. Cancer Cell 2016;30:108–19.
[183] Donkor MK, et al. Mammary tumor heterogeneity in the expansion of myeloid-derived suppressor cells. Int Immunopharmacol 2009;9:937–48.
[184] Fujita M, et al. COX-2 blockade suppresses gliomagenesis by inhibiting myeloid-derived suppressor cells. Cancer Res 2011;71:2664–74.
[185] Kalinski P. Regulation of immune responses by prostaglandin E2. J Immunol 2012;188:21–8.
[186] Obermajer N, Muthuswamy R, Lesnock J, Edwards RP, Kalinski P. Positive feedback between PGE2 and COX2 redirects the differentiation of human dendritic cells toward stable myeloid-derived suppressor cells. Blood 2011;118:5498–505.
[187] Obermajer N, Muthuswamy R, Odunsi K, Edwards RP, Kalinski P. PGE2-induced CXCL12 production and CXCR4 expression controls the accumulation of human MDSCs in ovarian cancer environment. Cancer Res 2011;71:7463–70.
[188] Rodriguez PC, et al. Arginase I production in the tumor microenvironment by mature myeloid cells inhibits T-cell receptor expression and antigen-specific T-cell responses. Cancer Res 2004;64:5839–49.
[189] Gao Q, et al. Cancer-cell-secreted CXCL11 promoted CD8+ T cells infiltration through docetaxel-induced-release of HMGB1 in NSCLC. J Immunother Cancer 2019;7:42.
[190] Liu E, et al. Use of CAR-transduced natural killer cells in CD19-positive lymphoid tumors. N Engl J Med 2020;382:545–53.
[191] Yang L, Pang Y, Moses HL. TGF-beta and immune cells: an important regulatory axis in the tumor microenvironment and progression. Trends Immunol 2010;31:220–7.
[192] Wikström P, Stattin P, Franck-Lissbrant I, Damber JE, Bergh A. Transforming growth factor beta1 is associated with angiogenesis, metastasis, and poor clinical outcome in prostate cancer. Prostate 1998;37:19–29.
[193] Krasagakis K, et al. Elevated plasma levels of transforming growth factor (TGF)-beta1 and TGF-beta2 in patients with disseminated malignant melanoma. Br J Cancer 1998;77:1492–4.
[194] Junker U, et al. Transforming growth factor beta 1 is significantly elevated in plasma of patients suffering from renal cell carcinoma. Cytokine 1996;8:794–8.
[195] Liu Z, Kuang W, Zhou Q, Zhang Y. TGF-β1 secreted by M2 phenotype macrophages enhances the stemness and migration of glioma cells via the SMAD2/3 signalling pathway. Int J Mol Med 2018;42:3395–403.
[196] Lazarova M, Steinle A. Impairment of NKG2D-mediated tumor immunity by TGF-β. Front Immunol 2019;10:2689.
[197] Dahmani A, Delisle J-S. TGF-β in T cell biology: implications for cancer immunotherapy. Cancers (Basel) 2018;10:194.
[198] Trotta R, et al. TGF-β utilizes SMAD3 to inhibit CD16-mediated IFN-γ production and antibody-dependent cellular cytotoxicity in human NK cells. J Immunol 2008;181:3784–92.
[199] Kopp H-G, Placke T, Salih HR. Platelet-derived transforming growth factor-beta down-regulates NKG2D thereby inhibiting natural killer cell antitumor reactivity. Cancer Res 2009;69:7775–83.
[200] Castriconi R, et al. Transforming growth factor β1 inhibits expression of NKp30 and NKG2D receptors: consequences for the NK-mediated killing of dendritic cells. Proc Natl Acad Sci U S A 2003;100:4120–5.
[201] Lee J-C, Lee K-M, Kim D-W, Heo DS. Elevated TGF-beta1 secretion and down-modulation of NKG2D underlies impaired NK cytotoxicity in cancer patients. J Immunol 2004;172:7335–40.
[202] Yvon ES, et al. Cord blood natural killer cells expressing a dominant negative TGF-β receptor: implications for adoptive immunotherapy for glioblastoma. Cytotherapy 2017;19:408–18.
[203] Yang B, et al. Blocking transforming growth factor-β signaling pathway augments antitumor effect of adoptive NK-92 cell therapy. Int Immunopharmacol 2013;17:198–204.
[204] Burga RA, et al. Engineering the TGFβ receptor to enhance the therapeutic potential of natural killer cells as an immunotherapy for neuroblastoma. Clin Cancer Res 2019;25:4400–12.
[205] Wang Z, et al. Augmented anti-tumor activity of NK-92 cells expressing chimeric receptors of TGF-βR II and NKG2D. Cancer Immunol Immunother 2017;66:537–48.

[206] Raffaghello L, et al. Downregulation and/or release of NKG2D ligands as immune evasion strategy of human neuroblastoma. Neoplasia 2004;6:558–68.
[207] Bi J, Tian Z. NK cell dysfunction and checkpoint immunotherapy. Front Immunol 2019;10, 1999.
[208] Dasgupta S, Bhattacharya-Chatterjee M, O'Malley BW, Chatterjee SK. Inhibition of NK cell activity through TGF-beta 1 by down-regulation of NKG2D in a murine model of head and neck cancer. J Immunol 2005;175:5541–50.
[209] Duan S, et al. Natural killer group 2D receptor and its ligands in cancer immune escape. Mol Cancer 2019;18:29.
[210] Mistry AR, O'Callaghan CA. Regulation of ligands for the activating receptor NKG2D. Immunology 2007;121:439–47.
[211] Harrison D, Phillips JH, Lanier LL. Involvement of a metalloprotease in spontaneous and phorbol ester-induced release of natural killer cell-associated fc gamma RIII (CD16-II). J Immunol 1991;147:3459–65.
[212] Liu Y, et al. Increased expression of programmed cell death protein 1 on NK cells inhibits NK-cell-mediated anti-tumor function and indicates poor prognosis in digestive cancers. Oncogene 2017;36:6143–53.
[213] Stojanovic A, Cerwenka A. Natural killer cells and solid tumors. J Innate Immun 2011;3:355–64.
[214] Parihar R, et al. NK cells expressing a chimeric activating receptor eliminate MDSCs and rescue impaired CAR-T cell activity against solid tumors. Cancer Immunol Res 2019;7:363–75.
[215] Chang Y-H, et al. A chimeric receptor with NKG2D specificity enhances natural killer cell activation and killing of tumor cells. Cancer Res 2013;73:1777–86.
[216] Xiao L, et al. Adoptive transfer of NKG2D CAR mRNA-engineered natural killer cells in colorectal cancer patients. Mol Ther 2019;27:1114–25.
[217] Navid F, Santana VM, Barfield RC. Anti-GD2 antibody therapy for GD2-expressing tumors. Curr Cancer Drug Targets 2010;10:200–9.
[218] Garnock-Jones KP, Keating GM, Scott LJ. Trastuzumab: a review of its use as adjuvant treatment in human epidermal growth factor receptor 2 (HER2)-positive early breast cancer. Drugs 2010;70:215–39.
[219] García-Foncillas J, Díaz-Rubio E. Progress in metastatic colorectal cancer: growing role of cetuximab to optimize clinical outcome. Clin Transl Oncol 2010;12:533–42.
[220] Winter MC, Ten Hancock BW. Years of rituximab in NHL. Expert Opin Drug Saf 2009;8:223–35.
[221] Albertini MR, Hank JA, Sondel PM. Native and genetically engineered anti-disialoganglioside monoclonal antibody treatment of melanoma. Cancer Chemother Biol Response Modif 2005;22:789–97.
[222] Yang RK, Sondel PM. Anti-GD2 strategy in the treatment of neuroblastoma. Drugs Future 2010;35:665.
[223] Parihar R, et al. A phase I study of interleukin 12 with trastuzumab in patients with human epidermal growth factor receptor-2-overexpressing malignancies: analysis of sustained interferon γ production in a subset of patients. Clin Cancer Res 2004;10:5027–37.
[224] Fleming GF, et al. A phase I trial of escalating doses of trastuzumab combined with daily subcutaneous interleukin 2: report of cancer and leukemia group B 9661. Clin Cancer Res 2002;8:3718–27.
[225] Snyder KM, et al. Expression of a recombinant high affinity IgG fc receptor by engineered NK cells as a docking platform for therapeutic mAbs to target cancer cells. Front Immunol 2018;9, 2873.
[226] Zhu H, et al. Pluripotent stem cell-derived NK cells with high-affinity noncleavable CD16a mediate improved antitumor activity. Blood 2020;135:399–410.
[227] Pomeroy EJ, et al. A genetically engineered primary human natural killer cell platform for cancer immunotherapy. Mol Ther 2020;28:52–63.
[228] Savan R, Chan T, Young HA. Lentiviral gene transduction in human and mouse NK cell lines. Methods Mol Biol 2010;612:209–21.
[229] Matosevic S. Viral and nonviral engineering of natural killer cells as emerging adoptive cancer immunotherapies. J Immunol Res 2018;2018, 4054815.
[230] Afolabi LO, Adeshakin AO, Sani MM, Bi J, Wan X. Genetic reprogramming for NK cell cancer immunotherapy with CRISPR/Cas9. Immunology 2019;158:63–9.
[231] Carlsten M, Childs RW. Genetic manipulation of NK cells for cancer immunotherapy: techniques and clinical implications. Front Immunol 2015;6:266.
[232] André P, et al. Anti-NKG2A mAb is a checkpoint inhibitor that promotes anti-tumor immunity by unleashing both T and NK cells. Cell 2018;175:1731–1743.e13.
[233] Gleason MK, et al. CD16xCD33 bispecific killer cell engager (BiKE) activates NK cells against primary MDS and MDSC CD33+ targets. Blood 2014;123:3016–26.

[234] Sarhan D, et al. 161533 TriKE stimulates NK-cell function to overcome myeloid-derived suppressor cells in MDS. Blood Adv 2018;2:1459–69.

[235] Vallera DA, et al. IL-15 trispecific killer engagers (TriKEs) make natural killer cells specific to CD33+ targets while also inducing persistence, in vivo expansion, and enhanced function. Clin Cancer Res 2016;22:3440–50.

[236] Labani-Motlagh A, Ashja-Mahdavi M, Loskog A. The tumor microenvironment: a milieu hindering and obstructing antitumor immune responses. Front Immunol 2020;11:940.

[237] Martens A, et al. Baseline peripheral blood biomarkers associated with clinical outcome of advanced melanoma patients treated with ipilimumab. Clin Cancer Res 2016;22:2908–18.

[238] Kurahara H, et al. Significance of M2-polarized tumor-associated macrophage in pancreatic cancer. J Surg Res 2011;167:e211–9.

[239] Liu H, et al. Role of NKG2D and its ligands in cancer immunotherapy. Am J Cancer Res 2019;9:2064–78.

[240] Sambi M, Bagheri L, Szewczuk MR. Current challenges in cancer immunotherapy: multimodal approaches to improve efficacy and patient response rates. J Oncol 2019;2019:4508794.

[241] Trombetta AC, et al. A circulating cell population showing both M1 and M2 monocyte/macrophage surface markers characterizes systemic sclerosis patients with lung involvement. Respir Res 2018;19:186.

[242] Zhu J, Huang X, Yang Y. Myeloid-derived suppressor cells regulate natural killer cell response to adenovirus-mediated gene transfer. J Virol 2012;86:13689–96.

[243] Hansson M, Asea A, Ersson U, Hermodsson S, Hellstrand K. Induction of apoptosis in NK cells by monocyte-derived reactive oxygen metabolites. J Immunol 1996;156:42–7.

[244] Akhiani AA, et al. Idelalisib rescues natural killer cells from monocyte-induced immunosuppression by inhibiting NOX2-derived reactive oxygen species. Cancer Immunol Res 2020. https://doi.org/10.1158/2326-6066.CIR-20-0055.

[245] Cong J, et al. Dysfunction of natural killer cells by FBP1-induced inhibition of glycolysis during lung cancer progression. Cell Metab 2018;28:243–255.e5.

[246] Caras I, et al. Evidence for immune defects in breast and lung cancer patients. Cancer Immunol Immunother 2004;53:1146–52.

[247] O'Brien KL, Finlay DK. Immunometabolism and natural killer cell responses. Nat Rev Immunol 2019;19:282–90.

[248] Chambers AM, Lupo KB, Matosevic S. Tumor microenvironment-induced immunometabolic reprogramming of natural killer cells. Front Immunol 2018;9, 2517.

[249] Terrén I, Orrantia A, Vitallé J, Zenarruzabeitia O, Borrego F. NK cell metabolism and tumor microenvironment. Front Immunol 2019;10, 2278.

[250] Wang J, Matosevic S. Adenosinergic signaling as a target for natural killer cell immunotherapy. J Mol Med 2018;96:903–13.

[251] Cekic C, Day Y-J, Sag D, Linden J. Myeloid expression of adenosine A2A receptor suppresses T and NK cell responses in the solid tumor microenvironment. Cancer Res 2014;74:7250–9.

[252] Dona AC. High-throughput metabolic screening. In: Holmes E, Nicholson JK, Darzi AW, Lindon JC, editors. Metabolic phenotyping in personalized and public healthcare. Academic Press; 2016. p. 111–36. https://doi.org/10.1016/B978-0-12-800344-2.00005-7 [Chapter 5].

[253] Draghiciu O, Nijman HW, Hoogeboom BN, Meijerhof T, Daemen T. Sunitinib depletes myeloid-derived suppressor cells and synergizes with a cancer vaccine to enhance antigen-specific immune responses and tumor eradication. Onco Targets Ther 2015;4, e989764.

CHAPTER

3

Natural killer cells as immunotherapeutic effectors for solid tumors

Sandro Matosevic

Department of Industrial and Physical Pharmacy, Purdue University, West Lafayette, IN, United States

Abstract

Among immune cells, natural killer (NK) cells have emerged as attractive effectors in cell-based therapy owing to their beneficial safety profile, ability to recognize and eliminate targets without prior antigen sensitization, and their potential for use in allogeneic adoptive transfer settings. Despite these advantages, antitumor responses of NK cells in solid tumors have been largely underwhelming. Multiple factors have contributed to the lack of responses, including heterogeneous tumor microenvironments, immunometabolic suppression driven by cancer cell metabolism and impaired activity of immune cells, and inhibition of NK cell cytotoxic activity due to an inconvenient balance of inhibitory to activating receptors. This has contributed to a generally poor infiltration of NK cells in solid tumors and limited antitumor activity. This chapter discusses the main challenges encountered with NK cell immunotherapies for solid tumors, and strategies under investigation and development which aim to improve and enhance these cells' activity in such therapeutic settings.

Abbreviations

ADO	adenosine
CAR	chimeric antigen receptor
CEACAM-1	carcinoembryonic antigen-related cell adhesion molecule 1
ECM	extracellular matrix
GvHD	graft versus host disease
IFN-γ	interferon
IL	interleukin
iPSC	induced pluripotent stem cell
KIR	killer immunoglobulin-like receptor
LAG-3	lymphocyte-activation gene 3
NK	natural killer
NKG2A/C/D	natural killer group 2A/C/D

https://doi.org/10.1016/B978-0-12-822620-9.00014-8

PD-1	programmed death protein 1
TIGIT	T cell immunoglobulin and ITIM domain
TIM-3	T cell immunoglobulin and mucin domain-containing protein 3
TME	tumor microenvironment

Conflict of interest

No potential conflicts of interest were disclosed.

Introduction

The use of natural killer (NK) cells in adoptive cell-based immunotherapy has emerged as a viable alternative to T cell therapy. NK cells have been touted for their ability to avoid or limit graft versus host disease (GvHD), thus enabling their allogeneic use, a much shorter persistence thus avoiding the risk of long-term toxicities and their innate ability to recognize and eliminate targets without prior antigen presentation, unlike T cells. In addition, NK cells do not require match between recipient MHC class I ligands and donor inhibitory killer immunoglobulin-like receptors [1–3]. These are remarkable characteristics that have helped NK cell-based immunotherapy advance significantly in recent years, so much so that the number of clinical programs with either native or genetically engineered NK cells is growing rapidly. Though NK cell immunobiology is unique and distinct to that of T cells or other immune cells, NK cell-based immunotherapy has benefited from the work done in the context of T cell therapy, particularly when it comes to genetic engineering.

However, this growth has not been easy, and several challenges are faced by the deployment of NK cell immunotherapies particularly in the setting of solid tumors [4]. They are, on one hand, related to the specific immunobiology of NK cells and, on the other, the characteristics of the tumors themselves. On the NK cell side, several unique properties of NK cells makes them, at once, remarkable innate antitumor effectors, and on the other, challenging to manipulate as such. For instance, NK cells are characterized by short persistence, usually about 2 weeks. This is advantageous because it ensures rapid elimination, beneficial particularly when these cells are engineered to target a cancer-associated antigen. However, such short persistence is not considered sufficient for durable antitumor responses. As a result, NK cells need continuous stimulation with cytokines, most commonly IL-15, although IL-2, IL-12, IL-18, and IL-21 have all been used either individually or in combination [5]. These cytokines have been provided either in soluble form or through engineering strategies, by genetically modifying NK cells to express one or more of them. And though genetically engineering NK cells to express IL-15 has been shown to enable their persistence in vivo in patients for at least 1 year [6], such an approach requires genetic manipulation, which is notoriously difficult with NK cells. Apart from short persistence, NK cells are averse to exogenous manipulation. Genetic engineering is, generally, difficult. Though viruses have had considerably more success than nonviral transfection approaches, NK cells are considered very difficult to engineer. This has limited the progress made in the context of chimeric antigen receptor (CAR)-NK cell therapy.

Tumors themselves pose particular challenges to NK cells. It is not surprising that most advances with NK cell immunotherapy have so far been made in the context of hematologic malignancies. These types of tumors lack many of the immunosuppressive characteristics of solid tumors. NK cells are particularly sensitive to mechanisms of immunosuppression induced by solid tumors. The overexpression of ligands to inhibitory receptors on NK cells is a proven mechanism that tumors utilize to avoid attack by NK cells. In many tumors, moreover, the expression of these receptors on NK cells is upregulated, while expression of NK cell-activating receptors, such as NKG2D, NKp46, and CD16, is downregulated [7,8]. NK cells traffic to solid tumors in response to gradients formed between chemokine receptors and their respective ligands. In solid tumors, these gradients are often altered in favor of dysfunctional or completely arrested NK cell migration. This occurs due to differential and altered expression of chemokines or their ligands compared to that on NK cells in healthy tissues. This is only part of the reason why many solid tumors are characterized by a very poor NK cell infiltrate. The poor intratumoral penetration of NK cells is considered one of the biggest obstacles to their effector function against solid tumors, particularly in light of clinical evidence that NK cell presence has been associated with improved survival of patients with a variety of cancers.

An emerging feature of solid tumors that has in recent years generated significant new insights into the immune behavior of NK cells in cancer is metabolism [9]. Is it well established that cancer cells have an altered metabolism that enables them to survive in settings of impaired glycolytic fueling as well as hypoxic environments. NK cells, on the other hand, have not evolved such an ability. The immunometabolic reprogramming of NK cells in the solid tumor microenvironment has become and extensive area of study, and novel immunotherapeutic strategies aimed at reversing immunometabolic dysfunction have appeared.

Collectively, NK cells are formidable immune effectors, with a proven ability to target cancer cells. The development of successful immunotherapies with these cells is tightly tied to the ability to overcome the severe immunosuppression that these cells experience in solid tumor microenvironments and restore impaired function. Because of the broad and serious immunosuppressive roles of the solid tumor microenvironment due to active metabolites such as adenosine, poor immune cell infiltration, and antigen escape, significant gaps in the ability to recapitulate immune cell function in favor of restored antitumor activity remain. This chapter discusses some of these challenges, including the associated strategies that have been evaluated with the aim of repairing dysfunctional NK cell immunity in solid tumors.

Sources of NK cells for immunotherapy (Table 1)

Traditionally, NK cells have been sourced from peripheral blood. In autologous adoptive transfer settings, peripheral blood has been sourced from patients. Approximately 10% of cells in peripheral blood are NK cells, making it a necessity to expand these cells substantially to achieved dosages suitable for infusion. Though peripheral blood is the most common

TABLE 1 Sources of NK cells.

Autologous patient derived
Cord blood
Induced pluripotent stem cells (iPSCs)
NK cell lines

source of NK cells for clinical use, multiple factors affect the number of cells that can be retrieved from peripheral blood [10]. Among them is the lower proportion of mature NK cells is in peripheral blood associated with pathological settings—NK cells in the peripheral blood of breast cancer patients, for instance, show diminished cytotoxicity, alongside a higher proportion of $CD56^{dim}CD16^{-}$ and $CD56^{bright}CD16^{-}$ cells [11,12], while a similar observation was also made in patients with colorectal cancer [13]. Additionally, obtaining high amounts of pure NK cells from leukapheresis, free from residual T and B cells, is often laborious and requires extensive manufacturing, potentially compromising the immunotherapy product.

Cord blood is another source of NK cells that contains a somewhat larger number of NK cells—up to 30%. Among the advantages of cord blood over peripheral blood is its relative ease of collection [14], with cord blood being often available in large amounts and this not requiring patient blood draws. Cord blood also contains lower number of T cells, which in turn reduces the risk of GvHD [15]. The ability of NK cells to be used allogeneically makes using cord blood attractive to achieve a truly allogeneic "off-the-shelf" cellular therapy product. Though functionally similar to peripheral blood in that cord blood NK cells include both $CD56^{dim}$ and $CD56^{bright}$ subsets, studies have indicated that there might be phenotypic and functional differences between NK cells sourced from cord blood and those from peripheral blood [16–18]. These include the presence of a $CD56^{-}/CD16^{+}$ subset in cord blood that is not present in peripheral blood [19]. These $CD56^{-}/CD16^{+}$ cells have been shown to be less cytolytically active than $CD56^{+}CD16^{+}$ NK cells, though they can be induced to express CD56 following activation with IL-2 or IL-15 [20]. This suggests that these cells may be less mature than $CD56^{+}/CD16^{+}$ NK cells.

As an alternative source of highly pure NK cells, induced pluripotent stem cells (iPSCs) have demonstrated to be able to be differentiated into functionally mature NK cells [21–24]. Due to the relative ease of genetically engineering iPSCs compared to NK cells from peripheral or umbilical cord blood, this approach may represent a promising strategy for generating off-the-shelf genetically-modified NK cells for use in immunotherapy of solid tumors. NK cells derived from CAR-engineered iPSCs are able to efficiently express CAR genes and target and kill more refractory tumors [25]. iPSC-derived stem cells have been successfully progressed to the clinic, with a number of trials evaluating these cells for a variety of cancers. One of the biggest hurdles to the widespread use of iPSC-derived NK cells is the laborious differentiation procedure used to obtain these cells. To date, most protocols utilize feeder layers or require peripheral blood and take many weeks to complete, often close to 2 months. Though new approaches have shortened this time and presented alternatives to feeder media in the differentiation process, they are still emerging, and differentiating NK cells from iPSCs is considered relatively difficult. Many of the protocols, moreover, achieve a fairly heterogeneous cell population with either a large proportion of immature NK cells, thus requiring sorting, or substantial contamination with NKT or T cells.

NK cell lines, such as NKG, NK-92, NKL, KHYG-1, YT, NK-YS, SNK-6, IMC-1, YTS, NKL, and HANK-1 [1], are able to circumvent blood isolation or differentiation procedures but either lack functional markers of blood-derived NK cells (such as, for example, NK-92 cells which lack CD16 among other receptors of mature NK cells) or they are proprietary and thus hard to obtain.

Expanding NK cells

Owing to the fact that only about 10% from peripheral blood are NK cells, and that doses in excess of 10^9 NK cells per patient are typically required, expansion of NK cells sourced from blood of donors is required prior to infusion [26,27]. To generate sufficient NK cells for infusion into patients, a number of strategies for expansion of cells sourced from blood have been employed. All of these strategies utilize various combination of cytokines, most commonly IL-15. Different cytokine combinations yield different activation patterns on NK cells, and they have been used for their specific capacity to promote proliferation or activation of NK cells, or both. Cytokines used for the expansion of NK cells include IL-2, IL-12, IL-15, IL-18, and IL-21. Traditionally, IL-2 and particularly IL-15 have been used to activate NK cells ex vivo and in vitro. Activating NK cells with a combination of IL-12, IL-15, and IL-18 followed by a 7- to 21-day rest period has the capacity to induce a "memory-like" phenotype on NK cells [28–30]. These memory-like NK cells have shown potent cytotoxic capacity in tumor challenge assays against acute myeloid leukemia and are being evaluated clinically. They enable the persistence of NK cell cytotoxicity and IFN-γ response in proliferating NK cells over an extended period of time. In $CD56^{dim}$ NK cells, the memory-like phenotype correlated to expression of NK cell receptors CD94, NKG2A, NKG2C, and CD69, alongside a lack of CD57 and killer immunoglobulin-like receptor (KIR). Despite their critical roles in promoting NK cell proliferation and activation, expansion rates with cytokines are not sufficient in reaching the high numbers of NK cells needed for adoptive transfer studies. For that reason, NK cells are typically expanded with cytokines in combination with feeder cells, most commonly K562 cells [31]. K562 is an erythroleukemic cell line that lacks HLA antigen expression. K562 feeder cells are often engineered to express the stimulatory molecule 4-1BBL and a gene encoding the cytokine IL-15, alongside the transmembrane domain of CD8α. Expansion rates of over 1000s of folds have been achieved in this way [32]. These 4-1BBL-K562 cells also upregulate activating NK cell surface receptors such as NKp30, NKp44, and NKp46 upon NK cell expansion. Membrane-bound IL-21 can be substituted for IL-15, a strategy that has been used clinically to expand genetically engineered NK cells for infusion into patients [33]. The use of soluble IL-21 and IL-15, alongside 4-1BBL, with K562 cells can also be used.

Challenges to NK cell immunotherapy in solid tumors (Table 2)

One feature of many solid tumors is that they are highly heterogeneous and characterized by multiple immunosuppressive mechanisms that impair the effectiveness of immunotherapies [34]. Such heterogeneity is commonly considered one of the major barrier to sustained effector function of NK cells. Though these cells have an innate ability recognize pathogens and migrate to sites of tumors, cells that eventually end up reaching those sites quickly lose potency and markers of activation, resulting in immune dysfunction and an inability to promote antitumor responses. This is one of the main reasons for the lack of success with trials in patients with solid tumors treated with adoptively transferred NK cells.

Solid tumor heterogeneity includes metabolic, functional, and phenotypic. The TME is a complex network on three-dimensional structures, which, together with TME-resident cells

TABLE 2 Tumor-related obstacles for NK function.

Tumor heterogeneity
Ligands to inhibitory ligands on NK cells
NKG2A, TIGIT, LAG-3, PD-1, CEACAM-1, KIRs, TIM-3
Hypoxia
Downregulation of activating receptors NKp46, NKp30, NKp44, and NKG2D
Increased activity of CD73 and generation of adenosine (ADO)
Poor migration to tumors because of altered chemokine expression

including cancer cells themselves, promote an environment of altered metabolic function, and resistance to immune attack. The TME consists of various cell types, including tumor-associated fibroblasts, endothelial cells, adipocytes, immune cells, and cancer cells, all connected by the extracellular matrix (ECM) [35,36]. The ECM is itself a complex network made up of collagen, glycoproteins, proteoglycans, and polysaccharides as well as a number of matricellular proteins, which are proteins that bind to matrix proteins, cell surface receptors, or cytokines. Together, these noncellular structures work to sustain the ECM and promote its various functions, ultimately creating a niche favorable for cancer homing and progression. Stem cells in the ECM niche are driven to the conversion into a cancer stem cell identity, resulting in a population of cells with phenotypic and functional potency to support cancer development and growth [37].

Solid tumors are characterized by the upregulation of ligands to inhibitory receptors on NK cells. NK cell inhibitory receptors include NKG2A, TIGIT, LAG-3, PD-1, and CEACAM-1 as well as inhibitory KIRs, among others [38], while other receptors such as TIM-3 have been described as having both activating and inhibitory roles [39]. Interaction between inhibitory receptors and ligands on NK cells drives NK cell inhibition and has been targeted via checkpoint blockade immunotherapy [40].

Another characteristic of the solid tumor TME is hypoxia. Hypoxia is associated with solid tumors due to inadequate vasculature and limited supply of oxygen. Oxygen levels in solid tumors are heterogeneous but generally lower than in healthy tissues, ranging from 0.5 to 4% O_2 saturation compared to 4%–14% O_2 saturation in healthy tissues [41]. Low oxygen reorganizes cellular functions and requires adaptation to such an environment. Cancer cells are able to adapt to conditions of low oxygen and carry out metabolic functions in an environment that is conducive to dysfunction of normal cells. Unfortunately, NK cells require oxygen for optimal metabolic and immune activity. A number of studies have demonstrated that although the role of hypoxia on NK cells is complex and varied, and it generally causes a reorganization of NK cell functions [42].

Low oxygen was shown to impair NK cell activity through the downregulation of activating receptors NKp46, NKp30, NKp44, and NKG2D [43]. Apart from impairing NK cell activation, other functional changes to NK cells have been reported to occur due to hypoxia, including altered cytotoxicity and impaired cytokine production [44]. In some cases, cytokine preactivation has been used as a mechanism to temper hypoxia-induced impairment of NK cell activity, and while this has been successful, it is not a universal strategy and hypoxia is more often than not considered detrimental to NK cell function [45]. It bears mentioning that the effects of hypoxia on NK cells also depend on the extent of hypoxia NK cells experience. It is generally thought that while extreme lack of oxygen is detrimental to NK cell function, mild hypoxia is less so, and NK cells are able to carry out their functions under such conditions, with enhancements possible with cytokine preconditioning [46–48].

CD73-adenosine

Hypoxia triggers the activity of CD73, an ectonucleotidase highly expressed on GBM and responsible for the generation of adenosine (ADO) from adenosine monophosphate [10]. Overexpression of CD73 is a clinically recognized prognostic indicator of the progression of many cancers, including brain cancer and lung cancer. The production of adenosine via CD73 is part of a large multienzyme cascade also involving CD39. CD39, the first enzyme in the cascade, converts extracellular adenosine triphosphate (ATP) into adenosine monophosphate (AMP), which is then taken up by CD73 to generate adenosine. Additionally, adenosine is also produced via a non-canonical cascade involving CD38, CD203a, and CD73 [49] and inhibits $CD8^+$ T cells. Though adenosine can also be produced through a pathway involving CD38, it is the CD39-CD73 axis that has received the most attention to date.

NK cells possess four adenosine receptors, namely A1, A2A, A2B, and A3. The A2 receptors (A and B) have been most widely studied in the context of cancer, with the A2A receptor having been shown to be involved in profound inhibitory binding of adenosine to NK cells [10]. It has been reported that adenosine severely impairs NK metabolism and cytotoxic functions, by altering NK cell glycolysis and oxidative phosphorylation [44]. More specifically, adenosine impairs the maturation of NK cells [50], the accumulation of cytotoxic $CD56^{dim}$ cells at tumor sites [50]. the expression of activating receptors, and NK effector function [21]. This occurs through the downregulation of metabolic control genes, including *LDHA* and *GAPD*, in the presence of adenosine. In addition, dysregulated cytokine responses via IFN-γ upregulation and impaired cytotoxicity against cancer targets in response to ADO have been observed. One interesting response to adenosine has been the upregulation of IFN-γ on NK cells, which has been observed with NK cells treated with adenosine. This was significantly upregulated when NK cells were stimulated with IL-15 or a combination of IL-12 and IL-15. Because the overall cytotoxic activity of NK cells was impaired in the presence of adenosine, the upregulation of IFN-γ not only signals altered cytokine signaling but also potentially impaired degranulation and activation processes that prevent mature cytotoxic responses.

Targeting the adenosine-CD73 axis has emerged as a viable a clinical strategy to rescue the solid tumor TME from immunometabolic dysfunction triggered by hypoxia [22]. Though targeting the immune checkpoint CD73 has resulted in a number of clinical programs, single antigen monotherapy has had limited effectiveness in targeting this receptor. It appears that anti-CD73 responses are more robust when CD73 is co-targeted with other immune checkpoints, such as PD-1. Multiple combination studies are ongoing which are targeting CD73 in combination with other receptors in solid tumors. Targeting CD73 was also shown to be able to restore impaired NK cell function in combination with blockade of the adenosine A2A receptor [23,24]. Similarly, combination immunotherapy with CD73 and NK cells engineered to express a CAR based on NKG2D with DAP10 and CD3ζ intracellular signaling was shown to improve anti-tumor responses in in vivo and in vitro models of lung adenocarcinoma [51].

The trafficking of NK cells to tumors is directed by a range of chemokine receptors and their respective ligands. This process is known as chemotaxis. NK cells express a number of these chemokine receptors, which include CXCR1, CXCR3, CXCR4, CXCR6 and CX3CR1 [52]. The ligands to these receptors include CXCL9, CXCL10, CXCL12, and CXCL16. In solid tumors, expression of both ligands and receptors is altered, dissuading NK cell migration to the tumor and promoting the recruitment of immunosuppressive immune cells. While many of these receptors are natively expressed on NK cells, their expression can also be modulated with cytokines. For example, expression of CXCR3 on human NK cells could be downregulated upon stimulation of NK cells for 6 or 24h with IL-2 and IL-12 either alone or in combination, while expansion of NK cells with IL-2 and feeder cells (EBV-LCLs) was able to upregulate the expression of CXCR3 on NK cells [53]. Modulating chemokine receptor-ligand axes is a relatively novel strategy aimed at enhancing NK cell penetration into solid tumors, achieved through a combination of genetic engineering or endogenous chemokine treatment, among other emerging strategies.

CD155-TIGIT

NK cells are subject to inhibition via the TIGIT-CD155 axis in many cancers. Cancer cells are characterized by the expression of poliovirus receptor (PVR or CD155). Cancer-associated CD155 promotes tumor progression through a variety of mechanisms that include cell invasion and migration, as well as cancer cell dispersal [54–58]. For instance, inhibition of CD155 on glioblastoma was shown to limit GBM dispersal and reduce the production of matrix metalloproteinase-2 (MMP-2), a known enhancer of GBM invasion. CD155 binds with high affinity to TIGIT on NK cells. TIGIT has been described as an immune checkpoint inhibitor, and its expression on tumor-infiltrating NK cells was shown to lead to cell exhaustion and impaired NK cell maturation [59]. TIGIT harbors an immunoglobulin tail tyrosine (ITT)-like phosphorylation motif and an ITIM (immunoreceptor tyrosine-based inhibition motif) on its cytoplasmic end. The interaction between TIGIT and CD155, primarily through the ITT-like motif which becomes phosphorylated at Tyr225, inhibits NK cell cytotoxicity, interferon-gamma (IFN-γ) production, granule polarization and antitumor functions [12,13]. Antibody-mediated blockade of TIGIT on NK cells was recently shown to promote NK cell effector functions, generate a potent anti-tumor memory response, and rescue NK cells from exhaustion [59]. In particular, NK cells were considered essential to a sustained antitumor response in the context of TIGIT blockade [59]. Interestingly, a number of studies contrasted this by showing that blockade of TIGIT resulted in no response when TIGIT was targeted alone. Stimulation with IL-15 was necessary to induce an antitumor response [60]. In relation to this, recent data have suggested that TIGIT might act as a co-receptor, targetable alongside other checkpoints, most notably PD-1. Indeed, combining TIGIT with anti-PD-L1 therapy is being explored on the wave of recent evidence that TIGIT expression is correlated to PD-L1 expression [61,62].

NKG2A

NKG2A is a recognized inhibitory receptor on NK cells. It has been studied extensively, with studies showing that blockade of NKG2A has the ability to restore NK cell cytotoxicity [63]. NKG2A heterodimerizes with CD94 on NK cells and binds HLA-E on tumor targets. This interaction induces inhibition through ITIM activation and SHP-1 recruitment on NK cells [64,65]. The surface expression of HLA-E is upregulated on cancer cells in response to IFN-γ produced by tumor cells [66]. NKG2A$^+$ NK cells as well as cancers expressing upregulated levels of HLA-E have been reported in ovarian, endometrial, CRC, and cervical cancer, while HLA-3 expression has been associated with poorer patient prognosis [67]. Co-expression of both NKG2A and PD-1 has also commonly been observed [68]. NKG2A blockade has been evaluated with the monoclonal antibody monalizumab, and it was shown that this can enhance NK cell effector functions, cytokine production and degranulation. In addition, co-blockade of NKG2A has also been proven effective in combination with PD-1- and EGFR-targeted therapies [68].

Genetic engineering of NK cells

In recent years, a growing strategy for enhancing the activation of NK cells in solid tumors has been the modulation of these cells' activities through genetic engineering [69]. Unlike exogenous stimulation of NK cells—which, in many ways is still needed even when NK cells are engineered—genetic engineering has the ability to intrinsically alter NK cells' activation status in a permanent manner. NK cells can and have been engineered using common approaches that have been used to engineer other cell types, including knocking down or out genes using shRNA or CRISPR/Cas9, or expressing CARs using nonviral or viral approaches [70]. These strategies have resulted in engineered NK cell variants with remarkable anti-tumor responses, some of which have advanced to clinical trials. Engineering NK cells is particularly attractive in the context of solid tumor immunotherapy, particularly considering the limited anti-tumor responses native, blood-derived NK cells have shown to have against these types of cancer, making genetic engineering almost a necessity [69].

However, engineering NK cells is considered very difficult [71]. There are many reasons for this, with one prominent line of thought surmising that NK cells have evolved sophisticated mechanisms of evasion form viral attack, thus translating such evasion mechanisms into resistance to viral transduction. As a result, genetic engineering of NK cells is cumbersome, nonstandardized, and highly heterogeneous both in terms of published procedures and in terms of outcomes. It bears mentioning that the NK cell line NK-92 is considered significantly easier to engineer than blood-derived NK cells. This is also evident in a larger number of studies utilizing this cell line over primary human NK cells.

NK cells have most commonly been engineered using viruses, either lenti- or retroviruses, or mRNA. Studies have shown that lentiviral transduction is superior to mRNA transfection. For instance, lentiviral transduction of cord blood-derived NK cells to express anti-CD19 and anti-CD20 CARs yielded 73% transduction efficiency, compared to less than 10% obtained with mRNA transfection [72]. This was, however, dependent on cells used, as peripheral blood NK cells yielded over 3× lower transduction efficiency. Transduction of NK cells can be enhanced via the use of transduction enhancers. These are typically small molecules or proteins that can enhance sensitization to viral entry. They include polybrene, protamine sulfate, DEAE-dextran, poly-L-lysine, or Retronectin [73–76]. Polybrene and promatine sulfate are cationic polymers and promote fusion of nucleic acids with the NK cell membrane to facilitate gene entry [77]. Polybrene is toxic to the NK cell line NK-92 so its use has mostly been with blood-derived NK cells. Retronectin is a truncated fragment of the extracellular matrix cell adhesion protein fibronectin, which acts by promoting colocalization of viral particles within its binding domains [78]. Vectofusin-1 has recently emerged as a new transduction enhancer, with a study suggesting it could serve as an alternative to Retronectin [79]. Despite a number of small molecule and protein-based transduction enhancer, there appears to be no standard nor is there a universally accepted strategy to ensure high efficiency transduction of NK cells. Similarly, virus pseudotyping, or the alterations in the host range of a virus, has been carried out to identify optimal viral host [80], most commonly with vesicular stomatitis virus GP (VSV-G). BX795, an inhibitor of TBK1/IKKε, was shown to have the ability to enhance transduction of NK cells with VSV-G-pseudotyped lentiviral vectors [81]. However, it is often thought that VSV-G-based transduction of NK cells does not work, despite it being a common approach for the engineering of T cells. Many of the transduction protocols described to date are very process specific and rely on highly variable conditions of cell line source and quality, donor variability, cell stimulation and expansion prior to transduction, cell fitness, and cell age. It could be for these reasons that NK cell engineering remains challenging regardless of approach used.

NK cells have so far been engineered to express receptors targeting a variety of tumor-associated antigens, including MUC1-CAR, HER2-CAR and NKG2D, GD2, PSCA, GPC3, EGFR/EGFRviii, EpCAM, and GPA7 [82–86]. Among these, MUC1, HER2, and NKG2D-CAR-NK cells have advanced to the clinic, and trials with these CAR-engineered NK cells are currently ongoing for a multitude of cancers. The majority of these trials utilize the NK cell line NK-92, largely due to its relative ease of engineering compared to blood-derived NK cells. An emerging strategy to engineer NK cells to avoid antigen escape has been to equip them with multitargeting CAR constructs. Such an example was described with iPSC-derived NK cells engineered to co-express CD19 and a noncleavable CD16, in addition to IL-15 and IL-15Rα [87]. Multitargeting CARs have been successfully demonstrated in the context of CAR T-cell therapy, with feasibility studies on NK cells still emerging.

Antigen escape and multitargeting CARs

The heterogeneity of GBM leads to outgrowth of antigen loss variants, which have been observed in all clinical studies to date with most GBM antigens. This is also called antigen escape and has been a problem for many antigen-targeting therapies, particularly those based on engineered immune cells that exert selection pressures on the targeted cancer-associated antigens over prolonged time. Antigen escape in CD19 CAR trials for acute lymphoblastic leukemia has been widely reported [88]. The potential conferral of a superior survival advantage to treated subjects by enhancing the duration of tumor suppression and eliminating recognition escape has been attempted with the incorporation of two antigens on a single immune cell that either share or have completely dissociated co-signaling domains [89]. These strategies—resulting in bispecific, dual or tandem targeting strategies, can enhance receptor activity while also selecting for specific signaling components to enhance CAR cell persistence and effector function. In this context, avoidance of activation-induced cell death [90] has been shown to be of particular concern for CAR-T-based therapies [91] and can be addressed by rational design of combinatorially incorporated signaling regions. Multiple CARs with T cells engineered to target two antigens, have been described preclinically [92,93]. Examples of multitargeting CAR-NK cells are limited. One such example involves multifunctional engineered NK cells co-targeting GD2 and ligands to NKG2D on cancer cells. The constructs encoding these two CARs was also engineered to trigger the release of an antibody fragment against CD73 on cancer cells in the TME and thus block the activity of CD73 on cancer cells, glioblastoma in this case, and limit the accumulation of extracellular adenosine [94]. In general, multitargeting engineered NK cells have not been extensively studied, and more work is necessary to establish the pre-clinical and clinical feasibility of such an approach when applied to NK cells.

NK cell engagers

A relatively strategy aimed at improving NK cell function involves the development of multispecific antibodies called engagers. These proteins are designed so that they bind and activate NK cells on one and cancer cells on the other. Additional moieties are often added to these structures, such as a cytokine-producing ability to sustain NK cell proliferation. Depending on the number of functional elements, engagers are classified as either bispecific or trispecific killer engagers (BiKEs and TriKEs). Typically, the NK cell-activating component has consisted of a ligand to one of NK cell-activating receptors, such as NKp30, NKp44 or NKp46, or antibody-dependent cell-mediated cytotoxicity (ADCC) ligand CD16. A specific receptor is typically picked in case its expression is unaltered in the observed cancer [95]. These engagers have promising results in studies targeting ligands on B-cell malignancies (CD19 and CD22), acute myeloid leukemia (AML; CD33), and colorectal cancer (CD133) [96–98]. Apart from targets on NK cells and cancer cells, NK cell engagers have commonly been engineered to secrete IL-15, which is a powerful cytokine that can aid NK cell activation, persistence and proliferation [99,100]. Among NK cell engagers, CD33-targeting BiKEs are being evaluated in the clinic to treat AML [101].

Conclusions

It is evident that NK cells are a particularly attractive proposition for immunotherapy of cancer, including solid tumors. Their relative safety compared to that of T cells, alloreactivity in the absence of recipient MHC class I ligands for donor inhibitory killer immunoglobulin-like receptors, and ability to attack cancer cells without prior antigen presentation makes them excellent candidates for the development of allogeneic immunotherapies. However, as this discussion has shown, NK cells are severely immunosuppressed in the microenvironment of solid tumors, and are victims to many of the cancer cells' mechanisms of immune evasion. Altered metabolism, downregulation of activating receptors, dysregulated chemokine receptor/ligand repertoire are some of the issues these cells experience in cancer. Efforts at overcoming such inhibitions have make substantial inroads into enhancing NK cell activity against solid tumors, however, much work still remains. The NK cells' aversion to exogenous manipulation, which includes their resistance to genetic modification, is an often-mentioned challenge that has yet to find a solution. New sources of NK cells, such as iPSCs, which promise easier and more controllable genetic manipulations to be carried out, could be part of a future wave of allogeneic NK cell immunotherapies which achieve effective and sustained elimination of solid tumors.

References

[1] Lupo KB, Matosevic S. Natural killer cells as allogeneic effectors in adoptive cancer immunotherapy. Cancers 2019;11:769. https://doi.org/10.3390/cancers11060769.

[2] Carotta S, Targeting NK. Cells for anticancer immunotherapy: clinical and preclinical approaches. Front Immunol 2016;7. https://doi.org/10.3389/fimmu.2016.00152.

[3] Geller MA, Miller JS. Use of allogeneic NK cells for cancer immunotherapy. Immunotherapy 2011;3: 1445–59. https://doi.org/10.2217/imt.11.131.

[4] Wang J, Matosevic S. Functional and metabolic targeting of natural killer cells to solid tumors. Cell Oncol 2020;43:577–600. https://doi.org/10.1007/s13402-020-00523-7.

[5] Zwirner NW, Domaica CI. Cytokine regulation of natural killer cell effector functions. Biofactors 2010;36: 274–88. https://doi.org/10.1002/biof.107.

[6] Liu E, Marin D, Banerjee P, Macapinlac HA, Thompson P, Basar R, et al. Use of CAR-transduced natural killer cells in CD19-positive lymphoid tumors. N Engl J Med 2020;382:545–53. https://doi.org/10.1056/NEJMoa1910607.

[7] Hu W, Wang G, Huang D, Sui M, Xu Y. Cancer immunotherapy based on natural killer cells: current progress and new opportunities. Front Immunol 2019;10. https://doi.org/10.3389/fimmu.2019.01205.

[8] Duan S, Guo W, Xu Z, He Y, Liang C, Mo Y, et al. Natural killer group 2D receptor and its ligands in cancer immune escape. Mol Cancer 2019;18:29. https://doi.org/10.1186/s12943-019-0956-8.

[9] DeBerardinis RJ. Tumor microenvironment, metabolism, and immunotherapy. N Engl J Med 2020;382: 869–71. https://doi.org/10.1056/NEJMcibr1914890.

[10] Wang J, Matosevic S. Adenosinergic signaling as a target for natural killer cell immunotherapy. J Mol Med 2018;96:903–13. https://doi.org/10.1007/s00109-018-1679-9.

[11] Poli A, Michel T, Thérésine M, Andrès E, Hentges F, Zimmer J. CD56bright natural killer (NK) cells: an important NK cell subset. Immunology 2009;126:458–65. https://doi.org/10.1111/j.1365-2567.2008.03027.x.

[12] de Jonge K, Ebering A, Nassiri S, Maby-El Hajjami H, Ouertatani-Sakouhi H, Baumgaertner P, et al. Circulating CD56 bright NK cells inversely correlate with survival of melanoma patients. Sci Rep 2019;9:4487. https://doi.org/10.1038/s41598-019-40933-8.

[13] Shinko D, McGuire HM, Diakos CI, Pavlakis N, Clarke SJ, Byrne SN, et al. Mass cytometry reveals a sustained reduction in CD16+ natural killer cells following chemotherapy in colorectal cancer patients. Front Immunol 2019;10. https://doi.org/10.3389/fimmu.2019.02584.

[14] Oran B, Shpall E. Umbilical cord blood transplantation: a maturing technology. Hematology Am Soc Hematol Educ Program 2012;2012:215–22. https://doi.org/10.1182/asheducation-2012.1.215.

[15] Nomura A, Takada H, Jin CH, Tanaka T, Ohga S, Hara T. Functional analyses of cord blood natural killer cells and T cells: a distinctive interleukin-18 response. Exp Hematol 2001;29:1169–76. https://doi.org/10.1016/s0301-472x(01)00689-0.

[16] Verneris MR, Miller JS. The phenotypic and functional characteristics of umbilical cord blood and peripheral blood natural killer cells. Br J Haematol 2009;147:185–91. https://doi.org/10.1111/j.1365-2141.2009.07768.x.

[17] Rutella S, Bonanno G, Marone M, De Ritis D, Mariotti A, Voso MT, et al. Identification of a novel subpopulation of human cord blood CD34-CD133-CD7-CD45+lineage- cells capable of lymphoid/NK cell differentiation after in vitro exposure to IL-15. J Immunol 2003;171:2977–88.
[18] Perez SA, Sotiropoulou PA, Gkika DG, Mahaira LG, Niarchos DK, Gritzapis AD, et al. A novel myeloid-like NK cell progenitor in human umbilical cord blood. Blood 2003;101:3444–50. https://doi.org/10.1182/blood-2002-05-1501.
[19] Sarvaria A, Jawdat D, Madrigal JA, Saudemont A. Umbilical cord blood natural killer cells, their characteristics, and potential clinical applications. Front Immunol 2017;8. https://doi.org/10.3389/fimmu.2017.00329.
[20] Gaddy J, Broxmeyer HE. Cord blood CD16+56- cells with low lytic activity are possible precursors of mature natural killer cells. Cell Immunol 1997;180:132–42. https://doi.org/10.1006/cimm.1997.1175.
[21] Raskovalova T, Lokshin A, Huang X, Jackson EK, Gorelik E. Adenosine-mediated inhibition of cytotoxic activity and cytokine production by IL-2/NKp46-activated NK cells: involvement of protein kinase A isozyme I (PKA I). Immunol Res 2006;36:91–9. https://doi.org/10.1385/IR:36:1:91.
[22] Chen S, Wainwright DA, Wu JD, Wan Y, Matei DE, Zhang Y, et al. CD73: an emerging checkpoint for cancer immunotherapy. Immunotherapy 2019;11:983–97. https://doi.org/10.2217/imt-2018-0200.
[23] Beavis PA, Divisekera U, Paget C, Chow MT, John LB, Devaud C, et al. Blockade of A2A receptors potently suppresses the metastasis of CD73+ tumors. Proc Natl Acad Sci U S A 2013;110:14711–6. https://doi.org/10.1073/pnas.1308209110.
[24] Qin L, Thompson LF, Kuzel TM, Zhang B. Requirement of NK cells for selective A2A receptor blockade to suppress CD73+ tumor metastasis. Immunotherapy 2014;6:19–21. https://doi.org/10.2217/imt.13.154.
[25] Li Y, Hermanson DL, Moriarity BS, Kaufman DS. Human iPSC-derived natural killer cells engineered with chimeric antigen receptors enhance anti-tumor activity. Cell Stem Cell 2018;23:181–192.e5. https://doi.org/10.1016/j.stem.2018.06.002.
[26] Fang F, Wang W, Chen M, Tian Z, Xiao W. Technical advances in NK cell-based cellular immunotherapy. Cancer Biol Med 2019;16:647–54. https://doi.org/10.20892/j.issn.2095-3941.2019.0187.
[27] Angelo LS, Banerjee PP, Monaco-Shawver L, Rosen JB, Makedonas G, Forbes LR, et al. Practical NK cell phenotyping and variability in healthy adults. Immunol Res 2015;62:341–56. https://doi.org/10.1007/s12026-015-8664-y.
[28] Romee R, Schneider SE, Leong JW, Chase JM, Keppel CR, Sullivan RP, et al. Cytokine activation induces human memory-like NK cells. Blood 2012;120:4751–60. https://doi.org/10.1182/blood-2012-04-419283.
[29] Leong JW, Chase JM, Romee R, Schneider SE, Sullivan RP, Cooper MA, et al. Pre-activation with IL-12, IL-15, and IL-18 induces CD25 and a functional high affinity IL-2 receptor on human cytokine-induced memory-like NK cells. Biol Blood Marrow Transplant 2014;20:463–73. https://doi.org/10.1016/j.bbmt.2014.01.006.
[30] Pahl JHW, Cerwenka A, Ni J. Memory-like NK cells: remembering a previous activation by cytokines and NK cell receptors. Front Immunol 2018;9. https://doi.org/10.3389/fimmu.2018.02796.
[31] Kundu S, Gurney M, O'Dwyer M. Generating natural killer cells for adoptive transfer: expanding horizons. Cytotherapy 2021. https://doi.org/10.1016/j.jcyt.2020.12.002.
[32] Lapteva N, Szmania SM, van Rhee F, Rooney CM. Clinical grade purification and expansion of natural killer cells. Crit Rev Oncog 2014;19:121–32.
[33] Ciurea SO, Schafer JR, Bassett R, Denman CJ, Cao K, Willis D, et al. Phase 1 clinical trial using mbIL21 ex vivo-expanded donor-derived NK cells after haploidentical transplantation. Blood 2017;130:1857–68. https://doi.org/10.1182/blood-2017-05-785659.
[34] Aum DJ, Kim DH, Beaumont TL, Leuthardt EC, Dunn GP, Kim AH. Molecular and cellular heterogeneity: the hallmark of glioblastoma. Neurosurg Focus 2014;37, E11. https://doi.org/10.3171/2014.9.FOCUS14521.
[35] Henke E, Nandigama R, Ergün S. Extracellular matrix in the tumor microenvironment and its impact on cancer therapy. Front Mol Biosci 2020;6. https://doi.org/10.3389/fmolb.2019.00160.
[36] Netti PA, Berk DA, Swartz MA, Grodzinsky AJ, Jain RK. Role of extracellular matrix assembly in interstitial transport in solid tumors. Cancer Res 2000;60:2497–503.
[37] Codd AS, Kanaseki T, Torigo T, Tabi Z. Cancer stem cells as targets for immunotherapy. Immunology 2018;153:304–14. https://doi.org/10.1111/imm.12866.
[38] Dao TN, Matosevic S. Immunometabolic responses of natural killer cells to inhibitory tumor microenvironment checkpoints. Immunometabolism 2019;1, e190003. https://doi.org/10.20900/immunometab20190003.

[39] Dao TN, Utturkar S, Atallah Lanman N, Matosevic S. TIM-3 expression is downregulated on human NK cells in response to cancer targets in synergy with activation. Cancers 2020;12:2417. https://doi.org/10.3390/cancers12092417.
[40] Cao Y, Wang X, Jin T, Tian Y, Dai C, Widarma C, et al. Immune checkpoint molecules in natural killer cells as potential targets for cancer immunotherapy. Signal Transduct Target Ther 2020;5:1–19. https://doi.org/10.1038/s41392-020-00348-8.
[41] Bhattacharya S, Calar K, Evans C, Petrasko M, de la Puente P. Bioengineering the oxygen-deprived tumor microenvironment within a three-dimensional platform for studying tumor-immune interactions. Front Bioeng Biotechnol 2020;8. https://doi.org/10.3389/fbioe.2020.01040.
[42] Matosevic S, Chambers AM. Immunometabolic dysfunction of natural killer cells mediated by the hypoxia-CD73 axis in solid tumors. Front Mol Biosci 2019;6:60. https://doi.org/10.3389/fmolb.2019.00060.
[43] Balsamo M, Manzini C, Pietra G, Raggi F, Blengio F, Mingari MC, et al. Hypoxia downregulates the expression of activating receptors involved in NK-cell-mediated target cell killing without affecting ADCC. Eur J Immunol 2013;43:2756–64. https://doi.org/10.1002/eji.201343448.
[44] Chambers AM, Wang J, Lupo KB, Yu H, Atallah Lanman NM, Matosevic S. Adenosinergic signaling alters natural killer cell functional responses. Front Immunol 2018;9. https://doi.org/10.3389/fimmu.2018.02533.
[45] Moon Y, Lim SA, Lee K-M, Park H. Abstract 122: hypoxia-enhanced proliferation and effector function of NK cells. Cancer Res 2018;78:122. https://doi.org/10.1158/1538-7445.AM2018-122.
[46] Fink T, Ebbesen P, Koppelhus U, Zachar V. Natural killer cell-mediated basal and interferon-enhanced cytotoxicity against liver cancer cells is significantly impaired under in vivo oxygen conditions. Scand J Immunol 2003;58:607–12.
[47] Loeffler DA, Juneau PL, Heppner GH. Natural killer-cell activity under conditions reflective of tumor microenvironment. Int J Cancer 1991;48:895–9.
[48] Lim SA, Moon Y, Park H, Lee K-M. NK cell effector functions are differentially regulated by hypoxic conditions (TUM9P.1014). J Immunol 2015;194:210.16.
[49] Horenstein AL, Chillemi A, Zaccarello G, Bruzzone S, Quarona V, Zito A, et al. A CD38/CD203a/CD73 ectoenzymatic pathway independent of CD39 drives a novel adenosinergic loop in human T lymphocytes. Onco Targets Ther 2013;2, e26246. https://doi.org/10.4161/onci.26246.
[50] Young A, Ngiow SF, Gao Y, Patch A-M, Barkauskas DS, Messaoudene M, et al. A2AR adenosine signaling suppresses natural killer cell maturation in the tumor microenvironment. Cancer Res 2018;78:1003–16. https://doi.org/10.1158/0008-5472.CAN-17-2826.
[51] Wang J, Lupo KB, Chambers AM, Matosevic S. Purinergic targeting enhances immunotherapy of CD73+ solid tumors with piggyBac-engineered chimeric antigen receptor natural killer cells. J Immunother Cancer 2018;6:136. https://doi.org/10.1186/s40425-018-0441-8.
[52] Yao X, Matosevic S. Chemokine networks modulating natural killer cell trafficking to solid tumors. Cytokine Growth Factor Rev 2021. https://doi.org/10.1016/j.cytogfr.2020.12.003.
[53] Wennerberg E, Kremer V, Childs R, Lundqvist A. CXCL10-induced migration of adoptively transferred human natural killer cells toward solid tumors causes regression of tumor growth in vivo. Cancer Immunol Immunother 2015;64:225–35. https://doi.org/10.1007/s00262-014-1629-5.
[54] Molfetta R, Zitti B, Lecce M, Milito ND, Stabile H, Fionda C, et al. CD155: a multi-functional molecule in tumor progression. Int J Mol Sci 2020;21. https://doi.org/10.3390/ijms21030922.
[55] Sloan KE, Stewart JK, Treloar AF, Matthews RT, Jay DG. CD155/PVR enhances glioma cell dispersal by regulating adhesion signaling and focal adhesion dynamics. Cancer Res 2005;65:10930–7. https://doi.org/10.1158/0008-5472.CAN-05-1890.
[56] Sloan KE, Eustace BK, Stewart JK, Zehetmeier C, Torella C, Simeone M, et al. CD155/PVR plays a key role in cell motility during tumor cell invasion and migration. BMC Cancer 2004;4:73. https://doi.org/10.1186/1471-2407-4-73.
[57] Lupo KB, Matosevic S. CD155 immunoregulation as a target for natural killer cell immunotherapy in glioblastoma. J Hematol Oncol 2020;13:76. https://doi.org/10.1186/s13045-020-00913-2.
[58] Liu L, You X, Han S, Sun Y, Zhang J, Zhang Y. CD155/TIGIT, a novel immune checkpoint in human cancers (review). Oncol Rep 2021;45:835–45. https://doi.org/10.3892/or.2021.7943.

[59] Zhang Q, Bi J, Zheng X, Chen Y, Wang H, Wu W, et al. Blockade of the checkpoint receptor TIGIT prevents NK cell exhaustion and elicits potent anti-tumor immunity. Nat Immunol 2018;19:723. https://doi.org/10.1038/s41590-018-0132-0.

[60] Judge SJ, Darrow MA, Thorpe SW, Gingrich AA, O'Donnell EF, Bellini AR, et al. Analysis of tumor-infiltrating NK and T cells highlights IL-15 stimulation and TIGIT blockade as a combination immunotherapy strategy for soft tissue sarcomas. J Immunother Cancer 2020;8, e001355. https://doi.org/10.1136/jitc-2020-001355.

[61] Hung AL, Maxwell R, Theodros D, Belcaid Z, Mathios D, Luksik AS, et al. TIGIT and PD-1 dual checkpoint blockade enhances antitumor immunity and survival in GBM. Onco Targets Ther 2018;7. https://doi.org/10.1080/2162402X.2018.1466769.

[62] Pal SK, Vanderwalde AM, Szeto C, Reddy S, Hamid O. 1183PPD-L1 expression is strongly associated with TIGIT, FOXP3 and LAG3 across advanced cancers, but not OX40, TIM3 and IDO. Ann Oncol 2018;29. https://doi.org/10.1093/annonc/mdy288.056.

[63] Kamiya T, Seow SV, Wong D, Robinson M, Campana D. Blocking expression of inhibitory receptor NKG2A overcomes tumor resistance to NK cells. J Clin Invest 2019;129. https://doi.org/10.1172/JCI123955.

[64] Rapaport AS, Schriewer J, Gilfillan S, Hembrador E, Crump R, Plougastel BF, et al. The inhibitory receptor NKG2A sustains virus-specific CD8+ T cells in response to a lethal poxvirus infection. Immunity 2015;43:1112–24. https://doi.org/10.1016/j.immuni.2015.11.005.

[65] Viant C, Fenis A, Chicanne G, Payrastre B, Ugolini S, Vivier E. SHP-1-mediated inhibitory signals promote responsiveness and anti-tumour functions of natural killer cells. Nat Commun 2014;5:5108. https://doi.org/10.1038/ncomms6108.

[66] Gustafson KS, Ginder GD. Interferon-gamma induction of the human leukocyte antigen-E gene is mediated through binding of a complex containing STAT1alpha to a distinct interferon-gamma-responsive element. J Biol Chem 1996;271:20035–46. https://doi.org/10.1074/jbc.271.33.20035.

[67] de Kruijf EM, Sajet A, van Nes JGH, Natanov R, Putter H, Smit VTHBM, et al. HLA-E and HLA-G expression in classical HLA class I-negative tumors is of prognostic value for clinical outcome of early breast cancer patients. J Immunol 2010;185:7452–9. https://doi.org/10.4049/jimmunol.1002629.

[68] André P, Denis C, Soulas C, Bourbon-Caillet C, Lopez J, Arnoux T, et al. Anti-NKG2A mAb is a checkpoint inhibitor that promotes anti-tumor immunity by unleashing both T and NK cells. Cell 2018;175:1731–1743.e13. https://doi.org/10.1016/j.cell.2018.10.014.

[69] Matosevic S. Viral and nonviral engineering of natural killer cells as emerging adoptive cancer immunotherapies. J Immunol Res 2018;, 4054815. https://doi.org/10.1155/2018/4054815.

[70] Mantesso S, Geerts D, Spanholtz J, Kučerová L. Genetic engineering of natural killer cells for enhanced antitumor function. Front Immunol 2020;11. https://doi.org/10.3389/fimmu.2020.607131.

[71] Carlsten M, Childs RW. Genetic manipulation of NK cells for cancer immunotherapy: techniques and clinical implications. Front Immunol 2015;6. https://doi.org/10.3389/fimmu.2015.00266.

[72] Boissel L, Betancur M, Lu W, Wels WS, Marino T, Etten RAV, et al. Comparison of mRNA and lentiviral based transfection of natural killer cells with chimeric antigen receptors recognizing lymphoid antigens. Leuk Lymphoma 2012;53:958–65. https://doi.org/10.3109/10428194.2011.634048.

[73] Swaney WP, Sorgi FL, Bahnson AB, Barranger JA. The effect of cationic liposome pretreatment and centrifugation on retrovirus-mediated gene transfer. Gene Ther 1997;4:1379–86. https://doi.org/10.1038/sj.gt.3300529.

[74] Jiang H, Zhang W, Shang P, Zhang H, Fu W, Ye F, et al. Transfection of chimeric anti-CD138 gene enhances natural killer cell activation and killing of multiple myeloma cells. Mol Oncol 2014;8:297–310. https://doi.org/10.1016/j.molonc.2013.12.001.

[75] Kruschinski A, Moosmann A, Poschke I, Norell H, Chmielewski M, Seliger B, et al. Engineering antigen-specific primary human NK cells against HER-2 positive carcinomas. Proc Natl Acad Sci U S A 2008;105:17481–6. https://doi.org/10.1073/pnas.0804788105.

[76] Huang L, Zhu P, Xia P, Fan Z. WASH has a critical role in NK cell cytotoxicity through Lck-mediated phosphorylation. Cell Death Dis 2016;7, e2301. https://doi.org/10.1038/cddis.2016.212.

[77] Davis HE, Morgan JR, Yarmush ML. Polybrene increases retrovirus gene transfer efficiency by enhancing receptor-independent virus adsorption on target cell membranes. Biophys Chem 2002;97:159–72.

[78] Hanenberg H, Xiao XL, Dilloo D, Hashino K, Kato I, Williams DA. Colocalization of retrovirus and target cells on specific fibronectin fragments increases genetic transduction of mammalian cells. Nat Med 1996;2:876. https://doi.org/10.1038/nm0896-876.

[79] Fenard D, Ingrao D, Seye A, Buisset J, Genries S, Martin S, et al. Vectofusin-1, a new viral entry enhancer, strongly promotes lentiviral transduction of human hematopoietic stem cells. Mol Ther Nucleic Acids 2013;2, e90. https://doi.org/10.1038/mtna.2013.17.

[80] Cronin J, Zhang X-Y, Reiser J. Altering the tropism of lentiviral vectors through pseudotyping. Curr Gene Ther 2005;5:387–98.

[81] Sutlu T, Nyström S, Gilljam M, Stellan B, Applequist SE, Alici E. Inhibition of intracellular antiviral defense mechanisms augments lentiviral transduction of human natural killer cells: implications for gene therapy. Hum Gene Ther 2012;23:1090–100.

[82] Wang W, Jiang J, Wu C. CAR-NK for tumor immunotherapy: clinical transformation and future prospects. Cancer Lett 2020;472:175–80. https://doi.org/10.1016/j.canlet.2019.11.033.

[83] Xie G, Dong H, Liang Y, Ham JD, Rizwan R, Chen J. CAR-NK cells: a promising cellular immunotherapy for cancer. EBioMedicine 2020;59. https://doi.org/10.1016/j.ebiom.2020.102975.

[84] Burger MC, Zhang C, Harter PN, Romanski A, Strassheimer F, Senft C, et al. CAR-engineered NK cells for the treatment of glioblastoma: turning innate effectors into precision tools for cancer immunotherapy. Front Immunol 2019;10. https://doi.org/10.3389/fimmu.2019.02683.

[85] Yilmaz A, Cui H, Caligiuri MA, Yu J. Chimeric antigen receptor-engineered natural killer cells for cancer immunotherapy. J Hematol Oncol 2020;13:168. https://doi.org/10.1186/s13045-020-00998-9.

[86] Rezvani K, Rouce R, Liu E, Shpall E. Engineering natural killer cells for cancer immunotherapy. Mol Ther 2017;25:1769–81. https://doi.org/10.1016/j.ymthe.2017.06.012.

[87] Goodridge JP, Mahmood S, Zhu H, Gaidarova S, Blum R, Bjordahl R, et al. FT596: translation of first-of-kind multi-antigen targeted off-the-shelf CAR-NK cell with engineered persistence for the treatment of B cell malignancies. Blood 2019;134:301. https://doi.org/10.1182/blood-2019-129319.

[88] Majzner RG, Mackall CL. Tumor antigen escape from CAR T-cell therapy. Cancer Discov 2018;8: 1219–26. https://doi.org/10.1158/2159-8290.CD-18-0442.

[89] Grada Z, Hegde M, Byrd T, Shaffer DR, Ghazi A, Brawley VS, et al. TanCAR: a novel bispecific chimeric antigen receptor for cancer immunotherapy. Mol Ther Nucleic Acids 2013;2, e105. https://doi.org/10.1038/mtna.2013.32.

[90] Gargett T, Yu W, Dotti G, Yvon ES, Christo SN, Hayball JD, et al. GD2-specific CAR T cells undergo potent activation and deletion following antigen encounter but can be protected from activation-induced cell death by PD-1 blockade. Mol Ther 2016;24:1135–49. https://doi.org/10.1038/mt.2016.63.

[91] Künkele A, Johnson AJ, Rolczynski LS, Chang CA, Hoglund V, Kelly-Spratt KS, et al. Functional tuning of CARs reveals signaling threshold above which CD8+ CTL antitumor potency is attenuated due to cell Fas-FasL-dependent AICD. Cancer Immunol Res 2015;3:368–79. https://doi.org/10.1158/2326-6066.CIR-14-0200.

[92] Zah E, Lin M-Y, Silva-Benedict A, Jensen MC, Chen YY. T cells expressing CD19/CD20 bispecific chimeric antigen receptors prevent antigen escape by malignant B cells. Cancer Immunol Res 2016;4:498–508. https://doi.org/10.1158/2326-6066.CIR-15-0231.

[93] Hegde M, Mukherjee M, Grada Z, Pignata A, Landi D, Navai SA, et al. Tandem CAR T cells targeting HER2 and IL13Rα2 mitigate tumor antigen escape. J Clin Invest 2016;126:3036–52. https://doi.org/10.1172/JCI83416.

[94] Wang J, Toregrosa-Allen S, Elzey BD, Utturkar S, Lanman NA, Bernal-Crespo V, et al. Tumor-responsive, multifunctional CAR-NK cells cooperate with impaired autophagy to infiltrate and target glioblastoma. BioRxiv 2020. https://doi.org/10.1101/2020.10.07.330043. 2020.10.07.330043.

[95] Gauthier L, Morel A, Anceriz N, Rossi B, Blanchard-Alvarez A, Grondin G, et al. Multifunctional natural killer cell engagers targeting NKp46 trigger protective tumor immunity. Cell 2019;177:1701–1713.e16. https://doi.org/10.1016/j.cell.2019.04.041.

[96] Cheng Y, Zheng X, Wang X, Chen Y, Wei H, Sun R, et al. Trispecific killer engager 161519 enhances natural killer cell function and provides anti-tumor activity against CD19-positive cancers. Cancer Biol Med 2020;17:1026–38. https://doi.org/10.20892/j.issn.2095-3941.2020.0399.

[97] Gleason MK, Ross JA, Warlick ED, Lund TC, Verneris MR, Wiernik A, et al. CD16xCD33 bispecific killer cell engager (BiKE) activates NK cells against primary MDS and MDSC CD33+ targets. Blood 2014;123: 3016–26. https://doi.org/10.1182/blood-2013-10-533398.

[98] Schmohl JU, Felices M, Oh F, Lenvik AJ, Lebeau AM, Panyam J, et al. Engineering of anti-CD133 trispecific molecule capable of inducing NK expansion and driving antibody-dependent cell-mediated cytotoxicity. Cancer Res Treat 2017;49:1140–52. https://doi.org/10.4143/crt.2016.491.
[99] Sahm C, Schönfeld K, Wels WS. Expression of IL-15 in NK cells results in rapid enrichment and selective cytotoxicity of gene-modified effectors that carry a tumor-specific antigen receptor. Cancer Immunol Immunother 2012;61:1451–61. https://doi.org/10.1007/s00262-012-1212-x.
[100] Romee R, Leong JW, Fehniger TA. Utilizing cytokines to function-enable human NK cells for the immunotherapy of cancer. Scientifica (Cairo) 2014;2014:205796. https://doi.org/10.1155/2014/205796.
[101] Guy D, Uy GL. Bispecific antibodies for the treatment of acute myeloid leukemia. Curr Hematol Malig Rep 2018;13:417–25. https://doi.org/10.1007/s11899-018-0472-8.

CHAPTER

4

Targeting NKG2D/NKG2D ligand axis for cancer immunotherapy

Jennifer Wu

Feinberg School of Medicine, Northwestern University, Chicago, IL, United States

Abstract

NKG2D is a major activation receptor on natural killer (NK) cells and a unique receptor on Cd8 T cells and subsets of NKT and γδ T cells. NKG2D is generally absent on CD4 T cells. The ligand of NKG2D (NKG2D-L) is generally absent on normal tissues, but prevalently induced on tumors cells as a response to genomic insults. Due to the unique nature of the NKG2D/NKG2D-L axis, attentions and efforts are emerging to harness the power of NKG2D/NKG2D-L for cancer immunotherapy. To support this development, this chapter will cover the basic biology of the NKG2D/NKG2D-L system and strategies to target this axis for cancer immunotherapy.

Abbreviations

ADAM a disintegrin and metalloproteinase
ATM ataxia telangiectasia mutated
ATR ATM and Rad3-related
GPI glycosylphosphatidylinositol
HCMV human cytomegalovirus
ITAM immunoreceptor tyrosine-based activation motif
KSHV Kaposi's sarcoma-associated herpesvirus
MDSC myeloid-derived suppressor cell
MGUS monoclonal gammopathy of undermined significance
MIC MHC I Chain related molecule
MULT1 murine ULBP-like transcript 1
NK Natural Killer
Syk spleen tyrosine kinase
TME tumor microenvironment
TCR T cell receptor
ULBPs UL16-binding proteins
UPR unfolder protein response
ZAP70 zeta-chain-associated protein kinase 70

https://doi.org/10.1016/B978-0-12-822620-9.00010-0

Conflict of interest

No potential conflicts of interest were disclosed.

Introduction

NKG2D (nature killer group 2, member D) is a C-type lectin-like activating receptor expressed by all human natural killer (NK) cells, most NKT cells, subsets of γδ T cells, and CD8 T cells. In mouse, NKG2D is also expressed by all NK cells and subsets of splenic γδ T cells and NKT cells, but only expressed by activated mouse CD8 T cells and activated mouse macrophages. NKG2D is located on a syntanic region of human choromosome 12 and on mouse chromosome 6, clustered with other NKG2 family members [1]. NKG2D serves as an invariant immune activating receptor upon engagement by ligands expressed on target cells, transformed or viral infected cells. Engagement of NKG2D by its ligands can activate NK cell and costimulate CD8 and γδ T cells [2,3]. The activation signals transmitted by NKG2D can override inhibitory signals transmitted by other NK receptors. NKG2D is therefore referred as the master activating receptor for NK cells to sense cells under abnormal physiological stress. The ligands for NKG2D are not commonly present in normal tissues but can be induced under abnormal physiological condition, such as cellular transformation or viral infection. The expression pattern of NKG2D ligands in tumor cells has been extensively studied. Emerging experimental evidence has indicated that NKG2D-mediated immunity can be very effective for tumor clearance by activating NK cells and co-stimulating CD8 T cells. However, it is widely accepted that NKG2D function is impaired in cancer patients, due to various mechanisms of tumor immunoediting and immune suppressive effect of tumor microenvironment. In this chapter, we will summarize the mechanisms of tumor immune editing NKG2D function and current development in harnessing the NKG2D-NKG2DL axis for cancer immunotherapy.

NKG2D

Molecular characteristics and signaling

NKG2D, encoded by *klrk1*, is a type II transmembrane glycoprotein, containing C-type lectin-like domains, similar to other known NKG2 family [4]. Although physically clustered with other NKG2 family members, NKG2D only shared 20%–30% sequence homology with other members of the NKG2 family. NKG2D is highly conserved between species. For instance, human NKG2D and mouse NKG2D share 70% amino acid identity [5]. NKG2D is constitutively expressed by all NK cells, NKT cells, and subsets of γδ T cells. NKG2D is expressed by all human CD8 T cells express NKG2D, but only activated mouse CD8 T cells. NKG2D can also be expressed by activated macrophages [6,7]. CD4 T cells generally do not express NKG2D. Under pathological conditions, a subset of CD4 T cell can be induced to express NKG2D [8–11].

The NKG2D molecule contains two β-sheets, two α-helices, four disulfate bonds, and a β-strand [7]. NKG2D forms homodimers on the cell membrane [5]. In both human and mouse lymphocytes, stable surface expression of NKG2D requires a complex formation of NKG2D homodimer with a Tyr-X-X-Met (YXXM) adaptor signaling molecule DAP10 at the cell membrane [12]. NKG2D itself does not contain intracellular signaling domain; it signals through the adaptor molecule DAP10. Each of the NKG2D transmembrane domain can bind to two DAP10 molecules as a dimer [13]. Thus, one assembled NKG2D molecule can signal through four DAP10 molecules. Mouse NK cells also express a splice variant NKG2D-S, which is 13 aa shorter than normal NKG2D and signals through either DAP10 or the immunoreceptor tyrosine-based activation motif (ITAM)-containing adaptor molecule DAP12 [14]. NKG2D in activated human NK and CD8 T cells may also signal through DAP12 to execute the stimulatory function [15]. Classically, upon ligand engagement of NKG2D, DAP 10 is phosphorylated by src-family kinases, which permits the recruitment of the PI3K subunit p85 and the signaling intermediate Grb2-Vav 1 to fully activate NK cell cytotoxic pathways. In activated mouse NK cells, NKG2D-s may also independently signal through DAP12, which, after phosphorylation, recruits ZAP70 (zeta- chain-associated protein kinase 70) and Syk (spleen tyrosine kinase).

NKG2D ligands

An array of genes encoding ligands for NKG2D have been identified in human and mice (Table 1). Expression of NKG2D is mostly restricted to tumors or certain viral-infected cells

TABLE 1 Identified human and mouse NKG2D ligands.

Name	Alternate name	Cell surface attachment
Human		
MICA	PERB11.1	Transmembrane
MICB	PERB11.2	Transmembrane
ULBP1	RAET1I	GPI anchor
ULBP2	RAET1H	GPI anchor or not
ULBP3	RAET1N	GPI anchor
ULBP4	RAET1E,LETAL	Transmembrane
ULBP5	RAET1G	Transmembrane or GPI anchor
ULBP6	RAET1L	GPI anchor
Mice		
Rae-1α	Raet 1a	GPI anchor
Rae-1β	Raet 1b	GPI anchor
Rae-1γ	Raet 1c	GPI anchor
Rae-1δ	Raet 1d	GPI anchor
Rae-1ε	Raet e	GPI anchor
H60-a	n/a	Transmembrane
H60-b	n/a	Transmembrane
H60-c	n/a	GPI anchor
MULT1	n/a	Transmembrane

where genomic insults occurred and rarely identified in normal tissues. The high divergent NKG2D expression is most likely attributed to evolutional process driven by viral infection [16].

Human NKG2D ligands

Two families of NKG2D ligands are identified thus far in humans: the MHC class I chain related family molecules A (MICA) and B (MICB) and the family of HCMV (human cytomegalovirus) UL16-binding proteins 1–6 (ULBPs 1–6) [17]. All these molecules are distant HLA class I homologs but not associated with β-2 microglobulin nor have roles in antigen presentation [4]. Although highly conserved within each family, members of the MIC family share little sequence or structural similarity with those of the ULBP family. The expression pattern of the MIC and ULBPs is also dissimilar.

The MIC(A/B) molecule is consisted of three extracellular domains (α1, α2, and α3), a transmembrane region, and a cytoplasmic tail [17,18]. Only the α1 and α2 domains of MICA/B interacts with the receptor NKG2D [19]. Human MIC genes are located within the MHC class I region of chromosome 6 [17,18]. Seven MIC loci exist, but only two loci encode translated genes (MICA and MICB) [4]. Although MICA and MICB transcripts are widely found in normal human tissues [20], they are predominantly found in tumors and rarely found in nontumor tissues. Inflamed human intestinal epithelium was found in some individual to express MIC, possibly due to the contact with intestinal microbes and chronic intestinal inflammation. MICA and MICB share over 80% amino acid identity. Both MICA and MICB are highly pleomorphic. It has been reported more than 110 alleles of MICA and more than 40 MICB alleles (http://hla.alleles.org/data/mica.html; http://hla.alleles.org/data/mica.html). Population in different geo-location or showed different frequency of these alleles. To some degree, this diversity may provide protection against rapidly evolving cancers.

The ULBPs were apparently named for their ability to bind to the human cytomegalovirus UL16 protein. To date, six members of human ULBP gene family are identified to encode functional proteins of two structural classes. ULBPs 1–3 and 6 are glycosylphosphatidylinositol (GPI)-linked proteins, whereas ULBPs 4 and 5 are type I transmembrane proteins. Unlike the MICs family, the ULBP family lacks the α3 domain and only have the MHC class I-like α1 and α2 domains [7,21]. The expression pattern of ULBP family members is not well defined. ULBP transcripts appear widely expressed in humans [22,23], not restricted to transformed tissues. ULBP family of ligands are far less polymorphic in comparison to the MICA/B family of ligands. Only a handful of alleles of the ULBP family of ligands were identified [24].

Mouse NKG2D ligands

Mouse genome does not contain human MIC genes or homologs. The identified mouse NKG2D ligands include family members of the murine MHC I-related family members of retinoic acid early transcript RAE-1(α, β, γ, δ, and ε) and H60 (a, b, c) and the murine ULBP-like transcript 1 (MULT1) [25–27]. All mouse NKG2D ligands only have the MHC class I-like α1 and α2 extracellular domains. To date, five members of the Rae-1 family, named Rae-1α, Rae-1β, Rae-1γ, Rae-1δ, and Rae-1ε, were identified. They are differentially expressed in various mouse strains but highly related to each other (>85% identity). The H60 family comprises three members, H60a, H60b, and H60c [27,28]. MULT1 is the unique member of the ULBP-like family of murine NKG2D ligands and was found by database searching for mouse sequences with similarities to human ULBP [29]. The association of murine NKG2D ligand expression in primary tumorigenesis was not well established. Transcripts of mouse NKG2D ligand was found to be expressed in a broad range of normal tissues. H60a mRNA was found in multiple tissues, including the spleen, cardiac, skeletal muscle, thymus, and skin, whereas H60b mRNA is limited to cardiac and skeletal muscles. H60c is transcribed largely in the skin [27,30]. H60a is productively expressed in BALB/c mice but not in C57BL/6 mice, whereas H60b and H60c transcripts are detected in both C57BL/6 and BALB/c mouse. MULT1 mRNA is found in the heart, thymus, lung, and kidney across most mice strains [27,29]. However, the expression level of NKG2D ligand on normal tissues seems to be below the threshold of inducing activate immune response to cause tissue injury.

Regulation of NKG2D ligands expression

NKG2D is viewed as the master activating receptor for immune responses. Thus, expression of NKG2D ligand has to be delicately regulated for controlling abnormal cells and protect normal tissue integrity. The regulation is achieved at multiple levels of regulatory mechanisms, each of which will be discussed below.

Transcriptional regulation

Expression of the transcripts of NKG2D ligands is known to be mainly regulated by cellular stress, DNA damage repair response, TLR stimulation, cytokine exposure, and activation of sting pathways. The promoter region of the MICA and MICB contains contain sequences that are highly homologous to the heat shock elements of HSP70 [31], a stress-induced gene. Viral oncoproteins, such as adenoviral E1A protein, or cellular stress-response-related products can bind to the promoter region of MICA and/or MICB to induce or upregulate its expression [31]. Treatment of hepatocellular carcinoma cells with RA was shown to induce the expression of MICA and MICB [32]. The transcription factor AP-1, which is involved in tumorigenesis and cellular stress responses, was found to regulate Rae-1 transcription through the JunB subunit [33].

The DNA damage response pathway is involved in maintaining the integrity of the genome. The PI3K-related protein kinases ATM (ataxia telangiectasia, mutated) and ATR (ATM and Rad3-related) sense DNA lesions, specifically double-strand breaks and stalled DNA replication, respectively. This sensing results in cell cycle arrest and DNA repair or cell apoptosis if the DNA damage is too extensive to be repaired. This pathway has been shown to be constitutively active in human cancer cells [34–36]. Both mouse and human cells upregulate NKG2D ligands expression following treatment with DNA-damaging agents or reagents in unfolder protein response (UPR) pathway [37]. The effect was dependent on ATR or ATM function, as inhibitors of ATR and ATM expression or kinase activity can prevent ligand upregulation.

Posttranslational regulation

Expression of NKG2D ligand can also be regulated posttranslationally via various mechanisms. The ubiquitination on the lysines in cytoplasmic tail of MULT1 was shown to mediate its rapid degradation [38]. Ubiquitination can be reduced in response to heat shock or ultraviolet irradiation through the MARCH family of E3 ligases and thus allow upregulation of NKG2D ligand expression, such as MULT1 in mice and MIC (A/B) in humans [39]. The presence of multiple lysines in the cytoplasmic tail of H60a, H60b, MICA, MICB, and RAET-1G suggests that this translational control mechanism might be used by other NKG2D ligands. KSHV (Kaposi's sarcoma-associated herpesvirus)-encoded E3 ubiquitin ligase K5 can downregulate cell surface expression of MICA and MICB [40]. The ubiquitination may also redistribute MICA to the plasma membrane, rather than target to degradation as observed with MULT1. Lastly, one of the most commonly described mechanism to regulate surface NKG2D ligand expression in human cancer cells is protease-mediated shedding and soluble NKG2D ligand containing exosome secretion [41].

NKG2D in antitumor immunity

NKG2D ligands stimulate tumor immunity

NKG2D-mediated tumor rejection has been demonstrated very effective in experimental animal models. The rejection was mediated primarily by NK cells or through a cooperation of NK cells with CD8 T cells. Overexpression of a high level of mouse NKG2D ligands Rae-1 or H60 in mouse tumor cells of various origin, including the thymoma cell line EL4, the T-cell lymphoma cell line RMA, and the poorly immunogenic and highly metastatic melanoma variant B16-BL6, induced in vivo rejection or retarded tumor growth when implanted into syngeneic mice [42,43]. It was also found that the rejection of a small dose of Rae-1 or H60-expressing tumors (e.g., 1×10^4 cells) could be achieved by NK cells or CD8 T cells alone, whereas inhibition the growth of large dose of Rae-1 or H60-expressing tumor cells (e.g., 1×10^6 cell) required a cooperation of NK cells and CD8 T cells [43].

The significance of NKG2D in controlling tumor growth was further emphasized by in vivo NKG2D neutralization in experimental models. When mice (B6 or balb/c background) were injected with antibody to neutralize NKG2D, these animals showed increased susceptibility to carcinogen MCA-induced fibrosarcoma in comparison to control IgG-treated mice [44]. Perhaps the most direct genetic evidence to demonstrate the role of NKG2D in tumor immunity comes from the NKG2D-deficient mice. When TRAMP mice were crossed with NKG2D-deficient mice, the progeny had 4- time increased frequency of developing poorly-differentiated tumors than NKG2D^{WT} counterparts [45]. In transgenic animal models, a mutant form of membrane-restricted human MICB overexpressed in the mouse prostate prevented oncogene-driven prostate tumor to progress [46].

Human tumors shed NKG2D ligands to evade immune response

Although NKG2D ligands are prevalently expressed in tumors of many types of human malignancies, direct evidence to demonstrate the role of NKG2D in controlling tumor growth or progression in human is still lacking. Understanding the significance of NKG2D in human cancer progression mainly comes from correlative observation in cancer patients. Massive clinical data demonstrating impaired NKG2D function in cancer patients was mediated by various mechanisms. Several studies elegantly demonstrate the positive correlation of impaired NKG2D function with cancer disease stages. Impaired NKG2D-mediated NK cell function was shown to correlate with cancer stages in prostate cancer patients [47]. In this study, circulating NK cells were isolated from prostate cancer patients with various stages of diseases. NKG2D expression and NK cell function were analyzed in vitro. The result showed a gradually loss of NKG2D+ NK population from patients with low grade to high grade of cancer, with complete loss of NKG2D expression on NK cells from patients with advanced diseases. As an obvious consequence, NKG2D-mediated cytotoxicity of these NK cells against tumor cells was severely subverted. Similar observations were demonstrated in the progression of other types of cancers [21]. Several studies have also described that dysfunction of NKG2D on $CD3^{+}CD56^{+}$ NK-like-T cells, and subsets of γδ T cells were associated with poor prognosis of certain cancers [48–50].

Expression of NKG2D ligands on tumors should effectively trigger both innate and adaptive immune response at the early stage of tumorigenesis to eradicate or control tumor progression tumors in human. However, in many types of established tumors of human malignancy, the NKG2D ligand MIC was highly expressed. The very paradoxical question is: how can human epithelial tumors develop and persist while the surface MIC molecule should identify them as abnormal and flag them for immune destruction? Clinical studies demonstrated that most of the human malignancies have developed mechanisms to evade NKG2D-mediated anti-tumor immunity. One of the common mechanisms by which human cancers evade NKG2D immunity is releasing the soluble form of the NKG2D ligands to the circulation by proteolytic shedding or exosomes in the case of specific GPI-linked NKG2D ligands [51]. This mechanism has been described in almost all human malignancies [21,51]. Elevated serum levels of soluble NKG2D ligands have been shown to be correlative with advanced cancer stages [21,37,39,52–56]. Some studies have suggested that serum levels of soluble MIC may be used as a valid prognosis factor for cancer progression [56]. Recent studies have shown that cancer patients with high levels of serum-soluble NKG2D ligands sULBP2 and sMICB responded unfavorably to immune checkpoint inhibitor therapy [57,58].

Tumor-derived sMIC can impose several negative imprints on host immune system (Fig. 1). First, shedding can reduce the density of membrane-bound NKG2D ligands, namely MIC on tumor cells and thus reduce the visibility of tumor cells by the immune surveillance. Second, sMIC in the circulation can not only mask NKG2D on effector NK, NKT, and T cells but also induce NKG2D internalization [47,51]. Since human MIC is not expressed by mouse, with tumor models that engineered human MIC into mouse tumors, it was shown that tumor secreted sMIC perturbed NK cell homeostatic renewal [46]. Tumor-secreted sMICB can also facilitate the differentiation of bone-marrow-derived myeloid cells to become a more immune suppressive phenotype [59]. Most recently, it was demonstrated that tumor secreted soluble MIC (A/B) can reprogram NK cell to alternative proinflammatory function with diminished cytotoxic function [60]. Together, these understandings provide the rationale that human tumor secreted sMIC may be a therapeutic target.

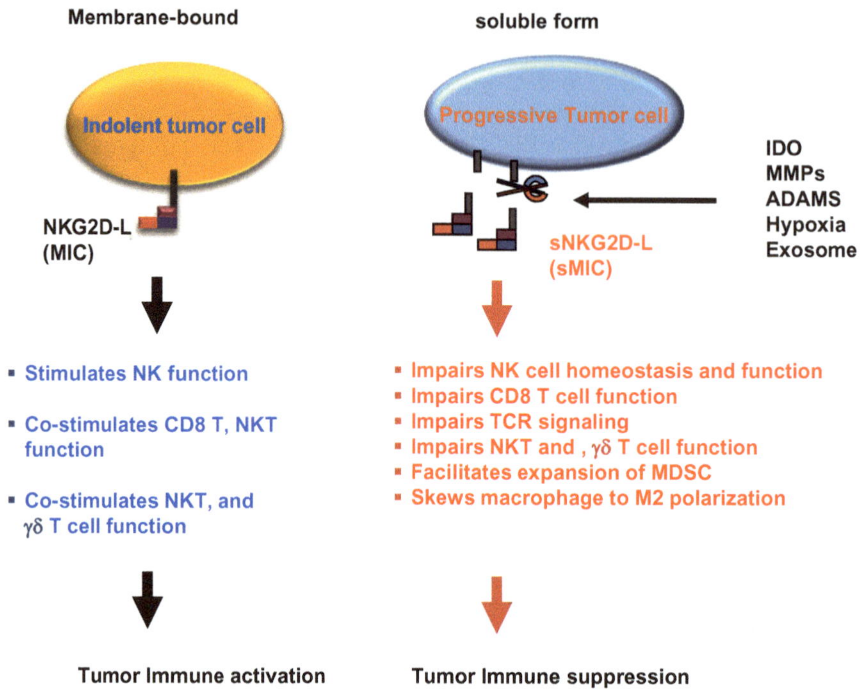

FIG. 1 Membrane-bound and soluble human NKG2D ligand (NKG2D-L) sMIC regulate tumor immunity in an opposite fashion.

Strategies to harness NKG2D/NKG2DL for cancer immunotherapy

NKG2D-based cell therapy such as NKG2D CAR-T cells and NKG2D CAR-NK cells are currently in development. In addition to these cell-based therapy, antibody-based therapy to harness the NKG2D/NKG2D ligand pathway to re-invigorate patients' endogenous NK and CD8 T cells are also in development. Two classes of antibodies targeting MIC have been reported in preclinical studies to effectively restore NK cell and CD8 T cell function: inhibition of MIC shedding and clearance of sMIC.

Studies have been done in many investigators to understand the mechanisms that regulate MIC shedding for potential therapeutic interventions. A diverse group of enzymes have recently been shown to be involved in MIC shedding. Studies from several groups have shown that the inhibition of cellular metalloproteinase activity by GM6001 markedly interferes with MIC shedding. Specific metalloproteinases, such as ADAM (a disintegrin and metalloproteinase)-10 and ADAM-17, were found contributing to MICA shedding [61], and ADAM-17 protease was found contributing to MICB shedding [62]. The type I membrane MMP (MT1-MMP, also called MMP14) also directly regulates MICA shedding independent of ADAMS [63]. The thiol isomerase ERp5, which catalyzes disulfide bond formation, reduction, and isomerization, was shown to be required for MIC shedding [64]. This was presumably accomplished by chaperoning conformational alterations of surface MIC through disulfide-bond exchange that render MIC susceptible for proteolytic cleavage.

ADAM-10 and -17 and the thioreductase ERp5 have been proposed to be potential cancer therapeutic targets for inhibiting MIC shedding. However, these enzymes are not only involved in pathology of diseases but also involved in many normal physiological functions. For instance, ADAM-17 is required for the generation of the active forms of epidermal growth factor receptor ligands that is essential for the development of epithelial tissues. In addition, although there are many examples of expression or upregulation of ADAMs in both tumor tissues and cell lines, the precise pattern of their expression within tumors is not always clear [65]. Furthermore, targeting ADAM-17 has been in clinical trials with a spectrum of inhibitors for over a decade. However, no single ADAM-17 inhibitor has passed a Phase II clinical trial because of high toxicity and nonspecific targeting [66]. As to the possibility of targeting ERp5, it has been suggested that disulfide bond exchange with cell surface molecule to enable the shedding may be a general mechanism by which ERp5 modulates cell signaling [67]. In addition, a wide role of ERp5 in cellular function has been implicated, such as in normal platelet activation [68]. These studies suggest that there are many facets of these enzymes that need to be understood before embarking with confidence on targeting them for cancer therapy. Therefore, a more specific and feasible target is needed for inhibiting MIC shedding for cancer therapy.

Targeting MIC shedding to sustain NKG2D immunity

By mass-spectrometry analyses, several independent studies have shown that MIC is cleaved at multiple sites in the near transmembrane region aa 253–289 in tumor cell lines [61,64,69]. Using genetic approach, a dispensable six-aa motif in the α3 ectodomain of MIC (A and B) was identified to be critical for regulating MIC shedding [69]. Mutation in the 6-aa motif completely prevented MIC shedding but did not interfere with MIC to be recognized by NKG2D. Because only the alpha1–alpha2 domain of MIC interacts with NKG2D, strategies to target the 6-aa motif or the entire alpha3 domain to inhibit MIC shedding has been sought. Preclinical studies demonstrated that monoclonal antibodies target the a3 domain of MIC to sustain surface MIC can inhibit tumor growth via stimulating NK cell cytotoxic activity [70]. However, one potential limitation of this strategy is that it may only be effective for controlling early stage of tumors where MIC is predominantly retained on cell surface.

For late stage of tumors, studies have shown that antibody clearance of sMIC as a single agent therapy can effectively restore and re-invigorate NK and CD8 T cell function in tumors [71]. It was demonstrated that the therapy efficacy involves multiple mechanisms: (1) restoring NKG2D expression on both NK cells and CD8 T cells; (2) restoring NK cell homeostatic self-renewal ability and homeostatic maintenance; (3) sustaining and expanding antigen-specific CD8 T cells in immune suppressive tumor microenvironment; (4) increase TCR clonality and complexity in TME; and (5) reducing arginase I producing MDSCs and alternatively activated macrophages in TME. It was further shown that antibody-targeting sMIC is effective to inhibit the growth of immune checkpoint nonresponsive tumors [72–74]. Co-targeting sMIC and PD1/PDL1 pathway provided dual co-stimulation of antigen-specific CD8 T cells via NKG2D and CD28 [73].

Considering the prevalence of MIC being expressed and sMIC being released by an array of advanced human malignant tumors, these preclinical studies provided the proof-of-concept that targeting soluble human NKG2D ligand sMIC holds a great promise as a therapy. Retrospective data have provided the ultimate clinical proof-of-concept. In a clinical trial with an anti-CTLA-4 antibody blockade or vaccines for melanoma therapy, patients who generated anti-MICA antibodies during the therapy showed significantly better clinical outcome than those who did not [75]. Studies by the same investigator also demonstrated that development of anti-sMIC autoantibody in monoclonal gammopathy of undetermined significance (MGUS) patients is associated with progression free to multiple myeloma [37]. The beneficial effects of anti-sMIC autoantibody were shown to act through antibody antagonizing sMICA-induced suppression of NK and CD8 T cells and stimulating DC for cross-presentation to generate antitumor responses.

Conclusions

NKG2D is a master activator of NK, TCR-engaged CD8 T cells, NKT cells, and subsets of γδ T cells. Expression of membrane-bound NKG2D ligands on transformed cells stimulates tumor immunity (Fig. 1). Conversely, tumor secreted soluble NKG2D ligands impairs tumor immunity. Preclinical studies and clinical correlative studies support the concept that targeting NKG2D ligands holds a great promise for cancer immunotherapy [76–78]. In particular, therapeutic concept of targeting the soluble NKG2D ligand sMIC to harness NKG2D immunity has been well proven in preclinical studies. However, the significance of this pathways has not been widely recognized for clinical development. With the emerging cancer immunology therapeutic regimens and FDA approvals and limitations of current therapies, it is time to consider co-targeting the NKG2D/NKG2D ligand pathway to improve clinical outcomes.

Note. My sincerely apologies to those whose important work not being cited.

References

[1] Glienke J, Sobanov Y, Brostjan C, Steffens C, Nguyen C, Lehrach H, Hofer E, Francis F. The genomic organization of NKG2C, E, F, and D receptor genes in the human natural killer gene complex. Immunogenetics 1998;48: 163–73.

[2] Bauer S, Groh V, Wu J, Steinle A, Phillips JH, Lanier LL, Spies T. Activation of NK cells and T cells by NKG2D, a receptor for stress-inducible MICA. Science 1999;285:727–9.

[3] Groh V, Rhinehart R, Randolph-Habecker J, Topp MS, Riddell SR, Spies T. Costimulation of CD8alphabeta T cells by NKG2D via engagement by MIC induced on virus-infected cells. Nat Immunol 2001;2:255–60.

[4] Eagle RA, Trowsdale J. Promiscuity and the single receptor: NKG2D. Nat Rev Immunol 2007;7:737–44.

[5] Raulet DH. Roles of the NKG2D immunoreceptor and its ligands. Nat Rev Immunol 2003;3:781–90.

[6] Diefenbach A, Jamieson AM, Liu SD, Shastri N, Raulet DH. Ligands for the murine NKG2D receptor: expression by tumor cells and activation of NK cells and macrophages. Nat Immunol 2000;1:119–26.

[7] Mistry AR, O'Callaghan CA. Regulation of ligands for the activating receptor NKG2D. Immunology 2007;121:439–47.

[8] Babic M, Dimitropoulos C, Hammer Q, Stehle C, Heinrich F, Sarsenbayeva A, Eisele A, Durek P, Mashreghi MF, Lisnic B, Van Snick J, Lohning M, Fillatreau S, Withers DR, Gagliani N, Huber S, Flavell RA, Polic B, Romagnani C. NK cell receptor NKG2D enforces proinflammatory features and pathogenicity of Th1 and Th17 cells. J Exp Med 2020;217.

[9] Groh V, Bruhl A, El-Gabalawy H, Nelson JL, Spies T. Stimulation of T cell autoreactivity by anomalous expression of NKG2D and its MIC ligands in rheumatoid arthritis. Proc Natl Acad Sci U S A 2003;100:9452–7.

[10] Hamada S, Caballero-Benitez A, Duran KL, Stevens AM, Spies T, Groh V. Soluble MICB in plasma and urine explains population expansions of NKG2D(+)CD4 T cells inpatients with juvenile-onset systemic lupus erythematosus. Open J Immunol 2017;7:1–17.
[11] Phoksawat W, Jumnainsong A, Sornkayasit K, Srisak K, Komanasin N, Leelayuwat C. IL-17 and IFN-gamma productions by CD4+ T cells and T cell subsets expressing NKG2D associated with the number of risk factors for cardiovascular diseases. Mol Immunol 2020;122:193–9.
[12] Ogasawara K, Lanier LL. NKG2D in NK and T cell-mediated immunity. J Clin Immunol 2005;25:534–40.
[13] Garrity D, Call ME, Feng J, Wucherpfennig KW. The activating NKG2D receptor assembles in the membrane with two signaling dimers into a hexameric structure. Proc Natl Acad Sci U S A 2005;102:7641–6.
[14] Long EO. Versatile signaling through NKG2D. Nat Immunol 2002;3:1119–20.
[15] Karimi M, Cao TM, Baker JA, Verneris MR, Soares L, Negrin RS. Silencing human NKG2D, DAP10, and DAP12 reduces cytotoxicity of activated CD8+ T cells and NK cells. J Immunol 2005;175:7819–28.
[16] Lanier LL. NKG2D receptor and its ligands in host defense. Cancer Immunol Res 2015;3:575–82.
[17] Bahram S, Inoko H, Shiina T, Radosavljevic M. MIC and other NKG2D ligands: from none to too many. Curr Opin Immunol 2005;17:505–9.
[18] Bahram S, Bresnahan M, Geraghty DE, Spies T. A second lineage of mammalian major histocompatibility complex class I genes. Proc Natl Acad Sci U S A 1994;91:6259–63.
[19] Groh V, Bahram S, Bauer S, Herman A, Beauchamp M, Spies T. Cell stress-regulated human major histocompatibility complex class I gene expressed in gastrointestinal epithelium. Proc Natl Acad Sci U S A 1996;93:12445–50.
[20] Schrambach S, Ardizzone M, Leymarie V, Sibilia J, Bahram S. In vivo expression pattern of MICA and MICB and its relevance to auto-immunity and cancer. PLoS One 2007;2, e518.
[21] Dhar P, Wu JD. NKG2D and its ligands in cancer. Curr Opin Immunol 2018;51:55–61.
[22] Cosman D, Mullberg J, Sutherland CL, Chin W, Armitage R, Fanslow W, Kubin M, Chalupny NJ. ULBPs, novel MHC class I-related molecules, bind to CMV glycoprotein UL16 and stimulate NK cytotoxicity through the NKG2D receptor. Immunity 2001;14:123–33.
[23] Radosavljevic M, Cuillerier B, Wilson MJ, Clement O, Wicker S, Gilfillan S, Beck S, Trowsdale J, Bahram S. A cluster of ten novel MHC class I related genes on human chromosome 6q24.2-q25.3. Genomics 2002;79:114–23.
[24] Romphruk AV, Romphruk A, Naruse TK, Raroengjai S, Puapairoj C, Inoko H, Leelayuwat C. Polymorphisms of NKG2D ligands: diverse RAET1/ULBP genes in northeastern Thais. Immunogenetics 2009;61:611–7.
[25] Cerwenka A, Bakker AB, McClanahan T, Wagner J, Wu J, Phillips JH, Lanier LL. Retinoic acid early inducible genes define a ligand family for the activating NKG2D receptor in mice. Immunity 2000;12:721–7.
[26] Diefenbach A, Hsia JK, Hsiung MY, Raulet DH. A novel ligand for the NKG2D receptor activates NK cells and macrophages and induces tumor immunity. Eur J Immunol 2003;33:381–91.
[27] Takada A, Yoshida S, Kajikawa M, Miyatake Y, Tomaru U, Sakai M, Chiba H, Maenaka K, Kohda D, Fugo K, Kasahara M. Two novel NKG2D ligands of the mouse H60 family with differential expression patterns and binding affinities to NKG2D. J Immunol 2008;180:1678–85.
[28] Malarkannan S, Shih PP, Eden PA, Horng T, Zuberi AR, Christianson G, Roopenian D, Shastri N. The molecular and functional characterization of a dominant minor H antigen, H60. J Immunol 1998;161:3501–9.
[29] Carayannopoulos LN, Naidenko OV, Fremont DH, Yokoyama WM. Cutting edge: murine UL16-binding protein-like transcript 1: a newly described transcript encoding a high-affinity ligand for murine NKG2D. J Immunol 2002;169:4079–83.
[30] Whang MI, Guerra N, Raulet DH. Costimulation of dendritic epidermal gammadelta T cells by a new NKG2D ligand expressed specifically in the skin. J Immunol 2009;182:4557–64.
[31] Venkataraman GM, Suciu D, Groh V, Boss JM, Spies T. Promoter region architecture and transcriptional regulation of the genes for the MHC class I-related chain A and B ligands of NKG2D. J Immunol 2007;178:961–9.
[32] Jinushi M, Takehara T, Tatsumi T, Kanto T, Groh V, Spies T, Kimura R, Miyagi T, Mochizuki K, Sasaki Y, Hayashi N. Expression and role of MICA and MICB in human hepatocellular carcinomas and their regulation by retinoic acid. Int J Cancer 2003;104:354–61.
[33] Nausch N, Florin L, Hartenstein B, Angel P, Schorpp-Kistner M, Cerwenka A. Cutting edge: the AP-1 subunit JunB determines NK cell-mediated target cell killing by regulation of the NKG2D-ligand RAE-1epsilon. J Immunol 2006;176:7–11.

[34] Gasser S, Raulet D. The DNA damage response, immunity and cancer. Semin Cancer Biol 2006;16:344–7.
[35] Gasser S, Raulet DH. The DNA damage response arouses the immune system. Cancer Res 2006;66:3959–62.
[36] Gorgoulis VG, Vassiliou LV, Karakaidos P, Zacharatos P, Kotsinas A, Liloglou T, Venere M, Ditullio Jr RA, Kastrinakis NG, Levy B, Kletsas D, Yoneta A, Herlyn M, Kittas C, Halazonetis TD. Activation of the DNA damage checkpoint and genomic instability in human precancerous lesions. Nature 2005;434:907–13.
[37] Jinushi M, Vanneman M, Munshi NC, Tai YT, Prabhala RH, Ritz J, Neuberg D, Anderson KC, Carrasco DR, Dranoff G. MHC class I chain-related protein A antibodies and shedding are associated with the progression of multiple myeloma. Proc Natl Acad Sci U S A 2008;105:1285–90.
[38] Nice TJ, Coscoy L, Raulet DH. Posttranslational regulation of the NKG2D ligand Mult1 in response to cell stress. J Exp Med 2009;206:287–98. https://www.ncbi.nlm.nih.gov/pubmed/19171762.
[39] Nice TJ, Deng W, Coscoy L, Raulet DH. Stress-regulated targeting of the NKG2D ligand Mult1 by a membrane-associated RING-CH family E3 ligase. J Immunol 2010;185:5369–76.
[40] Thomas M, Wills M, Lehner PJ. Natural killer cell evasion by an E3 ubiquitin ligase from Kaposi's sarcoma-associated herpesvirus. Biochem Soc Trans 2008;36:459–63. https://www.ncbi.nlm.nih.gov/pubmed/18481981.
[41] Baragano Raneros A, Suarez-Alvarez B, Lopez-Larrea C. Secretory pathways generating immunosuppressive NKG2D ligands: new targets for therapeutic intervention. Onco Targets Ther 2014;3, e28497.
[42] Cerwenka A, Baron JL, Lanier LL. Ectopic expression of retinoic acid early inducible-1 gene (RAE-1) permits natural killer cell-mediated rejection of a MHC class I-bearing tumor in vivo. Proc Natl Acad Sci U S A 2001;98:11521–6.
[43] Diefenbach A, Jensen ER, Jamieson AM, Raulet DH. Rae1 and H60 ligands of the NKG2D receptor stimulate tumour immunity. Nature 2001;413:165–71.
[44] Smyth MJ, Swann J, Cretney E, Zerafa N, Yokoyama WM, Hayakawa Y. NKG2D function protects the host from tumor initiation. J Exp Med 2005;202:583–8.
[45] Guerra N, Tan YX, Joncker NT, Choy A, Gallardo F, Xiong N, Knoblaugh S, Cado D, Greenberg NM, Raulet DH. NKG2D-deficient mice are defective in tumor surveillance in models of spontaneous malignancy. Immunity 2008;28:571–80.
[46] Liu G, Lu S, Wang X, Page ST, Higano CS, Plymate SR, Greenberg NM, Sun S, Li Z, Wu JD. Perturbation of NK cell peripheral homeostasis accelerates prostate carcinoma metastasis. J Clin Invest 2013;123:4410–22.
[47] Wu JD, Higgins LM, Steinle A, Cosman D, Haugk K, Plymate SR. Prevalent expression of the immunostimulatory MHC class I chain-related molecule is counteracted by shedding in prostate cancer. J Clin Invest 2004;114:560–8.
[48] Bilgi O, Karagoz B, Turken O, Kandemir EG, Ozturk A, Gumus M, Yaylaci M. Peripheral blood gamma-delta T cells in advanced-stage cancer patients. Adv Ther 2008;25:218–24.
[49] Marten A, von Lilienfeld-Toal M, Buchler MW, Schmidt J. Soluble MIC is elevated in the serum of patients with pancreatic carcinoma diminishing gammadelta T cell cytotoxicity. Int J Cancer 2006;119:2359–65.
[50] Wang H, Yang D, Xu W, Wang Y, Ruan Z, Zhao T, Han J, Wu Y. Tumor-derived soluble MICs impair CD3(+) CD56(+) NKT-like cell cytotoxicity in cancer patients. Immunol Lett 2008;120:65–71.
[51] Groh V, Wu J, Yee C, Spies T. Tumour-derived soluble MIC ligands impair expression of NKG2D and T-cell activation. Nature 2002;419:734–8.
[52] Doubrovina ES, Doubrovin MM, Vider E, Sisson RB, O'Reilly RJ, Dupont B, Vyas YM. Evasion from NK cell immunity by MHC class I chain-related molecules expressing colon adenocarcinoma. J Immunol 2003;171:6891–9.
[53] Holdenrieder S, Stieber P, Peterfi A, Nagel D, Steinle A, Salih HR. Soluble MICA in malignant diseases. Int J Cancer 2006;118:684–7.
[54] Holdenrieder S, Stieber P, Peterfi A, Nagel D, Steinle A, Salih HR. Soluble MICB in malignant diseases: analysis of diagnostic significance and correlation with soluble MICA. Cancer Immunol Immunother 2006;55:1584–9.
[55] Rebmann V, Schutt P, Brandhorst D, Opalka B, Moritz T, Nowrousian MR, Grosse-Wilde H. Soluble MICA as an independent prognostic factor for the overall survival and progression-free survival of multiple myeloma patients. Clin Immunol 2007;123:114–20.
[56] Tamaki S, Kawakami M, Ishitani A, Kawashima W, Kasuda S, Yamanaka Y, Shimomura H, Imai Y, Nakagawa Y, Hatake K, Kirita T. Soluble MICB serum levels correlate with disease stage and survival rate in patients with oral squamous cell carcinoma. Anticancer Res 2010;30:4097–101. https://www.ncbi.nlm.nih.gov/pubmed/21036725.

[57] Lopez-Soto A, Gonzalez S, Galluzzi L. Soluble NKG2D ligands limit the efficacy of immune checkpoint blockade. Onco Targets Ther 2017;6, e1346766.
[58] Maccalli C, Giannarelli D, Chiarucci C, Cutaia O, Giacobini G, Hendrickx W, Amato G, Annesi D, Bedognetti D, Altomonte M, Danielli R, Calabro L, Di Giacomo AM, Marincola FM, Parmiani G, Maio M. Soluble NKG2D ligands are biomarkers associated with the clinical outcome to immune checkpoint blockade therapy of metastatic melanoma patients. Onco Targets Ther 2017;6, e1323618.
[59] Xiao G, Wang X, Sheng J, Lu S, Yu X, Wu JD. Soluble NKG2D ligand promotes MDSC expansion and skews macrophage to the alternatively activated phenotype. J Hematol Oncol 2015;8:13.
[60] Dhar P, Basher F, Ji Z, Huang L, Qin S, Wainwright DA, Robinson J, Hagler S, Zhou J, MacKay S, Wu JD. Tumor-derived NKG2D ligand sMIC reprograms NK cells to an inflammatory phenotype through CBM signalosome activation. Commun Biol 2021;4:905.
[61] Waldhauer I, Goehlsdorf D, Gieseke F, Weinschenk T, Wittenbrink M, Ludwig A, Stevanovic S, Rammensee HG, Steinle A. Tumor-associated MICA is shed by ADAM proteases. Cancer Res 2008;68:6368–76.
[62] Boutet P, Aguera-Gonzalez S, Atkinson S, Pennington CJ, Edwards DR, Murphy G, Reyburn HT, Vales-Gomez M. Cutting edge: the metalloproteinase ADAM17/TNF-alpha-converting enzyme regulates proteolytic shedding of the MHC class I-related chain B protein. J Immunol 2009;182:49–53.
[63] Liu G, Atteridge CL, Wang X, Lundgren AD, Wu JD. The membrane type matrix metalloproteinase MMP14 mediates constitutive shedding of MHC class I chain-related molecule A independent of A disintegrin and metalloproteinases. J Immunol 2010;184:3346–50.
[64] Kaiser BK, Yim D, Chow IT, Gonzalez S, Dai Z, Mann HH, Strong RK, Groh V, Spies T. Disulphide-isomerase-enabled shedding of tumour-associated NKG2D ligands. Nature 2007;447:482–6.
[65] Edwards DR, Handsley MM, Pennington CJ. The ADAM metalloproteinases. Mol Aspects Med 2008;29:258–89.
[66] DasGupta S, Murumkar PR, Giridhar R, Yadav MR. Current perspective of TACE inhibitors: a review. Bioorg Med Chem 2009;17:444–59.
[67] Jordan PA, Gibbins JM. Extracellular disulfide exchange and the regulation of cellular function. Antioxid Redox Signal 2006;8:312–24.
[68] Jordan PA, Stevens JM, Hubbard GP, Barrett NE, Sage T, Authi KS, Gibbins JM. A role for the thiol isomerase protein ERP5 in platelet function. Blood 2005;105:1500–7.
[69] Wang X, Lundgren AD, Singh P, Goodlett DR, Plymate SR, Wu JD. An six-amino acid motif in the alpha3 domain of MICA is the cancer therapeutic target to inhibit shedding. Biochem Biophys Res Commun 2009;387:476–81.
[70] Ferrari de Andrade L, Tay RE, Pan D, Luoma AM, Ito Y, Badrinath S, Tsoucas D, Franz B, May Jr KF, Harvey CJ, Kobold S, Pyrdol JW, Yoon C, Yuan GC, Hodi FS, Dranoff G, Wucherpfennig KW. Antibody-mediated inhibition of MICA and MICB shedding promotes NK cell-driven tumor immunity. Science 2018;359:1537–42.
[71] Lu S, Zhang J, Liu D, Li G, Staveley-O'Carroll KF, Li Z, Wu JD. Nonblocking monoclonal antibody targeting soluble MIC revamps endogenous innate and adaptive antitumor responses and eliminates primary and metastatic tumors. Clin Cancer Res 2015;21:4819–30.
[72] Basher F, Dhar P, Wang X, Wainwright DA, Zhang B, Sosman J, Ji Z, Wu JD. Antibody targeting tumor-derived soluble NKG2D ligand sMIC reprograms NK cell homeostatic survival and function and enhances melanoma response to PDL1 blockade therapy. J Hematol Oncol 2020;13:74.
[73] Zhang J, Larrocha PS, Zhang B, Wainwright D, Dhar P, Wu JD. Antibody targeting tumor-derived soluble NKG2D ligand sMIC provides dual co-stimulation of CD8 T cells and enables sMIC(+) tumors respond to PD1/PD-L1 blockade therapy. J Immunother Cancer 2019;7:223.
[74] Zhang J, Liu D, Li G, Staveley-O'Carroll KF, Graff JN, Li Z, Wu JD. Antibody-mediated neutralization of soluble MIC significantly enhances CTLA4 blockade therapy. Sci Adv 2017;3, e1602133.
[75] Jinushi M, Hodi FS, Dranoff G. Therapy-induced antibodies to MHC class I chain-related protein A antagonize immune suppression and stimulate antitumor cytotoxicity. Proc Natl Acad Sci USA 2006;103:9190–5. https://www.ncbi.nlm.nih.gov/pubmed/16754847.
[76] Frazao A, Rethacker L, Messaoudene M, Avril MF, Toubert A, Dulphy N, Caignard A. NKG2D/NKG2-ligand pathway offers new opportunities in cancer treatment. Front Immunol 2019;10:661.
[77] Lazarova M, Steinle A. The NKG2D axis: an emerging target in cancer immunotherapy. Expert Opin Ther Targets 2019;23:281–94.
[78] Schmiedel D, Mandelboim O. NKG2D ligands-critical targets for cancer immune escape and therapy. Front Immunol 2018;9:2040.

CHAPTER

5

Chimeric antigen receptor-modified cells for the treatment of solid tumors: First steps in a thousand-mile march

Hind Rafei, Rafet Basar, Katayoun Rezvani, and May Daher

Department of Stem Cell Transplantation and Cellular Therapy, The University of Texas MD Anderson Cancer Center, Houston, TX, United States

Abstract

Chimeric antigen receptor (CAR)-engineered T-cell therapy is a promising cancer treatment strategy that harnesses the power of immunotherapy and exploits precision medicine. It consists of the genetic reprogramming of a patient's own T lymphocytes to express a synthetic receptor that recognizes a specific tumor antigen while transducing signals that activate the T cells and hence mediate tumor killing. CAR T-cells demonstrated unprecedented durable responses in otherwise refractory hematologic malignancies and have been approved by the Food and Drug Administration for the treatment of acute lymphoblastic leukemia, diffuse large B cell lymphoma, mantle cell lymphoma, and multiple myeloma. Despite early promising results, consistent and reproducible clinical gains from CAR T-cell therapy in solid tumors have not yet been demonstrated. In this chapter, we summarize the results of the clinical trials investigating the use of CAR T-cell therapy in the treatment of a variety of solid cancers and discuss the obstacles encountered while shedding light on innovative strategies that may lead to success. We also briefly highlight the ongoing research harnessing the potential of other immune cells in CAR engineering for the treatment of solid tumors.

Abbreviations

AACR	American Association for Cancer Research
ABA	5-amino-2-hydroxymethylphenyl boronic acid
AICD	activation-induced cell death
AML	acute myeloid leukemia
BATF	basic leucine zipper ATF-like transcription factor
BET	bromodomain and extra-terminal
bZIP	basic leucine zipper
CA	carbonic anhydrase
CAIX	carbonic anhydrase IX

https://doi.org/10.1016/B978-0-12-822620-9.00021-5

CAR	chimeric antigen receptor
CDX	cell-line-derived xenograft
CLDN6	claudin 6
CLTX	chlorotoxin
COX	cyclooxygenase
CRS	cytokine release syndrome
CTL	cytotoxic T-lymphocytes
CTLA-4	cytotoxic T lymphocyte-associated antigen 4
DLL3	delta-like ligand 3
ECM	extracellular matrix
EGFR	epidermal growth factor receptor
FAP	fibroblast activation protein
FRα	folate receptor alpha
GPC3	glypican 3
GSH	glutathione
GVHD	graft-vs-host disease
$HCO3^-$	bicarbonate
HOXA1	homeobox A1
HPSE	heparanase
iCARs	inhibitory CARs
IDO	indoleamine-2,3-dioxygenase
IL	interleukin
IL-13Rα2	IL-13 receptor alpha 2
iNKT cells	invariant natural killer T cells
iPSCs	induced pluripotent stem cells
IRF4	interferon-regulatory factor 4
LAG-3	lymphocyte-activation protein-3
LDHA	lactate dehydrogenase
MCT	monocarboxylate transporter
MDSCs	myeloid-derived suppressor cells
MSKCC	Memorial Sloan Kettering Cancer Center
MUC-4	mucin-4
NK cells	natural killer cells
NKT cells	natural killer T cells
NSCLC	nonsmall cell lung cancer
PCK1	phosphoenolpyruvate carboxykinase 1
PD-1	programmed death-1
PD-L1	PD-1 ligand 1
PD-L2	PD-1 ligand 2
PGE2	prostaglandin E2 PGE2
PKA	protein kinase A
PMSA	prostate-specific membrane antigen
PSCA	prostate stem-cell antigen
PTPN1	nonreceptor protein tyrosine phosphatase type 1
SCLC	small cell lung cancer
STING	stimulator of interferon gene
TAA	tumor-associated antigen
TAG-72	tumor-associated glycoprotein 72
TAMs	tumor-associated macrophages
TDO	tryptophan-2,3-dioxygenase
TGFβR2	TGF-β receptor 2
TIGIT	T cell Ig and ITIM domain
TILs	tumor-infiltrating lymphocytes

TIM-3	T-cell membrane protein-3
TME	tumor microenvironment
TP53I11	tumor protein p53-inducible protein 11
TSA	tumor-specific antigen
VEGFR	vascular endothelial growth factor receptor
γδ T cells	gamma-delta (γδ) T cells

Conflict of interest

R.B., M.D., K.R., and The University of Texas MD Anderson Cancer Center have an institutional financial conflict of interest with Takeda Pharmaceutical for the licensing of the technology related to CAR-NK cell research. MD Anderson has implemented an Institutional Conflict of Interest Management and Monitoring Plan to manage and monitor the conflict of interest with respect to MDACC's conduct of any other ongoing or future research related to this relationship. R.B., K.R., and The University of Texas MD Anderson Cancer Center have an institutional financial conflict of interest with Affirmed GmbH. Because MD Anderson is committed to the protection of human subjects and the effective management of its financial conflicts of interest in relation to its research activities, MD Anderson is implementing an Institutional Conflict of Interest Management and Monitoring Plan to manage and monitor the conflict of interest with respect to MD Anderson's conduct of any other ongoing or future research related to this relationship. K.R. participates on the Scientific Advisory Board for GemoAb, AvengeBio, Virogin, GSK, Caribou, and Bayer.

Introduction

Chimeric antigen receptor CAR T-cell therapy consists of the genetic engineering of a patient's own T lymphocytes to express a synthetic receptor against a specific tumor antigen (Fig. 1). It has shown impressive success in the treatment of relapsed and refractory hematologic malignancies; most prominently with the use of anti-CD19 CAR T-cells in the treatment of acute B-lymphoblastic leukemia and B-cell lymphoma [1,2] and more recently using anti-BCMA CAR T-cells in multiple myeloma [3]. This led to a growing number of preclinical studies and clinical trials that explore extending the use of CAR T-cell therapy in solid tumors, targeting a variety of antigens. Unfortunately, the efficacy in solid tumors has been less encouraging, but some recent positive results fuel hope for the potential application of this immunotherapeutic treatment in solid tumors [4,5].

The limited success of CAR T-cell therapy in solid tumors vs hematologic malignancies is multifactorial, primarily related to the physical and functional barriers preserving the different solid tumor compartments. Such factors include trafficking of CAR T-cells from the blood into the solid tumor sites, infiltration of the stromal elements of solid tumors, and overcoming the hostile tumor microenvironment (TME) characterized by hypoxia, acidic pH, nutrient depletion, and immunosuppression. Many elements contribute to the immunosuppressive nature of the TME, primarily (a) the abundance of immunosuppressive cells, (b) the T-cell intrinsic inhibitory mechanisms that regulate their function, and (c) the presence of immunosuppressive substances and cytokines. Another challenge has been the identification of specific tumor antigens that are both highly and uniformly expressed among solid tumors, while simultaneously showing low expression on normal tissues. Here we discuss the progress made thus far in the application of CAR T-cell therapy for the treatment of solid tumors as well as the barriers encountered, while proposing potential solutions and future directions for research.

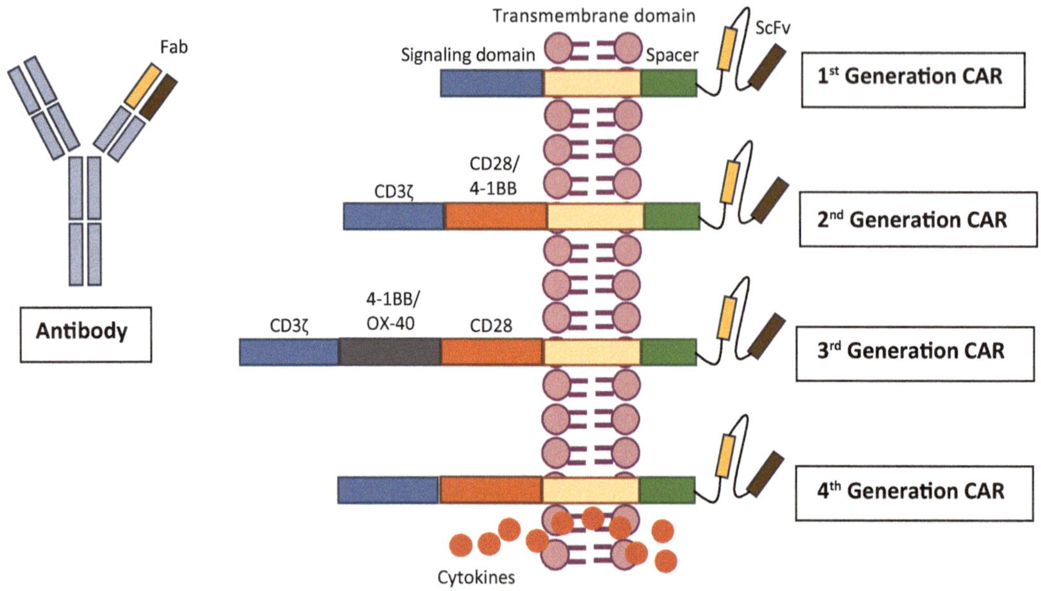

FIG. 1 The different chimeric antigen receptor (CAR) generations. The first CAR generation has only one intracellular signaling region. The second CAR generation incorporates costimulatory domains CD28 or 4-1BB. The third CAR generation incorporates more than one costimulatory domain in addition to CD3ζ. The fourth CAR generation involves cytokine overexpression, gene knockout and knock-in, targeting multiple antigens simultaneously, introduction of suicide genes, and other techniques.

CAR T-cell therapy experience in solid tumors

Trials as early as 2006 employed CAR T-cells lacking costimulatory endodomains for the treatment of solid tumors targeting a variety of antigens including FRα, CAIX, and L1-CAM in ovarian cancer, renal cancer, and neuroblastoma, with minimal efficacy [6,7]. Encouraging results, however, have been seen in early phase trials in patients with neuroblastoma treated with anti-GD2 CAR T-cells, and for ErbB2-positive sarcomas treated with HER2-directed CAR T-cells [8]. Recently, data from two phase I solid tumor CAR T-cell trials illustrate this modality's potential in mesothelioma and sarcoma [4,5] (Table 1).

Glioblastoma

Antigens most commonly targeted in clinical trials investigating CAR T-cell therapy for glioblastoma include EGFRvIII, HER2, and IL-13 receptor alpha 2 (IL-13Rα2). IL13Rα2 expression increases with the grade of glioma, with approximately 58% of grade IV gliomas overexpressing this receptor [9]. This receptor is not expressed at significant levels in normal brain tissue, making it an attractive target for CAR therapy [10–13]. First-generation autologous IL-13Rα2-targeting CAR T-cells were investigated in a pilot safety and feasibility trial, where three patients with recurrent glioblastoma received these CAR T-cells via intracavitary

TABLE 1 Selected published clinical trials of chimeric antigen receptor (CAR)-T cell therapy for solid tumors.

Reference	Publication year	Tumor type	Targeted antigen	Number of patients	CAR generation	Delivery method	Response	Adverse events
Brown et al.	2015	Glioblastoma	IL-13Rα2	3	First generation	Intracavitary/ intratumoral	Significant decrease in IL13Rα2 expression, 2/3 pts. with no recurrence at the border of resected cavity	Headaches and transient neurologic deficits
Louis et al.	2011	Neuroblastoma	GD2	11	First generation	Intravenous	CR 27%	Grades 1–3 localized pain
Feng et al.	2016	NSCLC	EGFR	11	Second generation	Intravenous	PR 18%, SD 45%	No severe toxicity
Beatty et al.	2014	Mesothelioma/ pancreatic cancer	Mesothelin	2	Second generation	Intravenous	PR 50%	Pt 1: Cardiac arrest, respiratory failure, DIC, CRS Pt 2: Jejunal obstruction
Adusumilli et al.	2019	Malignant pleural disease (mesothelioma, breast cancer, lung cancer)	Mesothelin	21	Second generation	Intrapleural	>50% decrease in mesothelin-related peptide in blood in 13 pts	No >grade 2 toxicity
Kershaw et al.	2006	Ovarian cancer	FRα	8	First generation	Intravenous	PD	Grade 3–4 IL-2 related toxicity

Continued

TABLE 1 Selected published clinical trials of chimeric antigen receptor (CAR)-T cell therapy for solid tumors—cont'd

Reference	Publication year	Tumor type	Targeted antigen	Number of patients	CAR generation	Delivery method	Response	Adverse events
Junghans et al.	2016	Prostate cancer	PSMA	5	First generation	Intravenous	PR 40%	No T cell attributed toxicities
Lamers et al.	2013	Renal cell carcinoma	CAIX	3	First generation	Intravenous	PD	Grade 2–4 liver toxicity
Navai et al.	2019	Sarcoma	HER2	10	Second generation	Intravenous	CR 20% SD 30% PD 50%	Grade 1–2 CRS

Pt, patient; *CR*, complete response; *PR*, partial response; *PD*, progressive disease; *SD*, stable disease; *NSCLC*, nonsmall cell lung cancer; *DIC*, disseminated intravascular coagulopathy; *CRS*, cytokine release syndrome.

administration following tumor resection. The study demonstrated the successful manufacture and delivery of this modality as well as the safety of administration. Adverse events, included headache and transient neurologic deficits, and were manageable. Promising early signs of activity were proven by a significant loss of IL13Rα2 tumor cell expression and improvement in overall survival [14]. Subsequently, a second-generation CAR adding a 4-1BB costimulatory domain (Fig. 1) was investigated and the patient treated with these CAR T-cells showed tumor regression after multiple intracranial (intracavitary and intraventricular) infusions. Interestingly, CAR T-cells with intracavitary administration stabilized only local tumor but did not control progression of the tumor at the metastatic sites. In contrast, intraventricular infusions resulted in significant tumor control in all intracranial and spinal tumors [15].

HER2 was also targeted in glioblastoma. Seventeen patients with progressive HER2-positive disease were treated on a phase I trial with peripheral infusions of HER2-targeted CAR-modified virus-specific T cells. No dose-limiting toxicity was noted, although two patients had grade 2 seizures/headaches. Of 16 evaluable patients, 1 had a partial response lasting for more than 9 months, and 7 had stable disease ranging between 8 weeks to 29 months (with three of these remaining progression free during follow-up of 24–29 months). No in vivo T-cell expansion was noted, but persistence up to 1 year at a low frequency was observed [16].

EGFRIII is another glioma antigen with approximately 50% of EGFR-amplified glioblastomas carrying constitutively active and oncogenic EGFRvIII [17,18]. Ten patients with recurrent glioblastoma were treated with a single dose of peripherally infused EGFRvIII-directed CAR T-cells in a first-in-human phase I trial. No evidence of off-tumor toxicity or cytokine release syndrome (CRS) was seen. The study demonstrated effective engraftment of the CAR T-cells in the peripheral blood but also significant trafficking and expansion of CAR T-cells in situ within active regions of glioblastoma [19]. These trials show early signs of feasibility, safety, and efficacy of CAR T-cell therapy in glioblastoma, though many challenges remain to be addressed.

Neuroblastoma

Targets explored in the development of CAR-based therapies for neuroblastoma include GD2 and CD171. GD2 is expressed on tumors of neuroectodermal origin such as melanoma and neuroblastoma [20]. First-generation anti-GD2 CAR T-cells were investigated for the treatment of neuroblastoma in a phase I trial and demonstrated a complete remission rate of 27% (3/11 patients). CAR T-cells were detected up to 192 weeks after administration [21]. CD171 is associated with many tumors and thought to enhance tumor activity [22]. CD171-specifc CAR T-cells exhibited robust antitumor activity in vitro [23]. However, in a clinical trial of first-generation CD171-targeting $CD8^+$ lymphocytes, disease control was not achieved and lymphocyte persistence was inversely correlated with disease burden [7]. The minimal antitumor activity was thought by the authors to be due to the lack of coadministration of IL-2, which is especially critical for the function of first-generation CARs. It is also important to note that a defined ratio of $CD4^+$:$CD8^+$ T cells in the product transferred is essential for the function and persistence of T cells which could perhaps partially explain the lack of efficacy seen in the adoptive transfer of exclusively $CD8^+$ T cells [24]. Recently, Moghimi et al. designed antihuman B7H3 CAR T-cells for the treatment of neuroblastoma by generating SynNotch gated CAR-T, GD2-B7H3, recognizing GD2 as the gate and B7H3 as the target. The cells generated demonstrated improved efficacy and high specificity in vitro and in metastatic xenograft mouse models of neuroblastoma [25].

Lung cancer

Epidermal growth factor receptor (EGFR) is frequently overexpressed in patients with nonsmall cell lung cancer (NSCLC) and EGFR kinase inhibitors constitute the backbone of treatment of patients with EGFR-overexpressing tumors [26]. Second-generation EGFR-targeting CAR T-cells (Fig. 1) that incorporate both CD137 and CD3ζ signaling domains were developed and investigated in a phase I trial where 2/11 patients with refractory NSCLC experienced a partial response. CAR T-cells were detected in both peripheral blood as well as tumor sites and posttreatment biopsies revealed eradication of EGFR-positive tumor cells [27].

In small cell lung cancer (SCLC), a phase I clinical trial is currently ongoing at our center (The University of Texas MD Anderson Cancer Center) investigating delta-like ligand 3 (DLL3)-targeted CAR T-cells in patients with relapsed/refractory SCLC [28]. In two separate studies, more than 80% of patients with SCLC were found to be positive for DLL3 expression, with 32%–47% of patients exhibiting high levels of expression [29,30].

Mesothelioma

Mesothelin is highly expressed in a number of solid tumors, including the vast majority of mesotheliomas and pancreatic adenocarcinomas, approximately 70% of ovarian cancers and 50% of lung adenocarcinomas [31–34]. Its expression on normal tissues is limited to the mesothelial cell lining of the pleura, pericardium, and peritoneum [35,36]. These characteristics have led to its targeting using CAR-modified T cells both in preclinical studies as well as clinical trials.

In a phase I clinical trial, four patients with advanced mesothelioma or pancreatic cancer were treated with multiple infusions of second-generation mesothelin-specific CAR T-cells. Antitumor responses were moderate, but CAR T-cell persistence and trafficking to the tumor site were detected. It is important to note that treatment with CAR T-cells in this study induced an antitumor humoral immune response, suggested by an elevated antibody response to a variety of tumor-associated proteins. This highlights that CAR T-cell therapy is potentially capable of eliciting a systemic immune response against a broader range of antigens mediated through epitope spreading. One patient experienced anaphylaxis and cardiac arrest after the third infusion, which was believed to be associated with the development of antibodies against the murine-derived scFv [37].

More recently, a mesothelin-targeting trial was presented at the AACR 2019 Annual Meeting involving 21 patients with malignant pleural disease (19 with malignant mesothelioma, 1 with lung cancer, and 1 with breast cancer) who received a single intrapleural dose of the CAR product with or without cyclophosphamide preconditioning. Preliminary results show CAR T-cell persistence in the peripheral blood of 13 patients during the 38-week evaluation period, a finding associated with a $\geq$50% decrease in the levels of mesothelin-related peptide in blood and tumor regression on imaging. The best objective responses (8/11 patients) occurred in patients who received a combination of cyclophosphamide lymphodepletion, CAR T-cells, and at least three doses of an anti-PD-1 agent. There was no evidence of CAR T-cell-related toxicity, such as neurotoxicity or CRS, greater than grade 2 [4].

Breast cancer

HER2 is an oncogene that is amplified in approximately 20% of breast cancers [38] and has a significant impact on the recurrence and survival of patients with breast cancer [39]. HER2-targeted therapies are currently the hallmark of treatment of HER2-overexpressing breast cancer [40,41]. Several clinical trials of HER2-specific CAR T-cell therapy in breast cancer are in progress including a phase I/II study in HER2-overexpressing breast cancer refractory to chemotherapy or HER2 antibody inhibitor (NCT01935843). Other breast cancer-associated antigens are also being investigated as targets of CAR T-cell therapy such as cMet (NCT03060356), mesothelin (NCT02580747), and MUC1 (NCT02587689).

Ovarian cancer

MUC16, FRα, mesothelin, and HER2 are the most commonly targeted antigens by CARs in ovarian cancer. MUC16 also known as CA-125 is a highly glycosylated mucin that is expressed by the majority of ovarian cancers and hence is a well-established tumor marker for the diagnosis and follow-up of patients with ovarian cancer [42,43]. Investigators at Memorial Sloan Kettering Cancer Center (MSKCC) engineered an IL-12-secreting MUC16 ecto-directed CAR T-cell that is currently being investigated in a phase I trial in refractory or relapsed ovarian cancer [44].

Folate receptor alpha (FRα) is a surface glycosylphosphatidylinositol-anchored protein that is overexpressed on a number of epithelial tumors including ovarian cancer, breast cancer, lung cancer, colorectal cancer, brain tumors, and other solid malignancies, while at the same time having a limited expression on normal tissues [45]. In ovarian cancer, FRα overexpression is associated with high-grade disease and poor prognosis [46]. In 2006, FRα-targeting CAR T-cells were shown in a clinical trial to have minimal antitumor activity. Modified T cells were detected in the circulation in large numbers for the first 2 days after transfer; however, their number rapidly declined thereafter, and their presence was barely detectable after 1 month in the majority of patients. This clinical trial established the safety of CAR T-cell administration but demonstrated limited persistence [6]. To improve their antitumor activity, CAR T-cells containing the FRα-specific scFv MOv19 coupled to the CD3ζ chain signaling module alone (MOv19-ζ) or in combination with the costimulatory motif CD137 (4-1BB) in tandem (MOv19-BBζ) were engineered and showed significant cytotoxic function in vitro. In vivo, only costimulated MOv19-BBζ CAR T-cell-mediated tumor regression in immunodeficient mice with FRα+human ovarian cancer [47].

Mesothelin is another promising target for the treatment of ovarian cancer with expression in greater than 80% of serous epithelial ovarian cancer cells [32]. Preliminary results of a phase I clinical trial investigating mesothelin-specific CAR T-cells revealed that the product is well tolerated without an off-tumor effect or CRS in patients with mesothelioma, ovarian cancer, or pancreatic cancer [48].

Finally, HER2 has been demonstrated to be expressed in approximately 40% of epithelial ovarian cancers [49]. A clinical trial of HER2-directed CAR T-cell therapy in ovarian cancer is currently ongoing (NCT02713984).

Prostate cancer

Prostate stem-cell antigen (PSCA) and prostate-specific membrane antigen (PSMA) are two of the most commonly used target antigens for CAR T-cell therapy for prostate cancer. Prostate stem-cell antigen is a glycosylphosphatidylinositol-anchored antigen expressed on the surface of prostate cells [50]. PSMA is a type II transmembrane protein normally present at low levels on the apical or cytosolic surface of prostate tissue [51] and that translocates to the extracellular or luminal epithelial surface upon transformation of prostate tissue into adenocarcinoma [52]. A phase I clinical trial reported a 40% response rate (2/5 patients), as measured by prostate-specific antigen (PSA) levels, with a first-generation PSMA-directed CAR after nonmyeloablative preconditioning and concurrent administration of IL-2 [53]. In another phase I trial, second-generation PSMA-directed CAR T-cells were shown to be tolerated and systemic persistence of up to 2 weeks was documented [54]. A major concern in prostate cancer is the immunosuppressive TME, and therefore TGFβ-insensitive PSMA-directed CAR T-cells were developed. This newer construct resulted in increased proliferation, enhanced cytokine secretion, decreased exhaustion, and long-term in vivo persistence in a human prostate cancer mouse model [55]. NCT03089203 is a phase I clinical trial currently being conducted at the University of Pennsylvania to assess the safety and preliminary efficacy of this lentivirally transduced PSMA-directed/TGFβ-insensitive CAR T-cell product in men with metastatic castrate-resistant prostate cancer [56].

Renal cell carcinoma

Carbonic anhydrase IX (CAIX) is a metalloprotease highly expressed in metastatic renal cell carcinoma and hence was exploited in CAR T-cell therapy [57,58]. However, it is also expressed on normal tissue, such as the epithelium of the gastric mucosa, small intestine, duodenum, and biliary tree [59,60]. Lamers et al. treated three patients with CAIX-positive metastatic renal cell carcinoma with first-generation anti-CAIX CAR T-cells in combination with IL-2, but no prior lymphodepletion. Two of these patients developed grade 2–4 liver toxicity and liver biopsies detected engineered T cells around the bile ducts causing cholangitis. CAIX was found to be overexpressed on the biliary ductal epithelium. Antibodies against the murine-derived scFv were detected in all three patients. To avoid liver toxicity, the investigators pre-administered unmodified antibody from which scFv was derived, to saturate the liver before CAR T-cell administration [57]. With this approach, no hepatotoxicity was observed in the four patients who received antibody pretreatment. No human antimouse antibodies against the cellular product were detected in patients who received the pretreatment, suggesting that the inflammation caused by the cholangitis possibly contributed to the generation of human antimouse antibodies. Unfortunately, no significant clinical responses occurred despite CAR T-cell persistence for 3–5 weeks [57,58].

Other antigens are being exploited for CAR T-cell therapy in renal cell carcinoma including CD70 that is significantly overexpressed in this disease [26]. Preclinical evaluation of CD70-targeting CD27-containing CAR in CD70-expressing renal cell carcinoma supported its safety and efficacy [61]. A clinical trial of anti-CD70 CAR in CD70-expressing solid tumors including renal cell carcinoma is currently recruiting (NCT02830724).

It should be noted that sunitinib, a multikinase inhibitor that is well established for the treatment of metastatic renal cell carcinoma, has been shown to decrease myeloid-derived suppressor cells (MDSCs) [62] that are involved in T-cell suppression [63,64], enhance type-I interferon responses, and decrease Treg function [65]. As noted by Yeku et al., it would be interesting to investigate the role of a vascular endothelial growth factor receptor (VEGFR) tyrosine kinase inhibitor, such as sunitinib, in preconditioning and maintenance after CAR T-cell therapy in renal cell carcinoma [66].

Sarcoma

Despite being a heterogeneous group of mesenchymal-derived tumors, there has been progress in identifying common tumor antigens in sarcoma for exploitation in CAR T-cell therapy. Recently, a group reported promising results at the AACR 2019 Annual Meeting with the use of HER2-directed CARs in the treatment of sarcoma [5]. The HEROS study is investigating a second-generation anti-HER2 CAR in pediatric and adult patients with advanced refractory or metastatic heavily pretreated HER2-positive sarcomas (NCT00902044). Ten patients were treated with up to three infusions of HER2-targeted CAR T-cells after lymphodepletion with either fludarabine or fludarabine and cyclophosphamide. The patients who had responses to this initial treatment received up to five additional infusions of CAR T-cells without lymphodepletion. CAR T-cells expanded in all but two patients, with a median peak expansion on day 7, and CAR T-cells were detected in all patients 6 weeks after infusion. Three patients achieved stable disease, and five had progressive disease. Overall, the patients experienced limited treatment-related toxicities [5].

Huang et al. targeted another antigen, IL-11Rα, through a second-generation CAR. IL-11Rα was shown to be markedly expressed in primary osseous and pulmonary metastatic osteosarcoma but absent from normal tibia and lung [67]. Signaling via the IL-11/IL-11Rα pathway is thought to promote osteoclastogenesis [68]. In a nude mouse model of osteosarcoma, anti-IL-11Rα CAR T-cells were effective against both primary tumors and pulmonary metastasis [69].

Overcoming challenges facing CAR therapy in solid tumors

The previously highlighted trials delineate promising early results but also major obstacles encountered in CAR T-cell therapy for solid tumors. These obstacles could be tumor-specific factors but often are limiting factors common to all solid tumors. Herein, we discuss these obstacles while shedding light on ongoing and future research to circumvent these obstacles (Fig. 2).

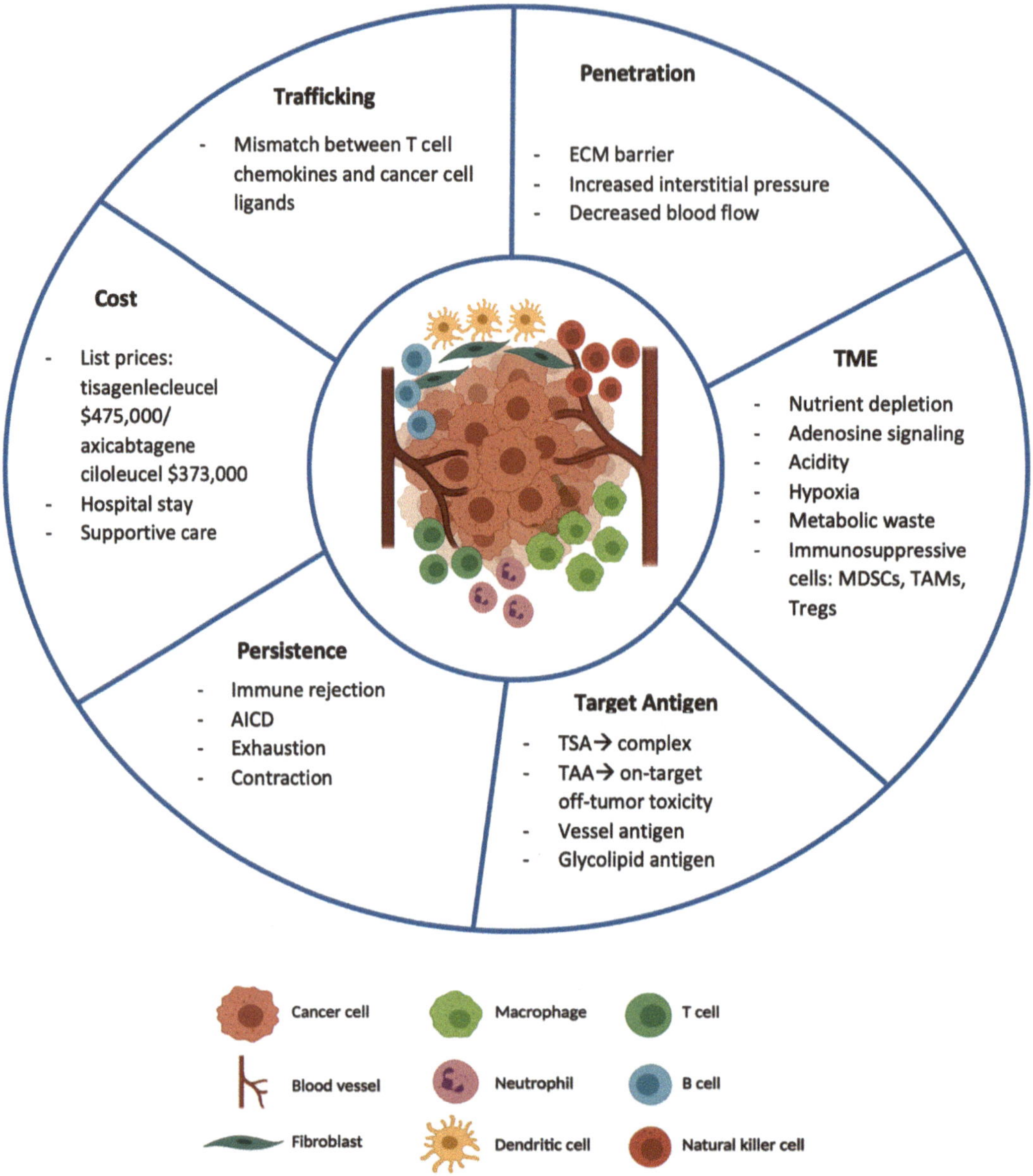

FIG. 2 Challenges for the application of chimeric antigen receptor (CAR)-T cell therapy for solid tumors. *ECM*, extracellular matrix; *TME*, tumor microenvironment; *MDSCs*, myeloid-derived suppressor cells; *TAMs*, tumor-associated macrophages; *Tregs*, regulatory T cells; *TSA*, tumor-specific antigen; *TAA*, tumor-associated antigen; *AICD*, activation-induced cell death.

Tumor microenvironment

Nutrient deprivation

The solid TME is characterized by its resistance to the infiltration and survival of immune cells, primarily due to the hypoxia, acidity, and nutrient deprivation that govern it; factors that are regulated by the metabolic activity of tumor cells [70,71]; as well as both intrinsic and extrinsic immunosuppressive mechanisms. The metabolic competition between cancer and immune cells for key nutrients such as glutamine, glucose, lipids, and amino acids was shown in a mouse sarcoma model to metabolically restrict T cells, leading to a decrease in mTOR pathway activity, glycolysis and cytokine production, and hence favoring an exhausted phenotype and setting the stage for cancer progression [70]. The competition takes place as both cancer and T effector cells primarily depend on the uptake of nutrients from the extracellular space as opposed to de novo synthetic pathways that are either defective or unable to fulfill the high metabolic demands. Tryptophan depletion in the TME, mediated by high expression of indoleamine-2,3-dioxygenase (IDO) and tryptophan-2,3-dioxygenase (TDO) by macrophages and cancer cells that convert tryptophan to kynurenine [72–74], leads to impairment of effector T cell survival and reciprocal development of immunosuppressive Tregs and MDSCs [75,76]. Cysteine depletion contributes to pathologic accumulation of reactive oxygen species (ROS), as it is among the substrates for the de novo synthesis of glutathione (GSH), the most abundant cellular antioxidant [77]. Cysteine depletion is particularly pronounced in the absence of cystathionase, which converts methionine to cysteine, putting T cells at a metabolic disadvantage compared to cancer cells [78]. Arginine is another essential amino acid, and its depletion such as that mediated by the production of the arginine-degrading enzyme, arginase, in the TME, leads to T cell anergy [79].

Immunosuppressive substances

In addition to nutrient deprivation in the TME, numerous substances, such as adenosine, produced by the tumor and immune cells mediate immunosuppression. Production of adenosine in the TME has recently gained attention as a potential target to reverse tumor-induced immunosuppression. Adenosine is generated from extracellular ATP by the ectoenzymes CD39 and CD73 [80] and is found at high concentrations in the TME [81]. CD73 is expressed on both cancer cells as well as immunosuppressive cells such as Tregs and MDSCs and is associated with a negative prognosis in certain types of cancers such as breast, prostate, and ovarian cancer [82–84]. Adenosine mediates intracellular signaling through binding to four identified receptors: A_1, A_{2A}, A_{2B}, and A_3, but the immunosuppressive function of adenosine is predominantly mediated by A_{2A}Rs expressed on activated T cells [81,85]. Indeed, A_{2A}R-deficient mice exhibit enhanced antitumor immune responses through resisting the immunosuppressive effects of adenosine [86,87].

Moreover, many cell types of the TME such as tumor cells, fibroblasts, and endothelial cells activate cyclooxygenase (COX)-2 and prostaglandin E synthase leading to the production of the lipid-signaling molecule prostaglandin E2 (PGE2) that mediates immunosuppression, angiogenesis and tumor progression [88]. Similar to adenosine, PGE2 activates protein kinase A (PKA) that inhibits antigen/receptor-triggered T cell activation and plays a role in attracting Tregs and MDSCs [89].

Acidity

Acidity is a universal feature of the solid TME. Hypoxic tumor cells upregulate lactate dehydrogenase (LDHA), which converts pyruvate into lactate, contributing to tumor acidosis [90,91]. However, tumor cells adopt mechanisms to survive in the acidic TME such as enhancing the export of lactate/H^+ through upregulation of the monocarboxylate transporter (MCT) family of genes [91–93]. A low pH in the TME not only contributes to immunosuppression but also promotes tumor proliferation, angiogenesis, and metastasis [91]. Besides its contribution to the TME acidosis, emerging data is showing the role of lactate as a signaling molecule in pathways involved in cancer proliferation and metastasis [94,95]. Lactate is also involved in signaling pathways leading to cytokine production, such as IL-23 and IL-6, mediating tumor-associated inflammation [96]. Lactate was found in multiple studies to be a prognostic marker of tumor progression and patient survival [97,98]. Other factors contributing to the acidosis of the TME include increased levels of CO_2 generated by oxidative metabolism, which is hydrated by carbonic anhydrases (CA) into bicarbonate (HCO_3^-) and protons (H^+) [99]. Previously mentioned CAIX, which has been targeted in CAR T-cell therapy for renal cell carcinoma, plays a key role in pH regulation in the hypoxic TME as its expression is induced by hypoxia via HIF-1α [100,101].

Hypoxia

It is currently well established that hypoxia plays a major role in regulating antitumor immune responses. HIF-1α, upregulated in hypoxic conditions, was shown to decrease the susceptibility of lung cancer cells to cytotoxic T-lymphocytes (CTL) through STAT3 phosphorylation in tumor cells [102]. In addition to this mechanism, hypoxia regulates NANOG, a stem cell self-renewal transcription factor which in turn is involved in the regulation of CTL-mediated cytotoxicity [103,104]. The hypoxia-induced microRNA-210 also regulates CTL-mediated tumor cytotoxicity. HIF-1 induces the expression of microRNA-210, which subsequently affects a number of target genes including homeobox A1 (HOXA1), nonreceptor protein tyrosine phosphatase type 1 (PTPN1), and tumor protein p53-inducible protein 11 (TP53I11), through which it decreases CTL tumor lysis [105]. It is also important to note that hypoxia upregulates autophagy, which is a catabolic cell degradation process that prevents the accumulation of altered cell components while providing more nutrients as an adaptation mechanism to metabolic stress and hypoxic conditions [106,107]. More recently, data showing the important link between autophagy and shaping antitumor immune responses are emerging [108]. It is now well established that hypoxia plays a key role in promoting an immunosuppressive TME while at the same time promoting tumor resistance to immunotherapy and adoptive cell therapy such as CAR T-cells.

Cell-mediated immunosuppression

Along with metabolic stress and hypoxia, the TME is characterized by the presence of immunosuppressive cells that counteract effector immune cells and adoptively transferred CAR T-cells. In some instances, removal of CAR T-cells from the TME restores their function [109]. A number of studies have suggested that modulating the TME by directly targeting immunosuppressive cells or engineering transferred T cells to counteract cell-mediated immunosuppression could improve CAR T-cell function [109–111].

Several immunosuppressive cells within the TME have been described. Tregs are a well-established type that act through multiple mechanisms to inhibit antitumor activity and hence promote tumor proliferation and metastasis. Such mechanisms include cell-cell contact-mediated inhibition, production of immunosuppressive cytokines such as TGF-β and IL-10 while depleting IL-2 [112]. Tumor-associated macrophages (TAMs) are another type of immunosuppressive cells commonly found in solid tumors. Macrophages can adopt a variety of phenotypes based on signals received from the surrounding niche and hence can be modulated to either be tumor-suppressive (M1 macrophages) or tumor-promoting (M2 macrophages) [113]. The TME often polarizes macrophages toward a tumor-promoting phenotype [114] that promotes angiogenesis, tumor proliferation, immune escape, and metastasis. MDSCs are also common in solid tumors and can dampen CAR T-cell function in the TME through a variety of mechanisms; including the production of regulatory cytokines and expression of inhibitory checkpoints [115], blocking lymphocyte homing [116], stimulation of other immunosuppressive cells such as Tregs [117], shifting macrophages to an immunosuppressive phenotype [118], depletion of metabolites essential for T-cell function through expression of enzymes metabolizing amino acids [119], expression of ectoenzymes promoting adenosine metabolism [120], and the production of reactive oxygen and nitrogen species [119,121,122].

In addition to immunosuppressive cytokines, tumor cells, and other cells in the TME could overexpress checkpoint molecules that block the function of CAR T-cells, such as programmed death-1 (PD-1), PD-1 ligands (PD-L1 and PD-L2), T-cell membrane protein-3 (TIM-3), lymphocyte-activation protein-3 (LAG-3), T cell Ig and ITIM domain (TIGIT), cytotoxic T lymphocyte-associated antigen 4 (CTLA-4), and others [123,124].

Prospects

Modulating immunometabolism to enhance immunotherapy is an active area of research. The different costimulatory endodomains incorporated into CARs predict different T-cell metabolic programs and mitochondrial biogenesis. 4-1BB signaling is associated with enhanced mitochondrial biogenesis and fatty acid oxidation compared with CD28 signaling that has enhanced aerobic glycolysis (Warburg phenomenon) [125]. Therefore, one strategy to resist the hostile TME is to specify the T-cell metabolic program through engineering them with different stimulatory domains. A group devised CAR T-cells that overexpress the antioxidant enzyme catalase to overcome the oxidative stress of the TME resulting in a decrease in the ROS while maintaining the antitumor activity [126]. Another group created a smart self-decision making CAR T-cell system that is primarily activated under hypoxic conditions through the fusion of oxygen-sensitive subdomain of HIF1α to a CAR scaffold [127]. This system as potential to decrease CAR T-cell associated on-target off-tumor toxicity by exclusively restricting its function to hypoxic microenvironments such as the TME.

Targeting adenosine signaling pathways has shown significant promise. Newick et al. targeted the downstream signaling of both adenosine and PGE2 by engineering CAR T-cells that produce a small peptide that inhibits the association of PKA with ezrin. Counteracting PKA immunosuppressive effect leads to increased cytokine production and T-cell cytotoxicity [128]. More recently, Beavis et al. combined CAR T-cells with A2AR antagonists and devised A2AR knockout CAR T-cells that showed enhanced cytokine production and antitumor activity both in vitro and in vivo [129].

Another immunometabolic modulatory approach consists of mitigating the effect of acidity of the TME on immune cells. Programming T cells to overexpress phosphoenolpyruvate carboxykinase 1 (PCK1), which increases gluconeogenesis from lactate and thus allows T cells to thrive in a glucose restricted and acidic TME, was shown to enhance antitumor T-cell function in melanoma murine models [130]. Potential strategies to modulate the TME acidosis and other ionic immunosuppression include (a) reducing the potassium levels through overexpression of potassium channels in T cells [131] and (b) supplementation with lactate-binding compounds such as 5-amino-2-hydroxymethylphenyl boronic acid (ABA), which binds lactate, accelerates the consumption of protons and normalizes pH [132].

To counteract the effect of immunosuppressive cells, a group designed CAR T-cells that target macrophages with a tumor-promoting phenotype. In a mouse model of Hodgkin's lymphoma, this CAR T-cell product led to long-term immunological memory [133]. Another way of offsetting cell-mediated immunosuppression is by targeting TGF-β, commonly produced in high amounts by Tregs, MDSCs, and tumor cells. Overexpression of a dominant-negative TGF-β receptor that blocks TGF-β signaling, on the surface of CAR T-cells resulted in enhanced T cell proliferation and persistence, cytokine production, and antitumor activity in aggressive prostate cancer mouse models [55]. The knockout of endogenous TGF-β receptor 2 (TGFβR2) from CAR T-cells by CRISPR/Cas9 technology was also investigated and shown to reduce the induced Treg conversion and prevent second-generation antimesothelin (M28z) CAR T-cell exhaustion. It also improved antitumor activity in vitro as well as in mesothelin-expressing CRL5826 cell-line derived xenograft (CDX) and in pancreatic carcinoma patient-derived xenograft (PDX) mouse models [134]. Another group showed that combining antimesothelin CAR T-cells with oncolytic adenovirus targeting TGF-β enhanced their activity against breast cancer xenograft mouse models [135].

Finally, different groups devised different strategies to block checkpoints on T cells. One strategy consisted of combining CAR T-cells with anti-PD-1 antibodies, which was shown to inhibit tumor growth and decrease the frequency of MDSC population of cells [136]. Another strategy entailed engineering the CAR T-cell itself to secrete checkpoint inhibitors [111,137]. Other investigators coexpressed antimesothelin CAR as well as a dominant-negative PD-1 receptor and demonstrated significant improvement in antitumor activity [123]. With the advent of CRISPR/CRISPR-Cas9 technology, CTLA-4 and PD-1 knockout CAR T-cells could be engineered to potentiate their effect [138]. A TME-resistant CAR using a triple-inhibitory receptor strategy, downregulating PD-1, TIM-3, and LAG-3 was also devised [139].

Extracellular matrix barrier

The presence of a strong surrounding barrier is a hallmark of solid tumors compared to hematologic malignancies, which poses a significant limitation to CAR T-cell penetration into the tumors. This barrier is made of stromal cells, immune cells, cancer cells, and extracellular matrix (ECM) components (proteins and glycans). Collagens, hyaluronan, proteoglycans, laminin, fibronectin, and others contribute to the proliferation of a fibrous connective tissue stroma of many solid tumors significantly hindering delivery of therapeutics [140–142]. This applies as well to the infiltration of CAR T-cells and other immune cells into the tumor stroma [143,144]. Moreover, the mass effect created by solid tumors compresses the lymphatics and the vasculature, which leads to elevation in the interstitial fluid pressure and hence reduced blood flow. This in turn leads to hypoxia, immunosuppression, and tumor progression [145]. In addition, the tumor-associated vasculature is often dysfunctional and might not express certain receptors compared to their normal counterparts, such as ICAM-1, CCAM-1, and P- and E-selectins, that are necessary for T-cell homing and infiltration [143,145]. Finally, it is important to note that ex vivo expanded CAR-transduced T cells often lack the enzyme heparanase (HPSE), which degrades heparin sulfate proteoglycans, essential components of ECM [146]. The physical barriers governing solid tumors remain a major hurdle in the accessibility and function of CAR T-cells in solid tumors compared to blood cancers.

Prospects

To enhance ECM breakdown and penetration of CAR T-cells into solid tumors, administration of collagenases and hyaluronidase could be coupled to the CAR T-cell infusion, such as is done with chemotherapy and antibody therapies in preclinical studies [147,148]. Alternatively, targeting ECM components such as cancer-associated fibroblasts (discussed later) could significantly impair solid tumor growth and weaken the ECM allowing for infiltration of CAR T-cells [149,150]. Heparanase-overexpressing CAR T-cells were shown to improve T-cell infiltration into tumors and antitumor effectiveness [146]. It is important to note, however, that breaking ECM could have a paradoxical deleterious effect as it could free the cancer cells from physical restraints and hence lead to cancer spread [151,152].

CAR-T cell trafficking

Not only do CAR-T cells have limited infiltration into the tumors, but also their homing to solid tumors sites after systemic administration has been a challenge. Trafficking of CAR T-cells into tumor beds is regulated by complex interactions between chemokines secreted by CAR T-cells with those secreted by tumor cells. Mature T cells express adhesion molecules and chemokines that allow them to interact with endothelial cells [143]. CXCR3 and CCR5 are two prominent G protein-coupled receptors that are often expressed on active tumor-infiltrating lymphocytes (TILs), as has been demonstrated for melanoma cells, colorectal cancer cells and breast cancer cells [153]. The lack of their corresponding ligands, CXCL9 and CXCL10, on tumor cells impedes the recruitment of those effector T cells to tumor beds, a phenomenon that has been demonstrated to hinder CAR T-cell trafficking [154].

Prospects

To circumvent the mismatch between the T-cell chemokines and cancer ligands, certain groups coexpressed CARs with better matched chemokines and have shown improved trafficking of CAR T-cells into tumor cells [155,156]. Others investigated CXCR2-expressing CAR T-cells in hepatocellular carcinoma and found that they exhibit significantly increased migration ability in vitro as well as accelerated trafficking, tumor-specific accumulation, and antitumor effect in xenograft tumor models [157]. Xu et al. used the stimulator of interferon gene (STING) agonists DMXAA and cGAMP in conjunction with CAR T-cell therapy and demonstrated that these combinations modulate the TME to favor CAR T-cell trafficking and persistence, which is supported by the generation of a chemokine milieu that enhances

CAR T-cell recruitment [158]. Other investigators directly infused the CAR T-cell products into tumor beds as opposed to peripheral infusion. They showed that intrapleural infused CAR T-cells elicited much stronger antitumor activity compared to intravenous infusion for malignant pleural disease in patients with primary malignant mesothelioma or pleural metastases from lung or breast cancer [159]. Based on these results, a clinical trial of intrapleural CAR T-cell infusion for patients with malignant pleural disease from mesothelioma, lung or breast cancer is currently underway (preliminary results discussed earlier) [4].

CAR T-cell persistence

Expansion and persistence of adoptively transferred CAR T-cells are essential prerequisites for their long-term efficacy [160,161]. CAR T-cell expansion differs from physiological T-cell expansion in several ways. First, T-cell transduction with CAR occurs ex vivo, and transduced T cells are subsequently expanded independent of TCR stimulation. Moreover, CAR T-cell expansion ex vivo depends on elements related to T-cell engineering such as CD3ζ/CD28 (or other costimulatory domains) stimulation in the presence of recombinant cytokines [162]. In vivo, CAR T-cell expansion is promoted by the lymphodepleted state created by the preconditioning regimen received and tumor burden [163]. Many factors contribute to lack of CAR T-cell maintenance after their infusion and hence loss of response. Such factors include (a) suboptimal initial expansion of the CAR T-cell product infused, (b) rejection of the infused CAR T-cell product through mounting of an endogenous immune response against immunogenic CAR sequences, (c) the previously discussed TME-immunosuppressive factors, (d) activation-induced cell death (AICD), (e) T-cell exhaustion following persistent CAR or TCR signaling, and (f) lack of stimulation by either antigen loss or decreased level of expression leading to contraction [164]. Exhaustion refers to irreversible epigenetic changes acquired by terminally differentiated T cells [165,166]. While exhaustion is an evolutionary mechanism that allows for the inactivation of repeatedly stimulated T cells by self MHC [167], it prevents a long-lasting antitumor activity despite antigen persistence for often long periods of time and leads to their susceptibility to AICD [166,168]. Kunkele et al. demonstrated that CAR T-cells are vulnerable to AICD through Fas-FasL-dependent pathway [168]. Additionally, T-cell contraction can occur independent of this pathway via cell intrinsic exhaustion, replicative senescence, growth factor withdrawal, and others.

Prospects

It is currently well established that the addition of different costimulatory domains promotes differential persistence capabilities of CAR T-cells [47,169]. Culture systems that prevent telomere loss and maintain endogenous costimulatory receptor expression have evolved to improve on the functional lifespan of transferred T cells [170,171]. Reducing the production time of CAR T-cells allows for the generation of less differentiated and hence more potent cells [172]. Other ways to decrease T-cell differentiation during in vitro expansion include inhibition of signaling pathways downstream of IL-2 such as PI3K [173], glycogen synthase kinase 3β or Wnt 3a [174], and protein kinase B (AKT) [175]. Moreover, replacing IL-2 with other cytokines such as IL-7 and IL-15 that regulate T-cell survival independent of TCR stimulation could enhance in vivo expansion and persistence of CAR T-cells [176]. Altering the culture milieu by the addition of IL-21 was found to improve CAR-dependent T-cell effector function [177]. IL-12 was also investigated in that regard. The constitutive secretion of IL-12 was shown to enhance T-cell cytotoxicity and promote resistance against Treg immunosuppression [178]. Batra et al. generated anti-glypican 3 (GPC3) CAR T-cells simultaneously encoding IL-15, IL-21, or both and incorporating 4-1BB costimulatory endodomains and found that T cells expressing both cytokines (21.15.GBBz. GPC3.CAR T-cells) have superior in vitro and in vivo expansion and persistence and the most robust antitumor activity in murine hepatocellular carcinoma xenograft mouse models [179].

CAR T-cells with a more naïve phenotype could be generated from genetic reprogramming of induced pluripotent stem cells (iPSCs) derived from somatic cells [180]. In addition, epigenetic modulation of CAR-T cells such as inhibition of bromodomain and extraterminal motif (BET) proteins [181] and disruption of the methylcytosine dioxygenase TET2 gene [182] promotes a stem cell- and a central memory T cell-like phenotype of T cells [183] and could be employed to enhance CAR T-cell persistence and potency. Lynn et al. studied the biology of CAR-T cell exhaustion and demonstrated an association between exhaustion and defect in the production of IL-2 by comparing T cells expressing a GD2 CAR with tonic signaling expressing profound features of exhaustion to T cells expressing a CD19 CAR without tonic signaling. They uncovered an imbalance of activating vs exhaustion-promoting AP-1 family members as a contributing factor to T-cell exhaustion and that the canonical AP-1 transcription factor, a heterodimer of c-Jun and c-Fos, is essential for T-cell activity and the production of IL-2. AP-1 family members are homo- or heterodimers of basic leucine zipper (bZIP) transcription factors, such as the activation factors c-Fos and c-Jun and the exhaustion-promoting factors JunB, basic leucine zipper ATF-like transcription factor (BATF), BATF3 and interferon-regulatory factor 4 (IRF4). Moreover, they overexpressed c-Jun in GD2 CAR-T cells which lead to a reduction in the expression of inhibitory receptors and improved cytokine production, cell expansion, and CAR-T cell activity in multiple mouse models of cancer, including the resistant 143B osteosarcoma tumor models and in models of tumors with low antigen expression. In agreement with these findings, the CRISPR-mediated deletion of the exhaustion-promoting factors JunB, BATF and IRF4 also increased cytokine production [184].

The progress made in the field of genome editing has allowed for the accelerated growth of strategies to overcome T-cell dysfunction and exhaustion such that some of these strategies have already moved to the clinic. The results of a first-in-human clinical trial utilizing CRISPR-edited T cells in the clinic were recently published [185]. T cells were engineered by removing the endogenous TCR and the checkpoint PD-1 and by introducing a synthetic cancer-specific TCR transgene (NY-ESO-1). The manufactured cell products were infused into three patients, two with advanced refractory myeloma and one with metastatic sarcoma. T-cell engraftment was shown in all three patients at stable levels for at least 9 months with no clinical toxicity. Trafficking of T cells to tumor sites was shown in all three patients with reduction in target antigens NY-ESO-1 and/or LAGE-1. The success shown in the clinical implementation of gene editing of TCR-based cell therapy opens the door for potential implementation of CAR-based therapies especially in an effort to expand their use for solid tumors.

Target antigens

One of the major limitations of CAR T-cell therapy in solid tumors has been the identification of ideal antigens specific to tumor cells that can be safely targeted such as CD19 and BCMA in blood cancers. Antigen targets that could be exploited in CAR T-cell therapy include proteins, carbohydrates, or glycolipids. The ideal target is one that is overexpressed on tumor tissues, with virtually no expression on normal tissues. Tumor-specific antigens (TSA) are ideal due to their unique and excessive expression on tumor cells [186]. EGFRvIII is one type of TSA that has been targeted in glioblastoma as we have previously discussed [19]. The challenge that has been encountered in the targeting of these types of antigens in the clinic is the fact that they are highly heterogeneous among patients with the same type of tumor. The generation of CAR-targeting TSAs would require the identification of TSAs for individual patients and the manufacture of personalized CARs, a process that is costly, complicated, and time-consuming and often nonaffordable [186].

Because of the limitations of TSAs, another type of tumor antigen called tumor-associated antigen (TAA) has been explored for targeting using CAR T-cell therapy. However, targeting TAAs is often associated with on-target off-tumor toxicity because they are often expressed on normal tissues at low levels [187]. We have previously discussed a myriad of TAAs that are the target of investigational CAR products such as HER2, mesothelin, and CEA.

Alternative antigens that could be targeted with CAR T-cell therapy are ones that are not expressed on the solid tumors themselves but instead on the tumor-associated vasculature such as VEGFR-2 [188] or tumor-associated fibroblasts such as fibroblast activation protein (FAP) [189]. Targeting these types of antigens impedes stromal formation and angiogenesis that are essential for tumor growth and metastasis while at the same time offering a wider spectrum of clinical applicability [190,191].

Finally, a distinct type of antigens that are candidates for targeting by CAR T-cells are glycolipid antigens. The most well-studied glycolipid antigen is GD2, which is overexpressed in neuroblastoma, astrocytoma, retinoblastoma, Ewing's sarcoma, breast cancer, melanoma and has been targeted by CAR-T [192,193] with good efficacy [194,195].

The success of CAR T-cell therapy relies in a significant part on the careful target antigen selection in accordance with the specific TME. A well-selected target antigen is key to reduce toxicity but also to determine the CAR-T antitumor efficacy.

Prospects

To overcome the paucity of target antigens as a major drawback in the expansion of the field of CAR T-cell therapy for solid tumors, researchers continue to explore novel targets. These include CD44v6 in lung and ovarian adenocarcinoma [196] which has been implicated in tumorigenesis, tumor cell invasion, and metastasis [197]. Other novel targets are GPC3 in hepatocellular carcinoma [179,198], PSCA in gastric cancer [199], B7-H3 in multiple tumors [200], AXL in triple-negative breast cancers [201], and Mucin-4 (MUC-4) in pancreatic adenocarcinoma [202].

An interesting strategy to overcome the limited number of cell-surface targets with high cancer-specific specificity was recently published [203]. Second-generation CAR T-cells targeting claudin 6 (CLDN6), a tetraspanin membrane protein involved in tight junction formation and with cancer-associated expression, and incorporating 4-1BB costimulatory domain were engineered with high specificity and binding affinity and were found to have high potency in lysing CLDN6-transfected lung cancer cell lines as well as in vivo in mice xenografted with ovarian cell lines. Moreover, they also introduced the concept of CARVac that aims to enhance engraftment and persistence of CAR T-cells through an RNA vaccine consisting of a liposomal CLDN6-encoding vaccine (CLDN6-LPX). CLDN6-LPX triggered increased display of CLDN6 on dendritic cell surface leading to increased proliferation and cytokine secretion of CLDN6 CAR T-cells both in vitro and in vivo. The adoptive transfer of subtherapeutic doses of CLDN6 CAR T-cells in combination with CLDN6-LPX led to complete regression of large tumors and increased survival of mice xenografted with large Lewis lung tumors and colon cancers.

Because tumor heterogeneity is a major concern in the development of CAR-T cells against glioblastoma, a recent study by Wang et al. sought to expand the repertoire of target antigens of glioblastoma by exploiting the glioblastoma-binding potential of chlorotoxin (CLTX), a peptide component of scorpion venom [204]. Demonstrating that CLTX binds a large proportion of tumor cells and tumor cell constituents, they developed CAR T cells using CLTX as the targeting domain (CLTX-CAR T cells) and showed a strong antitumor activity in orthotopic xenograft glioblastoma tumor models without observable off-target activity against normal cells. They also observed that matrix metalloproteinase-2 is required for targeting by CTLX-CAR T cells.

To minimize on-target off-tumor toxicity, dual antigen recognizing CAR-T cells have been designed. Two approaches are adopted; double tumor antigen-directed CARs and inhibitory CARs (iCARs) [205,206]. The expression of two CARs recognizing different epitopes and designed differentially to either incorporate co-stimulatory domains or deliver killing allows for a safer construct that restricts tandem CARs activation to tumor cells simultaneously expressing both epitopes [207]. The recognition of only one antigen expressed on a normal tissue leads to suboptimal activation of T cells in the absence of engagement of stimulatory domain-containing CAR [207]. On the other hand, iCARs combine two CARs, one of which is activating that recognizes tumor antigen and the other generates dominant negative signals. When iCARs recognize a specific antigen expressed only on normal tissues, the negative signals dampen the activating signals generated by the activating CAR and thus can distinguish cancer cells from normal cells and limit toxicity [205]. A dual CAR-T cell therapy strategy was recently developed that involves the coexpression of both an antitumor-associated glycoprotein 72 (TAG-72) CAR as well as an anti-CD47 CAR that is truncated so that it is devoid of intracellular signaling domains. This strategy was shown to enhance CAR-T cell killing in ovarian cancer both in vitro as well as in NSG mice bearing MESOV-derived tumors [208].

Beyond CAR T-cell therapy

With the high cost associated with extracting a patient's T cells and the severe toxicities associated with CAR T-cell therapy, other immune cells have recently gained interest for potential CAR-based therapies. Some of these studied cells include gamma-delta (γδ) T cells, natural killer (NK) cells, invariant natural killer T (iNKT) cells, and macrophages.

The short lifespan of NK cells allowed them to enter the picture of CAR-cell therapy as potentially good candidates [209]. Advantages of NK cells include (a) their strict regulation by signals from an array of activating and inhibitory receptors that helps in preventing on-target off-tumor side effects seen with CAR-T cells and (b) their intrinsic capacity of recognizing tumor cells, which allows them to kill cancer cells even when tumors evolve antigen escape mechanisms to evade targeting via CARs [210,211]. Importantly, CAR-NKs are not associated with the risk of graft-vs-host disease (GVHD) as demonstrated in a number of studies [212–215]. These unique characteristics fueled interest in harnessing the potential of NK cells for the generation of off-the-shelf, readily available, affordable, and safer CAR-NK cells for cancer immunotherapy [216]. Numerous preclinical studies have demonstrated the efficacy and safety of CAR-NK cells in cancer, including B-cell malignancies by targeting CD19 and CD20 [217–220], T-cell malignancies by targeting CD3 and CD5 [221,222], multiple myeloma via targeting CD138 and CS1 [223,224], HER2-positive tumors targeting HER2 [225,226], neuroblastoma targeting GD2 [227], and glioblastoma and breast cancer targeting EGFR [228,229]. Our group is currently leading a first-in-human phase I/II study of the safety and efficacy of CAR.CD19-CD28-zeta-2A-iC9-IL15-transduced cord blood NK cells in patients with relapsed/refractory B lymphoid malignancies (NCT03056339), preliminary results of which were recently published showing a 73% objective response rate in 11 treated patients, 7 of whom achieved complete responses with no major toxicity [230]. Other ongoing trials of CAR NK-cell therapy include targeting CD33 in acute myeloid leukemia (AML) (NCT02944162), CD7 in leukemia and lymphoma (NCT02742727), CD19 in B-cell malignancies (NCT02892695), and MUC1 in MUC1 positive tumors (NCT02839954).

Besides NK cells, natural killer T (NKT) cells have gained interest for CAR modification primarily due to emerging data showing that tumor infiltration with NKT cells is associated with improved outcome [231,232] and the CD1d restricted nature of NKT antigen recognition is thought to decrease potential off-target toxicity [232]. Multiple groups generated CAR-modified NKT cells such as anti-GD2 CAR-NKT cells for the treatment of neuroblastoma and anti-CD19 CAR-NKT cells for lymphoma and demonstrated good cytokine production and promising efficacy [233–235].

Both γδ T cells and macrophages have also been successfully transduced to express CARs and tested in preclinical models demonstrating targeted antitumor functions [236,237]. CAR-based therapies from each of these cell types have been faced with many challenges such as engineering efficacy and in vivo persistence. Nonetheless, they offer new promise for the growth of the field of CAR-based therapies and hope for less expensive and more widely adopted strategies.

Conclusion

Despite the numerous challenges encountered in advancing the field of CAR therapy for the treatment of solid tumors, the progress made thus far is exciting and the preclinical and early clinical data are promising. Many strategies have already been devised to overcome some of these challenges and others that could be further explored. Improving antigen selection and specificity of targeting, enhancing cell trafficking and homing to tumor beds, improving immune cell infiltration despite strong stromal components, potentiating immune cells to overcome the hostile TME, bolstering immune cell expansion and persistence, abrogating toxicities, and devising universal constructs, are prerequisites for the achievement of complete and durable remissions with CAR-based modalities in solid cancers. With the exploding advances in the fields of cellular therapy and cell engineering, synthetic biology, and genome editing, the future for CAR-cell therapy for solid tumors is bright and global efforts will continue to advance the field toward curative treatments for highly resistant cancers.

Acknowledgments

This work was supported in part by the generous philanthropic contributions to The University of Texas MD Anderson Cancer Center Moon Shots Program, by the generous support of Ann and Clarence Cazalot, Lyda Hill Philanthropies, and The Sally Cooper Murray endowment; by Grants from CPRIT (RP160693), by a Stand Up To Cancer Dream Team Research Grant (SU2C-AACR-DT-29-19), by Grants (1 R01 CA211044-01, 5 P01CA148600-03, and P50CA100632-16) from the National Institutes of Health (NIH), the Specialized Program of Research Excellence (SPORE) in Brain Cancer Grant (P50CA127001), the Specialized Program of Research Excellence (SPORE) in Leukemia grant (P50CA100632), and by a Grant (CA016672) to the MD Anderson Cancer Center from the NIH.

References

[1] Maude SL, Laetsch TW, Buechner J, Rives S, Boyer M, Bittencourt H, Myers GD. Tisagenlecleucel in children and young adults with B-cell lymphoblastic leukemia. N Engl J Med 2018;378(5):439–48.

[2] Neelapu SS, Locke FL, Bartlett NL, Lekakis LJ, Miklos DB, Jacobson CA, Lin Y. Axicabtagene ciloleucel CAR T-cell therapy in refractory large B-cell lymphoma. N Engl J Med 2017;377(26):2531–44.

[3] Munshi NC, Anderson LD, Shah N, Madduri D, Berdeja J, Lonial S, San-Miguel J. Idecabtagene vicleucel in relapsed and refractory multiple myeloma. N Engl J Med 2021;384(8):705–16. https://doi.org/10.1056/NEJMoa2024850.

[4] Adusumilli PS, Zauderer MG, Rusch VW, O'Cearbhaill RE, Zhu A, Ngai DA, Sadelain M. Abstract CT036: a phase I clinical trial of malignant pleural disease treated with regionally delivered autologous mesothelin-targeted CAR T cells: safety and efficacy. Cancer Res 2019;79(13 Suppl):CT036. https://doi.org/10.1158/1538-7445.Am2019-ct036.

[5] Navai S, Derenzo C, Joseph S. Administration of HER2-CAR T cells after lymphodepletion safely improves T cell expansion and induces clinical responses in patients with advanced sarcomas. In: Paper presented at the AACR Annual Meeting, Atlanta; 2019.

[6] Kershaw MH, Westwood JA, Parker LL, Wang G, Eshhar Z, Mavroukakis SA, Rogers Freezer L. A phase I study on adoptive immunotherapy using gene-modified T cells for ovarian cancer. Clin Cancer Res 2006;12(20):6106–15.

[7] Park JR, DiGiusto DL, Slovak M, Wright C, Naranjo A, Wagner J, Ostberg JR. Adoptive transfer of chimeric antigen receptor re-directed cytolytic T lymphocyte clones in patients with neuroblastoma. Mol Ther 2007;15(4):825–33.

[8] Ahmed N, Brawley VS, Hegde M, Robertson C, Ghazi A, Gerken C, Corder A. Human epidermal growth factor receptor 2 (HER2)-specific chimeric antigen receptor-modified T cells for the immunotherapy of HER2-positive sarcoma. J Clin Oncol 2015;33(15):1688.

[9] Brown CE, Warden CD, Starr R, Deng X, Badie B, Yuan Y-C, Barish ME. Glioma IL13Rα2 is associated with mesenchymal signature gene expression and poor patient prognosis. PLoS One 2013;8(10), e77769.

[10] Debinski W, Gibo DM, Hulet SW, Connor JR, Gillespie GY. Receptor for interleukin 13 is a marker and therapeutic target for human high-grade gliomas. Clin Cancer Res 1999;5(5):985–90.

[11] Jarboe JS, Johnson KR, Choi Y, Lonser RR, Park JK. Expression of interleukin-13 receptor α2 in glioblastoma multiforme: implications for targeted therapies. Cancer Res 2007;67(17):7983–6.

[12] Thaci B, Brown CE, Binello E, Werbaneth K, Sampath P, Sengupta S. Significance of interleukin-13 receptor alpha 2-targeted glioblastoma therapy. Neuro Oncol 2014;16(10):1304–12.

[13] Wykosky J, Gibo DM, Stanton C, Debinski W. Interleukin-13 receptor α2, EphA2, and Fos-related antigen 1 as molecular denominators of high-grade astrocytomas and specific targets for combinatorial therapy. Clin Cancer Res 2008;14(1):199–208.

[14] Brown CE, Badie B, Barish ME, Weng L, Ostberg JR, Chang W-C, Wright C. Bioactivity and safety of IL13Rα2-redirected chimeric antigen receptor CD8+ T cells in patients with recurrent glioblastoma. Clin Cancer Res 2015;21(18):4062–72.

[15] Brown CE, Alizadeh D, Starr R, Weng L, Wagner JR, Naranjo A, Simpson J. Regression of glioblastoma after chimeric antigen receptor T-cell therapy. N Engl J Med 2016;375(26):2561–9.

[16] Ahmed N, Brawley V, Hegde M, Bielamowicz K, Kalra M, Landi D, Wakefield A. Her2-specific chimeric antigen receptor-modified virus-specific T cells for progressive glioblastoma: a phase 1 dose-escalation trial. JAMA Oncol 2017;3(8):1094–101.

[17] Ekstrand AJ, Sugawa N, James CD, Collins VP. Amplified and rearranged epidermal growth factor receptor genes in human glioblastomas reveal deletions of sequences encoding portions of the N- and/or C-terminal tails. Proc Natl Acad Sci 1992;89(10):4309–13.

[18] Sugawa N, Ekstrand AJ, James CD, Collins VP. Identical splicing of aberrant epidermal growth factor receptor transcripts from amplified rearranged genes in human glioblastomas. Proc Natl Acad Sci 1990;87(21):8602–6.

[19] O'Rourke DM, Nasrallah MP, Desai A, Melenhorst JJ, Mansfield K, Morrissette JJ, Shen A. A single dose of peripherally infused EGFRvIII-directed CAR T cells mediates antigen loss and induces adaptive resistance in patients with recurrent glioblastoma. Sci Transl Med 2017;9(399), eaaa0984.

[20] Schulz G, Cheresh DA, Varki NM, Yu A, Staffileno LK, Reisfeld RA. Detection of ganglioside GD2 in tumor tissues and sera of neuroblastoma patients. Cancer Res 1984;44(12 Pt 1):5914–20.

[21] Louis CU, Savoldo B, Dotti G, Pule M, Yvon E, Myers GD, Liu E. Antitumor activity and long-term fate of chimeric antigen receptor–positive T cells in patients with neuroblastoma. Blood 2011;118(23):6050–6.

[22] Gavert N, Conacci-Sorrell M, Gast D, Schneider A, Altevogt P, Brabletz T, Ben-Ze'ev A. L1, a novel target of β-catenin signaling, transforms cells and is expressed at the invasive front of colon cancers. J Cell Biol 2005;168 (4):633–42.

[23] Gonzalez S, Naranjo A, Serrano LM, Chang WC, Wright CL, Jensen MC. Genetic engineering of cytolytic T lymphocytes for adoptive T-cell therapy of neuroblastoma. J Gene Med 2004;6(6):704–11.

[24] Turtle CJ, Hanafi L-A, Berger C, Gooley TA, Cherian S, Hudecek M, Budiarto TM. CD19 CAR-T cells of defined CD4+: CD8+ composition in adult B cell ALL patients. J Clin Invest 2016;126(6):2123–38.

[25] Moghimi B, Muthugounder S, Jambon S, Tibbetts R, Hung L, Bassiri H, Asgharzadeh S. Preclinical assessment of the efficacy and specificity of GD2-B7H3 SynNotch CAR-T in metastatic neuroblastoma. Nat Commun 2021;12(1):511. https://doi.org/10.1038/s41467-020-20785-x.

[26] Creighton CJ, Morgan M, Gunaratne PH, Wheeler DA, Gibbs RA, Gordon Robertson A, University of North Carolina at Chapel Hill. Comprehensive molecular characterization of clear cell renal cell carcinoma. Nature 2013;499(7456):43–9. https://doi.org/10.1038/nature12222.

[27] Feng K, Guo Y, Dai H, Wang Y, Li X, Jia H, Han W. Chimeric antigen receptor-modified T cells for the immunotherapy of patients with EGFR-expressing advanced relapsed/refractory non-small cell lung cancer. Sci China Life Sci 2016;59(5):468–79.

[28] Byers LA, Chiappori A, Smit M-AD. Phase 1 study of AMG 119, a chimeric antigen receptor (CAR) T cell therapy targeting DLL3, in patients with relapsed/refractory small cell lung cancer (SCLC). J Clin Oncol 2019;37 (15_suppl):TPS8576. https://doi.org/10.1200/JCO.2019.37.15_suppl.TPS8576.

[29] Furuta M, Sakakibara-Konishi J, Kikuchi H, Yokouchi H, Nishihara H, Minemura H, Fujita Y. Analysis of DLL3 and ASCL1 in surgically resected small cell lung cancer (HOT1702). Oncologist 2019;24(11):e1172–9.

[30] Tanaka K, Isse K, Fujihira T, Takenoyama M, Saunders L, Bheddah S, Okamoto I. Prevalence of Delta-like protein 3 expression in patients with small cell lung cancer. Lung Cancer 2018;115:116–20.

[31] Argani P, Iacobuzio-Donahue C, Ryu B, Rosty C, Goggins M, Wilentz RE, Yeo CJ. Mesothelin is overexpressed in the vast majority of ductal adenocarcinomas of the pancreas: identification of a new pancreatic cancer marker by serial analysis of gene expression (SAGE). Clin Cancer Res 2001;7(12):3862–8.

[32] Hassan R, Kreitman RJ, Pastan I, Willingham MC. Localization of mesothelin in epithelial ovarian cancer. Appl Immunohistochem Mol Morphol 2005;13(3):243–7.

[33] Ordóñez NG. Application of mesothelin immunostaining in tumor diagnosis. Am J Surg Pathol 2003;27 (11):1418–28.

[34] Ordóñez NG. Value of mesothelin immunostaining in the diagnosis of mesothelioma. Mod Pathol 2003;16 (3):192.

[35] Chang K, Pastan I. Molecular cloning of mesothelin, a differentiation antigen present on mesothelium, mesotheliomas, and ovarian cancers. Proc Natl Acad Sci 1996;93(1):136–40.
[36] Chang K, Pastan I, Willingham MC. Isolation and characterization of a monoclonal antibody, K1, reactive with ovarian cancers and normal mesothelium. Int J Cancer 1992;50(3):373–81.
[37] Beatty GL, Haas AR, Maus MV, Torigian DA, Soulen MC, Plesa G, June CH. Mesothelin-specific chimeric antigen receptor mRNA-engineered T cells induce anti-tumor activity in solid malignancies. Cancer Immunol Res 2014;2(2):112–20. https://doi.org/10.1158/2326-6066.cir-13-0170.
[38] Witton CJ, Reeves JR, Going JJ, Cooke TG, Bartlett JM. Expression of the HER1–4 family of receptor tyrosine kinases in breast cancer. J Pathol 2003;200(3):290–7.
[39] O'Shaughnessy JA. Molecular signatures predict outcomes of breast cancer. N Engl J Med 2006;355(6):615–7. Mass Medical Soc.
[40] Baselga J, Cortés J, Kim S-B, Im S-A, Hegg R, Im Y-H, Knott A. Pertuzumab plus trastuzumab plus docetaxel for metastatic breast cancer. N Engl J Med 2012;366(2):109–19.
[41] Slamon DJ, Leyland-Jones B, Shak S, Fuchs H, Paton V, Bajamonde A, Pegram M. Use of chemotherapy plus a monoclonal antibody against HER2 for metastatic breast cancer that overexpresses HER2. N Engl J Med 2001;344(11):783–92.
[42] Fritsche HA, Bast RC. CA 125 in ovarian cancer: advances and controversy. Clin Chem 1998;44(7):1379–80.
[43] Singh AP, Senapati S, Ponnusamy MP, Jain M, Lele SM, Davis JS, Batra SK. Clinical potential of mucins in diagnosis, prognosis, and therapy of ovarian cancer. Lancet Oncol 2008;9(11):1076–85.
[44] Koneru M, O'Cearbhaill R, Pendharkar S, Spriggs DR, Brentjens RJ. A phase I clinical trial of adoptive T cell therapy using IL-12 secreting MUC-16 ecto directed chimeric antigen receptors for recurrent ovarian cancer. J Transl Med 2015;13(1):102.
[45] Leamon CP, Low PS. Delivery of macromolecules into living cells: a method that exploits folate receptor endocytosis. Proc Natl Acad Sci U S A 1991;88(13):5572.
[46] Toffoli G, Cernigoi C, Russo A, Gallo A, Bagnoli M, Boiocchi M. Overexpression of folate binding protein in ovarian cancers. Int J Cancer 1997;74(2):193–8.
[47] Song D-G, Ye Q, Carpenito C, Poussin M, Wang L-P, Ji C, Powell DJ. In vivo persistence, tumor localization, and antitumor activity of CAR-engineered T cells is enhanced by costimulatory signaling through CD137 (4-1BB). Cancer Res 2011;71(13):4617–27.
[48] Tanyi JL, Haas AR, Beatty GL, Stashwick CJ, O'Hara MH, Morgan MA, June CH. Anti-mesothelin chimeric antigen receptor T cells in patients with epithelial ovarian cancer. J Clin Oncol 2016;34 (15_suppl):5511. https://doi.org/10.1200/JCO.2016.34.15_suppl.5511.
[49] Wong Y, Cheung T, Lam S, Lu H, Zhuang Y, Chan MY, Chung TK. Prevalence and significance of HER-2/neu amplification in epithelial ovarian cancer. Gynecol Obstet Invest 1995;40(3):209–12.
[50] Tricoli JV, Schoenfeldt M, Conley BA. Detection of prostate cancer and predicting progression: current and future diagnostic markers. Clin Cancer Res 2004;10(12):3943–53.
[51] Leek J, Lench N, Maraj B, Bailey A, Carr IM, Andersen S,. Prostate-specific membrane antigen: evidence for the existence of a second related human gene. Br J Cancer 1995;72(3):583–8. https://doi.org/10.1038/bjc.1995.377.
[52] DeMarzo AM, Nelson WG, Isaacs WB, Epstein JI. Pathological and molecular aspects of prostate cancer. Lancet 2003;361(9361):955–64. https://doi.org/10.1016/s0140-6736(03)12779-1.
[53] Junghans RP, Ma Q, Rathore R, Gomes EM, Bais AJ, Lo AS, Cohen SI. Phase I trial of anti-PSMA designer CAR-T cells in prostate cancer: possible role for interacting interleukin 2-T cell pharmacodynamics as a determinant of clinical response. Prostate 2016;76(14):1257–70. https://doi.org/10.1002/pros.23214.
[54] Rathkopf DE, Morris MJ, Fox JJ, Danila DC, Slovin SF, Hager JH, Scher HI. Phase I study of ARN-509, a novel antiandrogen, in the treatment of castration-resistant prostate cancer. J Clin Oncol 2013;31(28):3525–30. https://doi.org/10.1200/jco.2013.50.1684.
[55] Kloss CC, Lee J, Zhang A, Chen F, Melenhorst JJ, Lacey SF, June CH. Dominant-negative TGF-β; receptor enhances PSMA-targeted human CAR T cell proliferation and augments prostate cancer eradication. Mol Ther 2018;26(7):1855–66. https://doi.org/10.1016/j.ymthe.2018.05.003.
[56] Narayan V, Gladney W, Plesa G, Vapiwala N, Carpenter E, Maude SL, Haas NB. A phase I clinical trial of PSMA-directed/TGFβ-insensitive CAR-T cells in metastatic castration-resistant prostate cancer. J Clin Oncol 2019;37(7_Suppl):TPS347. https://doi.org/10.1200/JCO.2019.37.7_suppl.TPS347.

[57] Lamers CH, Sleijfer S, Van Steenbergen S, Van Elzakker P, Van Krimpen B, Groot C, Debets R. Treatment of metastatic renal cell carcinoma with CAIX CAR-engineered T cells: clinical evaluation and management of on-target toxicity. Mol Ther 2013;21(4):904–12.
[58] Lamers CHJ, Sleijfer S, Vulto AG, Kruit WHJ, Kliffen M, Debets R, Oosterwijk E. Treatment of metastatic renal cell carcinoma with autologous T-lymphocytes genetically retargeted against carbonic anhydrase IX: first clinical experience. J Clin Oncol 2006;24(13):e20–2. https://doi.org/10.1200/jco.2006.05.9964.
[59] Pastorekova S, Parkkila S, Parkkila AK, Opavsky R, Zelnik V, Saarnio J, Pastorek J. Carbonic anhydrase IX, MN/CA IX: analysis of stomach complementary DNA sequence and expression in human and rat alimentary tracts. Gastroenterology 1997;112(2):398–408.
[60] Tafreshi NK, Lloyd MC, Bui MM, Gillies RJ, Morse DL. Carbonic anhydrase IX as an imaging and therapeutic target for tumors and metastases. In: Carbonic anhydrase: mechanism, regulation, links to disease, and industrial applications. Springer; 2014. p. 221–54.
[61] Wang QJ, Yu Z, Hanada K-I, Patel K, Kleiner D, Restifo NP, Yang JC. Preclinical evaluation of chimeric antigen receptors targeting CD70-expressing cancers. Clin Cancer Res 2017;23(9):2267–76.
[62] Ko JS, Zea AH, Rini BI, Ireland JL, Elson P, Cohen P, Garcia J. Sunitinib mediates reversal of myeloid-derived suppressor cell accumulation in renal cell carcinoma patients. Clin Cancer Res 2009;15(6):2148–57.
[63] Finke JH, Rayman PA, Ko JS, Bradley JM, Gendler SJ, Cohen PA. Modification of the tumor microenvironment as a novel target of renal cell carcinoma therapeutics. Cancer J 2013;19(4).
[64] Walter S, Weinschenk T, Stenzl A, Zdrojowy R, Pluzanska A, Szczylik C, Mendrzyk R. Multipeptide immune response to cancer vaccine IMA901 after single-dose cyclophosphamide associates with longer patient survival. Nat Med 2012;18(8):1254.
[65] Finke JH, Rini B, Ireland J, Rayman P, Richmond A, Golshayan A, Dreicer R. Sunitinib reverses type-1 immune suppression and decreases T-regulatory cells in renal cell carcinoma patients. Clin Cancer Res 2008;14(20): 6674–82.
[66] Yeku O, Li X, Brentjens RJ. Adoptive T-cell therapy for solid tumors. Am Soc Clin Oncol Educ Book 2017;37:193–204. https://doi.org/10.1200/edbk_180328.
[67] Lewis VO, Ozawa MG, Deavers MT, Wang G, Shintani T, Arap W, Pasqualini R. The interleukin-11 receptor alpha as a candidate ligand-directed target in osteosarcoma: consistent data from cell lines, orthotopic models, and human tumor samples. Cancer Res 2009;69(5):1995–9. https://doi.org/10.1158/0008-5472.can-08-4845.
[68] Schwertschlag US, Trepicchio WL, Dykstra KH, Keith JC, Turner KJ, Dorner AJ. Hematopoietic, immunomodulatory and epithelial effects of interleukin-11. Leukemia 1999;13(9):1307–15.
[69] Huang G, Yu L, Cooper LJ, Hollomon M, Huls H, Kleinerman ES. Genetically modified T cells targeting interleukin-11 receptor alpha-chain kill human osteosarcoma cells and induce the regression of established osteosarcoma lung metastases. Cancer Res 2012;72(1):271–81. https://doi.org/10.1158/0008-5472.can-11-2778.
[70] Chang CH, Qiu J, O'Sullivan D, Buck MD, Noguchi T, Curtis JD, Pearce EL. Metabolic competition in the tumor microenvironment is a driver of cancer progression. Cell 2015;162(6):1229–41. https://doi.org/10.1016/j.cell.2015.08.016.
[71] Hensley CT, Wasti AT, DeBerardinis RJ. Glutamine and cancer: cell biology, physiology, and clinical opportunities. J Clin Invest 2013;123(9):3678–84. https://doi.org/10.1172/jci69600.
[72] Lob S, Konigsrainer A, Rammensee HG, Opelz G, Terness P. Inhibitors of indoleamine-2,3-dioxygenase for cancer therapy: can we see the wood for the trees? Nat Rev Cancer 2009;9(6):445–52. https://doi.org/10.1038/nrc2639.
[73] Munn DH, Mellor AL. Indoleamine 2,3-dioxygenase and tumor-induced tolerance. J Clin Invest 2007;117 (5):1147–54. https://doi.org/10.1172/jci31178.
[74] Pilotte L, Larrieu P, Stroobant V, Colau D, Dolusic E, Frederick R, Van den Eynde BJ. Reversal of tumoral immune resistance by inhibition of tryptophan 2,3-dioxygenase. Proc Natl Acad Sci U S A 2012;109(7):2497–502. https://doi.org/10.1073/pnas.1113873109.
[75] Mezrich JD, Fechner JH, Zhang X, Johnson BP, Burlingham WJ, Bradfield CA. An interaction between kynurenine and the aryl hydrocarbon receptor can generate regulatory T cells. J Immunol 2010;185(6):3190–8. https://doi.org/10.4049/jimmunol.0903670.
[76] Platten M, Wick W, Van den Eynde BJ. Tryptophan catabolism in cancer: beyond IDO and tryptophan depletion. Cancer Res 2012;72(21):5435–40. https://doi.org/10.1158/0008-5472.can-12-0569.

[77] Rashida Gnanaprakasam JN, Wu R, Wang R. Metabolic reprogramming in modulating t cell reactive oxygen species generation and antioxidant capacity. Front Immunol 2018;9(1075). https://doi.org/10.3389/fimmu.2018.01075.

[78] Srivastava MK, Sinha P, Clements VK, Rodriguez P, Ostrand-Rosenberg S. Myeloid-derived suppressor cells inhibit T-cell activation by depleting cystine and cysteine. Cancer Res 2010;70(1):68–77. https://doi.org/10.1158/0008-5472.can-09-2587.

[79] Feun L, You M, Wu CJ, Kuo MT, Wangpaichitr M, Spector S, Savaraj N. Arginine deprivation as a targeted therapy for cancer. Curr Pharm Des 2008;14(11):1049–57.

[80] Beavis PA, Stagg J, Darcy PK, Smyth MJ. CD73: a potent suppressor of antitumor immune responses. Trends Immunol 2012;33(5):231–7. https://doi.org/10.1016/j.it.2012.02.009.

[81] Ohta A, Gorelik E, Prasad SJ, Ronchese F, Lukashev D, Wong MK, Sitkovsky M. A2A adenosine receptor protects tumors from antitumor T cells. Proc Natl Acad Sci U S A 2006;103(35):13132–7. https://doi.org/10.1073/pnas.0605251103.

[82] Leclerc BG, Charlebois R, Chouinard G, Allard B, Pommey S, Saad F, Stagg J. CD73 expression is an independent prognostic factor in prostate cancer. Clin Cancer Res 2016;22(1):158–66. https://doi.org/10.1158/1078-0432.ccr-15-1181.

[83] Loi S, Pommey S, Haibe-Kains B, Beavis PA, Darcy PK, Smyth MJ, Stagg J. CD73 promotes anthracycline resistance and poor prognosis in triple negative breast cancer. Proc Natl Acad Sci U S A 2013;110(27):11091–6. https://doi.org/10.1073/pnas.1222251110.

[84] Turcotte M, Spring K, Pommey S, Chouinard G, Cousineau I, George J, Stagg J. CD73 is associated with poor prognosis in high-grade serous ovarian cancer. Cancer Res 2015;75(21):4494–503. https://doi.org/10.1158/0008-5472.can-14-3569.

[85] Sitkovsky MV, Hatfield S, Abbott R, Belikoff B, Lukashev D, Ohta A. Hostile, hypoxia-A2-adenosinergic tumor biology as the next barrier to overcome for tumor immunologists. Cancer Immunol Res 2014;2(7):598–605. https://doi.org/10.1158/2326-6066.cir-14-0075.

[86] Beavis PA, Divisekera U, Paget C, Chow MT, John LB, Devaud C, Darcy PK. Blockade of A2A receptors potently suppresses the metastasis of CD73+ tumors. Proc Natl Acad Sci U S A 2013;110(36):14711–6. https://doi.org/10.1073/pnas.1308209110.

[87] Beavis PA, Milenkovski N, Henderson MA, John LB, Allard B, Loi S, Darcy PK. Adenosine receptor 2A blockade increases the efficacy of anti-PD-1 through enhanced antitumor T-cell responses. Cancer Immunol Res 2015;3(5):
506–17. https://doi.org/10.1158/2326-6066.cir-14-0211.

[88] Kobayashi K, Omori K, Murata T. Role of prostaglandins in tumor microenvironment. Cancer Metastasis Rev 2018;37(2):347–54. https://doi.org/10.1007/s10555-018-9740-2.

[89] Mandapathil M, Szczepanski MJ, Szajnik M, Ren J, Jackson EK, Johnson JT, Whiteside TL. Adenosine and prostaglandin E2 cooperate in the suppression of immune responses mediated by adaptive regulatory T cells. J Biol Chem 2010;285(36):27571–80. https://doi.org/10.1074/jbc.M110.127100.

[90] Huber V, Camisaschi C, Berzi A, Ferro S, Lugini L, Triulzi T, Rivoltini L. Cancer acidity: an ultimate frontier of tumor immune escape and a novel target of immunomodulation. Semin Cancer Biol 2017;43:74–89. https://doi.org/10.1016/j.semcancer.2017.03.001.

[91] Parks SK, Cormerais Y, Marchiq I, Pouyssegur J. Hypoxia optimises tumour growth by controlling nutrient import and acidic metabolite export. Mol Aspects Med 2016;47–48:3–14. https://doi.org/10.1016/j.mam.2015.12.001.

[92] Halestrap AP. Monocarboxylic acid transport. Compr Physiol 2013;3(4):1611–43. https://doi.org/10.1002/cphy.c130008.

[93] Kim Y, Choi J-W, Lee J-H, Kim Y-S. Expression of lactate/H+ symporters MCT1 and MCT4 and their chaperone CD147 predicts tumor progression in clear cell renal cell carcinoma: immunohistochemical and The Cancer Genome Atlas data analyses. Hum Pathol 2015;46(1):104–12. https://doi.org/10.1016/j.humpath.2014.09.013.

[94] Hirschhaeuser F, Sattler UG, Mueller-Klieser W. Lactate: a metabolic key player in cancer. Cancer Res 2011;71(22):6921–5. https://doi.org/10.1158/0008-5472.can-11-1457.

[95] Marchiq I, Pouyssegur J. Hypoxia, cancer metabolism and the therapeutic benefit of targeting lactate/H(+) symporters. J Mol Med (Berl) 2016;94(2):155–71. https://doi.org/10.1007/s00109-015-1307-x.

[96] Husain Z, Huang Y, Seth P, Sukhatme VP. Tumor-derived lactate modifies antitumor immune response: effect on myeloid-derived suppressor cells and NK cells. J Immunol 2013;191(3):1486–95. https://doi.org/10.4049/jimmunol.1202702.
[97] Walenta S, Schroeder T, Mueller-Klieser W. Lactate in solid malignant tumors: potential basis of a metabolic classification in clinical oncology. Curr Med Chem 2004;11(16):2195–204.
[98] Walenta S, Wetterling M, Lehrke M, Schwickert G, Sundfor K, Rofstad EK, Mueller-Klieser W. High lactate levels predict likelihood of metastases, tumor recurrence, and restricted patient survival in human cervical cancers. Cancer Res 2000;60(4):916–21.
[99] Supuran CT. Carbonic anhydrases: novel therapeutic applications for inhibitors and activators. Nat Rev Drug Discov 2008;7(2):168–81. https://doi.org/10.1038/nrd2467.
[100] McDonald PC, Dedhar S. Carbonic anhydrase IX (CAIX) as a mediator of hypoxia-induced stress response in cancer cells. Subcell Biochem 2014;75:255–69. https://doi.org/10.1007/978-94-007-7359-2_13.
[101] Pastorek J, Pastorekova S. Hypoxia-induced carbonic anhydrase IX as a target for cancer therapy: from biology to clinical use. Semin Cancer Biol 2015;31:52–64. https://doi.org/10.1016/j.semcancer.2014.08.002.
[102] Noman MZ, Buart S, Van Pelt J, Richon C, Hasmim M, Leleu N, Chouaib S. The cooperative induction of hypoxia-inducible factor-1 alpha and STAT3 during hypoxia induced an impairment of tumor susceptibility to CTL-mediated cell lysis. J Immunol 2009;182(6):3510–21. https://doi.org/10.4049/jimmunol.0800854.
[103] Hasmim M, Noman MZ, Lauriol J, Benlalam H, Mallavialle A, Rosselli F, Chouaib S. Hypoxia-dependent inhibition of tumor cell susceptibility to CTL-mediated lysis involves NANOG induction in target cells. J Immunol 2011;187(8):4031–9.
[104] Hasmim M, Noman MZ, Messai Y, Bordereaux D, Gros G, Baud V, Chouaib S. Cutting edge: hypoxia-induced Nanog favors the intratumoral infiltration of regulatory T cells and macrophages via direct regulation of TGF-β 1. J Immunol 2013;191(12):5802–6.
[105] Noman MZ, Buart S, Romero P, Ketari S, Janji B, Mari B, Chouaib S. Hypoxia-inducible miR-210 regulates the susceptibility of tumor cells to lysis by cytotoxic T cells. Cancer Res 2012;72(18):4629–41.
[106] Mathew R, White E. Autophagy, stress, and cancer metabolism: what doesn't kill you makes you stronger. In: Paper presented at the Cold Spring Harbor symposia on quantitative biology; 2011.
[107] Viry E, Paggetti J, Baginska J, Mgrditchian T, Berchem G, Moussay E, Janji B. Autophagy: an adaptive metabolic response to stress shaping the antitumor immunity. Biochem Pharmacol 2014;92(1):31–42.
[108] Viry E, Baginska J, Berchem G, Noman MZ, Medves S, Chouaib S, Janji B. Autophagic degradation of GZMB/granzyme B: a new mechanism of hypoxic tumor cell escape from natural killer cell-mediated lysis. Autophagy 2014;10(1):173–5.
[109] Moon EK, Wang LC, Dolfi DV, Wilson CB, Ranganathan R, Sun J, Albelda SM. Multifactorial T-cell hypofunction that is reversible can limit the efficacy of chimeric antigen receptor-transduced human T cells in solid tumors. Clin Cancer Res 2014;20(16):4262–73. https://doi.org/10.1158/1078-0432.ccr-13-2627.
[110] Rafiq S, Yeku OO, Jackson HJ, Purdon TJ, van Leeuwen DG, Drakes DJ, Brentjens RJ. Targeted delivery of a PD-1-blocking scFv by CAR-T cells enhances anti-tumor efficacy in vivo. Nat Biotechnol 2018;36(9):847–56. https://doi.org/10.1038/nbt.4195.
[111] Suarez ER, Chang DK, Sun J, Sui J, Freeman GJ, Signoretti S, Marasco WA. Chimeric antigen receptor T cells secreting anti-PD-L1 antibodies more effectively regress renal cell carcinoma in a humanized mouse model. Oncotarget 2016;7(23):34341–55. https://doi.org/10.18632/oncotarget.9114.
[112] Sakaguchi S, Miyara M, Costantino CM, Hafler DA. FOXP3+ regulatory T cells in the human immune system. Nat Rev Immunol 2010;10(7):490–500. https://doi.org/10.1038/nri2785.
[113] Long KB, Collier AI, Beatty GL. Macrophages: key orchestrators of a tumor microenvironment defined by therapeutic resistance. Mol Immunol 2019;110:3–12. https://doi.org/10.1016/j.molimm.2017.12.003.
[114] Mantovani A, Sozzani S, Locati M, Allavena P, Sica A. Macrophage polarization: tumor-associated macrophages as a paradigm for polarized M2 mononuclear phagocytes. Trends Immunol 2002;23(11):549–55.
[115] Lu C, Redd PS, Lee JR, Savage N, Liu K. The expression profiles and regulation of PD-L1 in tumor-induced myeloid-derived suppressor cells. Oncoimmunology 2016;5(12), e1247135. https://doi.org/10.1080/2162402x.2016.1247135.
[116] Hanson EM, Clements VK, Sinha P, Ilkovitch D, Ostrand-Rosenberg S. Myeloid-derived suppressor cells down-regulate L-selectin expression on CD4+ and CD8+ T cells. J Immunol 2009;183(2):937–44. https://doi.org/10.4049/jimmunol.0804253.

[117] Huang B, Pan PY, Li Q, Sato AI, Levy DE, Bromberg J, Chen SH. Gr-1+CD115+ immature myeloid suppressor cells mediate the development of tumor-induced T regulatory cells and T-cell anergy in tumor-bearing host. Cancer Res 2006;66(2):1123–31. https://doi.org/10.1158/0008-5472.can-05-1299.

[118] Beury DW, Parker KH, Nyandjo M, Sinha P, Carter KA, Ostrand-Rosenberg S. Cross-talk among myeloid-derived suppressor cells, macrophages, and tumor cells impacts the inflammatory milieu of solid tumors. J Leukoc Biol 2014;96(6):1109–18. https://doi.org/10.1189/jlb.3A0414-210R.

[119] Raber PL, Thevenot P, Sierra R, Wyczechowska D, Halle D, Ramirez ME, Rodriguez PC. Subpopulations of myeloid-derived suppressor cells impair T cell responses through independent nitric oxide-related pathways. Int J Cancer 2014;134(12):2853–64. https://doi.org/10.1002/ijc.28622.

[120] Li J, Wang L, Chen X, Li L, Li Y, Ping Y, Zhang Y. CD39/CD73 upregulation on myeloid-derived suppressor cells via TGF-beta-mTOR-HIF-1 signaling in patients with non-small cell lung cancer. Oncoimmunology 2017;6 (6), e1320011. https://doi.org/10.1080/2162402x.2017.1320011.

[121] Groth C, Hu X, Weber R, Fleming V, Altevogt P, Utikal J, Umansky V. Immunosuppression mediated by myeloid-derived suppressor cells (MDSCs) during tumour progression. Br J Cancer 2019;120(1):16–25. https://doi.org/10.1038/s41416-018-0333-1.

[122] Liu Y, Wei J, Guo G, Zhou J. Norepinephrine-induced myeloid-derived suppressor cells block T-cell responses via generation of reactive oxygen species. Immunopharmacol Immunotoxicol 2015;37(4):359–65. https://doi.org/10.3109/08923973.2015.1059442.

[123] Cherkassky L, Morello A, Villena-Vargas J, Feng Y, Dimitrov DS, Jones DR, Adusumilli PS. Human CAR T cells with cell-intrinsic PD-1 checkpoint blockade resist tumor-mediated inhibition. J Clin Invest 2016;126(8): 3130–44. https://doi.org/10.1172/jci83092.

[124] Mirzaei HR, Rodriguez A, Shepphird J, Brown CE, Badie B. Chimeric antigen receptors T cell therapy in solid tumor: challenges and clinical applications. Front Immunol 2017;8:1850. https://doi.org/10.3389/fimmu.2017.01850.

[125] Kawalekar OU, O'Connor RS, Fraietta JA, Guo L, McGettigan SE, Posey AD, June CH. Distinct signaling of coreceptors regulates specific metabolism pathways and impacts memory development in CAR T cells. Immunity 2016;44(2):380–90. https://doi.org/10.1016/j.immuni.2016.01.021.

[126] Ligtenberg MA, Mougiakakos D, Mukhopadhyay M, Witt K, Lladser A, Chmielewski M, Kiessling R. Coexpressed catalase protects chimeric antigen receptor-redirected T cells as well as bystander cells from oxidative stress-induced loss of antitumor activity. J Immunol 2016;196(2):759–66. https://doi.org/10.4049/jimmunol.1401710.

[127] Juillerat A, Marechal A, Filhol JM, Valogne Y, Valton J, Duclert A, Poirot L. An oxygen sensitive self-decision making engineered CAR T-cell. Sci Rep 2017;7:39833. https://doi.org/10.1038/srep39833.

[128] Newick K, O'Brien S, Sun J, Kapoor V, Maceyko S, Lo A, Albelda SM. Augmentation of CAR T-cell trafficking and antitumor efficacy by blocking protein kinase A localization. Cancer Immunol Res 2016;4(6):541–51. https://doi.org/10.1158/2326-6066.cir-15-0263.

[129] Beavis PA, Henderson MA, Giuffrida L, Mills JK, Sek K, Cross RS, Darcy PK. Targeting the adenosine 2A receptor enhances chimeric antigen receptor T cell efficacy. J Clin Invest 2017;127(3):929–41. https://doi.org/10.1172/JCI89455.

[130] Ho PC, Bihuniak JD, Macintyre AN, Staron M, Liu X, Amezquita R, Kaech SM. Phosphoenolpyruvate is a metabolic checkpoint of anti-tumor T cell responses. Cell 2015;162(6):1217–28. https://doi.org/10.1016/j.cell.2015.08.012.

[131] Eil R, Vodnala SK, Clever D, Klebanoff CA, Sukumar M, Pan JH, Restifo NP. Ionic immune suppression within the tumour microenvironment limits T cell effector function. Nature 2016;537(7621):539–43. https://doi.org/10.1038/nature19364.

[132] Acharya AP, Rafi M, Woods EC, Gardner AB, Murthy N. Metabolic engineering of lactate dehydrogenase rescues mice from acidosis. Sci Rep 2014;4:5189. https://doi.org/10.1038/srep05189.

[133] Ruella M, Klichinsky M, Kenderian SS, Shestova O, Ziober A, Kraft DO, Gill S. Overcoming the immunosuppressive tumor microenvironment of hodgkin lymphoma using chimeric antigen receptor T cells. Cancer Discov 2017;7(10):1154–67. https://doi.org/10.1158/2159-8290.cd-16-0850.

[134] Tang N, Cheng C, Zhang X, Qiao M, Li N, Mu W, Wang H. TGF-β inhibition via CRISPR promotes the long-term efficacy of CAR T cells against solid tumors. JCI Insight 2020;5(4). https://doi.org/10.1172/jci.insight.133977.

[135] Li Y, Xiao F, Zhang A, Zhang D, Nie W, Xu T, Wang L. Oncolytic adenovirus targeting TGF-β enhances antitumor responses of mesothelin-targeted chimeric antigen receptor T cell therapy against breast cancer. Cell Immunol 2020;348:104041. https://doi.org/10.1016/j.cellimm.2020.104041.

[136] John LB, Devaud C, Duong CP, Yong CS, Beavis PA, Haynes NM, Darcy PK. Anti-PD-1 antibody therapy potently enhances the eradication of established tumors by gene-modified T cells. Clin Cancer Res 2013;19(20): 5636–46. https://doi.org/10.1158/1078-0432.ccr-13-0458.

[137] Li S, Siriwon N, Zhang X, Yang S, Jin T, He F, Wang P. Enhanced cancer immunotherapy by chimeric antigen receptor-modified T Cells engineered to secrete checkpoint inhibitors. Clin Cancer Res 2017;23(22): 6982–92. https://doi.org/10.1158/1078-0432.ccr-17-0867.

[138] Ren J, Zhang X, Liu X, Fang C, Jiang S, June CH, Zhao Y. A versatile system for rapid multiplex genome-edited CAR T cell generation. Oncotarget 2017;8(10):17002–11. https://doi.org/10.18632/oncotarget.15218.

[139] Zou F, Lu L, Liu J, Xia B, Zhang W, Hu Q, Jing S. Engineered triple inhibitory receptor resistance improves antitumor CAR-T cell performance via CD56. Nat Commun 2019;10(1):1–14.

[140] Cho A, Howell VM, Colvin EK. The extracellular matrix in epithelial ovarian cancer—a piece of a puzzle. Front Oncol 2015;5:245. https://doi.org/10.3389/fonc.2015.00245.

[141] Kaushik S, Pickup MW, Weaver VM. From transformation to metastasis: deconstructing the extracellular matrix in breast cancer. Cancer Metastasis Rev 2016;35(4):655–67. https://doi.org/10.1007/s10555-016-9650-0.

[142] Li J, Wientjes MG, Au JL. Pancreatic cancer: pathobiology, treatment options, and drug delivery. AAPS J 2010;12 (2):223–32. https://doi.org/10.1208/s12248-010-9181-5.

[143] Enblad G, Karlsson H, Loskog AS. CAR T-cell therapy: the role of physical barriers and immunosuppression in lymphoma. Hum Gene Ther 2015;26(8):498–505. https://doi.org/10.1089/hum.2015.054.

[144] Newick K, Moon E, Albelda SM. Chimeric antigen receptor T-cell therapy for solid tumors. Mol Ther Oncolytics 2016;3:16006. https://doi.org/10.1038/mto.2016.6.

[145] Stylianopoulos T, Martin JD, Chauhan VP, Jain SR, Diop-Frimpong B, Bardeesy N, Jain RK. Causes, consequences, and remedies for growth-induced solid stress in murine and human tumors. Proc Natl Acad Sci U S A 2012;109(38):15101–8. https://doi.org/10.1073/pnas.1213353109.

[146] Caruana I, Savoldo B, Hoyos V, Weber G, Liu H, Kim ES, Dotti G. Heparanase promotes tumor infiltration and antitumor activity of CAR-redirected T lymphocytes. Nat Med 2015;21(5):524–9. https://doi.org/10.1038/nm.3833.

[147] Eikenes L, Tari M, Tufto I, Bruland OS, de Lange Davies C. Hyaluronidase induces a transcapillary pressure gradient and improves the distribution and uptake of liposomal doxorubicin (Caelyx) in human osteosarcoma xenografts. Br J Cancer 2005;93(1):81–8. https://doi.org/10.1038/sj.bjc.6602626.

[148] Goodman TT, Olive PL, Pun SH. Increased nanoparticle penetration in collagenase-treated multicellular spheroids. Int J Nanomedicine 2007;2(2):265–74.

[149] Feig C, Jones JO, Kraman M, Wells RJ, Deonarine A, Chan DS, Fearon DT. Targeting CXCL12 from FAP-expressing carcinoma-associated fibroblasts synergizes with anti-PD-L1 immunotherapy in pancreatic cancer. Proc Natl Acad Sci U S A 2013;110(50):20212–7. https://doi.org/10.1073/pnas.1320318110.

[150] Wang LC, Lo A, Scholler J, Sun J, Majumdar RS, Kapoor V, Albelda SM. Targeting fibroblast activation protein in tumor stroma with chimeric antigen receptor T cells can inhibit tumor growth and augment host immunity without severe toxicity. Cancer Immunol Res 2014;2(2):154–66. https://doi.org/10.1158/2326-6066.cir-13-0027.

[151] Ozdemir BC, Pentcheva-Hoang T, Carstens JL, Zheng X, Wu CC, Simpson TR, Kalluri R. Depletion of carcinoma-associated fibroblasts and fibrosis induces immunosuppression and accelerates pancreas cancer with reduced survival. Cancer Cell 2015;28(6):831–3. https://doi.org/10.1016/j.ccell.2015.11.002.

[152] Rhim AD, Oberstein PE, Thomas DH, Mirek ET, Palermo CF, Sastra SA, Stanger BZ. Stromal elements act to restrain, rather than support, pancreatic ductal adenocarcinoma. Cancer Cell 2014;25(6):735–47. https://doi.org/10.1016/j.ccr.2014.04.021.

[153] Mikucki ME, Fisher DT, Matsuzaki J, Skitzki JJ, Gaulin NB, Muhitch JB, Evans SS. Non-redundant requirement for CXCR3 signalling during tumoricidal T-cell trafficking across tumour vascular checkpoints. Nat Commun 2015;6:7458. https://doi.org/10.1038/ncomms8458.

[154] Kershaw MH, Wang G, Westwood JA, Pachynski RK, Tiffany HL, Marincola FM, Hwu P. Redirecting migration of T cells to chemokine secreted from tumors by genetic modification with CXCR2. Hum Gene Ther 2002;13 (16):1971–80. https://doi.org/10.1089/10430340260355374.

[155] Craddock JA, Lu A, Bear A, Pule M, Brenner MK, Rooney CM, Foster AE. Enhanced tumor trafficking of GD2 chimeric antigen receptor T cells by expression of the chemokine receptor CCR2b. J Immunother 2010;33 (8):780–8. https://doi.org/10.1097/CJI.0b013e3181ee6675.

[156] Moon EK, Carpenito C, Sun J, Wang LC, Kapoor V, Predina J, Albelda SM. Expression of a functional CCR2 receptor enhances tumor localization and tumor eradication by retargeted human T cells expressing a mesothelin-specific chimeric antibody receptor. Clin Cancer Res 2011;17(14):4719–30. https://doi.org/10.1158/1078-0432.ccr-11-0351.

[157] Liu G, Rui W, Zheng H, Huang D, Yu F, Zhang Y, Lin X. CXCR2-modified CAR-T cells have enhanced trafficking ability that improves treatment of hepatocellular carcinoma. Eur J Immunol 2020;50(5):712–24. https://doi.org/10.1002/eji.201948457.

[158] Xu N, Palmer DC, Robeson AC, Shou P, Bommiasamy H, Laurie SJ, Serody JS. STING agonist promotes CAR T cell trafficking and persistence in breast cancer. J Exp Med 2020;218(2). https://doi.org/10.1084/jem.20200844.

[159] Adusumilli PS, Cherkassky L, Villena-Vargas J, Colovos C, Servais E, Plotkin J, Sadelain M. Regional delivery of mesothelin-targeted CAR T cell therapy generates potent and long-lasting CD4-dependent tumor immunity. Sci Transl Med 2014;6(261), 261ra151. https://doi.org/10.1126/scitranslmed.3010162.

[160] Fraietta JA, Lacey SF, Orlando EJ, Pruteanu-Malinici I, Gohil M, Lundh S, Melenhorst JJ. Determinants of response and resistance to CD19 chimeric antigen receptor (CAR) T cell therapy of chronic lymphocytic leukemia. Nat Med 2018;24(5):563–71. https://doi.org/10.1038/s41591-018-0010-1.

[161] Gardner RA, Finney O, Annesley C, Brakke H, Summers C, Leger K, Kelly-Spratt KS. Intent-to-treat leukemia remission by CD19 CAR T cells of defined formulation and dose in children and young adults. Blood 2017;129 (25):3322–31.

[162] Zhang C, Liu J, Zhong JF, Zhang X. Engineering CAR-T cells. Biomarker Res 2017;5:22. https://doi.org/10.1186/s40364-017-0102-y.

[163] Park JH, Geyer MB, Brentjens RJ. CD19-targeted CAR T-cell therapeutics for hematologic malignancies: interpreting clinical outcomes to date. Blood 2016;127(26):3312–20. https://doi.org/10.1182/blood-2016-02-629063.

[164] McLellan AD, Ali Hosseini Rad SM. Chimeric antigen receptor T cell persistence and memory cell formation. Immunol Cell Biol 2019. https://doi.org/10.1111/imcb.12254.

[165] Ladle BH, Li KP, Phillips MJ, Pucsek AB, Haile A, Powell JD, Gamper CJ. De novo DNA methylation by DNA methyltransferase 3a controls early effector CD8+ T-cell fate decisions following activation. Proc Natl Acad Sci U S A 2016;113(38):10631–6. https://doi.org/10.1073/pnas.1524490113.

[166] Wherry EJ, Kurachi M. Molecular and cellular insights into T cell exhaustion. Nat Rev Immunol 2015;15(8):486–99. https://doi.org/10.1038/nri3862.

[167] Leavy O. Benefits of exhaustion. Nat Rev Immunol 2015;15:468. https://doi.org/10.1038/nri3890.

[168] Kunkele A, Johnson AJ, Rolczynski LS, Chang CA, Hoglund V, Kelly-Spratt KS, Jensen MC. Functional tuning of CARs reveals signaling threshold above which CD8+ CTL antitumor potency is attenuated due to cell fas-fasl-dependent AICD. Cancer Immunol Res 2015;3(4):368–79. https://doi.org/10.1158/2326-6066.cir-14-0200.

[169] Krause A, Guo H-F, Latouche J-B, Tan C, Cheung N-KV, Sadelain M. Antigen-dependent CD28 signaling selectively enhances survival and proliferation in genetically modified activated human primary T lymphocytes. J Exp Med 1998;188(4):619–26.

[170] Roth A, Yssel H, Pene J, Chavez EA, Schertzer M, Lansdorp PM, Luiten RM. Telomerase levels control the lifespan of human T lymphocytes. Blood 2003;102(3):849–57. https://doi.org/10.1182/blood-2002-07-2015.

[171] Topp MS, Riddell SR, Akatsuka Y, Jensen MC, Blattman JN, Greenberg PD. Restoration of CD28 expression in CD28- CD8+ memory effector T cells reconstitutes antigen-induced IL-2 production. J Exp Med 2003;198 (6):947–55. https://doi.org/10.1084/jem.20021288.

[172] Ghassemi S, Nunez-Cruz S, O'Connor RS, Fraietta JA, Patel PR, Scholler J, Milone MC. Reducing ex vivo culture improves the antileukemic activity of chimeric antigen receptor (CAR) T cells. Cancer Immunol Res 2018;6 (9):1100–9. https://doi.org/10.1158/2326-6066.cir-17-0405.

[173] Bowers JS, Majchrzak K, Nelson MH, Aksoy BA, Wyatt MM, Smith AS, Paulos CM. PI3Kdelta inhibition enhances the antitumor fitness of adoptively transferred CD8(+) T cells. Front Immunol 2017;8:1221. https://doi.org/10.3389/fimmu.2017.01221.

[174] Gattinoni L, Zhong XS, Palmer DC, Ji Y, Hinrichs CS, Yu Z, Restifo NP. Wnt signaling arrests effector T cell differentiation and generates CD8+ memory stem cells. Nat Med 2009;15(7):808–13. https://doi.org/10.1038/nm.1982.
[175] Klebanoff CA, Scott CD, Leonardi AJ, Yamamoto TN, Cruz AC, Ouyang C, Restifo NP. Memory T cell-driven differentiation of naive cells impairs adoptive immunotherapy. J Clin Invest 2016;126(1):318–34. https://doi.org/10.1172/jci81217.
[176] Kaneko S, Mastaglio S, Bondanza A, Ponzoni M, Sanvito F, Aldrighetti L, Bonini C. IL-7 and IL-15 allow the generation of suicide gene-modified alloreactive self-renewing central memory human T lymphocytes. Blood 2009;113(5):1006–15. https://doi.org/10.1182/blood-2008-05-156059.
[177] Singh H, Figliola MJ, Dawson MJ, Huls H, Olivares S, Switzer K, Cooper LJ. Reprogramming CD19-specific T cells with IL-21 signaling can improve adoptive immunotherapy of B-lineage malignancies. Cancer Res 2011;71(10):3516–27. https://doi.org/10.1158/0008-5472.can-10-3843.
[178] Chmielewski M, Abken H. TRUCKs: the fourth generation of CARs. Expert Opin Biol Ther 2015;15(8):1145–54. https://doi.org/10.1517/14712598.2015.1046430.
[179] Batra SA, Rathi P, Guo L, Courtney AN, Fleurence J, Balzeau J, Heczey A. Glypican-3-specific CAR T cells coexpressing IL15 and IL21 have superior expansion and antitumor activity against hepatocellular carcinoma. Cancer Immunol Res 2020;8(3):309–20. https://doi.org/10.1158/2326-6066.Cir-19-0293.
[180] Loh Y-H, Hartung O, Li H, Guo C, Sahalie JM, Manos PD, Daley GQ. Reprogramming of T cells from human peripheral blood. Cell Stem Cell 2010;7(1):15–9. https://doi.org/10.1016/j.stem.2010.06.004.
[181] Kagoya Y, Nakatsugawa M, Yamashita Y, Ochi T, Guo T, Anczurowski M, Hirano N. BET bromodomain inhibition enhances T cell persistence and function in adoptive immunotherapy models. J Clin Invest 2016;126(9):3479–94. https://doi.org/10.1172/jci86437.
[182] Fraietta JA, Nobles CL, Sammons MA, Lundh S, Carty SA, Reich TJ, Melenhorst JJ. Disruption of TET2 promotes the therapeutic efficacy of CD19-targeted T cells. Nature 2018;558(7709):307–12. https://doi.org/10.1038/s41586-018-0178-z.
[183] Pace L, Goudot C, Zueva E, Gueguen P, Burgdorf N, Waterfall JJ, Amigorena S. The epigenetic control of stemness in CD8(+) T cell fate commitment. Science 2018;359(6372):177–86. https://doi.org/10.1126/science.aah6499.
[184] Lynn RC, Weber EW, Sotillo E, Gennert D, Xu P, Good Z, Mackall CL. c-Jun overexpression in CAR T cells induces exhaustion resistance. Nature 2019;576(7786):293–300. https://doi.org/10.1038/s41586-019-1805-z.
[185] Stadtmauer EA, Fraietta JA, Davis MM, Cohen AD, Weber KL, Lancaster E, June CH. CRISPR-engineered T cells in patients with refractory cancer. Science 2020;367(6481), eaba7365. https://doi.org/10.1126/science.aba7365.
[186] Schumacher TN, Schreiber RD. Neoantigens in cancer immunotherapy. Science 2015;348(6230):69–74. https://doi.org/10.1126/science.aaa4971.
[187] Morgan RA, Yang JC, Kitano M, Dudley ME, Laurencot CM, Rosenberg SA. Case report of a serious adverse event following the administration of T cells transduced with a chimeric antigen receptor recognizing ERBB2. Mol Ther 2010;18(4):843–51. https://doi.org/10.1038/mt.2010.24.
[188] Chen F, Zhang Y, Cai W. Molecular MRI of VEGFR-2 reveals intra-tumor and inter-tumor heterogeneity. Am J Nucl Med Mol Imaging 2013;3(4):312–6.
[189] Scanlan MJ, Raj BK, Calvo B, Garin-Chesa P, Sanz-Moncasi MP, Healey JH, Rettig WJ. Molecular cloning of fibroblast activation protein alpha, a member of the serine protease family selectively expressed in stromal fibroblasts of epithelial cancers. Proc Natl Acad Sci U S A 1994;91(12):5657–61. https://doi.org/10.1073/pnas.91.12.5657.
[190] Chinnasamy D, Yu Z, Kerkar SP, Zhang L, Morgan RA, Restifo NP, Rosenberg SA. Local delivery of interleukin-12 using T cells targeting VEGF receptor-2 eradicates multiple vascularized tumors in mice. Clin Cancer Res 2012;18(6):1672–83. https://doi.org/10.1158/1078-0432.ccr-11-3050.
[191] Santos AM, Jung J, Aziz N, Kissil JL, Pure E. Targeting fibroblast activation protein inhibits tumor stromagenesis and growth in mice. J Clin Invest 2009;119(12):3613–25. https://doi.org/10.1172/jci38988.
[192] Suzuki M, Cheung NK. Disialoganglioside GD2 as a therapeutic target for human diseases. Expert Opin Ther Targets 2015;19(3):349–62. https://doi.org/10.1517/14728222.2014.986459.
[193] Thomas S, Straathof K, Himoudi N, Anderson J, Pule M. An optimized GD2-targeting retroviral cassette for more potent and safer cellular therapy of neuroblastoma and other cancers. PLoS One 2016;11(3), e0152196. https://doi.org/10.1371/journal.pone.0152196.

[194] Mount CW, Majzner RG, Sundaresh S, Arnold EP, Kadapakkam M, Haile S, Mackall CL. Potent antitumor efficacy of anti-GD2 CAR T cells in H3-K27M(+) diffuse midline gliomas. Nat Med 2018;24(5):572–9. https://doi.org/10.1038/s41591-018-0006-x.
[195] Yu J, Wu X, Yan J, Yu H, Xu L, Chi Z, Guo J. Anti-GD2/4-1BB chimeric antigen receptor T cell therapy for the treatment of Chinese melanoma patients. J Hematol Oncol 2018;11(1):1. https://doi.org/10.1186/s13045-017-0548-2.
[196] Porcellini S, Asperti C, Corna S, Cicoria E, Valtolina V, Stornaiuolo A, Traversari C. CAR T cells redirected to CD44v6 control tumor growth in lung and ovary adenocarcinoma bearing mice. Front Immunol 2020;11:99. https://doi.org/10.3389/fimmu.2020.00099.
[197] Wang Z, Zhao K, Hackert T, Zöller M. CD44/CD44v6 a reliable companion in cancer-initiating cell maintenance and tumor progression. Front Cell Dev Biol 2018;6:97.
[198] Li D, Li N, Zhang Y-F, Fu H, Feng M, Schneider D, Ho M. Persistent polyfunctional chimeric antigen receptor T cells that target glypican 3 eliminate orthotopic hepatocellular carcinomas in mice. Gastroenterology 2020. https://doi.org/10.1053/j.gastro.2020.02.011.
[199] Wu D, Lv J, Zhao R, Wu Z, Zheng D, Shi J, Yao Y. PSCA is a target of chimeric antigen receptor T cells in gastric cancer. Biomarker Res 2020;8:3. https://doi.org/10.1186/s40364-020-0183-x.
[200] Huang B, Luo L, Wang J, He B, Feng R, Xian N, Huang G. B7-H3 specific T cells with chimeric antigen receptor and decoy PD-1 receptors eradicate established solid human tumors in mouse models. Oncoimmunology 2019;9(1), 1684127. https://doi.org/10.1080/2162402X.2019.1684127.
[201] Zhao Z, Li Y, Liu W, Li X. Engineered IL-7 receptor enhances the therapeutic effect of AXL-CAR-T cells on triple-negative breast cancer. Biomed Res Int 2020;2020:4795171. https://doi.org/10.1155/2020/4795171.
[202] Gautam SK, Kumar S, Dam V, Ghersi D, Jain M, Batra SK. MUCIN-4 (MUC4) is a novel tumor antigen in pancreatic cancer immunotherapy. Semin Immunol 2020;, 101391. https://doi.org/10.1016/j.smim.2020.101391.
[203] Reinhard K, Rengstl B, Oehm P, Michel K, Billmeier A, Hayduk N, Sahin U. An RNA vaccine drives expansion and efficacy of claudin-CAR-T cells against solid tumors. Science 2020;367(6476):446–53. https://doi.org/10.1126/science.aay5967.
[204] Wang D, Starr R, Chang W-C, Aguilar B, Alizadeh D, Wright SL, Brown CE. Chlorotoxin-directed CAR T cells for specific and effective targeting of glioblastoma. Sci Transl Med 2020;12(533), eaaw2672. https://doi.org/10.1126/scitranslmed.aaw2672.
[205] Fedorov VD, Sadelain M, Kloss CC. Novel approaches to enhance the specificity and safety of engineered T cells. Cancer J 2014;20(2):160–5. https://doi.org/10.1097/ppo.0000000000000040.
[206] Sadelain M. Chimeric antigen receptors: driving immunology towards synthetic biology. Curr Opin Immunol 2016;41:68–76. https://doi.org/10.1016/j.coi.2016.06.004.
[207] Grada Z, Hegde M, Byrd T, Shaffer DR, Ghazi A, Brawley VS, Ahmed N. TanCAR: a novel bispecific chimeric antigen receptor for cancer immunotherapy. Mol Ther Nucleic Acids 2013;2, e105. https://doi.org/10.1038/mtna.2013.32.
[208] Shu R, Evtimov VJ, Hammett MV, Nguyen N-YN, Zhuang J, Hudson PJ, Boyd RL. Engineered CAR-T cells targeting TAG-72 and CD47 in ovarian cancer. Mol Ther Oncolytics 2021;20:325–41. https://doi.org/10.1016/j.omto.2021.01.002.
[209] Leboeuf C, Mailly L, Wu T, Bour G, Durand S, Brignon N, Robinet E. In vivo proof of concept of adoptive immunotherapy for hepatocellular carcinoma using allogeneic suicide gene-modified killer cells. Mol Ther 2014;22(3):634–44. https://doi.org/10.1038/mt.2013.277.
[210] Daher M, Rezvani K. Next generation natural killer cells for cancer immunotherapy: the promise of genetic engineering. Curr Opin Immunol 2018;51:146–53. https://doi.org/10.1016/j.coi.2018.03.013.
[211] Mehta RS, Rezvani K. Chimeric antigen receptor expressing natural killer cells for the immunotherapy of cancer. Front Immunol 2018;9:283. https://doi.org/10.3389/fimmu.2018.00283.
[212] Curti A, Ruggeri L, D'Addio A, Bontadini A, Dan E, Motta MR, Lemoli RM. Successful transfer of alloreactive haploidentical KIR ligand-mismatched natural killer cells after infusion in elderly high risk acute myeloid leukemia patients. Blood 2011;118(12):3273–9. https://doi.org/10.1182/blood-2011-01-329508.
[213] Miller JS, Soignier Y, Panoskaltsis-Mortari A, McNearney SA, Yun GH, Fautsch SK, McGlave PB. Successful adoptive transfer and in vivo expansion of human haploidentical NK cells in patients with cancer. Blood 2005;105(8):3051–7. https://doi.org/10.1182/blood-2004-07-2974.

[214] Olson JA, Leveson-Gower DB, Gill S, Baker J, Beilhack A, Negrin RS. NK cells mediate reduction of GVHD by inhibiting activated, alloreactive T cells while retaining GVT effects. Blood 2010;115(21):4293–301. https://doi.org/10.1182/blood-2009-05-222190.
[215] Ruggeri L, Mancusi A, Burchielli E, Capanni M, Carotti A, Aloisi T, Velardi A. NK cell alloreactivity and allogeneic hematopoietic stem cell transplantation. Blood Cells Mol Dis 2008;40(1):84–90. https://doi.org/10.1016/j.bcmd.2007.06.029.
[216] Couzin-Frankel J. Supply of promising T cell therapy is strained. Science 2017;356(6343):1112–3. https://doi.org/10.1126/science.356.6343.1112.
[217] Chu Y, Hochberg J, Yahr A, Ayello J, van de Ven C, Barth M, Cairo MS. Targeting CD20+ aggressive B-cell non-Hodgkin lymphoma by anti-CD20 CAR mRNA-modified expanded natural killer cells in vitro and in NSG mice. Cancer Immunol Res 2015;3(4):333–44. https://doi.org/10.1158/2326-6066.cir-14-0114.
[218] Liu E, Tong Y, Dotti G, Shaim H, Savoldo B, Mukherjee M, Rezvani K. Cord blood NK cells engineered to express IL-15 and a CD19-targeted CAR show long-term persistence and potent antitumor activity. Leukemia 2018;32(2):520–31. https://doi.org/10.1038/leu.2017.226.
[219] Muller T, Uherek C, Maki G, Chow KU, Schimpf A, Klingemann HG, Wels WS. Expression of a CD20-specific chimeric antigen receptor enhances cytotoxic activity of NK cells and overcomes NK-resistance of lymphoma and leukemia cells. Cancer Immunol Immunother 2008;57(3):411–23. https://doi.org/10.1007/s00262-007-0383-3.
[220] Shimasaki N, Fujisaki H, Cho D, Masselli M, Lockey T, Eldridge P, Campana D. A clinically adaptable method to enhance the cytotoxicity of natural killer cells against B-cell malignancies. Cytotherapy 2012;14(7):830–40. https://doi.org/10.3109/14653249.2012.671519.
[221] Chen KH, Wada M, Firor AE, Pinz KG, Jares A, Liu H, Ma Y. Novel anti-CD3 chimeric antigen receptor targeting of aggressive T cell malignancies. Oncotarget 2016;7(35):56219–32. https://doi.org/10.18632/oncotarget.11019.
[222] Chen KH, Wada M, Pinz KG, Liu H, Lin KW, Jares A, Ma Y. Preclinical targeting of aggressive T-cell malignancies using anti-CD5 chimeric antigen receptor. Leukemia 2017;31(10):2151–60. https://doi.org/10.1038/leu.2017.8.
[223] Chu J, Deng Y, Benson DM, He S, Hughes T, Zhang J, Yu J. CS1-specific chimeric antigen receptor (CAR)-engineered natural killer cells enhance in vitro and in vivo antitumor activity against human multiple myeloma. Leukemia 2014;28(4):917–27. https://doi.org/10.1038/leu.2013.279.
[224] Jiang H, Zhang W, Shang P, Zhang H, Fu W, Ye F, Hou J. Transfection of chimeric anti-CD138 gene enhances natural killer cell activation and killing of multiple myeloma cells. Mol Oncol 2014;8(2):297–310. https://doi.org/10.1016/j.molonc.2013.12.001.
[225] Kruschinski A, Moosmann A, Poschke I, Norell H, Chmielewski M, Seliger B, Charo J. Engineering antigen-specific primary human NK cells against HER-2 positive carcinomas. Proc Natl Acad Sci U S A 2008;105(45):17481–6. https://doi.org/10.1073/pnas.0804788105.
[226] Liu H, Yang B, Sun T, Lin L, Hu Y, Deng M, Jiao S. Specific growth inhibition of ErbB2 expressing human breast cancer cells by genetically modified NK92 cells. Oncol Rep 2015;33(1):95–102. https://doi.org/10.3892/or.2014.3548.
[227] Esser R, Muller T, Stefes D, Kloess S, Seidel D, Gillies SD, Wels WS. NK cells engineered to express a GD2 -specific antigen receptor display built-in ADCC-like activity against tumour cells of neuroectodermal origin. J Cell Mol Med 2012;16(3):569–81. https://doi.org/10.1111/j.1582-4934.2011.01343.x.
[228] Chen X, Han J, Chu J, Zhang L, Zhang J, Chen C, Yu J. A combinational therapy of EGFR-CAR NK cells and oncolytic herpes simplex virus 1 for breast cancer brain metastases. Oncotarget 2016;7(19):27764–77. https://doi.org/10.18632/oncotarget.8526.
[229] Han J, Chu J, Keung Chan W, Zhang J, Wang Y, Cohen JB, Yu J. CAR-engineered NK cells targeting wild-type EGFR and EGFRvIII enhance killing of glioblastoma and patient-derived glioblastoma stem cells. Sci Rep 2015;5:11483. https://doi.org/10.1038/srep11483.
[230] Liu E, Marin D, Banerjee P, Macapinlac HA, Thompson P, Basar R, Rezvani K. Use of CAR-transduced natural killer cells in CD19-positive lymphoid tumors. N Engl J Med 2020;382(6):545–53. https://doi.org/10.1056/NEJMoa1910607.
[231] Metelitsa LS, Wu HW, Wang H, Yang Y, Warsi Z, Asgharzadeh S, Seeger RC. Natural killer T cells infiltrate neuroblastomas expressing the chemokine CCL2. J Exp Med 2004;199(9):1213–21. https://doi.org/10.1084/jem.20031462.

[232] Song L, Asgharzadeh S, Salo J, Engell K, Wu HW, Sposto R, Metelitsa LS. Valpha24-invariant NKT cells mediate antitumor activity via killing of tumor-associated macrophages. J Clin Invest 2009;119(6):1524–36. https://doi.org/10.1172/jci37869.

[233] Heczey A, Courtney AN, Montalbano A, Robinson S, Liu K, Li M, Metelitsa LS. Anti-GD2 CAR-NKT cells in patients with relapsed or refractory neuroblastoma: an interim analysis. Nat Med 2020;26(11):1686–90. https://doi.org/10.1038/s41591-020-1074-2.

[234] Heczey A, Liu D, Tian G, Courtney AN, Wei J, Marinova E, Metelitsa LS. Invariant NKT cells with chimeric antigen receptor provide a novel platform for safe and effective cancer immunotherapy. Blood 2014;124 (18):2824–33. https://doi.org/10.1182/blood-2013-11-541235.

[235] Rotolo A, Caputo VS, Holubova M, Baxan N, Dubois O, Chaudhry MS, Karadimitris A. Enhanced anti-lymphoma activity of CAR19-iNKT cells underpinned by dual CD19 and CD1d targeting. Cancer Cell 2018;34(4):596–610.e511. https://doi.org/10.1016/j.ccell.2018.08.017.

[236] Capsomidis A, Benthall G, Van Acker HH, Fisher J, Kramer AM, Abeln Z, Anderson J. Chimeric antigen receptor-engineered human gamma delta T cells: enhanced cytotoxicity with retention of cross presentation. Mol Ther 2018;26(2):354–65. https://doi.org/10.1016/j.ymthe.2017.12.001.

[237] Klichinsky M, Ruella M, Shestova O, Kenderian SS, Kim MY, O'Connor R, Gill S. Abstract 4575: chimeric antigen receptor macrophages (CARMA) for adoptive cellular immunotherapy of solid tumors. Cancer Res 2017;77(13 Suppl):4575. https://doi.org/10.1158/1538-7445.am2017-4575.

CHAPTER

6

Tumor-infiltrating lymphocyte (TIL) therapy

Kelly Mahuron and Yuman Fong

Department of Surgery, City of Hope Comprehensive Cancer Center, Duarte, CA, United States

Abstract

Adoptive cell therapy with autologous tumor-infiltrating lymphocytes (TIL ACT) is an emerging area of immunotherapy for cancer. This chapter will summarize the evolution of this therapy over the last 3 decades. Early studies utilized IL-2 to stimulate in situ T-cells to target cancer. The current approach involves isolating TILs from surgically harvested tumor, stimulating and expanding the TILs, and reinfusion as ACT. Many processes have been tried to make the TILs more active against tumor including engineering the cells to targeting neoantigens, stimulation of the TILs ex vivo to target tumor neoantigens. Methods have also been devised to enhance TIL engraftment by alteration of the tumor immune microenvironment and to enhance TIL longevity by augmenting TIL products for desired cellular phenotypes. The importance to lymphodepletion of the patient prior to TIL administration will also be summarized. The most important trials supporting these protocol steps in TIL generation and administration are summarized.

Abbreviations

ACT	adoptive cell therapy
CTL	cytolytic T-lymphocyte
CTLA-4	cytotoxic T-lymphocyte-associated protein 4
CXCL1	CXC-motif chemokine ligand 1
CXCL8	CXC-motif chemokine ligand 8
CXCR2	CXC chemokine receptor 2
HER2	human epidermal growth factor 2
ICI	immune checkpoint inhibitors
IL-2	interleukin 2
IL-15	interleukin 15
NK	natural killer cell
NMA	nonmyeloablative
PD-1	programmed cell death 1
PD-L1	programmed-death ligand 1
TCR	T-cell receptor
TGF-β	tissue growth factor-beta

https://doi.org/10.1016/B978-0-12-822620-9.00026-4

TIL tumor-infiltrating lymphocyte
TME tumor microenvironment
Treg T regulatory cells

Conflict of interest

No potential conflicts of interest were disclosed.

Introduction

T cells are the workhorses of tumor immune surveillance [1] and the pillar upon which modern cancer immunotherapy has been built [2]. Adoptive cell therapy with autologous tumor-infiltrating lymphocytes (TIL ACT) is an emerging area of immunotherapy that has had major advancements over the past decade. Multiple promising clinical trial programs involving TIL ACT are currently underway.

Cytotoxic T lymphocytes mediate antigen-specific antitumor immunity

Due to their antigen specificity and cytotoxic capabilities, $CD8^+$ T cells are well equipped for antitumor immunity as they develop into effector cells. Naïve $CD8^+$ T cells circulate throughout the body and recognize malignant cells when their T cell receptors (TCRs) encounter compatible short peptide sequences derived from tumor-associated antigens. These peptide sequences are displayed as MHC-peptide complexes by specialized antigen-presenting cells, such as dendritic cells, or by nucleated cells such as the tumor cells themselves. Successful TCR-peptide-MHC interactions lead to antigen-specific activation and clonal expansion of these cells into effector $CD8^+$ T cells [3]. These effector $CD8^+$ T cells, called cytotoxic T lymphocytes (CTL), gain potent cytolytic potential and migrate to tumor sites to directly facilitate cancer cell destruction in a tumor-antigen-specific fashion [4].

Tumor immune evasion and T cell exhaustion

Cancer cells utilize multiple immune evasion strategies to overcome the body's inherent antitumor immune defenses, many of which are the same natural processes responsible for tolerance and protection against autoimmunity [5]. These strategies include regulatory cells recruitment and suppressive cytokine production and results in an immunosuppressive tumor microenvironment that inhibits CTL antitumor functions [6].

Another tolerance-promoting process that impedes CTL function is T cell exhaustion, which refers to an acquired state of differentiation and dysfunction that occurs when T cells are exposed to chronic antigen. CTLs with sustained exposure to tumor-antigen have increased expression of inhibitory checkpoint receptors such as programmed cell death protein 1 (PD-1) and cytotoxic T-lymphocyte-associated antigen 4 (CTLA-4), decreased cytokine production, and decreased effector and cytolytic capability [7]. These CTLs become dysfunctional and have decreased ability to elicit antitumor responses. Overcoming T cell exhaustion and other tumor immune evasion mechanisms to enhance CTL-mediated antitumor responses is therefore the goal of many immunotherapy approaches.

Antigen nonspecific T cell immunotherapy approaches

Early immunotherapy approaches involved the systemic administration of the T cell growth factor and cytokine interleukin 2 (IL-2). IL-2 was discovered in 1976 [8] and received FDA approval for treatment of metastatic renal cell carcinoma in 1992 and metastatic melanoma in 1998. This approach nonspecifically activated CTLs and demonstrated durable complete responses for up to 10% of patients with metastatic disease [9]. However, IL-2 therapy has been limited by its toxicity profile and low-response rates.

By 2011, the first checkpoint inhibitor had been approved for the treatment of metastatic melanoma. Checkpoint blockade involves systemic administration of monoclonal antibodies that target the inhibitory receptors upregulated during the process of T cell exhaustion. A few years following the approval of ipilimumab, an anti-CTLA-4 monoclonal antibody, anti-PD-1 antibodies such as pembrolizumab and nivolumab were also approved for metastatic melanoma. These checkpoint inhibitors are often given together as combination therapy, and overall survival rates up to 52% have been demonstrated in patients with advanced melanoma [10]. Checkpoint blockade has revolutionized cancer therapy, and its use has expanded to other cancer types including nonsmall cell lung cancer, squamous cell head and neck cancer, and triple-negative breast cancer.

Although IL-2 therapy and checkpoint blockade demonstrate promising responses in select patient groups, the majority of patients and tumor types do not respond to these therapies. For these therapies to be successful, tumor-reactive CTLs must already be present at the tumor site, and their activation must overcome the suppressive tumor immune microenvironment promoted by the tumor itself [11]. Furthermore, these therapies act in an antigen nonspecific manner, which may result in lower efficiency and greater side effects. Adoptive cell therapy (ACT) has therefore emerged as an immunotherapy approach of interest as it involves administration of a large number of tumor-antigen specific T cells and allows for alteration of the host to overcome these tolerogenic mechanisms.

Adoptive cell therapy with autologous TILs (TIL ACT)

The practice of ACT involves the following main components: (1) identification and extraction of host tumor-antigen specific T cells, (2) expansion of these T cells to large numbers ex vivo, and (3) administration of the expanded autologous T cells back to the host to traffic to the tumor site for enhanced antitumor immunity (Fig. 1).

The first component can be achieved as tumor-antigen specific T cells arrest their migration when they encounter compatible MHC-peptide complexes. This leads to their accumulation at tumors site and within the surrounding stromal tissue, and these tumor-infiltrating lymphocytes (TILs) can be collected through excision of the primary tumor or metastatic sites [12]. The second component of this process is possible due to the discovery of IL-2; this cytokine

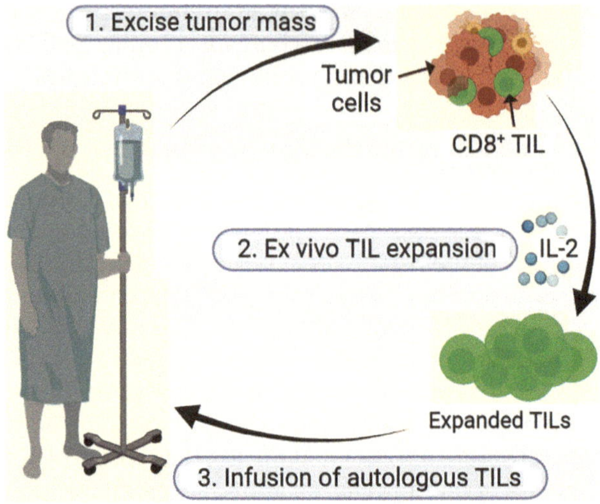

FIG. 1 Graphical depiction of the three main components of adoptive cell therapy with autologous TILs (TIL ACT).

TABLE 1 Advantages and challenges of TIL ACT.

Advantages	Challenges
Isolation of autologous tumor-antigen-specific TILs	Accessing tumor sites for excision/biopsy
Ability to screen TILs for desired characteristics prior to expansion	Obtaining adequate number of viable TILs
Ex vivo expansion to produce large numbers of activated TILs	Tumors must be immunogenic with TILs present
Peripheral infusion of autologous TIL products with migration to tumor site	Antigen-specific TILs must expand with ex vivo activation
Lymphodepleting immunotherapy to optimize host tumor immune microenvironment	Patients must tolerate lymphodepleting chemotherapy, IL-2, and timeframe for TIL expansion

permits the ex vivo expansion and growth of TILs in culture [13]. The efficacy of TIL expansion has since been improved with rapid expansion protocols utilizing anti-CD3 stimulation and irradiated feeder cells [14].

ACT using autologous TILs (TIL ACT) offers many advantages as an immunotherapy approach (Table 1). As described above, tumor-antigen specific TILs can be directly extracted from a host. Prior to expansion, TILs can be additionally selected for phenotypes such as tumor reactivity or expression of desired receptors. The expansion step also includes ex vivo activation, which in addition to the large quantify of infused TILs is thought to overcome the inhibition promoted by the in vivo immunosuppressive tumor microenvironment. Lastly, ACT allows for optimization of the host tumor immune microenvironment prior to

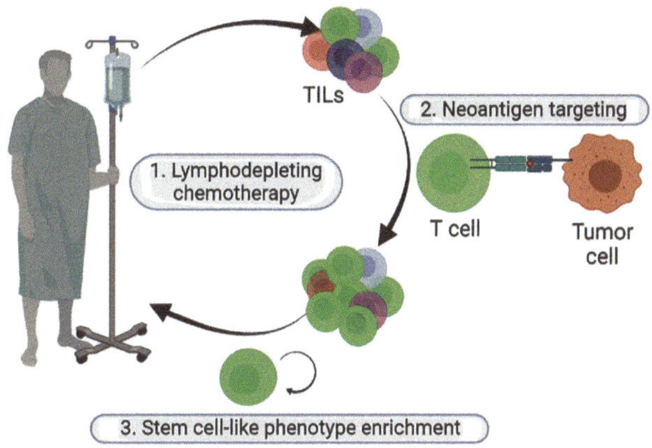

FIG. 2 Three strategies to optimize the efficacy and durability of TIL ACT.

TIL infusion to improve engraftment of the cells. This is performed with lymphodepleting chemotherapy and will be discussed in detail in the next section of this chapter.

Challenges to performing successful TIL ACT include that an adequate number of viable TILs must be present in the tumor immune microenvironment, which can be a limitation for small necrotic or nonimmunogenic tumors (Table 1). Tumor sites must be accessible for excision or biopsy, and patients must tolerate the conditioning process with lymphodepleting chemotherapy and high-dose IL-2. Additionally, initial protocols took up to 4–6 weeks for TIL production due to the selection process. In this timeframe, many patients developed disease progression and were no longer candidates for therapy. The complexity of TIL production also limits ACT to select centers. Therefore, not all patients are good candidates for or have had accessibility to TIL ACT, and initial successes have been with immunogenic and T-cell-rich tumors such as metastatic melanoma.

The initial studies utilizing TIL ACT for metastatic melanoma were pioneered by the lab of Dr. Steven Rosenberg of the National Cancer Institute. In 1986, Dr. Rosenberg and colleagues demonstrated that TILs isolated from mice and expanded ex vivo with IL-2 had cytotoxic potential and could be adoptively transferred back to the mouse host with significant antitumor response [15]. These findings led to the first phase I clinical trial of TIL ACT and included 86 patients with metastatic melanoma who received two cycles of autologous TILs plus high-dose IL-2 [16,17]. Overall, the authors found an objective response rate (ORR) of 34%. Their results demonstrated the clinical potential of adoptive immunotherapy but were significantly limited by low rates of TIL persistence and durable responses.

Therefore, the following three areas have been the focus of research strategies for the past 2 decades to improve the efficacy of TIL ACT as an immunotherapy approach, and these areas will be the focus of the remainder of this chapter (Fig. 2):

- Optimization of TIL engraftment by alteration of the tumor immune microenvironment.
- Optimization of TIL efficacy through targeting specific neoantigens.
- Optimization of TIL longevity by augmenting TIL products for desired cellular phenotypes.

Alteration of the immune tumor microenvironment to improves the engraftment of autologous TILs

While the first TIL ACT phase I clinical trials were encouraging and demonstrated a higher response rate than high-dose IL-2 alone, infused TILs had low rates of engraftment and short life spans. Follow-up studies utilizing retroviral insertion of the neomycin phosphotransferase to tag the infused TILs showed that less than 0.01% of circulating T cells were from the infused TILs at 1 week after administration [18]. Therefore, the ACT protocol was modified to include a nonmyeloablative (NMA) lymphodepleting regimen based upon success seen in murine models [19,20].

Lymphodepletion is thought to improve the lifespan and antitumor efficacy of transferred autologous T cells through multiple mechanisms. Murine studies have demonstrated that lymphodepletion eliminates suppressive T regulatory cells [21]. Additionally, removal of endogenous lymphocytes that typically act as "cytokine sinks" and consume IL-7 and IL-15 results in higher availability of these cytokines [22]. These cytokines are important for the survival and proliferation of $CD8^+$ T cells, and increased availability augments the antitumor effector functions of the infused autologous TILs.

Rosenberg and colleagues performed a series of three sequential clinical trials that included 93 patients with metastatic melanoma refractory to standard therapies that incorporated NMA lymphodepleting chemotherapy with cyclophosphamide and fludarabine in their ACT protocol. Definitive analysis of these trials was performed in 2011, with 19 of the 20 patients who achieved a complete response having ongoing regression beyond 3 years [23]. A randomized trial looked at the impact of augmenting lymphodepletion with total body irradiation, but no differences in ORRs or OS were observed [24]. NMA-lymphodepletion is now a standard component of modern-day TIL ACT protocols.

Additional efforts to overcome the immunosuppressive conditions of the tumor microenvironment include adoptive transfer with genetically modified TILs (Fig. 3). One example involves the retroviral transduction of a dominant-negative form of the TGF-β receptor into TILs. TGF-β is an inhibitory cytokine that downregulates T cell function and is central to tumor immune evasion. Genetically modified TILs expressing the dominant-negative form of its receptor can therefore resist TGF-β and its suppressive effects. Another approach involves retroviral transduction of the chemokine receptor CXCR2. Tumor cells are known to promote angiogenesis and autocrine by producing the chemokines CXCL1 and CXCL8. By overexpressing CXCR2, the ligand for these chemokines, TILs display enhanced tumor trafficking and antitumor immunity. Clinical trials are currently underway utilizing these approaches (Table 2). Clinical trials are also underway looking at gene editing TILs to be resistant to checkpoints, including cell-surface check points PD-1, PD-L1 [25], as well as intracellular check points such as CISH [26].

Improving antitumor efficacy of TIL ACT by targeting neoantigens

While the earliest TIL ACT clinical trials were initially performed with unselected TILs, protocols began screening TILs for tumor reactivity prior to ex vivo expansion. This was performed by coculturing TILs with autologous tumor or tumor cell lines and performing

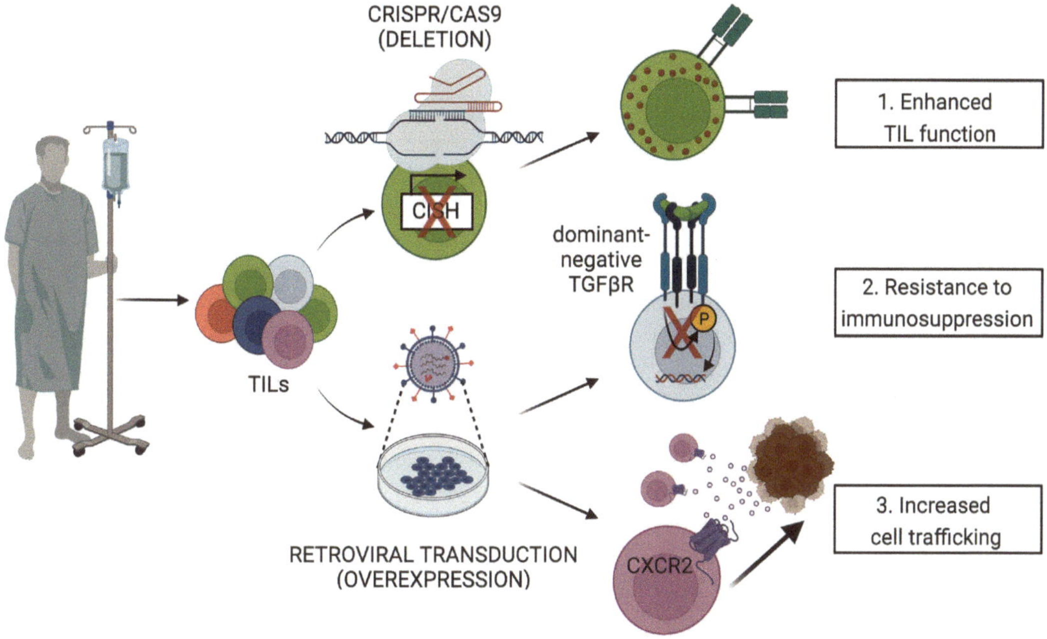

FIG. 3 Examples utilizing CRISPR/CAS9 or retroviral transduction for genetic modification of TILs. (1) TIL function and tumor reactivity are enhanced by CRISPR/CAS9-mediated deletion of internal checkpoint inhibitor CISH. (2) Overexpression of dominant-negative TGF- β receptor resulting in resistance to TGF-β and its immunosuppression. (3) Overexpression of CXCR2 to increase TIL trafficking into tumor sites.

TABLE 2 Current TIL ACT clinical trials that modulate the tumor immune microenvironment.

NCT identifier	TIL therapy	Description	Cancer	Trial phase
01740557	TILs with overexpressed CXCR2	CXCR2-transduced autologous TILs	Melanoma	1/2
01955460	TILs expressing TGFbDNRII	TGFbDNRII-transduced autologous TILs	NY-ESO-1-associated malignancies	1
02650986	TILs expressing TGFbDNRII	TGFbDNRII-transduced autologous TILs	Melanoma	1/2

TGFbDNRII, dominant-negative TGF-β receptor; *NY-ESO-1*, New York esophageal squamous cell carcinoma-1.

cytokine release assays [14]. However, even though TILs selected for tumor reactivity demonstrated strong antitumor potency ex vivo, the in vivo antigenic targets are unknown [27]. Therefore, there has been increased interest in the selection of TILs targeting specific tumor antigens to improve TIL ACT efficacy.

De novo somatic mutations called neoantigens are thought to be ideal targets for immunotherapy as these antigens are uniquely expressed by tumor cells and not by normal tissue. Studies characterizing patients with complete and durable responses to TIL ACT have confirmed the persistence of clonally expanded TILs that recognized neoantigens, supporting the concept that responses to TIL ACT are largely driven by neoantigen targets [28]. ACT utilizing TILs enriched for neoantigen-specific T cells may be especially advantageous for cancers with a low mutational burden as the TILs obtained from their biopsies may have a lower likelihood of tumor reactivity. Therefore, there has been significant interest in the development of personalized TIL infusion products targeting specific tumor mutations, but consistent identification of neoantigen-reactive TILs remains a challenge.

Initial methods to identify neoantigen-reactive TILs involved screening cDNA libraries, but this was time consuming and ineffective. Emerging approaches have relied on next-generation sequencings in conjunction with high-throughput screening methods [29]. An algorithm that screened for neoepitopes that could bind to major histocompatibility complexes was then utilized to filter the number of neoepitope candidates for T cell reactivity screens. These approaches have been used to identify neoantigen-targets for TIL ACT in a variety of nonmelanoma solid tumors such as cholangiocarcinoma and colorectal cancer [30,31]. Further advances have been made in the identification of tumor neoantigen through a minigene construct approach. This approach has enabled screening of all somatic mutations identified by whole-exome sequencing of cancerous tissue [32] and has led to an expanded number of identified unique neoantigens.

Although there are individual case reports of success using neoantigen-specific TIL ACT, it still remains unknown if TIL products with a high frequency of clonal populations or with a diverse TCR repertoire lead to better treatment efficacy. As discussed in the next section, the majority of TIL expansion protocols now omit the tumor reactivity selection step. Currently, few clinical trials utilize TILs with specific neoantigen targets (Table 3). TIL products that are selectively enriched for TILs that target clonal neoantigens has been developed by Achilles Therapeutics and by Intima Biosciences. Both are currently being tested in a phases 1 and 2 clinical trials. Whole-exome sequencing of blood and tumor samples is performed to identify clonal neoantigens, and dendritic cells loaded with

TABLE 3 Current TIL ACT clinical trials utilizing cancer neoantigen-specific TILs for the treatment of solid tumors.

NCT identifier	TIL therapy	Description	Cancer	Trial phase
02959905	TSA-CTL	Tumor-specific antigen-induced TILs	Melanoma	1
02789228	TAA-CTL	Tumor-associated antigen lymphocytes	Solid tumors	1
03171220	NRTs	Neoantigen-reactive T cells	Solid tumors	1/2
03658785	Neoantigen-specific TILs	Neoantigen-specific TILs	Cervical cancer	1/2
02584829	MCPyV TAg-specific TILs	Merkel cell polyoma virus T antigen-specific polyclonal autologous TILs	Merkel cell carcinoma	1/2
03997474	ATL001	TILs targeting clonal neoantigens	Melanoma	1/2

peptides correlating to these neoantigens are introduced into the expansion cultures to preferentially expand TILs that recognize the neoantigen. Although there has been much interest in neoantigen-specific TIL ACT, good candidate targets have been hard to identify in poorly immunogenic tumors, and immunogenic tumors such as melanoma have had good outcomes without TIL selection.

Impact of cellular phenotype upon TIL ACT longevity

T cell activation and differentiation follow different pathways depending on the strength and chronicity of the antigen exposure they receive. With acute infection, naïve T cells are activated and demonstrate a robust proliferation and differentiation into effector T cells. After antigen clearance, a majority of these effector T cells die with a subset differentiating into a pool of memory T cells that maintain stem cell-like regenerative qualities and can rapidly expand and regain effector functions upon restimulation [33].

In the setting of cancer, effector T cells are persistently activated by chronic antigen and undergo a process of T cell senescence and exhaustion resulting in a dysfunctional state with loss of cytotoxic effector functions [34]. This has implications for TIL ACT as a large subset of tumor reactive TILs experience sustained activation and are thought to have an exhaustive phenotype with high expression of inhibitory receptors. A study of TIL ACT in advanced melanoma patients demonstrated that a population of memory-progenitor T cells with a stem cell-like phenotype (CD39-CD69-) facilitates cancer regression and TIL persistence [35]. These responses were thought due to the ability of these TILs to self-renew and persist after infusion. There have been multiple investigations regarding the impact that cellular phenotype has upon TIL ACT efficacy and durability.

TIL expansion protocols have been observed to directly impact the cellular phenotype of the resultant TILs. Longer time in culture with additional cell cycle divisions results in TILs with decreased cellular functionality and activation potential. Therefore, the development of culture protocols with minimally cultured TILs, called young TILs, emerged. Young TILs are generated by foregoing the preexpansion cytokine measurement step and selection for tumor reactivity and by pooling multiple cultures together [36]. These changes have not only reduced TIL production time in half but they have been shown to generate TILs with greater cytotoxic potential. Shorter cultures times are now a standard component of TIL ACT protocols, and multiple clinical trials utilizing young TILs are underway (Table 4).

Another approach to impact the cellular phenotype of infused TILs can been to activate costimulatory signaling pathways during the expansion step. Current TIL expansion protocols, including the rapid expansion protocol, involve high concentration of anti-CD3/CD28 stimulation and do not optimally select for effector-memory CD8+ T cells that persist in vivo and rapidly expand upon restimulation. Therefore, studies have looked at activating costimulatory targets such as the tumor necrosis family receptor member 4-1BB (also known as CD137) during the expansion process to enhance TIL effector function and survival in vivo [37]. Clinical trials are currently ongoing to determine the impact of targeting this costimulatory pathway (Table 4).

In addition to improving survival with costimulatory agonists, CRISPR/Cas9 approaches have been used to genetically modify TILs and knockout genes contributing to

TABLE 4 Current TIL ACT clinical trials that augment cellular phenotype through augmentation of culture protocols, costimulatory pathways, or check point inhibition.

NCT identifier	TIL therapy	Description	Cancer	Trial phase
01319565, 01993719, 01174121, 02621021, 02133196	Young TILs	Shorter culture time for less senescent phenotype	Melanoma NSCLC Solid tumors	1/2
02652455	TILs + anti-CD137 antibody	Costimulation of CD137 pathway to promote cellular survival	Melanoma	1
03610490	MDA-TIL	Costimulation of CD137 pathway to promote cellular survival	Ovarian cancer Colorectal cancer Pancreatic cancer	2
04426669	CISH inactivated TIL	CRISPR/Cas9- mediated CISH inactivated TILs	Gastrointestinal epithelial tumors	2

immunosuppressive phenotypes. The gene PDCD1, which encodes PD-1, is thought to be an ideal target due to success with anti-PD-1 checkpoint blockade. A small phase 1 clinical trial of CRISPR-Cas9 PD-1 edited TILs in patients with advanced NSCLC demonstrated that CRISPR-Cas9 editing with TIL ACT is safe with minimal off-target effects [38]. Another gene target for TIL ACT is cytokine-induced SH2 protein (CISH), an intracellular checkpoint thought to inhibit TCR activation. A study from the NCI and University of Minnesota demonstrated that knockout of the CISH gene with CRISPR-Cas9 editing resulted in TILs with increased TCR avidity and cytotoxic potential [26]. A phase 2 clinical trial studying CISH knockout TIL ACT for patients with metastatic gastrointestinal epithelial tumors is underway (Table 4).

Improvement in TIL manufacturing and modern-day clinical trials

One of the initial major limitations to TIL ACT has been that many patients had disease progression during the 4–6 weeks required for TIL production under the original protocols. It is estimated that nearly one-third of enrolled patients progressed during this time and did not receive therapy. Therefore, clinical trials were developed to determine whether combining checkpoint inhibitors with TIL ACT could decrease the attrition rate. A small trial of 13 patients with metastatic melanoma was performed, and patients received ipilimumab prior to TIL ACT. The attribution rate was minimal, and 12 of the 13 (92%) received TIL ACT therapy [39]. This study supported the addition of immunotherapy to improve the feasibility of TIL ACT.

To streamline the TIL ACT manufacturing process, industry partners have collaborated with academic centers to develop new protocols that substantially reduced TIL expansion time. These protocols are modeled after the young TIL protocol that uses shorter culturing time and omits the preexpansion TIL selection for tumor reactivity. The company Iovance

TABLE 5 Current large-scale commercialized TIL ACT clinical trials in melanoma and other solid tumors.

NCT identifier	TIL therapy	Description	Cancer	Phase
02360579	Lifileucel (LN-144)	TIL Gen 2 process (Iovance)	Melanoma	2
03108495	LN-145	TIL Gen 2 process (Iovance)	Cervical cancer	2
03449108	LN-145/LN-145-S1	TIL Gen 2 process ± PD-1 selection (Iovance)	Ovarian cancer, anaplastic thyroid cancer, sarcoma	2
03645928	LN-144 (Lifileucel)/LN-145/LN-145-S1	TIL Gen 2 or 3 process ± PD-1 selection (Iovance)	Melanoma, NSCLC, HNSCC	2
04614103	LN-145	TIL Gen 2 process (Iovance)	NSCLC	2
04111510	LN-145	TIL Gen 2 process (Iovance)	TNBC	2
05050006	ITIL-168	(Instil Bio)	Melanoma	2

NSCLC, nonsmall cell lung cancer; *HNSCC*, head and neck squamous cell carcinoma; *TNBC*, triple-negative breast cancer.

Biotherapeutic has developed a Gen 2 process that reduces TIL production time to 22 days. Additionally, the cell product is cryopreserved for easier storage and administration. This process has been used in the clinical development of TIL products lifileucel (LN-144) and LN-145 (Table 5). The FDA granted Fast Track designation to lifileucel for advanced melanoma in 2017. Iovance is currently in the process of developing a Gen 3 process, which will further reduce production time to 16 days.

Lifileucel has been studied in the multicohort phase 2 clinical trial C-144-01 in patients with advanced melanoma who have progressed on immunotherapy. Follow-up on cohort 2, which included 66 patients, demonstrated an ORR of 36.4% and 2 CRs at a median follow-up of 18.7 months [40]. Follow-up data from cohort 4 showed an ORR of 32.4% and 1 CR in 68 patients at a median follow-up of 5.3 months. There is continued interest in large-scale production of TILs for improved public accessibility, and the cell therapy company Instil Bio received orphan drug designation for their TIL product ITIL-168, with phase 1 clinical trials in advanced melanoma set to begin in 2021 (Table 5).

While TIL ACT has mostly been used as savage therapy for patients that have progressed on immunotherapy and targeted BRAF/MEK inhibitors, studies are emerging that prior treatment may decrease the effectiveness of TIL ACT. Additional analysis of the phase 2 clinical trial C-144-01 examined the impact of prior anti-PD-1 therapy on duration of response of lifileucel [41]. Although no association was found between ORR and prior therapy, shorter duration of prior anti-PD-1 therapy was correlated with longer duration of response to lifileucel. Therefore, the investigators concluded that lifileucel should be given earlier at the time of initial progression on PD-1 to maximize response benefit.

A retrospective study from the Surgery Branch of NCI analyzed the impact of prior immunotherapy on the response rates of 226 patients with melanoma who underwent TIL ACT [42]. Patients naïve to anti-PD-1 therapy demonstrated an ORR of 56% and median melanoma-specific survival of 28.5 months compared with 24% and 11.6 months, respectively, in patients refractory to anti-PD-1. Forty-six of 48 anti-PD-1 native patients who had a CR have ongoing responses after a single treatment at a median follow-up of 89 months, as well as a 10-year melanoma-specific survival of 96%. This study gives additional support that TIL ACT should be considered earlier in the treatment plan for patients with advanced melanoma.

Lastly, another advantage of TIL ACT compared to other forms of immunotherapy is its lower risk for autoimmune side effects. Typical toxicities are related to lymphodepletion or IL-2 administration, and they are short lived as only single TIL ACT treatments are needed to obtain durable responses [43]. The infused cells in TIL ACT are mostly tumor antigen-specific with little to no off-target reactivity. Therefore, TIL ACT may be better tolerated than immune checkpoint blockade, which is associated with significant long-term immune-related adverse events in up to 40% of receiving patients [44].

TILs and solid tumors other than melanoma

Although TIL ACT is an effective therapy for refractory metastatic melanoma, its adoption to other solid cancer has been limited due to the lower immunogenicity of these tumors. [45]. However, there have been case reports demonstrating success in nonmelanoma cancers, such as cholangiocarcinoma [31], colon cancer [30], and breast cancer [46]. Additionally, in a phase I trial of patients with hepatocellular carcinoma who received TIL ACT following resection, 80% showed no evidence of recurrence at 14 months [47].

Clinical trials are advancing the utilization of TIL ACT for nonmelanoma solid tumors by including these patients in their studies. In a phase 1 clinical trial (NCT03215810), patients with metastatic NSCLC and progression on nivolumab received TILs produced at Moffitt Cancer Center. Of the 13 evaluable patients, 3 had confirmed responses and 11 had reduction in tumor burden, with a median best change of 35%. Two patients had CR that was ongoing at 1.5 years [48]. Iovance's LN-145, another TIL product utilizing the Gen 2 process, is currently being studied in the phase 2 clinical trial C-145-04 of patients with persistent or metastatic cervical cancer. Preliminary results included 27 patients and demonstrated an ORR of 44.4% with 3 CRs at a median follow-up of 7.4 months [49]. With these findings, the FDA granted LN-145 Breakthrough Therapy designation in 2019. LN-145 is also being studied in the clinical trial IOV-COM-202 (NCT04614103, Table 5), which includes patients with heavily pretreated metastatic NSCLC. Preliminary results were presented at SITC 2021 and demonstrated an evaluable ORR of 25% with 1 CR and 5 PRs at a median follow-up of 9.8 months. These trials demonstrate the feasibility of TIL ACT in nonmelanoma solid tumors and show efficacy even for treatment-refractory patients and further support the consideration of TIL ACT as an earlier-line therapy.

Summary

Cytotoxic T cells are the focus of modern immunotherapy due to their antigen specificity and potent antitumor capabilities. Adoptive cell therapy allows for the infusion of large numbers of autologous neoantigen-specific TILs after lymphodepleting chemotherapy to enhance engraftment. Recent clinical studies have demonstrated promising responses to TIL ACT, but only 25% of patients undergoing this therapy experience cure. Current investigations to optimize TIL ACT involve the alteration and selection of the TIL product to further module the tumor immune microenvironment, target-specific tumor neoantigens, and select TILs with favorable phenotypes. Advantages in TIL production have enabled the commercialization and large-scale application of TIL ACT, even for patients with nonmelanoma solid tumors. AS TIL ACT continues to evolve, its favorable adverse event profile compared to other forms of immunotherapy may support its use as a first-line therapy.

Acknowledgments

Y.F. has received funds as scientific advisor to Imugene, Eureka, Iovance, and Boehringer Ingelheim. Y.F. receives royalties from Imugene, Merck, and Xdemics.

References

[1] Gajewski TF, Schreiber H, Fu YX. Innate and adaptive immune cells in the tumor microenvironment. Nat Immunol 2013;14:1014–22.
[2] Raskov H, Orhan A, Christensen JP, Gögenur I. Cytotoxic CD8. Br J Cancer 2021;124:359–67.
[3] Chakraborty AK, Weiss A. Insights into the initiation of TCR signaling. Nat Immunol 2014;15:798–807.
[4] Maher J, Davies ET. Targeting cytotoxic T lymphocytes for cancer immunotherapy. Br J Cancer 2004;91:817–21.
[5] Caspi RR. Immunotherapy of autoimmunity and cancer: the penalty for success. Nat Rev Immunol 2008;8:970–6.
[6] Vinay DS, Ryan EP, Pawelec G, Talib WH, Stagg J, Elkord E, Lichtor T, Decker WK, Whelan RL, Kumara HMCS, Signori E, Honoki K, Georgakilas AG, Amin A, Helferich WG, Boosani CS, Guha G, Ciriolo MR, Chen S, Mohammed SI, Azmi AS, Keith WN, Bilsland A, Bhakta D, Halicka D, Fujii H, Aquilano K, Ashraf SS, Nowsheen S, Yang X, Choi BK, Kwon BS. Immune evasion in cancer: mechanistic basis and therapeutic strategies. Semin Cancer Biol 2015;35(Suppl):S185–98.
[7] Wherry EJ, Ha SJ, Kaech SM, Haining WN, Sarkar S, Kalia V, Subramaniam S, Blattman JN, Barber DL, Ahmed R. Molecular signature of CD8+ T cell exhaustion during chronic viral infection. Immunity 2007;27:670–84.
[8] Jiang T, Zhou C, Ren S. Role of IL-2 in cancer immunotherapy. Onco Targets Ther 2016;5, e1163462.
[9] Smith FO, Downey SG, Klapper JA, Yang JC, Sherry RM, Royal RE, Kammula US, Hughes MS, Restifo NP, Levy CL, White DE, Steinberg SM, Rosenberg SA. Treatment of metastatic melanoma using interleukin-2 alone or in conjunction with vaccines. Clin Cancer Res 2008;14:5610–8.
[10] Larkin J, Chiarion-Sileni V, Gonzalez R, Grob JJ, Rutkowski P, Lao CD, Cowey CL, Schadendorf D, Wagstaff J, Dummer R, Ferrucci PF, Smylie M, Hogg D, Hill A, Márquez-Rodas I, Haanen J, Guidoboni M, Maio M, Schöffski P, Carlino MS, Lebbé C, McArthur G, Ascierto PA, Daniels GA, Long GV, Bastholt L, Rizzo JI, Balogh A, Moshyk A, Hodi FS, Wolchok JD. Five-year survival with combined nivolumab and ipilimumab in advanced melanoma. N Engl J Med 2019;381:1535–46.
[11] Binnewies M, Roberts EW, Kersten K, Chan V, Fearon DF, Merad M, Coussens LM, Gabrilovich DI, Ostrand-Rosenberg S, Hedrick CC, Vonderheide RH, Pittet MJ, Jain RK, Zou W, Howcroft TK, Woodhouse EC, Weinberg RA, Krummel MF. Understanding the tumor immune microenvironment (TIME) for effective therapy. Nat Med 2018;24:541–50.
[12] Zippelius A, Batard P, Rubio-Godoy V, Bioley G, Liénard D, Lejeune F, Rimoldi D, Guillaume P, Meidenbauer N, Mackensen A, Rufer N, Lubenow N, Speiser D, Cerottini JC, Romero P, Pittet MJ. Effector function of human tumor-specific CD8 T cells in melanoma lesions: a state of local functional tolerance. Cancer Res 2004;64:2865–73.
[13] Morgan DA, Ruscetti FW, Gallo R. Selective in vitro growth of T lymphocytes from normal human bone marrows. Science 1976;193:1007–8.
[14] Dudley ME, Wunderlich JR, Shelton TE, Even J, Rosenberg SA. Generation of tumor-infiltrating lymphocyte cultures for use in adoptive transfer therapy for melanoma patients. J Immunother 2003;26:332–42.
[15] Rosenberg SA, Spiess P, Lafreniere R. A new approach to the adoptive immunotherapy of cancer with tumor-infiltrating lymphocytes. Science 1986;233:1318–21.
[16] Rosenberg SA, Packard BS, Aebersold PM, Solomon D, Topalian SL, Toy ST, Simon P, Lotze MT, Yang JC, Seipp CA. Use of tumor-infiltrating lymphocytes and interleukin-2 in the immunotherapy of patients with metastatic melanoma. A preliminary report. N Engl J Med 1988;319:1676–80.

[17] Rosenberg SA, Yannelli JR, Yang JC, Topalian SL, Schwartzentruber DJ, Weber JS, Parkinson DR, Seipp CA, Einhorn JH, White DE. Treatment of patients with metastatic melanoma with autologous tumor-infiltrating lymphocytes and interleukin 2. J Natl Cancer Inst 1994;86:1159–66.
[18] Rosenberg SA, Aebersold P, Cornetta K, Kasid A, Morgan RA, Moen R, Karson EM, Lotze MT, Yang JC, Topalian SL. Gene transfer into humans—immunotherapy of patients with advanced melanoma, using tumor-infiltrating lymphocytes modified by retroviral gene transduction. N Engl J Med 1990;323:570–8.
[19] Berenson JR, Einstein AB, Fefer A. Syngeneic adoptive immunotherapy and chemoimmunotherapy of a friend leukemia: requirement for T cells. J Immunol 1975;115:234–8.
[20] North RJ. Cyclophosphamide-facilitated adoptive immunotherapy of an established tumor depends on elimination of tumor-induced suppressor T cells. J Exp Med 1982;155:1063–74.
[21] Antony PA, Piccirillo CA, Akpinarli A, Finkelstein SE, Speiss PJ, Surman DR, Palmer DC, Chan CC, Klebanoff CA, Overwijk WW, Rosenberg SA, Restifo NP. CD8+ T cell immunity against a tumor/self-antigen is augmented by CD4+ T helper cells and hindered by naturally occurring T regulatory cells. J Immunol 2005;174:2591–601.
[22] Gattinoni L, Finkelstein SE, Klebanoff CA, Antony PA, Palmer DC, Spiess PJ, Hwang LN, Yu Z, Wrzesinski C, Heimann DM, Surh CD, Rosenberg SA, Restifo NP. Removal of homeostatic cytokine sinks by lymphodepletion enhances the efficacy of adoptively transferred tumor-specific CD8+ T cells. J Exp Med 2005;202:907–12.
[23] Rosenberg SA, Yang JC, Sherry RM, Kammula US, Hughes MS, Phan GQ, Citrin DE, Restifo NP, Robbins PF, Wunderlich JR, Morton KE, Laurencot CM, Steinberg SM, White DE, Dudley ME. Durable complete responses in heavily pretreated patients with metastatic melanoma using T-cell transfer immunotherapy. Clin Cancer Res 2011;17:4550–7.
[24] Goff SL, Dudley ME, Citrin DE, Somerville RP, Wunderlich JR, Danforth DN, Zlott DA, Yang JC, Sherry RM, Kammula US, Klebanoff CA, Hughes MS, Restifo NP, Langhan MM, Shelton TE, Lu L, Kwong ML, Ilyas S, Klemen ND, Payabyab EC, Morton KE, Toomey MA, Steinberg SM, White DE, Rosenberg SA. Randomized, prospective evaluation comparing intensity of lymphodepletion before adoptive transfer of tumor-infiltrating lymphocytes for patients with metastatic melanoma. J Clin Oncol 2016;34:2389–97.
[25] Beane JD, Lee G, Zheng Z, Mendel M, Abate-Daga D, Bharathan M, Black M, Gandhi N, Yu Z, Chandran S, Giedlin M, Ando D, Miller J, Paschon D, Guschin D, Rebar EJ, Reik A, Holmes MC, Gregory PD, Restifo NP, Rosenberg SA, Morgan RA, Feldman SA. Clinical scale zinc finger nuclease-mediated gene editing of PD-1 in tumor infiltrating lymphocytes for the treatment of metastatic melanoma. Mol Ther 2015;23:1380–90.
[26] Palmer DC, Webber BR, Patel Y, Johnson MJ, Kariya CM, Lahr WS, Parkhurst MR, Gartner JJ, Prickett TD, Lowery FJ, Kishton RJ, Gurusamy D, Franco Z, Vodnala SK, Diers MD, Wolf NK, Slipek NJ, McKenna DH, Sumstad D, Viney L, Henley T, Bürckstümmer T, Baker O, Hu Y, Yan C, Meerzaman D, Padhan K, Lo W, Malekzadeh P, Jia L, Deniger DC, Patel SJ, Robbins PF, McIvor RS, Choudhry M, Rosenberg SA, Moriarity BS, Restifo NP. Internal checkpoint regulates T Cell neoantigen reactivity and susceptibility to PD1 blockade. Available at SSRN: https://ssrn.com/abstract=3751760 or; 2020. https://doi.org/10.2139/ssrn.3751760.
[27] Restifo NP, Dudley ME, Rosenberg SA. Adoptive immunotherapy for cancer: harnessing the T cell response. Nat Rev Immunol 2012;12:269–81.
[28] Robertson J, Salm M, Dangl M. Adoptive cell therapy with tumour-infiltrating lymphocytes: the emerging importance of clonal neoantigen targets for next-generation products in non-small cell lung cancer. Immunooncol Technol 2019;3:1–7.
[29] Tran E, Robbins PF, Rosenberg SA. 'Final common pathway' of human cancer immunotherapy: targeting random somatic mutations. Nat Immunol 2017;18:255–62.
[30] Tran E, Robbins PF, Lu YC, Prickett TD, Gartner JJ, Jia L, Pasetto A, Zheng Z, Ray S, Groh EM, Kriley IR, Rosenberg SA. T-cell transfer therapy targeting mutant KRAS in cancer. N Engl J Med 2016;375:2255–62.
[31] Tran E, Turcotte S, Gros A, Robbins PF, Lu YC, Dudley ME, Wunderlich JR, Somerville RP, Hogan K, Hinrichs CS, Parkhurst MR, Yang JC, Rosenberg SA. Cancer immunotherapy based on mutation-specific CD4+ T cells in a patient with epithelial cancer. Science 2014;344:641–5.
[32] Lu YC, Yao X, Crystal JS, Li YF, El-Gamil M, Gross C, Davis L, Dudley ME, Yang JC, Samuels Y, Rosenberg SA, Robbins PF. Efficient identification of mutated cancer antigens recognized by T cells associated with durable tumor regressions. Clin Cancer Res 2014;20:3401–10.
[33] Dutton RW, Bradley LM, Swain SL. T cell memory. Annu Rev Immunol 1998;16:201–23.
[34] Wherry EJ, Kurachi M. Molecular and cellular insights into T cell exhaustion. Nat Rev Immunol 2015;15:486–99.
[35] Krishna S, Lowery FJ, Copeland AR, Bahadiroglu E, Mukherjee R, Jia L, Anibal JT, Sachs A, Adebola SO, Gurusamy D, Yu Z, Hill V, Gartner JJ, Li YF, Parkhurst M, Paria B, Kvistborg P, Kelly MC, Goff SL, Altan-Bonnet

G, Robbins PF, Rosenberg SA. Stem-like CD8 T cells mediate response of adoptive cell immunotherapy against human cancer. Science 2020;370:1328–34.

[36] Dudley ME, Gross CA, Langhan MM, Garcia MR, Sherry RM, Yang JC, Phan GQ, Kammula US, Hughes MS, Citrin DE, Restifo NP, Wunderlich JR, Prieto PA, Hong JJ, Langan RC, Zlott DA, Morton KE, White DE, Laurencot CM, Rosenberg SA. CD8+ enriched "young" tumor infiltrating lymphocytes can mediate regression of metastatic melanoma. Clin Cancer Res 2010;16:6122–31.

[37] Tavera RJ, Forget MA, Kim YU, Sakellariou-Thompson D, Creasy CA, Bhatta A, Fulbright OJ, Ramachandran R, Thorsen ST, Flores E, Wahl A, Gonzalez AM, Toth C, Wardell S, Mansaray R, Radvanyi LG, Gombos DS, Patel SP, Hwu P, Amaria RN, Bernatchez C, Haymaker C. Utilizing T-cell activation signals 1, 2, and 3 for tumor-infiltrating lymphocytes (TIL) expansion: the advantage over the sole use of Interleukin-2 in cutaneous and uveal melanoma. J Immunother 2018;41:399–405.

[38] Lu Y, Xue J, Deng T, Zhou X, Yu K, Deng L, Huang M, Yi X, Liang M, Wang Y, Shen H, Tong R, Wang W, Li L, Song J, Li J, Su X, Ding Z, Gong Y, Zhu J, Zou B, Zhang Y, Li Y, Zhou L, Liu Y, Yu M, Zhang X, Yin L, Xia X, Zeng Y, Zhou Q, Ying B, Chen C, Wei Y, Li W, Mok T. Safety and feasibility of CRISPR-edited T cells in patients with refractory non-small-cell lung cancer. Nat Med 2020;26:732–40.

[39] Mullinax JE, Hall M, Prabhakaran S, Weber J, Khushalani N, Eroglu Z, Brohl AS, Markowitz J, Royster E, Richards A, Stark V, Zager JS, Kelley L, Cox C, Sondak VK, Mulé JJ, Pilon-Thomas S, Sarnaik AA. Combination of Ipilimumab and adoptive cell therapy with tumor-infiltrating lymphocytes for patients with metastatic melanoma. Front Oncol 2018;8:44.

[40] Sarnaik AA, Hamid O, Khushalani NI, Lewis KD, Medina T, Kluger HM, Thomas SS, Domingo-Musibay E, Pavlick AC, Whitman ED, Martin-Algarra S, Corrie P, Curti BD, Oláh J, Lutzky J, Weber JS, Larkin JMG, Shi W, Takamura T, Jagasia M, Qin H, Wu X, Chartier C, Graf Finckenstein F, Fardis M, Kirkwood JM, Chesney JA. Lifileucel, a tumor-infiltrating lymphocyte therapy, in metastatic melanoma. J Clin Oncol 2021;39:2656–66.

[41] Larkin J, Sarnaik A, Chesney JA, Khushalani NI, Kirkwood JM, Weber JS, Lewis KD, Medina TM, Kluger HM, Thomas SS, Domingo-Musibay E, Olah J, Whitman ED, Martin-Algarra S, Corrie PG, Lutzky J, Shi W, Wu RX, Fardis M, Hamid O. Lifileucel (LN-144), a cryopreserved autologous tumor infiltrating lymphocyte (TIL) therapy in patients with advanced melanoma: evaluation of impact of prior anti-PD-1 therapy. J Clin Oncol 2021;39:9505.

[42] Seitter SJ, Sherry RM, Yang JC, Robbins PF, Shindorf ML, Copeland AR, McGowan CT, Epstein M, Shelton TE, Langhan MM, Franco Z, Danforth DN, White DE, Rosenberg SA, Goff SL. Impact of prior treatment on the efficacy of adoptive transfer of tumor-infiltrating lymphocytes in patients with metastatic melanoma. Clin Cancer Res 2021;27:5289.

[43] Yang JC. Toxicities associated with adoptive T-cell transfer for cancer. Cancer J 2015;21:506–9.

[44] Patrinely Jr JR, Johnson R, Lawless AR, Bhave P, Sawyers A, Dimitrova M, Yeoh HL, Palmeri M, Ye F, Fan R, Davis EJ, Rapisuwon S, Long GV, Haydon A, Osman I, Mehnert JM, Carlino MS, Sullivan RJ, Menzies AM, Johnson DB. Chronic immune-related adverse events following adjuvant anti-PD-1 therapy for high-risk resected melanoma. JAMA Oncol 2021;7:744–8.

[45] Morotti M, Albukhari A, Alsaadi A, Artibani M, Brenton JD, Curbishley SM, Dong T, Dustin ML, Hu Z, McGranahan N, Miller ML, Santana Gonzalez L, Seymour LW, Shi T, Van Loo P, Yau C, White H, Wietek N, Church DN, Wedge DC, Ahmed AA. Promises and challenges of adoptive T-cell therapies for solid tumours. Br J Cancer 2021;124:1759–76.

[46] Zacharakis N, Chinnasamy H, Black M, Xu H, Lu Y-C, Zheng Z, Pasetto A, Langhan M, Shelton T, Prickett T, Gartner J, Jia L, Trebska-McGowan K, Somerville RP, Robbins PF, Rosenberg SA, Goff SL, Feldman SA. Immune recognition of somatic mutations leading to complete durable regression in metastatic breast cancer. Nat Med 2018;24:724–30.

[47] Jiang SS, Tang Y, Zhang YJ, Weng DS, Zhou ZG, Pan K, Pan QZ, Wang QJ, Liu Q, He J, Zhao JJ, Li J, Chen MS, Chang AE, Li Q, Xia JC. A phase I clinical trial utilizing autologous tumor-infiltrating lymphocytes in patients with primary hepatocellular carcinoma. Oncotarget 2015;6:41339–49.

[48] Creelan BC, Wang C, Teer JK, Toloza EM, Yao J, Kim S, Landin AM, Mullinax JE, Saller JJ, Saltos AN, Noyes DR, Montoya LB, Curry W, Pilon-Thomas SA, Chiappori AA, Tanvetyanon T, Kaye FJ, Thompson ZJ, Yoder SJ, Fang B, Koomen JM, Sarnaik AA, Chen DT, Conejo-Garcia JR, Haura EB, Antonia SJ. Tumor-infiltrating lymphocyte treatment for anti-PD-1-resistant metastatic lung cancer: a phase 1 trial. Nat Med 2021;27:1410–8.

[49] Jazaeri AA, Zsiros E, Amaria RN, Artz AS, Edwards RP, Wenham RM, Slomovitz BM, Walther A, Thomas SS, Chesney JA, Morris R, Matsuo K, Gaillard S, Rose PG, Donas JG, Tromp JM, Tavakkoli F, Li H, Fardis M, Monk BJ. Safety and efficacy of adoptive cell transfer using autologous tumor infiltrating lymphocytes (LN-145) for treatment of recurrent, metastatic, or persistent cervical carcinoma. J Clin Oncol 2019;37:2538.

CHAPTER

7

Biology and status of chimeric antigen receptor-engineered T cell therapy

John P. Murad, Anthony K. Park, and Saul J. Priceman

Department of Hematology and Hematopoietic Cell Transplantation, City of Hope, Duarte, CA, United States

Abstract

The past few decades have witnessed striking advancements in our understanding of redirecting the immune system against cancer. Exciting research in this field has led to novel and clinically impactful immunotherapies that are improving outcomes for cancer patients. Specifically, the use of adoptive cell transfer of a patient's own T cells, engineered to express chimeric antigen receptors (CAR), has shown to be an effective method to redirect T cells to target known tumor antigens in both hematological and solid tumor malignancies. Over the past 5 years, multiple CAR T cell therapies targeting various B cell malignancies have gained Food and Drug Administration approval and have shown dramatic and durable responses in patients. The generation of these CAR T cells, and optimization of their modular synthetic components, can result in distinct functionalities that can have tremendous impact on clinical outcomes. Further, optimal engineering and manufacturing of these cellular products may lead to better accessibility, efficacy, and safety of future generations of CAR T cells. Additionally, to expand the overall efficacy of CAR T cells against solid tumors, a consideration of the critical role the tumor microenvironment plays in the suppression of immune responses, as well as the potential for antigen escape of heterogenous tumors, must also be addressed. In this chapter, we will detail the status of current FDA approved CAR T cell products and their clinical impact, highlight unique hurdles for CAR T cells for solid tumors, and discuss perspectives on the future of CAR-engineered cell therapies.

Abbreviations

A2AR	adenosine 2A receptor
ACT	adoptive cell therapy
ALL	acute lymphoblastic leukemia
BCMA	B-cell maturation antigen
CAR	chimeric antigen receptor
CAS9	CRISPR-associated protein 9
CCL17	C-C motif chemokine ligand 17
CCL18	C-C motif chemokine ligand 18
CCL22	C-C motif chemokine ligand 22

https://doi.org/10.1016/B978-0-12-822620-9.00018-5

CD137 cluster of differentiation 137
CD19 cluster of differentiation 19
CD27 cluster of differentiation 27
CD28 cluster of differentiation 28
CD4 cluster of differentiation 4
CD8 cluster of differentiation 8
CD80 cluster of differentiation 80
CD86 cluster of differentiation 86
COX-2 cyclooxygenase-2
CR complete response
CRISPR clustered regularly interspaced short palindromic repeats
CTLA-4 cytotoxic T-lymphocyte-associated antigen 4
CXCL1 CXC-motif chemokine ligand 1
CXCL12 CXC-motif chemokine ligand 12
DC dendritic cell
DLBCL diffuse large B-cell lymphoma
EGFRvIII epidermal growth factor receptor version III
FAP fibroblast activation protein
FDA Food and Drug Administration
FOXP3 forkhead box protein P3
GD2 disialoganglioside
GM-CSF granulocyte-macrophage colony-stimulating factor
H2O2 hydrogen peroxide
HER2 human epidermal growth factor 2
HIF hypoxia-inducible factor
ICOS inducible T cell co-stimulator
IDO indolamine 2,3-dioxygenase
IL-10 interleukin 10
IL-13Rα2 interleukin 13 receptor alpha 2
IL-2 interleukin 2
INF-γ interferon gamma
iNOS inducible nitric oxide synthase
LAG3 lymphocyte activation gene 3
MDSC myeloid-derived suppressor cells
MHC major histocompatibility complex
MM multiple myeloma
MMP2 matrix metallopeptidase 2
MMP9 matrix metallopeptidase 9
NK natural killer cell
ORR overall response rate
OX40 cluster of differentiation 134/tumor necrosis factor receptor 40
PBMC peripheral blood mononuclear cell
PD-1 programmed cell death 1
PD-L1 programmed-death ligand 1
PGE2 prostaglandin E2
PKA protein kinase A
PSCA prostate stem cell antigen
PSMA prostate specific membrane antigen
RIAD regulatory subunit I anchoring disruptor
ROS reactive oxygen species
scFv single-chain variable fragment
shRNA short hairpin ribonucleic acid
TAA tumor-associated antigen

TALEN	transcription activator-like effector nucleases
TAMs	tumor-associated macrophages
TANs	tumor-associated neutrophils
TCR	T-cell receptor
TGF-β	transforming growth factor beta
Th1	T helper type 1
Th2	T helper type 2
TIGIT	T cell immunoglobulin and ITIM domain
TIM3	T cell immunoglobulin and mucin 3
TLR9	toll-like receptor 9
TM	transmembrane domain
TME	tumor microenvironment
Treg	T regulatory cells
VEGF	vascular endothelial growth factor
ZFN	zinc-finger nuclease

Conflict of interest

J.P.M. has no conflicts to disclose. A.K.P. receives royalties from Imugene Ltd. S.J.P. is a scientific advisor to and has received royalties from Mustang Bio and Imugene Ltd.

Introduction

Over the past 20 years, a revolution in cancer therapy has occurred. The emergence of harnessing and retraining the immune system to engage in productive antitumor responses has opened the doors to novel and effective cancer immunotherapies. Where early attempts aimed at boosting immunity with cancer vaccines or other means may have been difficult to execute or translate, recent developments in our understanding in tumor immunology have helped in developing advanced engineering strategies to produce potent and durable anti-tumor immunity. Most recently, findings that cancer cells potentiate their survival by actively shutting down immune responses has led to the development of immune checkpoint pathway blockade, including those targeting PD-1/PD-L1 or CTLA-4 signaling. These molecules have shown great promise in re-activating tumor infiltrated immune T cells within a patient. Immune checkpoint blockade first gained FDA approval in 2011 and have since seen additional FDA approvals and remarkable responses in a variety of hematological and solid tumor indications, with active research ongoing in combination approaches. In 2013, the field of cancer immunology was named scientific breakthrough of the year by *Science Magazine*, and a wave of novel therapeutics to further exploit immune based therapies has been steadily building. In 2018 the Nobel Prize in Physiology or Medicine was awarded to Dr. James Allison and Dr. Tasuku Honjo for their contributions to the biology and targeting approaches for immune checkpoint proteins to enhance anticancer immune function.

Another novel approach in cancer immunotherapy has been re-engineering T cells to directly target known tumor antigens for both hematological and solid tumor indications. These T cells are engineered to express a chimeric antigen receptor (CAR) molecule that redirects T cells to identify and activate against tumor cells independent of peptide-MHC presentation. These CAR-engineered T cells can then be adoptively transferred back to patients generating a highly personalized immunotherapy strategy. CAR T cell therapy gained FDA approval in 2017 for the treatment of CD19-positive B-cell malignancies and has been expansive in terms of clinical investigations around other targets and disease indications.

Enhancing the presence of antigen-specific and CAR T cells to target both hematological malignancies and solid tumors is a major hurdle for effective cell-based immunotherapy. Immune escape mechanisms in the tumor microenvironment (TME) such as major histocompatibility complex (MHC) class I and II downregulation [1–3] as well as reduced presentation of neoantigens or tumor-associated antigens (TAAs) can lower the proliferation, activation and survival of helper and effector T cells in the TME [4,5]. This issue is intensified by the presence of suppressive immune populations including regulatory T cells (Tregs), tumor-associated macrophages (TAMs), and myeloid-derived suppressor cells (MDSCs) [6–8]. To overcome these barriers, next-generation engineering strategies of CAR T cells and the incorporation of CAR T cells with rational combination immunotherapies hold great promise as an effective cancer immunotherapy. The success of CAR T cell therapy in the hematological space has opened an extremely active area of preclinical research and development. Work is on-going to improve design and structure of these CAR T cell constructs to optimize tumor cytotoxicity, proliferation, and persistence within the suppressive TME. Furthermore, target validation and focus on improving their clinical impact in the treatment of solid tumors is the next frontier for CAR T cell therapy.

Generations of CAR T cells

CAR T cells are T cells that are engineered to express synthetic fusion proteins with an extracellular tumor-antigen-binding domain (often a single chain variable fragment or scFv) tethered to intracellular signaling domains to confer T cell activation. The scFv is linked to extracellular spacer and transmembrane (TM) domains derived from naturally occurring molecules, such as CD28 and CD8a, which then signals to an intracellular primary and secondary stimulation signaling motifs. A CD3ζ motif (as illustrated in Fig. 1) [9] generates traditional TCR-like activation, degranulation, and perforin-mediated T cell cytotoxicity toward antigen-positive tumor targets. First-generation CAR T cells featuring only CD3ζ activation signal displayed notable cytotoxic properties in vitro but lacked the optimal signaling for proliferation and persistence required for durable antitumor responses [10,11]. As such, second- and third-generation CARs include a variety of co-stimulatory secondary signals mediated by either single or combinations of CD28, 41BB, OX40, CD27, and/or ICOS, among others as illustrated in Fig. 2 [9,12]. These critical secondary signals allowed for superior CAR T cell production of cytokines (e.g., IL-2), which aids in proliferation and persistence, and subsequently yielded greater cytotoxic capacity both in vitro and in vivo [13]. Fourth-generation CAR T cells or "armored" T cells (Fig. 2) may incorporate a co-stimulatory secondary signal but also include coexpression of transgenes that allow for expression of CAR along with surface receptors or secreted factors (such as cytokines or proinflammatory ligands), which aid in T cell expansion, persistence, and/or durable antitumor activity [14].

Manufacturing of CAR T cells begins with leukapheresis of a patient, where bulk peripheral blood-mononuclear cells (PBMCs) or enriched T cells are isolated, activated, and engineered (by transduction or transfection) to express the CAR of interest. CAR T cells are then expanded ex vivo and cryopreserved preceding infusion back to the patient as an autologous product [15,16]. Recently, memory or effector phenotype of T cells prior to

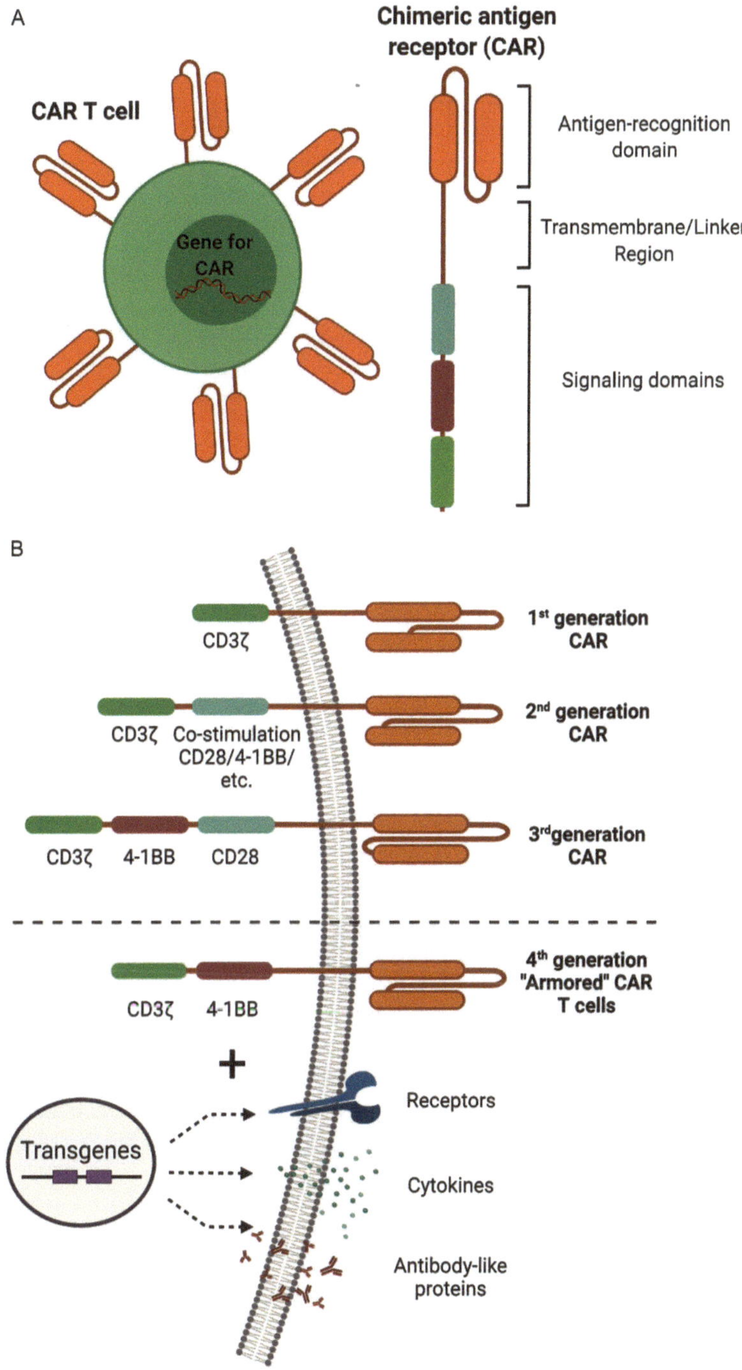

FIG. 1 (A) Description of chimeric antigen receptor (CAR) on T cell surface and illustration of traditional intra-cellular and extra-cellular domain positioning. (B) Illustration of multiple generations of CAR constructs and examples of modular design options used to incorporate unique cell signaling pathways or transgene integration.

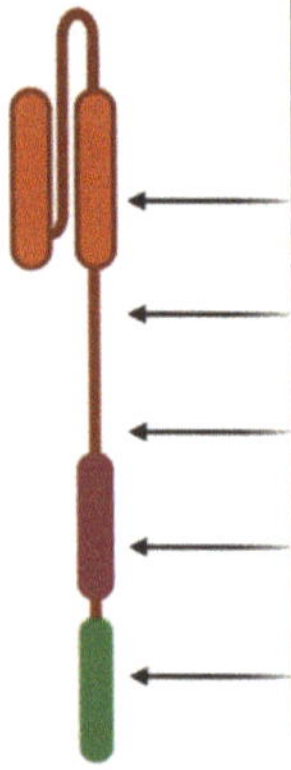

CARs	CD19-Targeting CAR T cells			BCMA-Targeting CAR T cells
	Kymriah	Yescarta	Breyanzi	Abecma
scFv	FMC63	FMC63	FMC63	anti-BCMA
Linker	CD8α	CD28	IgG4	CD8α
TM	CD8α	CD28	CD28	CD8α
Co-stimulatory domain	4-1BB	CD28	4-1BB	4-1BB
Cytolytic domain	CD3ζ	CD3ζ	CD3ζ	CD3ζ

FIG. 2 Table of the currently FDA approved CAR T cell therapies against hematological malignancies and their unique combination of antigen binding single-chain variable fragment (scFv), extracellular linkers, transmembrane (TM), co-stimulatory, and cytolytic domains.

transduction, as well as CD4 and CD8 ratios and other surface marker expression patterns in the final product have been shown to have predictive value in determining long-term persistence and overall antitumor efficacy of CAR T cells [17–19]. These findings suggest that the starting cell material as well as the final product are important components in optimizing effective CAR T cell therapies.

Building on the current approaches for the generation and manufacturing of CAR T cell products, numerous methodologies are being investigated to determine optimal engineering strategies, which may impact safety and therapeutic activity. Some of these approaches include using genome editing tools, such as CRISPR, zinc-finger nucleases (ZFN), or transcription activator-like effector nucleases (TALEN), to precisely knock-out or knock-in genes of interest with the potential for exquisite specificity and functional benefits, such as directly into the T-cell receptor locus [20,21]. The use of nonviral methods to introduce gene edits, such as transposon-derived vectors, may reduce manufacturing hurdles and costs presented by current viral transduction methods [22]. Furthermore, using nonviral RNA transfection to transiently express CAR may reduce potential safety concerns [23]. Finally, many groups are exploring CAR engineering of alternate immune cell types in addition to conventional T cells, including natural killer (NK) cells, regulatory T cells (Tregs), and gamma-delta T cells, to facilitate improved cellular therapy outcomes not only for a greater spectrum of cancer types but also for autoimmune diseases and other conditions [21].

FDA approved CAR T cells in hematologic malignancies

To date, the most impressive antitumor responses to CAR T cell-based therapy has been observed using CD19-specific CAR T cells for the treatment of patients with B-cell malignancies, which have compellingly demonstrated the potential of this approach [24]. Work in this area resulted in two landmark FDA approvals for patients with CD19+ hematological malignancies. Kymriah (Tisagenlecleucel), a CD19 targeted CAR T cell developed by Novartis, was the first ever approved CAR T cell therapy, closely followed by Yescarta (Axicabtagene ciloleucel) developed by Kite Pharma/Gilead. Kymriah is currently indicated for relapsed or refractory B-cell precursor acute lymphoblastic leukemia (ALL) as well as diffuse large B-cell lymphoma (DLBCL). Yescarta is currently indicated also for DLBCL and Non-Hodgkins lymphoma. Clinical successes for both are attributed to the nearly ubiquitous CD19 expression in these hematological malignancies, and its restriction to B-cells, which exemplifies the ideal CAR T cell target for efficacy and safety [25]. However, the fundamental ability for modular component switching for most CAR T cell designs has resulted in unique differences in Kymriah and Yescarta structure, independent of their antigen-binding domain, which result in different functional characteristics both preclinically and clinically.

Both Kymriah and Yescarta use the same CD19 antigen-binding scFv (clone FMC63), however, differ in extracellular linker, TM domain, and co-stimulatory signaling domain (Fig. 3) [26]. Kymriah uses a CD8-alpha (CD8α) linker and TM domain sequence in contrast to CD28 linker and TM domain used in Yescarta. In preclinical studies, it was found that use of CD28, versus CD8α, linker, and TM in multiple CAR designs have been implicated with improved proliferative capacity, a lowered threshold of antigen density for CAR T cell activation against CD19, and higher amounts of IFNγ secretion following stimulation [27–29]. The linker and TM domains established from endogenous T cell surface molecules such as CD8 and CD28 allow for potential dimerization or interaction with existing transmembrane domains, which may influence their signaling capacity or alter the clustering of multiple CAR molecules even prior to antigen engagement [29].

The second main difference between these two FDA-approved CAR T cells is in their secondary signal T cell activation molecules. Kymriah uses a 4-1BB (CD137) co-stimulatory domain, whereas Yescarta harbors a CD28 costimulatory domain. There has been significant characterization of the differences between these two co-stimulatory molecules in the CAR T cell field. Though there are fairly similar response rates between CD19 CAR T cells harboring either of these domains, the persistence of each is markedly different. Preclinical studies of persistence between 4-1BB and CD28 costimulatory domains in both hematological malignancies and solid tumors have highlighted the benefit of 4-1BB on persistence and long-term antitumor responses in vivo [30–32]. In contrast, CD28 co-stimulatory containing CAR T cells exhibit more rapid T cell activation, proliferation, and cytotoxicity as a tradeoff for their lack in persistence. Importantly, the context in which they are being used may dictate which construct gives the better clinical responses. If a strong initial antitumor response is necessary, then CD28 co-stimulation may be warranted, such as a bridge to hematopoietic cell transplant. However, if persistence is favored, then 4-1BB may be more appropriate. Clinically, CD28-based CD19-directed CARs are usually undetectable in patients after 3–6 months [33,34]. Alternatively, 4-1BB-containing counterparts have been documented to persist several years after initial therapy [35]. The fine tuning of activation in response to binding to target antigen is a feature that can be beneficial in mediating optimal target cytotoxicity, delivering desired clinical responses, and reducing potential for activation related toxicities.

With regard to clinical responses, the excitement regarding efficacy and future implications to the field are warranted. Based on results from phase II trials, Kymriah achieves an early objective response rate (ORR) of 52%, which normalizes to 33% at 6 months. As for complete response (CR), Kymriah achieves early CR of 40%, which sits at 29% at 6 months. Comparatively, Yescarta, with its CD28-linker/TM/co-stimulation design, has an early ORR of

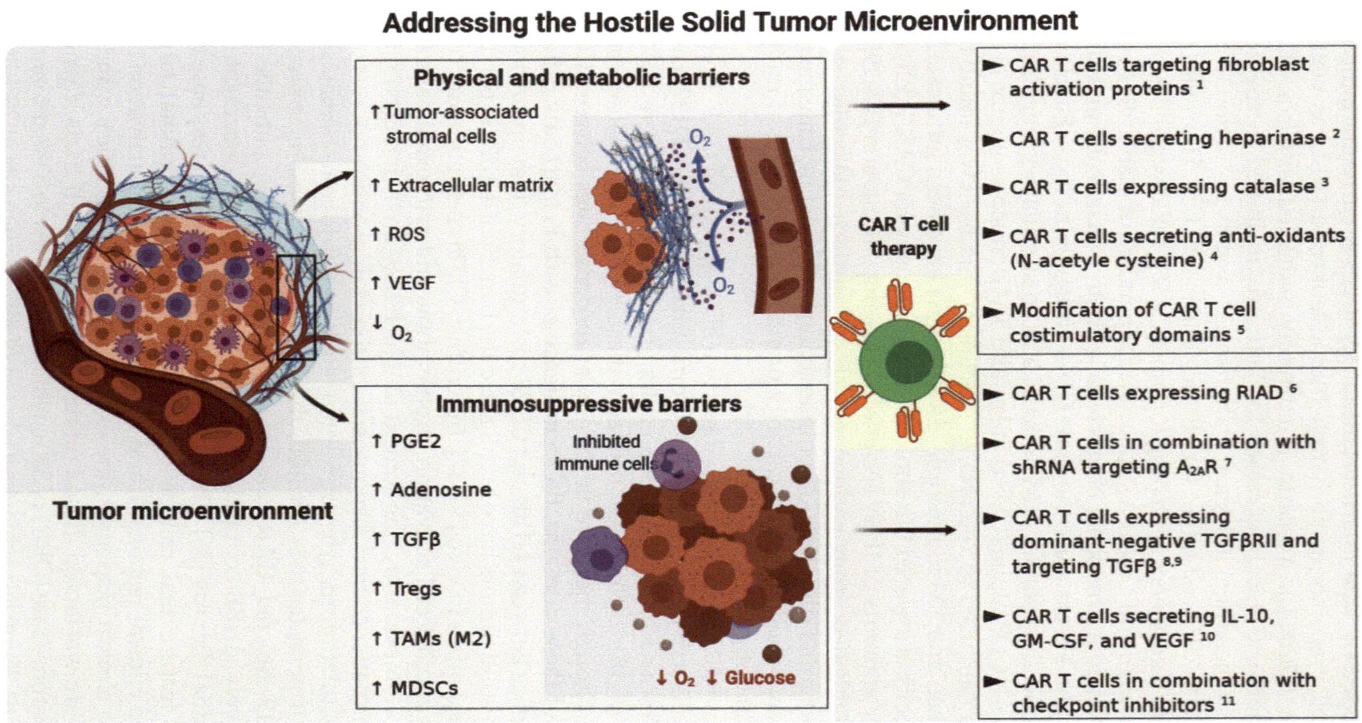

FIG. 3 Description of the physical and biochemical barriers imposed by the solid tumor microenvironment (TME) on CAR T cell therapy and novel CAR T cell engineering approaches proposed to help overcome these unique limitations.

82% (28% at 27 months) and early CR of 58% (37% at 27 months) [36]. Favorable results may appear to lean toward the CD28 costim-containing product in these settings; however, long-term results are still to be determined. The trade-off for enhanced activity may be the reported toxicities. Grade 3 cytokine release syndrome was reported in 22% and 11% of patients on Kymriah or Yescarta, respectively. However, the more dramatic, and possibly deadly, neurotoxicity associated with CD19-CAR T cells is at 12% and 32% of patients on Kymriah or Yescarta, respectively. Therefore, the design of these CAR T cells may hold some keys to determining impact of response and possible toxicities [36,37].

The successes in the CD19-targeing space have established that CAR T cell therapy can be optimized to induce durable and complete responses in patients, even under conditions of highly refractory disease [38]. Furthermore, current clinical trials are on-going in expanding CD19 targeting CAR T cells against multiple subtypes of non-Hodgkin's lymphomas [39,40]. Nearly 3 years after the original two products, the FDA approved the third CD19 targeting CAR T cell, Breyanzi (lisocabtagene maraleucel), developed by Bristol Myers Squibb, which uses a 4-1BB secondary signal but also fixed ratio of CD4 and CD8 in final product, has been reported to have less toxicities than Kymriah and Yescarta [36,41]. The last several years the immunotherapy field has seen an expansion of CAR T cell biology and function, concentrating on translating these impressive responses against CD19 malignancies to additional hematological diseases as well as toward solid cancers [42].

Concerns that the successes of CAR T cell therapy may have been a unique circumstance, having only gained clinical benefit in CD19+ B-cell malignancies, have been eased with recent approval of B-cell maturation antigen (BCMA) targeting product Abecma (idecabtagene vicleucel) for multiple myeloma (MM) [43,44]. Again, as with CD19, BCMA is nearly uniformly expressed on MM cells, with restriction to plasma or B cell lineages, making it an ideal target for this disease setting. The BCMA-targeting construct has a CD8 linker and TM domain followed by a 4-1BB co-stimulatory signaling molecule and CD3ζ primary signaling domain (Fig. 3) [45]. In phase II trials recently completed, Abecma achieved 73% partial response with 33% having CR or stringent complete response rate. Importantly, BCMA may be also expressed on recurrent B cell malignancy escape variants from relapsed patients treated originally with CD19-targeting therapies, and thus allowing for an additional avenue for these patients [46]. These results again highlight the potential of these CAR-engineered strategies in the hematological space and give insights to structure and function considerations for future solid tumor targeting CAR T cells.

Challenges in solid tumor targeting with CAR T cell therapy

Despite the promising clinical success demonstrated by CAR T cells against hematological malignancies, therapy against solid tumors have proven to be more difficult. To date, there have been an unsatisfying number of cases reporting positive therapeutic efficacy against solid tumors. For example, epidermal growth factor receptor variant III (EGFRvIII) targeting CAR T cell delivered intravenously demonstrated one patient with residual stable disease for over 18 months, while others showed detectable transient expansion and infiltration of CAR T cells at the tumor site without robust antitumor activity [47]. Disialoganglioside GD2 has also been targeted for the treatment of neuroblastoma using a first-generation CAR T cell without lymphodepletion, which resulted in 3 out of 11 patients with complete remissions [48]. Additionally, HER2 targeting second-generation CAR T cells treated without nonmyeloablative preconditioning against sarcoma showed 4 out of 17 patients with stable disease for up to 14 months [49]. There are currently numerous ongoing clinical trials targeting multiple tumor antigens that have failed to demonstrate comparable outcomes to CAR T cell therapies for B-cell malignancies. Therefore, the treatment of various solid cancer continues to be an exciting frontier for CAR T cell therapy, but the advancement in this landscape has been met with major challenges including the physical and immunosuppressive barriers in the solid tumor microenvironment and antigen escape mechanisms mediated by tumor antigen heterogeneity.

Physical and metabolic barriers in the solid tumor microenvironment

The tumor microenvironment is a remarkably dynamic and heterogeneous milieu composed of tumor cells, blood vessels, immune cells, fibroblasts, extracellular matrix, and signaling molecules. Better understanding of this complex network is necessary for the success of CAR T cell therapy. One of the major features of the hostile tumor microenvironment is the physical and metabolic barriers. Most solid tumors have originated from epithelial cells allowing them to maintain intercellular junction, which influences the protective mechanism against infiltration and dissemination of CAR T cells. The physical barriers surrounding solid tumors include tumor stroma and extracellular matrix that limit the penetration of CAR T cells into the tumor space. There are cancer-associated stroma cells that play an extensive role in the pathogenesis of solid tumors called cancer associated fibroblasts. These cells enhance immunosuppression in the tumor microenvironment by secreting cytokines and chemokines such as stromal cell derived factor 1 also known as CXCL12 and vascular endothelial growth factor (VEGF) to promote tumor growth, metastasis, and angiogenesis [50]. To combat the physical barrier, tumor stromal cells have been targeted to potentially promote increased efficacy of CAR T cell therapy. CAR T cells targeting a surface marker of cancer-associated fibroblasts called fibroblast activation protein (FAP) have been developed to reduce tumor fibroblasts by attacking the tumor promoting stromal cells to augment endogenous T cell anti-tumor responses [51]. CAR T cells that secrete heparinase, which degrade a major component of the extracellular matrix known as heparan sulfate proteoglycans to promote T cell infiltration into the tumor and increase antitumor activity have also been developed [52].

Metabolic barriers of the tumor microenvironment play a critical role in the downregulation of T cell activity. Hypoxia is a notable characteristic, which is a pathophysiologic consequence of structurally and functionally disrupted microcirculation and diffusion leading to an inadequate supply of oxygen. This unfavorable condition dampens lymphocytic activation in addition to decreasing proliferation since its metabolic activity is heavily dependent on oxygen [53]. Costimulatory domains of CAR T cells were compared to improve antitumor activity in hypoxic conditions. BBζ CAR T cells have been reported to increase mitochondrial spare respiratory capacity resulting in higher metabolic efficiency compared to 28ζ CAR T cells [54]. Hypoxia also initiates the Warburg effect allowing tumor cells to utilize glycolysis pathway rather than oxidative phosphorylation releasing lactate resulting in an acidic tumor microenvironment. The drop in pH levels has a detrimental effect on T cell function. Additionally, this generates increased levels of reactive oxygen species (ROS) produced by all cells as a metabolic product. In solid tumors, the increased ROS aids in proliferation of malignant cells by inhibiting anti-tumor immunity. Therefore, CAR T cells that co-express catalase, which is an antioxidant enzyme that degrades H_2O_2, have been developed to allow for increased resilience toward the ROS in the tumor microenvironment providing protection from oxidative damage [55]. CAR T cells that secrete antioxidants such as *N*-acetyl cysteine to overcome oxidative stress has been tested as another approach to potentially minimize the negative effects of metabolic stress [56]. Furthermore, the hypoxic tumor microenvironment leads to the production of hypoxia-inducible factor (HIF) that enhances tumor angiogenesis, proliferation, invasion, metastasis, and growth [57]. Recently, HIFs have been exploited to sensitize CAR T cells through hypoxia to increase tumor specificity [58].

The immunosuppressive TME

Another major component of the hostile tumor microenvironment includes the secreted tumor-derived soluble factors and cytokines. Multiple clinical trials have identified various immunosuppressive soluble factors in the serum, ascites fluid, and tissues extracted from cancer patients. Prostaglandin E2 (PGE2) is an important soluble factor in the tumor microenvironment. PGE2 is a small molecule secreted by tumor cells that contributes to cancer immunotherapy resistance by aiding in tumor promoting activities [59]. This molecule has the capacity to suppress the function of macrophages, neutrophils, and Th1 cells. Additionally, it can downregulate T cell proliferation and production of Th1 cytokines [60]. Adenosine is another molecule found in high levels within solid tumors due to its hypoxic conditions, which can promote tumor growth and progression at multiple pathophysiological levels and can act as a potent inhibitor of T cell function and proliferation [61]. The combination of PGE2 and adenosine found in the tumor microenvironment has demonstrated the inhibition of antitumor activity and cytokine production by activating protein kinase A (PKA) [62]. CAR T cells that express a small peptide called regulatory subunit I anchoring disruptor (RIAD) have been developed, which has shown to augment mesothelin targeting CAR T cell trafficking and efficacy by blocking PKA localization [63]. Furthermore, dampened CAR T cell activity by the upregulation of adenosine 2A Receptors ($A_{2A}R$) was reversed using either $A_{2A}R$ antagonists or genetic targeting of $A_{2A}R$ using shRNA [64].

Cytokines are found in the tumor microenvironment to regulate immunity. One of the most notable inhibitory cytokines is transforming growth factor β (TGF-β), which is released by cancer-associated fibroblasts, mesenchymal stem cells, blood endothelial cells, lymphatic epithelial cells, pericytes, and multiple lineages of leukocytes such as macrophages. TGF-β is known to promote tumor initiation and progression while suppressing the function of CD8 + T cells. Furthermore, it mediates the differentiation of T cells to become more immunosuppressive. Naïve T cells treated with TGF-β have shown to induce the expression of Forkhead box protein P3 (FOXP3) promoting its conversion into regulatory T cells (Treg), which suppresses the adaptive immune response in the tumor microenvironment [65,66]. In addition to the negative effects on the adaptive immune system, TGF-β has been reported to adversely impact the innate immune system. This cytokine significantly disrupts the recruitment, activation, and function of macrophages, neutrophils, natural killer cells, and dendritic cells by acting as an antagonist of major immune functions to decrease the recognition and clearance of tumor cells [67]. TGF-β induces M2 macrophage polarization, which is associated with the production of nitric oxide that suppresses monocyte-mediated cell death [68]. It is also known to inhibit the functional maturation of NK cells resulting in immature NK cells that do not effectively respond to foreign antigens. Neutrophils exposed to TGF-β lead to the emergence of N2 neutrophil expression profile that are associated with the expression of pro-tumorigenic factors such as MMP9 and CXCL1 [69]. Currently, there are multiple ongoing efforts to counteract the effects of this immunosuppressive cytokine. Prostate-specific membrane antigen (PSMA) targeting CAR T cells that co-express a dominant-negative TGF-βRII have demonstrated increased proliferation, persistence, cytokine secretion, resistance to exhaustion, and antitumor activity in mouse models [70]. CAR T cells have also been rewired to respond to soluble ligands such as TGF-β mediating CAR dimerization. Additionally, a bifunctional anti-PD-L1/TGF-β trap fusion protein has been engineered to show regression of the cytokine-dependent immunosuppressive activity [71].

The hostile tumor microenvironment is crowded with immunosuppressive immune populations including Tregs, myeloid-derived suppressor cells (MDSC), tumor-associated macrophages (TAM) with M2 phenotype, tumor-associated neutrophils (TAN) with N2 phenotype, and immature dendritic cells (iDC), which have the capacity to secrete immunosuppressive factors. Tregs suppress CD8+ T cell activity by downregulating CD80/CD86 signaling pathways associated with T cell activation to inhibit its antitumor activity leading to enhanced tumor development and progression [72]. They also secrete immunosuppressive cytokines TGF-β and IL-10 while intercepting IL-2, a cytokine that enhance CAR T cell proliferation and efficacy [73,74]. MDSCs play a major role in the immunosuppressive activity within the tumor microenvironment by secreting and expressing IL-10, TGF-β, inducible nitric oxide synthase (iNOS), arginine, cyclooxygenase-2 (COX-2), and indolamine 2,3-dioxygenase (IDO) [75]. They also have the potential to induce Treg activation. Thus, CAR T cell activity can be significantly impacted by the presence of MDSCs. TLR-9 targeting STAT3-siRNAs have been developed to alleviate MDSC-mediated immunosuppression [76]. MDSCs have also been targeted by an antibody drug conjugate called Gemtuzumab ozogamicin to enhance CAR T cell persistence and anti-tumor activity [77]. TAMs with M2 phenotype have a significant role in tumor development and immunosuppression by secreting IL-10, granulocyte-macrophage colony stimulating factor (GM-CSF), and VEGF [78]. TAMs secrete matrix metalloproteinases (MMP) such as MMP2 and MMP9 that break up the extracellular matrix of solid tumors to induce metastasis [79]. These cells also secrete chemokines such as CCL17, CCL18, and CCL22 that can attract Tregs to the tumor site to disrupt CAR T cell activity. TAMs express high levels of programmed death-ligand 1 (PD-L1) to inhibit CAR T cell function. Similarly, TANs promote tumor progression by inhibiting CD8 + T cell activity by secreting arginase 1 and modulating PD-L1/programmed cell death protein 1 (PD-1) signaling. In addition to the immunosuppressive immune populations, immune inhibitory checkpoints play a major role in suppressing CAR T cell activity. Currently, there are multiple inhibitory receptors that have been identified on the surface of T cells including PD-1, cytotoxic T lymphocyte antigen 4 (CTLA-4), lymphocyte activation gene 3 (LAG3), T cell immunoglobulin and mucin-3 (TIM3), T cell immunoglobulin and ITIM domain (TIGIT), B and T lymphocyte attenuator (CTLA), and $A_{2A}R$ [80]. The upregulation of these checkpoint inhibitors has been demonstrated on CAR T cells when reacting with its corresponding ligands found on the surface of tumor cells or immunosuppressive immune cells. Therefore, various checkpoint inhibitors have been combined with CAR T cells to improve antitumor activity against solid tumors [81].

Solid tumor antigen heterogeneity and antigen escape

Unlike hematological malignancies, solid tumors lack amenable tumor-specific antigens such as CD19 to be effectively and safely targeted by CAR T cells. Antigen escape is a prominent mechanism of relapse given the heterogeneous nature of solid tumors. Therefore, single-antigen CAR T cell therapy often leads to tumor resistance due to the outgrowth of target-less tumor cells. Even with majority of tumor cells expressing a target, these antigen expressions can be downregulated since they do not have a critical impact on its growth and maintenance. Many combinatorial approaches and modified CAR T cells are currently being tested to overcome this major hurdle for CAR T cell therapy against solid tumors (Fig. 4).

CAR T cells that have the capacity to target multiple antigens are currently being investigated to combat antigen heterogeneity. EGFRvIII-CAR T cells that secrete bispecific T cell engagers against EGFR have been developed to target not only the glioblastoma-specific tumor antigen but also the wild-type EGFR frequently overexpressed in glioblastoma [82]. Another approach is to target multiple antigens using dual transduced or tandem CARs. PSCA and PSMA dual transduced CAR T cells have been tested against prostate cancer while tandem CAR T cells targeting HER2 and IL13Rα2 have been developed to mitigate tumor antigen

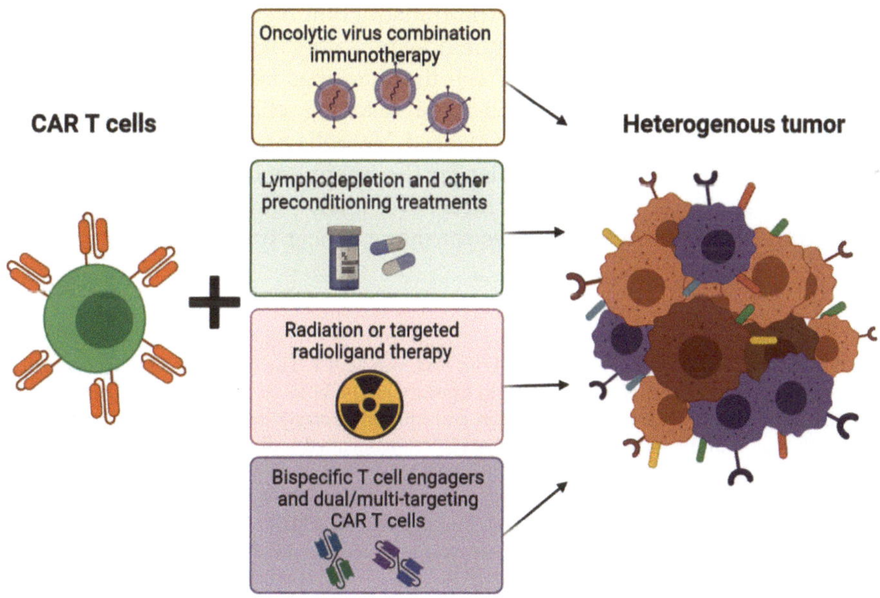

FIG. 4 Brief overview of the current state of combinatorial treatment or engineering strategies to incorporate with CAR T cells in attempts to stimulate endogenous immune function and help overcome the antigen heterogeneity present in a majority of solid tumors.

escape in glioblastoma [83,84]. Various modifications have been made to CAR T cells to enhance multitargeting capabilities. Recently, a synNotch receptor that recognizes a specific priming antigen has been expressed on CAR T cells to enable controlled tumor cell killing by targeting antigens that are expressed on both tumor and normal cells [85].

Future perspective of CAR T cell therapy approaches to tackle solid tumors

Established clinical efficacy of CAR T cell therapy approaches for B-cell malignancies, coupled with multiple pathways to overcome challenges facing solid tumors, has led to a burgeoning field of cell-based immunotherapies for targeting cancer. Next-generation CAR T cell approaches to overcome the immunosuppressive TME and antigen escape mechanisms that limit solid tumor CAR T cell therapies now being clinical investigated hold high hope. Beyond these novel strategies including intrinsic modifications to CAR T cell products as well as the incorporation of rational combination strategies, overall feasibility in treating advanced disease will still require manufacturing optimization and further clinical considerations. Enormous efforts are now being taken to develop allogeneic cell therapies, which have several potential benefits over the majority of current clinical trials involving autologous approaches. These benefits include (1) shorter time to treatment, (2) potential for resultant CAR T cell products that are neither corrupted by the cancer itself nor by the previous harsh cytotoxic therapies, and (3) more equivalent cell therapy products across multiple cancer patients [86]. Additional strategies include innovative engineering strategies including gene editing such as CRISPR-Cas9, ZFN, and TALEN. While these approaches are less well established than the more conventional viral transduction methods, specificity of CAR integration, accuracy of engineering, and the potential for safety have recently been reported [87]. Finally, while engineered conventional T cell therapies has been the dominant cell therapy approach to date, multiple cell therapies are reaching clinical development that may rival T cell therapies, including CAR-engineered NK cells and gamma-delta T cells, which hold promise for future cell-based therapies for the treatment of cancer [88]. Overall, the cell therapy field is expansive, and in the near future will likely offer multiple strategies that show effective and durable antitumor responses for many cancer types.

References

[1] Garrido F, Aptsiauri N, Doorduijn EM, Garcia Lora AM, van Hall T. The urgent need to recover MHC class I in cancers for effective immunotherapy. Curr Opin Immunol 2016;39:44–51.
[2] Bubenik J. Tumour MHC class I downregulation and immunotherapy (review). Oncol Rep 2003;10:2005–8.
[3] Mikyskova R, et al. Immune escape phenotype of HPV16-associated tumours: MHC class I expression changes during progression and therapy. Int J Oncol 2005;26:521–7.
[4] Reinis M, et al. Immunization with MHC class I-negative but not -positive HPV16-associated tumour cells inhibits growth of MHC class I-negative tumours. Int J Oncol 2007;30:1011–7.
[5] Paulson KG, et al. Downregulation of MHC-I expression is prevalent but reversible in Merkel cell carcinoma. Cancer Immunol Res 2014;2:1071–9.
[6] Tcyganov E, Mastio J, Chen E, Gabrilovich DI. Plasticity of myeloid-derived suppressor cells in cancer. Curr Opin Immunol 2018;51:76–82.
[7] Abrams SI, Netherby CS, Twum DY, Messmer MN. Relevance of interferon regulatory factor-8 expression in myeloid-tumor interactions. J Interferon Cytokine Res 2016;36:442–53.

[8] Gabrilovich DI. Myeloid-derived suppressor cells. Cancer Immunol Res 2017;5:3–8.
[9] Hartmann J, Schüßler-Lenz M, Bondanza A, Buchholz CJ. Clinical development of CAR T cells—challenges and opportunities in translating innovative treatment concepts. EMBO Mol Med 2017;, e201607485.
[10] Abate-Daga D, Davila ML. CAR models: next-generation CAR modifications for enhanced T-cell function. Mol Ther Oncolytics 2016;3.
[11] D'Aloia MM, Zizzari IG, Sacchetti B, Pierelli L, Alimandi M. CAR-T cells: the long and winding road to solid tumors. Cell Death Dis 2018;9:282.
[12] Chen L, Flies DB. Molecular mechanisms of T cell co-stimulation and co-inhibition. Nat Rev Immunol 2013;13:227–42.
[13] Chen N, Li X, Chintala NK, Tano ZE, Adusumilli PS. Driving CARs on the uneven road of antigen heterogeneity in solid tumors. Curr Opin Immunol 2018;51:103–10.
[14] Yeku OO, Brentjens RJ. Armored CAR T-cells: utilizing cytokines and pro-inflammatory ligands to enhance CAR T-cell anti-tumour efficacy. Biochem Soc Trans 2016;44:412–8.
[15] Jung I-Y, Lee J. Unleashing the therapeutic potential of CAR-T cell therapy using gene-editing technologies. Mol Cells 2018;41:717–23.
[16] Levine BL, Miskin J, Wonnacott K, Keir C. Global manufacturing of CAR T cell therapy. Mol Ther Methods Clin Dev 2016;4:92–101.
[17] Hurton LV, et al. Tethered IL-15 augments antitumor activity and promotes a stem-cell memory subset in tumor-specific T cells. Proc Natl Acad Sci U S A 2016;113:E7788–97.
[18] Danilo M, Chennupati V, Silva JG, Siegert S, Held W. Suppression of Tcf1 by inflammatory cytokines facilitates effector CD8 T cell differentiation. Cell Rep 2018;22:2107–17.
[19] Wang D, et al. Glioblastoma-targeted CD4+ CAR T cells mediate superior antitumor activity. JCI Insight 2018;3, e99048.
[20] Odé Z, Condori J, Peterson N, Zhou S, Krenciute G. CRISPR-mediated non-viral site-specific gene integration and expression in T cells: protocol and application for T-cell therapy. Cancer 2020;12:1704.
[21] Globerson Levin A, Rivière I, Eshhar Z, Sadelain M. CAR T cells: building on the CD19 paradigm. Eur J Immunol 2021;51:2151–63.
[22] Lukjanov V, Koutná I, Šimara P. CAR T-cell production using nonviral approaches. J Immunol Res 2021;2021:6644685.
[23] Choosing the right tool for genetic engineering: clinical lessons from chimeric antigen receptor-T cells. Hum Gene Ther 2021;32:1044–58.
[24] Gattinoni L, Klebanoff CA, Restifo NP. Paths to stemness: building the ultimate antitumour T cell. Nat Rev Cancer 2012;12:671–84.
[25] Wang K, Wei G, Liu D. CD19: a biomarker for B cell development, lymphoma diagnosis and therapy. Exp Hematol Oncol 2012;1:36.
[26] Gu R, et al. Efficacy and safety of CD19 CAR T constructed with a new anti-CD19 chimeric antigen receptor in relapsed or refractory acute lymphoblastic leukemia. J Hematol Oncol 2020;13:122.
[27] Alabanza L, et al. Function of novel anti-CD19 chimeric antigen receptors with human variable regions is affected by hinge and transmembrane domains. Mol Ther 2017;25:2452–65.
[28] Majzner RG, et al. Tuning the antigen density requirement for CAR T-cell activity. Cancer Discov 2020;10:702.
[29] Muller YD, et al. The CD28-transmembrane domain mediates chimeric antigen receptor heterodimerization with CD28. Front Immunol 2021;12.
[30] Priceman SJ, et al. Co-stimulatory signaling determines tumor antigen sensitivity and persistence of CAR T cells targeting PSCA+ metastatic prostate cancer. Onco Targets Ther 2018;7, e1380764.
[31] Milone MC, et al. Chimeric receptors containing CD137 signal transduction domains mediate enhanced survival of T cells and increased antileukemic efficacy in vivo. Mol Ther 2009;17:1453–64.
[32] Carpenito C, et al. Control of large, established tumor xenografts with genetically retargeted human T cells containing CD28 and CD137 domains. Proc Natl Acad Sci 2009;106:3360.
[33] Brentjens RJ, et al. Safety and persistence of adoptively transferred autologous CD19-targeted T cells in patients with relapsed or chemotherapy refractory B-cell leukemias. Blood 2011;118:4817–28.
[34] Park JH, et al. Long-term follow-up of CD19 CAR therapy in acute lymphoblastic leukemia. N Engl J Med 2018;378:449–59.

[35] Fraietta JA, et al. Determinants of response and resistance to CD19 chimeric antigen receptor (CAR) T cell therapy of chronic lymphocytic leukemia. Nat Med 2018;24:563–71.
[36] Halim L, Maher J. CAR T-cell immunotherapy of B-cell malignancy: the story so far. Ther Adv Vaccines Immunother 2020;8, 2515135520927164.
[37] Gust J, Taraseviciute A, Turtle CJ. Neurotoxicity associated with CD19-targeted CAR-T cell therapies. CNS Drugs 2018;32:1091–101.
[38] Pang Y, Hou X, Yang C, Liu Y, Jiang G. Advances on chimeric antigen receptor-modified T-cell therapy for oncotherapy. Mol Cancer 2018;17:91.
[39] Wang X, et al. Phase I studies of central-memory-derived CD19 CAR T cell therapy following autologous HSCT in patients with B-cell NHL. Blood 2016;127(24):2980-90.
[40] Till BG, et al. Adoptive immunotherapy for indolent non-Hodgkin lymphoma and mantle cell lymphoma using genetically modified autologous CD20-specific T cells. Blood 2008;112:2261–71.
[41] Abramson JS, et al. Lisocabtagene maraleucel for patients with relapsed or refractory large B-cell lymphomas (TRANSCEND NHL 001): a multicentre seamless design study. Lancet 2020;396:839–52.
[42] Priceman SJ, Forman SJ, Brown CE. Smart CARs engineered for cancer immunotherapy. Curr Opin Oncol 2015;27:466–74.
[43] Munshi NC, et al. Idecabtagene Vicleucel in relapsed and refractory multiple myeloma. N Engl J Med 2021;384:705–16.
[44] Raje N, et al. Anti-BCMA CAR T-cell therapy bb2121 in relapsed or refractory multiple myeloma. N Engl J Med 2019;380:1726–37.
[45] Friedman KM, et al. Effective targeting of multiple B-cell maturation antigen-expressing hematological malignances by anti-B-cell maturation antigen chimeric antigen receptor T cells. Hum Gene Ther 2018;29:585–601.
[46] Rodríguez-Lobato LG, et al. CAR T-cells in multiple myeloma: state of the art and future directions. Front Oncol 2020;10:1243.
[47] O'Rourke DM, et al. A single dose of peripherally infused EGFRvIII-directed CAR T cells mediates antigen loss and induces adaptive resistance in patients with recurrent glioblastoma. Sci Transl Med 2017;9.
[48] Louis CU, et al. Antitumor activity and long-term fate of chimeric antigen receptor-positive T cells in patients with neuroblastoma. Blood 2011;118:6050–6.
[49] Ahmed N, et al. Human epidermal growth factor receptor 2 (HER2)-specific chimeric antigen receptor-modified-T cells for the immunotherapy of HER2-positive sarcoma. J Clin Oncol Off J Am Soc Clin Oncol 2015;33:1688–96.
[50] Tahmasebi S, Elahi R, Esmaeilzadeh A. Solid tumors challenges and new insights of CAR T cell engineering. Stem Cell Rev Rep 2019;15:619–36.
[51] Wang LC, et al. Targeting fibroblast activation protein in tumor stroma with chimeric antigen receptor T cells can inhibit tumor growth and augment host immunity without severe toxicity. Cancer Immunol Res 2014;2:154–66.
[52] Caruana I, et al. Heparanase promotes tumor infiltration and antitumor activity of CAR-redirected T lymphocytes. Nat Med 2015;21:524–9.
[53] McNamee EN, Korns Johnson D, Homann D, Clambey ET. Hypoxia and hypoxia-inducible factors as regulators of T cell development, differentiation, and function. Immunol Res 2013;55:58–70.
[54] Kawalekar OU, et al. Distinct signaling of coreceptors regulates specific metabolism pathways and impacts memory development in CAR T cells. Immunity 2016;44:380–90.
[55] Ligtenberg MA, et al. Coexpressed catalase protects chimeric antigen receptor-redirected T cells as well as bystander cells from oxidative stress-induced loss of antitumor activity. J Immunol 2016;196:759.
[56] Scheffel MJ, et al. Efficacy of adoptive T-cell therapy is improved by treatment with the antioxidant N-acetyl cysteine, which limits activation-induced T-cell death. Cancer Res 2016;76:6006.
[57] Maybin JA, et al. Hypoxia and hypoxia inducible factor-1α are required for normal endometrial repair during menstruation. Nat Commun 2018;9:295.
[58] Juillerat A, et al. An oxygen sensitive self-decision making engineered CAR T-cell. Sci Rep 2017;7:39833.
[59] Park A, et al. Prostaglandin E2 secreted by thyroid cancer cells contributes to immune escape through the suppression of natural killer (NK) cell cytotoxicity and NK cell differentiation. Front Immunol 2018;9:1859.
[60] Nakanishi M, Rosenberg DW. Multifaceted roles of PGE2 in inflammation and cancer. Semin Immunopathol 2013;35:123–37.
[61] Vaupel P, Mayer A. Hypoxia-driven adenosine accumulation: a crucial microenvironmental factor promoting tumor progression. Adv Exp Med Biol 2016;876:177–83.

[62] Su Y, et al. Cooperation of adenosine and prostaglandin E2 (PGE2) in amplification of cAMP-PKA signaling and immunosuppression. Cancer Immunol Immunother 2008;57:1611–23.
[63] Newick K, et al. Augmentation of CAR T-cell trafficking and antitumor efficacy by blocking protein kinase a localization. Cancer Immunol Res 2016;4:541–51.
[64] Beavis PA, et al. Targeting the adenosine 2A receptor enhances chimeric antigen receptor T cell efficacy. J Clin Invest 2017;127:929–41.
[65] Chen W, et al. Conversion of peripheral CD4+CD25− naive T cells to CD4+CD25+ regulatory T cells by TGF-β induction of transcription factor Foxp3. J Exp Med 2003;198:1875–86.
[66] Tone Y, et al. Smad3 and NFAT cooperate to induce Foxp3 expression through its enhancer. Nat Immunol 2008;9:194–202.
[67] Flavell RA, Sanjabi S, Wrzesinski SH, Licona-Limón P. The polarization of immune cells in the tumour environment by TGFbeta. Nat Rev Immunol 2010;10:554–67.
[68] Gong D, et al. TGFβ signaling plays a critical role in promoting alternative macrophage activation. BMC Immunol 2012;13:31.
[69] Fridlender ZG, et al. Polarization of tumor-associated neutrophil phenotype by TGF-beta: "N1" versus "N2" TAN. Cancer Cell 2009;16:183–94.
[70] Kloss CC, et al. Dominant-negative TGF-β receptor enhances PSMA-targeted human CAR T cell proliferation and augments prostate cancer eradication. Mol Ther 2018;26:1855–66.
[71] Lan Y, et al. Enhanced preclinical antitumor activity of M7824, a bifunctional fusion protein simultaneously targeting PD-L1 and TGF-β. Sci Transl Med 2018;10(424)
[72] Scarfò I, Maus MV. Current approaches to increase CAR T cell potency in solid tumors: targeting the tumor microenvironment. J Immunother Cancer 2017;5:28.
[73] Ohue Y, Nishikawa H. Regulatory T (Treg) cells in cancer: can Treg cells be a new therapeutic target? Cancer Sci 2019;110:2080–9.
[74] Chinen T, et al. An essential role for the IL-2 receptor in T(reg) cell function. Nat Immunol 2016;17:1322–33.
[75] Gabrilovich DI, Nagaraj S. Myeloid-derived suppressor cells as regulators of the immune system. Nat Rev Immunol 2009;9:162–74.
[76] Kortylewski M, et al. In vivo delivery of siRNA to immune cells by conjugation to a TLR9 agonist enhances antitumor immune responses. Nat Biotechnol 2009;27:925–32.
[77] Fultang L, et al. MDSC targeting with Gemtuzumab ozogamicin restores T cell immunity and immunotherapy against cancers. eBioMedicine 2019;47:235–46.
[78] Solinas G, et al. Tumor-conditioned macrophages secrete migration-stimulating factor: a new marker for M2-polarization, influencing tumor cell motility. J Immunol 2010;185:642–52.
[79] Condeelis J, Pollard JW. Macrophages: obligate partners for tumor cell migration, invasion, and metastasis. Cell 2006;124:263–6.
[80] Zahavi DJ, Weiner LM. Targeting multiple receptors to increase checkpoint blockade efficacy. Int J Mol Sci 2019;20:158.
[81] Grosser R, Cherkassky L, Chintala N, Adusumilli PS. Combination immunotherapy with CAR T cells and checkpoint blockade for the treatment of solid tumors. Cancer Cell 2019;36:471–82.
[82] Choi BD, et al. CAR-T cells secreting BiTEs circumvent antigen escape without detectable toxicity. Nat Biotechnol 2019;37:1049–58.
[83] Kloss CC, Condomines M, Cartellieri M, Bachmann M, Sadelain M. Combinatorial antigen recognition with balanced signaling promotes selective tumor eradication by engineered T cells. Nat Biotechnol 2013;31:71–5.
[84] Hegde M, et al. Tandem CAR T cells targeting HER2 and IL13Ralpha2 mitigate tumor antigen escape. J Clin Invest 2016;126:3036–52.
[85] Choe JH, et al. SynNotch-CAR T cells overcome challenges of specificity, heterogeneity, and persistence in treating glioblastoma. Sci Transl Med 2021;13.
[86] Caldwell KJ, Gottschalk S, Talleur AC. Allogeneic CAR cell therapy—more than a pipe dream. Front Immunol 2020;11:618427.
[87] Stadtmauer EA, et al. CRISPR-engineered T cells in patients with refractory cancer. Science 2020;367.
[88] Morgan MA, Buning H, Sauer M, Schambach A. Use of cell and genome modification technologies to generate improved "off-the-shelf" CAR T and CAR NK cells. Front Immunol 2020;11:1965.

SECTION II

Process and trials optimization: Diagnostics, readouts, route, and production to optimize cell therapy

CHAPTER

8

Optimization of production for cell therapies

Xiuyan Wang[a,b,c] *and Isabelle Rivière*[a,b,c]

[a]Cell Therapy and Cell Engineering Facility, Memorial Sloan-Kettering Cancer Center, New York, NY, United States [b]Center for Cell Engineering, Memorial Sloan-Kettering Cancer Center, New York, NY, United States [c]Molecular Pharmacology Program, Memorial Sloan-Kettering Cancer Center, New York, NY, United States

Abstract

The development of cells as therapeutic agents to treat diseases has had unparalleled growth in recent years. Establishment and optimization of the manufacturing procedure are the prerequirement for the wide applications of such therapies. Determining the target product profile and the critical quality attributes, implementing the quality by design approach and understanding the mode of action are key steps for a successful process development endeavor. For a safe, cost-effective, and compliant cellular product manufacturing procedure, it is critical to integrate scientific knowledge, perform risk analysis, and connect measurable characteristics of the process to the final product release specifications during the optimization and process development phases.

Abbreviations

cGMP current good manufacturing practice
CPA cryoprotectant agents
CPPs critical process parameters
CQA critical quality attribute
CRF controlled rate freezer
DoE design of experiment
FDA Food and Drug Administration
GLP good laboratory practice
HPC hematopoietic progenitor cells
KPPs key process parameters
MOA mechanism of action
PAT process analytical technology
QbD quality by design
QC quality control
TPP target product profile

https://doi.org/10.1016/B978-0-12-822620-9.00004-5

Conflict of interest

I.R. reports research funding from Juno Therapeutics, Fate Therapeutics, Takeda Pharmaceuticals, and Atara Biotherapeutics, unrelated to the present manuscript.

Since the report of the first successful bone marrow transplantation for the treatment of leukemia in 1966 [1], cell therapies have been an active area of investigation aiming to cure diseases by adoptive cell transfer. In recent years, the relatively straightforward adoptive cell transfer approaches have been combined with genetic engineering to harness our immune system with additional powers. Cell therapies have been designed to treat prominent disorders such as cardiovascular, neurological, and autoimmune diseases as well as cancer using cells such as hematopoietic stem cells [2], mesenchymal stem cells [3], dendritic cells [4], T cells [5,6], and natural killer cells [7]. Up to March 2021, US Food and Drug Administration (FDA) has approved 17 Cellular Therapy Products (Table 1). A growing number of cell therapies are advancing from early to late clinical trial stages toward approval and FDA is anticipating the approval of 10–20 cell and gene therapy products per year by 2025.

Development of a new cell therapy typically encompasses three major stages, the preclinical proof-of-concept studies in an inventor's research laboratory, followed by technology transfer and process development. Many of the cell therapy products are initially developed in the academic setting. To translate a good laboratory practice (GLP) procedure from the research setting to a good manufacturing practice (GMP) procedure for clinical application, a gap analysis for GMP compliance is required at each step, ranging from incoming raw material quality, bill of materials for production, cell isolation/enrichment/genetic modification protocols, scale up or scale out manufacturing platform selection/implementation, formulation and cryopreservation scheme, up to the final product shipping containers (Fig. 1). At later stages, the product manufacturing process often needs to be transferred to industrial partners for commercialization purposes. To facilitate such tech transfer, it is helpful to standardize (1) the manufacturing process, (2) the parameters of the process, and (3) the characterization

TABLE 1 Approved cell therapy products [8,9].

Category	Trade name	Proper name	Manufacturer	Approval time	Indications
Cancer vaccine therapy	Provenge	Autologous cellular immunotherapy	Dendreon Corporation	April 29, 2010	Prostate cancer
Stem cell therapy	HEMACORD	HPC, cord blood	New York Blood Center, Inc	Nov. 10, 2011	Disorders affecting the hematopoietic system
	None	HPC, cord blood	Clinimmune Labs, University of Colorado Cord Blood Bank	May 24, 2012	Disorders affecting the hematopoietic system
	DUCORD	HPC, cord blood	Duke University School of Medicine	Oct. 8, 2012	Disorders affecting the hematopoietic system

TABLE 1 Approved cell therapy products [8,9]—cont'd

Category	Trade name	Proper name	Manufacturer	Approval time	Indications
	ALLOCORD	HPC, cord blood	SSM Cardinal Glennon Children's Medical Center	May 30, 2013	Disorders affecting the hematopoietic system
	None	HPC, cord blood (5th)	LifeSouth Community Blood Centers, Inc.	June 13, 2013	Disorders affecting the hematopoietic system
	CLEVECORD	HPC, cord blood	Cleveland Cord Blood Center	Sep 12, 2016	Disorders affecting the hematopoietic system
	None	HPC, cord blood	Bloodworks	January 28, 2016	Disorders affecting the hematopoietic system
	None	HPC, cord blood	MD Anderson Cord Blood Bank	June 21, 2018	Disorders affecting the hematopoietic system
Adult somatic cells	GINTUIT	Allogeneic cultured keratinocytes and fibroblasts in bovine collagen	Organogenesis Incorporated	March 12, 2012	Wounds of the oral soft tissue defects
	Laviv	Azficel-T	Fibrocell Technologies, Inc	June 22, 2011	Aesthetic therapy
	MACI	Autologous cultured chondrocytes on a porcine collagen membrane	Vericel Corporation	May 31, 2019	Cartilage defects of the knee
CAR-T cell therapy	Kymriah	Tisagenlecleucel	Novartis Pharmaceuticals Corporation	Aug. 30, 2017	B-cell precursor acute lymphoblastic leukemia (ALL)
	YESCARTA	Axicabtagene ciloleucel	Kite Pharma, Incorporated	Oct. 18, 2017	B-cell lymphoma
	TECARTUS	brexucabtagene autoleucel	Kite Pharma, Incorporated	July 24, 2020	Mantle cell lymphoma
	BREYANZI	lisocabtagene maraleucel	Juno Therapeutics, Inc., a Bristol-Myers Squibb Company	Feb. 5, 2021	B-cell lymphoma
	Abecma	Idecabtagene vicleucel	Juno Therapeutics, Inc., a Bristol-Myers Squibb Company	March 26, 2021	Multiple myeloma

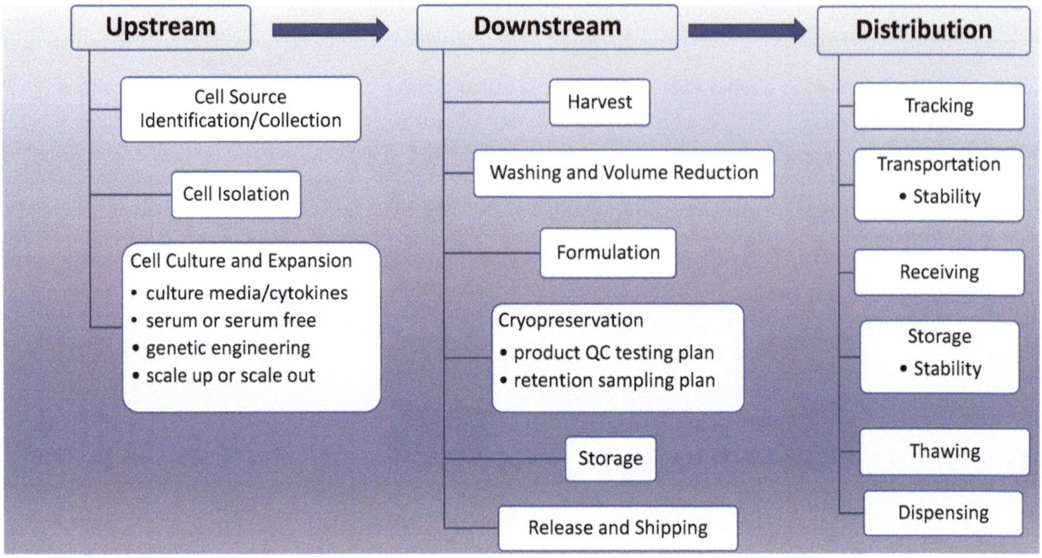

FIG. 1 Flowchart of a typical cell therapy manufacturing procedure and distribution.

of the product, starting from the initial process development to the optimization stage. It is also advisable to replace manual process with validated automated unit operations and manufacturing solutions when possible. These efforts will pay extra dividends at the time of process scaling up.

Keep the end goal in mind from early stage: Define the target product profile

The knowledge of mechanism of action (MOA) and target product profile (TPP) of a cellular product is a great starting point to develop a manufacturing process. The inception of the TPP concept aims at improving the workflow between the sponsor and the FDA during the drug development process. In 2007, FDA issued a draft guidance focusing on the use of TPP as a strategic development tool [10]. TPP defines the goals of the drug development and contains the overall plan for the drug development program, spanning nonclinical research activities, process development, clinical implementation, and marketing competitive strategies. It is a helpful in identifying desired product attributes and focusing team efforts on program implementation early in the process.

Table 2 lists some examples of product or process elements addressed in TPP. Each element is defined at three levels: the minimal acceptable level, the desired ideal specification in the best scenario, and the target specification. It is worthwhile to note that TPP is a "living" document that should be continuously updated to reflect changes in clinical development programs, outcomes of clinical trials, input from the FDA, or increasing knowledge of the product.

TABLE 2 Examples of product or process elements in TPP.

Parameters	Example elements	Acceptable process	Desired ideal process	Target process
Upstream	Starting cell population	Apheresis product	Defined master cell bank	Selected subsets of cells from apheresis product
	Culture time	Varies	Set manufacturing length	Set manufacturing length
	Culture technology	Flasks or cell bags	Bioreactor	Bioreactor
	In process product phenotype testing turn over time	24h	2h	4h
Downstream	Lot size	2×10^9	20×10^9	5×10^9
	End of process cell viability	60%	90%	75%
	Processing technology	Manual with open manipulation	Closed system, automatic	Semiclosed system, largely automatic
	Net volume per dose	varies	15mL	15mL
	Sterility testing turn over time	14days	7days	7–10days
Product storage condition	Storage condition	4°C	Liquid nitrogen	Liquid nitrogen
	Product stability	Overnight	Years	Months to years
Drug administration	Fresh or frozen product	Fresh	Frozen	Frozen
	Administration route	Intravenous	Intravenous	Intravenous
	Dosing strategy	Per patient weight	Flat dosing	Flat dosing

Product critical quality attributes (CQAs) and critical process parameters (CPPs), both part of TPP, are perhaps the most important aspects for production optimization and establishment of proper process controls (Fig. 2). CQAs refer to the physical, chemical, biological, and microbial characteristics of the cell therapy product that are informative for product safety, purity, identity, and efficacy [11]. CPPs are critical process parameters that have a direct impact on CQAs, and they cannot be identified in the absence of CQAs. For example, if the culture must be fed every 2days with a specific cytokine to generate the desired phenotype, then feeding every 2days with the specific cytokine is a CPP. Due to the inherent biological limitations and technical issues, it is challenging to define the CQAs and CPPs for cell therapy products. Several candidate CQAs based on product MOA could be identified first and systematically analyzed to correlate the final CQAs with product quality and clinical

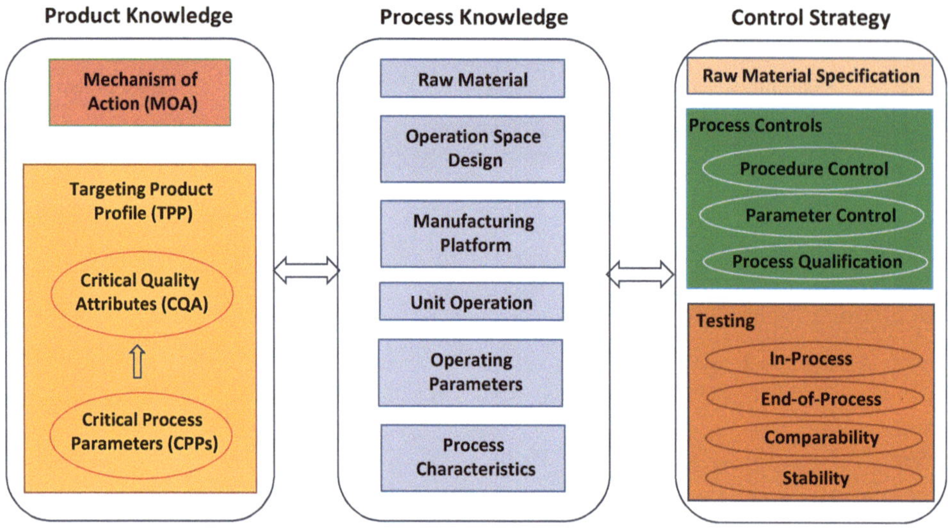

FIG. 2 Production process design and development for cell therapy products.

outcomes. There are other important process parameters that are not necessarily linked to CQAs. For example, if the quality of the cellular product is the same for a culture time of 7–10 days, then culture time of 7 days or 10 days is not a CPP but rather an economical consideration, given longer culture time may produce more cells or may require more resources. In the absence or only partial knowledge of CPPs, full knowledge of noncritical key process parameters (KPPs), such as the range of growth factor concentration needed during culture, would become equally important.

Process design: Take a quality by design approach

The norm that "the process is the product" is widely accepted in the cell therapy field. It is beneficial to address manufacturing concerns upfront as putting off or finding a suboptimal solution for a specific concern in the early stage of process development could lead to drastic, complicated and expensive remedies, such as bridging studies down the road [12,13]. Quality by design (QbD) is a quality system that integrates scientific knowledge and risk analysis into manufacturing process development and optimization. The biological complexity of cell therapy product makes this concept especially relevant. QbD is a tool that encompasses product and process description, characterization, design, monitoring, and continuous improvement (Fig. 2). A thorough understanding of the fundamental biology of a cellular product and the engineering requirement underlying its production is the basis for this approach.

The connection of measurable molecular and cellular characteristics of a cell product to its therapeutic efficacy through QbD is a crucial step in realizing the full potential of a cell therapy. Once TPP and the CQA have been established, a risk analysis could identify the most

FIG. 3 A schematic representation for quality by design. *CQA*, critical quality attributes; *PAT*, process analytical technology; *TPP*, targeting product profile.

influential CPPs and material attributes. It is important to remember that quality is not tested into the cellular product, but designed into the manufacturing process. A control strategy for the entire manufacturing process is required to ensure that product CQAs remain within an acceptable range. In addition, the manufacturing procedure should be validated at scale, monitored continually during manufacturing runs, and improved and optimized with increasing knowledge of the process (Fig. 3) [14]. Process design and process optimization are interactive and continuously being refined. Design of experiment (DoE) is another useful tool to help design an experiment to test multivariates at the same time for the rather sophisticated manufacturing process of cell therapy product [14].

Get to the goal: Many decisions to make

Many decisions need to be made during process development and optimization. It is important to remember that a robust and reproducible manufacturing process starts from the use of well-defined raw materials and should be conducted under current good manufacturing practice (cGMP). The manufacturing platform for a cellular product is dictated by the therapeutic design of the cellular product. For a product used in an autologous setting, a relatively small platform could be used since each production only needs to generate enough cells for one patient. In order to expand the manufacturing capacity for an autologous product, a "scale out" production model in which the same process will be replicated many times needs to be established. Production for an allogenic product, on the other hand, needs to support the generation of multiple doses for many patients in each individual lot, and a "scale-up" production model needs to be validated to increase manufacture throughput. We can arbitrarily simplify the design space for the process development and optimization into three sections: upstream processing, downstream processing, and final formulation/cryopreservation.

Upstream processing

The upstream process includes the entire process from early cell isolation and cultivation to cell banking and culture expansion, until termination of the culture, and collection of the live cell batch. Multiple cell types can be utilized for cell therapies [15]. The cell source used for a specific type of cell therapy and its isolation processes could vary greatly and need to be addressed early on. A range of technologies are available for cell isolation based on different mechanisms, such as density gradient centrifugation [16], plastic adhesion [17], counterflow centrifugation [18], and magnetic cell separation [19]. Culture optimization (for both media composition and culture condition) is a good example in which reliable empirical performance can leverage the DoE methodology. Culture media provide nutrients to the cells and could arguably be considered the most critical factor contributing to the health of the culture. The quality and consistency of the components used in the media are key factors for the performance of the medium. It is recommended to use high quality, such as cGMP grade, components and serum free, and chemically defined media whenever possible. Depending on the dose and technology requirements, the choice of a scalable production platform is a key decision in developing a cell-based therapy and need to be made in the very early stages of product development.

Downstream processing

Downstream refers to the steps where the cell mass from upstream is processed to meet purity and quality requirements. Downstream processing for the manufacturing of a cellular product can be divided into cell disruption, cell purification, and a polishing step. Pending on the particular application, one of the aforementioned steps may not be needed. For example, the cell disruption step will not be required for cells growing in suspension. Factors to be considered during downstream processing optimization include, but are not limited to, volume handling, operational restraints, processing time of each step, and number of maximal unit operations. Downstream processing methods for cellular therapy have been improved but are still relatively limited. Traditional centrifugation-based cell washing technology such as COBE2991, discontinuous flow system such as Cell Saver 5, acoustic cell processing separation such as Ekko platform or spinning membrane filtration-based technology such as LOVO have been either adapted from their conventional usage for blood product processing [20] or developed further for downstream cell washing, clarifying, and concentrating. These technologies are largely more suitable for a harvest scale of 1–10 L. For larger processing volume, bioprocessing technologies such as tangential-flow filtration and continuous or counterflow centrifugation have been adapted [21]. An ideal downstream process is a closed, scalable, well-regulated process that supports high recovery, high cell viability, and maintains functionality of the expanded cells.

Filling and cryopreservation

The filling step and cryopreservation step transform the formulated cellular drug substance into the final cellular drug product. It is a high-stake operation step that is inevitably required for commercial development of cellular products. Not only the final product but also likely the starting material and process intermediate would need to be cryopreserved. Suboptimal cryopreservation could not only lead to batch to batch variation but also lower cellular functionality and reduce cell yield. Optimization of cryopreservation is as important as optimization of the rest of the manufacturing procedure, and it is a key element in the successful delivery of cell-based therapies [22]. The major areas of consideration for this critical step include choice of a container, choice of filling systems, cell concentration, choice of cryoprotectant agents (CPA), freezing media composition, and the freezing program and instrument.

Conventional cryovials have been routinely used for master cell banks. However, storage in cryovials is not best practice for clinical and scaled-up process due to potential contamination hazard of stored cells [23]. Closed cryogenic vials have been developed, such as the CellSeal vials [24]. Although they provide an option for freezing smaller volumes of cell products, up to now the filling procedure for such vials is largely manual. In addition, cells cryopreserved in vials will need to be retrieved or transferred to syringes at time of infusion, which requires logistic coordination and trained staff. To freeze large quantities of cells, cryobags are needed. Owing to their relative fragility, cryobags need to be packaged in an overwrap bag and metallic casing at time of cryopreservation and storage [25] to guard against mechanical damage that might lead to material loss and compromised sterility. A fully automated fill-and-finish solution for large numbers of cryobags is still being developed. The recently developed FINIA fill and finish system has the limitation of output of only three cryobags at one time. CPAs predominantly protect cells from the damaging effects of freezing. There are two types of CPAs: penetrating small-molecular-weight CPAs such as DMSO and nonpenetrating large molecular weight CPAs such as hydroxyethyl starch. The degree of protection gained by CPAs depends on variables like freezing cell density, concentration of added CPA, duration of exposure to CPAs, and nature of the carrier solution for the CPA. In various cryopreservation buffer compositions, DMSO is undoubtedly the most widely used CPA along with vehicle components such as serum albumin. Nonetheless, DMSO-free and serum-free cryopreservation buffer have become available in recent years [26]. Two types of devices are available for cryopreservation: passive cooling device, such as CoolCell or Nalgene Mr. Frosty freezing containers, which are manual and relies on −80°C freezer as the cold source; and controlled rate freezer (CRF), which is automatic and programmable for multiple-step cooling profiles. Currently, there are two kinds of CRFs: the liquid nitrogen fed system, such as the Cryomed series [27] and the liquid nitrogen free, heat-pump-driven system, such as the VIA Freeze CRFs [28]. Product stability program should be established in early-stage process development to ensure the implementation of adequate cryopreservation procedures.

Understand the analytical needs

The advancement of new indications and therapeutic modalities is paralleled by the development of analytical technologies. Analytics is not only needed for product-release testing to demonstrate product quality and safety but also required for in-process testing to monitor the manufacturing procedure to ensure the process is carried on per design. Analytical testing methods and procedures developed for the cell therapy approach must be reproducible, accurate, sensitive, and rapid. They should only require small volumes of the product in order to maximize the therapeutic doses available for patients and to limit time in culture. They should also be cost effective. It is vital to develop analytical methods early in the product life cycle. It is highly desirable that product-specific potency methods are qualified and validated early in order to avoid dosing changes and dosing shifts during clinical stages (Fig. 4).

Process characterization is crucial for both the process development and clinical manufacturing stage. The correct application of process analytical technology (PAT) is the foundation to develop a robust, optimized, and reproducible process. The guidance of PAT was released by FDA in 2004 [29]. It is a system for designing, analyzing, and controlling manufacturing

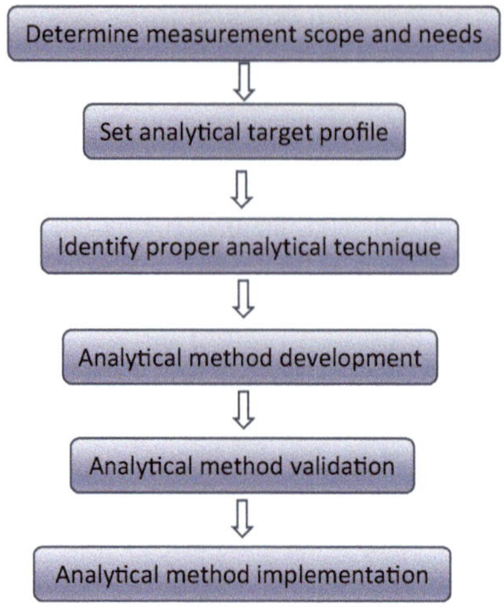

FIG. 4 Method development for PAT.

through in-process measurements to ensure final product quality. PAT provides the potential for continuous real-time quality assurance, resulting in improved operational control and compliance. It also plays a key role in creating control strategy and monitoring ongoing process to guarantee the CQAs of the cellular products.

A unique aspect of cell therapy is the starting cell population. Process development and optimization for cell therapy face the challenge of inherited donor variability. The evaluation of this special type of raw material alongside with other materials and process variables enables the generation of an overall risk assessment road map that should contribute to the overall success of process design and manufacturing strategy. A defined characterization profile for in-process and end of process samples ensures the comparability among different manufacturing runs. A broad screening of product profile early in process development and optimization is a good way to gain in-depth knowledge of the product CQAs. Down the road, the defined final signature profile of a cellular product should communicate its identity, purity and release criteria, and should ideally correlate with its therapeutic efficacy. Of note, the analytical techniques to be used for raw material, the process or product characterization could be restrained by technique availability and must be accepted by the regulatory agency.

Identify and overcome the distribution, thawing, and dispensing challenges

Along with the above considerations, storage conditions, packaging, and transportation should also be thoroughly investigated and validated. The product stability program should incorporate sampling and testing during distribution steps to identify and appropriately control critical parameters [30]. The optimization of thawing and post-thaw processing protocols for these "living drugs" as well as post-thaw cell characteristics are critical in defining the attributes of a robust manufacturing process [31].

Conclusion

The success of process development and process optimization depends on the thorough understanding of the cell-based product and sound analytics for the cell and process characterization. The implementation of DoE and PAT approaches at the early stages of process development will help fast-track the process optimization steps and facilitate safe, operable, reproducible, cost-effective, and regulatory compliant manufacturing processes for novel cell-based therapies.

References

[1] Pillow RP, Epstein RB, Buckner CD, Giblett ER, Thomas ED. Treatment of bone-marrow failure by isogeneic marrow infusion. N Engl J Med 1966;275:94–7.
[2] Chivu-Economescu M, Rubach M. Hematopoietic stem cells therapies. Curr Stem Cell Res Ther 2017;12:124–33.
[3] Andrzejewska A, Lukomska B, Janowski M. Concise review: mesenchymal stem cells: from roots to boost. Stem Cells 2019;37:855–64.
[4] Sadeghzadeh M, Bornehdeli S, Mohahammadrezakhani H, et al. Dendritic cell therapy in cancer treatment; the state-of-the-art. Life Sci 2020;254, 117580.
[5] Yeku O, Li X, Brentjens RJ. Adoptive T-cell therapy for solid tumors. Am Soc Clin Oncol Educ Book 2017;37:193–204.
[6] Nair R, Westin J. CAR T-cells. Adv Exp Med Biol 2020;1244:215–33.
[7] Rezvani K. Adoptive cell therapy using engineered natural killer cells. Bone Marrow Transplant 2019;54:785–8.
[8] FDA. Approved cellular and gene therapy products. FDA; 2021.
[9] Golchin A, Farahany TZ. Biological products: cellular therapy and FDA approved products. Stem Cell Rev Rep 2019;15:166–75.
[10] FDA. Target product profile—a strategic development process tool. FDA; 2007.
[11] FDA. Pharmaceutical development. International Council for Harmonisation of Technical Requirements for Pharmaceuticals for Human Use; 2009.
[12] Garber K. RIKEN suspends first clinical trial involving induced pluripotent stem cells. Nat Biotechnol 2015;33:890–1.
[13] Tyagarajan S, Spencer T, Smith J. Optimizing CAR-T cell manufacturing processes during pivotal clinical trials. Mol Ther Methods Clin Dev 2020;16:136–44.
[14] Lipsitz YY, Timmins NE, Zandstra PW. Quality cell therapy manufacturing by design. Nat Biotechnol 2016;34:393–400.
[15] Buzhor E, Leshansky L, Blumenthal J, et al. Cell-based therapy approaches: the hope for incurable diseases. Regen Med 2014;9:649–72.
[16] Yeo C, Saunders N, Locca D, et al. Ficoll-Paque versus Lymphoprep: a comparative study of two density gradient media for therapeutic bone marrow mononuclear cell preparations. Regen Med 2009;4:689–96.
[17] Cao Y, Sun H, Zhu H, et al. Allogeneic cell therapy using umbilical cord MSCs on collagen scaffolds for patients with recurrent uterine adhesion: a phase I clinical trial. Stem Cell Res Ther 2018;9:192.
[18] Powell Jr DJ, Brennan AL, Zheng Z, Huynh H, Cotte J, Levine BL. Efficient clinical-scale enrichment of lymphocytes for use in adoptive immunotherapy using a modified counterflow centrifugal elutriation program. Cytotherapy 2009;11:923–35.
[19] Miltenyi S, Muller W, Weichel W, Radbruch A. High gradient magnetic cell separation with MACS. Cytometry 1990;11:231–8.
[20] Lu M, Lezzar DL, Voros E, Shevkoplyas SS. Traditional and emerging technologies for washing and volume reducing blood products. J Blood Med 2019;10:37–46.
[21] Jacob Pattasseril HV, Lock LT, Rowley JA. Downstream technology landscape for large-scale therapeutic cell processing. Bioprocess Int 2013;11(Suppl):38–47.
[22] Meneghel J, Kilbride P, Morris GJ. Cryopreservation as a key element in the successful delivery of cell-based therapies—a review. Front Med (Lausanne) 2020;7, 592242.

[23] Andrews PW, Baker D, Benvinisty N, et al. Points to consider in the development of seed stocks of pluripotent stem cells for clinical applications: international stem cell banking initiative (ISCBI). Regen Med 2015;10:1–44.
[24] Woods EJ, Thirumala S. Packaging considerations for biopreservation. Transfus Med Hemother 2011;38:149–56.
[25] Sputtek A, Lioznov M, Kroger N, Rowe AW. Bioequivalence comparison of a new freezing bag (CryoMACS((R))) with the Cryocyte((R)) freezing bag for cryogenic storage of human hematopoietic progenitor cells. Cytotherapy 2011;13:481–9.
[26] Hunt CJ. Technical considerations in the freezing, low-temperature storage and thawing of stem cells for cellular therapies. Transfus Med Hemother 2019;46:134–50.
[27] Shu Z, Kang X, Chen H, et al. Development of a reliable low-cost controlled cooling rate instrument for the cryopreservation of hematopoietic stem cells. Cytotherapy 2010;12:161–9.
[28] Massie I, Selden C, Hodgson H, Fuller B, Gibbons S, Morris GJ. GMP cryopreservation of large volumes of cells for regenerative medicine: active control of the freezing process. Tissue Eng Part C Methods 2014;20:693–702.
[29] FDA. PAT—a framework for innovative pharmaceutical development, manufacturing, and quality assurance. FDA; 2004.
[30] Cell manufacturing roadmap to 2030., 2019, http://wwwcellmanufacturingusaorg/sites/default/files/Cell-Manufacturing-Roadmap-to-2030_ForWeb_110819pdf.
[31] Li R, Johnson R, Yu G, McKenna DH, Hubel A. Preservation of cell-based immunotherapies for clinical trials. Cytotherapy 2019;21:943–57.

CHAPTER

9

Lymphodepletion and cellular immunotherapy

Yuan Yuan[a], *Supriya Deshpande*[b], *Jamie Rand*[b], *and Yuman Fong*[b]

[a]Department of Medical Oncology, Cedars-Sinai Medical Center, Los Angeles, CA, United States
[b]Department of Surgery, City of Hope Comprehensive Cancer Center, Duarte, CA, United States

Abstract

Many adoptive cellular therapies (ACT) for cancer have advanced to the clinic for testing and treatment, including chimeric antigen receptor T-cell (CAR T-cell) therapy, tumor-infiltrating lymphocyte (TIL) therapy, and ex vivo expanded lymphokine-activated natural killer cells. Conditioning patients with lymphodepleting chemotherapy was a pivotal advance in enabling these therapies. In this review, we will discuss the various regimens that have been used and their associated toxicities. Only a limited number of studies have directly compared two or more regimens. Intermediate dose cyclophosphamide and fludarabine-based regimens are emerging as a popular approach that balances efficacy and safety. We will also discuss methods of avoiding lymphodepletion that are being developed to decrease toxicity of cellular immunotherapies.

Abbreviations

ACT	adoptive T cell transfer
ALL	acute lymphoblastic leukemia
Axi-Cel	Axicabtagene ciloleucel
B-ALL	B-cell lymphoblastic leukemia
BCL	B cell lymphoma
CAR T-cell	chimeric antigen receptor T-cell
CLL	chronic lymphocytic leukemia
CR	complete remission
CRR	complete response rate
CRS	cytokine release syndrome
Cy	cyclophosphamide
Flu	fludarabine
GBM	glioblastoma multiforme
ICANS	immune effector cell-associated neurotoxicity syndrome
IDO	indoleamine 2,3-dioxygenase
IFNγ	interferon gamma

https://doi.org/10.1016/B978-0-12-822620-9.00009-4

IL-12 interleukin-12
IL-13Ra2 interleukin-13 receptor antagonist 2
IL-15 interleukin-15
IL-18 interleukin-18
MSDC myeloid-derived suppressor cell
NMA nonmyeloablative
ORR overall response rate
PR partial response
R/R relapsed/refractory
ScFV single-chain antibody
TIL tumor-infiltrating lymphocytes
TMZ Temozolomide
T-reg regulatory T-cell

Conflict of interest

Dr. Yuman Fong is a scientific consultant for: Medtronics, Johnson and Johnson, Vergent Biologics, Eureka Therapeutics, Iovance Biotherapeutics, and Boehringer Ingelheim; and receives royalties from: Merck and Imugene.
Dr. Yuan Yuan has grants from: Merck, Pfizer, Novartis, Eisai, Imugene, Genentech; is a consultant for: Pfizer, Gilead, Novartis, Eisai, Genentech, AstraZeneca; and is on the speakers bureau for: Eisai, Genentech, Pfizer, Merck, Gilead, Daiichi Sankyo, AstraZeneca.
All other authors declare no conflicts of interest.

Introduction

The field of adoptive cell therapy (ACT), including tumor-infiltrating lymphocytes (TILs) or genetically modified chimeric antigen receptor T-cells (CAR T-cells), has progressed rapidly over the past few decades from the establishment of proof-of-principle preclinical studies to effective and durable U.S. Food and Drug Administration (FDA) approved therapies for treatment of leukemia and lymphoma. It is an emerging field with ongoing promise and challenges. The ACT approach involves isolation, ex vivo expansion, and activation of antigen-specific immune cells. It also involves genetic modification of these cells before reinfusion into patients for robust antitumor responses [1–4]. Lymphodepletion has routinely been used as a preconditioning regimen before ACT to create a conducive environment for the transferred T-cells (Table 1). However, there have been limited studies comparing lymphodepletion regimens, and there is no standard approach for the choice, dosage, and timing of these treatments.

TABLE 1 Goals of lymphodepletion.

Biologic	Clinical outcome
Increase sinks for ACT engraftment	Improved cancer killing
Increased proliferation and persistence of transferred cells	Improved persistence of therapeutic cells
Decreased Treg populations	Improved patient survival
Increased CD4+ and CD8+ proliferation	
Decrease competition by endogenous lymphocytes for IL-7 and IL-15	

Regimens for lymphodepletion

Whole-body radiation was used early on to achieve lymphodepletion [5,6]. Toxicities associated with such radiotherapy and the effectiveness of chemotherapeutic regimens have resulted in the latter becoming the lymphodepleting regimens of choice. The most common chemotherapy regimens are combinations of fludarabine (Flu) and cyclophosphamide (Cy).

Cy is one of the earliest cytotoxic agents identified and is an alkylating mustard [7]. It promotes DNA crosslinking and results in cell death [7]. Its preferential effects on lymphocytes are due to its inactivation by aldehyde dehydrogenase 1, an enzyme that is poorly expressed in lymphocytes [8]. Cy is the most potent lymphodepleting agent available. Flu is a purine analog prodrug that is converted by deoxycytidine kinase to its active triphosphate derivative (F-ara-ATP) [9]. Flu has dose-dependent neurotoxicity [10]. A consensus appears to be emerging around dosing levels of cy (300–500 mg/m^2 × 3 days) and flu (25–30 mg/m^2 × 3 days).

Tumor-infiltrating lymphocytes (TILs)

Tumor-infiltrating lymphocytes (TILs) are heterogenous lymphocytes, mainly T-cells and natural killer cells, that home to tumors and have been used as cellular immunotherapy for a long time [11]. In the early phase, a preclinical study by Dr Rosenberg and colleagues at NCI demonstrated significant tumor regression in murine sarcoma and colon adenocarcinoma transplanted in syngeneic mice after adoptive transfer of autologous TILs isolated and expanded ex vivo with IL-2 [12]. This study provided a rationale for the use of TILs to treat humans with advanced cancer and formed the basis for the pioneering TIL-ACT clinical trial in which metastatic melanoma patients were treated with autologous TILs in addition to high-dose IL-2 [13–15]. However, the benefits of infused TILs were short-lived, and a nonmyeloablative lymphodepleting regimen was added based on previous studies [16,17].

Preclinical studies provided a rationale for lymphodepletion with irradiation or Cy by showing that the host immune environment affected ACT and that immunosuppression before ACT increased the durability of infused T-cells (Table 2) [16,17,33]. Although there is variability in the doses used, Flu and Cy combination treatments are routinely used as lymphodepleting agents in TIL trials [34]. Dudley et al. demonstrated that ACT with TILs following host lymphodepletion with Cy and Flu and high-dose IL-2 therapy in metastatic melanoma patients resulted in the rapid growth of clonal populations of tumor antigen-specific T-cells and led to significant tumor regression, clearly establishing a favorable effect of lymphodepletion before ACT [18,19]. Increasing the durability of infused T-cells is essential for effective ACT. Further, these findings were supported by various groups that showed that preconditioning lymphodepletion regimens commonly consisting of Cy and Flu, to eliminate the host's regulatory mechanisms and allow accommodation of transferred T lymphocytes, boosting the efficacy of immunotherapies [1,35–38].

Gattinoni et al. demonstrated that transient removal of cellular sinks that compete for activation-inducing cytokines and increased availability of homeostatic cytokines are critical for enhancing the effectiveness of lymphodepletion [37]. Elimination of immunosuppressive cells such as Tregs and myeloid-derived suppressor cells (MSDCs) was shown to be a key mechanism that increases antitumor immunity following lymphodepletion. Other mechanisms for improved efficacy of ACT following lymphodepletion include better antigen presentation to

TABLE 2 Lymphodepletion regimens.

References	Disease	Regimen	Response rate
ACT—TIL			
[13]	Metastatic melanoma	Cy 25mg/kg × 1	ORR (55%) CR (5%)
[18]	Metastatic melanoma	Cy 60mg/kg × 2+Flu 25mg/m^2 × 5	PR 6 (46%)
[19]	Metastatic melanoma	Cy 60mg/kg × 2+Flu 25mg/m^2 × 5	OR (51%)
[20]	Metastatic melanoma	Cy 60mg/kg/d × 2+Flu 25mg/m^2/d × 5	CR (24%)
ACT—CAR T-CELL THERAPY			
[21] Tisagenlecleucel	B-cell lymphoblastic leukemia	Flu 30mg/m^2 × 4 Cy 500mg/m^2 × 2	ORR (81%) CR (60%)
[22] Tisagenlecleucel	Large B-cell lymphoma	Flu 25mg/m^2 × 2, Cy 250mg/m^2 × 3 Or Ben 90mg/m^2 × 2	ORR (52%) CR (40%) PR (12%)
[23] 19–28z 2G CAR	B-cell acute lymphoblastic leukemia	Cy 1.5–3.0g/m^2 × 1	ORR (88%)
[24] 19–28z 2G CAR	Diffuse large B-cell lymphoma	Flu 30mg/m^2 × 3 Cy 300–500mg/m^2 × 3	ORR (73%) CR (55%) PR (18%)
[25] 19-BB	B-cell acute lymphoblastic leukemia	Flu 25mg/m^2+Cy 30 OR 60mg/kg×2 OR 4days	MRD −ve CR (85%)
[26] 4–1BB based 2G CAR coexpressing EGFRt	R/R non-Hodgkin's lymphoma	Flu 25mg/m^2×2 or 4 Cy 30–60mg/kg×1	ORR (72%) CR (50%)
[27] 4–1BB based 2G CAR coexpressing EGFRt	B-cell non-Hodgkin lymphoma	**High intensity** Cy 60mg/kg × 1+Flu 25mg/m^2 × 3 Cy 60mg/kg × 1+Flu 25mg/m^2 × 5 **Low intensity** Cy 30mg/kg × 1+Flu 25mg/m^2 × 3 Cy 300mg/m^2 × 3+Flu 30mg/m^2 × 3 Cy 500mg/m^2 × 3+Flu 30mg/m^2 × 3	ORR (51%) CR (40%)
[28] Axi-Cil	Large B-cell lymphoma	Flu 30mg/m^2 × 3+Cy 500mg/m^2 × 3	ORR (82%) CRR (54%)

TABLE 2 Lymphodepletion regimens—cont'd

References	Disease	Regimen	Response rate
[29] Tisagenlecleucel	Chronic lymphocytic leukemia	Flu/cy Pen/Cy Ben (Dosage not reported)	ORR (57%) CR (29%) PR (29%)
[30] Lisa-Cel	Refractory large B-cell lymphoma	Flu 30 mg/m^2 × 3 + Cy 300 mg/m^2 × 3	ORR (73%) CR (53%)
[31] KTE-X19	Mantle cell lymphoma	Flu 30 mg/m^2 × 3 + Cy 500 mg/m^2 × 3	ORR (93%) CR (67%)
[32] KTE-X19	R/R B-precursor acute lymphoblastic leukemia	Flu 25 mg/m^2 × 3 + Cy 900 mg/m^2 × 1	CR (56%)

ACT, *adoptive cell therapy;* TIL, *tumor-infiltrating lymphocytes;* Cy, *cyclophosphamide;* Flu, *fludarabine;* Ben, *bendamustine;* R/R, *relapsed/refractory;* ORR, *overall response rate;* CRR, *complete response rate;* CR, *complete remission;* PR, *partial response.*

infused T cells, modification of the tumor microenvironment to enhance trafficking of infused T cells to tumors, increased immunogenic tumor cell death, decrease in the immunosuppressive enzyme, indoleamine 2,3-dioxygenase (IDO), in tumor cells, and bacterial translocation in gut microbiota leading to immunostimulatory effects [2,37,39,40].

Although inhibition of Tregs enhances antitumor immunity, studies have shown an increased percentage of Tregs following lymphodepletion in whole-body irradiated fibrosarcoma mouse models [2,39]. Studies have shown an increase in peripheral myeloid-derived suppressor cells (MDSCs) in melanoma and lung cancer patients following lymphodepletion and ACT that may contribute to ACT failure [41]. Not many studies have compared different lymphodepletion treatment regimens, and it is suggested that the intensities of these regimens could have different effects [2,40]. Factors other than lymphodepletion, such as the stage of CD8 T-cells differentiation, also affect the clinical response [42,43].

Radiation therapy has been used for lymphodepletion [1,44]. Preclinical data in a murine melanoma model showed that high-dose total-body irradiation (TBI) significantly increased antitumor effects of transferred tumor-specific CD8+ T cells. Increased intensity of TBI was associated not only with a greater tumor treatment efficacy but also with a release of proinflammatory cytokines [38]. Despite preclinical studies showing the correlation between intensity of lymphodepletion and therapeutic response, these results have not translated completely clinically [2,45] and have called attention to the toxicity of some of the preconditioning agents. Goff et al. conducted a randomized trial in metastatic melanoma patients receiving Cy and Flu with or without TBI prior to an infusion of autologous lymphocytes and high-dose IL-2. They reported that the addition of TBI did not improve patient outcomes but increased treatment-related toxicities [20]. A combination of lymphodepletion with TBI and Cy and Flu in metastatic melanoma patients receiving TIL with IL-2 following autologous CD34+ hematopoietic stem cells (HSC) transplantation showed increased objective response rates and complete response rates in the higher intensity radiation group compared to chemotherapy alone or lower intensity radiation [46]. Nissani et al. recently studied the effect of different nonmyeloablative (NMA) preconditioning regimens with different doses of Cy or TBI in combination with Flu on bone marrow suppression and recovery, cytokine serum levels, clinical response, and adverse events in metastatic melanoma patients. They concluded that among different cohorts tested (120Cy/125Flu and 60Cy/125Flu melanoma cohorts; 30Cy/75Flu ALL cohort; TBI/75Flu melanoma cohort), a dose of 60Cy/125Flu preconditioning achieved maximum effect with minimum toxicity [47].

Chimeric antigen receptor T-cells (CAR-T cells)

Another strategy for ACT is to use genetically modified T-cells expressing chimeric antigen receptors (CARs) with tumor-specific antigen recognition receptors and T cell signaling domains. CARs are known to redirect the specificity of T-cells and increase the potency and persistence of T-cells [48,49], putting them at the center stage for certain types of cancer immunotherapy. CAR-T cells therapy has evolved from the use of first-generation CARs with a single-chain antibody (ScFV) linked to an intracellular CD3ζ region of the T cell receptor (TCR) complex to second or third-generation CARs with one or more added intracellular costimulatory domains such as CD28, 4-1BB, or OX-40 molecules [4,50]. The later generation CARs have a more robust proliferation, cytokine secretion, effector function, and persistence of transduced cells [51,52].

Due to tumor cell-specific and homogenous expression of CD19, successful clinical outcomes have been observed with CD19-specific CAR T-cells in relapsed hematological malignancies [49,52,53]. The first two anti-CD19 CAR T-cell therapies that the FDA approved are Tisagenlecleucel (Kymriah), a second-generation CAR with a 4-1BB costimulatory domain for relapsed/refractory (R/R) acute lymphoblastic leukemia (ALL) and Axicabtagene ciloleucel (Yescarta), a second-generation CAR with a CD28 costimulatory domain for R/R large B cell lymphoma (BCL) [21,28].

Early clinical trials using CD-19-targeted CAR T-cells in B cell malignancies were not very promising and notably lacked the use of lymphodepletion regimens [54,55]. Currently, lymphodepletion is essential to CAR T-cell therapy (Table 2). It enhances treatment efficiency by decreasing immunoregulatory cells to reduce the risk of removing modified cells as non-self, decreasing the immunosuppressive IDO enzyme, increasing the release of stimulatory factors and homeostatic cytokines for expansion, proliferation, and improving the response of infused CAR T-cells [34,38,51,56]. CAR T-cell administration following lymphodepletion with Cy with Flu, Cy alone, and bendamustine (Ben) has been used in clinical trials [53].

Kochenderfer et al. demonstrated the first promising anti-CD-19 CAR T-cell approach for treating diffuse large B-cell lymphoma (DLBCL) patients receiving a lymphodepletion regimen with Flu and high-dose Cy. Eight of 25 patients achieved complete remissions (CRs). However, the high-dose regimen led to significant toxicity [57]. The intensity of the lymphodepletion regimen was reduced later. The reduced chemotherapy dosing resulted in depletion of blood lymphocytes and increased serum IL-15 levels. High serum IL-15 levels were associated with high peak blood CAR^+ cell levels and lymphoma remissions. A higher dose of Cy (500 mg/m^2) was given in an attempt to increase the CR rate; however, the higher dose caused notable toxicities when compared to a low dose of Cy (300 mg/m^2) in these advanced-stage lymphoma patients, thus the high-dose regimen was stopped prematurely after treating only four patients [24].

An early, promising clinical result demonstrating the importance of lymphodepletion intensity for CAR T-cell efficacy was shown with second-generation anti-CD19 CAR T-cells in chemotherapy-refractory chronic lymphocytic leukemia (CLL) and relapsed B-cell acute lymphoblastic leukemia (B-ALL) patients. A CD19 targeted 19-28z CAR in CLL and B-ALL patients showed enhanced persistence and efficacy of modified T cells in patients who received a Cy conditioning regimen compared to patients treated without prior Cy [36]. This same CAR T-cell therapy resulted in an 88% complete response rate in a phase 1 clinical trial with R/R B-ALL patients [23].

Dai et al. showed further support for CAR T-cell therapy in refractory B-ALL in a pilot clinical trial of T-cells modified to express a CAR targeting CD19 with the 4-1BB and CD3ζ domain. They demonstrated the benefits of lymphodepletion in patients who received prior conditioning therapy that included Cy, resulting in a more rapid in vivo expansion after cell infusion [58]. Further, Geyer et al. performed a phase 1 clinical trial with CD19-targeted CAR T-cells with a CD20 costimulatory domain (19-28z) in R/R CLL patients. They evaluated the safety and toxicity of 19-28z CAR T-cell therapy with or without conditioning therapy in these patients. They observed the lowest lymphocyte counts and greatest circulating CAR T-cells in R/R CLL patients treated with a Flu-containing regimen. Three of 12 CLL patients receiving conditioning therapy showed durable CR; 2 had minimal residual disease-negative CR [59]. The safety of CD19-specific CAR T-cell therapy for R/R B-ALL was further demonstrated by Curran et al. They showed a response rate of 94% in the high-dose Cy ($3\,g/m^2$) cohort without a negative impact on the toxicity profile. These results suggested that more intense lymphodepletion could reduce immune-mediated suppression or cell-mediated rejection of CAR T-cells, improving responses [60].

Hirayama et al. recently performed a multivariable analysis of various factors affecting response following anti-CD19 CAR T-cell therapy in aggressive B-cell non-Hodgkin lymphoma patients with low or high-intensity Cy/Flu lymphodepletion. They reported that a favorable cytokine profile was associated with a durable response, and this outcome was more likely with a higher dose of lymphodepletion than low-intensity lymphodepletion. However, multivariable modeling showed that in the high-dose lymphodepletion group, a favorable cytokine profile but not the intensity of lymphodepletion was associated with improved outcomes. These results highlight that the biological effects of conditioning regimens, in addition to the intensity of lymphodepletion, are crucial for improving the efficacy of immunotherapies [27].

A study with a second-generation 4-1BB CAR (CTL-019) in R/R CLL showed that a lymphodepleting regimen with either Flu/Cy, Pentostatin/Cy, or Ben was not associated with CART cell expansion and persistence [29], but a durable CR was dependent on the intrinsic T cell fitness [61]. This CAR (developed as Tisagenlecleucel) was subsequently used in the ELIANA nonrandomized multicenter trial following a lymphodepletion regimen of Flu and Cy in children and young adults with B-CLL. Tisagenlecleucel produced durable remission and long-term persistence without additional therapy in R/R B-cell ALL patients with an overall remission rate of 81% and a complete remission of 60% [21]. The JULIET trial with Tisagenlecleucel in adult patients with adult R/R DLBCL, used either Flu and Cy or Ben. An overall response rate of 52% and a complete response of 40% were reported [22]. A long-term median follow-up of 40.3 months of this trial continued to show a durable response. Post hoc analysis of progression-free survival by lymphodepleting chemotherapy showed that patients receiving Flu/Cy lymphodepletion had higher 24-month progression-free survival estimates than those receiving Ben conditioning or no conditioning [62].

A multicenter phase 2 trial in large B-cell lymphoma patients treated with Axicabtagene ciloleucel (Axi-Cel) anti-CD19 CAR T-cell therapy after low-dose Cy and Flu lymphodepletion showed an objective response rate of 82% and a complete response rate of 54% [28]. A 2-year follow-up of this trial indicated durable responses and a median overall survival of >2 years [63]. The results of the ZUMA-1 trial resulted in the approval of Axi-Cel for R/R DLBCL.

The TRANSCEND trial, the most extensive clinical study reported with CD-19-directed CAR T-cells Lisocabtagene maraleucel (liso-cel) in R/R large B-cell lymphomas showed durable responses with a low incidence of grade 3 or 4 toxicities. These patients received lymphodepletion with Flu (30 mg/m^2)+Cy (300 mg/m^2) daily for 3 days. An objective response of 73% and a complete response of 53% were reported [30].

Further studies that have used conditioning lymphodepletion regimens have demonstrated increased CAR T-cell persistence and improved outcomes of lymphodepletion treatment vs nontreatment in B-cell malignancies. In a first attempt to assess the feasibility of using selected T-cell subsets with a CD19 CAR, Turtle et al. conducted a clinical trial of CD19 CAR T-cells prepared from defined CD4+ and CD8+ T-cell subsets and administered these in a defined composition to adult B-cell ALL patients. A remarkable 93% of patients achieved bone marrow remission. They showed that the addition of Flu to lymphodepletion delayed anti-CAR immune responses and increased CD4+ and CD8+ CAR T-cell proliferation and persistence in these patients [64]. In a long-term follow-up study, Hay et al. showed that incorporation of Flu in the lymphodepletion regimen and allogeneic hematopoietic cell transplantation (HCT) after CAR T-cell therapy improved event-free survival. They also reported that patients who received Flu-containing lymphodepletion had CD19-negative relapses, while CD19-positive relapses were seen in patients who did not receive Flu. Additionally, a higher CAR T-cell area under the curve was associated with Flu-containing lymphodepletion [25]. Turtle et al. treated non-Hodgkin's lymphoma patients with CD19 CAR T-cells prepared from defined CD4+ and CD8+ T-cell subsets and administered them in a controlled ratio after Cy-based lymphodepletion with or without Flu. The authors reported increased CAR T-cell expansion and persistence and higher response rates (72% overall response rate and 50% complete remission) in patients who had received Cy/Flu treatment compared to only Cy-based treatment. It was postulated that increased homeostatic cytokines and decreased anti-CAR immune response, or decreased expression of IDO in the TME might be likely mechanisms for better persistence of CAR T-cells and improved outcomes by adding Flu to Cy [26].

A recent phase 1 trial of autologous CD19.28ζ-CAR T-cells in children and young adults with R/R B-ALL, showed the superiority of Flu/Cy regimens over alternate lymphodepletion regimens. A complete response of 69% was reported in Flu/Cy-based lymphodepletion compared to 25% in non-Flu/Cy-based lymphodepletion. Also, Flu/Cy regimens were associated with improved event-free survival and overall survival [65].

Ramos et al. conducted a phase 1 dose-escalation study to assess the safety of second-generation CD28ζ CAR-Ts targeting CD30 (CD30.CAR-Ts) in Hodgkin lymphoma and anaplastic large cell lymphoma patients without a lymphodepletion preconditioning treatment. They showed that CD30.CAR-Ts were safe and observed an overall response rate of only 33% [66]. Recently, they reported results from 2 parallel phase 1/2 studies in R/R HL patients. The CD30.CAR T-cells were infused after lymphodepletion with either Ben alone, Ben and Flu, or Cy and Flu. They showed an overall response rate of 72% and a complete response of 59% in patients who received Flu-based lymphodepletion compared to no objective clinical responses in the patients who received Ben alone. In addition to showing a significant improvement in PFS with Flu-based lymphodepletion, this study also highlighted that durable clinical responses without significant toxicity from CAR T-cell therapies are also possible in diseases other than B-cell malignancies [67].

Wang et al. performed a phase 2 trial with KTE-X19 in patients with R/R mantle-cell lymphoma. Durable remission was observed in majority of patients with an objective response of 93% and a complete response of 67%. They, however, concluded that the benefit provided by conditioning therapy, including Flu and Cy was marginal compared to that of KTE-X19 based on prolonged responses in patients and peak level and area under the curve for CAR T cells [31]. The phase 2 results of the ZUMA-3 trial of KTE-X19 CAR T-cell therapy in adults with R/R B-precursor acute lymphoblastic leukemia showed high and durable responses, with 71% of patients showing complete remission or complete remission with incomplete hematological recovery and 56% patients reaching complete remission. As a result of the promising outcomes of the ZUMA-3 trial that used Cy and Flu, Brexucabtagene autoleucel, the anti-CD19 CAR T-cell therapy, was recently approved by the FDA for B-ALL [32].

Solid tumors

While early-phase clinical trials using CAR T-cells against CD-19-positive hematological malignancies have shown promising results, there is still relapse in some patients due to nondurable effects of therapies [68,69]. The use of CAR T-cells for solid tumors presents distinctive obstacles such as lack of lineage-specific tumor-specific antigens, inefficient tumor localization, an immunosuppressive tumor microenvironment leading to T cell exhaustion, tumor heterogeneity, risk of on-target, off-tumor toxicity, inefficient tumor infiltration, and inadequate CAR T-cell expansion and persistence [70]. Currently, there are more CAR T-cell clinical trials for brain tumors compared to other solid tumors [50]. CAR T-cells targeting various tumor antigens such as CEA in colorectal cancers, breast cancers, gastric cancers, adenocarcinoma, and liver metastases, and others targeting CAIX, CD133, EGFRvIII, IL-13Ra2, ErB2/Her2, GD2 (neuroblastoma or sarcoma), and GPC3 (hepatocellular carcinoma) have been tested in early phase clinical studies [52]. Initially, first-generation CARs were used, which did not show very efficient results, mainly due to the absence of costimulatory endodomains and lymphodepletion regimens [71]. Further, even second- and third-generation CAR T-cell therapies often did not lead to durable responses for the treatment of solid tumors. The importance of lymphodepletion in a CAR T-cell therapy targeting a $CEACAM5^+$ malignancy was shown by Thistlethwaite et al. This study showed that patients treated with an increased intensity preconditioning regimen of Cy/Flu had significantly higher CAR T-cell engraftment compared to patients treated with Flu alone, but this effect was short-lived with a rapid decline of systemic CAR T-cells within 14 days. This study also highlighted the on-target off-tissue toxicity associated with CAR T-cell therapy leading to the premature closing of the trial [72]. In a case report by Hegde et al., durable remission of sarcoma was seen in a child with metastatic rhabdomyosarcoma with Flu/Cy conditioning resulting in subsequent improvement of autologous HER2 CAR T-cell expansion and increased levels of circulating IL-15, without any on-target, off-tumor toxicities, or adverse events [73].

Challenges associated with targeting brain tumors were demonstrated in a dose-escalation study of third-generation CAR T-cells targeting the EGFR variant III (EGFRvIII) in glioblastoma following lymphodepletion with Flu/Cy. While lymphodepletion induced the transient cytopenias and CAR$^+$ cells correlated with cell dose, ACT with CAR T-cells targeting EGFRvIII did not induce any objective clinical responses. There was also one treatment-related fatality associated with the highest dose of infused T-cells highlighting the challenges associated with targeting brain tumors [74]. Brown et al. demonstrated remarkable clinical response following CAR T-cell therapy targeting intracranial glioma-associated antigen IL13Rα2. It was the first report showing regression of all intracranial and spinal tumors along with increased cytokine and immune cells in the cerebrospinal fluid of a glioblastoma patient with manageable therapy-related toxic effects [75]. This was early evidence showing the potential of CAR T-cell immunotherapy in solid tumors. Combination therapies of CAR T-cells and immune checkpoint inhibition are currently being tested. CAR T-cells targeting disialoganglioside (GD2) in R/R-neuroblastoma showed a beneficial effect of lymphodepletion with Flu/Cy by enhancing CAR T-cell persistence. This study also showed that combining PD-1 inhibition with lymphodepletion did not further enhance the expansion or persistence of CAR T-cells [76].

Toxicity of lymphodepletion

Lymphodepletion using cytotoxic chemotherapy and/or total body irradiation produces significant toxicity. The most significant and consistent complications are related to (1) cytokine release syndrome (CRS) and (2) immune effector cell-associated neurotoxicity syndrome (ICANS). Each of these can result from lymphodepletion, effector cell administration, or the combination. The reported incidences of CRS and ICANS associated with various studies of Flu-Cy administration are shown in Table 3. Other adverse effects of lymphodepletion are listed in Table 4. While respiratory failure, pneumonia, and death are some of the most severe adverse events, the most common are anemia, lymphopenia, neutropenia, and thrombocytopenia. The intensity of the lymphodepletion regimen seems to be related to the risk of CRS and ICANS.

Lymphodepletion with Flu-Cy had a significantly higher chance of CRS compared to conditioning with bendamustine combinations [67]. Flu is associated with ICANS, especially at higher doses [87]. Guidelines for assessing, grading, and treating CRS and ICANS have been published [77].

Alternatives to lymphodepletion

Many approaches are under investigation to avoid lymphodepletion and the associated toxicities.

Regional delivery of cell therapies: For cancers restricted to single anatomic regions, investigators have attempted to deliver CAR T-cells locally without lymphodepletion. The concept underlying this approach is that dose intensification by regional delivery may overcome the local immunosuppressive and antiproliferative factors. In this way, Brown et al. achieved a complete response to CAR T-therapy in a case of advanced glioblastoma with multifocal leptomeningeal metastases using regional ventricular delivery [75]. IL13Rα2-targeted CAR T-cells were infused via Omiya reservoir. Whether delivery by peritoneal infusion can successfully treat peritoneal cancers or whether infusion by hepatic infusion may treat liver cancers is under active investigation.

TABLE 3 Incidence of cytokine release syndrome (CRS) and immune cell neurotoxicity syndrome (ICANS) associated with lymphodepletion prior to adoptive cell transfer.

Study	CAR-T	Cancer	Lymphodepletion regimen	≥Grade 3 CRS	≥Grade 3 ICANS
Brudno et al. [87]	Anti-BCMA	Multiple myeloma	Flu 30 mg/m^2 Cy 300 mg/m^2 × 3 days	55%	1%
Turtle et al. [26]	4–1BB 2G CAR expressing EGFRt	Non-Hodgkin's lymphoma	Cy 60 mg/kg × 1 Flu 25 mg/m^2 × 3–5 days	20%	35%
Kochenderfer [24]	19–28z 2G CAR	Non-Hodgkin's lymphoma	Flu 30 mg/m^2 Cy 300–500 mg/m^2	–	55%
Abramson et al. [30]	Liso-Cel (TRANSCEND)	Diffuse large B-cell lymphoma Transformed follicular lymphoma	Flu 30 mg/m^2 Cy 300 mg/m^2 × 3 days	2%	10%
Maude et al. [21]	Tisagenlecleucel (JULIET)	B-cell acute lymphoblastic leukemia	Flu 30 mg/m^2 × 4 Cy 500 mg/m2 × 2	47%	13%
Locke et al. [63]	Axi-Cil (ZUMA-1)	Diffuse large B-cell lymphoma Mediastinal B-cell lymphoma	Flu 30 mg/m^2 Cy 500 mg/m^2	11%	32%
Hay et al. [25]	19-BBz	B-cell acute lymphoblastic leukemia	Flu 25 mg/m^2 × 2 Cy 30–60 mg/kg × 2	19%	23%
Schuster et al. [22]	Tisagenlecleucel (JULIET)	Diffuse large B-cell lymphoma Follicular lymphoma	Flu 25 mg/m^2 × 2 Cy 250 mg/m^2 × 3	23%	12%

Cytokine-secreting CARs: A number of groups are examining the possibility of using CARs secreting cytokines that enhance CAR expansion or persistence as an alternative to lymphodepletion. The cytokines under investigation in this role include IL-12 [78], IL-18 [79], and IL-15 [80]. CAR-Ts secreting cytokines also act by recruitment of host immunity [81].

Lymphodepletion targeting CD45: CD45 is a receptor tyrosine phosphatase, a protein expressed on the cell surface of white blood cells and involved in B and T cell activation and signaling [82]. Work is underway to determine if anti-CD45 monoclonal antibodies may be an alternative to traditional lymphodepletion in conditioning patients prior to cytotoxic T-cell therapy [83]. A radioimmune approach to conditioning for cell therapy is attempting to use ^{131}I-labeled apamistamab (anti-CD45) in such a role [84].

Another approach for lymphodepletion involves using chemotherapies that also have direct cytotoxic effects on solid tumors. A current phase I clinical trial is attempting to use Temozolomide (TMZ), the standard treatment for GBM, which also induces lymphopenia, as a preconditioning regimen before CAR T-cell therapy in EGFvIII-positive GBM patients [85].

TABLE 4 Other adverse effects associated with lymphodepletion prior to adoptive cell transfer.

Common (>30%)	Less common (10%–30%)	Infrequent but severe
Neutropenia	Headache	Death
Anemia	Decreased appetite	Respiratory distress
Fatigue	Diarrhea	Pneumonia
Nausea	Dizziness	
Thrombocytopenia	Hypotension	
	Vomiting	
	Hypokalemia	
	Pyrexia	
	Leucopenia	
	Abdominal pain	
	Peripheral edema	
	Sinus tachycardia	
	Tremor	
	Confusion	
	Hypogammaglobulinemia	
	Hypertension	
	Dyspnea	
	Insomnia	
	Back pain	
	Chills	

Conclusion

Lymphodepletion before ACT-based immunotherapies can enhance antitumor responses through improving innate immunity, increasing access to homeostatic cytokines, and reducing the numbers of T-regs and MDSCs. Lymphodepletion significantly enhances the effects of most ACT therapy. For systemic delivery of CAR-T or TILs, Cy with Flu appears to have the best utility and acceptable toxicities. Low-dose Flu/Cy lymphodepletion regimens have been shown to reduce the risk of infection and severe neutropenia in CAR T-therapy. The next-generation CAR-Ts, which secrete immunostimulatory cytokines, may further improve efficacy and obviate the need for lymphodepletion.

References

[1] Muranski P, Boni A, Wrzesinski C, Citrin DE, Rosenberg SA, Childs R, et al. Increased intensity lymphodepletion and adoptive immunotherapy—how far can we go? Nat Clin Pract Oncol 2006;3(12):668–81.

[2] Watanabe S, Arita M, Takahashi M, Saida Y, Koya T, Kikuchi T. Effect of lymphodepletion on donor T cells and the role of recipient cells persisting after cytotoxic treatments in cancer immunotherapies. Crit Rev Immunol 2017;37(1):59–73.

[3] Dudley ME, Rosenberg SA. Adoptive-cell-transfer therapy for the treatment of patients with cancer. Nat Rev Cancer 2003;3(9):666–75.

[4] June CH, Sadelain M. Chimeric antigen receptor therapy. N Engl J Med 2018;379(1):64–73.

[5] Pinnix CC, Gunther JR, Dabaja BS, Strati P, Fang P, Hawkins MC, et al. Bridging therapy prior to Axicabtagene ciloleucel for relapsed/refractory large B-cell lymphoma. Blood Adv 2020;4(13):2871–83.

[6] Sim AJ, Jain MD, Figura NB, Chavez JC, Shah BD, Khimani F, et al. Radiation therapy as a bridging strategy for CAR T cell therapy with axicabtagene ciloleucel in diffuse large B-cell lymphoma. Int J Radiat Oncol Biol Phys 2019;105(5):1012–21.

[7] Pinto N, Ludeman SM, Dolan ME. Drug focus: pharmacogenetic studies related to cyclophosphamide-based therapy. Pharmacogenomics 2009;10(12):1897–903.

[8] Ahlmann M, Hempel G. The effect of cyclophosphamide on the immune system: implications for clinical cancer therapy. Cancer Chemother Pharmacol 2016;78(4):661–71.

[9] Gandhi V, Plunkett W. Cellular and clinical pharmacology of fludarabine. Clin Pharmacokinet 2002;41 (2):93–103.

[10] Lowe KL, Mackall CL, Norry E, Amado R, Jakobsen BK, Binder G. Fludarabine and neurotoxicity in engineered T-cell therapy. Gene Ther 2018;25(3):176–91.

[11] Jin Y, Dong Y, Zhang J, Sun J, Liu Y, Chen Y. The toxicity of cell therapy: mechanism, manifestations, and challenges. J Appl Toxicol 2021;41(5):659–67.

[12] Rosenberg SA, Spiess P, Lafreniere R. A new approach to the adoptive immunotherapy of cancer with tumor-infiltrating lymphocytes. Science (New York, NY) 1986;233(4770):1318–21.

[13] Rosenberg SA, Packard BS, Aebersold PM, Solomon D, Topalian SL, Toy ST, et al. Use of tumor-infiltrating lymphocytes and interleukin-2 in the immunotherapy of patients with metastatic melanoma. A preliminary report. N Engl J Med 1988;319(25):1676–80.

[14] Rosenberg SA, Yang JC, Topalian SL, Schwartzentruber DJ, Weber JS, Parkinson DR, et al. Treatment of 283 consecutive patients with metastatic melanoma or renal cell cancer using high-dose bolus interleukin 2. JAMA 1994;271(12):907–13.

[15] Rosenberg SA, Yannelli JR, Yang JC, Topalian SL, Schwartzentruber DJ, Weber JS, et al. Treatment of patients with metastatic melanoma with autologous tumor-infiltrating lymphocytes and interleukin 2. J Natl Cancer Inst 1994;86(15):1159–66.

[16] North RJ. Cyclophosphamide-facilitated adoptive immunotherapy of an established tumor depends on elimination of tumor-induced suppressor T cells. J Exp Med 1982;155(4):1063–74.

[17] Berenson JR, Einstein Jr AB, Fefer A. Syngeneic adoptive immunotherapy and chemoimmunotherapy of a Friend leukemia: requirement for T cells. J Immunol (Baltimore, Md: 1950) 1975;115(1):234–8.

[18] Dudley ME, Wunderlich JR, Robbins PF, Yang JC, Hwu P, Schwartzentruber DJ, et al. Cancer regression and autoimmunity in patients after clonal repopulation with antitumor lymphocytes. Science (New York, NY) 2002;298(5594):850–4.

[19] Dudley ME, Wunderlich JR, Yang JC, Sherry RM, Topalian SL, Restifo NP, et al. Adoptive cell transfer therapy following non-myeloablative but lymphodepleting chemotherapy for the treatment of patients with refractory metastatic melanoma. J Clin Oncol Off J Am Soc Clin Oncol 2005;23(10):2346–57.

[20] Goff SL, Dudley ME, Citrin DE, Somerville RP, Wunderlich JR, Danforth DN, et al. Randomized, prospective evaluation comparing intensity of lymphodepletion before adoptive transfer of tumor-infiltrating lymphocytes for patients with metastatic melanoma. J Clin Oncol Off J Am Soc Clin Oncol 2016;34(20):2389–97.

[21] Maude SL, Laetsch TW, Buechner J, Rives S, Boyer M, Bittencourt H, et al. Tisagenlecleucel in children and young adults with B-cell lymphoblastic leukemia. N Engl J Med 2018;378(5):439–48.

[22] Schuster SJ, Bishop MR, Tam CS, Waller EK, Borchmann P, McGuirk JP, et al. Tisagenlecleucel in adult relapsed or refractory diffuse large B-cell lymphoma. N Engl J Med 2019;380(1):45–56.

[23] Davila ML, Riviere I, Wang X, Bartido S, Park J, Curran K, et al. Efficacy and toxicity management of 19-28z CAR T cell therapy in B cell acute lymphoblastic leukemia. Sci Transl Med 2014;6(224), 224ra25.

[24] Kochenderfer JN, Somerville RPT, Lu T, Shi V, Bot A, Rossi J, et al. Lymphoma remissions caused by anti-CD19 chimeric antigen receptor T cells are associated with high serum interleukin-15 levels. J Clin Oncol Off J Am Soc Clin Oncol 2017;35(16):1803–13.
[25] Hay KA, Gauthier J, Hirayama AV, Voutsinas JM, Wu Q, Li D, et al. Factors associated with durable EFS in adult B-cell ALL patients achieving MRD-negative CR after CD19 CAR T-cell therapy. Blood 2019;133(15):1652–63.
[26] Turtle C.J., Hanafi L.A., Berger C., Hudecek M., Pender B., Robinson E., et al. Immunotherapy of non-Hodgkin's lymphoma with a defined ratio of CD8+ and CD4+ CD19-specific chimeric antigen receptor-modified T cells. Sci Transl Med. 2016;8(355):355ra116.
[27] Hirayama AV, Gauthier J, Hay KA, Voutsinas JM, Wu Q, Gooley T, et al. The response to lymphodepletion impacts PFS in patients with aggressive non-Hodgkin lymphoma treated with CD19 CAR T cells. Blood 2019;133 (17):1876–87.
[28] Neelapu SS, Locke FL, Bartlett NL, Lekakis LJ, Miklos DB, Jacobson CA, et al. Axicabtagene ciloleucel CAR T-cell therapy in refractory large B-cell lymphoma. N Engl J Med 2017;377(26):2531–44.
[29] Porter DL, Hwang WT, Frey NV, Lacey SF, Shaw PA, Loren AW, et al. Chimeric antigen receptor T cells persist and induce sustained remissions in relapsed refractory chronic lymphocytic leukemia. Sci Transl Med 2015;7 (303), 303ra139.
[30] Abramson JS, Palomba ML, Gordon LI, Lunning MA, Wang M, Arnason J, et al. Lisocabtagene maraleucel for patients with relapsed or refractory large B-cell lymphomas (TRANSCEND NHL 001): a multicentre seamless design study. Lancet (London, England) 2020;396(10254):839–52.
[31] Wang M, Munoz J, Goy A, Locke FL, Jacobson CA, Hill BT, et al. KTE-X19 CAR T-cell therapy in relapsed or refractory mantle-cell lymphoma. N Engl J Med 2020;382(14):1331–42.
[32] Shah BD, Ghobadi A, Oluwole OO, Logan AC, Boissel N, Cassaday RD, et al. KTE-X19 for relapsed or refractory adult B-cell acute lymphoblastic leukaemia: phase 2 results of the single-arm, open-label, multicentre ZUMA-3 study. Lancet (London, England) 2021;398(10299):491–502.
[33] North RJ. Gamma-irradiation facilitates the expression of adoptive immunity against established tumors by eliminating suppressor T cells. Cancer Immunol Immunother 1984;16(3):175–81.
[34] Tsimberidou AM, Van Morris K, Vo HH, Eck S, Lin YF, Rivas JM, et al. T-cell receptor-based therapy: an innovative therapeutic approach for solid tumors. J Hematol Oncol 2021;14(1):102.
[35] Radvanyi LG, Bernatchez C, Zhang M, Fox PS, Miller P, Chacon J, et al. Specific lymphocyte subsets predict response to adoptive cell therapy using expanded autologous tumor-infiltrating lymphocytes in metastatic melanoma patients. Clin Cancer Res 2012;18(24):6758–70.
[36] Brentjens RJ, Riviere I, Park JH, Davila ML, Wang X, Stefanski J, et al. Safety and persistence of adoptively transferred autologous CD19-targeted T cells in patients with relapsed or chemotherapy refractory B-cell leukemias. Blood 2011;118(18):4817–28.
[37] Gattinoni L, Finkelstein SE, Klebanoff CA, Antony PA, Palmer DC, Spiess PJ, et al. Removal of homeostatic cytokine sinks by lymphodepletion enhances the efficacy of adoptively transferred tumor-specific CD8+ T cells. J Exp Med 2005;202(7):907–12.
[38] Wrzesinski C, Paulos CM, Kaiser A, Muranski P, Palmer DC, Gattinoni L, et al. Increased intensity lymphodepletion enhances tumor treatment efficacy of adoptively transferred tumor-specific T cells. J Immunother (Hagerstown, Md: 1997) 2010;33(1):1–7.
[39] Baba J, Watanabe S, Saida Y, Tanaka T, Miyabayashi T, Koshio J, et al. Depletion of radio-resistant regulatory T cells enhances antitumor immunity during recovery from lymphopenia. Blood 2012;120(12):2417–27.
[40] Bechman N, Maher J. Lymphodepletion strategies to potentiate adoptive T-cell immunotherapy—what are we doing; where are we going? Expert Opin Biol Ther 2021;21(5):627–37.
[41] Innamarato P, Kodumudi K, Asby S, Schachner B, Hall M, Mackay A, et al. Reactive myelopoiesis triggered by lymphodepleting chemotherapy limits the efficacy of adoptive T cell therapy. Mol Ther J Am Soc Gene Ther 2020;28(10):2252–70.
[42] Fang L, Lonsdorf AS, Hwang ST. Immunotherapy for advanced melanoma. J Invest Dermatol 2008;128 (11):2596–605.
[43] Gattinoni L, Klebanoff CA, Palmer DC, Wrzesinski C, Kerstann K, Yu Z, et al. Acquisition of full effector function in vitro paradoxically impairs the in vivo antitumor efficacy of adoptively transferred CD8+ T cells. J Clin Invest 2005;115(6):1616–26.
[44] Dudley ME, Yang JC, Sherry R, Hughes MS, Royal R, Kammula U, et al. Adoptive cell therapy for patients with metastatic melanoma: evaluation of intensive myeloablative chemoradiation preparative regimens. J Clin Oncol Off J Am Soc Clin Oncol 2008;26(32):5233–9.

[45] Rosenberg SA, Restifo NP. Adoptive cell transfer as personalized immunotherapy for human cancer. Science (New York, NY) 2015;348(6230):62–8.
[46] Rosenberg SA, Yang JC, Sherry RM, Kammula US, Hughes MS, Phan GQ, et al. Durable complete responses in heavily pretreated patients with metastatic melanoma using T-cell transfer immunotherapy. Clin Cancer Res 2011;17(13):4550–7.
[47] Nissani A, Lev-Ari S, Meirson T, Jacoby E, Asher N, Ben-Betzalel G, et al. Comparison of non-myeloablative lymphodepleting preconditioning regimens in patients undergoing adoptive T cell therapy. J Immunother Cancer 2021;9(5).
[48] June CH, O'Connor RS, Kawalekar OU, Ghassemi S, Milone MC. CAR T cell immunotherapy for human cancer. Science (New York, NY) 2018;359(6382):1361–5.
[49] Amini L, Silbert SK, Maude SL, Nastoupil LJ, Ramos CA, Brentjens RJ, et al. Preparing for CAR T cell therapy: patient selection, bridging therapies and lymphodepletion. Nat Rev Clin Oncol 2022;19(5):342–55.
[50] Akhavan D, Alizadeh D, Wang D, Weist MR, Shepphird JK, Brown CE. CAR T cells for brain tumors: lessons learned and road ahead. Immunol Rev 2019;290(1):60–84.
[51] Dasyam N, George P, Weinkove R. Chimeric antigen receptor T-cell therapies: optimising the dose. Br J Clin Pharmacol 2020;86(9):1678–89.
[52] Hartmann J, Schussler-Lenz M, Bondanza A, Buchholz CJ. Clinical development of CAR T cells-challenges and opportunities in translating innovative treatment concepts. EMBO Mol Med 2017;9(9):1183–97.
[53] Pietrobon V, Todd LA, Goswami A, Stefanson O, Yang Z, Marincola F. Improving CAR T-cell persistence. Int J Mol Sci 2021;22(19).
[54] Jensen MC, Popplewell L, Cooper LJ, DiGiusto D, Kalos M, Ostberg JR, et al. Antitransgene rejection responses contribute to attenuated persistence of adoptively transferred CD20/CD19-specific chimeric antigen receptor redirected T cells in humans. Biol Blood Marrow Transplant J Am Soc Blood Marrow Transplant 2010;16 (9):1245–56.
[55] Savoldo B, Ramos CA, Liu E, Mims MP, Keating MJ, Carrum G, et al. CD28 costimulation improves expansion and persistence of chimeric antigen receptor-modified T cells in lymphoma patients. J Clin Invest 2011;121(5):1822–6.
[56] Ninomiya S, Narala N, Huye L, Yagyu S, Savoldo B, Dotti G, et al. Tumor indoleamine 2,3-dioxygenase (IDO) inhibits CD19-CAR T cells and is downregulated by lymphodepleting drugs. Blood 2015;125(25):3905–16.
[57] Kochenderfer JN, Dudley ME, Kassim SH, Somerville RP, Carpenter RO, Stetler-Stevenson M, et al. Chemotherapy-refractory diffuse large B-cell lymphoma and indolent B-cell malignancies can be effectively treated with autologous T cells expressing an anti-CD19 chimeric antigen receptor. J Clin Oncol Off J Am Soc Clin Oncol 2015;33(6):540–9.
[58] Dai H, Zhang W, Li X, Han Q, Guo Y, Zhang Y, et al. Tolerance and efficacy of autologous or donor-derived T cells expressing CD19 chimeric antigen receptors in adult B-ALL with extramedullary leukemia. Onco Targets Ther 2015;4(11), e1027469.
[59] Geyer MB, Riviere I, Senechal B, Wang X, Wang Y, Purdon TJ, et al. Safety and tolerability of conditioning chemotherapy followed by CD19-targeted CAR T cells for relapsed/refractory CLL. JCI Insight 2019;5.
[60] Curran KJ, Margossian SP, Kernan NA, Silverman LB, Williams DA, Shukla N, et al. Toxicity and response after CD19-specific CAR T-cell therapy in pediatric/young adult relapsed/refractory B-ALL. Blood 2019;134 (26):2361–8.
[61] Fraietta JA, Lacey SF, Orlando EJ, Pruteanu-Malinici I, Gohil M, Lundh S, et al. Determinants of response and resistance to CD19 chimeric antigen receptor (CAR) T cell therapy of chronic lymphocytic leukemia. Nat Med 2018;24(5):563–71.
[62] Schuster SJ, Tam CS, Borchmann P, Worel N, McGuirk JP, Holte H, et al. Long-term clinical outcomes of tisagenlecleucel in patients with relapsed or refractory aggressive B-cell lymphomas (JULIET): a multicentre, open-label, single-arm, phase 2 study. Lancet Oncol 2021;22(10):1403–15.
[63] Locke FL, Ghobadi A, Jacobson CA, Miklos DB, Lekakis LJ, Oluwole OO, et al. Long-term safety and activity of axicabtagene ciloleucel in refractory large B-cell lymphoma (ZUMA-1): a single-arm, multicentre, phase 1-2 trial. Lancet Oncol 2019;20(1):31–42.
[64] Turtle CJ, Hanafi LA, Berger C, Gooley TA, Cherian S, Hudecek M, et al. CD19 CAR-T cells of defined CD4+:CD8+ composition in adult B cell ALL patients. J Clin Invest 2016;126(6):2123–38.
[65] Shah NN, Lee DW, Yates B, Yuan CM, Shalabi H, Martin S, et al. Long-term follow-up of CD19-CAR T-cell therapy in children and young adults with B-ALL. J Clin Oncol Off J Am Soc Clin Oncol 2021;39(15):1650–9.
[66] Ramos CA, Ballard B, Zhang H, Dakhova O, Gee AP, Mei Z, et al. Clinical and immunological responses after CD30-specific chimeric antigen receptor-redirected lymphocytes. J Clin Invest 2017;127(9):3462–71.

[67] Ramos CA, Grover NS, Beaven AW, Lulla PD, Wu MF, Ivanova A, et al. Anti-CD30 CAR-T cell therapy in relapsed and refractory Hodgkin lymphoma. J Clin Oncol Off J Am Soc Clin Oncol 2020;38(32):3794–804.
[68] Nie Y, Lu W, Chen D, Tu H, Guo Z, Zhou X, et al. Mechanisms underlying CD19-positive ALL relapse after anti-CD19 CAR T cell therapy and associated strategies. Biomarker Res 2020;8:18.
[69] Xu X, Sun Q, Liang X, Chen Z, Zhang X, Zhou X, et al. Mechanisms of relapse after CD19 CAR T-cell therapy for acute lymphoblastic leukemia and its prevention and treatment strategies. Front Immunol 2019;10:2664.
[70] Kakarla S, Gottschalk S. CAR T cells for solid tumors: armed and ready to go? Cancer J (Sudbury, Mass) 2014;20(2):151–5.
[71] Wagner J, Wickman E, DeRenzo C, Gottschalk S. CAR T cell therapy for solid tumors: bright future or dark reality? Mol Ther J Am Soc Gene Ther 2020;28(11):2320–39.
[72] Thistlethwaite FC, Gilham DE, Guest RD, Rothwell DG, Pillai M, Burt DJ, et al. The clinical efficacy of first-generation carcinoembryonic antigen (CEACAM5)-specific CAR T cells is limited by poor persistence and transient pre-conditioning-dependent respiratory toxicity. Cancer Immunol Immunother 2017;66(11):1425–36.
[73] Hegde M, Joseph SK, Pashankar F, DeRenzo C, Sanber K, Navai S, et al. Tumor response and endogenous immune reactivity after administration of HER2 CAR T cells in a child with metastatic rhabdomyosarcoma. Nat Commun 2020;11(1):3549.
[74] Goff SL, Morgan RA, Yang JC, Sherry RM, Robbins PF, Restifo NP, et al. Pilot trial of adoptive transfer of chimeric antigen receptor-transduced T cells targeting EGFRvIII in patients with glioblastoma. J Immunother (Hagerstown, Md: 1997) 2019;42(4):126–35.
[75] Brown CE, Alizadeh D, Starr R, Weng L, Wagner JR, Naranjo A, et al. Regression of glioblastoma after chimeric antigen receptor T-cell therapy. N Engl J Med 2016;375(26):2561–9.
[76] Heczey A, Louis CU, Savoldo B, Dakhova O, Durett A, Grilley B, et al. CAR T cells administered in combination with lymphodepletion and PD-1 inhibition to patients with neuroblastoma. Mol Ther J Am Soc Gene Ther 2017;25(9):2214–24.
[77] Schmidts A, Wehrli M, Maus MV. Toward better understanding and management of CAR-T cell-associated toxicity. Annu Rev Med 2021;72:365–82.
[78] Pegram HJ, Lee JC, Hayman EG, Imperato GH, Tedder TF, Sadelain M, et al. Tumor-targeted T cells modified to secrete IL-12 eradicate systemic tumors without need for prior conditioning. Blood 2012;119(18):4133–41.
[79] Huang Y, Li D, Zhang PF, Liu M, Liang X, Yang X, et al. IL-18R-dependent and independent pathways account for IL-18-enhanced antitumor ability of CAR-T cells. FASEB J 2020;34(1):1768–82.
[80] Lanitis E, Rota G, Kosti P, Ronet C, Spill A, Seijo B, et al. Optimized gene engineering of murine CAR-T cells reveals the beneficial effects of IL-15 coexpression. J Exp Med 2021;218(2).
[81] Avanzi MP, Yeku O, Li X, Wijewarnasuriya DP, van Leeuwen DG, Cheung K, et al. Engineered tumor-targeted T cells mediate enhanced anti-tumor efficacy both directly and through activation of the endogenous immune system. Cell Rep 2018;23(7):2130–41.
[82] Rheinlander A, Schraven B, Bommhardt U. CD45 in human physiology and clinical medicine. Immunol Lett 2018;196:22–32.
[83] Louis CU, Straathof K, Bollard CM, Gerken C, Huls MH, Gresik MV, et al. Enhancing the in vivo expansion of adoptively transferred EBV-specific CTL with lymphodepleting CD45 monoclonal antibodies in NPC patients. Blood 2009;113(11):2442–50.
[84] Iomab-ACT: A Pilot Study of 131-I Apamistamab Followed by CD19-Targeted CAR T-Cell Therapy for Patients With Relapsed or Refractory B-Cell Acute Lymphoblastic Leukemia or Diffuse Large B-Cell Lymphoma. [cited 2022]. Available from: https://clinicaltrials.gov/ct2/show/NCT04512716?term=04512716&cntry=US&draw=2&rank=1.
[85] Suryadevara CM, Desai R, Abel ML, Riccione KA, Batich KA, Shen SH, et al. Temozolomide lymphodepletion enhances CAR abundance and correlates with antitumor efficacy against established glioblastoma. Onco Targets Ther 2018;7(6), e1434464.
[86] Brudno JN, Maric I, Hartman SD, Rose JJ, Wang M, Lam N, et al. T Cells Genetically Modified to Express an Anti-B-Cell Maturation Antigen Chimeric Antigen Receptor Cause Remissions of Poor-Prognosis Relapsed Multiple Myeloma. Journal of clinical oncology : official journal of the American Society of Clinical Oncology 2018;36(22):2267–80.
[87] Santomasso BD, Park JH, Salloum D, Riviere I, Flynn J, Mead E, et al. Clinical and biological correlates of neurotoxicity associated with CAR T-cell therapy in patients with B-cell acute lymphoblastic leukemia. Cancer Discov 2018;8(8):958–71.

CHAPTER

10

Imaging the immune cell in immunotherapy

Alessia Volpe[a], *Ronald Blasberg*[a,b,c,d], *Inna Serganova*[b,e,f], *and Vladimir Ponomarev*[a,c]

[a]Department of Radiology, Memorial Sloan Kettering Cancer Center, New York, NY, United States [b]Department of Neurology, Memorial Sloan Kettering Cancer Center, New York, NY, United States [c]Molecular Pharmacology and Chemistry Program, Memorial Sloan Kettering Cancer Center, New York, NY, United States [d]Memorial Sloan Kettering Cancer Center, New York, NY, United States [e]Department of Medicine, Division of Hematology and Medical Oncology, Weill Cornell Medicine, New York, NY, United States [f]Human Oncology and Pathogenesis Program, Memorial Sloan Kettering Cancer Center, New York, NY, United States

Abstract

The clinical success of cancer immunotherapy and the challenges currently limiting its further application to treat more cancer types have highlighted the pivotal role played by molecular imaging.

In this chapter, we will discuss the following topics:

(1) Imaging objectives from an immunological perspective
(2) Imaging modalities and labeling strategies for cancer immunotherapy
(3) Application of the Imaging Toolbox toward cancer immunotherapy
(4) What the future holds.

Abbreviations

BLI	bioluminescence imaging
CAR	chimeric antigen receptor
CBLuc	click beetle luciferase
CCD	charge-coupled device
CEST	chemical-exchange saturation transfer
CEU	contrast-enhanced ultrasound
CFP	cyan fluorescent protein
CFSE	carboxyfluorescein succinimidyl ester

https://doi.org/10.1016/B978-0-12-822620-9.00008-2

CLI	Cherenkov luminescence imaging
CLuc	Cypridina luciferase
CMV	cytomegalovirus
CTL	cytolytic T cell
CTLA-4	cytotoxic T lymphocyte-associated antigen 4
D_2R	dopamine 2 receptor
DC	dendritic cells
EBV	Epstein-Barr virus
eDHFR	*Escherichia coli* dihydrofolate reductase enzyme
FDA	Food and Drug Administration
FLI	fluorescence imaging
FLuc	firefly luciferase
FMT	fluorescence-mediated tomography
GFP	green fluorescence protein
GLuc	gaussian luciferase
hCEA	human carcinoembryonic antigen
hdCK	human deoxycytidine kinase
hdGK	human deoxyguanosine kinase
hNET	human norepinephrine transporter
hNIS	human sodium iodide symporter
hSSTR	human somatostatin receptor subtype 2
HSV1-*tk*	herpes simplex virus 1 thymidine kinase
hTK2	human mitochondrial thymidine kinase 2
ICOS	inducible T-cell co-stimulator
IFN-gamma	interferon-gamma
IL-2	interleukin 2
LAG-3	lymphocyte activation gene 3 protein
LE	labeling efficiency
LRP	lysine-rich-protein
LTR	long terminal repeat
MHC	major histocompatibility complex
MRI	magnetic resonance imaging
NFAT	nuclear factor of activated T cells
NIR	near-infrared
NK	natural killer
OI	optical imaging
OX40	tumor necrosis factor receptor family, member 4 (TNFRSF40)
PD-1	programmed cell death receptor 1
PD-L1	programmed cell death ligand 1
PEG	polyethylene glycol
PET	positron emission tomography
PFC	perfluorocarbon emulsion
PGK	3-phosphoglycerate kinase
PSMA	prostate specific membrane antigen
RF	radiofrequency
RFP	red fluorescence protein
Rluc	Renilla luciferase
RSV	respiratory syncytial virus
SFFV	spleen focus forming virus
SPECT	single photon emission computed tomography
SPIO	superparamagnetic iron oxide
TCR	T-cell receptor
TfR	transferrin
TIL	tumor-infiltrating lymphocyte

TIM-3	T cell immunoglobulin and mucin domain-containing protein 3
TME	tumor microenvironment
TNBC	triple-negative breast cancer
Treg	regulatory T cell
TYR	tyrosinase
US	ultrasound
USPIO	ultrasmall supermagnetic iron oxide
VLuc	vargula luciferase
VSPIO	very small paramagnetic iron oxide

Conflict of interest

No potential conflicts of interest were disclosed.

Imaging objectives from an immunological perspective

The mechanism of action of cell-based immunotherapies is more diverse and complex than conventional therapy (chemo and radiation therapy) and targeted molecular therapy (e.g., anti-EGFR). Successful targeting, expansion, infiltration, and effector function are only some of the many aspects influencing the way patients respond to immunotherapy and most are still largely unknown. Outstanding questions include

(1) What immune cell types are critical for a positive response to immunotherapy?
(2) What are the kinetics of distribution of cell-based immunotherapies in the human body?
(3) Do they successfully traffic to the tumor?
(4) What is their activation status?
(5) How long do they survive, persist, and retain their function in the tumor?
(6) What is the relation between persistence and response?
(7) What determines the heterogeneous response in patients and how can we predict it?
(8) Can we better predict and/or identify the side effects associated with immunotherapy (including cytokine release syndrome (CRS), on-target off-tumor toxicity, and neurotoxicity)?

As part of the so-called "Immunoimaging Toolbox" (Fig. 1), noninvasive imaging methodologies can help us better answer these questions and better understand the processes involved in successful immunotherapy and, more importantly, its failure.[a]

Imaging modalities and labeling strategies for cancer immunotherapy

In this section, we provide a comprehensive overview of the available imaging modalities and labeling strategies that can be tailored toward cancer immunotherapy. The imaging

[a] All aspects listed above will be discussed in "3. Application of the Imaging Toolbox toward Cancer Immunotherapy," where appropriate examples from the past and recent literature and across imaging modalities and immune cell types will also be provided.

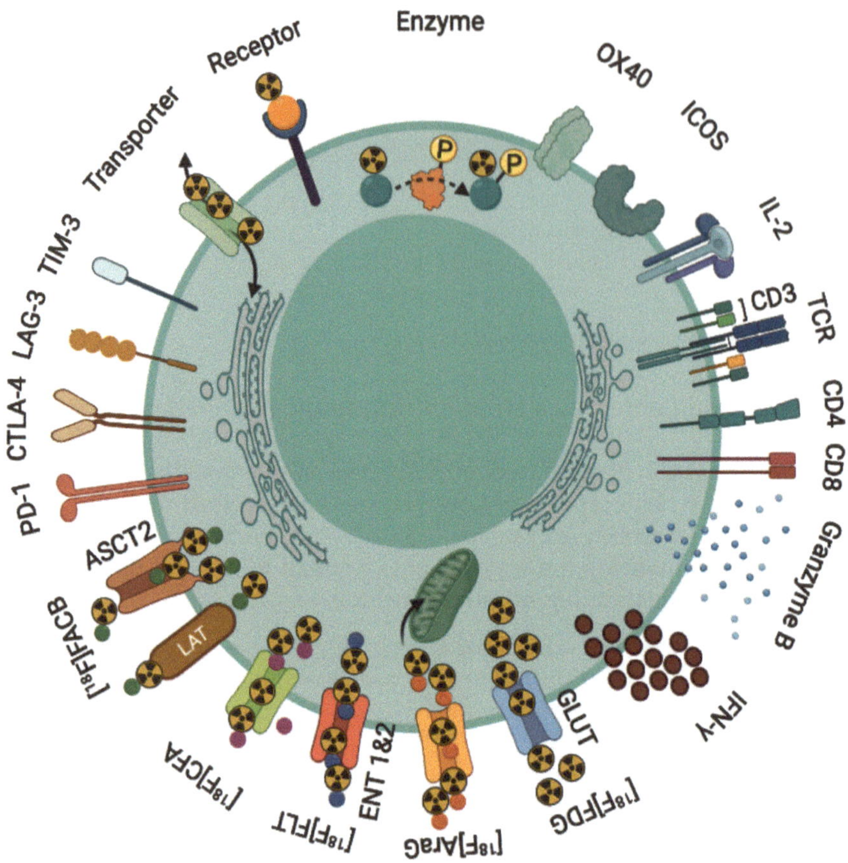

FIG. 1 The "Imaging Toolbox": immune-related molecular imaging targets.

modalities include (1) optical, (2) ultrasound, (3) magnetic resonance, and (4) nuclear-based imaging.

Optical imaging (OI)

Optical imaging (OI) is a noninvasive imaging technology using light (including visible, ultraviolet, and infrared light) and the special properties of photons to acquire biological information. As it is not involving exposure to ionizing radiation, it can be safely and reliably used for repeated imaging and monitoring of disease progression and therapy. The so-called "optical window" for optimal optical imaging is set between 650 and 950 nm, while at wavelengths longer than 950 nm, the increased absorption by water, hemoglobin, and lipids renders it not ideal [1,2]. Among optical imaging technologies, fluorescence (FLI) and bioluminescence (BLI) imaging have emerged as powerful modalities for molecular imaging.

Fluorescence imaging (FLI)

The technique is based on an external light of the appropriate wavelength exciting a target fluorescent molecule, followed by the release of longer wavelength, lower energy light, which can be ultimately detected and imaged. FLI relies on the use of light-activated fluorescent molecules, including dyes, conjugated fluorophores, fluorescent proteins (e.g., green fluorescent protein GFP) and optical contrast agents [3,4]. Since fluorescence imaging is based on nonionizing emissions, the reporters are suitable for repeated administration. As each fluorescence reporter emits at a different wavelength, more of them can be imaged simultaneously. Most are widely available at a low price and can be easily incorporated in DNA constructs, cells, and tissues. Intraoperative fluorescence imaging can provide anatomical, molecular, and functional information either through direct real-time imaging of the surgical field or by intraoperative optical specimen [5,6].

FLI encompasses a wide range of resolution and imaging depths, from subcellular analysis ($< 400\,\mu m$ with intravital microscopy) to small-animal imaging (between 1 and 3mm spatial resolution at <10 cm with fluorescence-mediated tomography (FMT)) [7].

Below are listed the most relevant examples of reporters for fluorescence molecular imaging:

- Artificial cell surface molecules

The anti-PEG reporter NIR797 belongs to this class. This reporter has been used for both in vitro and in vivo imaging [8]. Authors have cloned and expressed a membrane-anchored antipolyethylene glycol (PEG) reporter consisting of the Fab fragment of a mouse anti-PEG monoclonal antibody (AGF3), fused to the C-like extracellular-transmembrane-cytosolic domains of the mouse B7-1 receptor. Immunogenicity, specificity, and probe toxicity have been evaluated alongside with the already established HSV1-tk reporter gene and then successfully tested in a metastatic HeLa xenograft model. The same reporter is also suitable for PET and MRI imaging when paired with its corresponding probe (for PET: ^{124}I-PEG-SHPP; for MRI: SPIO-PEG). Below are listed some of the main fluorescent proteins that can be used for FLI.

- Monomeric fluorescent proteins (mFPs)

Examples of mFPs are eGFP A206K (Ex. 488/Em. 507nm) [9]; mCherry (Ex. 587/Em. 610nm) [10]; TagRFP (Ex. 555/Em. 584nm) [11]; mPlum (Ex. 590/Em. 649nm) [12]; mNeptune (Ex. 600/Em. 650nm) [13]; All the above fluorescent proteins do not have a clinical use and can be fused with other reporter genes allowing multimodal multiscale imaging.

- Fluorescent proteins

Far red tetrameric E2-Crimson (Ex. 611/Em. 646) [14]; tetrameric DsRed (Ex. 551/Em. 583) [15].

- Near-infrared (NIR) fluorescent proteins

IFP1.4 (Ex. 684/Em. 708nm; dimer) [16]; iRFP 670 (Ex. 643/Em. 713nm) [17,18]; iRFP 713 (Ex. 690/Em/713nm) [16]. All the above NIR reporter requires the co-factor biliverdin to be added exogenously to generate detectable fluorescence. The only two near-infrared dyes currently approved by the FDA for intraoperative use are indocyanine green (ICG) and methylene blue (MB) allowing signal detection up to several millimeters of depth under the tissue surface [19].

- Photoactivatable proteins

Kaede (Ex. 518/Em. 580nm) [20]; IrisFP (Ex. 516, Em. 580) [21].

- Photoconvertible proteins

Dendra2 [22] has an excitation-emission at 490–507nm, similar to enhanced-GFP and other green fluorescent proteins. It can therefore be detected with the commonly used filter sets EGFP and FITC. The unique feature of this protein is the ability to photoconvert to red fluorescent state upon intense-blue-light irradiation (at 460–500nm).

Bioluminescence imaging (BLI)

BLI is an optical molecular imaging approach based on the detection of visible light emitted during the enzyme-mediate oxidation of a molecular substrate and has a very favorable signal-to-noise ratio. The enzyme (e.g., a luciferase) is introduced in the desired cells by genetic engineering and serves as an imaging reporter gene. As with many other imaging techniques involving the use of reporter genes, it allows in vivo longitudinal imaging and serial quantification of biological processes, without having to sacrifice the study animal. This can significantly reduce the number of animals required to conduct a study, particularly when studying a biological process sequentially over time, and minimizes interanimal variability. Cells expressing BLI reporters can be imaged using sensitive charge-coupled device (CCD) cameras for the detection of light, after injecting the enzyme's substrate. Optimum imaging time is typically influenced by the route of administration of the substrate (e.g., retro-orbital, intraperitoneal, or intravenous) and amount of substrate injected.

A variety of bioluminescence systems do exist in nature, each having its unique luciferase enzyme and corresponding substrate. Luciferases derived from terrestrial organisms primarily use D-luciferin as a substrate, while luciferases coming from marine organisms primarily use coelenterazine Here we report the most used BLI enzymes and related characteristics.

- Firefly luciferase (FLuc)

Cloned in 1985 [23], FLuc derives from the North American Firefly *Photinus pyralis* and represents an excellent marker for gene expression since it lacks of posttranslational modifications and has an in vivo half-life of about 3h. The firefly luciferase needs ATP, magnesium and a benzothiazolyl-tiazole luciferin (or derivatives) as a substrate in order to catalyze the reaction responsible for producing emitted photons [24,25]. The corresponding light emission is at 530–640nm (peaking at 562nm). This emission spectrum, together with the intrinsic optical properties of tissues, allows light to penetrate in the tissues for several centimeters, subsequently allowing deep-tissue imaging. There exist also other variants of FLuc, such as the Italian firefly luciferase, from *Luciola italica red* [26], and the North American firefly luciferase from *Photinus pyralis red* [27].

- Click beetle luciferase (including CBred, CBG68, CBG99)

CBLuc derives from the *Pyrophorus plagiophtalamus*. Similar to FLuc, the substrate for CBLuc is luciferin. The emission varies depending on the different subspecies: the emission for green (CBG68 and CBG99) is 537nm; the emission for red (CBred) is 611nm [28–30]. Although the emission at 611nm is advantageous for in vivo imaging, the firefly luciferase still remains the best candidate for bioluminescence molecular imaging because its reaction with luciferin produces more light.

- Renilla luciferase (RLuc)

Renilla reniformis is from a sea pansy, and the luciferase uses coelenterazine as a substrate [31]. RLuc emission is at 480nm, while RLuc8 and RLuc8.6 variants emit at 480 and 535nm, respectively.

- Gaussia luciferase (GLuc)

GLuc derives from *Gaussia princeps* and has an emission at 480nm. The corresponding substrate is coelenterazine [32,33].

- Cypridina luciferase (CLuc)

CLuc is from Cypridina noctiluca, with emission at 460nm and uses Cypridina luciferin (Vargulin) as a substrate [34]. A recent study from Moroz et al. demonstrated the applicability of a newly generated membrane-anchored Cypridina (maCLuc) in combination with other established luciferase-based reporter systems for the multiplexed in vivo imaging of four different targets in the same animal [35].

- Vargula luciferase (VLuc)

This sea firefly luciferase is derived from *Vargula hilgendorfii* and emits at 460nm. The paired substrate is Cypridina luciferin (Vargulin) [36].

An active area of research in bioluminescence imaging is represented by the identification of new luciferase reporter genes and mutated variants. One example is represented by the engineered firefly luciferase-inspired enzyme, also known as PLG2. PLG2 has a greater in vitro sensitivity (threefold) than FLuc. This chimera contains the N-domain of the *Photinus pyralis* luciferase, the C-domain of the *Luciola italica* luciferase, and five modified amino acids to enhance its bioluminescence properties [37]. NanoLuc, first isolated from a deep-sea shrimp *Oplophorus gracilirostris* [38], also falls into this new category. It has been further engineered to work optimally with a novel coelenterazine analog, named furimazine, where the phenol group was replaced with a furane. Another analog, fluorofurimazine, was recently tested in vivo; it was demonstrated that the NanoLuc-fluorofurimazine combination has the highest bioluminescence intensity post intravenous administration, thereby upgrading the NanoLuc system to deep tissue imaging [39]. The small molecular size of NanoLuc (19kDa compared to the 61.5kDa of firefly) and bright luminescence (10-fold brighter than firefly or renilla luciferase) allow it to be more easily incorporated into genetic-transduction reporter constructs and provides exquisite sensitivity for bioluminescence imaging. Renilla luciferase has also been fused to the iRFP near-infrared proteins in order to create a new near-infrared bioluminescence resonance energy transfer (BRET) system for deep-tissue imaging [40].

Considerations for optical imaging

Innovations in both molecular biology and chemistry have led to the development of optical methods for imaging a variety of cellular and molecular processes, in both healthy and diseased models. In the past, optical imaging has been primarily involved in preclinical research in small animal models, but there are several areas where this modality could potentially translate to the clinic. Key challenges for moving optical imaging into clinical application are (i) the tissue autofluorescence responsible for producing a high background noise and (ii) the low tissue penetration of light. For light in the visible spectrum, absorption by hemoglobin, water, and other molecules is responsible for the overall reduction of optical signal by approximately 10-fold per centimeter of tissue [41]. Both aspects restrict the application of most current optical methods to small animals or to the study of surface objects in the human body. Imaging and detection of deeper objects is only possible by using the light within the near-infrared range [19]. At these wavelengths, the absorption of light by hemoglobin, water, and lipid is low, autofluorescence is reduced, and tissue penetration is improved [42,43]. The many limitations of the optical imaging strategy are also reflected in the ability to obtain only a qualitative (relative quantification) output signal, while radionuclide-based imaging (such as PET) allows for absolute quantification.

Ultrasound imaging (US)

Ultrasound imaging is the second most used technique in clinical imaging as it is noninvasive, widely available, allows real-time imaging without the use of ionizing radiation, and is relatively inexpensive compared to other imaging modalities described in this chapter [44–46]. Ultrasound is based on the propagation of mechanical (ultrasonic) waves from a transducer into the tissue and the interaction of these waves with the mechanical discontinuity characteristics of different tissue. US generated echoes that can be detected and resolved to depict tissue anatomy in the so-called brightness mode (B-mode). Contrast-enhanced ultrasound (CEU) molecular imaging is a technique that relies on the detection of (i) encapsulated gas microbubbles or (ii) nonmicrobubble contrast agents (other nano- and micro-particles binding to target molecules and/or cells).

Here we describe the most relevant US contrast agents as well their advantages and disadvantages for ultrasound molecular imaging.

Encapsulated gas microbubbles

Microbubbles are small size (usually between 1 and 4 μm) gas-liquid emulsions, with a gaseous core encapsulated in a shell. The small size allows them to easily traffic through the vascular compartment and avoids leakage into the extravascular space. The gas type can vary from room air, to nitrogen, or heavy inert gases (e.g., perfluoropropane, perfluorobutane, perfluorohexane, and sulfur hexafluoride) [47]. Heavy gasses are preferred as they stabilize the microbubble by retaining a poor solubility in the blood (and therefore a longer half-life during circulation in the blood stream). The shell, on the other hand, can be composed of different materials (including phospholipids, biocompatible polymers, and proteins such as albumin). Coating with a polyethylene glycol (PEG) polymer can provide the microbubble with additional stearic protection, prevent absorption of macromolecules to the microbubble surface, and prevent their in vivo aggregation. A "targeting ligand" can also be attached to the shell, which provides targeting/recognition motif to direct the microbubble to localize at specific molecules or cells. The targeting ligand can be an antibody, a peptide or an engineered scaffold directly incorporated into the shell during, or immediately after synthesis. Alongside the many clinical developments, a wide range of preclinical seminal studies have employed microbubbles in the context of inflammation [48–52], angiogenesis [53,54], plaque formation [55,56], and thrombosis [57,58].

Nonmicrobubble contrast agents

Nanobubbles are gas-liquid emulsions encapsulated in a biodegradable polymer and generating a high ultrasound signal. Solid nanoparticles, on the other hand, can be silica-based or iron oxide particles containing gas pockets in their pores. Belonging to this class are all agents with a submicron or nanosize (between 10 and 1000 nm) [59]. Based on both composition and size, a wide range of particles have been synthesized for US and include echogenic liposomes, perfluorocarbon emulsion (PFC) nanodroplets, nanobubbles, and solid nanoparticles. Echogenic liposomes display an aqueous core surrounded by a lipidic bilayer. The air pockets within the bilayer are responsible for generating acoustic reflectivity [60]. PFC nanodroplets are liquid emulsions consisting of a perfluorocarbon core encapsulated in a phospholipid monolayer. The high PFC nanodroplets accumulation at the target leads to acoustic detection.

Considerations for ultrasound imaging

Despite the variety of contrast agents readily available for routine preclinical and clinical US, microbubbles are still the most commonly used contrast agents for US. Circulating unbound particles do generate a very low background signal that rapidly decreases. The efficient in vivo distribution and clearance of the background signal allow the operator to perform the ultrasound imaging within a few minutes following administration of the probe. This compares favorably to other molecular imaging techniques (including SPECT and PET, where the imaging session frequently starts 1-h post contrast injection). The rapid clearance can be leveraged for longitudinal imaging by allowing repetitive injection of the microbubbles in the desired recipient.

Ultrasound molecular imaging is cheap, easy to perform, and does not require complex infrastructure; most important, it provides good spatial and temporal resolution. Importantly, US is also quantitative and has both high sensitivity and a high signal-to-noise ratio. Areas of current development are mainly related to microbubble targeting optimization, identification of new disease targets, and development of new detection algorithms. Another area of development is that of acoustic reporter genes for noninvasive ultrasound molecular imaging. In a recent study published in *Nature*, Bourdeau and colleagues reported on the first reporter gene system for ultrasound, showing the ability to visualize genetically modified *E. coli* and *S. typhimurium* bacteria in living animals [61].

Magnetic resonance imaging (MRI)

Magnetic resonance imaging (MRI) is a noninvasive tomographic imaging technology avoiding the use of ionizing radiation and excelling in soft tissue contrast. MRI is based on the manipulation of the nuclear magnetic moment of endogenous nuclei (^1H in H_2O) in a magnetic field. A comprehensive overview of the basic principles of this modality can be found elsewhere [62]. Following a magnetic perturbation, nuclei relax using two co-dependent relaxation mechanism, namely T1 (spin-lattice relaxation) and T2 (spin-spin relaxation) and realign with the magnetic field. In T1 relaxation, protons return to their original position and the energy is transferred to their surroundings (an enhanced "bright" signal is obtained with positive contrast agents, such as Gadolinium-DTPA). T2 relaxation refers to a progressive dephasing of the protons spinning synchronously and resulting in the transfer of energy in the transverse plane. The most relevant contrast agents tailored for magnetic resonance imaging and here discussed are the so-called negative contrast agents (resulting in a reduction or absence of signal).

Negative contrast agents

T2 contrast agents are iron-oxide nanoparticles, which in turn can be (i) superparamagnetic iron oxide (SPIO), (ii) very small paramagnetic iron oxide (VSPIO), or (iii) ultrasmall supermagnetic oxide (USPIO). The material used for the coating of these magnetic particles is nontoxic, biocompatible allows the targetable delivery with particle localization within a specific area. For example, the SPIOs are often coated with silicon material [63–65], polyacrylic acid [66,67], dextran [68,69], dopamine [70], deferoxamine [71,72], and other alternative organic polymers. Below the polymer coating is the crystalline iron-oxide core. Each of the particles contains thousands of iron atoms, thus resulting in high T2 relaxivities (up to 200 $(mMs)^{-1}$) [73] and the detection possible even at low concentrations of contrast agent (μmol to nmol range). Virtually any cell population of interest can be labeled with these magnetic nanoparticles and followed by in vivo MRI. Frank and colleagues have extensively described the existing methodologies for magnetic labeling of cells, including stem cells and other mammalian cells, and their translation into clinical setting [74]. Those particles are typically smaller than 300 nm (between 50 and 100 nm) and can be easily transported across the cell membrane. A major problem with the SPIOs is that they effectively are a "1-time use" for labeling cells (including immune cells). Two issues are limiting: (1) if a SPIO-labeled cell divides the SPIOs are split between the two cells and become diluted and (2) once the labeled cells die, the SPIOs usually remain (and are taken up by macrophages) and continue to be visualized as "black spots" by T2 MRI.

Reporter genes also belong to the class of negative contrast agents. They are employed to allow noninvasive imaging of gene expression, particularly of those genes that have been artificially introduced into the target cells to report on a specific molecular event. MRI reporter genes are categorized into three classes based on the nature of the encoded gene: (i) enzymes; (ii) receptors; (iii) endogenously expressed reporter genes. An advantage of MR-based reporter gene imaging is excellent spatial resolution; two major disadvantages is low sensitivity and the absence of good quantification of reporter expression.

- Enzyme-based reporter genes

Belonging to this class are tyrosinase (TYR) and β-galactosidase. TYR is the primary enzyme responsible for melanin production. Melanin chelates paramagnetic ions (such as Fe^{3+}). The high iron-chelating capacity depends on the channels within the melanin granules, resulting in enhanced MR signal in the transfected cells [75]. Weissleder and Enochs were the first ones to develop TYR as an MRI imaging reporter, demonstrating that the high concentration of metallo-melanin is the reason for the reduced T1 relaxivity [75,76]. As the uncontrolled production of melanin would most likely be harmful to the cells, the TYR reporter gene must be regulated by specific promoters [77]. The fact that its expression can be inducible is a great advantage to minimize the risk of toxicity due to the melanin-driven production of highly reactive oxygen species [78], making it way superior to the expression of constituent reporter genes. However, its high molecular weight and low expression limit its utility in MRI imaging.

β-Galactosidase is a glycoside hydrolase [79] catalyzing the reaction converting X-gal substrate into a Prussian Blue-like product. This reporter has a large number of both chromogenic and fluorogenic different substrates for optical imaging. It is encoded by the bacterial LacZ gene and mainly utilized for the detection of proliferation and cell survival. As an MRI reporter, it requires an extra contrast agent, one of which is the widely used EgadMe (sort name for (1-(2-(b-galactopyranosyloxy)propyl)-4,10-tris(carboxymethyl)-1,4,7,10-tetraazacyclododecne)gadolinium(III)). EgadMe is a paramagnetic Gd^{3+} chelate with a high binding affinity for gadolinium. A major disadvantage of EgadMe is that this probe is rapidly cleared through renal excretion after systemic administration. It also gets diluted while cells are dividing, thus leading to a reduction of MR contrast over-time.

- Receptor-based reporter genes

The transferrin reporter gene (TfR) belongs to this category. It encodes for the cell-surface transferrin receptor, a homodimer transmembrane glycoprotein almost ubiquitously expressed in all mammalian cells and binding to transferrin (a protein mediating the transport of iron in the blood). TfR binds to two iron-loaded transferrin to form the so-called "TfR-Tf-Fe" complex and internalizes it via endocytosis, allowing iron release from the complex. The high expression of TfR determines a higher intake of iron from the cells and decreases T2 relaxation time [80]. The receptor has been used as a reporter gene for MRI in multiple studies displaying excellent contrast [81–83]. One example is represented by transferrin-conjugated SPIO particles that were successfully internalized by gliosarcoma cells ectopically expressing TfR [83].

- Endogenously expressed reporter genes

Ferritin is an intracellular protein ubiquitously expressed mainly in liver, spleen, and brain. It plays a critical role in iron metabolism by storing it and releasing it only upon need. Ferritin is characterized by 24 heavy and light subunits, encapsulated with an iron oxide core. The ratio between heavy and light subunits varies depending on the different tissue, but all have a similar structure of 5 A-helix, just with a different function. The ability of ferritin to sequester iron makes it a suitable candidate as a reporter gene for MRI [84]. In fact, cells transfected to overexpress the ferritin reporter gene allow efficient capture of extracellular iron, resulting in a high superparamagnetism, significantly impacting on MRI relaxation time and producing detectable contrast. Unlike the SPIOs or similar nanoparticle-based approaches, a ferritin reporter has the advantage of not diluting its signal with cell division; the reporter is endogenously and ubiquitously expressed in dividing cells. A second advantage is that ferritin does not need ectopic probes compared to previously discussed TYR, β-galactosidase, and TfR. However, a major limitation is seen in the poor imaging capability (poor sensitivity) when body iron levels are low (as the MRI image quality solely depends on iron loading index [85]) and the inability to obtain meaningful measures of reporter expression (iron level dependent).

A new class of artificial proteins, also named chemical-exchange saturation transfer (CEST) agents, and technically classified as negative contrast agents, are emerging as reporters for MRI imaging. Lysine-rich protein (LRP) [86] is one of them and can be used to image a variety of physiological and pathological parameters, including pH, metabolites, cancer, and stroke. The principle behind them is that many molecules in the body exchange protons with bulk water via the selective radiofrequency (RF) irradiation of exchanging protons. The RF irradiation component allows this type of reporter to switch on and off, changing the image contrast via an RF presaturation pulse agent.

Nuclear-based imaging

Imaging using radionuclide-based probes offers whole-body noninvasive tracking capabilities for both preclinical and clinical studies and can be achieved using direct or indirect imaging approaches [87]. Here we discuss the existing nuclear-based imaging strategies, from basic principles to advantages and limitations.

Basic principles of direct labeling

Direct labeling is achieved by ex vivo labeling of cells prior administration into the desired recipient (animal models and/or humans). Labeled cells, preincubated with ready-to-use contrast agents, are employed to track a variety of cell types, including immune cells [88–90]. Lipophilic chelators can help the label enter the cells, and once the latter has successfully bound its intracellular target, the chelator is removed from the cell. Transfection agents and electroporation are alternative approaches employed to facilitate the label entry into the target cells. Transfection agents are non- or low-toxic cationic molecules, polymers, liposomes, and lipoplexes with positively charged head groups that neutralize the electrostatic charge of the plasma membrane by interacting with the negatively charged phospholipids in the inner layer. Electroporation is based on intense electric pulses that are responsible for changes in the membrane permeability, thereby allowing the introduction of the label into the cell. Another approach for ex vivo labeling directly takes advantage of their cellular processes (e.g., receptor-mediated internalization, transport-mediated mechanism, phagocytosis, etc.).

The standard protocol for ex vivo labeling involves three critical experimental steps: (i) incubate of the cells of interest with the desired imaging agent (time and temperature of incubation vary and need to be determined by the operator), (ii) perform multiple washes to remove unbound or noninternalized radiotracer, and (iii) measure the radioactivity present in the cells and the washes using an automated or manual gamma counter. Cell labeling efficiency (LE(%)) may vary depending on labeling conditions and cell type [91].

Once cells are successfully labeled in vitro, and the corresponding labeling efficiency is calculated, they can be administered into a living animal and followed over-time by radionuclide-based imaging.

Considerations for direct labeling

Although indirect labeling enables a straight-forward labeling of any cell of interest, it has multiple drawbacks that need to be factored in the experimental design. In fact, the imaging agent used to label cells can be lost over time due to two biological processes: (i) efflux/loss of the label from the cell and (ii) a dilution effect due to cell division. When efflux occurs, the isotope leaks out of the target cells and can circulate in the blood stream and redistribute in other sites of the body (if it is not immediately excreted). Both circulation and redistribution generate background, rendering it more difficult to discriminate between labeled cells and free isotope. As the directly labeled cells continue to actively divide, the label is distributed to the daughter cells and is progressively diluted. Depending on the radiotracer of choice (and its corresponding half-life), the overall signal strength will be retained until radiotracer decay. However, the limitation described above makes direct labeling an unsuitable choice for the long-term monitoring of rapidly dividing cells, such as cancer cells or T cells, as the number of individually labeled cells would decrease over time [92]. Signal loss due to efflux, dilution, and ultimately decay can be compensated by using radioisotopes with longer half-life. When using long half-life tracers, a few more considerations need to be made, particularly in light of a potential translation of direct labeling into a clinical setting: in fact, tracers with a long half-life (and thus taking more time to decay) will have a dramatic impact on the management of the patient and his body excreta. Moreover, the longer decay prolongs the time the radioactivity is emitted in the patient's body and may therefore result in more damage to the healthy tissues. After prelabeled cells are systemically administered, they will generate a signal regardless of their viability status. In fact, dead cells will still lead to signal, resulting in the generation of unreliable sets of data and a potential misinterpretation of the therapeutic response. Also, some clinically used radiotracers employed for ex vivo labeling can impact on cell viability and/or function, despite displaying a great labeling efficiency. One example is that of ^{111}In-oxine and ^{18}F-FDG, both responsible for the progressive reduction of T cell proliferative ability, impaired cytotoxic function, and ultimately cell death when used at high concentrations [93].

Indirect cell labeling can overcome many of the above challenges associated with direct labeling and will be discussed in the following section. Despite its known limitations, the ex vivo labeling process described above is simple in its execution and does not involve genetic manipulation of the cells, meaning that very few regulatory hurdles for clinical application have to be overcome, thereby rendering this approach easily translatable into patients for short-term monitoring of immune cells and therapeutics.

Basic principles of indirect labeling with reporter genes

Indirect labeling is achieved through genetic engineering of cells to ectopically express a reporter gene, ultimately leading to its genomic integration and long-term expression. A recent review of reporter gene imaging outlines its historical development and explores whether the "promise" has been delivered [94].

The versatility of reporter gene imaging is due to the flexibility to tailor the expression cassette to the experimental objectives. The typical components of a reporter system are the so-called "regulatory complex" and the "reporter gene." The regulatory complex includes the promoter, which can either be (i) a constitutive regulatory element, always "on," hence leading to continuous transcription (e.g., CMV, RSV, SFFV, PGK, LTR, etc.) or (ii) an inducible regulatory element, also described as a "molecular genetic sensor" and leading to a controlled/induced expression [95]. The latter is placed under a transcriptional control and responds to endogenous cell signals (e.g., signaling molecules involved in a signaling pathway, and/or tissue- or tumor-specific transcription factors), thereby driving the expression of the reporter gene and linking it to a particular biological process or mechanism. Because of their intrinsic properties, reporters based on constitutive promoters are best suited for immune cell trafficking, persistence, and viability, while those based on inducible promoters are more suitable for monitoring activation and cell function.

Once a suitable cDNA expression cassette is designed, reporter genes are ectopically introduced into cells using (i) viral vectors (retro-, adeno-, and lenti-viral based and randomly integrating the transgene into the host genome) [96], (ii) episomal plasmids (transposons) [97,98], or through (iii) gene editing [99–101]. After the generation of reporter gene-expressing cells, those cells are expanded in vitro, and selected to isolate positive clones (only reporter expressing cells). The selection can either happen through flow cytometric sorting or by taking advantage of existing selection markers. Importantly, reporter-expressing cells can then be tested for radioactive uptake, where cellular radiotracer uptake is expressed as % Uptake [91].

Cells are then injected into the desired subject (including humans), followed by the intravenous injection of a radiolabeled probe (complimentary to the specific reporter gene) and radionuclide imaging (PET or SPECT) is performed for localization and quantitation of reporter expression at selected times postprobe injection. The indirect labeling and imaging paradigm involves the coupling of the "reporter-transgene" with its complementary "reporter probe." The probe accumulation reflects the endogenous signaling/transcription factors that drive the reporter gene expression, stabilization of mRNA and subsequent translation of the mRNA into the gene product, thereby informing on levels of reporter gene expression. The probes accumulate where certain endogenous receptors or transporters are highly expressed, and can therefore be used to image human-derived reporter genes (i.e., [^{131}I]MIBG for NET [102,103]; 3-(2′[^{18}F]-fluoroethyl)spiperone (FESP) for D_2R [104]; [^{111}In] DTPA-octreotide for SSTR2 [105]; iodide and $^{99m}TcO_4^-$ for NIS [106]). The reporter transgenes can encode for: (i) transporters, (ii) receptors, or (iii) enzymes. Here we provide a description of the above classes and their main features.

- Transport-based human reporter genes

Reporter genes encoding for a transporter enable a radiolabeled probe to be transferred into the cell where it accumulates and generates a signal. Here we describe two main transporters that have specific advantages for human studies, namely, they involve human genes and are less likely to induce a host immune reaction (this possibility exists for viral reporter genes such as HSV1-tk).

The sodium iodide symporter (NIS) is a known imaging and therapeutic target in thyroid cancer [107,108]. It has been successfully employed as a reporter gene in a number of preclinical studies [109–113]. NIS is endogenously expressed in a limited number of tissues, including thyroid gland [114] and stomach [115], and mediates the transport of several anions with an electroneutral stoichiometry [116]. NIS accurately reports on cell viability as the transport relies on an intact Na^+/K^+ gradient [117,118]. The [^{18}F]BF_4^- tracer for NIS has emerged as a more desirable radiolabeled probe (compared to $^{123/124/131}I$) due to its faster clearance from the circulation, reaching lower background levels at the time of imaging, and has been used safely in humans [119]. NIS represents an excellent reporter gene candidate for future regulatory approval in humans.

The human norepinephrine transporter (NET) is exclusively expressed in organs with sympathetic innervation (heart, brain) [120] and has been employed as a reporter gene in different cell types [121–123]. Significant effort has been placed into the development of radiolabeled norepinephrine analogs for NET imaging of brain tumors, exploiting its endogenous expression in the bran and overcoming the limitation of the currently available tracers (MIBG and HED) in passing the blood-brain barrier (BBB) [124].

Transporters such as NIS and NET typically provide signal amplification due to their ability to transport multiple radiotracer molecules into the cell. Another transporter is also the dopamine transporter (DAT).

- Receptor-based human reporter genes

A reporter gene can also encode for a receptor that irreversibly binds to a radiolabeled probe, allowing for visualization by molecular imaging. One of them is the somatostatin receptor type 2 (SSTR2), endogenously expressed in several tissues and tumor cells, including pituitary, endocrine, medullary thyroid and pheochromocytomas. There are several paired radiotracers available, some of which are already used in clinical practice (i.e., [^{111}In] DTPA-octreotide [125,126]). Tracers with the high affinity for SSTR2 and excellent image quality (low background) include ^{68}Ga-DOTA-TATE [127] and DOTA-TOC [128]. The dopamine receptor D_2R and its mutant D_2R80A can be used for noninvasive reporter-gene imaging [129,130], and are mainly expressed in the striatum of the brain and pituitary gland. Compared to transport-based reporters, most receptors bind irreversibly to their complementary radiolabeled probe with a 1:1 stoichiometry, and therefore cannot provide any signal amplification.

- Enzyme-based reporter genes

Reporter genes can encode for enzymes that chemically modify the corresponding radiolabeled probe, such that it can no longer exit the cell. Thus, radioactivity becomes "trapped" and accumulates within the cell and the level of accumulation reflects the level of reporter gene expression in the transduced cells. Endogenous enzymes used as transporters include the human deoxycytidine kinase (hdCK) [131,132] and the human mitochondrial thymidine kinase 2 (hTK2) and its mutants [133,134].

The human deoxycytidine kinase (dCK) converts deoxy-cytydine, deoxy-adenosine, deoxy-guanosine, and pyrimidine-based cytotoxic drugs into their monophosphate forms [135]. Likar and colleagues described the endogenous hdCK and mutants as reporter genes suitable for PET imaging with ^{18}F-FEAU. Importantly, the hdKC reporter also showed prosuicidal activity in cells treated with pyrimidine-based antineoplastic prodrugs (gemcitabine and Ara-C), allowing for the safe elimination of reporter-expressing cells following their administration [132].

hmTK2 is typically expressed in the mitochondria and phosphorylates deoxy-cytidine, deoxy-thymidine, deoxy-uridine, and various anticancer and antiviral nucleoside analogs. However, the nucleoside analogs of thymidine (such as 2′-fluoro-2-deoxy-1-beta-D-arabinofuranosyl-5-iodouracil (FIAU) and 2-fluoro-5-ethyl-1-beta-D-arabinofuranosyluracil

(FEAU)) are not efficiently entrapped due to their poor accessibility to the reporter protein in the cell [133]. Therefore, a truncated version of the N terminus of the human-derived reporter gene has been developed (hΔTK2), allowing for expression in the cell cytoplasm and subsequent entrapment of [$^{123/124}$I]FIAU or [^{18}F]FEAU for noninvasive SPECT and PET imaging, respectively [133].

Alternatively, a viral thymidine kinase can be used as a reporter gene for PET imaging. The herpes simplex virus 1 thymidine kinase (HSV1-tk) and its mutant derivatives (e.g., HSV1-sr39tk and HSV1-A167Ytk) [136] catalyze the phosphorylation of nucleoside analogs (e.g., FHBG or FIAU), and entrap them in the cytoplasm by adding a phosphate (an extra negative charge) to them. In contrast, endogenous mammalian tk has limited or no ability to phosphorylate these compounds. HSV1-tk and HSV1-sr39tk are susceptible to GCV-induced cell death, thus displaying a dual reporter-suicide feature that can be exploited in case of immune rejection, as they are both non mammalian [137]. Several radiolabeled probes have been developed for HSV1-tk PET imaging [136,138–143], including 9-[4-fluoro-3-(hydroxymethyl)butyl]guanine (FHBG). In a comparative study, Tjuvajev et al. showed that FIAU has a greater sensitivity and a lower abdominal noise over FHPG and FHBG for the in vivo imaging of HSV1-tk expression by PET [144]. The pharmacokinetics and dosimetry of [^{18}F]FHBG proved to be safe in healthy volunteers [145].

Considerations for indirect labeling

Reporter genes are a powerful and versatile tool for the noninvasive and repetitive observation of biological events and the development and monitoring of new therapeutics (for more information, please refer to the review from Volpe et al. [91]). Expression and function of the reporters are two exclusive prerequisites of viable cells. Therefore, the resulting radioactive signal can only be attributed to them. So-called "foreign" reporters are potentially more suited for the imaging of systemically administered adoptively transferred immune cells, as they are uniquely expressed in them and do not generate any background signal. Alternatively, host-compatible reporters (such as NIS) can also be employed and have the advantage of not eliciting a host immune response.

An advantage of indirect cell labeling is the permanent integration of the reporter gene into the genome of the host cell and does not result in label dilution during cell division. This becomes particularly relevant for the long-term monitoring of cellular immunotherapeutics, since the reporter gene is passed onto the daughter cells. Indirect labeling allows for imaging and quantifying a population of rapidly expanding cells (including immune cells) with no time limit [89]. While the observation time of directly labeled cells is typically dictated by the radiotracer half-life, this is not the case for reporter gene-expressing cells, where short-lived isotopes can be used and are recommended to avoid any potential associated radio damage and/or toxicity.

However, the (i) potential for immunogenicity, the (ii) very costly and complex protocols, the (iii) need for a well-established nuclear medicine infrastructure, and the (iv) regulatory concerns relating to the use of viral vectors, have slowed their clinical translation.

Application of the "Imaging Toolbox" toward cancer immunotherapy

Immunotherapy seeks to enhance a patient's immune response against cancer. The main classes of immunotherapies being developed are monoclonal antibodies, recombinant cytokines, cancer-specific vaccines and adoptive cell-based therapies (including tumor-infiltrating lymphocytes (TILs), transgenic endogenous T cell receptor (TCR)-T cells and chimeric antigen receptor (CAR)-T cells, regulatory T cells (Tregs), natural killer (NK) cells, and dendritic cells (DCs)).

Traditional imaging methodologies can provide important anatomic information on the tumor mass but fail to give a complete representation of treatment response or tumor progression. This is particularly important since many immunotherapies often result in "pseudo-progression" in many patients. It is increasingly important to know how immune cells distribute in the human body and the kinetics of that distribution. Do immune cells continue to expand? And do they continue to expand and remain activated at the tumor site, as well as in the regional lymph nodes, spleen, and thymus? How long do they exert their cytotoxicity function and how long to they survive/persist at the tumor site? Can a subset of memory T cells be identified, localized, and monitored by noninvasive imaging? These are all unmet questions.

Current clinical trials involving cell- or antibody-based immunotherapies are performed without knowledge of the whole-body in vivo distribution of immune cells during treatment. Accurate molecular imaging approaches are needed and could play a critical role in monitoring and assessing immunotherapy success and failure. Importantly, cell-based therapeutic cells can potentially redistribute and lead to so-called "on-target off-site" toxic effects that could be detrimental or fatal for the patient. Molecular imaging can be exploited as a method to predict and/or promptly identify these side effects. The development of new more effective anticancer immunotherapies is also very much dependent on the identification and in vivo monitoring of existing and novel biomarkers to evaluate immune responses in patients undergoing immunotherapy. The imaging toolbox (Fig. 1) can inform not only on the presence of and activation status of therapeutic immune cells but also on biomarkers as a readout of adoptive cell-based therapy response, thus contributing to the identification of the patients most likely to benefit from immunotherapy.

Imaging T cell trafficking and persistence

The assessment of immunotherapy treatment dramatically benefits from the noninvasive long-term imaging methodologies. The spatiotemporal in vivo dynamics of distribution, persistence at the tumor site, and overall survival of systemically infused cytotoxic T lymphocytes and CAR-T cell immunotherapies remains largely unknown. In the paragraphs below, we summarize the most relevant preclinical and clinical studies using molecular imaging for the noninvasive monitoring of T-cell-based therapies using different labeling approaches and modalities.

Imaging cytolytic T cells (CTLs)

Direct imaging strategies. Nuclear imaging of directly labeled CTLs has been successfully performed in a number of studies [132,146–149]. Pittet et al. monitored CTLs trafficking at the tumor site and demonstrated that lympho-depletion prior to adoptive transfer of CTLs increases their intratumoral delivery and overall antitumor effect in a CT26 model of murine colon carcinoma [146]. The authors employed the FDA-approved ^{111}In-oxine clinical probe for this study and were able to assess whether immunizations would lead to more effective therapies. In a study from Grimfors et al., ex vivo labeling of CD4+ T cells with ^{111}In-oxine revealed trafficking and sequestration of the cells at the lesion site in 35 patients with Hodgkin's lymphoma [147]. A study from Sato et al. used ^{89}Zr-oxine to monitor the accumulation of antigen-specific activated cytotoxic T cells within the tumor [150]. Importantly, the antitumor activity resulting in shrinkage of tumor mass indicated that the radiolabeling did not have an impact on CTLs in vivo function.

Indirect imaging strategies. The indirect strategy for labeling T cells usually involves bioluminescence and nuclear imaging reporter genes. The engineering of T lymphocytes with luciferase reporter genes for bioluminescence has been extensively reported [151–154]. The high sensitivity, methodological simplicity and very low cost make BLI imaging the most suited to study the trafficking of T cells in preclinical mouse models.

For imaging T cell trafficking in patients, different types of nuclear-based reporter genes are used. The in vivo imaging of T cells has been tested by Moroz and colleagues [155]. The authors compared several reporters (of foreign derivation and host) and relative imaging probes (including hNET/[^{18}F]MFBG and [^{123}I/^{124}I]MIBG, hNIS/^{124}I, dCKDM/^{18}F-FEAU, and HSV1-tk/^{124}I-FIAU) for the imaging of reporter transduced T cells. Results demonstrate that hNET/[^{18}F]MFBG is the most sensitive reporter/probe combination for the repeated imaging of hNET-expressing T cells (reporting on less than 1×10^5 T cells/cm^3). Likar et al. developed and tested the mutated human deoxycytidine kinase (hdCK) reporter gene for the in vivo imaging of systemically administered Pz1-transduced PSMA-specific T cells [132]. They also demonstrated that the cytolytic effect against target cells was not impaired by the PET reporter. The mutant dCK was also used by McCracken and colleagues for the whole-body detection of tumor-infiltrating adoptive T cells using [^{18}F]FMAU [156]. Koehne et al. applied the same reporter gene strategy involving HSV1-tk and showed for the first time successful long-term in vivo tracking of CTLs with [^{124}I]FIAU and [^{131}I]FIAU [148]. Dotti et al. detected the in vivo trafficking of intravenously administered autologous polyclonal macaque CTLs in nonhuman primates using the mutant version of HSV1-tk reporter gene (HSV1-sr39tk), paving the way for in-human studies [157]. Kenanova et al. employed a human carcinoembryonic antigen (hCEA) reporter for the molecular imaging of adoptive T cell therapies; however, its application is restricted to the imaging of target tissues not endogenously expressing hCEA [158].

The advantage of using reporter gene imaging is the possibility to track more than one T cell population of interest simultaneously. In fact, interactions between different T cell populations do have a large impact on kinetics, antitumor activity, and persistence of adoptive immune responses. Doubrovin et al. independently tracked the fate of two EBV-specific T cell subpopulations in the same mouse model and using dual-modality SPECT and PET imaging [159]. The two T cell populations were expressing different reporters and corresponding probes: (i) CD4+ T cells were expressing hNET and were visualized by [^{123}I]FIAU-SPECT and (ii) CD8+ T cells were expressing HSV1-tk and were visualized by [^{124}I]FIAU-PET.

Imaging chimeric antigen receptor T (CAR-T) cells

Direct labeling approaches for the in vivo imaging of CAR-T cells are widely used.

Weist et al. optimized the ^{89}Zr-oxine labeling protocol for the in vivo tracking of PSCA-targeted CAR-T cell distribution in two xenograft models of glioblastoma and prostate tumor [160]. Authors showed feasibility of this approach, with a 75% radiolabeling yield and over 60% of the tracer still retained by the CAR-T cells after 6 days post administration. Lee et al. assessed the ^{89}Zr-DFO-labeled CAR-T suitability to investigate present and future cell-based immunotherapies kinetics and safety profile [161]. Another study from Bhatnagar et al. allowed in vivo distribution analysis of CD19-specific CAR-T cells also engineered with gold nanoparticles functionalized with ^{64}Cu (using a DOTA chelator) and the firefly luciferase by PET and BLI imaging [162].

Synthetic direct-labeling, site-specific constructs have also be developed for the noninvasive tracking of CAR-T therapy [163,164]. Krebs et al. used a single-chain variable fragment (scFv) of the murine antilanthanide-DOTA IgG1 antibody 2D12.5 (developed by Goodwin and colleagues using high-affinity DOTA-chelates of lantanoids), that were fused to the human IgG4-CH2-CH3 spacer and trans-membrane domain CD4. This reporter construct was used to study CD19-targeted CAR-T cell distribution and persistence at the site of the tumor in real time [165].

Indirect genetic labeling provides a much greater advantage for long-term monitoring of adoptive cell therapies and assessment of their functional status. Volpe et al. recently applied the NIS reporter for imaging CAR-T cell trafficking by PET in a model of triple-negative breast cancer (TNBC) [112]. They revealed CAR-T retention differences in MDA-MB-231 and MDA-MD-436 TNBC models. The loss of CAR-T cell in vivo detection at the tumor 14 days post administration was inversely correlated with PD-L1 immune checkpoint expression on the surface of MDA-MB-231 cancer cells, suggesting that this could reflect CAR-T cells exhaustion and death. Here, $[^{18}F]BF_4^-$ was used as the imaging probe due to its faster clearance from circulation and lower background levels in the PET images compared to radioactive iodide tracers for NIS [155,166] and to its proven safety and low toxicity in humans [119]. The detection sensitivity of CAR-T cells in this study was as low as 3000 cells/million non hNIS-expressing cells, and the authors provided a sensitive and clinically compliant toolkit for the in vivo monitoring of administered CAR-T cell therapies via NIS. Vedvyas et al. used the SSTR2 reporter to engineer ICAM-1-targeting CAR-T cells by ^{68}Ga-DOTATOC-PET imaging [167]. Real-time PET imaging revealed biphasic CAR-T cell expansion and contraction at the tumor, which could be critical information to estimate therapy success in cancer patients. A strategy involving a triple-reporter comprising of the *Escherichia coli* dihydrofolate reductase enzyme (eDHFR), the yellow fluorescent protein (YFP), and the Renilla luciferase (RLuc) was used to evaluate anti-GD2 CAR-T cell trafficking by $[^{18}F]$TMP-PET, fluorescence, and bioluminescence imaging [168]. Minn et al. tracked CD19-mediated CAR-T cells engineered to express the $tPSMA^{N9Del}$ reporter gene variant in an acute lymphoblastic model [169]. Interestingly, this study showed that CAR-T signal at the tumor did not correlate with either peripheral CAR-T blood counts or CAR-T presence in the bone marrow, thus highlighting the importance of tracking the administered CAR-Ts in order to assess homing, retention, and therapy success.

Najjar et al. reported on T cells expressing the CD19-specific CAR together with the Firefly luciferase (FLuc) and HSV1-tk reporter genes engineered using a nonviral based sleeping beauty trasposone/trasposase system and imaged persistence and retention of cytotoxic function at the tumor by bioluminescent and PET imaging [170]. Dobrenkov et al. constructed a triple reporter to be expressed by the CAR-T cells; it contained the HSV1-tk PET reporter, the green fluorescent protein, and the click beetle red luciferase BLI reporter [171]. The PET strategy enabled imaging of adoptively transferred T cells providing spatial and temporal information on their trafficking and persistence.

The first documented clinical studies of nuclear imaging based on reporter gene expression are from Yaghoubi et al. [172] and Keu et al. [173]. The authors administered CD8+ cytolytic T cells genetically engineered to express the interleukin-13 (IL-13) zetakine CAR and the PET reporter HSV1-tk into the medial left frontal lobe tumor of patients with recurrent glioma and were able to noninvasively monitor the trafficking and homing to the tumor by [^{18}F]FHBG-afforded PET imaging. MRI was used to evaluate tumor mass before and after intratumoral administration of therapeutic cells [173].

Available strategies for the in vivo detection of CAR-T cell therapies by reporter gene-based bioluminescence and nuclear imaging are still expanding and will increasingly contribute to maximize our understanding of immunotherapy fate.

Imaging T cell activation

Upon antigen recognition and engagement, naïve T cells become activated and start increasing the expression of several types of activation markers (surface or metabolic) and release specific antitumor cytokines. Regardless of the cellular expression (membrane, cytosolic, or nuclear), those activation markers can be easily targeted by molecular imaging and are here listed.

Direct imaging T cell activation biomarkers

OX40 (CD134) is a cell-surface marker of T cell activation and can be imaged by directly labeling anti-OX40 antibodies with radionuclide-based tracers. Alam et al. validated a ^{64}Cu-DOTA-AbOX40 immuno-PET tracer to monitor the dynamics of T cell activation following in situ administration of a CpG oligodeoxynucleotide vaccine [174] and the diagnosis of acute graft-versus-host disease [175]. The ^{64}Cu-DOTA-AbOX40 is a promising candidate for the successful monitoring of OX40-based cancer immunotherapies.

The T cell inducible COStimulator (ICOS, CD278) is another example of cell-surface marker and an ideal target for the in vivo monitoring of CAR-T cell therapy [176]. Simonetta et al. employed a ^{89}Zr-DFO-ICOS antibody for the noninvasive tracking of murine CD19-28z CAR-T cells in a syngeneic model of systemic B-cell lymphoma [177,178]. To do so, an anti-ICOS mAb was conjugated to the bifunctional chelator deferoxamine (DFO), and the immunocomplex was radiolabeled with ^{89}Zr. The ICOS immune-PET can be tailored to all available CAR-T cell products on the market and contribute to predict CAR-T cell therapy success in cancer patients.

Hartimath et al. targeted the cytokine IL-2 using the [^{18}F]FB-IL-2 (N-(4-[^{18}F]fluorobenzoyl)-interleukin-2) to detect activated $CD4^+$ and $CD8^+$ T cells by PET imaging [179]. The tracer distribution has also being tested in a phase 1 clinical trial in five patients to validate its feasibility for T cell activation monitoring (ClinicalTrial.gov identifier NCT02922283).

Other examples of cell surface markers associated with T cell lymphocytes activation are PD-1, CTLA-4, and LAG3, also classified as immune checkpoint molecules and discussed in "Imaging the tumor microenvironment (TME)" section.

Markers of T cell activation also include lineage defining molecules. Murine CD3 was successfully targeted using a ^{89}Zr-DFO-anti-CD3 antibody in a syngeneic model of bladder cancer [180]. Larimer et al. used a ^{89}Zr-isothiocyaanatobenzyl-deferozamine-CD3 PET probe to assess T cell infiltration post CTLA-4 blockade in a model of colon cancer [181]. Fraise et al. used the *cys*-diabody ^{89}Zr-malDFO-GK1.5, previously tested in a murine model of colitis [182], to detect in vivo $CD4^+$ T cell viability, proliferation, and function [183].

The murine and the human CD8 was targeted using a *cys*-diabody ^{64}Cu-169 [184], a *cys*-diabody ^{89}Zr-malDFO-169 [185], and a minibody ^{89}Zr-Df-IAB22M2C [186] immuno-PET probes, respectively. The latter is also being tested in several clinical trials (ClinicalTrial.gov identifiers NCT03802123, NCT03107663, and NCT04029181).

Approaches involving CD3-, CD4-, and CD8-based in vivo imaging are less invasive and still very quantitative ways of monitoring immune responses and T cell infiltration that can replace the current bioptic and genetic methodologies.

Direct imaging of targets related to T cell activation and effector function

Granzyme B can be used as a downstream effector of both cytotoxicity and an early predictive biomarker for response to immunotherapy. Granzyme B can be imaged preclinically using PET imaging to increase the detection of effective immune responses rather than simply the presence of T cells. Larimer et al. developed a PET imaging agent (a peptide called GZP) designed to recognize murine granzyme B and allowing labeling of a several PET radiometals, including ^{68}Ga [187,188].

IFN-gamma is another attractive target for molecular imaging. Gibson et al. tested an ^{89}Zr-labeled IFN-gamma immuno-PET antibody for the monitoring of therapeutic outcome [189]. In an orthotopic mouse mammary tumor model, PET imaging detected high levels of the cytokine and was able to correlate it with response to therapy.

Indirect imaging NFAT-mediated T cell activation

Indirect approaches based on reporter gene imaging can be applied for the monitoring of TCR-induced T cell activation. A seminal study is from Ponomarev and colleagues who described a novel dual-reporter strategy using the herpes simplex virus type 1 thymidine kinase paired with green fluorescent protein (HSV1-tk/GFP or TKGFP) and allowing to inform on NFAT-mediated activation of T lymphocytes by optical fluorescence imaging and positron emission tomography (PET) [190]. The authors constructed an TCR-dependent NFAT-TKGFP inducible reporter in a retroviral cassette and under the control of an artificial *cis*-acting NFAT-specific enhancer. Upon TCR engagement with MHC-complex on antigen-presenting cells, NFAT-TKGFP starts reporting on NFAT-mediated transcription of various target genes, including IL-2 and other cytokines [191], thereby providing information on the activation status of T cells in vivo. The system can be easily translated by tailoring the reporter gene/probe to meet the clinical needs (e.g., use of nonimmunogenic reporter).

NFAT fused to enhanced CFP (eCFP) and GFP (eGFP) can also be used for the small- and large-scale CAR scFv screening for the rapid testing and identification of the leading CAR on the basis of the maximum NFAT-mediated T cell activation and antitumor activity [192].

Imaging metabolic targets

Activated T cells typically undergo metabolic reprogramming, allowing them to survive the hostile tumor microenvironment and keep their antitumor function also in peripheral tissues [193]. The major metabolic pathways that appear upregulated in activated T cells are glycolysis [194], DNA synthesis [195], and nucleic acid metabolism [196]. The change in metabolism can be imaged, using direct labeling strategies. These strategies are often linked to both therapeutic T cells activation and expansion (e.g., if the magnitude of the signal increases overtime).

The well-established strategy to target the glycolysis pathway is based on [^{18}F]FDG imaging, particularly in the context of immunotherapy treatment [197]. [^{18}F]FDG imaging is used to assess metabolic changes related to glycolysis and to the prediction of an effective anticancer immune response in patients and for a stratification of patients being assessed [198]. Response to PD-1 and CTLA-4 blocking antibodies has been assessed in a number of studies using both [^{18}F]FDG [199–203] and [^{18}F]FLT [204]. To target DNA synthesis as a method for detection of quantity and kinetics of antigen-specific T cells, several ^{18}F-based PET tracers have been employed: (i) [^{18}F]FLT for both checkpoint inhibition [204] and T cell response [195]; and (ii) [^{18}F]F-AraG (accumulation in T cells via the dGK pathway) for the visualization of activated T cells [205], CD8-cytotoxic T cells post chemotherapy [206], and as a predictor of immune response post anti-PD-1 therapy [207]; and (iii) [^{18}F]CFA for T cell activation [208,209]. The same applies to amino acid metabolism: [^{18}F]FACBC was used in several studies for the visualization of activated T cells in the tumor microenvironment [210] and the discrimination between activated T and B cells compared to other immune cell types (e.g., granulocytes and macrophages) [211].

The main clinical trials involving the use of PET-afforded metabolic imaging use as a target the DNA synthesis and their ClinicalTrial.gov identifiers are here listed: NCT03142204, NCT03007719, NCT03409419, and NCT03409419.

Imaging the tumor microenvironment (TME)

The clinical success of immunotherapy also depends on the influence of the tumor microenvironment (TME). The latter is an extracellular fluid milieu containing cancer cells, extracellular matrix, immune cells, blood vessels, and signaling molecules. The TME influences both tumor initiation and response to therapy. The TME frequently has many nonnormal characteristics (including a low pH, hypoxia, and nutrient limitations—such as low glucose levels). The TME can be extremely harsh on T cells and limit their function leading to tumor escape. This immunosuppression is often accompanied by upregulation of a number of checkpoint pathways, including the programmed cell death protein/protein ligand 1 (PD-1/PD-L1), cytotoxic T lymphocyte-associated protein 4 (CTLA-4), T cell immunoglobulin and mucin domain-containing protein 3 (TIM3), lymphocyte activation gene 3 protein (LAG3), and the surface enzyme CD73. The use of immune checkpoint inhibitors targeting these pathways has given extraordinary results in several malignancies [212–214], demonstrating the ability to overcome immune escape by enhancing T cell anticancer response in some patients. Clinical trials based on immune checkpoint inhibition in renal cancer [215], nonsmall cell lung cancer [216,217], and melanoma [218] effectively determined endogenous anticancer immune responses.

Currently approved by the American Food and Drug Administration (FDA) [219] for clinical treatment and utilized in the daily practice are

(1) PD-1 inhibitors Nivolumab, Pembrolizumab
(2) PD-L1 inhibitors Atezolizumab, Avelumab, Durvalumab
(3) CTLA-4 inhibitor Ipilimumab (human IgG1 anti-CTLA-4 monoclonal antibody)

Numerous agents for molecular imaging have already been developed and used to assist scientists in noninvasively interrogate the heterogeneous expression and dynamics of immune checkpoint molecules in in vivo preclinical models. The goal is to establish a clinically translatable system for the stratification of patients before immunotherapy treatment. All the listed studies below employ direct labeling molecular imaging strategies.

Imaging of programmed cell death protein 1 (PD-1/CD279)

PD-1 is expressed on the surface of T and B cells as well as NK cells and plays a key role in negatively regulating T cell activity [220]. Because of this, imaging techniques targeting PD-1 only allow the in vivo visualization of immune cells. Natarajan et al. developed a novel immunoPET radiotracer for the assessment of PD-1 status in tumor-infiltrating T lymphocytes (TILs) and is based on a ^{64}Cu-labeled PD-1 antimouse antibody (Keytruda; humanized IgG4) [221]. This was the first reported study in a preclinical model of melanoma. The already approved humanized IgG4 anti-PD-1 Pembrolizumab was also radiolabeled with ^{89}Zr-deferoxamine (Df) and its pharmacokinetics, biodistribution, and dosimetry were evaluated in two rodent models and a humanized murine one by PET imaging [222]. The same has been done also with Nivolumab [223]. Hettich and colleagues used a ^{64}Cu-NOTA-labeled anti-PD-1 radiotracer for PET imaging to correlate the high CD8-positive TILs levels in the tumor with a high expression of PD-1 [224]. Natarajan also labeled the murine-derived ^{64}Cu-DOTA-anti-PD-1 for preclinical application [221]. IRDye800cw and ^{64}Cu were conjugated to liposomes with PD-1 antibody labeling, while doxorubicin was loaded into the liposomes, and successfully permitted PD-1 expression visualization using a dual-modality near-infrared fluorescence (NIRF) and PET imaging [225].

All the above strategies can be paired with [^{18}F]FDG [197]. The latter can be used for the evaluation of tumor response post Ipilimumab [199–201,226], Nivolumab [202,227,228] or Pembrolizumab [203,227]. Ongoing clinical trial for PD-1 antibody ^{89}Zr-labeled Pembrolizumab is looking at safety and treatment response assessment (ClinicalTrials.gov identifiers NCT02760225).

Imaging of programmed cell death protein ligand 1 (PD-L1/CD274)

The most relevant examples of successful in vivo monitoring of PD-L1 blockade in cancer cells include the following:

Heskamp et al. developed the first noninvasive imaging technique to monitor tumor PD-L1 expression by radiolabeling the monoclonal antibody PD-L1.3.1 (murine IgG1) with

^{111}In-SPECT in a xenograft model of MDA-MB-231 human triple-negative breast cancer [229]. A murine anti-PD-L1 antibody was also conjugated with ^{111}In-DTPA, and the biodistribution and dosimetry were evaluated [230]. Broos et al. developed several nanobodies raised against PD-L1 and radiolabeled with ^{99m}Tc for SPECT imaging [231]. A number of PET-compatible strategies have also been used, including engineered peptides [231–233] and labeled monoclonal antibodies [234,235]. Chatterjee et al. employed a highly specific ^{64}Cu-labeled PET-afforded peptide (WL12) for the rapid detection of PD-L1 expression in tumors [236]. Similarly, Lesniak et al. quantified ^{64}Cu-WL12 accumulation and its subsequent use in the detection of PD-L1 in MDA-MB-231, SUM149 and 4T1 orthotopic models of breast cancer [237]. Another PET-afforded anti-PD-L1 antibody was conjugated to ^{64}Cu-NOTA as a therapeutic in vivo checkpoint antibody [224]. A ^{89}Zr-DFO anti-PDL-1 antibody [238] as well as ^{64}Cu-NOTA [239,240] and a ^{68}Ga-NOTA [239] have been developed for preclinical use. ^{18}F-conjugated examples are a tumor-targeting anti-PD-L1 adnectin (^{18}F-BMS-986192) [241,242], a NOTA-ZPDL-1_1 affibody [243], and a single-domain antibody (VHHs) against mouse PD-L1 [244]. The atezolizumab antibody against PD-L1 has been also conjugated with Licor-800, a near-infrared dye also suitable for fluorescence optical imaging and tested in multiple cancer lines [236]. On the basis of this work, Ren et al. have recently developed a dual-label imaging agent ^{89}Zr-DFO-PD-L1-IRDye800cw for the noninvasive detection of PD-L1 by PET, followed by near-infrared fluorescence (NIRF) imaging to guide in vivo surgical resection of the tumor and evaluation of treatment response [245]. Another dual-label dual-modal system is based on nanohybrid liposomal nanoparticles loaded with paclitaxel, decorated with PD-L1 antibody to enable tumor-specific targeting, and labeled with IRDye800cw and Gd-DOTA. The system allowed dual-modality fluorescence and MRI imaging and assessment of therapeutic efficacy in 4T1 breast and CT26 colon tumor xenografts [246]. Zhao et al published the first study to successfully measure by noninvasive Cherenkov luminescence imaging (CLI) the PD-L1 dynamics in a human CRC subcutaneous xenograft tumor model using a ^{131}I-labeled anti-PD-L1 monoclonal antibody [247].

^{111}In-, ^{64}Cu-, and ^{89}Zr-Atezolizumab humanized IgG1 antibodies for both SPECT and PET imaging are being tested in clinical trials. [^{89}Zr]MPDL3280 (humanized PD-L1 mAb; ClinicalTrials.gov identifiers NCT02453984 and NCT02478099) is a PD-L1 antibody under clinical evaluation for the treatment of advanced and metastatic melanoma [248], triple-negative breast cancer [249], nonsmall cell lung cancer [250], renal [251], urothelial [252], and bladder cancer [253]. Preliminary results have already been published and showed correlation between clinical response and pretreatment assessed by PET imaging, confirming the validity of molecular imaging for the assessment and prediction of clinical response to anti-PD-L1 checkpoint blockade [235]. ^{89}Zr-Durvalumab (ClinicalTrial.gov identifier NCT03610061 and NCT03829007) and serial ^{89}Zr-Atezolizumab (ClinicalTrials.gov identifier NCT03850028) are also running to evaluate tumor saturation by atezolizumab treatment. Another multicenter trial started with ^{89}Zr-labeled DFO-REGN3504, a nonregistered anti-PD-L1 antibody (ClinicalTrail.gov identifier NCT03746704).

Imaging of cytotoxic T lymphocyte-associated protein 4 (CTLA-4/CD152)

CTLA-4 targeted antibodies augment endogenous T cell response against malignant cells. This has attracted the molecular imaging community's attention as a target molecule for imaging in cancer immunotherapy. Higashikawa et al. developed a [^{64}Cu]Cu-DOTA-anti-CTLA-4 antibody by introducing a DOTA group to the antimouse CTLA-4 mAb and subsequent radiolabeling with the PET tracer ^{64}Cu [254]. A ^{64}Cu-DOTA-Ipilimumab proved to be a reliable tool for the noninvasive monitoring of CTLA-4-expressing tissues in three different human nonsmall cell lung cancers [255]. The same author developed and validated two tracers targeting CTLA-4 (one a full antibody and the other a $F(ab')_2$ fragment) and radiolabeled with ^{64}Cu to localize CTLA-4-positive tissues in humanized mouse models [256]. Another way of monitoring anti-CTLA-4 therapy is based on Tremelimumab administration and detection of the changes in glucose metabolism by [^{18}F]FDG or [^{18}F]FLT PET imaging in advanced melanoma patients [204]. This specific study did not show significant changes in tumor size. However, the increase of [^{18}F]FLT uptake in the spleen was correlated to immunotherapy effect.

A currently running phase 2 clinical trial involves the assessment of [^{89}Zr]Zr-Ipilimumab uptake and its correlation to treatment response in patients with metastatic melanoma (ClinicalTrial.gov identifier NCT03313323).

Imaging of lymphocyte activation gene 3 protein (LAG3)

Prolonged tumor antigen exposure and the inflammatory environment are also responsible for the expression of LAG-3 on the surface of T lymphocytes. Therapeutic antibodies blocking LAG-3 produce a durable response as a single agent and in combination with other immune checkpoint inhibitors. Lecocq et al. used $^{99m}TcO_4^-$-labeled nanobodies (recombinant variable domains VHH of heavy-chain only antibodies from the *Camelidae* family) to noninvasively evaluate mouse LAG-3 expression using SPECT imaging [257]. Kelly and colleagues labeled the fully human anti-LAG-3 antibody REGN3767 with ^{89}Zr using the bifunctional chelator p-SCN-Bn-Deferoxamine (DFO) and tested it in a preclinical model of MC38 murine colon adenocarcinoma expressing human LAG-3 [258]. The same full anti-LAG antibody labeled with [^{89}Zr]Zr-DFO ([^{89}Zr]Zr-REGN3767), alone or in combination with REGN2810 (anti PD-1), is being tested in clinical trial (ClinicalTrial.gov identifier NCT03005782) as a tool to monitor and predict response to anti-LAG3 therapy in combination with anti-PD-1 treatment. A phase 1 clinical trial is investigating the biodistribution and tumor uptake of radiolabeled BI-754091 anti-LAG-3 monoclonal antibody with ^{89}Zr in patients with advanced nonsmall cell lung cancer and head and neck squamous cell carcinoma (ClinicalTrial.gov identifier NCT03780725).

Imaging of T cell immunoglobulin and mucin domain-containing protein 3 (TIM3)

TIM-3 is another immune checkpoint playing an important role in T cell exhaustion and with several clinical trials currently testing the efficacy of TIM-3 blocking antibodies. There is currently one clinical trial testing the feasibility of ^{18}F-Clofarabine post anti-TIM-3 monoclonal antibody treatment with and without anti-PD-1 antibody (ClinicalTrial.gov identifier NCT03409419).

Imaging of T cell immunoreceptor with Ig and ITIM domains (TIGIT)

^{64}Cu- and ^{89}Zr-labeled antibodies for TIGIT were developed for the preclinical quantification of TIGIT expression on tumor-infiltrating lymphocytes [259].

Imaging regulatory T cells (Tregs)

Regulatory T cells (Tregs) have been in vivo monitored using bioluminescence reporter genes in preclinical models [260–263]. As the field of regulatory T cells is quickly evolving from polyclonal to antigen-specific Tregs, more CARs are being developed and need testing both preclinically and clinically. A study from Dawson et al. lentivirally transduced HLA-A2 CAR-Treg cells with a luciferase-GFP fusion protein to in vivo track them and described the peak of CAR-Tregs infiltration to the A2+ skin graft at 7 days post intravenous infusion [263]. However, the bioluminescence methodology and the use of foreign reporters (nonhuman) does not allow clinical translation (please refer to "Imaging modalities and labeling strategies for cancer immunotherapy" section for details). The most promising platform for CAR-Tregs imaging is that of the host reporter genes paired with clinically compatible imaging modalities (radionuclide-based imaging, e.g., SPECT, PET). As a proof of concept of the feasibility of host reporter genes for the in vivo monitoring of Tregs (and CAR-Tregs) dynamics, Sharif-Paghaleh et al. retrovirally engineered NIS-expressing Tregs and detect their presence 24 h post infusion by $^{99m}TcO_4^-$ SPECT imaging. The authors showed that the radiolabeling of adoptively transferred Tregs was not impairing their viability or function, thereby allowing noninvasive tracking in both preclinical and clinical settings. Jacob et al. isolated human Tregs using a GMP-compatible protocol and lentivirally transduced them to express NIS reporter [264]. The authors investigated Tregs kinetics of distribution and homing within a human skin transplant in a humanized model by SPECT and showed traceable Tregs in the skin graft for up to 40 days post intravenous administration. Moreover, HLA-A2 CAR-Tregs were engineered to express IL-10 giving them an additional suppressive advantage, and NIS reporter allowing their in vivo monitoring by nuclear imaging [265]. These studies using radionuclide-based reporter gene imaging of polyclonal Tregs demonstrate the applicability of the methodology toward the development and clinical translation of adoptive Treg therapy.

Imaging NK cells

The main advances in monitoring NK cell-based immunotherapies by molecular imaging have been made in the development of in vivo tracking methodologies. Visualization of NK cells by optical imaging has been achieved in multiple studies involving bioluminescence (mainly FLuc as a BLI reporter) [266–271], fluorescence imaging (using a near-infrared dye 1,1′-dioctadecyl-3,3,3′,3′-tetramethylindodiccarbocyanine (DiD) or quantum dots emitting in the near-infrared region (QD705)) [272,273], and intravital fluorescence microscopy (using either GFP fluorescent protein or carboxyfluorescein succinimidyl ester CFSE) [274–277] in preclinical models of cancer. Microscopy is particularly ideal to study the molecular and cellular interactions of NK cells in vitro, but rather unsuitable for noninvasive in vivo tracking of the cells in both animals and humans. For this purpose, magnetic resonance and nuclear imaging are better suited and are mostly achieved by direct cell labeling.

NK cells can be directly labeled with superparamagnetic iron oxide (SPIO) and ultrasmall superparamagnetic iron oxide (USPIO) nanoparticles [278–281] and monitored for up to 4 days in vivo as shown by Mallett et al [280]. In recent years, also ^{19}F has emerged as a novel agent to label NK cell-based immunotherapies for MRI. Two independent studies proved that this nonradioactive tracer is safe and effective in monitoring the in vivo distribution, homing and therapeutic potential of NK cells [282,283]. Despite the various drawbacks intrinsic to the direct labeling with paramagnetic particles (including label dilution and label persistence after cell death), the approach can be directly translated into the clinic for short-term monitoring of engineered NK cells. A few iron-oxide nanoparticles (namely ferumoxide, ferumoxytol, and ferucarbotran) have already been approved by the FDA [278,279]. Notably, Daldrup-Link et al. labeled HER2-targeting NK cells with both ferumoxide and ferucarbotran clinically applicable contrast agents to monitor the dynamics of NK distribution and accumulation at the tumor site [278]. Similarly, Meier et al. showed increased prostate cancer targeting of CAR-NK cells directly labeled with ferumoxide [279].

However, MRI is less sensitive in detecting labeled NK cells compared to nuclear imaging, and nuclear imaging is more easily translatable into the clinic by using already FDA-approved probes. Short-lived isotopes were used to directly label NK cells, including [^{18}F]FDG [284] (half-life: 109.8 min) and ^{11}C-methyl-iodide (half-life: 20 min) [285], both for PET imaging. Meier et al. studied the accumulation of HER2/neu-specific CAR-NK cells prelabeled with [^{18}F]FDG, compared to nonengineered naïve NK cells in a mouse HER2-positive breast cancer model [284]. Melder et al. studied the localization of activated and ^{11}C-methyl-iodide NK cells at the tumor site after 1 h post intravenous delivery in a murine sarcoma model [285]. Long-lived radioisotopes (such as ^{89}Zr for PET, ^{111}In-oxine for SPECT, and ^{99m}Tc for gamma camera) were also employed to allow long-term repetitive imaging (until the radioactive isotope is decayed). A clinically relevant study from Sato et al. employed ^{89}Zr-oxine for the in vivo tracking of adoptively transferred rhesus macaque NK cells [286]. The authors demonstrated that the labeling does not have an impact on cell phenotype, viability nor function. Shaffer et al. also tested the radiolabeling of human NK cells with ^{64}Cu and ^{89}Zr-NKp30Ab and demonstrated that the latter had a superior in vivo contrast at their respective optimal time points [287]. ^{111}In-oxine was used to label autologous [288] and allogeneic [289] NK cells that allowed detection up to 72 h due to the half-life of the radiotracer. Moreover, Galli et al. used a ^{99m}Tc-anti-CD56 monoclonal antibody to label primary reinfused NK cells and monitor intratumor infiltration in a xenograft model of anaplastic thyroid cancer [290]. The same author also proved that labeling NK cells with ^{99m}Tc was much safer than labeling with ^{111}In-oxine [291]. It is already well established that despite being able to achieve a high cell labeling, ^{111}In-oxine is responsible for the progressive reduction of cell viability and function [93]. In fact, the exposure to radionuclide isotopes can be particularly toxic for NK cells and is very much influenced by both the radionuclide and the dose used. While the in vivo monitoring of CAR-NK cells is possible using direct labeling strategies, the challenges posed by the NK cells properties (e.g., poor transduction using retro- and lentiviral approaches) have dramatically delayed the development of long-term and clinically compatible reporter gene-based strategies [292].

Imaging dendritic cell vaccines

Dendritic cell vaccines are typically imaged using direct labeling approaches, where the antigen and the label are ex vivo loaded into the cells prior to infusion into cancer patients. Radionuclide imaging and MRI are the preferred methodologies to label adoptively transferred DCs. A number of studies have been successfully conducted with ^{111}In radiotracer for SPECT as it displays a high radiolabeling yield (due to the binding to cytoplasmatic proteins within the cells), as well as good labeling stability at 24 h [93,293–296]. Molecular imaging has proven that the route of administration of DCs is relevant for their distribution, localization and therefore function [297–299]. Data involving ^{111}In-labeled DCs intravenously administered in patients show in vivo detection in the liver and spleen for over 7 days [296]. In another clinical study in melanoma patients, ^{111}In-labeled DCs were injected either intradermally or in the proximity of or inside regional lymph nodes. Micro-autoradiography of resected lymph nodes anatomically distant from the ones injected revealed migration of the radioactive mature DC population. As for the intradermally injected patients, they also showed a high percentage of mature DCs migrating at the neighboring lymph node. The study demonstrated that the DCs have the ability to determine an immune response that directly correlates with their ability to migrate in vivo [298]. Ridolfi et al. used ^{99m}Tc-HMPAO- and ^{111}In-labeled DCs; he/she compared their efficiency to migrate and estimated the number of cells migrated to the lymph nodes following their administration by two different routes (subcutaneous versus intradermal) in a clinical vaccination trial [300]. The study revealed a higher migration capacity of the cell-based therapy when administered intradermally, as more of the cell suspension drained into the afferent lymphatic vessels due to elasticity of the skin. Aarntzen et al. showed that pretreatment with an extra dose of DCs, Toll-like receptor ligands or proinflammatory cytokines (e.g., TNF-alpha, IL-1-alpha) does not improve migration of intradermally administered ^{19}F-labeled DCs (imaged by MRI) (ClinicalTrial.gov identifier NCT00243594) [297]. Furthermore, the patients in this study displayed evident hypoxia at the site of injection. In fact, when a high density of cells is injected, this may prevent them to access sufficient nutrients and oxygen at that site, thereby hampering migration and overall therapeutic response. Another study from de Vries et al. monitored intranodally injected DCs labeled with superparamagnetic iron oxide nanoparticles and tracked their in vivo distribution in melanoma patients. Notably, the authors were able to identify which were the patients that underwent immunotherapy misinjection [301].

Notable examples of reporter gene-based imaging of dendritic cell migration are also present in literature for PET and MRI. Lee and colleagues engineered DCs to express the sodium iodide symporter (NIS) and tracked their migration by ^{124}I-PET and FLuc BLI imaging [302]. Others have used the novel [^{18}F]BF_4 radiotracer for NIS for this purpose [303]. Kim et al. employed the ferritin reporter for noninvasive imaging of DCs by MRI [304].

Several studies reporting in vitro or in vivo tracking of DC-based vaccines using optical imaging [305–310] have been published but are not relevant for clinical application.

Areas of research where molecular imaging is playing a key role mainly relate to the following aspects: (i) establishment of solid clinical protocols with optimal number of DCs to be administered to achieve an adequate immune response; (ii) in vivo distribution and migration; and (iii) successful homing in secondary lymphoid organs to improve or predict therapeutic success.

What the future holds

The advent of immunotherapy has led to durable responses in cancer patients, including those with an initial poor prognosis. However, still too few patients (less than 50%) are benefitting from current immunotherapies, and additional treatments and treatment combinations need to be developed. With the continuous expansion of the immunotherapy arsenal, the Immunoimaging Toolbox is evolving too. Its main focus is to find ways to inform on the dynamic changes of the immune system following administration of cell-based immunotherapies in order to monitor and predict therapeutic response.

Areas of development include improved radiotracer design, the engineering of new and more powerful imaging devices, and the combination of immunoimaging with radiomics. While most of the existing probes for direct and indirect nuclear-based imaging are already clinically approved, the portfolio will likely continue to grow with the development of new and better radiotracers for both direct and indirect imaging. Current efforts are focused on the improvement of existing imaging technologies: two recent examples in this direction are represented by the advent of total-body PET [311] and hybrid scanners (such as PET-MRI) [312,313].

The first human imaging study using a total-body PET described by Badawi and colleagues generates dynamic images with a high temporal resolution and shows an unprecedented sensitivity [314]. The combination of a more sensitive PET system with new available radiotracers will lead to the identification of new molecular targets and biomarkers for immunotherapy. Thanks to mathematic modeling and machine learning the way molecular imaging studies will be interpreted is likely to change too and shift toward automation.

Immunoimaging needs to be seen as a companion tool, to be paired with existing and new blood-based biomarkers as well as new assay "omics" (including epigenetics, genomics, proteomics, and metabolomics). "Omics" are important for the identification of the tumor intrinsic properties, but lack of spatial-temporal information. Combining those with immunoimaging is a necessary step to advance our knowledge about the mechanism of action of cell-based immunotherapies and to provide a way to select the best candidates for therapeutic success. The same applies to "radiomics": a new, and not yet fully exploited, imaging processing and analysis technology enabling the conversion of radiological images into quantitative data [315]. The ultimate goal of radiomics is to help establish imaging-based biomarkers, which will play a critical role in interpreting and even predicting response in patients treated with immunotherapies [316,317].

The Immunoimaging Toolbox is already becoming increasingly relevant in providing preclinical support and justification for moving toward implementation in clinical trials. In the coming years, we expect to see the Immunoimaging Toolbox play a central role in guiding the FDA approval of newly generated off-the-shelf cell-based immunotherapies and will provide a diagnostic and monitoring tool to improve patient management and support clinical decision making.

Acknowledgments

This work was supported in part by NIH grants R01 CA163980, R01 CA204924, R01 CA215136 (to R.G.B.); P30 CA008748 (MSK Cancer Center Support Grant/Core Grant); R50 CA221810 (to I.S.); R01 CA220524-01A1 and R21 CA213139 01A1 (to V.P.), the EIO Fellowship Award (FP00001443) from the MSKCC Center for Experimental Immuno-Oncology and the Tow Foundation Fellowship Award (FP00004141) from the MSKCC Center for Molecular Imaging and Nanotechnology (FP00004141) (to A.V.).

References

[1] Weissleder R. A clearer vision for in vivo imaging. Nat Biotechnol 2001;19(4):316–7.
[2] Smith AM, Mancini MC, Nie S. Bioimaging: second window for in vivo imaging. Nat Nanotechnol 2009;4 (11):710–1.
[3] Davidson MW, Campbell RE. Engineered fluorescent proteins: innovations and applications. Nat Methods 2009;6(10):713–7.
[4] Lichtman JW, Conchello JA. Fluorescence microscopy. Nat Methods 2005;2(12):910–9.
[5] Zhang RR, et al. Beyond the margins: real-time detection of cancer using targeted fluorophores. Nat Rev Clin Oncol 2017;14(6):347–64.
[6] Nagaya T, et al. Fluorescence-guided surgery. Front Oncol 2017;7:314.
[7] Graves EE, Weissleder R, Ntziachristos V. Fluorescence molecular imaging of small animal tumor models. Curr Mol Med 2004;4(4):419–30.
[8] Chuang KH, et al. Development of a universal anti-polyethylene glycol reporter gene for noninvasive imaging of PEGylated probes. J Nucl Med 2010;51(6):933–41.
[9] Ormo M, et al. Crystal structure of the *Aequorea victoria* green fluorescent protein. Science 1996;273(5280):1392–5.
[10] Shaner NC, et al. Improved monomeric red, orange and yellow fluorescent proteins derived from *Discosoma* sp. red fluorescent protein. Nat Biotechnol 2004;22(12):1567–72.
[11] Merzlyak EM, et al. Bright monomeric red fluorescent protein with an extended fluorescence lifetime. Nat Methods 2007;4(7):555–7.
[12] Lin MZ, et al. Autofluorescent proteins with excitation in the optical window for intravital imaging in mammals. Chem Biol 2009;16(11):1169.
[13] Kremers GJ, et al. Photoconversion in orange and red fluorescent proteins. Nat Methods 2009;6(5):355–8.
[14] Liu M, et al. In vivo three dimensional dual wavelength photoacoustic tomography imaging of the far red fluorescent protein E2-crimson expressed in adult zebrafish. Biomed Opt Express 2013;4(10):1846–55.
[15] Matz MV, et al. Fluorescent proteins from nonbioluminescent Anthozoa species. Nat Biotechnol 1999;17 (10):969–73.
[16] Shcherbakova DM, Verkhusha VV. Near-infrared fluorescent proteins for multicolor in vivo imaging. Nat Methods 2013;10(8):751–4.
[17] Filonov GS, et al. Bright and stable near-infrared fluorescent protein for in vivo imaging. Nat Biotechnol 2011;29 (8):757–61.
[18] Deliolanis NC, et al. Deep-tissue reporter-gene imaging with fluorescence and optoacoustic tomography: a performance overview. Mol Imaging Biol 2014;16(5):652–60.
[19] van Manen L, et al. A practical guide for the use of indocyanine green and methylene blue in fluorescence-guided abdominal surgery. J Surg Oncol 2018;118(2):283–300.
[20] Ando R, et al. An optical marker based on the UV-induced green-to-red photoconversion of a fluorescent protein. Proc Natl Acad Sci U S A 2002;99(20):12651–6.
[21] Adam V, et al. Structural characterization of IrisFP, an optical highlighter undergoing multiple photo-induced transformations. Proc Natl Acad Sci U S A 2008;105(47):18343–8.
[22] Gurskaya NG, et al. Engineering of a monomeric green-to-red photoactivatable fluorescent protein induced by blue light. Nat Biotechnol 2006;24(4):461–5.
[23] de Wet JR, et al. Cloning of firefly luciferase cDNA and the expression of active luciferase in *Escherichia coli*. Proc Natl Acad Sci U S A 1985;82(23):7870–3.
[24] Mezzanotte L, et al. In vivo molecular bioluminescence imaging: new tools and applications. Trends Biotechnol 2017;35(7):640–52.
[25] Ugarova NN. Luciferase of *Luciola mingrelica* fireflies. Kinetics and regulation mechanism. J Biolumin Chemilumin 1989;4(1):406–18.
[26] Michelini E, et al. Spectral-resolved gene technology for multiplexed bioluminescence and high-content screening. Anal Chem 2008;80(1):260–7.
[27] Branchini BR, et al. Red-emitting luciferases for bioluminescence reporter and imaging applications. Anal Biochem 2010;396(2):290–7.
[28] Mezzanotte L, et al. A new multicolor bioluminescence imaging platform to investigate NF-kappaB activity and apoptosis in human breast cancer cells. PLoS ONE 2014;9(1), e85550.

[29] Mezzanotte L, et al. Sensitive dual color in vivo bioluminescence imaging using a new red codon optimized firefly luciferase and a green click beetle luciferase. PLoS ONE 2011;6(4), e19277.
[30] Wood KV, et al. Complementary DNA coding click beetle luciferases can elicit bioluminescence of different colors. Science 1989;244(4905):700–2.
[31] Lorenz WW, et al. Isolation and expression of a cDNA encoding *Renilla reniformis* luciferase. Proc Natl Acad Sci U S A 1991;88(10):4438–42.
[32] Tannous BA. Gaussia luciferase reporter assay for monitoring biological processes in culture and in vivo. Nat Protoc 2009;4(4):582–91.
[33] Tannous BA, et al. Codon-optimized Gaussia luciferase cDNA for mammalian gene expression in culture and in vivo. Mol Ther 2005;11(3):435–43.
[34] Nakajima Y, et al. cDNA cloning and characterization of a secreted luciferase from the luminous Japanese ostracod, *Cypridina noctiluca*. Biosci Biotechnol Biochem 2004;68(3):565–70.
[35] Moroz MA, et al. Introducing a new reporter gene, membrane-anchored *Cypridina* luciferase, for multiplex bioluminescence imaging. Mol Ther Oncolytics 2021;21:15–22.
[36] Thompson EM, Nagata S, Tsuji FI. Cloning and expression of cDNA for the luciferase from the marine ostracod *Vargula hilgendorfii*. Proc Natl Acad Sci U S A 1989;86(17):6567–71.
[37] Branchini BR, et al. An enhanced chimeric firefly luciferase-inspired enzyme for ATP detection and bioluminescence reporter and imaging applications. Anal Biochem 2015;484:148–53.
[38] Hall MP, et al. Engineered luciferase reporter from a deep sea shrimp utilizing a novel imidazopyrazinone substrate. ACS Chem Biol 2012;7(11):1848–57.
[39] Gaspar N, et al. Evaluation of NanoLuc substrates for bioluminescence imaging of transferred cells in mice. J Photochem Photobiol B 2021;216, 112128.
[40] Rumyantsev KA, Turoverov KK, Verkhusha VV. Near-infrared bioluminescent proteins for two-color multimodal imaging. Sci Rep 2016;6:36588.
[41] Contag CH, et al. Photonic detection of bacterial pathogens in living hosts. Mol Microbiol 1995;18(4):593–603.
[42] Shah K, Weissleder R. Molecular optical imaging: applications leading to the development of present day therapeutics. NeuroRx 2005;2(2):215–25.
[43] Ntziachristos V, et al. Looking and listening to light: the evolution of whole-body photonic imaging. Nat Biotechnol 2005;23(3):313–20.
[44] Kiessling F, et al. Recent advances in molecular, multimodal and theranostic ultrasound imaging. Adv Drug Deliv Rev 2014;72:15–27.
[45] Wilson SR, Burns PN. Microbubble-enhanced US in body imaging: what role? Radiology 2010;257(1):24–39.
[46] Pysz MA, Willmann JK. Targeted contrast-enhanced ultrasound: an emerging technology in abdominal and pelvic imaging. Gastroenterology 2011;140(3):785–90.
[47] Klibanov AL. Ligand-carrying gas-filled microbubbles: ultrasound contrast agents for targeted molecular imaging. Bioconjug Chem 2005;16(1):9–17.
[48] Laschinger M, Engelhardt B. Interaction of alpha4-integrin with VCAM-1 is involved in adhesion of encephalitogenic T cell blasts to brain endothelium but not in their transendothelial migration in vitro. J Neuroimmunol 2000;102(1):32–43.
[49] Lindner JR, et al. Ultrasound assessment of inflammation and renal tissue injury with microbubbles targeted to P-selectin. Circulation 2001;104(17):2107–12.
[50] Christiansen JP, et al. Noninvasive imaging of myocardial reperfusion injury using leukocyte-targeted contrast echocardiography. Circulation 2002;105(15):1764–7.
[51] Hauff P, et al. Molecular targeting of lymph nodes with L-selectin ligand-specific US contrast agent: a feasibility study in mice and dogs. Radiology 2004;231(3):667–73.
[52] Bachmann C, et al. Targeting mucosal addressin cellular adhesion molecule (MAdCAM)-1 to noninvasively image experimental Crohn's disease. Gastroenterology 2006;130(1):8–16.
[53] Ellegala DB, et al. Imaging tumor angiogenesis with contrast ultrasound and microbubbles targeted to alpha(v) beta3. Circulation 2003;108(3):336–41.
[54] Leong-Poi H, et al. Assessment of endogenous and therapeutic arteriogenesis by contrast ultrasound molecular imaging of integrin expression. Circulation 2005;111(24):3248–54.
[55] Hamilton AJ, et al. Intravascular ultrasound molecular imaging of atheroma components in vivo. J Am Coll Cardiol 2004;43(3):453–60.

[56] Villanueva FS, et al. Microbubbles targeted to intercellular adhesion molecule-1 bind to activated coronary artery endothelial cells. Circulation 1998;98(1):1–5.
[57] Unger EC, et al. In vitro studies of a new thrombus-specific ultrasound contrast agent. Am J Cardiol 1998;81 (12a):58g–61g.
[58] Schumann PA, et al. Targeted-microbubble binding selectively to GPIIb IIIa receptors of platelet thrombi. Investig Radiol 2002;37(11):587–93.
[59] Lanza GM, Wickline SA. Targeted ultrasonic contrast agents for molecular imaging and therapy. Prog Cardiovasc Dis 2001;44(1):13–31.
[60] Buchanan KD, et al. Echogenic liposome compositions for increased retention of ultrasound reflectivity at physiologic temperature. J Pharm Sci 2008;97(6):2242–9.
[61] Bourdeau RW, et al. Acoustic reporter genes for noninvasive imaging of microorganisms in mammalian hosts. Nature 2018;553(7686):86–90.
[62] Jackson EF. Principles of magnetic resonance imaging and magnetic resonance spectroscopy. In: Kim EE, Yang DJ, editors. Targeted Molecular Imaging in Oncology. New York, NY: Springer New York; 2001. p. 30–61.
[63] Lien YH, Wu TM. Preparation and characterization of thermosensitive polymers grafted onto silica-coated iron oxide nanoparticles. J Colloid Interface Sci 2008;326(2):517–21.
[64] Momenbeik F, Yazdani E. Application of methyl silane coated iron oxide magnetic nanoparticles for solid-phase extraction and determination of fat-soluble vitamins by high performance liquid chromatography. J AOAC Int 2015;98(2):336–44.
[65] Tsiapa I, et al. (99m)Tc-labeled aminosilane-coated iron oxide nanoparticles for molecular imaging of ανβ3-mediated tumor expression and feasibility for hyperthermia treatment. J Colloid Interface Sci 2014;433:163–75.
[66] Couto D, et al. Polyacrylic acid coated and non-coated iron oxide nanoparticles are not genotoxic to human T lymphocytes. Toxicol Lett 2015;234(2):67–73.
[67] Couto D, et al. Polyacrylic acid-coated and non-coated iron oxide nanoparticles induce cytokine activation in human blood cells through TAK1, p38 MAPK and JNK pro-inflammatory pathways. Arch Toxicol 2015;89 (10):1759–69.
[68] Barbaro D, et al. Glucose-coated superparamagnetic iron oxide nanoparticles prepared by metal vapour synthesis are electively internalized in a pancreatic adenocarcinoma cell line expressing GLUT1 transporter. PLoS ONE 2015;10(4), e0123159.
[69] Borny R, et al. Nucleophilic cross-linked, dextran coated iron oxide nanoparticles as basis for molecular imaging: synthesis, characterization, visualization and comparison with previous product. Contrast Media Mol Imaging 2015;10(1):18–27.
[70] Wu M, et al. Nanocluster of superparamagnetic iron oxide nanoparticles coated with poly (dopamine) for magnetic field-targeting, highly sensitive MRI and photothermal cancer therapy. Nanotechnology 2015;26(11), 115102.
[71] Leung KC, et al. Increased efficacies in magnetofection and gene delivery to hepatocellular carcinoma cells with ternary organic-inorganic hybrid nanocomposites. Chem Asian J 2013;8(8):1760–4.
[72] Leung KC, et al. Ternary hybrid nanocomposites for gene delivery and magnetic resonance imaging of hepatocellular carcinoma cells. Quant Imaging Med Surg 2013;3(6):302–7.
[73] Gimi B, et al. Molecular imaging of cancer: applications of magnetic resonance methods. Proc IEEE Inst Electr Electron Eng 2005;93(4):784–99.
[74] Frank JA, et al. Methods for magnetically labeling stem and other cells for detection by in vivo magnetic resonance imaging. Cytotherapy 2004;6(6):621–5.
[75] Weissleder R, et al. MR imaging and scintigraphy of gene expression through melanin induction. Radiology 1997;204(2):425–9.
[76] Enochs WS, et al. Paramagnetic metal scavenging by melanin: MR imaging. Radiology 1997;204(2):417–23.
[77] Alfke H, et al. In vitro MR imaging of regulated gene expression. Radiology 2003;228(2):488–92.
[78] Gilad AA, et al. Developing MR reporter genes: promises and pitfalls. NMR Biomed 2007;20(3):275–90.
[79] Louie AY, et al. In vivo visualization of gene expression using magnetic resonance imaging. Nat Biotechnol 2000;18(3):321–5.
[80] Zhang R, et al. Real-time specific light-up sensing of transferrin receptor: image-guided photodynamic ablation of cancer cells through controlled cytomembrane disintegration. Anal Chem 2016;88(9):4841–8.

[81] Pereira SM, et al. Overexpression of the MRI reporter genes ferritin and transferrin receptor affect iron homeostasis and produce limited contrast in mesenchymal stem cells. Int J Mol Sci 2015;16(7):15481–96.
[82] Pereira SM, et al. Evaluating the effectiveness of transferrin receptor-1 (TfR1) as a magnetic resonance reporter gene. Contrast Media Mol Imaging 2016;11(3):236–44.
[83] Weissleder R, et al. In vivo magnetic resonance imaging of transgene expression. Nat Med 2000;6(3):351–5.
[84] Genove G, et al. A new transgene reporter for in vivo magnetic resonance imaging. Nat Med 2005;11(4):450–4.
[85] Ono K, et al. Ferritin reporter used for gene expression imaging by magnetic resonance. Biochem Biophys Res Commun 2009;388(3):589–94.
[86] Gilad AA, et al. Artificial reporter gene providing MRI contrast based on proton exchange. Nat Biotechnol 2007;25(2):217–9.
[87] Blasberg RG, Gelovani J. Molecular-genetic imaging: a nuclear medicine-based perspective. Mol Imaging 2002;1(3):280–300.
[88] Adonai N, et al. Ex vivo cell labeling with 64Cu-pyruvaldehyde-bis(N4-methylthiosemicarbazone) for imaging cell trafficking in mice with positron-emission tomography. Proc Natl Acad Sci U S A 2002;99(5):3030–5.
[89] Kircher MF, Gambhir SS, Grimm J. Noninvasive cell-tracking methods. Nat Rev Clin Oncol 2011;8(11):677–88.
[90] Read EJ, et al. In vivo traffic of indium-111-oxine labeled human lymphocytes collected by automated apheresis. J Nucl Med 1990;31(6):999–1006.
[91] Volpe A. Applications of nuclear-based imaging in gene and cell therapy: probe considerations. Mol Ther Oncolytics 2021;20:447–58.
[92] Grimm J, Kircher MF, Weissleder R. Cell tracking. Principles and applications. Radiologe 2007;47(1):25–33.
[93] Botti C, et al. Comparison of three different methods for radiolabelling human activated T lymphocytes. Eur J Nucl Med 1997;24(5):497–504.
[94] Serganova I, Blasberg RG. Molecular imaging with reporter genes: has its promise been delivered? J Nucl Med 2019;60(12):1665–81.
[95] Minn I, et al. Molecular-genetic imaging of cancer. Adv Cancer Res 2014;124:131–69.
[96] Goswami R, et al. Gene therapy leaves a vicious cycle. Front Oncol 2019;9:297.
[97] Ronald JA, et al. Development and validation of non-integrative, self-limited, and replicating minicircles for safe reporter gene imaging of cell-based therapies. PLoS ONE 2013;8(8), e73138.
[98] Lufino MM, Edser PA, Wade-Martins R. Advances in high-capacity extrachromosomal vector technology: episomal maintenance, vector delivery, and transgene expression. Mol Ther 2008;16(9):1525–38.
[99] Maggio I, Gonçalves MA. Genome editing at the crossroads of delivery, specificity, and fidelity. Trends Biotechnol 2015;33(5):280–91.
[100] Bressan RB, et al. Efficient CRISPR/Cas9-assisted gene targeting enables rapid and precise genetic manipulation of mammalian neural stem cells. Development 2017;144(4):635–48.
[101] Ashmore-Harris C, Fruhwirth GO. The clinical potential of gene editing as a tool to engineer cell-based therapeutics. Clin Transl Med 2020;9(1):15.
[102] Glowniak JV, et al. Evaluation of metaiodobenzylguanidine uptake by the norepinephrine, dopamine and serotonin transporters. J Nucl Med 1993;34(7):1140–6.
[103] Shulkin BL, et al. Iodine-123-4-amino-3-iodobenzylguanidine, a new sympathoadrenal imaging agent: comparison with iodine-123 metaiodobenzylguanidine. J Nucl Med 1986;27(7):1138–42.
[104] Barrio JR, et al. 3-(2′-[18F]fluoroethyl)spiperone: in vivo biochemical and kinetic characterization in rodents, nonhuman primates, and humans. J Cereb Blood Flow Metab 1989;9(6):830–9.
[105] Hemminki A, et al. An adenovirus with enhanced infectivity mediates molecular chemotherapy of ovarian cancer cells and allows imaging of gene expression. Mol Ther 2001;4(3):223–31.
[106] Penheiter AR, Russell SJ, Carlson SK. The sodium iodide symporter (NIS) as an imaging reporter for gene, viral, and cell-based therapies. Curr Gene Ther 2012;12(1):33–47.
[107] Choudhury PS, Gupta M. Differentiated thyroid cancer theranostics: radioiodine and beyond. Br J Radiol 2018;91(1091):20180136.
[108] Ravera S, et al. The sodium/iodide symporter (NIS): molecular physiology and preclinical and clinical applications. Annu Rev Physiol 2017;79:261–89.
[109] Che J, et al. hNIS-IRES-eGFP dual reporter gene imaging. Mol Imaging 2005;4(2):128–36.

[110] Terrovitis J, et al. Ectopic expression of the sodium-iodide symporter enables imaging of transplanted cardiac stem cells in vivo by single-photon emission computed tomography or positron emission tomography. J Am Coll Cardiol 2008;52(20):1652–60.
[111] Fruhwirth GO, et al. A whole-body dual-modality radionuclide optical strategy for preclinical imaging of metastasis and heterogeneous treatment response in different microenvironments. J Nucl Med 2014;55 (4):686–94.
[112] Volpe A, et al. Spatiotemporal PET imaging reveals differences in CAR-T tumor retention in triple-negative breast cancer models. Mol Ther 2020;28(10):2271–85.
[113] Volpe A, et al. Radionuclide-fluorescence reporter gene imaging to track tumor progression in rodent tumor models. J Vis Exp 2018;133.
[114] Jhiang SM, et al. An immunohistochemical study of Na^+/I^- symporter in human thyroid tissues and salivary gland tissues. Endocrinology 1998;139(10):4416–9.
[115] Kotani T, et al. Characterization of gastric Na^+/I^- symporter of the rat. Clin Immunol Immunopathol 1998;89 (3):271–8.
[116] Eskandari S, et al. Thyroid Na^+/I^- symporter. Mechanism, stoichiometry, and specificity. J Biol Chem 1997;272 (43):27230–8.
[117] Dohan O, et al. The sodium/iodide symporter (NIS): characterization, regulation, and medical significance. Endocr Rev 2003;24(1):48–77.
[118] Portulano C, Paroder-Belenitsky M, Carrasco N. The Na^+/I^- symporter (NIS): mechanism and medical impact. Endocr Rev 2014;35(1):106–49.
[119] O'Doherty J, et al. (18)F-Tetrafluoroborate, a PET probe for imaging sodium/iodide symporter expression: whole-body biodistribution, safety, and radiation dosimetry in thyroid cancer patients. J Nucl Med 2017;58 (10):1666–71.
[120] Axelrod J, Kopin IJ. The uptake, storage, release and metabolism of noradrenaline in sympathetic nerves. Prog Brain Res 1969;31:21–32.
[121] Altmann A, et al. Increased MIBG uptake after transfer of the human norepinephrine transporter gene in rat hepatoma. J Nucl Med 2003;44(6):973–80.
[122] Moroz MA, et al. Imaging hNET reporter gene expression with 124I-MIBG. J Nucl Med 2007;48(5):827–36.
[123] Anton M, et al. Use of the norepinephrine transporter as a reporter gene for non-invasive imaging of genetically modified cells. J Gene Med 2004;6(1):119–26.
[124] Ding YS, Fowler J. New-generation radiotracers for nAChR and NET. Nucl Med Biol 2005;32(7):707–18.
[125] Bakker WH, et al. [111In-DTPA-D-Phe1]-octreotide, a potential radiopharmaceutical for imaging of somatostatin receptor-positive tumors: synthesis, radiolabeling and in vitro validation. Life Sci 1991;49(22):1583–91.
[126] Bakker WH, et al. In vivo application of [111In-DTPA-D-Phe1]-octreotide for detection of somatostatin receptor-positive tumors in rats. Life Sci 1991;49(22):1593–601.
[127] FDA Letter of Approval for [68Ga]Ga-DOTA-TATE. Available from: https://www.fda.gov/news-events/press-announcements/fda-approves-new-diagnostic-imaging-agent-detect-rare-neuroendocrine-tumors.
[128] FDA Letter of Approval for [68Ga]Ga-DOTA-TOC. Available from: https://www.accessdata.fda.gov/drugsatfda_docs/appletter/2019/210828Orig1s000ltr.pdf.
[129] MacLaren DC, et al. Repetitive, non-invasive imaging of the dopamine D2 receptor as a reporter gene in living animals. Gene Ther 1999;6(5):785–91.
[130] Liang Q, et al. Noninvasive, quantitative imaging in living animals of a mutant dopamine D2 receptor reporter gene in which ligand binding is uncoupled from signal transduction. Gene Ther 2001;8(19):1490–8.
[131] Lee JT, et al. Comparative analysis of human nucleoside kinase-based reporter systems for PET imaging. Mol Imaging Biol 2017;19(1):100–8.
[132] Likar Y, et al. A new pyrimidine-specific reporter gene: a mutated human deoxycytidine kinase suitable for PET during treatment with acycloguanosine-based cytotoxic drugs. J Nucl Med 2010;51(9):1395–403.
[133] Ponomarev V, et al. A human-derived reporter gene for noninvasive imaging in humans: mitochondrial thymidine kinase type 2. J Nucl Med 2007;48(5):819–26.
[134] Campbell DO, et al. Structure-guided engineering of human thymidine kinase 2 as a positron emission tomography reporter gene for enhanced phosphorylation of non-natural thymidine analog reporter probe. J Biol Chem 2012;287(1):446–54.
[135] Arnér ES, Eriksson S. Mammalian deoxyribonucleoside kinases. Pharmacol Ther 1995;67(2):155–86.

[136] Gambhir SS, et al. A mutant herpes simplex virus type 1 thymidine kinase reporter gene shows improved sensitivity for imaging reporter gene expression with positron emission tomography. Proc Natl Acad Sci U S A 2000;97(6):2785–90.
[137] Yaghoubi SS, et al. Imaging progress of herpes simplex virus type 1 thymidine kinase suicide gene therapy in living subjects with positron emission tomography. Cancer Gene Ther 2005;12(3):329–39.
[138] Alauddin MM, et al. 9-[(3-[18F]-fluoro-1-hydroxy-2-propoxy)methyl]guanine ([18F]-FHPG): a potential imaging agent of viral infection and gene therapy using PET. Nucl Med Biol 1996;23(6):787–92.
[139] Alauddin MM, Conti PS. Synthesis and preliminary evaluation of 9-(4-[18F]-fluoro-3-hydroxymethylbutyl) guanine ([18F]FHBG): a new potential imaging agent for viral infection and gene therapy using PET. Nucl Med Biol 1998;25(3):175–80.
[140] Gambhir SS, et al. Imaging adenoviral-directed reporter gene expression in living animals with positron emission tomography. Proc Natl Acad Sci U S A 1999;96(5):2333–8.
[141] Haberkorn U, et al. Ganciclovir uptake in human mammary carcinoma cells expressing herpes simplex virus thymidine kinase. Nucl Med Biol 1998;25(4):367–73.
[142] Alauddin MM, et al. Synthesis of 2′-deoxy-2′-[18F]fluoro-5-bromo-1-beta-D-arabinofuranosyluracil ([18F]-FBAU) and 2′-deoxy-2′-[18F]fluoro-5-chloro-1-beta-D-arabinofuranosyl-uracil ([18F]-FCAU), and their biological evaluation as markers for gene expression. Nucl Med Biol 2004;31(4):399–405.
[143] Fuchigami T, et al. Synthesis and characterization of 9-(4-[(18)F]fluoro-3-(hydroxymethyl)butyl)-2-(phenylthio)-6-oxopurine as a novel PET agent for mutant herpes simplex virus type 1 thymidine kinase reporter gene imaging. Mol Imaging Biol 2020;22(5):1151–60.
[144] Tjuvajev JG, et al. Comparison of radiolabeled nucleoside probes (FIAU, FHBG, and FHPG) for PET imaging of HSV1-tk gene expression. J Nucl Med 2002;43(8):1072–83.
[145] Yaghoubi S, et al. Human pharmacokinetic and dosimetry studies of [(18)F]FHBG: a reporter probe for imaging herpes simplex virus type-1 thymidine kinase reporter gene expression. J Nucl Med 2001;42(8):1225–34.
[146] Pittet MJ, et al. In vivo imaging of T cell delivery to tumors after adoptive transfer therapy. Proc Natl Acad Sci U S A 2007;104(30):12457–61.
[147] Grimfors G, et al. Tumour imaging of indium-111 oxine-labelled autologous lymphocytes as a staging method in Hodgkin's disease. Eur J Haematol 1989;42(3):276–83.
[148] Koehne G, et al. Serial in vivo imaging of the targeted migration of human HSV-TK-transduced antigen-specific lymphocytes. Nat Biotechnol 2003;21(4):405–13.
[149] Dubey P, et al. Quantitative imaging of the T cell antitumor response by positron-emission tomography. Proc Natl Acad Sci U S A 2003;100(3):1232–7.
[150] Sato N, et al. (89)Zr-Oxine complex PET cell imaging in monitoring cell-based therapies. Radiology 2015;275 (2):490–500.
[151] Zhang W, et al. Rapid in vivo functional analysis of transgenes in mice using whole body imaging of luciferase expression. Transgenic Res 2001;10(5):423–34.
[152] Rabinovich BA, et al. Visualizing fewer than 10 mouse T cells with an enhanced firefly luciferase in immunocompetent mouse models of cancer. Proc Natl Acad Sci U S A 2008;105(38):14342–6.
[153] Hardy J, et al. Bioluminescence imaging of lymphocyte trafficking in vivo. Exp Hematol 2001;29(12):1353–60.
[154] Costa GL, et al. Adoptive immunotherapy of experimental autoimmune encephalomyelitis via T cell delivery of the IL-12 p40 subunit. J Immunol 2001;167(4):2379–87.
[155] Moroz MA, et al. Comparative analysis of T cell imaging with human nuclear reporter genes. J Nucl Med 2015;56(7):1055–60.
[156] McCracken MN, et al. Noninvasive detection of tumor-infiltrating T cells by PET reporter imaging. J Clin Invest 2015;125(5):1815–26.
[157] Dotti G, et al. Repetitive noninvasive monitoring of HSV1-tk-expressing T cells intravenously infused into nonhuman primates using positron emission tomography and computed tomography with 18F-FEAU. Mol Imaging 2009;8(4):230–7.
[158] Kenanova V, et al. Recombinant carcinoembryonic antigen as a reporter gene for molecular imaging. Eur J Nucl Med Mol Imaging 2009;36(1):104–14.
[159] Doubrovin MM, et al. In vivo imaging and quantitation of adoptively transferred human antigen-specific T cells transduced to express a human norepinephrine transporter gene. Cancer Res 2007;67(24):11959–69.

[160] Weist MR, et al. PET of adoptively transferred chimeric antigen receptor T cells with (89)Zr-oxine. J Nucl Med 2018;59(10):1531–7.
[161] Lee SH, et al. Feasibility of real-time in vivo 89Zr-DFO-labeled CAR T-cell trafficking using PET imaging. PLoS ONE 2020;15(1), e0223814.
[162] Bhatnagar P, et al. Imaging of genetically engineered T cells by PET using gold nanoparticles complexed to Copper-64. Integr Biol (Camb) 2013;5(1):231–8.
[163] Corneillie TM, et al. Converting weak binders into infinite binders. Bioconjug Chem 2004;15(6):1389–91.
[164] Goodwin DA, et al. Pharmacokinetics of pretargeted monoclonal antibody 2D12.5 and 88Y-Janus-2-(p-nitrobenzyl)-1,4,7,10-tetraazacyclododecanetetraacetic acid (DOTA) in BALB/c mice with KHJJ mouse adenocarcinoma: a model for 90Y radioimmunotherapy. Cancer Res 1994;54(22):5937–46.
[165] Krebs S, et al. Antibody with infinite affinity for in vivo tracking of genetically engineered lymphocytes. J Nucl Med 2018;59(12):1894–900.
[166] Diocou S, et al. [18F]tetrafluoroborate-PET/CT enables sensitive tumor and metastasis in vivo imaging in a sodium iodide symporter-expressing tumor model. Sci Rep 2017;7(1):946.
[167] Vedvyas Y, et al. Longitudinal PET imaging demonstrates biphasic CAR T cell responses in survivors. JCI Insight 2016;1(19), e90064.
[168] Sellmyer MA, et al. Imaging CAR T cell trafficking with eDHFR as a PET reporter gene. Mol Ther 2020;28(1):42–51.
[169] Minn I, et al. Imaging CAR T cell therapy with PSMA-targeted positron emission tomography. Sci Adv 2019;5 (7), eaaw5096.
[170] Najjar AM, et al. Imaging of sleeping beauty-modified CD19-specific T cells expressing HSV1-thymidine kinase by positron emission tomography. Mol Imaging Biol 2016;18(6):838–48.
[171] Dobrenkov K, et al. Monitoring the efficacy of adoptively transferred prostate cancer-targeted human T lymphocytes with PET and bioluminescence imaging. J Nucl Med 2008;49(7):1162–70.
[172] Yaghoubi SS, et al. Noninvasive detection of therapeutic cytolytic T cells with 18F-FHBG PET in a patient with glioma. Nat Clin Pract Oncol 2009;6(1):53–8.
[173] Keu KV, et al. Reporter gene imaging of targeted T cell immunotherapy in recurrent glioma. Sci Transl Med 2017;9(373).
[174] Alam IS, et al. Imaging activated T cells predicts response to cancer vaccines. J Clin Invest 2018;128(6):2569–80.
[175] Alam IS, et al. Visualization of activated T cells by OX40-immunopet as a strategy for diagnosis of acute graft-versus-host disease. Cancer Res 2020;80(21):4780–90.
[176] Xiao Z, et al. ICOS is an indicator of T-cell-mediated response to cancer immunotherapy. Cancer Res 2020;80 (14):3023–32.
[177] Simonetta F, et al. Molecular imaging of chimeric antigen receptor T cells by ICOS-immunoPET. Clin Cancer Res 2021;27(4):1058–68.
[178] Volpe A, et al. Predicting CAR-T cell immunotherapy success through immunoPET. Clin Cancer Res 2021;27 (4):911–2.
[179] Hartimath SV, et al. Noninvasive monitoring of cancer therapy induced activated T cells using [(18)F]FB-IL-2 PET imaging. Onco Targets Ther 2017;6(1), e1248014.
[180] Beckford Vera DR, et al. Immuno-PET imaging of tumor-infiltrating lymphocytes using zirconium-89 radiolabeled anti-CD3 antibody in immune-competent mice bearing syngeneic tumors. PLoS ONE 2018;13(3), e0193832.
[181] Larimer BM, et al. Quantitative CD3 PET imaging predicts tumor growth response to anti-CTLA-4 therapy. J Nucl Med 2016;57(10):1607–11.
[182] Freise AC, et al. Immuno-PET in inflammatory bowel disease: imaging CD4-positive T cells in a murine model of colitis. J Nucl Med 2018;59(6):980–5.
[183] Freise AC, et al. ImmunoPET imaging of murine CD4(+) T cells using anti-CD4 Cys-diabody: effects of protein dose on T cell function and imaging. Mol Imaging Biol 2017;19(4):599–609.
[184] Seo JW, et al. CD8(+) T-cell density imaging with (64)Cu-labeled Cys-diabody informs immunotherapy protocols. Clin Cancer Res 2018;24(20):4976–87.
[185] Tavaré R, et al. An effective immuno-PET imaging method to monitor CD8-dependent responses to immunotherapy. Cancer Res 2016;76(1):73–82.

[186] Pandit-Taskar N, et al. First-in-humans imaging with (89)Zr-Df-IAB22M2C anti-CD8 minibody in patients with solid malignancies: preliminary pharmacokinetics, biodistribution, and lesion targeting. J Nucl Med 2020;61 (4):512–9.
[187] Larimer BM, et al. The effectiveness of checkpoint inhibitor combinations and administration timing can be measured by granzyme B PET imaging. Clin Cancer Res 2019;25(4):1196–205.
[188] Larimer BM, et al. Granzyme B PET imaging as a predictive biomarker of immunotherapy response. Cancer Res 2017;77(9):2318–27.
[189] Gibson HM, et al. IFNγ PET imaging as a predictive tool for monitoring response to tumor immunotherapy. Cancer Res 2018;78(19):5706–17.
[190] Ponomarev V, et al. Imaging TCR-dependent NFAT-mediated T-cell activation with positron emission tomography in vivo. Neoplasia 2001;3(6):480–8.
[191] Rao A, Luo C, Hogan PG. Transcription factors of the NFAT family: regulation and function. Annu Rev Immunol 1997;15:707–47.
[192] Rydzek J, et al. Chimeric antigen receptor library screening using a novel NF-κB/NFAT reporter cell platform. Mol Ther 2019;27(2):287–99.
[193] Buck MD, et al. Mitochondrial dynamics controls T cell fate through metabolic programming. Cell 2016;166 (1):63–76.
[194] Fox CJ, Hammerman PS, Thompson CB. Fuel feeds function: energy metabolism and the T-cell response. Nat Rev Immunol 2005;5(11):844–52.
[195] Aarntzen EH, et al. Early identification of antigen-specific immune responses in vivo by [18F]-labeled 3′-fluoro-3′-deoxy-thymidine ([18F]FLT) PET imaging. Proc Natl Acad Sci U S A 2011;108(45):18396–9.
[196] Iurescia S, Fioretti D, Rinaldi M. Targeting cytosolic nucleic acid-sensing pathways for cancer immunotherapies. Front Immunol 2018;9:711.
[197] Aide N, et al. FDG PET/CT for assessing tumour response to immunotherapy: report on the EANM symposium on immune modulation and recent review of the literature. Eur J Nucl Med Mol Imaging 2019;46 (1):238–50.
[198] Schwenck J, et al. Cancer immunotherapy is accompanied by distinct metabolic patterns in primary and secondary lymphoid organs observed by non-invasive in vivo (18)F-FDG-PET. Theranostics 2020;10(2):925–37.
[199] Sachpekidis C, et al. Predictive value of early 18F-FDG PET/CT studies for treatment response evaluation to ipilimumab in metastatic melanoma: preliminary results of an ongoing study. Eur J Nucl Med Mol Imaging 2015;42(3):386–96.
[200] Sachpekidis C, et al. The role of interim (18)F-FDG PET/CT in prediction of response to ipilimumab treatment in metastatic melanoma. Eur J Nucl Med Mol Imaging 2018;45(8):1289–96.
[201] Cho SY, et al. Prediction of response to immune checkpoint inhibitor therapy using early-time-point (18)F-FDG PET/CT imaging in patients with advanced melanoma. J Nucl Med 2017;58(9):1421–8.
[202] Goldfarb L, et al. Monitoring anti-PD-1-based immunotherapy in non-small cell lung cancer with FDG PET: introduction of iPERCIST. EJNMMI Res 2019;9(1):8.
[203] Castello A, et al. (18)F-FDG PET/CT for response assessment in *Hodgkin lymphoma* undergoing immunotherapy with checkpoint inhibitors. Leuk Lymphoma 2019;60(2):367–75.
[204] Ribas A, et al. Imaging of CTLA4 blockade-induced cell replication with (18)F-FLT PET in patients with advanced melanoma treated with tremelimumab. J Nucl Med 2010;51(3):340–6.
[205] Ronald JA, et al. A PET imaging strategy to visualize activated T cells in acute graft-versus-host disease elicited by allogenic hematopoietic cell transplant. Cancer Res 2017;77(11):2893–902.
[206] Levi J, et al. (18)F-FAraG PET for CD8 profiling of tumors and assessment of immunomodulation by chemotherapy. J Nucl Med 2020.
[207] Levi J, et al. Imaging of activated T cells as an early predictor of immune response to anti-PD-1 therapy. Cancer Res 2019;79(13):3455–65.
[208] Kim W, et al. [18F]CFA as a clinically translatable probe for PET imaging of deoxycytidine kinase activity. Proc Natl Acad Sci U S A 2016;113(15):4027–32.
[209] Radu CG, et al. Molecular imaging of lymphoid organs and immune activation by positron emission tomography with a new [18F]-labeled 2′-deoxycytidine analog. Nat Med 2008;14(7):783–8.

[210] Kanagawa M, et al. Comparison of trans-1-amino-3-[18F]fluorocyclobutanecarboxylic acid (anti-[18F]FACBC) accumulation in lymph node prostate cancer metastasis and lymphadenitis in rats. Nucl Med Biol 2014;41 (7):545–51.

[211] Oka S, et al. Differences in transport mechanisms of trans-1-amino-3-[18F]fluorocyclobutanecarboxylic acid in inflammation, prostate cancer, and glioma cells: comparison with L-[methyl-11C]methionine and 2-deoxy-2-[18F]fluoro-D-glucose. Mol Imaging Biol 2014;16(3):322–9.

[212] Topalian SL, et al. Safety, activity, and immune correlates of anti-PD-1 antibody in cancer. N Engl J Med 2012;366(26):2443–54.

[213] Larkin J, et al. Combined nivolumab and ipilimumab or monotherapy in untreated melanoma. N Engl J Med 2015;373(1):23–34.

[214] Hodi FS, et al. Improved survival with ipilimumab in patients with metastatic melanoma. N Engl J Med 2010;363(8):711–23.

[215] Motzer RJ, et al. Nivolumab plus ipilimumab versus sunitinib in advanced renal-cell carcinoma. N Engl J Med 2018;378(14):1277–90.

[216] Gettinger S, et al. Nivolumab monotherapy for first-line treatment of advanced non-small-cell lung cancer. J Clin Oncol 2016;34(25):2980–7.

[217] Herbst RS, et al. Pembrolizumab versus docetaxel for previously treated, PD-L1-positive, advanced non-small-cell lung cancer (KEYNOTE-010): a randomised controlled trial. Lancet 2016;387(10027):1540–50.

[218] Postow MA, et al. Nivolumab and ipilimumab versus ipilimumab in untreated melanoma. N Engl J Med 2015;372(21):2006–17.

[219] Vaddepally RK, et al. Review of indications of FDA-approved immune checkpoint inhibitors per NCCN guidelines with the level of evidence. Cancers (Basel) 2020;12(3).

[220] Dolan DE, Gupta S. PD-1 pathway inhibitors: changing the landscape of cancer immunotherapy. Cancer Control 2014;21(3):231–7.

[221] Natarajan A, et al. Novel radiotracer for immunoPET imaging of PD-1 checkpoint expression on tumor infiltrating lymphocytes. Bioconjug Chem 2015;26(10):2062–9.

[222] England CG, et al. Preclinical pharmacokinetics and biodistribution studies of 89Zr-labeled pembrolizumab. J Nucl Med 2017;58(1):162–8.

[223] England CG, et al. (89)Zr-labeled nivolumab for imaging of T-cell infiltration in a humanized murine model of lung cancer. Eur J Nucl Med Mol Imaging 2018;45(1):110–20.

[224] Hettich M, et al. High-resolution PET imaging with therapeutic antibody-based PD-1/PD-L1 checkpoint tracers. Theranostics 2016;6(10):1629–40.

[225] Du Y, et al. Nuclear and fluorescent labeled PD-1-liposome-DOX-(64)Cu/IRDye800CW allows improved breast tumor targeted imaging and therapy. Mol Pharm 2017;14(11):3978–86.

[226] Breki CM, et al. Fractal and multifractal analysis of PET/CT images of metastatic melanoma before and after treatment with ipilimumab. EJNMMI Res 2016;6(1):61.

[227] Kong BY, et al. Residual FDG-PET metabolic activity in metastatic melanoma patients with prolonged response to anti-PD-1 therapy. Pigment Cell Melanoma Res 2016;29(5):572–7.

[228] Kaira K, et al. Metabolic activity by (18)F-FDG-PET/CT is predictive of early response after nivolumab in previously treated NSCLC. Eur J Nucl Med Mol Imaging 2018;45(1):56–66.

[229] Heskamp S, et al. Noninvasive imaging of tumor PD-L1 expression using radiolabeled anti-PD-L1 antibodies. Cancer Res 2015;75(14):2928–36.

[230] Josefsson A, et al. Imaging, biodistribution, and dosimetry of radionuclide-labeled PD-L1 antibody in an immunocompetent mouse model of breast cancer. Cancer Res 2016;76(2):472–9.

[231] Broos K, et al. Non-invasive assessment of murine PD-L1 levels in syngeneic tumor models by nuclear imaging with nanobody tracers. Oncotarget 2017;8(26):41932–46.

[232] Kumar D, et al. Peptide-based PET quantifies target engagement of PD-L1 therapeutics. J Clin Invest 2019;129 (2):616–30.

[233] Lv G, et al. PET imaging of tumor PD-L1 expression with a highly specific nonblocking single-domain antibody. J Nucl Med 2020;61(1):117–22.

[234] Christensen C, et al. Quantitative PET imaging of PD-L1 expression in xenograft and syngeneic tumour models using a site-specifically labelled PD-L1 antibody. Eur J Nucl Med Mol Imaging 2020;47(5):1302–13.

[235] Bensch F, et al. (89)Zr-atezolizumab imaging as a non-invasive approach to assess clinical response to PD-L1 blockade in cancer. Nat Med 2018;24(12):1852–8.

[236] Chatterjee S, et al. Rapid PD-L1 detection in tumors with PET using a highly specific peptide. Biochem Biophys Res Commun 2017;483(1):258–63.

[237] Lesniak WG, et al. PD-L1 detection in tumors using [(64)Cu]Atezolizumab with PET. Bioconjug Chem 2016;27 (9):2103–10.

[238] Kikuchi J, et al. The novel orally active proteasome inhibitor K-7174 exerts anti-myeloma activity in vitro and in vivo by down-regulating the expression of class I histone deacetylases. J Biol Chem 2013;288(35):25593–602.

[239] Mayer AT, et al. Practical immuno-PET radiotracer design considerations for human immune checkpoint imaging. J Nucl Med 2017;58(4):538–46.

[240] Maute RL, et al. Engineering high-affinity PD-1 variants for optimized immunotherapy and immuno-PET imaging. Proc Natl Acad Sci U S A 2015;112(47):E6506–14.

[241] Donnelly DJ, et al. Synthesis and biologic evaluation of a novel (18)F-labeled adnectin as a PET radioligand for imaging PD-L1 expression. J Nucl Med 2018;59(3):529–35.

[242] Stutvoet TS, et al. Molecular imaging of PD-L1 expression and dynamics with the adnectin-based PET tracer (18)F-BMS-986192. J Nucl Med 2020;61(12):1839–44.

[243] González Trotter DE, et al. In vivo imaging of the programmed death ligand 1 by (18)F PET. J Nucl Med 2017;58 (11):1852–7.

[244] Ingram JR, et al. PD-L1 is an activation-independent marker of brown adipocytes. Nat Commun 2017;8(1):647.

[245] Ren J, et al. Preclinical study of a dual mode PD-L1 target imaging probe facilitates patient screening for immunotherapy. J Nucl Med 2020;61(supplement 1):1027.

[246] Du Y, et al. Liposomal nanohybrid cerasomes targeted to PD-L1 enable dual-modality imaging and improve antitumor treatments. Cancer Lett 2018;414:230–8.

[247] Zhao S, et al. Cerenkov luminescence imaging is an effective preclinical tool for assessing colorectal cancer PD-L1 levels in vivo. EJNMMI Res 2020;10(1):64.

[248] Hamid O, et al. Clinical activity, safety, and biomarkers of MPDL3280A, an engineered PD-L1 antibody in patients with locally advanced or metastatic melanoma (mM). J Clin Oncol 2013;31(15_suppl):9010.

[249] Emens LA, et al. Abstract 2859: inhibition of PD-L1 by MPDL3280A leads to clinical activity in patients with metastatic triple-negative breast cancer (TNBC). Cancer Res 2015;75(15_Supplement):2859.

[250] Spigel DR, et al. Clinical activity, safety, and biomarkers of MPDL3280A, an engineered PD-L1 antibody in patients with locally advanced or metastatic non-small cell lung cancer (NSCLC). J Clin Oncol 2013;31 (15_suppl):8008.

[251] Cho DC, et al. Clinical activity, safety, and biomarkers of MPDL3280A, an engineered PD-L1 antibody in patients with metastatic renal cell carcinoma (mRCC). J Clin Oncol 2013;31(15_suppl):4505.

[252] Petrylak DP, et al. Atezolizumab (MPDL3280A) monotherapy for patients with metastatic urothelial cancer: long-term outcomes from a phase 1 study. JAMA Oncol 2018;4(4):537–44.

[253] Powles T, et al. MPDL3280A (anti-PD-L1) treatment leads to clinical activity in metastatic bladder cancer. Nature 2014;515(7528):558–62.

[254] Higashikawa K, et al. 64Cu-DOTA-anti-CTLA-4 mAb enabled PET visualization of CTLA-4 on the T-cell infiltrating tumor tissues. PLoS ONE 2014;9(11), e109866.

[255] Ehlerding EB, et al. ImmunoPET imaging of CTLA-4 expression in mouse models of non-small cell lung cancer. Mol Pharm 2017;14(5):1782–9.

[256] Ehlerding EB, et al. Antibody and fragment-based PET imaging of CTLA-4+ T-cells in humanized mouse models. Am J Cancer Res 2019;9(1):53–63.

[257] Lecocq Q, et al. Noninvasive imaging of the immune checkpoint LAG-3 using nanobodies, from development to pre-clinical use. Biomolecules 2019;9(10).

[258] Kelly MP, et al. Abstract 3033: immuno-PET detection of LAG-3 expressing intratumoral lymphocytes using the zirconium-89 radiolabeled fully human anti-LAG-3 antibody REGN3767. Cancer Res 2018;78(13 Supplement):3033.

[259] Shaffer T, Natarajan A, Gambhir SS. PET imaging of TIGIT expression on tumor-infiltrating lymphocytes. Clin Cancer Res 2021;27(7):1932–40.

[260] Suffner J, et al. Dendritic cells support homeostatic expansion of Foxp3+ regulatory T cells in Foxp3.LuciDTR mice. J Immunol 2010;184(4):1810–20.

[261] Cheng HY, et al. Bioimaging of alloantigen-stimulated regulatory T cells in rat vascularized composite allotransplantation. PLoS ONE 2018;13(9), e0203624.
[262] Nguyen VH, et al. In vivo dynamics of regulatory T-cell trafficking and survival predict effective strategies to control graft-versus-host disease following allogeneic transplantation. Blood 2007;109(6):2649–56.
[263] Dawson NA, et al. Systematic testing and specificity mapping of alloantigen-specific chimeric antigen receptors in regulatory T cells. JCI Insight 2019;4(6).
[264] Jacob J, et al. Spatiotemporal in vivo tracking of polyclonal human regulatory T cells (Tregs) reveals a role for innate immune cells in Treg transplant recruitment. Mol Ther Methods Clin Dev 2021;20:324–36.
[265] Mohseni YR, et al. Chimeric antigen receptor-modified human regulatory T cells that constitutively express IL-10 maintain their phenotype and are potently suppressive. Eur J Immunol 2021;51(10):2522–30.
[266] Knorr DA, et al. Engineered human embryonic stem cell-derived lymphocytes to study in vivo trafficking and immunotherapy. Stem Cells Dev 2013;22(13):1861–9.
[267] Zhu L, et al. Natural killer cell (NK-92MI)-based therapy for pulmonary metastasis of anaplastic thyroid cancer in a nude mouse model. Front Immunol 2017;8:816.
[268] Swift BE, et al. Natural killer cell lines preferentially kill clonogenic multiple myeloma cells and decrease myeloma engraftment in a bioluminescent xenograft mouse model. Haematologica 2012;97(7):1020–8.
[269] Edinger M, et al. Advancing animal models of neoplasia through in vivo bioluminescence imaging. Eur J Cancer 2002;38(16):2128–36.
[270] May RM, et al. Murine natural killer immunoreceptors use distinct proximal signaling complexes to direct cell function. Blood 2013;121(16):3135–46.
[271] Chandrasekaran S, et al. Super natural killer cells that target metastases in the tumor draining lymph nodes. Biomaterials 2016;77:66–76.
[272] Mukherjee S, et al. Identification of EpCAM as a molecular target of prostate cancer stroma. Am J Pathol 2009;175(6):2277–87.
[273] Lim YT, et al. Near-infrared emitting fluorescent nanocrystals-labeled natural killer cells as a platform technology for the optical imaging of immunotherapeutic cells-based cancer therapy. Nanotechnology 2009;20(47), 475102.
[274] Tavri S, et al. Optical imaging of cellular immunotherapy against prostate cancer. Mol Imaging 2009;8(1):15–26.
[275] Deguine J, et al. Intravital imaging reveals distinct dynamics for natural killer and CD8(+) T cells during tumor regression. Immunity 2010;33(4):632–44.
[276] Beuneu H, et al. Dynamic behavior of NK cells during activation in lymph nodes. Blood 2009;114(15):3227–34.
[277] Bajénoff M, et al. Natural killer cell behavior in lymph nodes revealed by static and real-time imaging. J Exp Med 2006;203(3):619–31.
[278] Daldrup-Link HE, et al. In vivo tracking of genetically engineered, anti-HER2/neu directed natural killer cells to HER2/neu positive mammary tumors with magnetic resonance imaging. Eur Radiol 2005;15(1):4–13.
[279] Meier R, et al. Depicting adoptive immunotherapy for prostate cancer in an animal model with magnetic resonance imaging. Magn Reson Med 2011;65(3):756–63.
[280] Mallett CL, et al. Migration of iron-labeled KHYG-1 natural killer cells to subcutaneous tumors in nude mice, as detected by magnetic resonance imaging. Cytotherapy 2012;14(6):743–51.
[281] Sheu AY, et al. MRI-monitored transcatheter intra-arterial delivery of SPIO-labeled natural killer cells to hepatocellular carcinoma: preclinical studies in a rodent model. Investig Radiol 2013;48(6):492–9.
[282] Bouchlaka MN, et al. (19)F-MRI for monitoring human NK cells in vivo. Onco Targets Ther 2016;5(5), e1143996.
[283] Su Z, et al. MRI-guided interventional natural killer cell delivery for liver tumor treatment. Cancer Med 2018;7(5):1860–9.
[284] Meier R, et al. Tracking of [18F]FDG-labeled natural killer cells to HER2/neu-positive tumors. Nucl Med Biol 2008;35(5):579–88.
[285] Melder RJ, et al. A method for labeling cells for positron emission tomography (PET) studies. J Immunol Methods 1994;175(1):79–87.
[286] Sato N, et al. In vivo tracking of adoptively transferred natural killer cells in rhesus macaques using (89)zirconium-oxine cell labeling and PET imaging. Clin Cancer Res 2020;26(11):2573–81.
[287] Shaffer TM, et al. PET imaging of the natural killer cell activation receptor NKp30. J Nucl Med 2020;61(9):1348–54.

[288] Matera L, et al. In vivo migration of labeled autologous natural killer cells to liver metastases in patients with colon carcinoma. J Transl Med 2006;4:49.
[289] Meller B, et al. Monitoring of a new approach of immunotherapy with allogenic (111)In-labelled NK cells in patients with renal cell carcinoma. Eur J Nucl Med Mol Imaging 2004;31(3):403–7.
[290] Galli F, et al. In vivo imaging of natural killer cell trafficking in tumors. J Nucl Med 2015;56(10):1575–80.
[291] Galli F, Histed S, Aras O. NK cell imaging by in vitro and in vivo labelling approaches. Q J Nucl Med Mol Imaging 2014;58(3):276–83.
[292] Varani M, et al. State of the art of natural killer cell imaging: a systematic review. Cancers (Basel) 2019;11(7).
[293] Davis 2nd HH, et al. 111Indium-labeled human alveolar macrophages and monocytes: function and ultrastructure. J Immunol Methods 1980;36(2):99–107.
[294] Fisher B, et al. Tumor localization of adoptively transferred indium-111 labeled tumor infiltrating lymphocytes in patients with metastatic melanoma. J Clin Oncol 1989;7(2):250–61.
[295] Marienhagen J, et al. 111In-oxine labelling of tumour-cytotoxic macrophages generated in vitro from circulating blood monocytes: an in vitro evaluation. Nucl Med Commun 1995;16(5):357–61.
[296] Mackensen A, et al. Homing of intravenously and intralymphatically injected human dendritic cells generated in vitro from CD34+ hematopoietic progenitor cells. Cancer Immunol Immunother 1999;48(2–3):118–22.
[297] Aarntzen EH, et al. Targeting of 111In-labeled dendritic cell human vaccines improved by reducing number of cells. Clin Cancer Res 2013;19(6):1525–33.
[298] De Vries IJ, et al. Effective migration of antigen-pulsed dendritic cells to lymph nodes in melanoma patients is determined by their maturation state. Cancer Res 2003;63(1):12–7.
[299] Lesterhuis WJ, et al. Route of administration modulates the induction of dendritic cell vaccine-induced antigen-specific T cells in advanced melanoma patients. Clin Cancer Res 2011;17(17):5725–35.
[300] Ridolfi R, et al. Evaluation of in vivo labelled dendritic cell migration in cancer patients. J Transl Med 2004;2(1):27.
[301] de Vries IJ, et al. Magnetic resonance tracking of dendritic cells in melanoma patients for monitoring of cellular therapy. Nat Biotechnol 2005;23(11):1407–13.
[302] Lee HW, et al. Tracking of dendritic cell migration into lymph nodes using molecular imaging with sodium iodide symporter and enhanced firefly luciferase genes. Sci Rep 2015;5(1):9865.
[303] Lee SB, et al. Tracking dendritic cell migration into lymph nodes by using a novel PET probe (18)F-tetrafluoroborate for sodium/iodide symporter. EJNMMI Res 2017;7(1):32.
[304] Kim HS, et al. In vivo tracking of dendritic cell using MRI reporter gene, ferritin. PLoS ONE 2015;10(5), e0125291.
[305] Pham W, Xie J, Gore JC. Tracking the migration of dendritic cells by in vivo optical imaging. Neoplasia 2007;9 (12):1130–7.
[306] Christian NA, et al. In vivo dendritic cell tracking using fluorescence lifetime imaging and near-infrared-emissive polymersomes. Mol Imaging Biol 2009;11(3):167–77.
[307] Lee SB, et al. Radionuclide-embedded gold nanoparticles for enhanced dendritic cell-based cancer immunotherapy, sensitive and quantitative tracking of dendritic cells with PET and *Cerenkov luminescence*. NPG Asia Mater 2016;8(6):e281.
[308] Piao YJ, Kim HS, Moon WK. Noninvasive photoacoustic imaging of dendritic cell stimulated with tumor cell-derived exosome. Mol Imaging Biol 2020;22(3):612–22.
[309] Schimmelpfennig CH, et al. Ex vivo expanded dendritic cells home to T-cell zones of lymphoid organs and survive in vivo after allogeneic bone marrow transplantation. Am J Pathol 2005;167(5):1321–31.
[310] Stoll S, et al. Dynamic imaging of T cell-dendritic cell interactions in lymph nodes. Science 2002;296 (5574):1873–6.
[311] Cherry SR, et al. Total-body PET: maximizing sensitivity to create new opportunities for clinical research and patient care. J Nucl Med 2018;59(1):3–12.
[312] Shao Y, et al. Simultaneous PET and MR imaging. Phys Med Biol 1997;42(10):1965–70.
[313] Slates RB, et al. A study of artefacts in simultaneous PET and MR imaging using a prototype MR compatible PET scanner. Phys Med Biol 1999;44(8):2015–27.
[314] Badawi RD, et al. First human imaging studies with the EXPLORER total-body PET scanner. J Nucl Med 2019;60 (3):299–303.

[315] Wei H, Jiang H, Song B. Role of medical imaging for immune checkpoint blockade therapy: from response assessment to prognosis prediction. Cancer Med 2019;8(12):5399–413.
[316] Sun R, et al. A radiomics approach to assess tumour-infiltrating CD8 cells and response to anti-PD-1 or anti-PD-L1 immunotherapy: an imaging biomarker, retrospective multicohort study. Lancet Oncol 2018;19(9):1180–91.
[317] Durot C, et al. Metastatic melanoma: pretreatment contrast-enhanced CT texture parameters as predictive biomarkers of survival in patients treated with pembrolizumab. Eur Radiol 2019;29(6):3183–91.

CHAPTER

11

Radiologic assessment of tumor response to immunotherapy and its complications

Leonid Roshkovan[a] *and Sharyn I. Katz*[b,*]

[a]Division of Abdominal Imaging, Department of Radiology, University of Pennsylvania Perelman School of Medicine, Philadelphia, PA, United States [b]Division of Thoracic Radiology, Department of Radiology, University of Pennsylvania Perelman School of Medicine, Philadelphia, PA, United States

Abstract

Imaging plays a pivotal role in oncologic management. The introduction of immunotherapeutics into routine cancer care has resulted in major therapeutic breakthroughs, however, requires special consideration in use and interpretation of medical imaging.

Abbreviations

ARDS	adult respiratory distress syndrome
CT	computed tomography
CTLA4	cytotoxic T-lymphocyte-associated protein 4
CRS	cytokine release syndrome
FDG-PET	^{18}F-fluoro-deoxyglucose-positron emission tomography
HLH	hemophagocytic lymphohistiocytosis
ICI	immune-checkpoint inhibitor
irRC	immune-related response criteria
irRECIST	immune-related response criteria
iRECIST	immune response evaluation criteria in solid tumors, version 1.1
LYRIC	lymphoma response to immunomodulatory therapy criteria
MAS	macrophage activation syndrome

*Present address: Department of Radiology, Hospital of the University of Pennsylvania, 1 Silverstein Bldg., 3400 Spruce Street, Philadelphia, PA 19104, United States.

https://doi.org/10.1016/B978-0-12-822620-9.00019-7

mRECIST1.1	modified response evaluation criteria in solid tumors, version 1.1
MRI	magnetic resonance imaging
PD-1	programmed death 1
MTV	metabolic tumor volume
PDL-1	programmed death ligand 1
PERCIST	PET response criteria in solid tumors
imPERCIST	immunotherapy-modified PET RECIST
RECIST1.1	response evaluation criteria in solid tumors, version 1.1
PSMA	prostate-specific membrane antigen
SPECT	single-photon emission computed tomography
SUL	standardized uptake value corrected for lean body mass
SUV	standard uptake value

Conflict of interest

No potential conflicts of interest were disclosed.

Introduction

Medical imaging plays a central role in cancer management including in the setting of immunotherapy (Fig. 1). Different imaging modalities have relative strengths and weakness that result in their allocation to specific roles in oncology. At diagnosis, imaging assessment of disease extent allows optimal treatment planning. Once therapy is initiated, imaging is used to assess tumor response and clinically suspected complications. Immunotherapy incites an inflammatory response that is therapeutic at the sites of disease but can confound conventional imaging interpretation and may result in off-target inflammation. This characteristic of immunotherapy has led to current exploration of novel approaches to assess tumor response on imaging.

Immune checkpoint inhibitors are by far the most widely used immunotherapeutics in clinical use for cancer management. The widespread integration of these agents into the management of nearly every malignancy has provided much information on patterns of response and potential pitfalls of imaging interpretation. At the time of this writing only a handful of CAR T-cell therapies have been FDA approved for use in human cancer; however, this is likely to increase in the future. While we have learned much from the clinical trials of CAR T-cell therapy, our widespread experience with ICI therapy may provide insights into the strengths and weakness of commonly used imaging modalities and unconventional patterns of response that may be associated with favorable responses to immunotherapeutics.

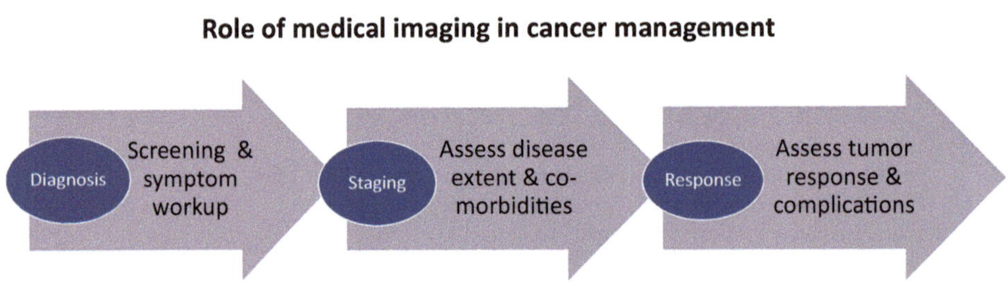

FIG. 1 **Role of medical imaging in cancer management.** Imaging plays a key role in cancer detection, staging, and during the course of therapy.

Uses of medical imaging in immunotherapy

Upon diagnosis of malignancy, patients undergo an assessment of the degree of local tumor invasion, potential lymph node involvement by malignancy, and presence of distant metastasis. This diagnostic process is called "staging" and serves the primary purpose of guiding patient therapy to determine if the patient is a surgical candidate or would likely benefit more from systemic therapy, such as immunotherapy, potentially in combination with radiation therapy. In addition, imaging is used to guide optimal target selection for tissue biopsy allowing histological and genetic expression analysis.

Computed tomography (CT) is the primary imaging tool for staging due to high resolution, speed of imaging acquisition and widespread availability [1]. CT images are generated by complex reconstruction algorithms that convert the degree of tissue absorption of ionizing radiation into an image that is essentially a map of tissue densities. However even given the high degree of spatial resolution on CT, subtle metastasis that do not generate enough architectural distortion in the surrounding tissues may be missed. For this reason, staging CT is often supplemented with ^{18}F-fluoro-deoxyglucose-positron emission tomography (FDG-PET) providing a synergy between the high resolution anatomic imaging on CT and highly sensitive metabolic imaging on PET [2]. FDG, which is an analog of glucose, accumulated rapidly in most cancers due to a high avidity for glucose, a result of metabolic reprogramming known as the Warburg effect [3]. This results in high sensitivity of FDG-PET for most tumors and allows detection of metastasis that are occult on CT alone.

While FDG is the primary tracer used for routine PET imaging in oncology, there are other tracers available and under development that are designed to target the activity of other aspects of tumor metabolism. For example, Gallium -PSMA-11 is used as an imaging agent for targeting prostate cancer and Gallium 68-dotatate is used for the detection of neuroendocrine tumors [4,5].

Magnetic resonance imaging (MRI) is typically used to assess anatomic structures that are limited on CT or FDG-PET (such as the brain and soft tissues) or to assist in characterizing indeterminate findings, though some have advocated for the use of whole body MRI, potentially fused with FDG-PET, as an alternative approach to oncologic staging [6]. MRI uses the differential magnetism of tissues that results from tissue content of water, protein, fat and metals (for example in blood products) to generate anatomical imaging. This results in a major advantage in tissue contrast and the ability to detect subtle differences in the composition of tissues. The disorganized, vascular nature of most malignancies results in differential magnetism due to abnormal features such as densely packed cells, surrounding edema and immature blood vessels. MRI is routinely used in staging of the brain for primary or metastatic brain neoplasms [7]. This is because the brain is normally diffusely, intensely FDG avid [8], which renders FDG-PET limited while the tissue contrast of the brain is moderate on CT (Fig. 2).

In addition to staging malignancy, medical imaging is instrumental in assessing the response of cancer to therapy including immunotherapy. This is primarily done on CT imaging with FDG-PET and MRI acting as adjunct imaging to characterize indeterminate tissue suspected to represent recurrent or metastatic disease. Baseline imaging acquired during the staging process is compared to subsequent imaging obtained at selected timepoints in the course of therapy. Change in tumor lesion size and the presence of new lesions are

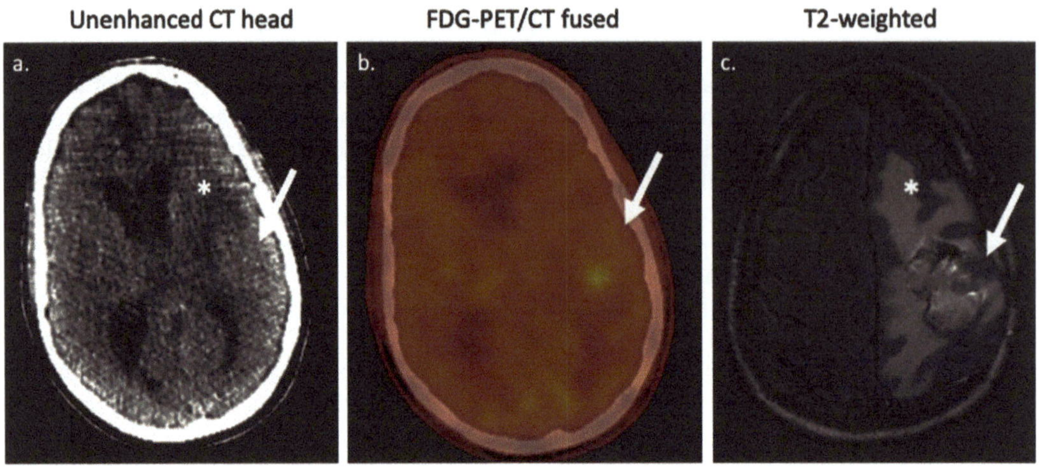

FIG. 2 **Multimodality approach to medical imaging.** The relative strengths and weakness of the available imaging modalities lead to differing roles in the management of cancer. For example, in this patient with a malignancy in the left parietal lobe (*white arrow*), the mass is less well characterized on unenhanced CT (A) since the attenuation of the mass is similar to that of the adjacent normal brain tissue. The conspicuity of this mass on CT could be improved with the administration of intravenous contrast. By comparison fused images of unenhanced CT and FDG-PET (B) in the same patient taking advantage of the elevated FDG avidity of the tumor relative to the normal adjacent brain. However, the brain is diffusely FDG avid which limits its utility in the brain. MRI (C) more clearly depicts the mass and surrounding tissue edema (*white asterisk*) due to the excellent tissue contrast inherent in this imaging modality.

assessed and the data is used to decide whether the patient is likely to benefit most from either continued therapy or a switch in therapy. A systematic approach to this practice of therapy response assessment on imaging has been standardized in the setting of clinical trials.

Challenges to medical imaging presented by immunotherapy

Cancer immunotherapeutics have a common endpoint of stimulating the host immune antitumor activity. While there are a number of approaches to achieve this common goal, including adoptive T-cell transfer therapy, in the last decade immune checkpoint inhibitors (ICIs) have become dominant in the field. ICIs have been rapidly integrated into the standard of care in multiple types of malignancies either as a standalone therapy or in combination with chemotherapy and/or radiation therapy. Anti-CTLA4 therapy, which works by amplifying the T cell response to stimulation by antigen in the context of MHC, was the first ICI to demonstrate marked success in melanoma with remarkable and durable responses, even in advanced disease [9]. This was followed by successes in a variety of malignancies with the anti-PD1/PDL1 class of ICIs which work by disabling the PDL1-PD1 interaction thereby unmasking the tumors to the host immune system [9]. Both approaches have been integrated in routine oncologic management. However, the inflammatory-producing mechanism of these drugs also presents unique challenges to imaging assessment of tumor therapy response including unconventional patterns of responses (Table 1). Since we have the most clinical experience with ICIs as immunotherapeutics in routine cancer management, we will

TABLE 1 Unconventional patterns of response to immunotherapy.

Unconventional patterns of response to immunotherapy	
Pseudo progression	Initial significant enlargement and/or development of new lesions with subsequent regression of disease on sustained therapy
Hyper progression	Rapid progression
Dissociated response	Two distinct patterns of response of tumor lesions with both increase and decrease in size

discuss what is known about the ICI therapy profile noting that there is much overlap with the challenges in pitfalls with this class of agents and other immunotherapeutics. We will then discuss the approaches to imaging adoptive T cell therapeutics.

Unconventional patterns of response in immunotherapy

Pseudo progression

A major potential confounder to imaging of therapy response is radiologic pseudo progression where patients exhibit an unconventional pattern of response characterized by initial tumor enlargement and even development of new lesions, with subsequent regression of tumor on sustained therapy [10,11]. This is thought to occur due to therapy-induced inflammatory infiltration of tumor tissue causing initial apparent enlargement on imaging (Fig. 3). With eventual tumor lysis, the tumor decreases in size demonstrating clear morphological

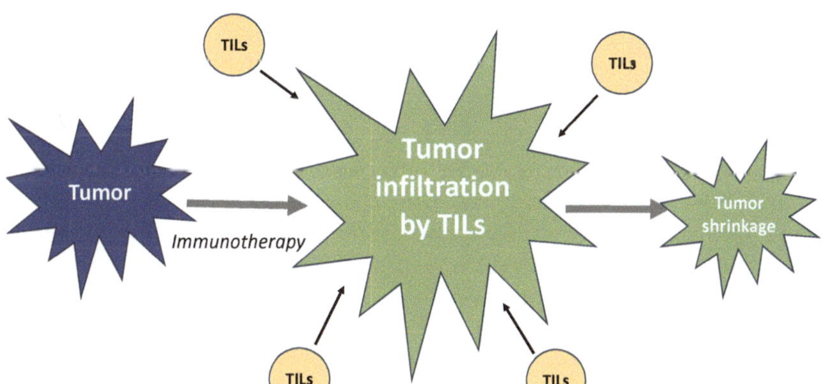

FIG. 3 **Radiologic pseudo progression during immunotherapy.** The successful immune targeting of the tumor by the host immune system results in infiltration of the tumor by tumor-associated infiltrating lymphocytes (TILs). In a subset of patients, this inflammatory response can result in morphological enlargement of the tumor that can be misinterpreted as progression. Sustained therapy results in tumor cytolysis and morphological decrease in size of the tumor lesion.

response. This unconventional pattern of response was first described in the setting of melanoma being treated with anti-CTLA4 therapy but is also observed with anti-PD1/PDL1 and other classes of immunotherapy including CAR T cell therapeutics. The incidence of pseudo progression varies by tumor type and ICI therapeutic with an incidence of 5%–13% of cancer patients receiving checkpoint inhibitor therapy [11,12] and may be more frequent in the younger population. Most commonly pseudo progression appears in early stages of therapy, usually occurring within the first 4 months after initiation of therapy though can be observed later [12]. Some literature suggests that patients experiencing pseudo progression may show overall better outcomes [13]. It must be emphasized, however, that the majority of patients showing increase in tumor burden on follow-up imaging are far more likely to have true progressive disease rather than pseudo progression [14].

Pseudo progression has also been reported in CAR T-cell therapy where it is typically more pronounced and rapid in development compared to pseudo progression in response to ICI therapy [15]. This may be due to the type of cancer being treated by current CAR T-cell therapies, currently primarily hematologic malignancies, or potentially a characteristic of CAR T-cell therapy. Differentiation between pseudo progression and true progression is particularly important for CAR T-cell therapy since salvage would include chemotherapies that may impede the efficacy of CAR T-cell therapy. As yet there is no reliable way to differentiate between true progression and pseudo progression based on imaging done at a single timepoint.

Hyperprogression

Hyperprogression is another unconventional pattern observed in the setting of immunotherapy and is characterized by a rapid increase in tumor growth after initiation of immunotherapy [10] (Fig. 4). The mechanism(s) of this devastating response to immunotherapy is currently unknown. Hyperprogression has been described in the setting of anti-PDL1 therapy though is not a unique phenomenon to immunotherapy and has been identified with other types of interventions. Hyperprogression in the setting of immune checkpoint inhibition is observed to be generally uncommon with the exception of reported incidence of up to 10% in some therapeutic settings. As yet there is no way to predict which patients will undergo hyperprogression [16].

Dissociated response

Another atypical pattern of response, first observed in nonsmall cell lung cancer patients treated with anti-PD-1/PDL-1 therapy, is called dissociated response. The term dissociated response refers to a pattern of mixed response characterized by the co-occurrence of two distinct response patterns where some lesions respond favorably to the therapy while others do not. The exact incidence of a dissociative response varies in the literature between 3% and 48% depending on tumor type and precise definition of dissociative response employed. Identification of the presence of a dissociative response is important since there is evidence that this response type may have a greater likelihood of a favorable prognosis, including increased survival and treatment efficacy, relative to other types of progression. When dissociative response is identified, patients may be benefit from being maintained on ICI therapy with potential local ablative therapy or surgical resection of the progressing lesions [17].

Finally, ICI therapy has been observed to result in persistent durable tumor response that can be maintained for up to years after stopping therapy in a subset of patients. Durable

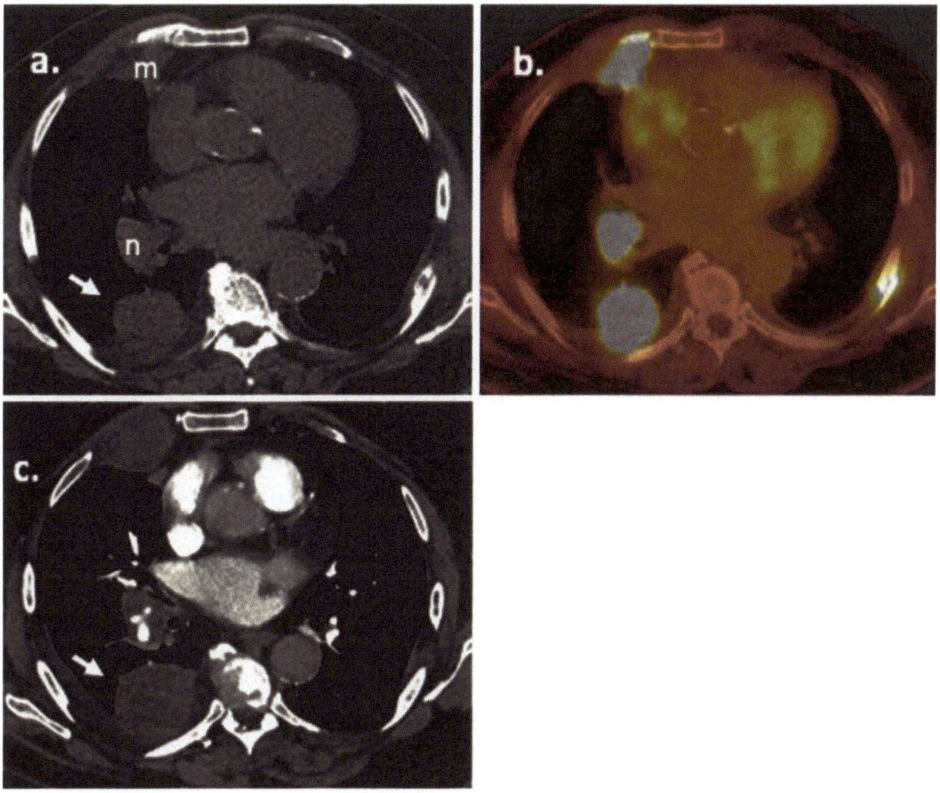

FIG. 4 **Hyper progression on immune checkpoint inhibitor therapy.** Unenhanced axial CT (A) and axial fused CT and FDG-PET (B) images obtained at baseline in a 72-year-old male with nonsmall cell lung cancer depict an FDG avid mass (*white arrow*) in the right lower lobe with pleural metastasis (*m*) and ipsilateral nodal metastasis (*n*). At first follow-up axial contrast enhanced CT image (C) after starting anti-PD1 therapy with pembrolizumab, there has been progression of disease with significantly increased size of the primary malignancy and metastatic disease (*white arrow*).

response (DR) was recently defined as "a progression-free survival that exceeded three times the median progression-free survival of the whole population." While a durable response is not unique to immunotherapy, it occurs more frequently with immunotherapy than with other systemic regimens. Among ICIs, durable responses are seen more frequently in anti-PD-1/PD-L1 therapies than anti-CTLA-4 therapies [18]. Durable responses are observed at a high incidence with some CAR T-cell therapies, for example, refractory Hodgkin's lymphoma, B-cell lymphoma, follicular lymphoma, and acute lymphoblastic leukemia [19–22].

Immune-related adverse events

Another aspect of immunotherapy that can confound radiologic interpretation is off-target activity of the amplified immune response. The tissues most affected by this off-target activity typically share an expression profile with the intended target of therapy though any tissue can be affected by this elevated immune activity. ICIs are in widespread use in the clinical practice of oncology, and immune-related adverse events are commonly observed with these agents. Immune-related adverse events with CAR T-cell therapy has shown a distinct pattern of side effects relative to other immunotherapies such as ICI therapy. However, this profile is likely to evolve as these agents become more prevalent and the number of CAR T-cell targets increase. Currently, there are only four CAR T-cell therapies approved for clinical use and they all target a single antigen, CD19, for application in extrahemopoietic tumor in hematologic malignancies [15].

Immune checkpoint inhibitors

The augmented immune response in anti-CTLA4 therapy or disabled immune tolerance to host tissue in anti-PDL1/PD1 therapy result in inflammation in normal tissues such as skin, lung, and bowel which are the most affected organs with this class of therapeutics. The incidence of adverse events with these agents is not insignificant with some degree of off-target inflammatory activity seen in up to 80% of patients on anti-CTLA4 therapy, particularly if in combination with another ICI, and 20% of patients on anti-PDL1/PD-1 agents [23]. In the setting of ICI therapy, approximately three-quarters of radiologically evident adverse events occur within the first 3 months of therapy though can manifest later in therapy [24]. This off-target inflammation can hinder detection of new metastasis and assessment of response of existing disease on CT. In addition, since inflammation results in an increased glucose avidity in activated immune cells, these off-target foci of inflammation can mimic tumor on FDG-PET [25]. For example, lymph node inflammation incited by immune checkpoint inhibition can be mistaken for metastasis. If clinically significant, these adverse events can be treated with immunosuppression, and in more severe disease with discontinuation of the immunotherapy.

Immune-related adverse effects related to ICI therapy can essentially involve any organ system with a predilection for skin, gastrointestinal tract, and lung. In the gastrointestinal tract, this manifest as colitis on CT (Fig. 5) or MRI with findings of diffuse or segmental wall thickening, vascular engorgement, mucosal hyperemia, and surrounding fat stranding. These

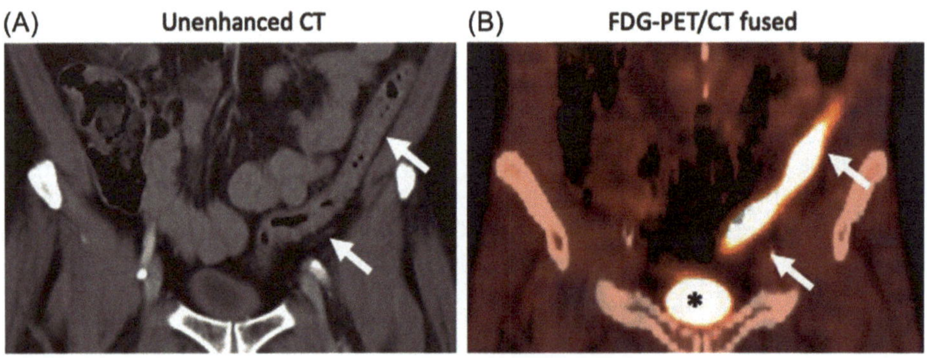

FIG. 5 **Immune checkpoint inhibitor-induced colitis.** Coronal unenhanced CT (A) and fused FDG-PET and CT images (B) reveal segmental thickening associated with intense FDG avidity in the sigmoid colon in this 53-year-old patient with poorly differentiated adenocarcinoma of the lung on durvalumab following therapy with cisplatin and docetaxel and radiation to the lung. This is in keeping with durvalumab induced colitis, a common adverse event associated with ICI therapy.

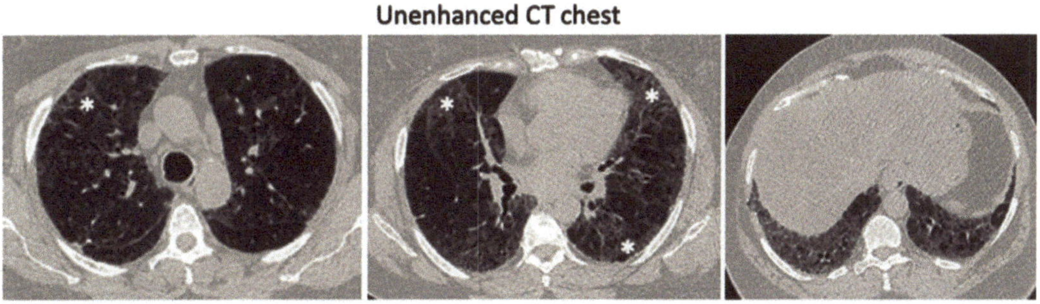

FIG. 6 **Immune checkpoint inhibitor-induced pneumonitis.** Unenhanced CT images of a 62-year-old female with renal cell carcinoma at 10 weeks of anti-PD1 therapeutic pembrolizumab. Images depict ill-defined patchy areas of ground glass opacity in the lungs. This is a common pattern of ICI therapy-induced pneumonitis on CT.

imaging findings can create a diagnostic dilemma since they are nonspecific in nature and can be the result of infection or ischemia in addition to immune-mediated colitis in patients undergoing immunotherapy [26]. Immune-mediated lung toxicity occurs in up to 3%–5% of patients treated with anti-PD1/anti-PDL1 therapy monotherapy [14]. Imaging presentation ranges between minimal inflammatory groundglass opacities (Fig. 6), more extensive evidence of organizing pneumonia pattern of disease with classically peripheral distribution of groundglass opacities, alveolar consolidations, and peribronchial interstitial opacities (Fig. 7). Another pattern of ICI-induced pneumonitis can manifest as a sarcoid-like reaction, which can occur in up to 5% of patients on these agents [27]. Imaging findings include mediastinal and bilateral hilar lymphadenopathy, with or without associated perilymphatic nodularity which is best appreciated on CT. Sarcoid-like reaction to ICI therapy can also result

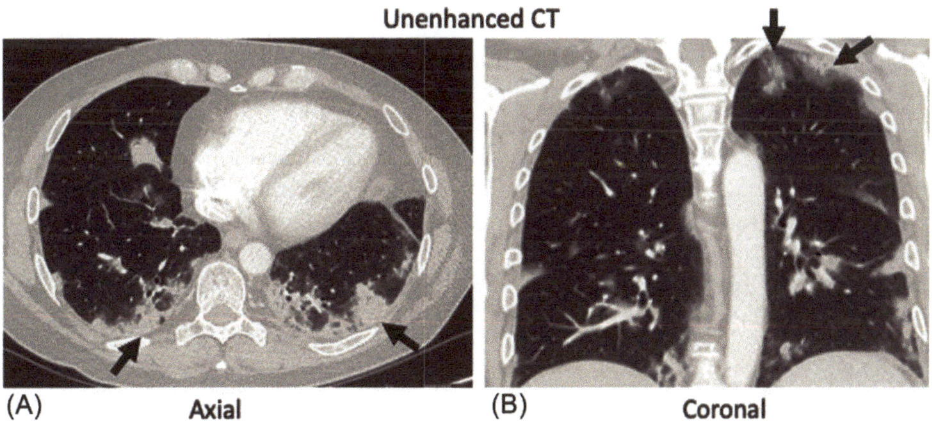

FIG. 7 **Immune checkpoint inhibitor-induced pneumonitis.** Intravenous contrast-enhanced CT images of the chest were obtained in a 58-year-old patient with a squamous cell carcinoma of the floor of the mouth on pembrolizumab therapy. Axial (A) and coronal (B) images reveal a pattern of peripherally distributed foci of consolidation and ground glass typical of organizing pneumonia. This is a manifestation of ICI drug toxicity.

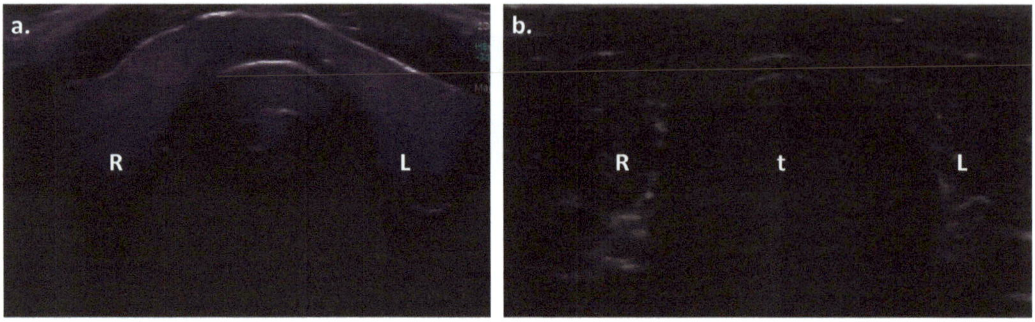

FIG. 8 **Immune checkpoint inhibitor-induced thyroiditis.** Axial images of the thyroid gland are obtained with gray-scale ultrasound imaging in a healthy patient (A) and 51-year-old female with follicular lymphoma and renal cell carcinoma on therapy with ICIs ipilimumab and nivolumab (B). In the healthy thyroid gland, there is homogenous high echogenicity. In the case of thyroiditis, there is diffusely low echogenicity of the thyroid gland, which is typically seen in the setting of inflammation. *R*, right thyroid lobe; *L*, left thyroid gland; *t*, trachea.

in increased nodal size and FDG avidity on PET imaging, which can present a challenge in differentiating drug reaction from progressive disease.

Less commonly, the off-target activity of ICIs can induce inflammation in the endocrine system such as thyroiditis, which is most frequently observed in patients on combination therapy of two ICIs [28]. Thyroiditis can be difficult to appreciate on CT but can appear as diffusely increased uptake on FDG PET/CT and heterogenous or diffusely decreased echogenicity and hypervascularity on ultrasound [29] (Fig. 8). Similarly, immune-related hypophysitis appears as focally increased hypophyseal avidity on FDG-PET/CT and as an enlarged, enhancing pituitary gland with pituitary stalk thickening on MRI (Fig. 9). Immune-related hypophysitis is uncommon though most described with anti-CTLA4 therapy, The pituitary gland often returns to its baseline size within several months [30].

Finally, immune activation incited by ICIs can potentiate the toxicity of other therapies, notably radiation therapy. Patients treated with concurrent immunotherapy and radiotherapy can result in increased rates of radiation-induced pneumonitis [31]. In addition, subclinical changes of previous radiation-induced lung injury can be enhanced by the subsequent administration of immunotherapy, a phenomenon known as radiation-recall pneumonitis [32]. There is also evidence that immunotherapy can enhance the treatment effect of radiation therapy to tumor tissue not in the field of therapy which is known as the abscopal effect [33]. This is thought to be mediated through an enhanced immunotherapeutic response primed by the release of tumor antigens during radiation therapy.

CAR T cell therapies

There are several potentially severe adverse effects that can occur with CAR T-cell therapy (Table 2). One such commonly observed adverse effect with CAR T-cell therapy, not typically seen with ICI therapy, is cytokine release syndrome (CRS). While CRS can occur with any immunotherapy, it is most common with CAR T-cell therapy and also observed with

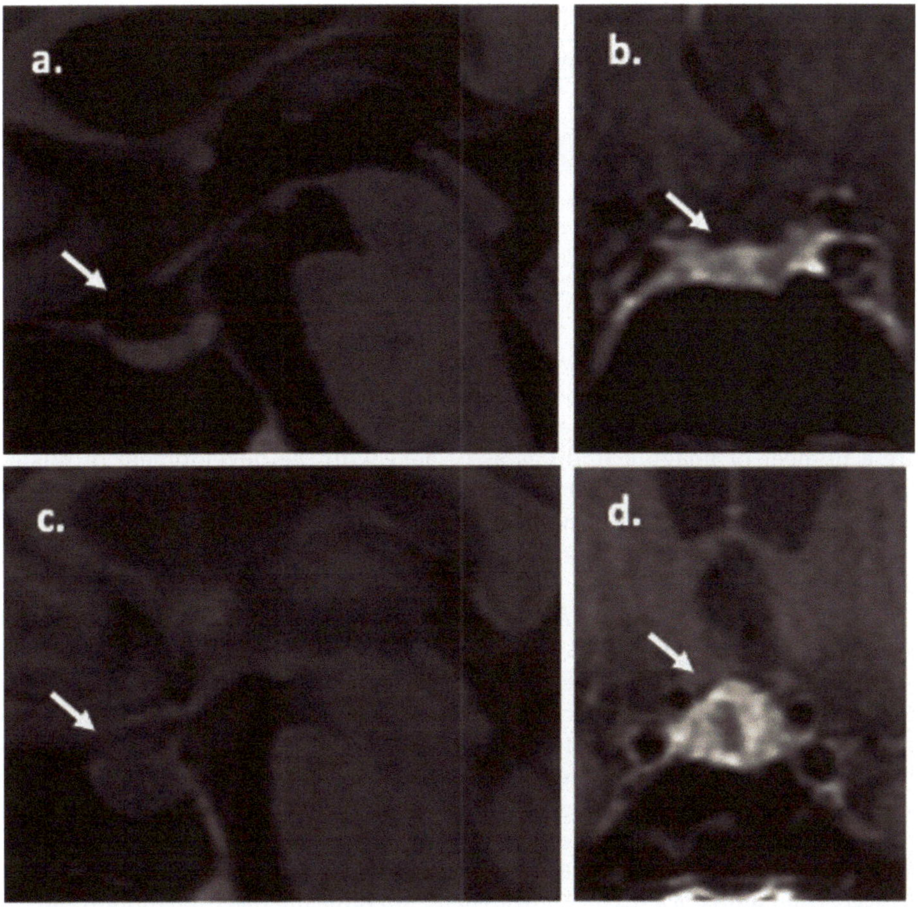

FIG. 9 **Immune checkpoint inhibitor-induced hypophysitis.** 77-year-old woman with metastatic melanoma treated with Ipilimumab and nivolumab. Pretreatment MRI demonstrated normal size and enhancement pattern of the hypophysis on sagittal unenhanced T1-weighted (A) and coronal IV contrast enhanced T1-weighted (B) images. Follow-up MRI sagittal unenhanced T1-weighted (C) and coronal IV contrast enhanced T1-weighted (D) images obtained after 3 cycles of ICI immunotherapy revealed an enlargement and heterogeneously hyperenhancement of the hypophysis (*white arrows*) consistent with hypophysitis. ICI therapy was subsequently halted and the inflammation resolved.

bispecific T-cell engager antibodies. CRS is due to a release of a burst of proinflammatory cytokines, triggered by IL-6 and IL-1, which occurs in response to T cell activation, proliferation, and cytokine release. This cytokine release leads to further stimulation of an exaggerated inflammatory response, potentially leading to hypotension, shock, acute renal failure, ARDS, cardiac arrhythmias, capillary leak syndrome, disseminated intravascular coagulation, and death [34]. Some degree of CRS occurs in 60%–93% of all patients receiving CAR T-cell therapy and is usually mild with approximately 13%–14% patients experiencing severe CRS (grade 3 or higher) [34]. CRS typically occurs within the first 7 days of therapy though can

TABLE 2 Classes of cancer response to therapy.

Class of cancer response to therapy	
Complete response (CR)	Complete resolution of disease
Durable response (DR)	Sustained complete resolution of disease
Partial response (PR)	Significant ($\geq$30%) decrease in tumor lesions
Stable disease (SD)	Disease change in between PR and PD
Progressive disease (PD)	Significant ($\geq$20%) increase in tumor lesions

potentially occur up to 14 days following the administration of CAR T-cells. Risks factors for developing CRS include a large tumor burden, large doses of CAR T-cells, and acute lymphoblastic leukemia as the underlying malignancy [35]. Since CRS can have a multiorgan impact, there are a number of imaging findings associated with the development of this adverse event. In the lungs, increased capillary permeability results in pulmonary edema and pleural effusions and can progress to ARDS with more severe grades of CRS (Figs. 10 and 11). In the kidneys, delayed clearance of IV contrast may be observed due to acute renal failure and along with soft tissue edema [15].

Neurotoxicity is associated with CAR T-cell therapy and is relatively common, occurring in up to 64% of patients receiving CAR T-cell therapy [35] and typically occurs within the first week after administration of CAR T-cell therapy though has been reported 3–4 weeks following CAR T-cell infusion [35]. This mechanism for this neurotoxicity is not well understood but is thought to result from breakdown of the blood-brain barrier and activation of microglial cells resulting in robust inflammation which appears to be instigated by proinflammatory cytokines such as IL-6 [15]. These changes can manifest as mental status changes, seizures, impaired attention, memory loss, and expressive aphasia. Symptoms are typically reversible, however, severe cases can progress to cerebral edema, coma, and death [15,35,36]. Imaging findings associated with CAR T-cell neurotoxicity are typically negative though can manifest as cerebral edema (Fig. 11). Other imaging findings that have been reported include leptomeningeal enhancement and microhemorrhages [37] and signal abnormalities in the basal ganglia, cingulate gyrus, hippocampus, and splenium of the corpus callosum [38].

Finally, an uncommon but serious adverse effect of CAR T cell therapy is termed Hemophagocytic Lymphohistiocytosis (HLH)/Macrophage Activation Syndrome (MAS) and can occur in up to 3% of patients undergoing CAR T-cell therapy [39]. This is characterized by immune hyperactivation, primarily involving macrophages resulting in excess cytokine release, and phagocytization of host cells which are often hematologic in origin. When this syndrome is associated with autoimmunity, it is called macrophage activation syndrome (MAS) [15]. Clinically, HLH/MAS can be difficult to differentiate from CRS since both can result in a systemic inflammatory-like response, though tends to occur slightly later than CRS. Imaging findings most commonly include hepatosplenomegaly, pulmonary opacities, pleural effusions and brain parenchymal changes detectable on MRI including cerebral edema, hemorrhage, and white matter changes [15].

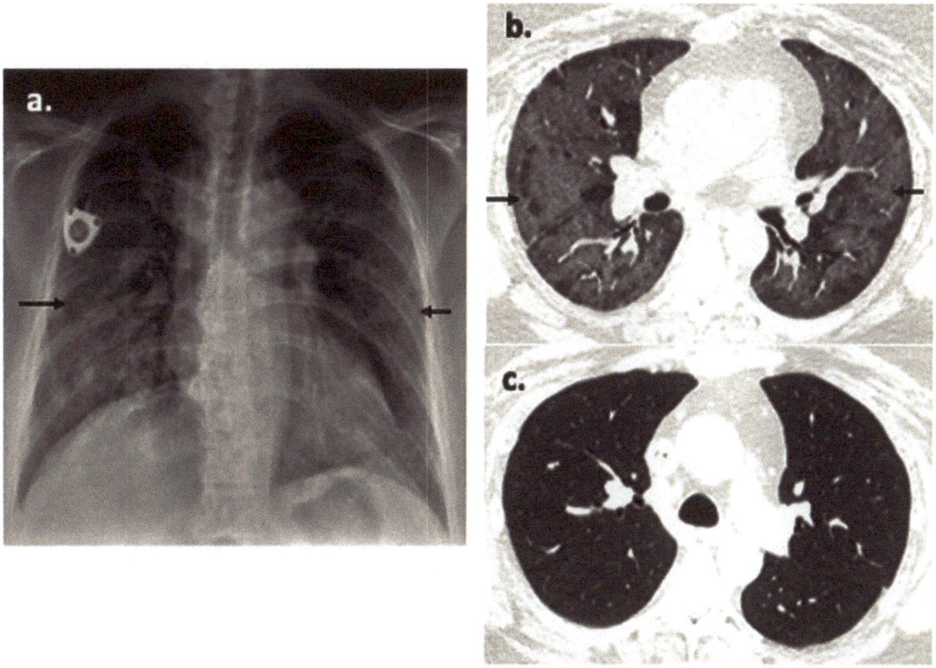

FIG. 10 **CAR T cell-mediated cytokine release syndrome.** 59-year-old patient with stage IIIC recurrent papillary serous ovarian cancer treated with CART therapy. The patient subsequently developed fever and hypoxia and was diagnosed with grade 3 cytokine release syndrome (CRS) at day 5 following CAR T-cell infusion. Chest radiograph at 5 day following the 1st CART infusion (A) demonstrates subtle diffuse pulmonary opacities suggestive of pulmonary edema. Axial unenhanced CT images obtained the same day (B) confirmed bilateral diffuse ground glass opacities typical of cytokine release syndrome. The symptoms of CRS resolved on Tocilizumab and dexamethasone and the patient was discharged home 1 week later. A follow-up axial unenhanced CT (C) demonstrated complete resolution of lung opacities.

Methods of measurement of tumor response

Accurate and consistent approaches to measurement of disease are critical to achieve the best patient outcomes in oncology. For clinical trials designed to assess the effectiveness of systemic therapy, there has been development of a number of standardized systems for tumor measurement of disease and are referred to as tumor response criteria. These tumor response criteria are tailored to assess behavior of tumor growth and response to therapy providing metrics used to assess the benefit of a given therapy. This is done by taking measurements of specific tumor lesions at baseline and comparing the size of these lesions over time noting the appearance of new lesions. While these criteria are not typically employed formally in every day oncologic practice, the general principles of measurement are used routinely by radiologists and considered best practice. Therefore, an understanding these criteria provides insight into how tumors are measured and therapy response assessed. Immunotherapeutics

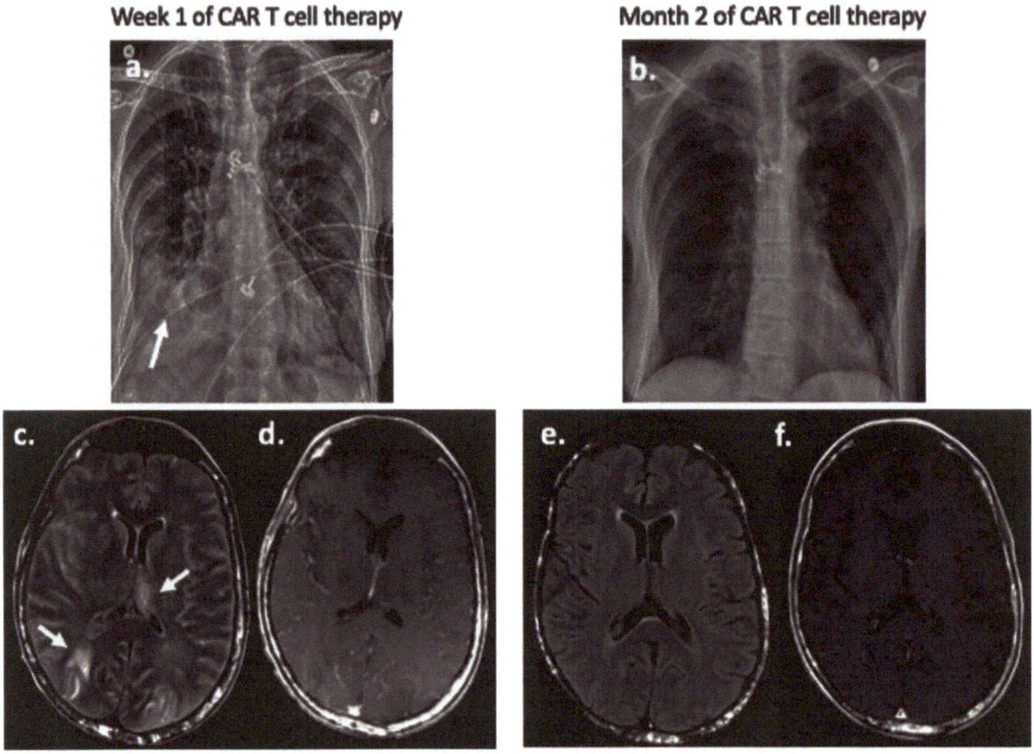

FIG. 11 **CAR T cell-mediated cytokine release syndrome.** 61-year-old female with history of multiple myeloma treated with CART therapy developed high fevers, tachycardia, and hypoxia at day 2 of CAR T-cell infusion. This is followed by development of immune effector cell-associated neurotoxicity syndrome (ICANS) complicated by seizures and status epilepticus corresponding to peripheral lymphocytosis. Chest X-ray obtained 2 days after first CAR T-cell infusion (A) demonstrates a right basilar pulmonary opacity (*white arrow*) with small right pleural effusion in keeping with the clinical symptoms of CRS with clearance of the pulmonary opacities and pleural effusion at 2 months following CAR T-cell infusion (B). MRI images obtained during the first week following CAR T-cell infusion demonstrates diffuse sulcal, cortical, subcortical signal abnormality with increasing sulcal effacement on FLAIR (C) and T1-weighted post contrast-enhanced image (D). Patient was treated with Cytoxan and a dexamethasone taper with full recovery. Follow-up MRI FLAIR (E) and T1-weighted post contrast-enhanced image (F) demonstrate resolution of FLAIR abnormality and edema corresponding to resolution of symptoms. There is heterogeneity in the marrow on all MRI sequences corresponding to the known multiple myeloma.

present specific challenges to these tumor assessments and as a result tumor response criteria adapted to immunotherapeutics have been developed though are still exploratory in nature.

Conventional tumor response criteria

The most commonly used of these imaging tumor response criteria is Response Evaluation Criteria in Solid Tumors, version 1.1 (RECIST1.1) [40]. In this tumor response criteria, longitudinal measurements of "target" lesions, selected based on size and distribution in the body, are used to approximate tumor burden. The sum of up to five target lesion measurements is

TABLE 3 Adverse effects associated with CAR T-cell therapy.

CAR T cell adverse effect	Clinical findings	Imaging findings
Cytokine Release Syndrome	Fever, malaise, hypotension, hypoxia, consumptive coagulopathy, and death	Pulmonary edema, ground-glass opacities, atelectasis, and pleural effusion
Immune Effector Cell-Associated Neurotoxicity Syndrome	Headache, confusion, aphasia, memory loss, seizures, and focal weakness	Most commonly no imaging findings If present: Vasogenic edema, leptomeningeal enhancement, micro-hemorrhages
Hemophagocytic Lymphohistiocytosis (HLH)/ Macrophage Activation Syndrome	Fever, hepatosplenomegaly, and cytopenia. Can involve CNS, lungs, liver, and spleen	Cerebral atrophy, white matter hyperintense lesions on MRI FLAIR, and demyelination. Pulmonary edema, pleural effusions. Hepatosplenomegaly, periportal edema, and ascites

calculated at each timepoint and the change in that sum over time is determined. Using percentage of change between imaging timepoints, disease is considered in four response categories (Table 3): complete response (CR, no detectable tumor) (Fig. 12), partial response (PR, greater than 30% decrease in tumor target lesion sum) (Fig. 13), progression disease (PD, 20% or greater increase in tumor target lesion sum or the presence of new lesions), or stable disease (SD, disease that does not meet criteria for PR or PD). Typically, the assessment of PD results in a change in therapy. Lesions that do not meet the criteria to be included in the assessment as

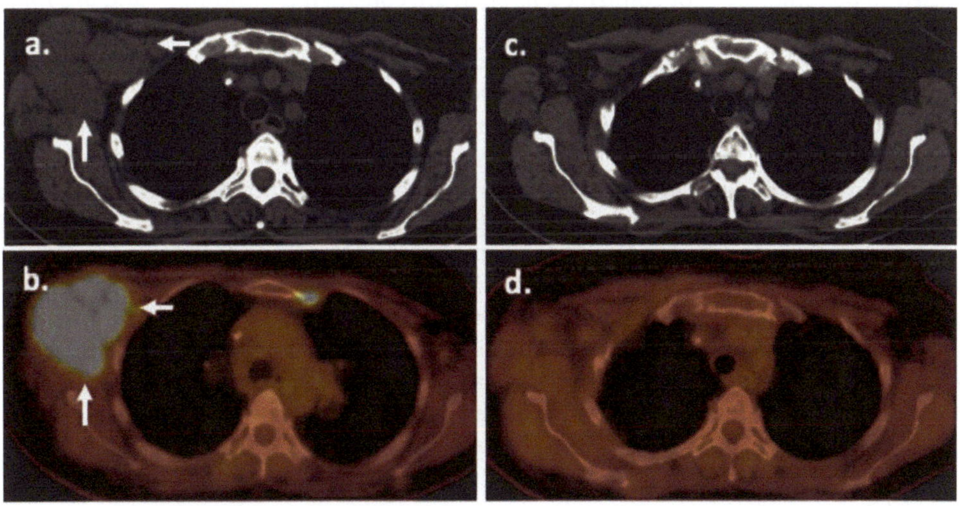

FIG. 12 **Therapy response to CAR T-cell therapy**. A 56-year-old female with diffuse large B-cell lymphoma transformed from follicular lymphoma with progression on prior chemotherapy regimen. Baseline FDG PET-CT prior to CAR T-cell therapy demonstrates bulky lymphadenopathy in the right axilla (*white arrows*) on unenhanced axial CT images (A) which are shown to be intensely avid on fused axial FDG-PET/CT images (B). Images obtained at 2 months of CAR T-cell therapy reveals a marked decrease in the size of right axillary lymph nodes on unenhanced axial CT (C) with a corresponding marked decrease in nodal FDG avidity on fused axial FDG-PET/CT images (D).

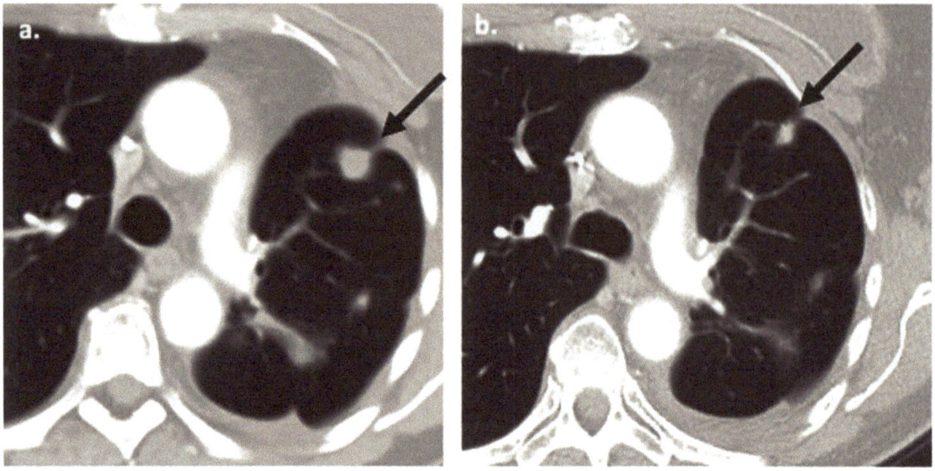

FIG. 13 **Tumor response to immunotherapy with immune checkpoint inhibition**. This 53-year-old female patient with metastatic leiomyosarcoma underwent therapy with a regimen of anti-CTLA4 and anti-PD1 therapy. At baseline (A), there is a metastatic lung nodule in the left upper lobe (*black arrow*). At 2 months of therapy, there is a significant decrease in the size of this metastatic nodule in keeping with treatment response (B).

"target" lesions can be included as "nontargets" and can impact the assessment. Lesions are considered nontargets if they are not large enough or have characteristics that limit accurate measurement such as position in pleural fluid, bone lesions that can partially calcify or coalescent lesions that may separate with subsequent disease response. Lesions that are suspected to be tumor but not confirmed as such are considered "findings" and are followed over time.

While RECIST1.1 is the most commonly used tumor response criteria in solid tumors, there are many others that are in use and are tailored to specific diseases and therapeutics, for example, Lugano [41] and Cheson [42] criteria for lymphoma and modified RECIST (mRECIST) for hepatocellular carcinoma [43]. There are also a number of tumor response criteria that are under development to address shortcomings in conventional tumor response criteria. For example, the CHOI criteria that measures changes in tumor tissue attenuation as a marker of tumor response in addition to changes in tumor size [44] and may have particular utility in following response to antiangiogenic therapy. Immunotherapeutics present unique challenges to tumor response assessment that has spurred the development of a number of tools to assess disease burden which seeks to address these potential confounders. One of the first directions was to create tumor response criteria tailored to the unconventional patterns of response that can be observed with immunotherapy, specifically ICI therapy.

Immune tumor response criteria

The immune-related response criteria (irRC) was the first tumor response criteria to address of challenges of response assessment in the setting of immunotherapy (Table 4). These criteria were introduced in 2007 in response to the observation of pseudoprogression in

TABLE 4 Tumor response criteria used in clinical trials of immunotherapy in cancer.

	RECIST1.1	iRECIST	PERCIST1.0	imPERCIST
Type of clinical trial endpoint	Primary	Exploratory	Primary	Exploratory
Target lesions	Up to five lesions that meet size criteria	Up to five lesions that meet size criteria	Single most avid lesion	Single most avid lesion
Index of response	% change of target sums	% change of target sums	% in SUL of most avid lesion	% in SUL of most avid lesion
Progression				
Target size increase	Assigned when target sum increases ≥20% or new lesions	Initially assigned iUPD, requires confirmation of PD at next timepoint	Assigned when SUL increases ≥30%, absolute increase > 0.8 or new lesions	Initially assigned iUPD, requires confirmation of PD at next timepoint
New lesions	Automatic PD	Initially assigned iUPD, requires >5 mm overall growth at next timepoint to confirm PD	Automatic PD	Initially assigned iUPD requires confirmation of PD at next timepoint
Nontarget	Unequivocal progression	Unequivocal progression	Unequivocal progression	Unequivocal progression

patients with melanoma on ipilimumab. The irRC was a modification of the World Health Organization criteria calling for bidimensional measurements for up to 10 tumor lesions. These criteria differed significantly from the conventional WHO since the irRC allowed for initial growth of lesion without calling progressive disease [45]. Later, the immune-related response criteria (irRECIST) were introduced in 2013, these set of criteria were based on the more widely used RECIST 1.1 criteria, utilizing five target lesions and assessing change by using a sum of the lesions unidimensional measurements [46]. These criteria were shown to be more reproducible and allowed an easier comparison to the widely accepted RECIST criteria. The immune RECIST1.1 (iRECIST 1.1) introduced in 2017 [47] is the most widely accepted of these immune tumor response criteria. The major difference between iRECIST1.1 and RECIST1.1 lies in the definition of progressive disease.

In RECIST1.1, tumor target lesion sums increasing by greater than 20% between timepoints or the demonstration of new metastasis is considered progressive disease. RECIST1.1 is widely used in clinical trials, however, can underestimate response to therapy in immunotherapeutics in up to 15% of patients since it cannot differentiate between pseudo progression and true progression [22]. iRECIST introduced the term unconfirmed progression (iUPD) where the increased target lesion size, nontarget lesion size or new lesions meeting the conventional RECIST1.1 criteria for progressive disease is called iUPD when first observed [23]. Comparison is then made to subsequent follow-up CT in 4–8 weeks and if further increases are observed, then it is labeled confirmed progressive disease (iCPD). In this way, the patient is maintained on therapy for an observation period after initial evidence of potential progression to ensure differentiation from pseudo progression. In the case of pseudo progression, tumor will begin to regress by the follow-up timepoint 4–8 weeks later.

Therapy response assessment with FDG-PET/CT is an emerging tool for monitoring immunotherapy response with several published criteria. PET Response Criteria in Solid tumors (PERCIST) published in 2009 used standardized uptake value (SUV) corrected for the lean body mass, the SUL, the imaging tool for assessing response. SUL peak value is assessed within a lesion at the site of highest metabolic activity and complete response to therapy was defined as reduction of 18F-FDG uptake to the level of the background blood pool [48,49]. Immunotherapy-modified PET RECIST (imPERCIST) published in 2019 is based on PERCIST and uses SUL for response quantification but introduced the concept of iUPD requiring a 4–8 week confirmatory scan in order to call progressive disease similar to iRECIST [50].

Finally, Lymphoma Response to Immunomodulatory Therapy Criteria—LYRIC published in 2016 are based on the previously published LUGANO criteria utilizing either a 5-point score based on FDG uptake from the PET portion of the exam or the sum of lesion diameter from the CT portion of the exam to assess activity of lymphoma. Progressive disease without clinical deterioration based on either PET or CT metrics needs to be confirmed with a 12-week follow-up scan or biopsy [51].

These newly developed immune-specific tumor response criteria are still exploratory in nature, and it is still recommended that conventional response criteria such as RECIST1.1 be used as a primary imaging endpoint in clinical trials, whereas the immune criteria can be considered for exploratory endpoints.

FDG-PET for assessment of CAR T-cell therapy

FDG-PET/CT is the most commonly used imaging tool for both baseline and therapy response assessment of lymphoproliferative disorders, as such it plays a central role in assessing response of CAR T-cell therapy. Recent publications have shown that baseline lower metabolic tumor volume (MTV), which reflects the volume of the tumor with pathologic uptake on PET correlates with better overall survival and progression-free survival in patients with relapsed/refractory large B-cell lymphoma [52]. Higher values of SUV and MTV obtained 1 month after CAR T-cell therapy were shown to correlate with a greater risk of death and progressive disease and greater percent change in these parameters was shown to correlate with greater progression-free survival [53,54].

Future directions for imaging tumor response to immunotherapy

A major goal of research in cancer imaging is centered on methods to increase the specificity of differentiating malignant tumor from benign tissue. This can be very challenging particularly in the posttreatment setting where adjacent tissue may be distorted by therapy. In addition, methods to determine response of cancer to therapy early, and with precision, are also needed to prevent prolonged exposure to the toxicity and expense of therapies that will have no clinical benefit. Time on an ineffective therapy also gives tumors more time to grow and disseminate unchecked. A number of approaches to imaging cancer with precision are currently underway.

Radiomics and artificial intelligence

Radiomics is a field combining computer science, artificial intelligence, and radiology to enhance the precision of medical imaging. The advent of ICI therapy has raised interest in the development of novel radiomic tools for therapy response assessment on medical imaging in cancer patients receiving immunotherapeutics. The field of radiomics leverages the thousands of features imbedded in the digital image that is not visible to the naked eye. These features are a manifestation of tumor histologic type [55,56], phenotypic expression [57–59], and pathophysiology and so radiomic signatures that correspond with tumor type and that can predict response to therapy [60] are currently under development. Early success of this strategy has been observed in the setting of tumors being treated with immunotherapeutics. For example, radiomic signatures that correlate with tumor infiltration with lymphocytes or that predict benefit from ICI therapy have been described [61]. If translation to the clinic is ultimately successful, this work could result in an algorithm applied to conventional medical imaging that could provide diagnostic or prognostic data on tumors planned for or undergoing immunotherapy. This strategy has the potential to identify patients unlikely to benefit from immunotherapy, saving them unwanted toxicity, and to correctly differentiate between progression and pseudo progression.

Targeted radiopharmaceuticals

Beyond conventional imaging, there are also molecular imaging approaches that are under development that directly target the immune cells that migrate to the tumor after successful activation during immunotherapy. An example of this is anti-CD8 imaging where minibodies (truncated antibodies) targeting CD8 are labeled with a radioisotope and then injected in trace amounts to track the migration of activated CD8+ lymphocytes. The accumulation of these labeled immune cells then results in a signal that is detectable on PET imaging [62]. These agents are currently in human clinical trials [63] and are an exciting example of precision medicine approaches to imaging immunotherapy response in cancer. In addition to directly targeting T cells for imaging, there are also a number of emerging PET radiotracers that may have value in oncology, including in the setting of immunotherapy.

Surveillance of infused CAR T-cell activity can be accomplished through conventional measurements of tumor size and peripheral blood monitoring; however, these approaches do not yield information on the spatial distribution of CAR T-cell activity. The development of imaging tools that can serve as a dynamic, noninvasive monitoring of this living drug is urgently sought. Due to the superior sensitivity to small numbers of cells and availability of numerous imaging probes, PET imaging, either combined with CT or MRI, or single-photon emission computed tomography (SPECT) paired with CT are the favored imaging modalities for the development of dynamic imaging strategies for CAR T-cell therapy [64]. The two primary approaches to imaging CAR T cells using these modalities are through direct labeling of CAR T cells prior to infusion, or through use of gene reporters [65]. While direct labeling of the infused T cells is the simplest strategy, the gene reporter approach has the advantage of only being produced by live cells and so the imaging signal is indicative of live CAR T-cells [66]. The only reporter gene currently used for CAR-T imaging in human patients is an enzyme-based reporter gene, herpes simplex virus 1 thymidine kinase (HSV1-tk). The

HSV1-tk gene reporter strategy allows for PET imaging with multiple radiolabeled probes [64]. These dynamic imaging strategies are still exploratory and not for routine patient cancer imaging.

Conclusions

The introduction of immunotherapy into routine clinical practice has resulted in groundbreaking advances in patient survival in a wide range of malignancies. Common to these agents are mechanisms of action that lead to inflammation in the tumor resulting in tumor lysis. This differs significantly from conventional chemotherapeutics or agents targeted to facets of tumor metabolism which aim to shut down tumor cell growth. As a result, the use of immunotherapeutic in oncology has resulted in unconventional patterns of responses and adverse events many of which we are still striving to fully understand and which impact patient management. The implementation and interpretation of medical imaging of oncology patients receiving these agents require careful consideration of these potential factors in order to avoid common pitfalls in image interpretation in the setting of immunotherapy. These challenges have spurred development in a number of directions in the field of imaging including the development of novel tumor response criteria tailored to immunotherapy, radiomic, and artificial intelligence imaging approaches to imaging analysis and therapy response prediction, targeted imaging agents to immune cells allowing for tracking of activated immune cell populations and novel PET agents to measure activity through various metabolic pathways. In addition to the immunotherapeutics discussed above, novel immunotherapies are under development including bispecific antibodies to recruit immune effectors and may show promise in the future.

Acknowledgments

The authors would like to thank Corey Langer, MD, and Melina Marmarelis, MD, for their assistance with this chapter.

References

[1] Hopper KD, Singapuri K, Finkel A. Body CT and oncologic imaging. Radiology 2000;215(1):27–40.
[2] Seemann MD. PET/CT: fundamental principles. Eur J Med Res 2004;9(5):241–6.
[3] Koppenol WH, Bounds PL, Dang CV. Otto Warburg's contributions to current concepts of cancer metabolism. Nat Rev Cancer 2011;11(5):325–37.
[4] Fendler WP, Calais J, Eiber M, Flavell RR, Mishoe A, Feng FY, et al. Assessment of 68Ga-PSMA-11 PET accuracy in localizing recurrent prostate cancer: a prospective single-arm clinical trial. JAMA Oncol 2019;5(6):856–63.
[5] Hofman MS, Lau WF, Hicks RJ. Somatostatin receptor imaging with 68Ga DOTATATE PET/CT: clinical utility, normal patterns, pearls, and pitfalls in interpretation. Radiographics 2015;35(2):500–16.
[6] Schmidt GP, Reiser MF, Baur-Melnyk A. Whole-body MRI for the staging and follow-up of patients with metastasis. Eur J Radiol 2009;70(3):393–400.
[7] Kim M, Suh CH, Lee SM, Kim HC, Aizer AA, Yanagihara TK, et al. Diagnostic yield of staging brain MRI in patients with newly diagnosed non-small cell lung cancer. Radiology 2020;297(2):419–27.
[8] Shreve PD, Anzai Y, Wahl RL. Pitfalls in oncologic diagnosis with FDG PET imaging: physiologic and benign variants. Radiographics 1999;19(1):61–77. quiz 150-1.

[9] Lutje S, Feldmann G, Essler M, Brossart P, Bundschuh RA. Immune checkpoint imaging in oncology: a game changer toward personalized immunotherapy? J Nucl Med 2020;61(8):1137–44.

[10] Billan S, Kaidar-Person O, Gil Z. Treatment after progression in the era of immunotherapy. Lancet Oncol 2020;21 (10):e463–76.

[11] Chiou VL, Burotto M. Pseudoprogression and immune-related response in solid tumors. J Clin Oncol 2015;33 (31):3541–3.

[12] Park HJ, Kim KW, Pyo J, Suh CH, Yoon S, Hatabu H, et al. Incidence of pseudoprogression during immune checkpoint inhibitor therapy for solid tumors: a systematic review and meta-analysis. Radiology 2020;297 (1):87–96.

[13] Fujimoto D, Yoshioka H, Kataoka Y, Morimoto T, Hata T, Kim YH, et al. Pseudoprogression in previously treated patients with non-small cell lung cancer who received nivolumab monotherapy. J Thorac Oncol 2019;14(3): 468–74.

[14] Hammer M, Bagley S, Aggarwal C, Bauml J, Nachiappan AC, Simone 2nd CB, et al. Thoracic imaging of non-small cell lung cancer treated with anti-programmed death receptor-1 therapy. Curr Probl Diagn Radiol 2019;48 (2):142–7.

[15] Yoon JG, Smith DA, Tirumani SH, Caimi PF, Ramaiya NH. CAR T-cell therapy: an update for radiologists. Am J Roentgenol 2021;217(6):1461–74.

[16] Kurman JS, Murgu SD. Hyperprogressive disease in patients with non-small cell lung cancer on immunotherapy. J Thorac Dis 2018;10(2):1124–8.

[17] Guan Y, Feng D, Yin B, Li K, Wang J. Immune-related dissociated response as a specific atypical response pattern in solid tumors with immune checkpoint blockade. Ther Adv Med Oncol 2022;14, 17588359221096877.

[18] Pons-Tostivint E, Latouche A, Vaflard P, Ricci F, Loirat D, Hescot S, et al. Comparative analysis of durable responses on immune checkpoint inhibitors versus other systemic therapies: a pooled analysis of phase III trials. JCO Precis Oncol 2019;3:1–10.

[19] Ramos CA, Grover NS, Beaven AW, Lulla PD, Wu MF, Ivanova A, et al. Anti-CD30 CAR-T cell therapy in relapsed and refractory Hodgkin lymphoma. J Clin Oncol 2020;38(32):3794–804.

[20] Ying Z, Huang XF, Xiang X, Liu Y, Kang X, Song Y, et al. A safe and potent anti-CD19 CAR T cell therapy. Nat Med 2019;25(6):947–53.

[21] Roddie C, Dias J, O'Reilly MA, Abbasian M, Cadinanos-Garai A, Vispute K, et al. Durable responses and low toxicity after fast off-rate CD19 chimeric antigen receptor-T therapy in adults with relapsed or refractory B-cell acute lymphoblastic leukemia. J Clin Oncol 2021;39(30):3352–63.

[22] Hirayama AV, Gauthier J, Hay KA, Voutsinas JM, Wu Q, Pender BS, et al. High rate of durable complete remission in follicular lymphoma after CD19 CAR-T cell immunotherapy. Blood 2019;134(7):636–40.

[23] Martins F, Sofiya L, Sykiotis GP, Lamine F, Maillard M, Fraga M, et al. Adverse effects of immune-checkpoint inhibitors: epidemiology, management and surveillance. Nat Rev Clin Oncol 2019;16(9):563–80.

[24] Tirumani SH, Ramaiya NH, Keraliya A, Bailey ND, Ott PA, Hodi FS, et al. Radiographic profiling of immune-related adverse events in advanced melanoma patients treated with Ipilimumab. Cancer Immunol Res 2015;3 (10):1185–92.

[25] Aide N, Hicks RJ, Le Tourneau C, Lheureux S, Fanti S, Lopci E. FDG PET/CT for assessing tumour response to immunotherapy: report on the EANM symposium on immune modulation and recent review of the literature. Eur J Nucl Med Mol Imaging 2019;46(1):238–50.

[26] Som A, Mandaliya R, Alsaadi D, Farshidpour M, Charabaty A, Malhotra N, et al. Immune checkpoint inhibitor-induced colitis: a comprehensive review. World J Clin Cases 2019;7(4):405–18.

[27] Gkiozos I, Kopitopoulou A, Kalkanis A, Vamvakaris IN, Judson MA, Syrigos KN. Sarcoidosis-like reactions induced by checkpoint inhibitors. J Thorac Oncol 2018;13(8):1076–82.

[28] Haanen J, Carbonnel F, Robert C, Kerr KM, Peters S, Larkin J, et al. Management of toxicities from immunotherapy: ESMO clinical practice guidelines for diagnosis, treatment and follow-up. Ann Oncol 2017;28(suppl_4): iv119–42.

[29] Alessandrino F, Shah HJ, Ramaiya NH. Multimodality imaging of endocrine immune related adverse events: a primer for radiologists. Clin Imaging 2018;50:96–103.

[30] Lasocki A, Iravani A, Galligan A. The imaging of immunotherapy-related hypophysitis and other pituitary lesions in oncology patients. Clin Radiol 2021;76(5):325–32.

[31] Jang JY, Kim SS, Song SY, Kim YJ, Kim SW, Choi EK. Radiation pneumonitis in patients with non-small-cell lung cancer receiving chemoradiotherapy and an immune checkpoint inhibitor: a retrospective study. Radiat Oncol 2021;16(1):231.

[32] Teng F, Li M, Yu J. Radiation recall pneumonitis induced by PD-1/PD-L1 blockades: mechanisms and therapeutic implications. BMC Med 2020;18(1):275.
[33] Ngwa W, Irabor OC, Schoenfeld JD, Hesser J, Demaria S, Formenti SC. Using immunotherapy to boost the abscopal effect. Nat Rev Cancer 2018;18(5):313–22.
[34] Gutierrez C, McEvoy C, Mead E, Stephens RS, Munshi L, Detsky ME, et al. Management of the critically ill adult chimeric antigen receptor-T cell therapy patient: a critical care perspective. Crit Care Med 2018;46(9):1402–10.
[35] Schubert ML, Schmitt M, Wang L, Ramos CA, Jordan K, Muller-Tidow C, et al. Side-effect management of chimeric antigen receptor (CAR) T-cell therapy. Ann Oncol 2021;32(1):34–48.
[36] Rivera AM, May S, Lei M, Qualls S, Bushey K, Rubin DB, et al. CAR T-cell-associated neurotoxicity: current management and emerging treatment strategies. Crit Care Nurs Q 2020;43(2):191–204.
[37] Gust J, Taraseviciute A, Turtle CJ. Neurotoxicity associated with CD19-targeted CAR-T cell therapies. CNS Drugs 2018;32(12):1091–101.
[38] Santomasso BD, Park JH, Salloum D, Riviere I, Flynn J, Mead E, et al. Clinical and biological correlates of neurotoxicity associated with CAR T-cell therapy in patients with B-cell acute lymphoblastic leukemia. Cancer Discov 2018;8(8):958–71.
[39] Sandler RD, Tattersall RS, Schoemans H, Greco R, Badoglio M, Labopin M, et al. Diagnosis and management of secondary HLH/MAS following HSCT and CAR-T cell therapy in adults; a review of the literature and a survey of practice within EBMT centres on behalf of the autoimmune diseases working party (ADWP) and transplant complications working party (TCWP). Front Immunol 2020;11:524.
[40] Eisenhauer EA, Therasse P, Bogaerts J, Schwartz LH, Sargent D, Ford R, et al. New response evaluation criteria in solid tumours: revised RECIST guideline (version 1.1). Eur J Cancer 2009;45(2):228–47.
[41] Cheson BD, Fisher RI, Barrington SF, Cavalli F, Schwartz LH, Zucca E, et al. Recommendations for initial evaluation, staging, and response assessment of Hodgkin and non-Hodgkin lymphoma: the Lugano classification. J Clin Oncol 2014;32(27):3059–68.
[42] Cheson BD, Pfistner B, Juweid ME, Gascoyne RD, Specht L, Horning SJ, et al. Revised response criteria for malignant lymphoma. J Clin Oncol 2007;25(5):579–86.
[43] Lencioni R, Llovet JM. Modified RECIST (mRECIST) assessment for hepatocellular carcinoma. Semin Liver Dis 2010;30(1):52–60.
[44] Choi H, Charnsangavej C, Faria SC, Macapinlac HA, Burgess MA, Patel SR, et al. Correlation of computed tomography and positron emission tomography in patients with metastatic gastrointestinal stromal tumor treated at a single institution with imatinib mesylate: proposal of new computed tomography response criteria. J Clin Oncol 2007;25(13):1753–9.
[45] Wolchok JD, Hoos A, O'Day S, Weber JS, Hamid O, Lebbe C, et al. Guidelines for the evaluation of immune therapy activity in solid tumors: immune-related response criteria. Clin Cancer Res 2009;15(23):7412–20.
[46] Nishino M, Giobbie-Hurder A, Gargano M, Suda M, Ramaiya NH, Hodi FS. Developing a common language for tumor response to immunotherapy: immune-related response criteria using unidimensional measurements. Clin Cancer Res 2013;19(14):3936–43.
[47] Seymour L, Bogaerts J, Perrone A, Ford R, Schwartz LH, Mandrekar S, et al. iRECIST: guidelines for response criteria for use in trials testing immunotherapeutics. Lancet Oncol 2017;18(3):e143–52.
[48] Wahl RL, Jacene H, Kasamon Y, Lodge MA. From RECIST to PERCIST: evolving considerations for PET response criteria in solid tumors. J Nucl Med 2009;50(Suppl 1):122S–50S.
[49] O JH, Lodge MA, Wahl RL. Practical PERCIST: a simplified guide to PET response criteria in solid tumors 1.0. Radiology 2016;280(2):576–84.
[50] Goldfarb L, Duchemann B, Chouahnia K, Zelek L, Soussan M. Monitoring anti-PD-1-based immunotherapy in non-small cell lung cancer with FDG PET: introduction of iPERCIST. EJNMMI Res 2019;9(1):8.
[51] Cheson BD, Ansell S, Schwartz L, Gordon LI, Advani R, Jacene HA, et al. Refinement of the Lugano classification lymphoma response criteria in the era of immunomodulatory therapy. Blood 2016;128(21):2489–96.
[52] Dean EA, Mhaskar RS, Lu H, Mousa MS, Krivenko GS, Lazaryan A, et al. High metabolic tumor volume is associated with decreased efficacy of axicabtagene ciloleucel in large B-cell lymphoma. Blood Adv 2020;4(14):3268–76.
[53] Breen WG, Hathcock MA, Young JR, Kowalchuk RO, Bansal R, Khurana A, et al. Metabolic characteristics and prognostic differentiation of aggressive lymphoma using one-month post-CAR-T FDG PET/CT. J Hematol Oncol 2022;15(1):36.

[54] Sesques P, Tordo J, Ferrant E, Safar V, Wallet F, Dhomps A, et al. Prognostic impact of 18F-FDG PET/CT in patients with aggressive B-cell lymphoma treated with anti-CD19 chimeric antigen receptor T cells. Clin Nucl Med 2021;46(8):627–34.

[55] Beig N, Khorrami M, Alilou M, Prasanna P, Braman N, Orooji M, et al. Perinodular and intranodular radiomic features on lung CT images distinguish adenocarcinomas from granulomas. Radiology 2018;180910.

[56] Linning E, Lu L, Li L, Yang H, Schwartz LH, Zhao B. Radiomics for classifying histological subtypes of lung cancer based on multiphasic contrast-enhanced computed tomography. J Comput Assist Tomogr 2019;43 (2):300–6.

[57] Aerts HJ, Velazquez ER, Leijenaar RT, Parmar C, Grossmann P, Carvalho S, et al. Decoding tumour phenotype by noninvasive imaging using a quantitative radiomics approach. Nat Commun 2014;5:4006.

[58] He B, Dong D, She Y, Zhou C, Fang M, Zhu Y, et al. Predicting response to immunotherapy in advanced non-small-cell lung cancer using tumor mutational burden radiomic biomarker. J Immunother Cancer 2020;8(2).

[59] Jia TY, Xiong JF, Li XY, Yu W, Xu ZY, Cai XW, et al. Identifying EGFR mutations in lung adenocarcinoma by noninvasive imaging using radiomics features and random forest modeling. Eur Radiol 2019;29(9):4742–50.

[60] Chaddad A, Desrosiers C, Toews M, Abdulkarim B. Predicting survival time of lung cancer patients using radiomic analysis. Oncotarget 2017;8(61):104393–407.

[61] Sun R, Limkin EJ, Vakalopoulou M, Dercle L, Champiat S, Han SR, et al. A radiomics approach to assess tumour-infiltrating CD8 cells and response to anti-PD-1 or anti-PD-L1 immunotherapy: an imaging biomarker, retrospective multicohort study. Lancet Oncol 2018;19(9):1180–91.

[62] Tavare R, Escuin-Ordinas H, Mok S, McCracken MN, Zettlitz KA, Salazar FB, et al. An effective immuno-PET imaging method to monitor CD8-dependent responses to immunotherapy. Cancer Res 2016;76(1):73–82.

[63] Pandit-Taskar N, Postow MA, Hellmann MD, Harding JJ, Barker CA, O'Donoghue JA, et al. First-in-humans imaging with (89)Zr-Df-IAB22M2C anti-CD8 minibody in patients with solid malignancies: preliminary pharmacokinetics, biodistribution, and lesion targeting. J Nucl Med 2020;61(4):512–9.

[64] Shao F, Long Y, Ji H, Jiang D, Lei P, Lan X. Radionuclide-based molecular imaging allows CAR-T cellular visualization and therapeutic monitoring. Theranostics 2021;11(14):6800–17.

[65] Kircher MF, Gambhir SS, Grimm J. Noninvasive cell-tracking methods. Nat Rev Clin Oncol 2011;8(11):677–88.

[66] Massoud TF, Singh A, Gambhir SS. Noninvasive molecular neuroimaging using reporter genes: part I, principles revisited. Am J Neuroradiol 2008;29(2):229–34.

CHAPTER

12

Novel cell delivery systems: Intracranial and intrathecal

Lisa Feldman and Behnam Badie

Division of Neurosurgery, City of Hope National Medical Center, Duarte, CA, United States

Abstract

Brain tumors remain extraordinarily difficult to treat and remain a significant source of morbidity and mortality despite standard of care interventions such as surgery, chemotherapy, and radiation. The blood brain barrier (BBB) serves as a major obstacle to the delivery of antitumoral treatments. In recent years, new treatments have been developed that overcome the BBB through direct methods such as intranasal delivery, intracranial, or lumbar intrathecal delivery to the cerebrospinal fluid including implanted devices to allow serial injections of a variety of treatments, as well as convection-enhanced delivery. Alternative methods for drug delivery to the central nervous system involve the use of chemical and mechanical BBB disruption, focused ultrasound, and laser interstitial thermotherapy. This study also deals with the history, evolution, and examples of therapies.

Abbreviations

AADC	aromatic L-amino acid decarboxylase
AAV	adeno-associated firus
ALL	acute lymphocytic leukemia
Ara-C	cytarabine
BBB	blood brain barrier
BTB	blood tumor barrier
CAR T cell	chimeric antigen receptor T cell
CED	convection-enhanced delivery
CSF	cerebrospinal fluid
CT	computed tomography
EVD	external ventricular drain
FUS	focused ultrasound
GBM	glioblastoma
hAADC	human aromatic L-amino acid decarboxylase
HER2	human epidermal growth factor receptor 2
IL-13	interleukin-13
LITT	laser interstitial thermotherapy
MRI	magnetic resonance imaging

https://doi.org/10.1016/B978-0-12-822620-9.00007-0

MTX	methotrexate
PAMPs	pathogen-associated molecular pattern receptors
PD	Parkinson's disease
PET	positron emission tomography
WHO	World Health Organization

Conflict of interest

Lisa Feldman has no financial relationships to disclose.
Behnam Badie receives licensing (IP and royalties) payments from Mustang Bio., Inc.

Tumors of the central nervous system

Tumors of the central nervous system (CNS) pose a magnificent challenge in neuro-oncology, and is the 10th leading cause of death for men and women [1]. CNS tumors are the most common solid tumors and second most common cancer in children, resulting in an estimated 26% of childhood cancers [2]. In adults, 85%–90% of all primary CNS tumors arise in the brain [3]. In 2012, approximately 256,213 new CNS tumors were diagnosed worldwide, which resulted in about 189,382 deaths [4]. In the United States, an estimated 23,890 new cases (13,590 men and 10,300 women) [1] are expected to be diagnosed in 2020 which will result in 18,020 (10,190 men and 7830 women) deaths [1,5]. While tumors originating in the CNS are referred to as primary tumors, tumors in the brain and spinal cord that arise from tumors cells from outside of the CNS are referred to as secondary malignancies. The most common metastatic tumors to the CNS include bladder, breast, kidney, lung, melanoma, and blood born cancers such as leukemia and lymphoma [1].

CNS tumors are classified by the World Health Organization (WHO) into four main grades in which grade I is benign and grade IV is the most aggressive. In 2016, the WHO revised their classification system to define tumors not only by histological features but also by molecular parameters [6]. Across tumors from all four WHO grades, the 5-year survival rate for CNS tumors is about 36% and 10-year survival is approximately 31% [1]. Not surprisingly, survival rates decrease with age, but overall prognosis also depends on tumor histology, the WHO grade, the location of tumor, and the functional status of a person at diagnosis.

Unfortunately, high-grade gliomas, including glioblastoma (GBM), are the most common brain tumors in adults with an average incidence of 4.67–5.73 per 100,000 people [7,8]. These tumors are extraordinarily difficult to treat and confer a 5-year survival of only approximately 5% of patients [9]. Because GBMs are incredibly invasive into surrounding brain tissue, complete surgical resection becomes impossible. After surgical debulking, additional treatments such as chemotherapy, radiation therapy, and alternating electric fields therapy are often offered to patients. However, as these treatments fail to result in complete remission, scientists have been working hard to develop novel, efficacious treatments. Throughout the course of novel therapy development, a variety of methods to deliver drugs, peptides, electricity, immunotherapies, to name a few, have been designed to treat cancers of the CNS.

Blood brain barrier (BBB)

The blood brain barrier (BBB) is a network of continuous tight junctions between brain capillary endothelial cells that prevent access of the majority of therapeutics to the brain [10,11]

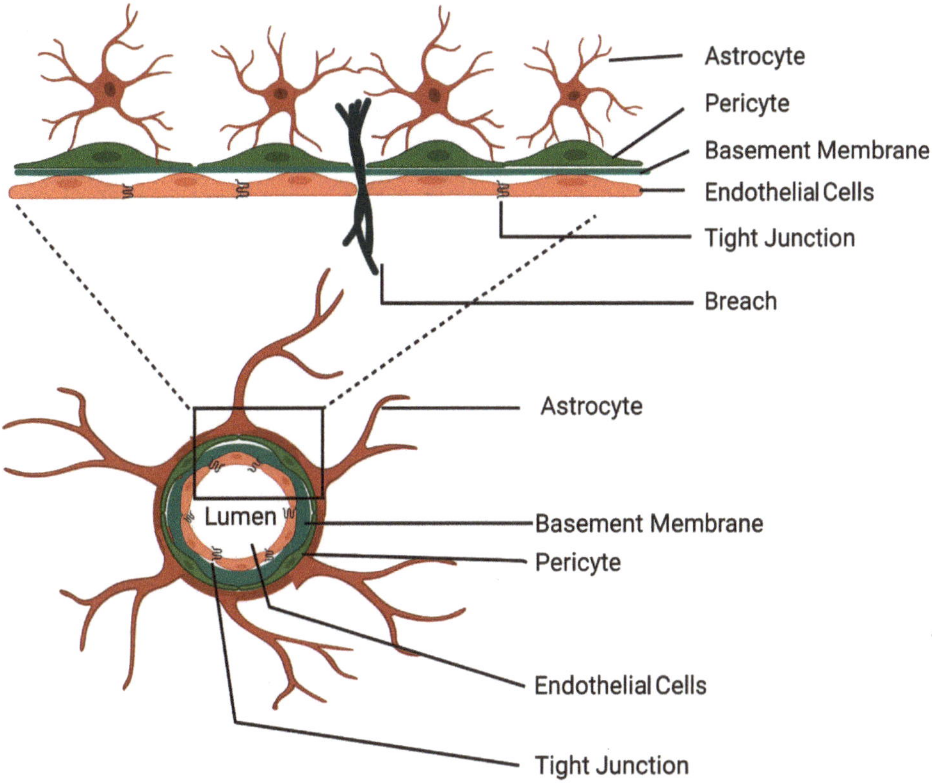

FIG. 1 The blood–brain barrier is made of endothelial cells that are linked by tight junctions, which share a basement membrane with pericytes. Astrocyte end-feet generate the outermost layer of the BBB.

(Fig. 1). The BBB is created by microvascular endothelial cells that line the blood vessels in the CNS, and the network they create through tight junctions and astrocytic end feet [12,13]. The tight junctions from endothelial cells, in concert with a variety of enzymes, receptors, and transporters [14], severely restrict access of intravascular molecules to the brain. This complex system was designed to effectively block out toxins to the brain, but it also restricts 100% large-molecule drugs and 98% of small-molecule drugs [15]. The BBB is so effective that only small molecules and proteins less than 400 Da successfully cross it [16]. The interendothelial junctions latch endothelial cells into such a strict barrier that strongly reduces the penetration of even water-soluble substances in the CNS [16]. Generally speaking, lipid-soluble molecules cross the BBB through transmembrane diffusion, but because of diffusion, they are restricted in terms of distribution throughout the brain [17].

With regard to CNS cancers, the majority of chemotherapies are bound to plasma proteins with molecular weights that far surpass 400 Da [18]. The BBB also expresses high concentrations of efflux transmembrane pumps, such as P-glycoprotein, which actively remove chemotherapies from the CNS [19,20]. One innovative strategy to improve the delivery of chemotherapeutic medications to the CNS involves blocking these transmembrane protein transporters [21]. Other delivery techniques have been developed to either weaken or bypass the BBB to augment CNS drug delivery.

Brain tumors themselves pose further challenges to CNS drug delivery. Many brain tumors harbor increased numbers of abnormal microvessels [22], which include distended capillaries with sluggish flow, and leaky walls that are more permeable than healthy brain vessels [23]. In fact, this unique blood-tumor barrier (BTB) allows tumors to be visualized with contrast agents on computed tomography or magnetic resonance imaging and might lead to inconsistent or reduced diffusion of therapeutics to tumors [24]. This highly abnormal blood flow in these vessels often results in areas of intratumoral hypoxia, rendering chemotherapies and radiation therapies less effective [25]. Altogether, the BTB presents an additional, daunting challenge to optimized delivery of therapeutics to CNS tumors [26].

A wide variety of approaches to circumvent the BBB have been developed over time with the gross aim of improving CNS drug delivery. For example, techniques to open the BBB include the manipulation of CNS capillary permeability, while other techniques involve delivering medications directly to cerebrospinal fluid (CSF) and bypass the BBB altogether. The older methods include intra-arterial injections of medications, such as hyperosmolar therapy; however, newer strategies include sophisticated novel pharmaceuticals that chemically pass the BBB, or utilization of technology that inject therapeutics directly into tumor or tumor cavity [23] (see Table 1). Overall, direct delivery to the CSF, either via intracerebroventricular, intracisternal, or intrathecal injection, is the most common method of administration of therapeutics to bypass the BBB.

TABLE 1 Various methods to circumvent BBB for therapeutic delivery to CNS.

Method	Advantages	Disadvantages
Drugs engineered to utilize BBB transport mechanisms	Noninvasive administration, delivered throughout CNS	Expensive drug development, manufacturing. Requires systemic administration, ongoing clinical development
Transnasal delivery	Noninvasive, easy to administer outside clinical setting, repeatable	Small volume of drug delivered, relatively restricted delivery, interindividual variability
Arterial injection of osmotic solutions or other agents	Effectively delivers targeted drugs throughout CNS, strong clinical experience	Invasive, requires general anesthesia, inconsistent results, may result in local neurotoxicity
Intrathecal, intraventricular injection	Effectively delivers drugs within CSF and to brain surface	Little drug penetration beyond brain surface, invasive, risk of infection with implanted catheter/system
Focused ultrasound, with or without microbubbles	Noninvasive, repeatable, allows target drug to reach high concentrations of drug in CNS, can be improved with use of microbubbles or magnetic particles for additional targeting	Requires systemic administration, risk of infection, technically challenging, ongoing clinical development

Modified from Aryal M, et al. Ultrasound-mediated blood-brain barrier disruption for targeted drug delivery in the central nervous system. Adv Drug Deliv Rev 2014;72:94–109.

Direct delivery

Intranasal delivery

Intranasal application bypasses the BBB to deliver medications to the brain and spinal cord noninvasively and within minutes. Drug absorption throughout the olfactory pathway involves drug crossing the nasal epithelium, traveling to the olfactory bulb, where it is dispersed through the brain parenchyma and into the CSF [27]. Even though the olfactory pathway was initially thought to be the exclusive channel for intranasal delivery to the CNS [28], it is now understood that the trigeminal nerve and pathways also are involved, particularly in the delivery to the brain stem and spinal cord [29]. Because the olfactory and trigeminal nerves are unique connections between the outside environment and the CNS, drug side effects that are often seen in systemic delivery are often avoided [30]. The nasal mucosa serves as a model pharmaceutical absorptive conduit due to its large surface area, high blood flow, porous endothelial membrane, and circumvention of hepatic first pass metabolism [31]. A wide range of therapeutics, including micro- and macromolecules, growth factors, viral vectors, and stem cells [32], have been delivered intranasally to treat Alzheimer's disease, stroke, neurodegenerative diseases, and cognitive functions such as memory, attention, and mood in both animal and human models [30].

Benefits of intranasal delivery include a painless, noninvasive, self-delivery of medications with a fast onset of action and avoidance of chemical/enzymatic degradation, which may occur in the gastrointestinal tract [31]. Limitations to this delivery, however, include limited clearance of drug by mucociliary removal, enzymatic degradation, and relatively low permeability for some drugs through the nasal epithelium [32] such as hydrophilic molecules, peptides, proteins, and nucleotides [33,34].

Intracranial delivery to CSF

Intracerebrovascular and intracisternal delivery of medications bypasses the BBB and delivers therapeutics directly into CSF in lateral ventricles or cisterna magna. This administration method has been used for decades to deliver treatments for a broad range of infections and a wide variety of cancers in pediatric and adult patients [35]. Please refer to Table 2 for antineoplastic agents currently administered via CSF delivery.

Generally, direct delivery into the CNS requires less medication and causes fewer side effects as compared to systemic or oral drug administration. Depending on the expected duration of drug administration, a variety of intraventricular devices may be placed. For short-term intraventricular drug administration, such as antibiotics or antivirals to treat meningitis or encephalitis, an external ventricular drain (EVD) may be placed. An EVD is a flexible plastic, often antibiotic-coated catheter that is passed through a small hole in the skull, into the lateral ventricle to allow drainage of CSF, and delivery of medications into the CSF. An EVD also provides added benefits of allowing intracranial pressure measurements, or to sample CSF regularly to assess therapeutic benefit over time. EVDs are typically placed at ICU bedside, or intraoperatively, and are temporary conduits to the ventricles. Delivery of medications into the cisterna magna involves the placement of a catheter between the first and second cervical vertebrae (C1–C2 interspace) under radiographic guidance.

TABLE 2 Neoplastic agents delivered into CSF.

Therapeutic agents	Neoplasms tested	Adverse effects
Methotrexate (MTX)	First-line treatment for CNS lymphoma and meningeal leukemia; breast, lung, osteosarcoma [36]	Infection [37]; bone marrow, hepatic, pulmonary, or renal toxicities [36]; neurotoxicity [38]; aseptic meningitis or chemical arachnoiditis [39]
Cytarabine (Ara-C)	CNS prophylaxis for acute lymphocytic leukemia (ALL), lymphoma, neoplastic meningitis [40]. Also administered in liposomal formula for slow-release [39]	Neurotoxicity [40], bone marrow, febrile neutropenia, cardiomyopathy [41], transverse myelopathy, aseptic meningitis, chemical arachnoiditis [39]
Hydrocortisone	Given in conjunction with MTX and/or Ara-C for CNS prophylaxis for B-progenitor acute lymphoblastic leukemia [42], non-Hodgkin lymphoma [43]. Coadministered to reduce chemotherapy side effects	Toxicity has not been well-documented in controlled studies, otherwise psychiatric adverse effects [39], headache, nausea, and confusion have been described [44]
Thiotepa (Thioplex)	Monoclonal anti-CD20 antibody to treat leptomeningeal carcinomatosis related to B-cell non-Hodgkin lymphoma, primary CNS lymphoma	Typically, well-tolerated. Myelosuppression, aseptic meningitis [45], polyneuropathy [46]
Rituximab (Rituxan)	Monoclonal anti-CD20 antibody to treat leptomeningeal carcinomatosis related to B-cell non-Hodgkin lymphoma, primary CNS lymphoma	Hypertension, nausea, vomiting, and double vision [47]
Trastuzumab (Herceptin)	Monoclonal antibody to treat leptomeningeal carcinomatosis from human epidermal growth factor receptor 2 positive (HER2+) breast brain metastasis	Very well tolerated with few side effects [48,49]; aseptic meningitis [50]
Oncolytic viral therapy	Wide variety including high-grade glioma, melanoma, squamous cell carcinoma, breast (review [51]). Utilizes viruses that selectively target and induce lysis of tumor cells (review [52,53]); also sensitizes tumor cells to chemotherapy and radiotherapy [54]	Wide variety specific to viral therapy, includes flu-like symptoms [51] nausea, vomiting, fever, chills, asthenia, leukopenia [53]
Immunotherapies	Wide variety including GBM, pediatric high-grade glioma, melanoma, breast, lymphoma, lung. Includes the use of immunostimulants (cytokines, interferons, pathogen-associated molecular pattern (PAMPs) receptors), chimeric antigen receptor (CAR) T-cell [55] T- and B-cell based immunotherapies and vaccines; checkpoint inhibitors (review [56])	Wide variety specific to immunotherapy, includes CNS-related immune toxicity (hypophysitis [57], hypothyroidism [58]), encephalitis, demyelinating polyneuropathy, encephalomyelitis [56]

Patients are evaluated for this type of delivery through use of noninvasive brain scans such as computed tomography (CT) or magnetic resonance imaging (MRI) to assess ventricle size. Typically, patients undergo a trial process involving temporary catheter placements, whereby patients are monitored in an intensive care unit for adverse reactions including cognitive changes, hallucinations, seizures, respiratory depression, or coma [59]. Moreover, patients are monitored as medication dosing and schedule are optimized for each individual [60]. If the trial is successful and the patient may need longer duration of treatment, he or she can undergo placement of a permanent implanted drug-delivery device. Drug reservoir devices, such as Ommaya or Rickham reservoirs, are implanted under the skin and contain a catheter that terminates in the lateral ventricle. Neurosurgeons may wish to utilize intraoperative neuro-navigation during placement to improve accuracy of placement of reservoir catheter. Patients are then able to receive delivery of therapeutics via small gauge needle administration through the skin, directly into the device reservoir. Although a trained health care provider administers the drug through reservoirs, repeated drug administration through these reservoirs is straight forward and well-tolerated by patients. This method is so efficacious that a number of clinical trials have been designed to deliver medications for late-infantile neuronal ceroid lipofuscinosis type 2 disease, mucopolysaccharidosis II, amyotrophic lateral sclerosis, and Parkinson's disease [35].

Implanted reservoirs can also be used for delivery of therapeutic cells into the brain. Our team at City of Hope designs and manufactures cell-based antiglioma therapies. Cellular therapies, which include engineered neural stem cells [61–63] and chimeric antigen receptor (CAR) cytotoxic T-cells [55,64,65], are injected into the CNS to target tumor cells. We can safely intracranially infuse as many as 10^8 engineered cells through an implanted reservoir/catheter system after tumor resection or biopsy. We have discovered that directly delivering CAR T cells into the ventricular system leads to their dispersion throughout the CNS and even spinal cisterns.

Although relatively rare, risks associated with placement of reservoirs are not benign and include intracerebral hemorrhage, infection, poor positioning of catheter, CSF leak, device malfunction (i.e., clotting), and risks associated with undergoing general anesthesia. A meta-analysis reviewed 35 published papers on the risks associated with reservoir placement and reported an overall infection rates between 0% and 27% and noninfectious complication rates ranging from 1% to 33%, with CSF leak, hemorrhage, catheter malposition, and catheter obstruction being the most common [35]. An additional risk of intraventricular administration drug administration is the increase in intracranial pressure, particularly with administration of large volumes [66]. To avoid this potential risk, most medical care providers withdraw CSF volumes equal to that being delivered prior to administration [67,68]. While the vast majority of patients retain their intraventricular reservoirs beyond the duration of their therapies [35], neurosurgeons must be readily available to repair or remove a device should a complication arise or should the patient wish to have it removed.

Lumbar intrathecal delivery

In 1899, a German surgeon named Dr August Bier, published the first well-recognized report on the use of spinal intrathecal medication, after he injected cocaine into his own intrathecal space, as well as in that of six other patients, to assess a novel anesthetic technique [69]. Subsequently, physicians added morphine to cocaine for intrathecal anesthetic to counteract adverse side effects of cocaine [70]. Since then, a substantial number of medications have been used for CNS analgesia including ziconotide, morphine, fentanyl, bupivacaine, sufentanil, all of which may be administered with or without clonidine for intractable pain, and baclofen for spasticity [71].

The major advantage of intrathecal delivery is the use of smaller doses of medications, with significant reduction in systemic side effects, such as respiratory depression and constipation with narcotics, and delirium, coma, and seizures with baclofen. Therefore, as opposed to when medications are given orally or intravenously, patients are much less likely to overdose from medication delivered intrathecally.

Intrathecal infusion pumps were developed in the 1970s to implement a steady, well-controlled delivery of opiates in patients [72], and since then, pumps have only become smaller and more efficient. Currently, intrathecal pumps are placed either temporarily with an external pump or permanently with a fully implanted device. A percutaneous catheter with an external pump is less invasive to place, and may be used in patients too sick to undergo surgery, or with a short life-expectancy, and may be used in patients for a brief drug trial [71]. The implanted pumps currently on the market are sophisticated and allow substantial adjustments per individual patient's needs (Fig. 2). For example, pumps might delivery drug in a fixed-rate or a variable-rate fashion, may allow the patient to self-delivery bolus as needed, may have a battery or be re-chargeable, and come in different reservoir chamber sizes to accommodate different volumes of medications. Many of these pumps are MRI-compatible and are interrogated or deactivated noninvasively with a hand-held wand. Patients with implanted pumps only require intermittent visits with medical care providers for pump refill, making this a convenient option for many patients.

Although there are no established guidelines for selecting patients for intrathecal pain pump placement, most physicians consider this option for patients with chronic pain (either cancer-related or not) or intractable spasticity that do not respond to maximum oral medications. Most physicians advocate for a trial period of intrathecal medication administration prior to recommending the placement of a permanent implant. An intrathecal drug trial may include a single or small number of injections of therapeutics into the lumbar intrathecal

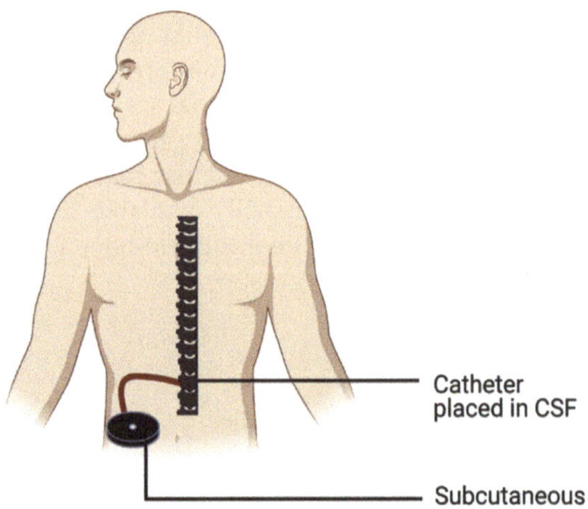

FIG. 2 The intrathecal pain pump has a catheter that terminates in the cerebrospinal fluid at the lumbar spine. The proximal end of the catheter is tunneled under the skin and is connected to a reservoir that is placed in a subcutaneous pocket.

space through a lumbar puncture, or as described above, through a temporary percutaneous pump placed under radiographic guidance.

Permanent device implantation takes place in the operating room, under general anesthesia. Two separate incisions are made to place the device: one on the lower back at midline, typically between the lumbar fourth and fifth vertebral bodies, and the second on the abdomen, with the side preoperatively selected by the patient. The posterior incision is used to deliver the pump catheter in between the spinous processes, into the dura, and is tunneled rostrally well-into the thecal space. The distal catheter is tunneled under the skin from the posterior incision to the anterior incision, where a subcutaneous pocket is created to hold the drug reservoir chamber. Risks associated with this procedure include bleeding, infection, reported at about 2%–5% [73,74], CSF leak, seroma or granuloma formation [71], long-term risk of pump malfunction or obstruction, and risk from general anesthesia. Should patients wish to have their pumps subsequently removed, a second surgery under general anesthesia is warranted, so this should be explained to patients prior to implantation.

Transplantation of stem cells and other therapeutic cells has proven promising for spinal cord injury in rodents. Neural precursor cells delivered via lumbar puncture to rats with thoracic spinal cord injuries reduced the size of spinal cord injury as well as provided functional recovery in motor and bladder function [75]. Lumbar intrathecal hematopoietic stem cell injections have also shown promise for animals with cervical spinal cord injury [76,77]. A human study enrolled 10 spinal cord injury patients who received 5 days of recombinant human granulocyte colony-stimulating factor, and then underwent aspiration of 100mL of bone marrow cells. 100×10^6 autologous mononuclear cells were collected from the aspirate and injected via lumbar puncture. This initial feasibility study showed delivering autologous bone marrow precursor cells via lumbar puncture was achievable, safe, and well tolerated by patients [78]. In another study, 16 patients with chronic spinal cord injury received either intrathecal autologous CD34+ cells labeled with magnetic nanoparticles or magnetic nanoparticles alone. These patients were then followed with MRI imaging, which showed CD34+ cells, but not magnetic beads alone, traveled to the site of spinal cord injury [79], suggesting this delivery modality shows clinical promise for this patient population.

Convection-enhanced delivery

Convection-enhanced delivery (CED), otherwise referred to as high-flow microinfusion, was developed in the 1990s by Dr Edward Oldfield and his team at the National Institute of Health [36,80]. This delivery system allows for the delivery of highly concentrated solutions, including nanoparticles, to intricately accurate target brain regions. The precision behind this technique delivers drug only to target brain tissue, such that healthy surrounding brain tissue is exposed to little drug, while the BBB restricts the infused agent from entering the systemic circulation. CED plays an important theoretical role in treating malignant gliomas, which have a high propensity to recur within centimeters of the first tumor [81], as well as carries applications to treat Parkinson's disease and Alzheimer's disease [82].

CED consists of an external, automated pump that continuously drives the flow of the desired therapeutic agent through a cannula or catheter that is stereotactically placed directly into the brain or tumor parenchyma, or just adjacent to the target region of infusion [36,80].

The inner cannula can then be delivered through the outer cannula such that its tip ends at the desired infusion point. In doing so, this two-cannulae technique maintains the integrity of the predetermined trajectory of the cannula placement [83]. Once the pump is activated, the pressure gradient at the tip of the catheter generates bulk flow of drug, rather than diffusion, through the interstitial space. Diffusive flow as described by Fick's law is dependent on tissue diffusivity, as well as drug concentration, gradient, and molecular weight [82]. This law generally describes why large molecular weight pharmaceuticals take longer to diffuse in brain tissue and may require greater and even dangerous concentrations to deliver them to target brain tissue.

With the use of sophisticated technology, such as MRI and sonograms, more complex CED models have been developed which allow for real-time tracking of infusing material during intraparenchymal delivery. For example, an exquisitely accurate computer-generated CED system was developed to use both anatomical and diffusion tensor magnetic resonance imaging as a source for data input. These data points are supplemented with a description of infusate movement using a stochastic differential equation that assesses both advective and diffusive terms. Predication of drug concentration at particular times and locations relative to the delivery site are either confirmed or manipulated using feedback from the imaging device. By following the movement of traceable material in the targeted region of the brain, these researchers were able to collect an array of information pertaining to varying concentration gradients at neighboring points in the tissue. Not only does this model allow for anisotropic variations but it also incorporates potential sinks for medication such as leakage into the subarachnoid space, drainage into cavities, and fluctuations in tissue clearance rates [84]. Further human trials conducted by these researchers have shown that treatment plans can be adjusted during the administration of the traceable material due to real-time feedback [85]. With sufficient intraoperative time, deviations from the predicted flow pattern and desired distribution can be immediately adjusted by manipulating delivery parameters such as injection site and infusion rate.

By controlling the variables that govern the efficacy of delivery, including drug particle size, surface area of particles, and fluid carrying the particles, CED enables a slow, controlled infusion of a desired agent. The drug delivery occurs in a manner that expands the extracellular space, similar to that seen in vasogenic edema, while preserving the integrity of the surrounding tissue. Indeed, CED has been used successfully deliver nanoparticles, particularly virus-sized particles as would be utilized in gene therapy to the brain [86]. To date, numerous human clinical trials have successfully demonstrated the delivery of nanoparticles using CED. Though a variety of particles including enzymes, antibodies, and viruses have been convected and delivered via CED. Most of these clinical trials have examined the delivery of receptor-directed toxins. Receptor-directed toxins consist of one protein that targets a particular feature on the cell surface and another protein that causes cytotoxicity [86]. The first reported clinical trial using CED was performed by Oldfield and utilized a chemotherapeutic agent selective for transferrin receptor Tf-CR107, a transferrin conjugate bound to a point-mutated diphtheria toxin, to treat malignant brain tumors [87].

In a phase I study, the delivery of Tf-CRM107 was well tolerated in patients, reduced tumor volume by more than 50% in 9 out of 10 patients, and resulted in a median survival of 75 weeks, compared to the 36 weeks in nonresponders [87]. In a phase II study, complete and partial tumor response without severe toxicity occurred in 35% of patients treated with

Tf-CRM107 [88]. Another immunotoxin clinical trial used cintredekin besudotox, a recombinant protein consisting of interleukin-13 (IL-13) and a truncated form of Pseudomonas exotoxin, PE38QQR. The initial phase I trial showed promising clinical results for the treatment of recurrent glioblastoma multiforme [89]; however, a phase III study (PRECISE) of 296 participants who received either CED of cintredekin besudotox postoperatively or a gliadel wafer implanted at the time of surgery demonstrated no survival benefit [90]. Unfortunately, it was found that only about 50% catheters were properly placed for this study, suggesting that proper catheter placement might have improved this clinical trial's overall outcome [91].

Viral particles for treatment of Parkinson's disease (PD) have also been successfully delivered through CED [92,93]. PD is characterized by loss of dopaminergic neurons, and ultimately severe decrease of dopamine in the putamen. Aromatic L-amino acid decarboxylase (AADC) is an enzyme in the dopamine pathway that converts L-dopa to dopamine. Adeno-associated virus (AAV) vector containing the human aromatic L-amino acid decarboxylase (hAADC) gene was convected into the putamen of 10 humans and 3 nonhuman primates with PD [93]. The primary endpoints of the study were safety and transgene expression after CED. Imaging with both MRI imaging and positron emission tomography (PET) scans showed appropriate targeting of drug to the putamen. Nonhuman primates also had immunohistochemical studies demonstrating that hAADC expression occurred in these regions. This result strongly supports the feasibility of using CED of nanoparticles such as virus for clinical gene therapy.

In summary, CED, which has been described as molecular neurosurgery, is a technique that can be used to deliver nanoparticles and various other therapeutics to target areas in the CNS. However, CED's utility for cellular therapies has not been investigated yet. It is likely that the slow infusion rate that is inherent to this technique (typically a few hours) may result in cell clumping or loss of cell viability that could potentially impact the efficacy of cell-based therapies. With advancements in in vivo cell imaging technologies, however, researchers should be able to optimize the infusion parameters to facilitate the application of this technique for CNS cellular therapies.

Methods to disrupt the BBB

It is well recognized that many CNS therapeutics are considerably more efficacious when delivered while the BBB is disrupted, however, permanently destroying the BBB can lead to permanent structural brain damage and death. Therefore, several invasive and noninvasive neurosurgical techniques have been developed to temporarily disrupt the BBB to provide a therapeutic window for intervention.

Chemical disruption of BBB

A number of endogenous physiological substances, including neurotransmitters, inflammatory markers, a variety of hormones [12], as well as physiological states of hypertension and hypercapnia, can all transiently open the BBB [94]. Unfortunately, the transient disruption to the BBB by these factors results in unpredictable barrier opening and inconsistent duration of opening [95]. Ideally, BBB disruption for therapeutic purposes should be transient and reversible. Hypertonic solutions, particularly by the use of intra-arterial mannitol, but also lactamide, saline, urea, and radiographic contrast agents, have been used for this purpose with success [94]. The intra-arterial injection of mannitol results in dramatic diffusion of fluid out of cells, dehydration, and shrinking of endothelial cells, ultimately opening the BBB tight junction for at least a number of hours [96] (Fig. 3). The delivery of chemotherapeutic agents following the disruption of BBB shows significant increase in medial survival time in patients with CNS lymphoma [97], as well as in patients with other chemo-responsive tumors [98]. Intra-arterial injection of osmotic agents, however, comes with inherent risks that patients must be fully informed about. First of all, the procedure requires general anesthesia and therefore adopts all associated anesthetic risks, which is considered an invasive procedure with possible damage to arteries and bleeding and results in inconsistent delivery of goal medications to the CNS [99].

A major limitation of disrupting BBB prior to chemotherapeutic agent delivery is neurotoxicity. For example, although doxorubicin and cisplatin are well-tolerated systemically, they can result in significant neurotoxicity when administrated with BBB disruption in canine and rodent models [100,101]. Moreover, additional systemic toxic risks associated with mannitol-induced BBB disruption include hyperkalemic ventricular tachycardia [102], seizures [103], and vasovagal episodes with bradycardia and hypotension [104].

Mechanical BBB disruption

To circumvent the toxic side effects of osmotic agents, scientists and physicians developed focused ultrasound (FUS) methods to disrupt BBB. FUS utilizes transcranially directed low-frequency ultrasound waves to briefly open the tight junctions in BBB endothelial cells [105]. One of the earliest uses of this technique was to deliver liposome-encapsulated doxorubicin across an FUS-disrupted membrane in a rat glioma model [106]. Because the FUS systems at

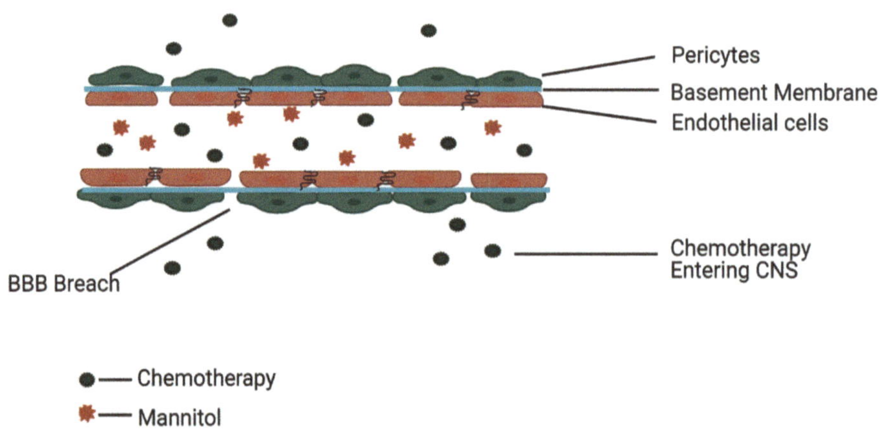

FIG. 3 Intra-arterial hyperosmolar injections result in endothelial cell dehydration and contraction. This disrupts the endothelial cell tight junctions, and allows for a leaky blood-brain barrier.

this time generated excessive heat on the skull, this technique required a craniotomy [107], and therefore was largely restricted to investigational animal studies.

Newer FUS ablation systems, however, have been developed to overcome this problem through active cooling of the scalp during treatment, as well as utilizing transducers with large apertures to distribute ultrasound energy over larger skull surface [108]. When used in conjunction with acoustic simulation calculated by skull CT scans to optimize phase and amplitude corrections, as well as MR temperature imaging to measure heat, these systems are able to achieve high intensities of energy to deliver thermal ablation through the human skull for clinical use [99]. FUS results in transient BBB opening, which can close within 6h of the procedure [109], and once closed, resumes normal physiological function [110,111].

More recently, the FUS technique has been enhanced to include the use of circulating microbubbles (Fig. 4). Intravenous administration of microbubble contrast agent allows the BBB to remain consistently open, without resulting in harm to underlying brain [112]. The oscillation of microbubbles during FUS leads to greater disruption of the BBB with lower frequencies of ultrasound [99], hence resulting in significantly less heat damage to skull or underlying brain [113]. This combination allowed a novel way to disrupt the BBB in a targeted, repeatable noninvasive fashion [99].

Laster interstitial thermotherapy (LITT), also known as MRI-guided laser ablation, is a minimally invasive surgical procedure in which a small hole is made in the skull under general anesthesia and a laser fiber is introduced into a brain tumor under image guidance. The laser transmits energy at the fiber tip, which is monitored in real-time MRI live temperature-mapping of the patient's brain. This precisely controlled heat generated at the fiber tip causes rapid, irreversible tissue damage to precisely destroy tumor and not surrounding, healthy tissue [114]. This thermotherapy results in cell membrane destruction in endothelial cells

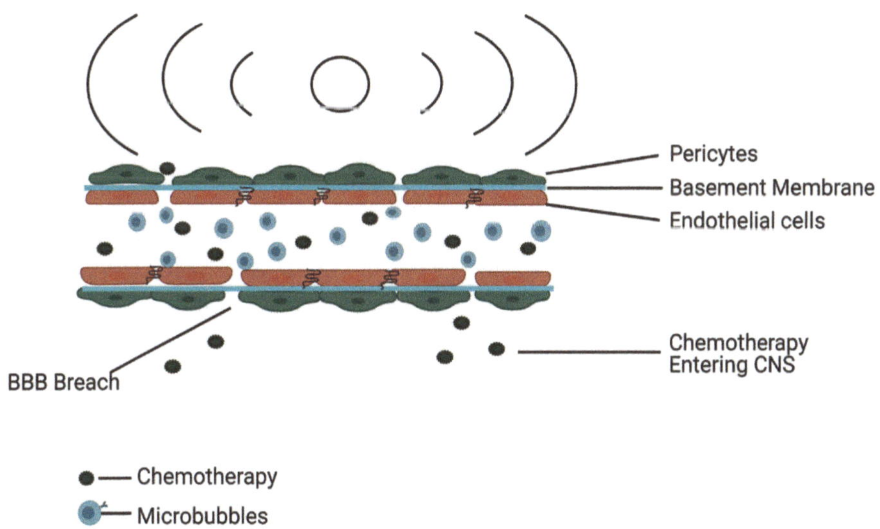

FIG. 4 During focused ultrasound delivery, microbubbles injected into the vasculature result in mechanical disruption of the endothelial cell tight junctions. This leads to breaches in the blood-brain barrier, allowing increased permeability of drugs into the central nervous system.

to disrupt the BBB and allow chemotherapeutic entry into the CNS [115]. Contrasted-brain MRI after LITT procedures show an area of enhancement around the thermal ablation area, reflecting the leaky BBB state [116], which has been shown to remain open for several days following treatment in a nonhuman primate model [117]. Indeed, mice with GBM (orthotopically implanted GL261 tumor cells) underwent LITT and had disrupted BBB for 4 weeks following treatment. Tumor-implanted mice then showed an increase in Dextran 10 and 70 kDa and human IgG extravasation after intravenous injection of tracers and authors showed by both immunohistochemistry and electron microscopy that LITT increases brain endothelial transcytosis to increase BBB permeability [118]. This technology to increase BBB permeability is currently only investigational but has the potential to be exploited for human therapy benefit.

Conclusion

In conclusion, although there have been improved adjuvant therapies for treatment of CNS tumors, such as immunotherapies and chemotherapies, there is considerable effort in designing creative ways to deliver high concentrations of novel-targeted agents directly to brain tumors. Directed treatments may reduce unwanted side effects seen in systemic delivery of therapeutic agents, and intracranial delivery of higher concentration of therapeutic agents would likely result in more efficacious treatments. A major limitation to development of new cell-based treatments involves circumventing the BBB. As unique ways of bypassing the BBB are developed such as via physical or chemical BBB disruption, or through the development of new engineered cells that utilize the barrier's endogenous endothelial receptors, novel CNS tumor therapeutics are on the horizon.

References

[1] Cancer.Net for American Society of Clinical Oncology. Brain tumor: statistics, 2020. [cited 2020]; Available from: https://www.cancer.net/cancer-types/brain-tumor/statistics.
[2] The American Cancer Society Medical and Editorial Content Team. Types of cancer that develop in children, 2019. [cited 2020]; Available from: https://www.cancer.org/cancer/cancer-in-children/types-of-childhood-cancers.html.
[3] PDQ Adult Treatment Editorial Board. Adult central nervous system tumors treatment (PDQ®): health professional version. PDQ cancer information summaries [Internet]. [Online] 2020 August 14, 2020. [cited 2020]; Available from: https://www.ncbi.nlm.nih.gov/books/NBK65982/.
[4] Ferlay J, Shin HR, Bray F, Forman D, Parkin M. Cancer incidence and mortality worldwide; 2013 [cited 2020; IARC CancerBase No. 11].
[5] American Cancer Society. American Cancer Society: cancer facts and figures 2020; 2020 [cited 2020].
[6] Louis DN, et al. The 2016 World Health Organization classification of tumors of the central nervous system: a summary. Acta Neuropathol 2016;131(6):803–20.
[7] Gousias K, et al. Descriptive epidemiology of cerebral gliomas in northwest Greece and study of potential predisposing factors, 2005-2007. Neuroepidemiology 2009;33(2):89–95.
[8] Larjavaara S, et al. Incidence of gliomas by anatomic location. Neuro Oncol 2007;9(3):319–25.
[9] Ostrom QT, et al. The epidemiology of glioma in adults: a "state of the science" review. Neuro Oncol 2014;16 (7):896–913.
[10] Oberoi RK, et al. Strategies to improve delivery of anticancer drugs across the blood-brain barrier to treat glioblastoma. Neuro Oncol 2016;18(1):27–36.

[11] Theodorakis PE, et al. Physical insights into the blood-brain barrier translocation mechanisms. Phys Biol 2017;14(4), 041001.
[12] Neuwelt EA. Mechanisms of disease: the blood-brain barrier. Neurosurgery 2004;54(1):131–40 [discussion 141–142].
[13] Guerra M, Blazquez JL, Rodriguez EM. Blood-brain barrier and foetal-onset hydrocephalus, with a view on potential novel treatments beyond managing CSF flow. Fluids Barr CNS 2017;14(1):19.
[14] Neuwelt EA, et al. Engaging neuroscience to advance translational research in brain barrier biology. Nat Rev Neurosci 2011;12(3):169–82.
[15] Abbott NJ, Romero IA. Transporting therapeutics across the blood-brain barrier. Mol Med Today 1996;2(3):106–13.
[16] Dong X. Current strategies for brain drug delivery. Theranostics 2018;8(6):1481–93.
[17] Banks WA. Characteristics of compounds that cross the blood-brain barrier. BMC Neurol 2009;9(Suppl 1):S3.
[18] Banks WA. Physiology and pathology of the blood-brain barrier: implications for microbial pathogenesis, drug delivery and neurodegenerative disorders. J Neurovirol 1999;5(6):538–55.
[19] Zhou SF. Structure, function and regulation of P-glycoprotein and its clinical relevance in drug disposition. Xenobiotica 2008;38(7–8):802–32.
[20] Choong E, et al. The permeability P-glycoprotein: a focus on enantioselectivity and brain distribution. Expert Opin Drug Metab Toxicol 2010;6(8):953–65.
[21] Miller DS, Bauer B, Hartz AM. Modulation of P-glycoprotein at the blood-brain barrier: opportunities to improve central nervous system pharmacotherapy. Pharmacol Rev 2008;60(2):196–209.
[22] Schlageter KE, et al. Microvessel organization and structure in experimental brain tumors: microvessel populations with distinctive structural and functional properties. Microvasc Res 1999;58(3):312–28.
[23] Groothuis DR. The blood-brain and blood-tumor barriers: a review of strategies for increasing drug delivery. Neuro Oncol 2000;2(1):45–59.
[24] Jain RK. Barriers to drug delivery in solid tumors. Sci Am 1994;271(1):58–65.
[25] Rockwell S. Use of hypoxia-directed drugs in the therapy of solid tumors. Semin Oncol 1992; 19(4 Suppl 11):29–40.
[26] Bidros DS, Vogelbaum M, Liu JK. Barriers to delivery of therapeutics to brain tumors. In: Winn HR, editor. Neurological surgery. Elsevier; 2011. p. 175–89.
[27] Khan S, et al. Formulation of intranasal mucoadhesive temperature-mediated in situ gel containing ropinirole and evaluation of brain targeting efficiency in rats. J Drug Target 2010;18(3):223–34.
[28] Chen XQ, et al. Delivery of nerve growth factor to the brain via the olfactory pathway. J Alzheimers Dis 1998;1 (1):35–44.
[29] Thorne RG, et al. Delivery of insulin-like growth factor-I to the rat brain and spinal cord along olfactory and trigeminal pathways following intranasal administration. Neuroscience 2004;127(2):481–96.
[30] Hanson LR, Frey 2nd WH. Intranasal delivery bypasses the blood-brain barrier to target therapeutic agents to the central nervous system and treat neurodegenerative disease. BMC Neurosci 2008;9(Suppl 3):S5.
[31] Jadhav KR, Gambhire M, Shaikh IM, Kadam VJ, Pisal SS. Nasal drug delivery system—factors affecting and applications. Curr Drug Ther 2007;2:27–38.
[32] Mittal D, et al. Insights into direct nose to brain delivery: current status and future perspective. Drug Deliv 2014;21(2):75–86.
[33] Lim ST, Forbes B, Brown MB, Martin GP. Physiological factors affecting nasal drug delivery. In: Touitou E, Barry BW, editors. Enhancement in drug delivery. Boca Raton: CRC Press; 2007.
[34] Pillon DJ, Arnold J, Meezan E. Nasal delivery of peptides. In: Touitou E, Barry BW, editors. Enhancement in drug delivery. Boca Raton: CRC Press; 2007.
[35] Cohen-Pfeffer JL, et al. Intracerebroventricular delivery as a safe, long-term route of drug administration. Pediatr Neurol 2017;67:23–35.
[36] Bobo RH, et al. Convection-enhanced delivery of macromolecules in the brain. Proc Natl Acad Sci U S A 1994;91 (6):2076–80.
[37] Pompe RS, et al. Intraventricular methotrexate as part of primary therapy for children with infant and/or metastatic medulloblastoma: feasibility, acute toxicity and evidence for efficacy. Eur J Cancer 2015;51(17):2634–42.
[38] Brugnoletti F, et al. Recurrent intrathecal methotrexate induced neurotoxicity in an adolescent with acute lymphoblastic leukemia: serial clinical and radiologic findings. Pediatr Blood Cancer 2009;52(2):293–5.
[39] Olmos-Jimenez R, et al. Stability of four standardized preparations of methotrexate, cytarabine, and hydrocortisone for intrathecal use. J Oncol Pharm Pract 2016;22(5):659–65.

[40] Jabbour E, et al. Neurologic complications associated with intrathecal liposomal cytarabine given prophylactically in combination with high-dose methotrexate and cytarabine to patients with acute lymphocytic leukemia. Blood 2007;109(8):3214–8.
[41] BC Cancer Agency Cancer Drug Manual. Cytarabine, 2014. Available from: http://www.bccancer.bc.ca/drug-database-site/Drug%20Index/Cytarabine_monograph_1May2014.pdf.
[42] Pullen J, et al. Extended triple intrathecal chemotherapy trial for prevention of CNS relapse in good-risk and poor-risk patients with B-progenitor acute lymphoblastic leukemia: a pediatric oncology group study. J Clin Oncol 1993;11(5):839–49.
[43] Tomita N, et al. Prophylactic intrathecal methotrexate and hydrocortisone reduces central nervous system recurrence and improves survival in aggressive non-Hodgkin lymphoma. Cancer 2002;95(3):576–80.
[44] Patten SB, Neutel CI. Corticosteroid-induced adverse psychiatric effects: incidence, diagnosis and management. Drug Saf 2000;22(2):111–22.
[45] Berg SL, Chamberlain MC. Systemic chemotherapy, intrathecal chemotherapy, and symptom management in the treatment of leptomeningeal metastasis. Curr Oncol Rep 2003;5(1):29–40.
[46] Martin Algarra S, et al. Severe polyneuropathy and motor loss after intrathecal thiotepa combination chemotherapy: description of two cases. Anticancer Drugs 1990;1(1):33–5.
[47] Rubenstein JL, et al. Phase I study of intraventricular administration of rituximab in patients with recurrent CNS and intraocular lymphoma. J Clin Oncol 2007;25(11):1350–6.
[48] Pluchart H, et al. Long-term survivor with intrathecal and intravenous trastuzumab treatment in metastatic breast cancer. Target Oncol 2016;11(5):687–91.
[49] Figura NB, et al. Intrathecal trastuzumab in the management of HER2+ breast leptomeningeal disease: a single institution experience. Breast Cancer Res Treat 2018;169(2):391–6.
[50] Freyer CW, Yaghmour G, Jennings K, Dhanapal V. Drug-induced aseptic meningitis associated with intrathecal trastuzumab. J Pharm Technol 2014;30(2):43–7.
[51] Malhotra A, et al. Oncolytic virotherapy for head and neck cancer: current research and future developments. Oncolytic Virother 2015;4:83–93.
[52] Kane JR, et al. Sui generis: gene therapy and delivery systems for the treatment of glioblastoma. Neuro Oncol 2015;17(Suppl 2):ii24–36.
[53] Shilpa P, et al. Oncolytic viruses in head and neck cancer: a new ray of hope in the management protocol. Ann Med Health Sci Res 2014;4(Suppl 3):S178–84.
[54] Mullen JT, Tanabe KK. Viral oncolysis. Oncologist 2002;7(2):106–19.
[55] Brown CE, et al. Regression of glioblastoma after chimeric antigen receptor T-cell therapy. N Engl J Med 2016;375(26):2561–9.
[56] McGranahan T, Li G, Nagpal S. History and current state of immunotherapy in glioma and brain metastasis. Ther Adv Med Oncol 2017;9(5):347–68.
[57] Blansfield JA, et al. Cytotoxic T-lymphocyte-associated antigen-4 blockage can induce autoimmune hypophysitis in patients with metastatic melanoma and renal cancer. J Immunother 2005;28(6):593–8.
[58] Downey SG, et al. Prognostic factors related to clinical response in patients with metastatic melanoma treated by CTL-associated antigen-4 blockade. Clin Cancer Res 2007;13(22 Pt 1):6681–8.
[59] Sheth S. Intraventricular or intracisternal drug delivery, 2016. Available from: https://www.neuromodulation.com/drug-delivery.
[60] Lee DJ, et al. Intracerebroventricular opiate infusion for refractory head and facial pain. World J Clin Cases 2014;2(8):351–6.
[61] Aboody KS, et al. Neural stem cell-mediated enzyme/prodrug therapy for glioma: preclinical studies. Sci Transl Med 2013;5(184), 184ra59.
[62] Portnow J, et al. Feasibility of intracerebrally administering multiple doses of genetically modified neural stem cells to locally produce chemotherapy in glioma patients. Cancer Gene Ther 2021;28(3–4):294–306.
[63] Portnow J, et al. Neural stem cell-based anticancer gene therapy: a first-in-human study in recurrent high-grade glioma patients. Clin Cancer Res 2017;23(12):2951–60.
[64] Mirzaei HR, et al. Chimeric antigen receptors T cell therapy in solid tumor: challenges and clinical applications. Front Immunol 2017;8:1850.
[65] Wang D, et al. Chlorotoxin-directed CAR T cells for specific and effective targeting of glioblastoma. Sci Transl Med 2020;12(533).
[66] Patel MM, et al. Getting into the brain: approaches to enhance brain drug delivery. CNS Drugs 2009;23(1):35–58.

[67] Blaney SM, et al. A phase-1 pharmacokinetic optimal dosing study of intraventricular topotecan for children with neoplastic meningitis: a pediatric brain tumor consortium study. Pediatr Blood Cancer 2013;60(4):627–32.

[68] Rubenstein JL, et al. Multicenter phase 1 trial of intraventricular immunochemotherapy in recurrent CNS lymphoma. Blood 2013;121(5):745–51.

[69] Bier A. Attempts over Cocainisirung of the Ruckenmarkers. Langenbecks Arch Klin Chir Ver Dtsch Z Chir 1899;51:361–9.

[70] Onofrio BM, Yaksh TL, Arnold PG. Continuous low-dose intrathecal morphine administration in the treatment of chronic pain of malignant origin. Mayo Clin Proc 1981;56(8):516–20.

[71] Bottros MM, Christo PJ. Current perspectives on intrathecal drug delivery. J Pain Res 2014;7:615–26.

[72] Blackshear PJ, et al. The implantable infusion pump: a new concept in drug delivery. Med Prog Technol 1979;6 (4):149–61.

[73] Kumar K, Nath R, Wyant GM. Treatment of chronic pain by epidural spinal cord stimulation: a 10-year experience. J Neurosurg 1991;75(3):402–7.

[74] Lang P. The treatment of chronic pain by epidural spinal cord stimulation—a 15 year follow up; present status. Axone 1997;18(4):71–3.

[75] Neuhuber B, et al. Stem cell delivery by lumbar puncture as a therapeutic alternative to direct injection into injured spinal cord. J Neurosurg Spine 2008;9(4):390–9.

[76] Bakshi A, et al. Minimally invasive delivery of stem cells for spinal cord injury: advantages of the lumbar puncture technique. J Neurosurg Spine 2004;1(3):330–7.

[77] Lepore AC, et al. Neural precursor cells can be delivered into the injured cervical spinal cord by intrathecal injection at the lumbar cord. Brain Res 2005;1045(1–2):206–16.

[78] Callera F, do Nascimento RX. Delivery of autologous bone marrow precursor cells into the spinal cord via lumbar puncture technique in patients with spinal cord injury: a preliminary safety study. Exp Hematol 2006;34(2):130–1.

[79] Callera F, de Melo CM. Magnetic resonance tracking of magnetically labeled autologous bone marrow CD34+ cells transplanted into the spinal cord via lumbar puncture technique in patients with chronic spinal cord injury: CD34+ cells' migration into the injured site. Stem Cells Dev 2007;16(3):461–6.

[80] Morrison PF, et al. High-flow microinfusion: tissue penetration and pharmacodynamics. Am J Physiol 1994;266 (1 Pt 2):R292–305.

[81] Barker 2nd FG, et al. Survival and functional status after resection of recurrent glioblastoma multiforme. Neurosurgery 1998;42(4):709–20 [discussion 720–723].

[82] Mehta AM, Sonabend AM, Bruce JN. Convection-enhanced delivery. Neurotherapeutics 2017;14(2):358–71.

[83] Lonser RR, et al. Real-time image-guided direct convective perfusion of intrinsic brainstem lesions [Technical note]. J Neurosurg 2007;107(1):190–7.

[84] Raghavan R, et al. Convection-enhanced delivery of therapeutics for brain disease, and its optimization. Neurosurg Focus 2006;20(4):E12.

[85] Sampson JH, et al. Induction of hyperintense signal on T2-weighted MR images correlates with infusion distribution from intracerebral convection-enhanced delivery of a tumor-targeted cytotoxin. Am J Roentgenol 2007;188(3):703–9.

[86] Chen MY, et al. Surface properties, more than size, limiting convective distribution of virus-sized particles and viruses in the central nervous system. J Neurosurg 2005;103(2):311–9.

[87] Laske DW, Youle RJ, Oldfield EH. Tumor regression with regional distribution of the targeted toxin TF-CRM107 in patients with malignant brain tumors. Nat Med 1997;3(12):1362–8.

[88] Weaver M, Laske DW. Transferrin receptor ligand-targeted toxin conjugate (Tf-CRM107) for therapy of malignant gliomas. J Neurooncol 2003;65(1):3–13.

[89] Kunwar S, et al. Direct intracerebral delivery of cintredekin besudotox (IL13-PE38QQR) in recurrent malignant glioma: a report by the Cintredekin Besudotox Intraparenchymal study group. J Clin Oncol 2007;25(7):837–44.

[90] Kunwar S, et al. Phase III randomized trial of CED of IL13-PE38QQR vs Gliadel wafers for recurrent glioblastoma. Neuro Oncol 2010;12(8):871–81.

[91] Sampson JH, et al. Poor drug distribution as a possible explanation for the results of the PRECISE trial. J Neurosurg 2010;113(2):301–9.

[92] Richardson RM, et al. Interventional MRI-guided putaminal delivery of AAV2-GDNF for a planned clinical trial in Parkinson's disease. Mol Ther J Am Soc Gene Ther 2011;19(6):1048–57.

[93] Valles F, et al. Qualitative imaging of adeno-associated virus serotype 2-human aromatic L-amino acid decarboxylase gene therapy in a phase I study for the treatment of Parkinson disease. Neurosurgery 2010;67 (5):1377–85.

[94] Kroll RA, Neuwelt EA. Outwitting the blood-brain barrier for therapeutic purposes: osmotic opening and other means. Neurosurgery 1998;42(5):1083–99 [discussion 1099–1100].
[95] Neuwelt EA. In: Neuwelt EA, editor. Implications of the blood-brain barrier and its manipulation, vols 1 and 2. New York: Plenum Press; 1989.
[96] Rapoport SI, Robinson PJ. Tight-junctional modification as the basis of osmotic opening of the blood-brain barrier. Ann N Y Acad Sci 1986;481:250–67.
[97] Dahlborg SA, et al. Non-AIDS primary CNS lymphoma: first example of a durable response in a primary brain tumor using enhanced chemotherapy delivery without cognitive loss and without radiotherapy. Cancer J Sci Am 1996;2(3):166–74.
[98] Doolittle ND, et al. Blood-brain barrier disruption for the treatment of malignant brain tumors: the National Program. J Neurosci Nurs 1998;30(2):81–90.
[99] Aryal M, et al. Ultrasound-mediated blood-brain barrier disruption for targeted drug delivery in the central nervous system. Adv Drug Deliv Rev 2014;72:94–109.
[100] Neuwelt EA, et al. Pharmacology and neurotoxicity of cis-diamminedichloroplatinum, bleomycin, 5-fluorouracil, and cyclophosphamide administration following osmotic blood-brain barrier modification. Cancer Res 1983;43(11):5278–85.
[101] Neuwelt EA, et al. Neurotoxicity of chemotherapeutic agents after blood-brain barrier modification: neuropathological studies. Ann Neurol 1983;14(3):316–24.
[102] Seto A, et al. Ventricular tachycardia caused by hyperkalemia after administration of hypertonic mannitol. Anesthesiology 2000;93(5):1359–61.
[103] Doolittle ND, et al. Safety and efficacy of a multicenter study using intraarterial chemotherapy in conjunction with osmotic opening of the blood-brain barrier for the treatment of patients with malignant brain tumors. Cancer 2000;88(3):637–47.
[104] Bellavance MA, Blanchette M, Fortin D. Recent advances in blood-brain barrier disruption as a CNS delivery strategy. AAPS J 2008;10(1):166–77.
[105] Hynynen K, et al. Focal disruption of the blood-brain barrier due to 260-kHz ultrasound bursts: a method for molecular imaging and targeted drug delivery. J Neurosurg 2006;105(3):445–54.
[106] Etame AB, et al. Focused ultrasound disruption of the blood-brain barrier: a new frontier for therapeutic delivery in molecular neurooncology. Neurosurg Focus 2012;32(1):E3.
[107] Heimburger RF. Ultrasound augmentation of central nervous system tumor therapy. Indiana Med 1985;78 (6):469–76.
[108] Hynynen K, McDannold N. MRI guided and monitored focused ultrasound thermal ablation methods: a review of progress. Int J Hyperthermia 2004;20(7):725–37.
[109] Burgess A, et al. Focused ultrasound-mediated drug delivery through the blood-brain barrier. Expert Rev Neurother 2015;15(5):477–91.
[110] McDannold N, et al. Temporary disruption of the blood-brain barrier by use of ultrasound and microbubbles: safety and efficacy evaluation in rhesus macaques. Cancer Res 2012;72(14):3652–63.
[111] McDannold N, et al. MRI-guided targeted blood-brain barrier disruption with focused ultrasound: histological findings in rabbits. Ultrasound Med Biol 2005;31(11):1527–37.
[112] Hynynen K, et al. Noninvasive MR imaging-guided focal opening of the blood-brain barrier in rabbits. Radiology 2001;220(3):640–6.
[113] McDannold N, King RL, Hynynen K. MRI monitoring of heating produced by ultrasound absorption in the skull: in vivo study in pigs. Magn Reson Med 2004;51(5):1061–5.
[114] Sloan AE, et al. Results of the NeuroBlate system first-in-humans phase I clinical trial for recurrent glioblastoma: clinical article. J Neurosurg 2013;118(6):1202–19.
[115] Schober R, et al. Fine structure of zonal changes in experimental Nd:YAG laser-induced interstitial hyperthermia. Lasers Surg Med 1993;13(2):234–41.
[116] Hawasli AH, et al. Stereotactic laser ablation of high-grade gliomas. Neurosurg Focus 2014;37(6):E1.
[117] Nakagawa M, et al. Acute effects of interstitial hyperthermia on normal monkey brain—magnetic resonance imaging appearance and effects on blood-brain barrier. Neurol Med Chir (Tokyo) 1994;34(10):668–75.
[118] Salehi A, Paturu M, Caine M, Klein R, Kim A. Laser therapy enhances blood-brain barrier and blood-tumor barrier permeability in a mouse model of glioblastoma. Neurosurgery 2019;66(Suppl 1):135.

CHAPTER

13

Diagnostic methods to assess the numbers, phenotype, and function of primary and engineered NK cells: Methods to predict prognosis and treatment outcome

Meng-Wei Ko[a], *Kawaljit Kaur*[a], *Po-Chun Chen*[a], *Barbara Breznik*[a,e], *Emanuela Senjor*[a,c,d], *Nishant Chovatiya*[a], *Paul Wong*[a], *Tamara Lah Turnsek*[e], *Janko Kos*[c,d], and *Anahid Jewett*[a,b]

[a]Division of Oral Biology and Medicine, The Jane and Jerry Weintraub Center for Reconstructive Biotechnology, University of California Los Angeles (UCLA), Los Angeles, CA, United States [b]The Jonsson Comprehensive Cancer Center, UCLA School of Dentistry and Medicine, Los Angeles, CA, United States [c]Department of Biotechnology, Jožef Stefan Institute, Ljubljana, Slovenia [d]Faculty of Pharmacy, University of Ljubljana, Ljubljana, Slovenia [e]Department of Genetic Toxicology and Cancer Biology, National Institute of Biology, Ljubljana, Slovenia

Abstract

Natural killer (NK) cells are the major immune effectors known to reduce tumor burden by inducing direct killing or differentiation of a number of different cancer stem cells/undifferentiated or poorly differentiated tumor cells. Differentiated tumor cells, in comparison to their stem-like counterparts, are less capable to grow and metastasize, so NK-induced differentiation leads to tumor growth inhibition and curtailment of tumor metastasis. In this chapter, we present an overview of phenotypic characteristics and functions of NK cells that play a crucial role in targeting tumors, and methodologies to test these characteristics and functions. In addition, we discuss the potential loss of NK functions in cancer patients, and how this loss of NK cell functions could be one of the contributing factors for the induction and progression of cancer. Moreover, we report the

https://doi.org/10.1016/B978-0-12-822620-9.00013-6

methodologies to test the surface and functional characteristics of NK cells. Finally, we discuss the significance of incorporating NK cell function testing as one of the diagnostic approaches to assess the functions of NK cells in high-risk cancer group and in cancer patients, as this assessment could be an ideal approach to optimize treatment strategies.

Abbreviations

ADCC	antibody-dependent cellular cytotoxicity (ADCC)
BLI	bioluminescence imaging
CD	cluster of differentiation
CRA	chromium release assay
CSCs	cancer stem cells
GM-CSF	granulocyte-macrophage colony-stimulating factor
HSCs	hematopoietic stem cells
IFN-γ	interferon-gamma
LDH	lactate dehydrogenase
mAb	monoclonal antibody
MCP	monocyte chemoattractant protein
MHC-class I	major histocompatibility complex molecule class I
NK cells	natural killer cells
OC	osteoclasts
PBMCs	peripheral blood mononuclear cells
RANTES	regulated on activation, normal T cell expressed and secreted
rhIL-2	recombinant human IL-2
RLU	relative light unit
TNF-α	tumor necrosis factor-α
TVA	target visualization assay

Conflict of interest

No potential conflicts of interest were disclosed.

Introduction

Natural killer cells: Basic biology and background

Peripheral blood mononuclear cells (PBMCs) are derived from hematopoietic stem cells through hematopoiesis and are composed of T cells, B cells, natural killer (NK) cells, and monocytes. Each cell type exhibits a certain percentage within the total PBMC population, with NK cells accounting for 5%–10% and can serve as an indicator of individuals' health [1]. As a key effector in the immune system, NK cells mediate selection and differentiation of cancer stem cells (CSCs) through lysis and secreted IFN-γ and/or membrane-bound TNF-α, respectively. Unlike cytotoxic CD8+ T cells, NK cells do not require priming with antigen prior to launching their cytotoxic function. Instead, their functions are conducted by the sum of the signals received through both activating and inhibitory receptors on their surface. Also, NK cells are responsible for shaping the tumor microenvironment and differentiating tumors through cytokine and chemokine secretion. As a result of these actions from NK cells, cancer cells lose their capabilities of self-renewal and proliferation and gain sensitivities to chemotherapeutic drugs, ultimately leading to the curtailment of tumor growth and metastasis.

Dysfunctional NK cells at the preneoplastic stage of tumorigenesis and in cancer patients; defects in multiple NK functions.

However, when functions of NK cells are suppressed or defective, CSCs escape from the boundaries of tumor surveillance and start proliferating, which results in tumor development. It was previously reported that NK cell functions were suppressed at the preneoplastic and neoplastic stages of tumorigenesis [2]. Therefore, incorporating NK cell functional evaluations into routine clinical examinations for individuals, who have genetic predispositions to developing cancer due to the inheritance of an abnormal gene or a combination of genetic and environmental factors, could help in early detection of immune dysfunction. Early detection of NK cell dysfunction could help to tailor NK cell-based immunotherapies to prevent and control cancer before the disease becomes uncontrollable. In this chapter, we present an overview of the evaluations of major parameters of NK cell function, along with tools and methodologies to assess those parameters. Furthermore, we provide the characteristics of NK cells in cancer patients and discuss why would the assessment of NK cell function in clinical setting potentially assist in customizing cancer immunotherapies to improve the treatment outcomes and prognostic factors in cancer patients.

It is well known that both cytotoxic function and IFN-γ secretion by the NK cells are compromised in cancer patients [3], and these are important underlying mechanisms for the survival and maintenance of tumors in cancer patients since NK cells are the primary effectors to target CSCs and aggressive tumors. Many mechanisms have been attributed to the loss of NK cell function, including downmodulation and decrease in important receptors, such as CD16 [4], NKG2D [5], zeta chain [6], decreased survival and expansion of NK cells [7], and decreased secretion of important cytokines such as IFN-γ [7]. Loss of NK cells not only is the key factor in decreased lysis of CSCs/undifferentiated tumors but it is also important in decreased secretion of IFN-γ that contributes to the increase in the frequency and load of undifferentiated tumors, since we have shown previously that IFN- γ is an important cytokine mediating differentiation of the tumor cells [8]. In addition, when using the same amounts of IFN-γ secreted from healthy and cancer patients' NK cells, those from cancer patients had much lower ability to differentiate tumors when compared to IFN-γ secreted from healthy individuals [7]. This finding is of particular interest since it has translational significance. Not only restoration of IFN-γ secretion by patient NK cells is important for the effective control of tumors in cancer patients but also the functional competence of secreted IFN-γ in differentiation of tumors is equally important.

We have recently demonstrated that feeding high-fat calorie diet (HFCD) to mice with pancreatic KRAS mutation severely inhibited NK cell function at the preneoplastic stage of pancreatic cancer [9,10]. NK cytotoxicity was decreased in the peripheral blood, spleen, pancreas and peripancreatic tissue in mice with KRAS mutation that were fed with HFCD, whereas in bone marrow, NK cytotoxicity was increased when compared to wild-type (WT) mice fed with lean control diet (CD) [10]. Mice with KRAS mutation fed with HFCD demonstrated the highest reduction in the expansion and function of NK cells, followed by mice with KRAS mutation which were fed with CD, and WT mice fed with HFCD as compared to WT mice fed with CD [10]. Interestingly, NK cells cultured with autologous monocytes from mice with KRAS mutation (KC) and WT mice fed with HFCD exhibited decreased expansion, cytotoxicity, and IFN-γ secretion. However, when NK cells from WT mice fed with CD were cultured with osteoclasts from KC mice fed with HFCD, NK cells exhibited no/lower cytotoxicity in the presence of increased cytokine secretion [10]. Thus, although NK cytotoxicity was always suppressed, depending on the extent of defect either in NK cells or in their interacting cells or both within the microenvironment, distinct profiles of IFN-γ secretion could be obtained [10]. These results suggested that NK functional loss could be selective depending on the extent and levels of suppression, which NK cells receive from the microenvironment. Loss of expansion and decreased cytotoxicity could precede loss of IFN-γ secretion, and under the most significant inactivation of NK cells all the key functions of NK cells can be compromised.

Rationale for establishing comprehensive diagnostic and prognostic tests to monitor patient NK cells before and after treatment

To be successful in eradicating cancer, NK cells have to withstand the suppressive nature of the tumor microenvironment, therefore, expanded NK cells should be able to maintain their viability and function when interacting with the tumor cells and other suppressive cells in TME. Such assessments, including those mentioned above, can be made along with the NK treatment strategies by using 3D tumor pieces from the patients to culture with the engineered NK cells, as well as the 2D assessments of the rate of expansion, cytotoxicity, and secretion and function of IFN-γ, to ensure that these tumors do not induce either functional inactivation or cell death in the expanded NK cells and that these cells are capable of eliminating tumors in vivo. The effectiveness of such 3D tests was shown in our recent work when functionally competent NK cells were cultured with tumorospheres, which were shown to penetrate and lyse the tumors by confocal microscopy. Therefore, using a set of diagnostic and prognostic tests are extremely important and should be designed to test the expandability and function of NK cells before, after and along with the patient treatment to safeguard against unwanted outcomes. The diagnostic tests listed in this review should be able to provide a comprehensive view of patient NK cells, and therefore, are valuable tools for diagnosis and prognosis of the disease (Fig. 1).

Major parameters to assess NK cell functions

Surface biomarkers of NK cells

Based on their surface receptors, NK cells are characterized into two major subsets: cytotoxic and regulatory. The majority of circulating NK cells are of the cytotoxic type, with the surface phenotype of $CD3^{-}CD16^{+}CD56^{dim}$. The regulatory NK cells are predominantly found in the mucosa, with a surface profile of $CD3^{-}CD16^{-}CD56^{bright}$, and are known for their capacity to secrete high levels of cytokines [11]. A minor subset of peripheral blood NK cells also exhibits $CD3^{-}CD16^{-}CD56^{bright}$ phenotype. Our laboratory has previously introduced the term "split anergy" in NK cells. NK cells become split anergized after being treated with IL-2 and anti-CD16 monoclonal antibody (mAb) or interacting with sensitive but not resistant NK tumors. Split anergized NK cells express a similar surface profile, as seen in tissue-associated NK cell subsets ($CD3^{-}CD16^{-/low}CD56^{bright}$) and exhibit increased IFN-γ secretion in the presence of no or low cytotoxicity [12].

Human NK cell surface profile in healthy individuals is found to be negative for CD3, CD127/IL-7Rα and positive for CD56/NCAM-1, CD94, CD122/IL-2Rβ, FcγRIII/CD16, KIR

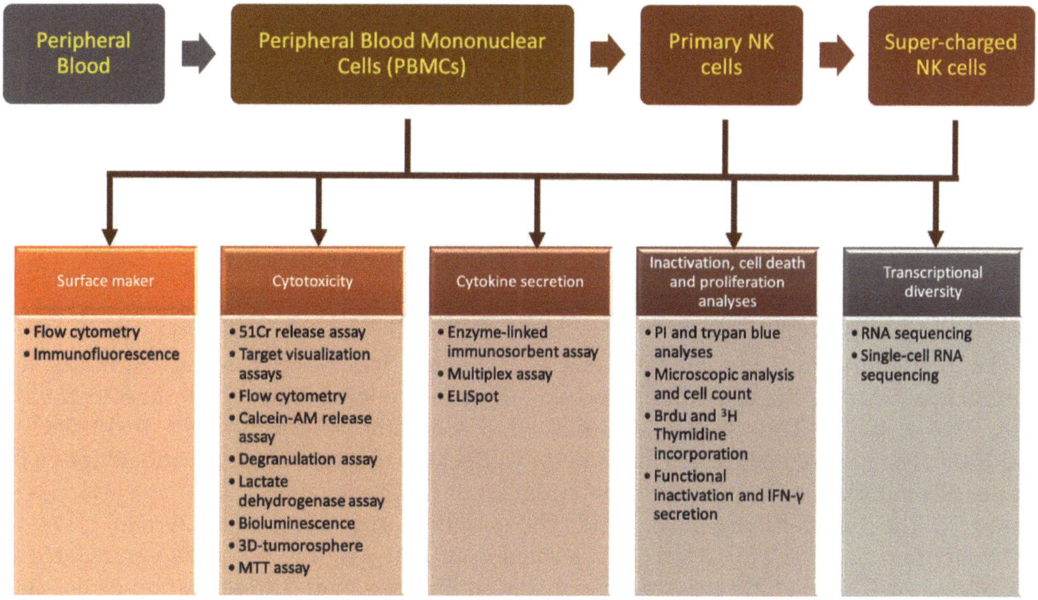

FIG. 1. Diagram of NK cell expansion and major parameters and methodologies to assess NK cell functions. Peripheral blood mononuclear cells (PBMCs) are isolated from periperal blood. Primary NK cells are subsquently purified from PBMCs. With oscteoclasts and sonicated bacteria, the number and the functions of NK cells are increased and enhanced. Five major parameters of NK cell function and their respective methodogies are listed, including surface marker profile, cytotoxicity, cytokine secretion profile, inactivation, cell death and proliferation analyses, and transcriptional diversity.

family receptors, NKG2A, NKG2D, NKp30, NKp44, NKp46, and NKp80. Aside from these markers, NK cells mediate antibody-dependent cellular cytotoxicity (ADCC) through the Fc region of the CD16 receptor. Higher levels of CD16 expression on NK cell surfaces are associated with greater capacity for ADCC [13]. The interplay between these activating and inhibitory receptors regulates and determines the functional fate of NK cells [14,15]. Therefore, a number of predictions in regards to NK function can be made through measuring NK cell surface receptors. Thus, the measurement of NK cells' surface markers may play a very important role in NK cell functional assessment.

Cytotoxic function of NK cells

Cytotoxic activity of NK cells is mediated through two distinct mechanisms: NK cell mediated direct killing and antibody-dependent cellular cytotoxicity (ADCC). NK cell-mediated direct killing is executed by the release of cytotoxic granules containing perforin and granzymes, following the formation of the immune synapse between NK and target cells. Perforin permeates the membrane of the target cell, allowing the entry of granzymes which activate a cascade of reactions that causes the apoptosis of the target cells [16]. Important regulators of this cytotoxic pathway that activate granzymes and perforin from their precursor forms are cathepsins C and L, respectively, and the activity of these cathepsins is regulated by cystatins [17]. Hence, not only the levels of perforin and granzymes but also their regulators, cathepsins and cystatins are important in optimally driving the NK cell-mediated cytotoxicity.

The second mode of cytotoxicity in NK cells is mediated through ADCC, in which NK cells through the binding to the Fc region of an antibody bound to the tumor cells mediate lysis of the tumor cells. Upon the ligation of these receptors on tumor cells by CD16 receptors on the NK cells, the cell death cascade is activated, leading to the lysis of the target cells [13,18]. Direct killing and ADCC are the two mechanisms of NK cell-mediated cytotoxicity that possess crucial roles in eliminating both CSCs/undifferentiated and differentiated tumors [19–21].

Cytokine, chemokine, and growth factors secreted by NK cells

The regulatory function of NK cells is mediated by the release of cytokines and chemokines. During the activation process, NK cells are known to secrete IL-6, IFN-γ, IL-1RA, TNF-α, IL-1B, IL-10, IL-12p70, G-CSF, IL-8, GM-CSF, RANTES, IP-10, IL-13, MCP-1, and IL-17. The released cytokines may not only affect the function of innate and adaptive immune cells but can also impact the differentiation of cancer stem cells. The key cytokines released by the NK cells are IFN-γ and TNF-α which together are known to drive the differentiation of CSCs [16]. Moreover, induction of IFN-γ by the NK cells is regulated by the release of IL-10, which plays a crucial role in limiting over-differentiation of the cells, which may result in cell death [22,23].

Transcriptional diversity in NK cells

Since NK cells are heterogeneous, and each NK cell subset behaves differently during activation, standard methodology can be complemented by single-cell analysis to determine the level of heterogeneity within the NK cells. The goal of exploring the diversity of NK cells is to define the baseline heterogeneity of circulating NK cells, in order to determine how NK cells from different population groups respond to stimulations, especially IL-2, and also in response to diseases.

Activation and expansion of NK cells

In addition to surface phenotype, cytotoxicity, and cytokine secretion in primary NK cells, it is pivotal to determine whether NK cells from cancer patients are capable of optimal activation and expansion, and whether the expanded NK cells from patients are functionally activated to the same degree as in healthy individuals, and are fully competent in carrying out their functions. This approach is valuable in assessing whether the allogenic or autologous platforms are appropriate for cancer immunotherapies.

Induction of differentiation of tumors by NK cells

We have previously shown that the supernatants from IL-2 and anti-CD16 mAb-treated NK cells, along with paraformaldehyde fixed NK cells, have the potential to induce differentiation in tumors [24]. The inability of patient-derived NK cells to differentiate tumors would be detrimental in curtailing the tumor growth, and thus such assessments are important to perform for NK cells obtained from patients.

Methodologies to assess NK cell surface markers, cytotoxicity, and secretion

Analysis of surface markers on NK cells

NK cells express CD16 and CD56 surface receptors in the absence of CD3 receptors and are further divided based on the levels of CD16 and CD56 expression. NK cells also express a number of activating and inhibitory receptors that play a crucial role in the functional balance of NK cells. Flow cytometry is the standard procedure in detecting surface receptor expression, although other procedures are also available to measure surface expressions. Given the plethora of different methodologies, each procedure has its own advantages and limitations (Table 1).

(a) Flow cytometry

Flow cytometry is a well-established methodology that is utilized to analyze the surface receptor expression on different types of cells. Using flow cytometry, we have developed a limited informative list of surface receptors on NK cells, which is a great tool for NK cell diagnostics in clinical settings.

(b) Immunofluorescence

Immunofluorescence is a technology used to detect the localization, relative expression, and also the activation state of target proteins and other biomolecules within fixed cells or tissue samples, which provides better insights of molecule expressions in NK cells.

Evaluation of NK cell cytotoxicity

Several distinct cytotoxicity assays have been established to assess the killing ability of NK cells, including ^{51}Cr release assay, target visualization assays, flow cytometry-based assays, calcein release assay, degranulation assay, lactate dehydrogenase assay, and bioluminescence-based assay. Each of these procedures has its own advantages and limitations (Table 1).

(a) ^{51}Cr release assay

The 4-h chromium release assay (CRA) is commonly used for precise and accurate quantification of cytotoxicity and is still considered as the "gold standard" in evaluating the cytotoxic activity of NK cells. In CRA, target cells are labeled with ^{51}Cr before their co-culture with NK cells in a series of effector-to-target (E:T) ratios. Percentages of lysed tumors in each E:T ratios are calculated based on the radioactivity released in the co-culture supernatants, and the percentages of cytotoxicity at different E:T ratio are often used to calculate the "lytic units," which represent the cytotoxic potential of a given NK cell population [43]. For instance, the lytic units $30/10^6$ cells are determined using the inverse number of NK cells required to lyse 30% of target cells $\times$ 100.

(b) Target visualization assays

Target visualization assays (TVA) enumerate live and dead cells by using high-throughput, single-cell imaging, and the counting is performed using the ImmunoSpot TVA software. In TVA, fluorescent dye-labeled target cells before and after interaction with the NK cells are imaged within the wells of a 96-well plate, and the image of the well

TABLE 1 Advantages and limitation of methodologies involved in NK cells functional evaluation.

Evaluation	Methodology	Advantages	Limitations	References
Surface markers	Flow cytometry	– Multiparameter data acquisition and multivariant data analysis – High-speed and allows subpopulation analysis of heterogeneous populations	– Does not provide quantitative metrics on proteins or cell surface markers – Expensive	[25]
	Immunofluorescence	– Provide clear subcellular localization of molecules – High specificity and sensitivity, and multiple targets can be detected in the same sample – Easy to operate	– Lower specificity	[26]
Cytotoxic function	^{51}Cr release assay	– Low number of effector cells required – Precise quantification	– Radioactive exposure – Limited to single time point readout only	[27]
	Target visualization assays	– No radioactive exposure – Measurements of cytolytic activity at the single-cell level	– Low specificity and sensitivity	[28]
	Flow cytometry	– Assessment of specific subsets of target cells – Precise panel designed for desired subsets distinguishing respective sensitivities to cell death from the whole populations	– Chances of high background – Hard to detect cells those lost their surface receptors	[29]
	Calcein-AM release assay	– Good retention in target cells and low pH sensitivity, and no stain transfer among cells – Effector cells can be re-used for additional analysis	– High spontaneous release	[30]
	Degranulation assay	– Provide direct information about effector cells – Allows discrimination of multiple NK cell subsets based on their level of activation	– Additional assays are required to achieve more profound knowledge of immune mechanisms taking placing against the given stimulus	[31]
	Lactate dehydrogenase assay	– No radioactive exposure – Highly reliable, high speed, and simple evaluation	– Limited to serum free or low-serum conditions	[32]

TABLE 1 Advantages and limitation of methodologies involved in NK cells functional evaluation—cont'd

Evaluation	Methodology	Advantages	Limitations	References
	Bioluminescence imaging	– No radioactive exposure – Shorter labeling time of target cells – Can be quantified quickly and at multiple time points – Low error in calculations of cell lysis	– Only luciferase expressing cells can be used – Transduction of luciferase gene is time consuming, and can be toxic to cells	[33]
	3D-tumorosphere model	– Allows measurement of several endpoints, in addition to cytotoxicity one can detect cytokine secretions and surface receptors	– Not standardized yet – Dependable on tumor cell type	[34–36]
	MTT assay	– High throughput screening	– Dependable on metabolic rate and number of mitochondria – Multiple wash steps	[37]
Cytokine secretions	Enzyme-linked immunosorbent assay	– Easy to run, safe and eco-friendly – Cost-effective	– Labor intensive – High chances of false positive or false negative – Only measures one analytes per time	[38]
	Multiplex assay	– Less sample required for more data points – Easy to perform and reproduce – High throughput, time effective	– Chances of cross-reactivity	[39]
	ELISpot	– Allows cell examination at single-cell level – Very high sensitivity	– Difficult to establish assay for new cytokines due to scarce availability of matching antibody pairs – Expensive	[40]
Transcriptional diversity	Single-cell RNA sequencing	– High accuracy in quantification of expression – Platform to detect novel transcript, splice variants, or allele-specific expression	– Expensive – Additional filtering is required to avoid noise in the output	[41,42]

containing the cells that has retained the label is captured in a tamper-proof format, thus keeping an audit trail for the raw data, then the fully automated image analysis is conducted, as live and dead target cells are recognized and counted.

(c) Flow cytometry

The flow cytometric approach has also been developed to detect cell death in target cells. With methods of specific monoclonal antibody staining in combination with annexin V and propidium iodide, cell death induced in the target cells can be detected before and after culture with the NK cells.

(d) Calcein-AM release assay

In this approach, target cells are labeled with calcein-AM. The incubation of effector and target cells at different E:T ratios, and testing is performed in the same manner as in CRA. However, the samples are measured using the spectrofluorimeter. Data are presented as arbitrary fluorescent units (AFU), and the percent lysis is calculated using the same formula used for CRA [30].

(e) Degranulation assay

During the degranulation process, the outer membrane of the granules merges with the NK cell plasma membrane, leading to the surface expression of CD107a molecules, which has been proposed as the mechanism responsible for protecting the effector cell from degranulation-associated suicide. Based on the detection of CD107a surface expression in NK cells by flow cytometry, degranulation assay can measure the cytotoxicity of immune cells [31].

(f) Lactate dehydrogenase assay

The lactate dehydrogenase (LDH) assay detects LDH enzyme released from the damaged and dying cells. LDH is not limited only to test cell-mediated cytotoxicity but also detects compound or toxic chemical-mediated cytotoxicity. The released LDH is measured with a coupled enzymatic reaction and the assay quantifies cytotoxicity based on the measurement of LDH released from damaged cells into the supernatant [32].

(g) Bioluminescence imaging

Bioluminescence imaging (BLI) has been applied to cytotoxic functional assessment, with its concept being similar to TVA. It detects signals from light-emitting enzymes in the cells, in which luciferase is most frequently used. Target cells are either transfected with luciferase or are given the substrate which would be converted into luminescence molecules. Since it requires ATP to synthesize bioluminescence, the lack of BLI serves as the indication of intracellular ATP exhaustion of target cells, which serves as a proxy for the viability of target cells in cytotoxicity assays. Furthermore, the cytotoxic function of the effector cells can be calculated from the acquired signals from different effector-to-target ratio and the no effector control group [33].

(h) 3D-tumorosphere model to assess cytotoxicity and penetration of NK cells

Tumorosphere models of cancers are used to assess NK cell penetration and cytotoxicity. These 3D tumorospheres are established by self-assembly of CSCs through external centrifugation force after the addition of methylcellulose in growth medium. Tumoropsheres are imaged during their interactions with the NK cells, and the number of live and dead cells are quantified with the spot detection tool of the Imaris software or by PI staining and flow cytometry after tumorosphere dissociation. Organoid cultures can also be made for such purpose.

(i) 3-[4,5-Dimethylthiazol-2-yl]-2,5-diphenyltetrazolium bromide, MTT assay

MTT assay is usually used in cell proliferation and cytotoxicity assessments. It quantifies the conversion of yellow salt MTT (3-[4,5-dimethylthiazol-2-yl]-2,5-diphenyltetrazolium bromide) to a purple insoluble formazan precipitation in organic solvent using spectrophotometry. Given that the MTT-formazan production is proportional to the number of viable cells, the cytotoxicity function can be evaluated through the comparison of different effector-to-target ratio and the target cells only control groups [44,45].

Evaluation of cytokine and chemokine release by the NK cells

Given that NK cells regulate the immune system and induce cell differentiation through cytokines, quantification of cytokines and chemokines should be regarded as one of the vital measurements for the evaluation of the NK cell function. The known methodologies to assess the secreted factors are enzyme-linked immunosorbent assay (ELISA), multiplex assay, and ELISpot. Each procedure has its own advantages and limitations (Table 1).

(a) Enzyme-linked immunosorbent assay (ELISA)

ELISA is the most commonly used and cost-effective assay for cytokine and chemokine measurements. This assay involves specific primary and secondary antibodies that binds to a particular cytokine in the sample. Upon recognition of the target protein, the secondary antibody is then linked to a reporter enzyme, of which streptavidin is commonly utilized. When the substrate to the enzyme is added, the detectable signals are determined by absorbance and can be subsequently converted into the concentration of the target proteins based on the established standard curve [46].

(b) Multiplex assay

Multiplex platforms have been developed for high-throughput analysis. In these multiplex platforms, several pooled beads are coated with different antibodies and can be identified by either the size or fluorescence of beads using flow-based detection instruments. By measuring the intensity of the signals on the beads, a larger panel of cytokine and chemokine secretion profile is achieved simultaneously, ultimately providing a broader perspective on NK cell secretion.

(c) Enzyme-Link Immune absorbent spot (ELISpot)

Besides measuring the secretion profile of NK cells in a total population, there are also other techniques for detecting cytokine secretion in a single-cell manner. Similar to ELISA, ELISpot requires specific antibodies for the particular target cytokine. Unlike ELISA, multiple analytes can be detected in one sample simultaneously in the ELISpot assay. However, instead of measuring the quantity of protein analyte in the supernatant, the ELISpot assay detects the relative number of cells producing the protein analyte as spot counts [47,48].

Evaluation of transcriptional diversity in NK cells

Study of NK cells' transcriptional diversity helps to characterize NK cell subsets with diverse functions, as well as different maturational status, and this allows researchers to discover the purpose and specific function of each of these subsets of cells. The understanding of NK cell function dynamics within the individual patient can aid diagnosis and cancer therapeutic strategies. RNA sequencing is often applied to assess the phenotypes of NK cells at the molecular level and to investigate protein expression and gene transcription, respectively. This procedure has its own advantages and limitations (Table 1).

(a) Single-cell RNA sequencing of NK cells

Single-cell RNA sequencing of NK cells is utilized to compare the gene expression in unstimulated and stimulated NK cells in different population groups. Thus, this methodology could be used to discover more characteristics of NK cell function besides cytokine secretion profiles. The application of these technologies in NK cells obtained from cancer patients and healthy individuals allows scientists to identify the mechanisms of the underlying immunopathogenesis in cancer patients. For instance, the inhibition of the transcription factor HIF-1α, which enhances NK cell antitumor activity, in cancer patients has been identified in the comparison of single-cell RNA-sequencing of NK cells from cancer patients and healthy individuals [49]. This particular finding, along with the potential for other similar discoveries, is helpful in strategizing new approaches in translational medicine for immunological intervention as part of cancer therapy in the future.

Evaluation of NK cells activation and expansion

As mentioned above, evaluating surface receptors, cytotoxic function, and IFN-γ secretion are essential to determine the functional activity of NK cells. In addition, there is one more parameter for the assessment of the NK cell potential: NK cell expansion. Our laboratory has previously established a methodology to expand and optimize the function of NK cells [7]. In this section, we discuss this methodology of NK cell expansion using osteoclasts (OCs) and probiotic bacteria.

(a) Methodology to expand NK cells

To induce the expansion of NK cells, NK cells need to be treated with rh-IL-2 (1000 U/mL) and anti-CD16 mAb (3 μg/mL) for 18–20 h before they are co-cultured with osteoclasts (OCs) and sonicated probiotic bacteria, sAJ2 (OCs:NK:sAJ2; 1:2:4). The medium is then refreshed every 3 days, and functional assay and surface phenotypical analysis are performed on the NK cells [7].

(b) Methodology to evaluate numbers and function of expanded NK cells

OC-expanded NK cells, or supercharged NK cells, can be evaluated by cell counts, expansion rate, and surface expression levels of CD16, in addition to other NK cell surface receptors, cytotoxic activity and cytokine secretion levels. We have previously shown that cancer patient NK cells exhibit a lower capacity of expansion, and this can be determined by direct microscopic cell count [7]. Second, we have previously reported that contaminating small fraction of T cells within purified NK cultures expand much faster if cancer patients' NK cells are used for expansion. This is monitored through flow cytometric analysis using CD16/CD56 and CD3 antibodies to determine the numbers of NK and T cells in the cell culture [7]. Third, cancer patients' expanded NK cells mediate lower cytotoxic activity secretion of cytokines, when compared to healthy individuals' expanded NK cell. Therefore, CRA and ELISA or multiplex assays need to be conducted to determine cytotoxic activity and cytokine secretion potentials, respectively.

Evaluation of NK cells mediated differentiation of tumors

We have previously reported that NK cells induce differentiation of tumors via secreted and membrane-bound forms of IFN-γ and TNF-α, respectively, with IFN-γ having a more dominant role [12,50–54]. Decreased IFN-γ secretion from the NK cells contributes to decreased NK-induced CSCs differentiation, which ultimately results in increased growth and load of undifferentiated tumors [8]. In this section, we discuss the methodology for and evaluation of tumor differentiation induction.

(a) Methodology to induce differentiation of tumors

To evaluate NK cell-mediated differentiation, we first need to treat the NK cells with rh-IL-2 (1000 U/mL) and anti-CD16 mAb (3 μg/mL) overnight, after which the supernatants need to be harvested to determine the levels of IFN-γ secretion by ELISA. The amounts of supernatants used to treat the tumor cells for differentiation are calculated based on the IFN-γ concentration in pg/mL by ELISA, and the total amount of supernatant required for differentiation can subsequently be divided into 3–5 days period of treatment to induce differentiation over a period of time [24].

(b) Methodology to evaluate differentiation levels of tumors

We have previously demonstrated that the same amounts of IFN-γ secreted from cancer patients' NK cells, when compared to those from healthy individuals, are less capable of differentiation of CSCs [7]. To evaluate NK cell-mediated differentiation, the differentiated phenotypes would be assessed, including cytotoxicity, cytokine secretion, and surface marker expression using CRA, ELISA, and flow cytometry, respectively. We have previously demonstrated that NK cells primarily target CSCs, whereas differentiated tumors are targeted less by NK cells [8,55–63]. In other words, CSCs become resistant to NK cell-mediated cytotoxicity once they become differentiated [8,12,24,64–66]. Therefore, the sensitivity of tumor cells to NK cell-mediated cytotoxicity can be predicted through the assessment of their differentiation level. Moreover, it has been reported that activation of the cytokine secretion of NK cells is not induced upon the interaction of differentiated tumor cells [67,68]. Tumor-induced activation of NK cells is tested by conducting ELISA or multiplex assay on the supernatants harvested from the co-cultures of NK cells and tumors cells. Also, the differentiation status of tumor cells is determined by measuring the surface expressions of CD54, MHC-class I and PD-L1 on tumor cells which are known to be upregulated and CD44 which is shown to be downregulated during cell differentiation [12,24,52,65].

Significance of NK cell diagnostics in the field of cancer

Altered immune cell subsets in cancer patients were shown to correlate with cancer progression and poor prognosis [69–71]. Monitoring the frequency and function of NK cells in PBMCs can serve as a pivotal parameter in tracking cancer progression, especially in high-risk populations. The altered proportions of circulating NK cells and their association with the clinical outcome of cancer therapeutics highlight the potential diagnostic application of biomarkers for NK cells and other immune cells using PBMCs of cancer patients. We have previously demonstrated that decreased numbers of PBMCs, along with decreased percentages of CD3+ T cells and B cells, but increased percentages of NK cells, monocytes, and CD11b+immune cells were seen in a number of distinct cancer patients including colon and pancreatic cancer patients, in comparison to those in healthy individuals [72]. Regarding surface markers of NK cells, studies have revealed lower expressions of activating, inhibitory, and co-stimulatory receptors on cancer patients' NK cells [7].

The most crucial functions of NK cells are cytotoxicity and cytokine secretion, which were shown to be compromised in cancer patients [3]. We have also observed reduced surface expression of CD16, along with lower levels of NK cell-mediated ADCC in cancer patients [72]. In addition, the CD16 signaling pathway during interaction of NK cells with monocytes is functionally defective in cancer patients [72]. In the NK cell expansion methodology mentioned above, cancer patients' NK cells were incapable of expanding to the same extent as those of healthy individuals. Nevertheless, cancer patient-expanded NK cells using either autologous or allogeneic OCs from healthy individuals exhibited lower cytokine secretions and cytotoxic function when compared to those expanded from NK cells of healthy individuals [7].

Based on current findings of suppressed NK cell functions in cancer patients, defects in NK cells' functional activity could result in tumor progression and metastasis. It has been demonstrated that both genetic and environmental factors play important roles in the functions of NK cells [9,10]. There is an urgent need to understand if the suppression of NK cell functions is due to the induction and progression of cancer, or whether the suppressed functions of NK cells precede the establishment of cancer and could be an underlying mechanism for the generation of tumors. Therefore, investigating the phenotype and functional profiles of NK cells using NK cell diagnostics in high-risk population would serve to be a promising tool in detecting immunosuppression and may aid in preventing and treating cancer at the early stage.

One other major advantage of NK cell-mediated differentiation of tumors is to make the tumors more sensitive to chemotherapeutic drugs. Thus, cytokine secretion from NK cells can serve as evidence to support the effectiveness of chemotherapies. Therefore, including NK cell diagnostics into routine examinations would serve as a potential indicator for early cancer detection, in addition to an avenue of predicting and evaluating prognoses and clinical outcomes following chemotherapy.

NK cell diagnostics: Future of cancer diagnostics

Currently, there are limited tests available in the field of cancer diagnostics to detect the levels of cytokines or cancer biomarkers. In acknowledging the important role that NK cells hold in halting the development and progression of cancer, it is pivotal to evaluate NK cell functions in the clinical setting to better understand the status of NK cells in cancer patients, in order to strategize approaches to optimize these cells' functions and halt the progression of tumors. Indeed, as mentioned above, NK cells are the main effectors that lyse and remove the most aggressive and metastatic clones of the cancer; CSCs and poorly differentiated tumors. Thus, implementing NK cell diagnostics would be useful for monitoring the functions of NK cells in cancer patients, along with high-risk population groups. Evaluating the phenotype and functional characteristics of NK cells in cancer patients, both pre- and posttreatment, could provide insights toward the effectiveness of treatments, as well as to predict the prognosis of patients following their treatment.

References

[1] van Meir H, et al. Impact of (chemo)radiotherapy on immune cell composition and function in cervical cancer patients. Onco Targets Ther 2017;6(2), e1267095.
[2] Jewett A, et al. Natural killer cells: diverse functions in tumor immunity and defects in pre-neoplastic and neoplastic stages of tumorigenesis. Mol Ther Oncolytics 2020;16:41–52.
[3] Imai K, et al. Natural cytotoxic activity of peripheral-blood lymphocytes and cancer incidence: an 11-year follow-up study of a general population. Lancet 2000;356(9244):1795–9.
[4] Gogali F, et al. CD3(−)CD16(−)CD56(bright) immunoregulatory NK cells are increased in the tumor microenvironment and inversely correlate with advanced stages in patients with papillary thyroid cancer. Thyroid 2013;23(12):1561–8.
[5] Lopez-Cobo S, et al. Impaired NK cell recognition of vemurafenib-treated melanoma cells is overcome by simultaneous application of histone deacetylase inhibitors. Onco Targets Ther 2018;7(2), e1392426.
[6] Ciszak L, et al. Impaired zeta chain expression and IFN-gamma production in peripheral blood T and NK cells of patients with advanced lung cancer. Oncol Rep 2009;21(1):173–84.
[7] Kaur K, et al. Novel strategy to expand super-charged NK cells with significant potential to lyse and differentiate cancer stem cells: differences in NK expansion and function between healthy and cancer patients. Front Immunol 2017;8:297.
[8] Bui VT, et al. Augmented IFN-gamma and TNF-alpha induced by probiotic bacteria in NK cells mediate differentiation of stem-like tumors leading to inhibition of tumor growth and reduction in inflammatory cytokine release; regulation by IL-10. Front Immunol 2015;6:576.
[9] Kaur K, et al. Suppression of gingival NK cells in precancerous and cancerous stages of pancreatic cancer in KC and BLT-humanized mice. Front Immunol 2017;8:1606.
[10] Kaur K, et al. Deficiencies in natural killer cell numbers, expansion, and function at the pre-neoplastic stage of pancreatic cancer by KRAS mutation in the pancreas of obese mice. Front Immunol 2018;9:1229.
[11] De Maria A, et al. Revisiting human natural killer cell subset function revealed cytolytic CD56(dim)CD16+ NK cells as rapid producers of abundant IFN-gamma on activation. Proc Natl Acad Sci U S A 2011;108(2):728–32.
[12] Tseng HC, Cacalano N, Jewett A. Split anergized natural killer cells halt inflammation by inducing stem cell differentiation, resistance to NK cell cytotoxicity and prevention of cytokine and chemokine secretion. Oncotarget 2015;6(11):8947–59.
[13] Hatjiharissi E, et al. Increased natural killer cell expression of CD16, augmented binding and ADCC activity to rituximab among individuals expressing the FcγRIIIa-158 V/V and V/F polymorphism. Blood 2007;110(7):2561–4.
[14] Pegram HJ, et al. Activating and inhibitory receptors of natural killer cells. Immunol Cell Biol 2011;89(2):216–24.
[15] Thanapati S, Ganu MA, Tripathy AS. Differential inhibitory and activating NK cell receptor levels and NK/NKT-like cell functionality in chronic and recovered stages of chikungunya. PLoS One 2017;12(11), e0188342.
[16] Abel AM, et al. Natural killer cells: development, maturation, and clinical utilization. Front Immunol 2018;9:1869.
[17] Perisic Nanut M, et al. Cysteine cathepsins as regulators of the cytotoxicity of NK and T cells. Front Immunol 2014;5:616.
[18] Jewett A, Giorgi JV, Bonavida B. Antibody-dependent cellular cytotoxicity against HIV-coated target cells by peripheral blood monocytes from HIV seropositive asymptomatic patients. J Immunol 1990;145(12):4065–71.
[19] Thompson TW, et al. Endothelial cells express NKG2D ligands and desensitize antitumor NK responses. eLife 2017;6.
[20] Pazina T, et al. Regulation of the functions of natural cytotoxicity receptors by interactions with diverse ligands and alterations in splice variant expression. Front Immunol 2017;8:369.
[21] Ochoa MC, et al. Enhancement of antibody-dependent cellular cytotoxicity of cetuximab by a chimeric protein encompassing interleukin-15. Onco Targets Ther 2018;7(2), e1393597.

[22] D'Andrea A, et al. Interleukin 10 (IL-10) inhibits human lymphocyte interferon gamma-production by suppressing natural killer cell stimulatory factor/IL-12 synthesis in accessory cells. J Exp Med 1993;178(3):1041–8.
[23] Jewett A, et al. Differential secretion of TNF-alpha and IFN-gamma by human peripheral blood-derived NK subsets and association with functional maturation. J Clin Immunol 1996;16(1):46–54.
[24] Tseng HC, et al. Induction of split anergy conditions natural killer cells to promote differentiation of stem cells through cell-cell contact and secreted factors. Front Immunol 2014;5:269.
[25] Davey HM, Kell DB. Flow cytometry and cell sorting of heterogeneous microbial populations: the importance of single-cell analyses. Microbiol Rev 1996;60(4):641–96.
[26] Tan WCC, et al. Overview of multiplex immunohistochemistry/immunofluorescence techniques in the era of cancer immunotherapy. Cancer Commun 2020;40(4):135–53.
[27] van der Haar Àvila I, et al. Evaluating antibody-dependent cell-mediated cytotoxicity by chromium release assay. Methods Mol Biol 2019;1913:167–79.
[28] Welter A, et al. High-throughput GLP-capable target cell visualization assay for measuring cell-mediated cytotoxicity. Cell 2018;7(5).
[29] Xu J, et al. Resveratrol reverses Doxorubicin resistance by inhibiting epithelial-mesenchymal transition (EMT) through modulating PTEN/Akt signaling pathway in gastric cancer. J Exp Clin Cancer Res 2017;36(1):19.
[30] Neri S, et al. Calcein-acetyoxymethyl cytotoxicity assay: standardization of a method allowing additional analyses on recovered effector cells and supernatants. Clin Diagn Lab Immunol 2001;8(6):1131–5.
[31] Lorenzo-Herrero S, et al. CD107a degranulation assay to evaluate immune cell antitumor activity. Methods Mol Biol 2019;1884:119–30.
[32] Weidmann E, et al. Lactate dehydrogenase-release assay: a reliable, nonradioactive technique for analysis of cytotoxic lymphocyte-mediated lytic activity against blasts from acute myelocytic leukemia. Ann Hematol 1995;70(3):153–8.
[33] Karimi MA, et al. Measuring cytotoxicity by bioluminescence imaging outperforms the standard chromium-51 release assay. PLoS One 2014;9(2):e89357.
[34] Hoogstad-van Evert JS, et al. Umbilical cord blood CD34+ progenitor-derived NK cells efficiently kill ovarian cancer spheroids and intraperitoneal tumors in NOD/SCID/IL2Rgnull mice. OncoImmunology 2017;6(8), e1320630.
[35] Lanuza PM, et al. Activated human primary NK cells efficiently kill colorectal cancer cells in 3D spheroid cultures irrespectively of the level of PD-L1 expression. OncoImmunology 2018;7(4), e1395123.
[36] Christakou AE, Ohlin M, Önfelt B, Wiklund M. Ultrasonic three-dimensional on-chip cell culture for dynamic studies of tumor immune surveillance by natural killer cells. Lab Chip 2015;15(15):3222–31.
[37] Hamid R, et al. Comparison of alamar blue and MTT assays for high through-put screening. Toxicol In Vitro 2004;18(5):703–10.
[38] Sakamoto S, et al. Enzyme-linked immunosorbent assay for the quantitative/qualitative analysis of plant secondary metabolites. J Nat Med 2018;72(1):32–42.
[39] Elshal MF, McCoy JP. Multiplex bead array assays: performance evaluation and comparison of sensitivity to ELISA. Methods 2006;38(4):317–23.
[40] Janetzki S. Overview, in elispot for rookies (and experts too). Cham: Springer International Publishing; 2016. p. 1–10.
[41] Hebenstreit D. Methods, challenges and potentials of single cell RNA-seq. Biology 2012;1(3):658–67.
[42] Eraslan G, et al. Single-cell RNA-seq denoising using a deep count autoencoder. Nat Commun 2019;10(1):390.
[43] Pross HF, Maroun JA. The standardization of NK cell assays for use in studies of biological response modifiers. J Immunol Methods 1984;68(1–2):235–49.
[44] Shpakova AP, Pavlova KS, Bulycheva TI. MTT-colorimetric method for detection the cytotoxic activity of human natural killer cells. Klin Lab Diagn 2000;2:20–3.
[45] Hussain RF, Nouri AM, Oliver RT. A new approach for measurement of cytotoxicity using colorimetric assay. J Immunol Methods 1993;160(1):89–96.
[46] Engvall E, Perlmann P. Enzyme-linked immunosorbent assay (ELISA). Quantitative assay of immunoglobulin G. Immunochemistry 1971;8(9):871–4.
[47] Scrivo R, et al. Evaluation of degranulation and cytokine production in natural killer cells from spondyloarthritis patients at single-cell level. Cytometry B Clin Cytom 2011;80(1):22–7.

[48] Derby EG, et al. Correlation of human CD56+ cell cytotoxicity and IFN-gamma production. Cytokine 2001; 13(2):85–90.
[49] Ni J, et al. Single-cell RNA sequencing of tumor-infiltrating NK cells reveals that inhibition of transcription factor HIF-1alpha unleashes NK cell activity. Immunity 2020;52(6):1075–1087.e8.
[50] Woroniecka K, et al. T cell exhaustion signatures vary with tumor type and are severe in glioblastoma. Clin Cancer Res 2018;24:4175–86.
[51] Schwartz RH. T cell anergy. Annu Rev Immunol 2003;21:305–34.
[52] Bui VT, et al. Augmented IFN-γ and TNF-α induced by probiotic bacteria in NK cells mediate differentiation of stem-like tumors leading to inhibition of tumor growth and reduction in inflammatory cytokine release; regulation by IL-10. Front Immunol 2015;6:576.
[53] Jewett A, et al. Rapid and potent induction of cell death and loss of NK cell cytotoxicity against oral tumors by F(ab')2 fragment of anti-CD16 antibody. Cancer Immunol Immunother 2008;57(7):1053–66.
[54] Jewett A, Cavalcanti M, Bonavida B. Pivotal role of endogenous TNF-alpha in the induction of functional inactivation and apoptosis in NK cells. J Immunol 1997;159(10):4815–22.
[55] Jewett A, et al. Strategies to rescue mesenchymal stem cells (MSCs) and dental pulp stem cells (DPSCs) from NK cell mediated cytotoxicity. PLoS One 2010;5(3), e9874.
[56] Ichise H, et al. NK cell alloreactivity against KIR-ligand-mismatched HLA-haploidentical tissue derived from HLA haplotype-homozygous iPSCs. Stem Cell Rep 2017;9(3):853–67.
[57] Jewett A, et al. Natural killer cells preferentially target cancer stem cells; role of monocytes in protection against NK cell mediated lysis of cancer stem cells. Curr Drug Deliv 2012;9(1):5–16.
[58] Ames E, et al. NK cells preferentially target tumor cells with a cancer stem cell phenotype. J Immunol 2015;195 (8):4010–9.
[59] Jewett A, Man YG, Tseng HC. Dual functions of natural killer cells in selection and differentiation of stem cells; role in regulation of inflammation and regeneration of tissues. J Cancer 2013;4(1):12–24.
[60] Kozlowska AK, et al. Differentiation by NK cells is a prerequisite for effective targeting of cancer stem cells/ poorly differentiated tumors by chemopreventive and chemotherapeutic drugs. J Cancer 2017;8(4):537–54.
[61] Kozlowska AK, et al. Resistance to cytotoxicity and sustained release of interleukin-6 and interleukin-8 in the presence of decreased interferon-gamma after differentiation of glioblastoma by human natural killer cells. Cancer Immunol Immunother 2016;65(9):1085–97.
[62] Wang B, et al. Metastatic consequences of immune escape from NK cell cytotoxicity by human breast cancer stem cells. Cancer Res 2014;74(20):5746–57.
[63] Jewett A, Tseng HC. Tumor induced inactivation of natural killer cell cytotoxic function; implication in growth, expansion and differentiation of cancer stem cells. J Cancer 2011;2:443–57.
[64] Kaur K, et al. Super-charged NK cells inhibit growth and progression of stem-like/poorly differentiated oral tumors in vivo in humanized BLT mice; effect on tumor differentiation and response to chemotherapeutic drugs Onco Targets Ther 2018;7(5), e1426518.
[65] Tseng HC, et al. Differential targeting of stem cells and differentiated glioblastomas by NK cells. J Cancer 2015; 6(9):866–76.
[66] Kaur K, et al. Probiotic-treated super-charged NK cells efficiently clear poorly differentiated pancreatic tumors in Hu-BLT mice. Cancers 2019;12(1).
[67] Jewett A, Bonavida B. Target-induced inactivation and cell death by apoptosis in a subset of human NK cells. J Immunol 1996;156(3):907–15.
[68] Jewett A, Bonavida B. Target-induced anergy of natural killer cytotoxic function is restricted to the NK-target conjugate subset. Cell Immunol 1995;160(1):91–7.
[69] Krijgsman D, et al. Characterization of circulating T-, NK-, and NKT cell subsets in patients with colorectal cancer: the peripheral blood immune cell profile. Cancer Immunol Immunother 2019;68(6):1011–24.
[70] Yang J, et al. Low percentage of CD24hiCD27^{+}CD19^{+} B cells decelerates gastric cancer progression in XELOX-treated patients. Int Immunopharmacol 2015;26(2):322–7.
[71] Hirata F, et al. Effect of bevacizumab plus XELOX (CapeOX) chemotherapy on liver natural killer cell activity in colorectal cancer with resectable liver metastasis. Ann Gastroenterol Surg 2018;2(5):383–93.
[72] Ko MW, Kaur K, Safaei T, Chen W, Sutanto C, Wong P, Jewett A. Defective patient NK function is reversed by AJ2 probiotic bacteria or addition of allogeneic healthy monocytes. Cells 2022;11(4):697.

SECTION III

Patient trials and combinational strategies in cellular immunotherapy; Successful cell therapy may depend on the selection of treatment strategies synergistic with cell therapy

CHAPTER

14

Combination of NK cell immunotherapy with chemotherapy and radiation enhances NK cell therapy and provides improved prognosis in cancer patients and in humanized BLT mouse model system

Saeed Sadeghi[a], *Po-Chun Chen*[b], *Anahid Jewett*[b,c], *and Kawaljit Kaur*[b]

[a]Division of Hematology-Oncology, Department of Medicine, UCLA David Geffen School of Medicine, Los Angeles, CA, United States [b]Division of Oral Biology and Medicine, The Jane and Jerry Weintraub Center for Reconstructive Biotechnology, University of California Los Angeles (UCLA), Los Angeles, CA, United States [c]The Jonsson Comprehensive Cancer Center, UCLA School of Dentistry and Medicine, Los Angeles, CA, United States

Abstract

Recent advancements in the field of immunology have provided a platform to develop immunotherapies for cancer therapeutics, and there is now a great interest in the development of natural killer cell (NK cell)-based therapies. NK cells originate from the bone marrow and are a key component of innate immune system. Through production of key chemokines and cytokines, NK cells are known to regulate the functions of other immune cells and are also known to mediate direct cytotoxicity as well as antibody-dependent cellular cytotoxicity. Recent advances in the production and expansion of NK cells have shown effectiveness to fight against cancer at both preclinical and clinical stages. In addition, efforts in combining NK cell-based adoptive transfer with chemotherapy and radiation are of growing interest. In this book chapter, we present an overview of recent studies on cancer immunotherapies and provide the rationale for the combination of chemotherapy, targeted therapy, and radiotherapy with NK cell-based immunotherapy. We will also highlight

NK Cells in Cancer Immunotherapy: Successes and Challenges
https://doi.org/10.1016/B978-0-12-822620-9.00022-7

the outcome of completed as well as ongoing clinical trials demonstrating the role of NK cells in enhancement of the effectiveness of cancer therapeutics. We further demonstrate and discuss the effectiveness of combined NK cell and chemotherapy treatments in tumor-bearing humanized-BLT mice.

Abbreviations

ADCC	antibody-dependent cellular cytotoxicity
CD	cluster of differentiation
CSCs	cancer stem-like cells
CTLA-4	cytotoxic T-lymphocyte-associated protein 4
hu-BLT	humanized-bone marrow/liver/thymus
IFN-γ	interferon-gamma
IL	interleukin
KIR	killer cell immunoglobulin-like receptors
MHC	major histocompatibility complex
NF-κB	nuclear factor kappa-B
NK cells	natural killer cells
sNK	supercharged NK
TAMs	tumor-associated macrophages
TGF	transforming growth factor
TME	tumor microenvironment
TNF	tumor necrosis factor
TRAIL	TNF-related apoptosis-inducing ligand

Conflict of interest statement

The authors declare no conflict of interest.

Introduction

Cancer is the second leading cause of mortality globally [1]. There is an urgent need to improve clinical outcome in cancer patients as current cancer therapeutics are limited in efficacy and may also have undesirable toxicities [2–4]. Despite intense research and improvement in therapeutic regimens, the 5-year survival rate in the metastatic cancer remains low. This is in part due to late presentation of symptoms or rapid progression which does not permit cure through surgical resection. Recent advances in understanding the molecular pathways have resulted in development of targeted therapies that have improved patient outcomes in various malignancies. More recently, clinical trials on cancer immunotherapies have demonstrated that immunotherapy is an effective treatment modality for many types of malignancies including metastatic melanoma, lung cancer, and bladder cancer [5–9]. Effectors of the immune system are thought to shape the survival and maturation of tumor cells and to participate in the elimination of cancer. Hence, while surgery in combination with chemotherapy and radiotherapy is considered a fundamental therapeutic strategy in many solid tumors, immunotherapy alone or in combination with other therapies is playing a larger role in treatment of various malignancies. The ultimate goal of immunotherapies is to assist immune system to eradicate the cancer cells and is already found to transform terminal cancer to perhaps a more chronic disease [10].

With the rapid advances in the immunotherapy approaches in cancer, there is now greater focus on development of cell-based immunotherapies. For instance, in a mouse model of brain tumor (glioblastoma), NK cells were found to reduce tumor growth and increase apoptosis, resulting in animal survival [11]. Clinical trials have also demonstrated the safety and decreased tumor relapse rate with natural killer (NK) cell-based immunotherapy [12–15] (Table 1). NK cells are cytotoxic lymphocytes and are well known to mediate direct natural cytotoxicity as well as antibody-dependent cellular cytotoxicity (ADCC). By producing key cytokines and chemokines, NK cells are known to regulate the functions of other immune cells [16,17]. NK cells' function is also mediated by the balance between activating and inhibitory signals which NK cells receive from their surface receptors [18]. Two major NK cells' surface expression are CD56 and CD16. In addition, NK cells express NKG2A [12], p58 [13], KIR2D [14] as inhibitory and CD16, NKp46 [15], NKp30 [16], NKp44 [17], NKG2D [18] as activating surface receptors. NK cell-mediated ADCC is known to be induced by the binding of NK cells' CD16 surface receptor to Fc proportion of antibody on target cells [19]. The threshold for activation of NK cells is lowered in a suitable cytokine microenvironment provided by IL-2, IL-12, IL-15, IL-18, IL-21, and type I interferons [20].

NK cells are known to exhibit two very important functions: cytotoxicity and release of cytokines. The cytotoxic function of NK cells is either via granule-mediated cytotoxicity or cytotoxicity mediated by death receptors. In granule-mediated cytotoxicity, NK cells mediate cytotoxicity against transformed tumor cells, as well as infected or stressed healthy cells, by releasing preformed granules of proteins, known as perforin and granzyme B, which can induce necrosis as well as apoptosis, or programmed cell death in target cells [20–22]. Both perforin and granzymes are synthesized in their precursor forms and are required to be cleaved by cathepsins to their active forms. Processing of perforin is catalyzed by cathepsin L [22], whereas cathepsins C and H activate the granzymes A and B [23–25]. The activity of these cathepsins is further regulated by cystatin F [26]. In NK cell-mediated cytotoxicity by death receptors, NK cells through cell surface ligands such as Fas ligand, TNF-α, and TRAIL (Apo2 L) induce death signals in target cells [27]. Both pathways can also be induced via antibody-dependent cellular cytotoxicity (ADCC) through binding of NK cells' CD16 surface receptors to the Fc portion of antibodies on target cells [19].

NK cell-mediated cytotoxicity leads to direct killing of cancer stem cells (CSCs), and NK cell-secreted cytokines mediate the differentiation of tumors [28]. CSCs sustain tumor growth because of their self-renewal capacity and their high rates of proliferation and exhibit generalized profile of surface receptor expression demonstrating a lack or a decrease in the levels of MHC-class I, CD54, and PD-L1 and an increase in CD44, CD133, and ALDH [19,23–27]. In addition, CSCs/undifferentiated or poorly differentiated tumors are known to have aggressive phenotype with a potential to give rise to metastatic growth after implantation of few tumor cells in mice [24,29–31]. During NK cell interaction with CSCs or activation through cytokines and receptor cross linking, NK cells secrete an array of cytokines, chemokines, and growth factors including IFN-γ [29]. IFN-γ plays a crucial role in differentiating cancer stem cells (CSCs)/undifferentiated tumors [32]. IFN-γ-mediated differentiation of CSCs results in the inhibition of tumor proliferation [33] and upregulation of MHC class I and adhesion molecules that increases the sensitivity of tumors to CD8+ T cell cytotoxic effects [34]. In addition, IFN-γ induces the activation of dendritic cells and monocytes, which are critical steps in the activation of adaptive immune responses [30,31].

TABLE 1 Clinical trial of NK cell-based immunotherapy.

Treatment approach	NK source	Cancer type	Status	ClinicalTrials.gov identifier
NK+SC IL-2	UCB (allogeneic)	Acute myeloid leukemia	Recruiting	NCT04347616
NK	UCB (allogeneic)	Gastric and gastroesophageal cancer	Recruiting	NCT04385641
BiCAR-NK cells		Pancreatic cancer/solid tumor expressing ROBO1	Recruiting	NCT03941457/ NCT03940820
Ex vivo expanded NK cells	PB (autologous)	Pediatric solid tumors	Complete	NCT01875601
NK+SC IL-2	PB	Advanced malignant tumors	Recruiting	NCT03619954
CAR-pNK	PB (allogeneic)	MUC1-positive relapsed or refractory solid tumor	Recruiting	NCT02839954
Ex vivo IL-2 cultured NK	Liver (allogeneic)	Hepatocellular carcinoma	Complete	NCT01147380
NK cell	PB (autologous)	Nonsmall cell lung cancer with and without EGFR mutation	Recruiting	NCT03662477
NK cell	PB (autologous)	Small-cell lung cancer	Recruiting	NCT03410368
Ex vivo IL-21-expanded NK	PB (allogeneic)	Acute myeloid leukemia	Recruiting	NCT04220684
NK+SC IL-2	PB (alloreactive)	Acute myeloid leukemia	Recruiting	NCT03955848
NK+SC IL-2	PB (allogeneic)	Hematologic malignancies	Complete	NCT00697671
Ex vivo expanded NK	PB (allogeneic)	Hematologic malignancies and solid tumors	Complete	NCT00640796
Ex vivo IL-21+K562-expanded NK+S·C IL-2	PB (allogeneic)	Acute myeloid leukemia and acute lymphoblastic leukemia	Recruiting	NCT04327037
Ex vivo IL-2 activated NK cells	PB (alloreactive)	Acute myeloid leukemia and hematologic malignancies	Recruiting	NCT01220544
Genetically modified NK cell	PB (allogeneic)	Hematologic malignancies	Recruiting	NCT04093622
Ex vivo expanded NK	PB (autologous)	pH+acute lymphoblastic leukemia	Recruiting	NCT02185781
Ex vivo IL-21+K562-expanded NK	PB (autologous)	Chronic myeloid leukemia	Recruiting	NCT03348033
Genetically modified NK cell	NK92	CD33-positive acute myeloid leukemias	Complete	NCT02944162

UCB, umbilical cord blood-derived; *SC*, subcutaneous; *PB*, peripheral blood.

It is known that differentiated tumors are favorable targets of CD8+T cells and chemotherapy, so NK cells assist CD8+T cells and chemotherapy in eradication of tumors [32,33]. Given that NK cell-based immunotherapies can represent a synergistic approach to ensure enhanced antitumor activity in conjunction with T cells and chemotherapy, it is important to optimize therapeutic strategies to preserve or enhance the function of NK cells in cancer patients [34]. In this book chapter, we will briefly review NK cells as a treatment modality for cancer and discuss emerging evidence for combination of NK immunotherapy with chemotherapy and radiation.

Preclinical and clinical development of NK cell-based immunotherapies

There are three widely used methodologies to develop NK cell-based immunotherapies. First one is the in vivo expansion of haploidentical NK cells alone or in combination with hematopoietic cell transplantation as cancer therapeutic [28]. Miller et al. showed that in acute myeloid leukemia (AML) patients, infusion of NK cells from haploidentical donors in combination with subcutaneous IL-2 after chemotherapy resulted in successful in vivo expansion of donor NK cells and complete remission of in 5 out of 19 AML patients [28]. Other studies have demonstrated that cytokines such as IL-2, IL-12, IL-15, IL-18, and IL-21 can induce expansion in NK cells [35,36].

Second approach is lenti-viral or retro-viral transduction in NK cells to express CAR. Efficacy of CAR-NK using NK92 cell lines as well as human peripheral blood, cord-blood, iPSCs-derived NK cells to inhibit the growth of hematological and solid tumors in preclinical studies have been shown in several studies [37–42]. In addition to direct cytotoxicity against the tumors, CAR-NK cells were also found to eliminate immunosuppressive cells in the TME including myeloid-derived suppressive cells (MDSCs) and M2 tumor-associated macrophages (TAMs) [43,44].

Third approach is the ex vivo expansion of autologous and allogeneic NK cells prior to infusion in cancer patients. Several in vitro, in vivo, and clinical studies including our studies have demonstrated the effectiveness of NK cells in inhibiting the tumor growth and metastasis [45–49]. A number of in vitro NK expansion techniques have been developed to allow for a higher therapeutic cell dose [50,51]. Our laboratory previously developed a novel efficient ex vivo human NK cells expansion methodology to generate large numbers of NK cells (supercharged NK cells (sNK cells)) with potent function for adoptive NK cell immunotherapy [47]. In our methodology, osteoclasts were used as feeder cells since they mediated potent expansion and activation of NK cells and triggered higher levels of NK cell cytotoxicity and secretion of cytokines and chemokines [52]. Osteoclasts produce a number of cytokines such as IL-15, IL-12, IL-18, and IFN-α, which are well-known activators of NK cells [52]. In addition, lower expression of MHC-class I and higher surface expression of NKG2D ligands such as ULBPs and MICA/B may represent other potential mechanisms by which osteoclasts can expand functionally potent NK cells [52]. In addition to osteoclasts, we have also used probiotics (combination of 7–8 gram-positive bacterial strains) to increase expansion, survival, and secretion of a variety of cytokines including IFN-γ by the NK cells. Therefore, combination of signals provided by both probiotics and osteoclasts participates in higher expansion of NK cells, and these expanded NK cells were found to be highly potent in lysis and in differentiation of CSCs both in in vitro and in vivo studies, and hence the designation of supercharged NK cells for such osteoclast-expanded NK cells [45–47,53,54].

Among various approaches of NK cell-based immunotherapies, ex vivo expansion and NK cell activation are the two most widely known mechanism that increase NK cell number and cytotoxicity, respectively.

Rationale for NK cell-based immunotherapy

Peripheral blood derived-NK cell-mediated cytotoxicity and, levels of NK cells infiltration of the tumor microenvironment (TME) are contributing factors associated with survival of cancer patients, with higher NK cell-mediated cytotoxicity and NK cells infiltration to TME being associated with improved patient prognosis [55–57]. NK cell numbers, NK cells' surface receptors (CD16, NKG2D, and zeta chain), cytokine secretions, NK cell survival, and NK cell-mediated cytotoxic function have been found to be suppressed both in peripheral blood of cancer patients as well as in TME [47,57–83]. In our studies, we have also demonstrated suppressed NK cell-mediated cytotoxicity and IFN-γ secretion in cancer patients when compared to healthy individuals [57]. In addition, we have also shown that IFN-γ secreted by patient NK cells was less capable to differentiate tumors when compared to IFN-γ secreted by healthy individuals' NK cells [47]. The suppression of NK cells' function contributes to reduced ability of NK cells in fighting cancer and thereby resulting in growth and metastasis of cancer [19].

We have previously demonstrated that NK cells from cancer patients were defective in both cytotoxic function and IFN-γ secretion [57]. Loss of these two important NK cell functions may be the underlying mechanisms for tumor survival in cancer patients, since NK cells are the primary effectors which target CSCs and aggressive tumors. Mechanisms that lead to the loss of NK cell function include downmodulation and decrease in important receptors, such as CD16 [81], NKG2D [83], zeta chain [82], decreased survival and expansion of NK cells [47], and decreased secretion of important cytokines [47]. NK cells in cancer patients have been found to secrete lower amounts of IFN-γ and, IFN-γ secreted by patient NK cells were found to be less capable to differentiate tumors when compared to IFN-γ secreted by healthy individuals' NK cells [47]. Loss of NK cells not only is the key factor in decreased lysis of CSCs/undifferentiated tumors but it is also important in decreased secretion of IFN-γ which contributes to the increase in the frequency and load of undifferentiated tumors [19].

Our studies have previously also demonstrated significantly lower NK cell expansion with lower cytokine secretions and cytotoxic function in cancer patients compared to healthy individuals [47]. Furthermore, faster expansion of contaminating patient T cells was observed in cultures of patient expanded NK cells which correlated with decreased NK cell number and cytotoxic function [47]. Other studies have also shown that adoptive transfer of autologous NK cells in melanoma and renal cell carcinoma patients did not result in meaningful clinical outcome [84]. Despite limitations of using autologous NK cells for cancer immunotherapies, several clinical trials have demonstrated benefits of autologous NK cells alone or in combination with other therapeutics to treat several cancers [85], including, multiple myeloma [51], hepatocellular carcinoma [86], AML [28], digestive cancer [87], and osteosarcoma [88] (Tables 1 and 2).

TABLE 2 Clinical trial of NK cell-based immunotherapy in combination with other cancer therapeutics.

Treatment approach	NK source	Cancer type	Status	ClinicalTrials.gov identifier
Ex vivo expanded NK cells +Trastuzumab	PB (autologous)	HER2-positive breast or gastric cancer	Recruiting	NCT02030561
Anti-HER2 NK cells	PB (allogeneic)	Advanced or metastatic HER2-expressing tumors	Recruiting	NCT04319757
NK+cryosurgery		Recurrent tongue cancer/esophageal cancer/laryngeal cancer/pharyngeal cancer/renal cancer/breast cancer/ovarian cancer/ovarian cancer/liver cancer	Complete	NCT02849379/NCT02843581/NCT02849314/NCT02849327/NCT02843607/NCT02844335/NCT02849353/NCT02849340/NCT02843802/NCT02849015
NK+Bevacizumab	PB (allogeneic)	Recurrent solid tumors	Complete	NCT02857920
NK+HLA-haploidentical transplant	PB (allogeneic)	High-risk solid tumors	Recruiting	NCT02100891
Ex vivo expanded NK+nimotuzumab	PB (autologous)	Late-stage malignancies	In process	NCT03554889
NK+irreversible electroporation	PB (autologous)	Recurrent liver cancer	Complete	NCT03008343
Ex vivo expanded NK+anti-GD2 antibodies	PB (allogeneic)	Recurrent neuroblastoma	Recruiting	NCT03242603
Ex vivo expanded NK+antibody ch14.18 +lenalidomide	PB (autologous)	Refractory or recurrent neuroblastoma	Recruiting	NCT02573896
NK+Trastuzumab		Recurrent breast cancer	Complete	NCT02843126
NK+Cetuximab		Recurrent nonsmall cell lung cancer with EGFR mutation	Complete	NCT02845856
NK+CD133 positively-selected autologous hematopoietic stem cells+SC IL-2	PB (allogeneic)	High-risk solid tumors or lymphomas	Complete	NCT02130869
NK+autologous stem cells	PB (allogeneic)	Hematologic malignancies	Complete	NCT00660166

Continued

TABLE 2 Clinical trial of NK cell-based immunotherapy in combination with other cancer therapeutics—cont'd

Treatment approach	NK source	Cancer type	Status	ClinicalTrials.gov identifier
Ex vivo expanded NK + chemotherapy + IL-2	PB (allogeneic)	Relapsed or refractory acute leukemia	Complete	NCT02074657
Expanded NK + chemotherapy + lenalidomide	UCB (allogeneic)	Chronic lymphocytic leukemia	Recruiting	NCT02280525
NK + Rituximab + Rhu-GMCSF	PB (allogeneic)	Chronic lymphocytic leukemia	Complete	NCT00383994
Anti-CD19 CAR-NK	NK92	CD19-positive leukemia	Recruiting	NCT02892695
KIR-ligand mismatched haploidentical NK + autologous stem cell transplant	PB (allogeneic)	Multiple myeloma	Complete	NCT00089453
Expanded NK + rituximab + chemotherapy + stem cell transplant	UCB (allogeneic)	B-cell non-Hodgkin's lymphoma	Recruiting	NCT03019640
NK cells + rituximab	PB (allogeneic)	B cell lymphoma	Complete	NCT02843061

UCB, umbilical cord blood-derived; *SC*, subcutaneous; *PB*, peripheral blood.

As mentioned above, NK cells mediate direct killing of CSCs/poorly differentiated tumors and induce differentiation of tumors via secreted cytokines. Increase in tumor differentiation is one of the important mechanisms in assisting CD8+ T cells, and chemotherapy drugs in targeting and eradication of tumors since differentiated tumors express higher levels of MHC class I and are well-known targets of CD8+ T cells and are more susceptible to chemotherapy-mediated killing. As such, NK cell-based immunotherapies can represent a synergistic approach to ensure enhanced antitumor activity in conjunction with T cells and chemotherapy, in order to optimize therapeutic strategies to preserve or enhance the function of NK cells in cancer patients [34].

Rationale for combination of chemotherapy with NK-based therapy

Despite advances in chemotherapy regimens for various malignancies, there remains a critical need to improve patient outcomes in both localized and metastatic cancer. In the past decade, addition of check point inhibitors to chemotherapy or other targeted therapies has resulted in improved disease survival rate for many different disease states, suggesting that perhaps addition of NK cell-based therapy to existing systemic therapies may also result in improved outcome. Here we will review some of the evidence that suggests chemotherapy or targeted therapy may enhance innate NK cell function and thereby provide rationale for combination of such therapies with NK cell-based therapies.

Preclinical work has shown that chemotherapy can impact both the number and functionality of NK cells. For example, maintenance chemotherapy used to maintain remission in acute lymphoblastic leukemia (ALL) can significantly suppress NK cell levels [89]. NK cell activity can also be affected by cytostatic drugs [90]. Using cell survival assays, Markesz et al. evaluated 29 commonly used cytostatic drugs on NK cell-mediated cytotoxicity. Drugs such as bevacizumab, bleomycin, doxorubicin, epirubicin, vinorelbine, carboplatin, methotrexate, ifosfamide, etoposide, hydroxyurea, asparginase, 5-fluorouraicl, 6-mercaptopurine, streptozocin, and cyclosphamide did not affect NK cell-mediated killing, whereas proteasome inhibitors or antimicrotubule drugs interfered with NK cell-mediated killing [91]. Other groups have shown that chemotherapy agents such as doxorubicin can enhance NK cell's cytotoxicity by increasing NKG2D ligand expression [92]. Chemotherapy agents such as 5-fluorouracil (5-FU) may regulate expression of NK cell receptor ligands on tumor cells, thereby sensitizing them to lysis by NK cells [93]. Cisplatin can also activate NK cells in hepatocellular carcinoma cell lines via increase in NKG2D ligand [94]. However, in prostate cancer, treatment with docetaxel increases androgen receptor expression which in turn leads to increased production of LLT1, an NK cell inhibitory ligand, resulting in reduced NK cell-mediated cytotoxicity against castration resistant prostate cancer cells [95]. Such studies highlight the complexities of the interplay between chemotherapy and NK cell activity in both in vitro and in vivo models.

There is some data on changes in immune cells in clinical patients undergoing chemotherapy. In a small clinical study, peripheral blood samples of 26 cancer patients who were undergoing various chemotherapy regimens were analyzed for changes in immune cell populations. While there was a marked decrease in B cells, there was no significant changes in the T cells or NK cells before and after chemotherapy [96]. This suggests that chemotherapy does not negatively impact the NK cell population and therefore can be considered in combination with NK cell therapy.

Chemotherapy may further enhance NK cell-based therapy as it can inhibit the immune suppressive nature of tumor microenvironment (TME) via reduction of immunosuppressive factors [97,98]. Zheng et al. also evaluated the role of chemotherapy on immune microenvironment in a cohort of 109 cervical cancer patients which were exposed to neoadjuvant chemotherapy (NACT). While there was variability in the immune effects of NACT, expansion of immune cells was demonstrated in over 50% of patients and there was frequent gain in CD56+ NK cells. Interestingly, a cluster of patients with highest levels of NK cells had a progression-free survival (PFS) advantage compared to patients who were NK deficient [99]. Similarly, in a study of oral squamous cell carcinoma (OSCC), patients who had received neoadjuvant chemotherapy had a higher level of CD56+ NK cells in the stroma. Moreover, neoadjuvant chemotherapy induced CD4+ T cell, CD8+ T cell, and CD56+ NK cell infiltration into the TME, suggesting that perhaps strategies in sequential use of chemotherapy and NK cell-based immunotherapy may enhance anti-tumor effect in OSCC patients [100]. Other groups have also demonstrated that low or intermittent dose of chemotherapy can maintain the drug cytotoxicity and enhance NK cell function [101,102]. Taking together, these studies provide further support for combination of chemotherapy and NK therapy to enhance immune response.

The interplay of chemotherapy and NK cell-based immunotherapy may be dose and time dependent. Feng et al. explored the interaction of epirubicin and NK cells in breast cancer cell lines. When epirubicin was combined with breast cancer and NK cell lines, there was time and NK cell dose-dependent growth inhibition in all breast cancer cell lines. Alternatively, if the breast cancer cells were pretreated with epirubicin followed by expanded NK cells, the cytotoxic effects were potentiated compared to either treatment alone. Epirubicin enhances NK cell cytotoxicity through several mechanisms including upregulation of NKG2D ligands, increased cytokine secretion by NK cells, and also by activation of the granzyme B/perforin pathway [103].

Preclinical evidence discussed in last section has led to a few clinical trials of combination of chemotherapy and NK cell-based therapy. Li et al. proposed that NK cell infusion in combination with chemotherapy following surgery for locally advanced colon cancer would significantly improve the prognosis of patients compared to chemotherapy alone. Among a total of 60 patients with stage III colon cancer, 33 received chemotherapy alone while 27 received chemotherapy with at least 2 course of autologous NK cell infusions. In vitro studies demonstrated that following exposure to 5-Fluorouracil (5-FU) or oxaliplatin, the expression of some NK activating receptor ligands was significantly increased in both cancer CSCs and non-CSCs. Moreover, not only 5FU and oxaliplatin sensitized both CSCs and non-CSCs to NK cell-mediated cytotoxicity, the susceptibility of the cells correlated positively with the length of exposure to chemotherapy agents. Clinically, patients in the NK cell therapy group had a higher 5-year PFS rate (51% vs 35%) with a median PFS of 34.3 months (19.4 months in non-NK group). Among subset of patients with poorly differentiated colon cancer the median PFS was still improved from 12.1 to 23.5 months with the additional NK cell infusion. In another study, colorectal cancer patients were offered 6 cycles of chemotherapy with 5FU and oxaliplatin (FOLFOX) regimen, followed by five infusions of dendritic cell cytokine-induced NK cells (DC-CIK). Compared to patients who received chemotherapy alone, those who received DC-CIK had no decrease in the CD4+ T cell, CD8+ T cell, and NK cell populations and demonstrated a prolonged progression-free survival and OS compared to chemotherapy only patients (PFS 8.8 vs 5.8 months; OS 14.7 vs 10.8 months) [104]. The synergistic benefit of DC-CIK with chemotherapy has also been demonstrated in patients with late-stage nonsmall cell lung cancer [105]. These evidence-based studies demonstrate the efficacy of combination cancer therapy using chemotherapy and NK cells (Table 2).

Combination of cisplatin with supercharged NK cell immunotherapy enhances NK cell mediated killing and increases the secretion of IFN-γ in humanized-BLT mice

In agreement with the data obtained from patients treated with the combination of NK cell immunotherapy and chemotherapy, we also observed similar outcomes when tumor implanted humanized-BLT (hu-BLT) mice received both NK cells and chemotherapy drugs. Briefly, oral squamous carcinoma stem cells (OSCSCs) were implanted in the oral cavity of hu-BLT mice and allowed the tumors to grow before they were injected supercharged NK cells (sNK cells). After a week of injection of sNK cells, cisplatin (CDDP) was administered, the mice were monitored and euthanized when tumor alone implanted mice exhibited signs of morbidity (Fig. 1A). Compared to tumor-alone implanted mice, those which received NK cells or CDDP had smaller tumors in general. However, mice which received both the NK cells and the CDDP exhibited the smallest tumors (Fig. 1B). Accordingly, when assessing

NK cell-mediated cytotoxicity and IFN-γ secretion, tumor-bearing mice that received no treatment exhibited suppression of both NK cell cytotoxicity and secretion of IFN-γ in splenocytes, bone marrow derived cells, and peripheral blood mononuclear cells when compared to control mice (no tumor) or those injected with sNK cells in the absence of tumor implantation (Fig. 1C–G). Mice that received tumor implantation and were injected with CDDP alone exhibited restoration of NK cell cytotoxicity in splenocytes (Fig. 1D), but the levels of IFN-γ secretion by splenocytes and PBMCs remained similar to those seen in tumor implanted mice in the absence of CDDP injection (Fig. 1C, E, and F). CDDP injection in tumor implanted mice exhibited higher secretion of IFN-γ by bone marrow derived cells when compared to those obtained from tumor alone implanted mice (Fig. 1G). The more striking differences can be seen in mice that received either sNK cells or the combination of sNK cells with the injection of CDDP (Fig. 1B–G). Although sNK cell injection increased cytotoxicity and secretion of IFN-γ in all three tissue compartments of mice when compared to those implanted with tumors in the absence of sNK injection, the highest increase in the NK cell cytotoxicity and secretion of IFN-γ were seen in tumor bearing mice injected with sNK cells and CDDP (Fig. 1C–G). As indicated above, differentiation of tumor cells by the membrane bound and secreted forms of IFN-γ and TNF-α by the NK cells was largely responsible in rendering the tumor cells susceptible to chemotherapeutic drugs as shown in our previous publications [33,53]. In addition, differentiation of tumor cells by the NK cell supernatants increased the surface expression levels of MICA/B on tumor cells, providing the ligands for the NKG2D-mediated activation of NK cells [33,106]. Likewise, it has been hypothesized that chemotherapy drug-mediated increase in NKG2D ligands could also be responsible for the increase in NK activation [33,106]. Thus, similar outcomes seen in the patients treated with the chemotherapeutic drugs in the presence of NK cell therapy, and in tumor-bearing hu-BLT mice injected with the sNK and treated with chemotherapy drugs provides the rationale for the combinatorial treatment of NK cell immunotherapy with chemotherapy drugs. Thus, collectively, these results suggest that patients can benefit from the combination of NK cell immunotherapy with the chemotherapy (Fig. 2).

Rationale for combination of targeted therapy with NK-based therapy

Beyond traditional chemotherapy, there is now evidence for potential immunomodulatory effect of targeted therapies. Sorafenib, a multi-kinase inhibitor used in the treatment of hepatocellular carcinoma (HCC), may exert its antitumor activity partially by enhancing NK cell recruitment via increase in macrophage related cytokines such as IL-18 and CCL5. In addition, sorafenib may trigger NK cell response via activation of pyroptotic macrophages, suggesting perhaps that HCC patients may benefit from NK directed therapies [107].

NK cells are a primary effector of antibody-dependent cellular cytotoxicity (ADCC) and as such addition of NK cells may potentially augment clinical benefit seen with existing cytotoxic therapies. Targeted therapy such as monoclonal antibodies may also enhance ADCC. Previously, it has been shown that the anti-CD20 antibody rituximab can enhance ADCC activity and improve NK cell killing ability significantly [108]. Elotuzumab has also been shown to enhance NK cell-mediated ADCC in myeloma cells which express SLAM-7 [109]. Daratumumab, an anti-CD38 antibody, has been shown to enhance NK cell-mediated

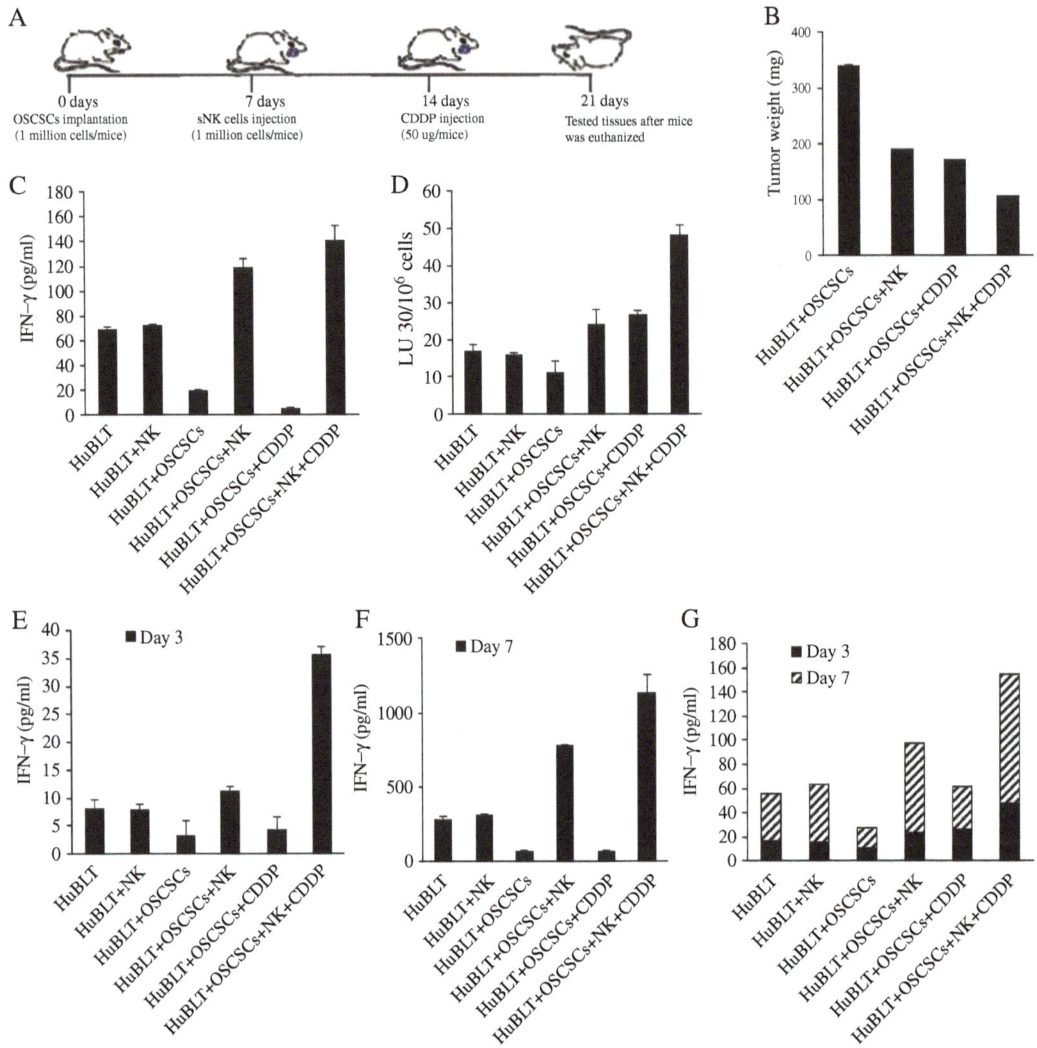

FIG. 1 **Immunotherapy with supercharged NK cells prior to injection of cisplatin inhibited tumor growth, and increased IFN-γ secretion, and NK cell-mediated cytotoxicity in different tissue compartments of hu-BLT mice.** Hu-BLT mice were orthotopically injected with 1×10^6 human OSCSCs into the floor of the mouth. One week after the tumor implantation, mice received sNK cells via tail-vein injection, and 1 week after sNK cells injection, mice received CDDP (50 μg/mice) via tail vein injection. The disease progression and weight loss were monitored for another week (A). Hu-BLT mice were implanted with OSCSC tumors and were injected with sNK cells and CDDP as described in (A). At the end of the experiment, hu-BLT mice were sacrificed; the tumors were harvested and weighed (B). Hu-BLT mice were implanted with OSCSC tumors and were injected with sNK cells and CDDP as described in (A). At the end of experiment, hu-BLT mice were sacrificed and spleens (C and D), peripheral blood (E and F), and bone marrow (G) were harvested, and single cell suspensions were obtained and cultured with IL-2 (10,000 U/mL) for 3 and 7 days. The supernatants were then harvested from the cultures on day 7 of spleen culture, and the secretions of IFN-γ were determined using single ELISA (C). Cytotoxicity of day 7 spleen cultures was determined against OSCSCs using standard 4-h ^{51}Cr release assay. The lytic units (LU) 30/10^6 cells were determined using the inverse number of NK cells required to lyse 30% of OSCSCs × 100 (D). The supernatants were harvested from the cultures on day 3 (E) and day 7 (F) of peripheral blood mononuclear cells and the secretions of IFN-γ were determined using single ELISA. The supernatants were harvested from the cultures on day 3 and day 7 of bone marrow derived cells and the secretions of IFN-γ were determined using single ELISA (G).

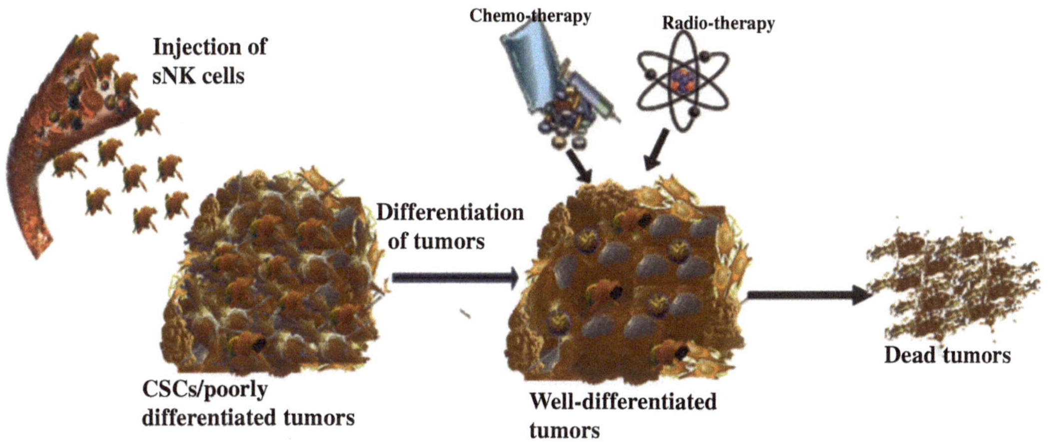

FIG. 2 **Schematic representation of supercharged NK cell-mediated differentiation of CSCs/poorly differentiated tumors and increased sensitization to the effects of chemotherapy and radiotherapy.**
The delivery of super-charged NK cells (sNK cells) through tail-vein injection in hu-BLT mice mobilizes sNK cells to the tumor microenvironment. sNK cells upon interaction with the tumor cells mediate tumor killing through increased NK cell-mediated cytotoxicity and tumor differentiation through the IFN-γ and TNF-γ secreted by the NK cells. Differentiated tumors in turn become susceptible to the effect mediated by the chemotherapy and radiotherapy. In addition, chemotherapy is likely to increase NK cell function through increase in stress ligands such as MICA/B. Thus, there is a bidirectional effect in not only increasing the potential of chemotherapy to target tumors, but chemotherapy in turn can activate NK cell function and mediate the enhancement of both NK cell function and increase in the function of chemotherapeutic drugs.

cytotoxicity and trigger ADCC, though this was limited to CD38 high myeloma cell lines. Furthermore, the NK cell activity can be further potentiated by using KIR-ligand mismatch donor NK cells, suggesting synergistic activity between daratumumab and NK cell therapy [110]. In follicular lymphoma, studies on lenalidomide enhances ADCC by NK cells via a cereblon-dependent mechanism and combination of lenalidomide with rituximab enhances formation of NK cell lytic immune synapses more than either drug alone, confirming that the combination therapy results in immune enhancement [111].

There are also examples of NK-mediated ADCC in solid tumors. In breast cancer, monoclonal antibody targeting HER2 such as trastuzumab exerts its therapeutic activity by activating NK cell-mediated ADCC [112]. In a preclinical mouse model of HER2+ breast cancer, NK cell depletion abolished anti-HER2 monoclonal antibody therapeutic activity [113,114]. Finally, a nude mouse colorectal cancer xenograft model has recently been used to demonstrate that cetuximab, a monoclonal antibody against epidermal growth factor receptor (EGFR) can enhance NK cell-mediated ADCC and combination of cetuximab and NK cells had a stronger inhibitory effect on the growth of xenograft which had higher EGFR expression [115,116]. A small pilot study explored the combination of Hu14.18K322A, an anti-GD2 monoclonal antibody with cytotoxic therapy, IL2, GM-CSF and haploidentical NK cells in patients with recurrent or refractory neuroblastoma. A total of 11 patients received the NK cells. The overall response rate was 61.5% with complete response rate of 38.5%, a superior outcome compared to historical response rates of 15%–50% for this disease. The results are intriguing as it provides clinical evidence for combination of NK cells with existing monoclonal antibodies [117] (Table 2).

Rationale for combination of radiation therapy and NK cell therapy

There has been growing interest in combining radiation and NK cell-based therapy. Radiation can enhance immune response by increasing the antigenicity of malignant cells as well as interaction with the TME [118]. In addition, radiation therapy may directly impact NK cellular function through several mechanisms. In vitro low dose ionizing radiation (LDIR) at 75–150 mGy can result in the proliferation of NK cells, enhancement of NK cell cytotoxicity, secretion NK cell effector proteins such as IFN-γ and TNF-α, and also lead to increased expression of perforin and Fas ligands in NK cells [119]. Alternatively, high dose ionizing radiation (HIDR) at dose greater or equal to 1.0 Gy decreased NK cell-mediated cytotoxicity though this could be mitigated by pretreatment with IL-2 [120,121]. Barsoumian et al. evaluated both low and high dose radiation along with systemic antiprogrammed cell death protein 1 (PD-1) and anticytotoxic T-lymphocyte associated protein 4 (CTLA-4) in a lung adenocarcinoma mice model. This in vivo model demonstrated improved antitumor activity by enhancing NK cell infiltration of tumor and shifting tumor associated macrophages (TAM) to M1 phenotype [122]. Of note, M1 macrophage phenotype can trigger cytotoxic NK cell function [123].

There is some clinical evidence demonstrating the effects of radiation therapy on NK cells in cancer patients. In one study, the effect of 20 Gy of intraoperative radiation therapy on immune system of 13 breast cancer patients was evaluated. Flow cytometry analysis showed a significant increase in the CD56+CD16+ NK cell populations at 3 weeks following radiation [124]. Murana et al. looked at the use of stereotactic body radiotherapy (SBRT) at dose of 10 Gy in 21 oligometastatic breast cancer patients and noted rapid activation of NK cells in more than half of the patients with single dose of SBRT. In addition, the group noted a positive correlation between higher number of activated NK cells at diagnosis and longer PFS, suggesting that perhaps a combined approach of radiation therapy and NK cell activity can result in better outcomes for patients [125]. In another study, the peripheral blood of 31 patients who were scheduled to undergo SBRT to any organ for a primary or metastatic disease was analyzed showing a significant decline in the percentage of total and cytotoxic $CD56^{dim}CD16^{high}$ NK cells following SBRT to parenchymal sites (lung or liver but not brain or bone). Of note there was no change in percentage of cytokine producing $CD56^{high}CD16^{dim}$ NK cells in periphery of any patient [126]. The authors proposed that this decrease in NK cells in peripheral blood may be due to increasing migration of NK cells to the irradiated tumor site given that prior xenograft model showed that tumor irradiation leads to increased upregulation of stress ligands such as NKG2D ligands, resulting in infiltration of tagged NK cells into the tumors [126,127].

Such work suggests that perhaps the combination of radiotherapy and autologous NK cell therapy may result in enhanced efficacy. A preclinical trial was done in dogs with canine sarcoma in which focal radiation therapy to the tumor bed was followed by intravenous transfer of previously expanded canine NK cells. The authors found significant increase in NK homing to tumors as well as an increased activation of circulating NK cells in vivo. In addition, of the 10 dogs studied, 5 remained metastatic free at the 6-month primary endpoint and with resolution of metastatic disease in 1 dog [128]. Recently, a randomized phase II clinical trial evaluated the benefit of adoptive NK cell-based therapy in nonsmall cell lung cancer patients with membrane bound heat shock protein 70 (mHsp70) who had received combined radiation and chemotherapy (RCT). mHsp70 expression is associated with more aggressive and metastatic disease. Patients who had received 60–70 Gy of radiation therapy along with platinum-based chemotherapy were randomized to receive up to four doses of autologous NK cells every 2–6 weeks following RCT. Clinical outcome in the intervention arm was significantly better with one complete and one partial responder and two patients had stable disease and two had progressive disease. On the contrary, in the control group, five patients had progressive disease. In addition, the clinical responses in the intervention arm were associated with increased prevalence of activated NK cells in peripheral blood [129].

Conclusion

Recent significant advancements in the pathobiology of immune cells, in particular, NK cells and the role which this immune cell subset plays in controlling cancer growth and metastasis, provide the platform to use NK cells in immunotherapy of cancer patients. Preclinical studies demonstrated promising results when NK cells alone, or in combination with chemotherapy and radiation therapies were used to treat various malignant tumors. Similar trends were also seen in our humanized mouse model system treated with the combination of NK cell immunotherapy and chemotherapy (Figs. 1 and 2). Recent clinical trials using NK cells alone or in combination with other cancer therapeutics support the efficacy of these therapies for improved prognosis and longer survival of patients. However, some of the clinical trials are still at an early phase of investigation, and therefore, further investigations are needed to establish efficacy based on the type of the cancer, and patient-related factors. In addition, defining the timing of NK cell infusion and optimal dose of these therapies in combination with other therapeutics is still being explored. Work presented in this book chapter suggests there is a potential in harnessing the power of NK cell therapy in combination with radiation and chemotherapy to treat various malignancies in the near future.

Acknowledgments

The authors are grateful to funding agencies and donors for supporting the work.

Author contributions

K.K. and S.S. wrote, reviewed, and edited the book chapter. P.C.C. reviewed and edited the book chapter. A.J. oversaw the conceptualization of the book chapter, reviewed, and edited the chapter and acquired funding.

Statement of originality

The authors declare that the work published in this review is original.

References

[1] Fitzmaurice C, et al. Global, regional, and national cancer incidence, mortality, years of life lost, years lived with disability, and disability-adjusted life-years for 32 cancer groups, 1990 to 2015: a systematic analysis for the global burden of disease study. JAMA Oncol 2017;3(4):524–48.

[2] Maroof H, et al. *Lactobacillus acidophilus* could modulate the immune response against breast cancer in murine model. J Clin Immunol 2012;32(6):1353–9.

[3] Hassan Z. Anti-cancer and biotherapeutic potentials of probiotic bacteria. J Cancer Sci Ther 2019;11.
[4] Górska A, et al. Probiotic bacteria: a promising tool in cancer prevention and therapy. Curr Microbiol 2019; 76(8):939–49.
[5] Liu E, et al. Use of CAR-transduced natural killer cells in CD19-positive lymphoid tumors. N Engl J Med 2020; 382(6):545–53.
[6] Terranova-Barberio M, et al. Exhausted T cell signature predicts immunotherapy response in ER-positive breast cancer. Nat Commun 2020;11(1):3584.
[7] Chen EX, et al. Effect of combined immune checkpoint inhibition vs best supportive care alone in patients with advanced colorectal cancer: the Canadian Cancer Trials Group CO.26 study. JAMA Oncol 2020;6(6):831–8.
[8] Rosenberg SA, Restifo NP. Adoptive cell transfer as personalized immunotherapy for human cancer. Science 2015;348(6230):62–8.
[9] Sharma P, Allison JP. The future of immune checkpoint therapy. Science 2015;348(6230):56–61.
[10] Emens LA, et al. Cancer immunotherapy: opportunities and challenges in the rapidly evolving clinical landscape. Eur J Cancer 2017;81:116–29.
[11] Poli A, et al. Targeting glioblastoma with NK cells and mAb against NG2/CSPG4 prolongs animal survival. Oncotarget 2013;4(9):1527–46.
[12] Ciurea SO, et al. Phase 1 clinical trial using mbIL21 ex vivo-expanded donor-derived NK cells after haploidentical transplantation. Blood 2017;130(16):1857–68.
[13] Hallner A, et al. The HLA-B -21 dimorphism impacts on NK cell education and clinical outcome of immunotherapy in acute myeloid leukemia. Blood 2019;133(13):1479–88.
[14] Björklund AT, et al. Complete remission with reduction of high-risk clones following haploidentical NK-cell therapy against MDS and AML. Clin Cancer Res 2018;24(8):1834–44.
[15] Ishikawa T, et al. Phase I clinical trial of adoptive transfer of expanded natural killer cells in combination with IgG1 antibody in patients with gastric or colorectal cancer. Int J Cancer 2018;142(12):2599–609.
[16] Fildes JE, Yonan N, Leonard CT. Natural killer cells and lung transplantation, roles in rejection, infection, and tolerance. Transpl Immunol 2008;19(1):1–11.
[17] Farag SS, Caligiuri MA. Human natural killer cell development and biology. Blood Rev 2006;20(3):123–37.
[18] Pegram HJ, et al. Activating and inhibitory receptors of natural killer cells. Immunol Cell Biol 2011;89(2):216–24.
[19] Bui VT, et al. Augmented IFN-gamma and TNF-alpha induced by probiotic bacteria in NK cells mediate differentiation of stem-like tumors leading to inhibition of tumor growth and reduction in inflammatory cytokine release; regulation by IL-10. Front Immunol 2015;6:576.
[20] Trinchieri G. Biology of natural killer cells. Adv Immunol 1989;47:187–376.
[21] Lanier LL. NK cell recognition. Annu Rev Immunol 2005;23:225–74.
[22] Tseng HC, et al. Increased lysis of stem cells but not their differentiated cells by natural killer cells; de-differentiation or reprogramming activates NK cells. PLoS One 2010;5(7), e11590.
[23] Todaro M, et al. Colon cancer stem cells dictate tumor growth and resist cell death by production of interleukin-4. Cell Stem Cell 2007;1(4):389–402.
[24] Ricci-Vitiani L, et al. Identification and expansion of human colon-cancer-initiating cells. Nature 2007; 445(7123):111–5.
[25] Toden S, et al. Oligomeric proanthocyanidins (OPCs) target cancer stem-like cells and suppress tumor organoid formation in colorectal cancer. Sci Rep 2018;8(1):3335.
[26] Melsen JE, et al. Human circulating and tissue-resident CD56(bright) natural killer cell populations. Front Immunol 2016;7:262.
[27] Kozlowska AK, et al. Resistance to cytotoxicity and sustained release of interleukin-6 and interleukin-8 in the presence of decreased interferon-gamma after differentiation of glioblastoma by human natural killer cells. Cancer Immunol Immunother 2016;65(9):1085–97.
[28] Miller JS, et al. Successful adoptive transfer and in vivo expansion of human haploidentical NK cells in patients with cancer. Blood 2005;105(8):3051–7.
[29] Guo Z, Hardin H, Lloyd RV. Cancer stem-like cells and thyroid cancer. Endocr Relat Cancer 2014;21(5):T285–300.
[30] Prince ME, et al. Identification of a subpopulation of cells with cancer stem cell properties in head and neck squamous cell carcinoma. Proc Natl Acad Sci U S A 2007;104(3):973–8.
[31] Hu Z, et al. Targeting tissue factor as a novel therapeutic oncotarget for eradication of cancer stem cells isolated from tumor cell lines, tumor xenografts and patients of breast, lung and ovarian cancer. Oncotarget 2017; 8(1):1481–94.

[32] Nandi M, et al. CD8(+)CD28(−) T cells: key cytotoxic players impacting disease pathogenesis in chronic HBV infection. Clin Sci (Lond) 2019;133(17):1917–34.
[33] Kozlowska AK, et al. Differentiation by NK cells is a prerequisite for effective targeting of cancer stem cells/ poorly differentiated tumors by chemopreventive and chemotherapeutic drugs. J Cancer 2017;8(4):537–54.
[34] Zingoni A, et al. Natural killer cell response to chemotherapy-stressed cancer cells: role in tumor immunosurveillance. Front Immunol 2017;8:1194.
[35] Denman CJ, et al. Membrane-bound IL-21 promotes sustained ex vivo proliferation of human natural killer cells. PLoS One 2012;7(1), e30264.
[36] Granzin M, et al. Shaping of natural killer cell antitumor activity by ex vivo cultivation. Front Immunol 2017;8:458.
[37] Imai C, Iwamoto S, Campana D. Genetic modification of primary natural killer cells overcomes inhibitory signals and induces specific killing of leukemic cells. Blood 2005;106(1):376–83.
[38] Liu E, et al. Cord blood NK cells engineered to express IL-15 and a CD19-targeted CAR show long-term persistence and potent antitumor activity. Leukemia 2018;32(2):520–31.
[39] Li Y, et al. Human iPSC-derived natural killer cells engineered with chimeric antigen receptors enhance antitumor activity. Cell Stem Cell 2018;23(2):181–192.e5.
[40] Wang J, et al. Purinergic targeting enhances immunotherapy of CD73(+) solid tumors with piggyBac-engineered chimeric antigen receptor natural killer cells. J Immunother Cancer 2018;6(1):136.
[41] Chang YH, et al. A chimeric receptor with NKG2D specificity enhances natural killer cell activation and killing of tumor cells. Cancer Res 2013;73(6):1777–86.
[42] Romanski A, et al. CD19-CAR engineered NK-92 cells are sufficient to overcome NK cell resistance in B-cell malignancies. J Cell Mol Med 2016;20(7):1287–94.
[43] Parihar R, et al. NK cells expressing a chimeric activating receptor eliminate MDSCs and rescue impaired CAR-T cell activity against solid tumors. Cancer Immunol Res 2019;7(3):363–75.
[44] Zhang P, et al. Effects of CSF1R-targeted chimeric antigen receptor-modified NK92MI & T cells on tumor-associated macrophages. Immunotherapy 2018;10(11):935–49.
[45] Kaur K, et al. Super-charged NK cells inhibit growth and progression of stem-like/poorly differentiated oral tumors in vivo in humanized BLT mice; effect on tumor differentiation and response to chemotherapeutic drugs. Onco Targets Ther 2018;7(5), e1426518.
[46] Kaur K, et al. Probiotic-treated super-charged NK cells efficiently clear poorly differentiated pancreatic tumors in Hu-BLT mice. Cancers 2019;12(1).
[47] Kaur K, et al. Novel strategy to expand super-charged NK cells with significant potential to lyse and differentiate cancer stem cells: differences in NK expansion and function between healthy and cancer patients. Front Immunol 2017;8:297.
[48] Morvan MG, Lanier LL. NK cells and cancer: you can teach innate cells new tricks. Nat Rev Cancer 2016;16(1):7–19.
[49] Sharma P, et al. Primary, adaptive, and acquired resistance to cancer immunotherapy. Cell 2017;168(4):707–23.
[50] Fujisaki H, et al. Expansion of highly cytotoxic human natural killer cells for cancer cell therapy. Cancer Res 2009;69(9):4010–7.
[51] Leivas A, et al. Novel treatment strategy with autologous activated and expanded natural killer cells plus anti-myeloma drugs for multiple myeloma. Onco Targets Ther 2016;5(12), e1250051.
[52] Tseng HC, et al. Bisphosphonate-induced differential modulation of immune cell function in gingiva and bone marrow in vivo: role in osteoclast-mediated NK cell activation. Oncotarget 2015;6(24):20002–25.
[53] Bui VT, et al. Augmented IFN-γ and TNF-α induced by probiotic bacteria in NK cells mediate differentiation of stem-like tumors leading to inhibition of tumor growth and reduction in inflammatory cytokine release; regulation by IL-10. Front Immunol 2015;6:576.
[54] Dong H, Rowland I, Yaqoob P. Comparative effects of six probiotic strains on immune function in vitro. Br J Nutr 2012;108(3):459–70.
[55] Burke S, et al. New views on natural killer cell-based immunotherapy for melanoma treatment. Trends Immunol 2010;31(9):339–45.
[56] Larsen SK, Gao Y, Basse PH. NK cells in the tumor microenvironment. Crit Rev Oncog 2014;19(1–2):91–105.
[57] Imai K, et al. Natural cytotoxic activity of peripheral-blood lymphocytes and cancer incidence: an 11-year follow-up study of a general population. Lancet 2000;356(9244):1795–9.
[58] Bruno A, et al. A think tank of TINK/TANKs: tumor-infiltrating/tumor-associated natural killer cells in tumor progression and angiogenesis. J Natl Cancer Inst 2014;106(8):dju200.

[59] Gross E, Sunwoo JB, Bui JD. Cancer immunosurveillance and immunoediting by natural killer cells. Cancer J 2013;19(6):483–9.
[60] Mirjacic Martinovic KM, et al. Decreased expression of NKG2D, NKp46, DNAM-1 receptors, and intracellular perforin and STAT-1 effector molecules in NK cells and their dim and bright subsets in metastatic melanoma patients. Melanoma Res 2014;24(4):295–304.
[61] Gubbels JA, et al. MUC16 provides immune protection by inhibiting synapse formation between NK and ovarian tumor cells. Mol Cancer 2010;9:11.
[62] Balsamo M, et al. Melanoma-associated fibroblasts modulate NK cell phenotype and antitumor cytotoxicity. Proc Natl Acad Sci U S A 2009;106(49):20847–52.
[63] Castriconi R, et al. Transforming growth factor beta 1 inhibits expression of NKp30 and NKG2D receptors: consequences for the NK-mediated killing of dendritic cells. Proc Natl Acad Sci U S A 2003;100(7):4120–5.
[64] Pietra G, et al. Melanoma cells inhibit natural killer cell function by modulating the expression of activating receptors and cytolytic activity. Cancer Res 2012;72(6):1407–15.
[65] Krockenberger M, et al. Macrophage migration inhibitory factor contributes to the immune escape of ovarian cancer by down-regulating NKG2D. J Immunol 2008;180(11):7338–48.
[66] Igarashi T, et al. Enhanced cytotoxicity of allogeneic NK cells with killer immunoglobulin-like receptor ligand incompatibility against melanoma and renal cell carcinoma cells. Blood 2004;104(1):170–7.
[67] White D, et al. Natural killer (NK) activity in peripheral blood lymphocytes of patients with benign and malignant breast disease. Br J Cancer 1982;46(4):611–6.
[68] Jewett A, Tseng HC. Tumor induced inactivation of natural killer cell cytotoxic function; implication in growth, expansion and differentiation of cancer stem cells. J Cancer 2011;2:443–57.
[69] Santos MF, et al. Comparative analysis of innate immune system function in metastatic breast, colorectal, and prostate cancer patients with circulating tumor cells. Exp Mol Pathol 2014;96(3):367–74.
[70] Matsumoto Y, et al. Abdominal infection suppresses the number and activity of intrahepatic natural killer cells and promotes tumor growth in a murine liver metastasis model. Ann Surg Oncol 2016;23(Suppl 2):S257–65.
[71] Baskic D, et al. Suppression of natural killer-cell and dendritic-cell apoptotic tumoricidal activity in patients with head and neck cancer. Head Neck 2013;35(3):388–98.
[72] Tseng HC, Cacalano N, Jewett A. Split anergized natural killer cells halt inflammation by inducing stem cell differentiation, resistance to NK cell cytotoxicity and prevention of cytokine and chemokine secretion. Oncotarget 2015;6(11):8947–59.
[73] Mickel RA, et al. Natural killer cell cytotoxicity in the peripheral blood, cervical lymph nodes, and tumor of head and neck cancer patients. Cancer Res 1988;48(17):5017–22.
[74] Ghoneum M, Gill G, Perry L. Natural killer cell activity in patients with carcinoma of the larynx and hypopharynx. Laryngoscope 1986;96(11):1300.
[75] Tartter PI, et al. The prognostic significance of natural killer cytotoxicity in patients with colorectal cancer. Arch Surg 1987;122(11):1264–8.
[76] Izawa S, et al. H2O2 production within tumor microenvironment inversely correlated with infiltration of CD56dim NK cells in gastric and esophageal cancer: possible mechanisms of NK cell dysfunction. Cancer Immunol Immunother 2011;60(12):1801–10.
[77] Nolibe D, Poupon MF. Enhancement of pulmonary metastases induced by decreased lung natural killer cell activity. J Natl Cancer Inst 1986;77(1):99–103.
[78] Vitale M, et al. Effect of tumor cells and tumor microenvironment on NK-cell function. Eur J Immunol 2014; 44(6):1582–92.
[79] Gallois A, et al. Reversal of natural killer cell exhaustion by TIM-3 blockade. Onco Targets Ther 2014;3(12), e946365.
[80] Hersey P, et al. Low natural-killer-cell activity in familial melanoma patients and their relatives. Br J Cancer 1979;40(1):113–22.
[81] Gogali F, et al. CD3(−)CD16(−)CD56(bright) immunoregulatory NK cells are increased in the tumor microenvironment and inversely correlate with advanced stages in patients with papillary thyroid cancer. Thyroid 2013;23(12):1561–8.
[82] Ciszak L, et al. Impaired zeta chain expression and IFN-gamma production in peripheral blood T and NK cells of patients with advanced lung cancer. Oncol Rep 2009;21(1):173–84.
[83] Lopez-Cobo S, et al. Impaired NK cell recognition of vemurafenib-treated melanoma cells is overcome by simultaneous application of histone deacetylase inhibitors. Onco Targets Ther 2018;7(2), e1392426.

[84] Parkhurst MR, et al. Adoptive transfer of autologous natural killer cells leads to high levels of circulating natural killer cells but does not mediate tumor regression. Clin Cancer Res 2011;17(19):6287–97.
[85] Liem NT, et al. Phase I clinical trial using autologous ex vivo expanded NK cells and cytotoxic T lymphocytes for cancer treatment in Vietnam. Int J Mol Sci 2019;20(13).
[86] Lee JH, et al. Adjuvant immunotherapy with autologous cytokine-induced killer cells for hepatocellular carcinoma. Gastroenterology 2015;148(7):1383–1391.e6.
[87] Sakamoto N, et al. Phase I clinical trial of autologous NK cell therapy using novel expansion method in patients with advanced digestive cancer. J Transl Med 2015;13:277.
[88] Judge SJ, et al. Blood and tissue biomarker analysis in dogs with osteosarcoma treated with palliative radiation and intra-tumoral autologous natural killer cell transfer. PLoS One 2020;15(2), e0224775.
[89] Komada Y, et al. Cellular immunosuppression in children with acute lymphoblastic leukemia: effect of consolidation chemotherapy. Cancer Immunol Immunother 1992;35(4):271–6.
[90] Ujhazy P, Babusikova O. NK-cell activity affected by some cytostatic drugs and their additives. Neoplasma 1991;38(3):303–12.
[91] Markasz L, et al. Effect of frequently used chemotherapeutic drugs on the cytotoxic activity of human natural killer cells. Mol Cancer Ther 2007;6(2):644–54.
[92] Soriani A, et al. ATM-ATR-dependent up-regulation of DNAM-1 and NKG2D ligands on multiple myeloma cells by therapeutic agents results in enhanced NK-cell susceptibility and is associated with a senescent phenotype. Blood 2009;113(15):3503–11.
[93] Ghiringhelli F, Apetoh L. Enhancing the anticancer effects of 5-fluorouracil: current challenges and future perspectives. Biom J 2015;38(2):111–6.
[94] Shi L, et al. Cisplatin enhances NK cells immunotherapy efficacy to suppress HCC progression via altering the androgen receptor (AR)-ULBP2 signals. Cancer Lett 2016;373(1):45–56.
[95] Tang M, et al. Docetaxel suppresses immunotherapy efficacy of natural killer cells toward castration-resistant prostate cancer cells via altering androgen receptor-lectin-like transcript 1 signals. Prostate 2020;80(10):742–52.
[96] Waidhauser J, et al. Chemotherapy markedly reduces B cells but not T cells and NK cells in patients with cancer. Cancer Immunol Immunother 2020;69(1):147–57.
[97] Lake RA, Robinson BW. Immunotherapy and chemotherapy—a practical partnership. Nat Rev Cancer 2005; 5(5):397–405.
[98] Kershaw MH, et al. Enhancing immunotherapy using chemotherapy and radiation to modify the tumor microenvironment. Onco Targets Ther 2013;2(9), e25962.
[99] Zhang Y, et al. Baseline immunity and impact of chemotherapy on immune microenvironment in cervical cancer. Br J Cancer 2021;124(2):414–24.
[100] Takakura H, et al. The immunological impact of chemotherapy on the tumor microenvironment of oral squamous cell carcinoma. Acta Med Okayama 2017;71(3):219–26.
[101] Wu J, Waxman DJ. Immunogenic chemotherapy: dose and schedule dependence and combination with immunotherapy. Cancer Lett 2018;419:210–21.
[102] Ghiringhelli F, et al. Metronomic cyclophosphamide regimen selectively depletes CD4+CD25+ regulatory T cells and restores T and NK effector functions in end stage cancer patients. Cancer Immunol Immunother 2007;56(5):641–8.
[103] Feng H, et al. Epirubicin pretreatment enhances NK cell-mediated cytotoxicity against breast cancer cells in vitro. Am J Transl Res 2016;8(2):473–84.
[104] Lin T, et al. Clinical effects of autologous dendritic cells combined with cytokine-induced killer cells followed by chemotherapy in treating patients with advanced colorectal cancer: a prospective study. Tumour Biol 2016;37(4):4367–72.
[105] Zhong R, et al. Dendritic cells combining with cytokine-induced killer cells synergize chemotherapy in patients with late-stage non-small cell lung cancer. Cancer Immunol Immunother 2011;60(10):1497–502.
[106] Kaur K, et al. ADCC against MICA/B is mediated against differentiated oral and pancreatic and not stem-like/ poorly differentiated tumors by the NK cells; loss in cancer patients due to down-modulation of CD16 receptor. Cancers 2021;13(2).
[107] Hage C, et al. Sorafenib induces pyroptosis in macrophages and triggers natural killer cell-mediated cytotoxicity against hepatocellular carcinoma. Hepatology 2019;70(4):1280–97.
[108] Shiokawa M, et al. In vivo assay of human NK-dependent ADCC using NOD/SCID/gammac(null) (NOG) mice. Biochem Biophys Res Commun 2010;399(4):733–7.

[109] Balasa B, et al. Elotuzumab enhances natural killer cell activation and myeloma cell killing through interleukin-2 and TNF-alpha pathways. Cancer Immunol Immunother 2015;64(1):61–73.
[110] Mahaweni NM, et al. Daratumumab augments alloreactive natural killer cell cytotoxicity towards CD38+ multiple myeloma cell lines in a biochemical context mimicking tumour microenvironment conditions. Cancer Immunol Immunother 2018;67(6):861–72.
[111] Chiu H, et al. Combination lenalidomide-rituximab immunotherapy activates anti-tumour immunity and induces tumour cell death by complementary mechanisms of action in follicular lymphoma. Br J Haematol 2019;185(2):240–53.
[112] De P, Hasmann M, Leyland-Jones B. Molecular determinants of trastuzumab efficacy: what is their clinical relevance? Cancer Treat Rev 2013;39(8):925–34.
[113] Jaime-Ramirez AC, et al. IL-12 enhances the antitumor actions of trastuzumab via NK cell IFN-gamma production. J Immunol 2011;186(6):3401–9.
[114] Stagg J, et al. Anti-ErbB-2 mAb therapy requires type I and II interferons and synergizes with anti-PD-1 or anti-CD137 mAb therapy. Proc Natl Acad Sci U S A 2011;108(17):7142–7.
[115] Chen S, et al. Cetuximab intensifies the ADCC activity of adoptive NK cells in a nude mouse colorectal cancer xenograft model. Oncol Lett 2016;12(3):1868–76.
[116] Veluchamy JP, et al. Combination of NK cells and cetuximab to enhance anti-tumor responses in RAS mutant metastatic colorectal cancer. PLoS One 2016;11(6), e0157830.
[117] Federico SM, et al. A pilot trial of humanized anti-GD2 monoclonal antibody (hu14.18K322A) with chemotherapy and natural killer cells in children with recurrent/refractory neuroblastoma. Clin Cancer Res 2017; 23(21):6441–9.
[118] Wennerberg E, et al. Immune recognition of irradiated cancer cells. Immunol Rev 2017;280(1):220–30.
[119] Yang G, et al. Low-dose ionizing radiation induces direct activation of natural killer cells and provides a novel approach for adoptive cellular immunotherapy. Cancer Biother Radiopharm 2014;29(10):428–34.
[120] Zhou L, et al. Validating the pivotal role of the immune system in low-dose radiation-induced tumor inhibition in Lewis lung cancer-bearing mice. Cancer Med 2018;7(4):1338–48.
[121] Zarcone D, et al. Radiation sensitivity of resting and activated nonspecific cytotoxic cells of T lineage and NK lineage. Blood 1989;73(6):1615–21.
[122] Barsoumian HB, et al. Low-dose radiation treatment enhances systemic antitumor immune responses by overcoming the inhibitory stroma. J Immunother Cancer 2020;8(2).
[123] Bellora F, et al. TLR activation of tumor-associated macrophages from ovarian cancer patients triggers cytolytic activity of NK cells. Eur J Immunol 2014;44(6):1814–22.
[124] Linares-Galiana I, et al. Changes in peripheral immune cells after intraoperative radiation therapy in low-risk breast cancer. J Radiat Res 2021;62(1):110–8.
[125] Muraro E, et al. Local high-dose radiotherapy induces systemic immunomodulating effects of potential therapeutic relevance in oligometastatic breast cancer. Front Immunol 2017;8:1476.
[126] McGee HM, et al. Stereotactic ablative radiation therapy induces systemic differences in peripheral blood immunophenotype dependent on irradiated site. Int J Radiat Oncol Biol Phys 2018;101(5):1259–70.
[127] Ames E, et al. Enhanced targeting of stem-like solid tumor cells with radiation and natural killer cells. Onco Targets Ther 2015;4(9), e1036212.
[128] Canter RJ, et al. Radiotherapy enhances natural killer cell cytotoxicity and localization in pre-clinical canine sarcomas and first-in-dog clinical trial. J Immunother Cancer 2017;5(1):98.
[129] Multhoff G, et al. Targeted natural killer cell-based adoptive immunotherapy for the treatment of patients with NSCLC after radiochemotherapy: a randomized phase II clinical trial. Clin Cancer Res 2020;26(20):5368–79.

CHAPTER

15

Combining oncolytic viruses with immune cell therapy as treatments for cancer: OV, CAR T-cell, and NK combinations

Shyambabu Chaurasiya, Yanghee Woo, Kevin Choong, Supriya Deshpande, and Yuman Fong

Department of Surgery, City of Hope Comprehensive Cancer Center, Duarte, CA, United States

Abstract

Two of the most promising anticancer therapies to emerge in the past 20 years are oncolytic viral therapy and engineered immune cell therapy. Oncolytic viral therapy (OV) relies on oncolytic viruses, which specifically infect and kill cancers and also activate local and systemic immune signals to generate a lasting immune response to cancer. Talimogene laherparepvec, a herpes simplex OV, has been approved for use in the treatment of melanoma. Other OVs are being tested in various cancers, and target cancer by direct killing and by amplifying local cytokine and checkpoint signals to attract effector immune cells to tumor sites. The field of engineered immune cell therapies has also seen much progress. Chimeric antigen receptor T cells and natural killer cells designed to target many tumor antigens are in human trials. One directed against B-cell malignancies has been approved for use as human medicine. This review summarizes some of the emerging work seeking to combine OVs and immune cell therapy. This combination therapy is an exciting area with the promise of harnessing both innate and adaptive immunity for effective cancer treatment.

Abbreviations

BiTE	bispecific T-cell engager
CAd-VEC*PDL1*	CAd-VEC expressing a PD-L1 blocking mini-body
CAR	chimeric antigen receptor
CAR-T	chimeric antigen receptor T cells
CCL20	C-C motif chemokine ligand 20
CCL5	chemokine (C-C motif) ligand 5
CCR5	chemokine receptor type 5

https://doi.org/10.1016/B978-0-12-822620-9.00006-9

CD19	cluster of differentiation 19
CD19t	truncated CD19
CEA	carcinoembryonic antigen
CIK	cytokine-induced killer
EGFR	epidermal growth factor receptor
FR-α	folate receptor alpha
GM-CSF	granulocyte-macrophage colony-stimulating factor
HDPDL1	helper-dependent PD-L1 blocking mini-body
HER2	human epidermal growth factor receptor 2
HNSCC	head and neck squamous cell carcinoma
IFN-γ	interferon gamma
IL-12	interleukin 12
IL-15	interleukin 15
IL-18	interleukin 18
IL-2	interleukin 2
MHC	major histocompatibility complex
NK	natural killer
oHSV	oncolytic herpes simplex virus
OV	oncolytic virus
PD-L1	programmed death-ligand 1
PGE2	prostaglandin E2
RANTES	regulated on activation, normal T cell expressed and secreted
TAG-72	tumor-associated glycoprotein 72
TILs	tumor-infiltrating lymphocytes
TME	tumor microenvironment
TNF	tumor necrosis factor
T-Vec	talimogene laherparepvec
VSV	vesicular stomatitis virus

Conflict of interest

Dr Fong is a paid scientific consultant for Imugene, Boehringer Ingelheim, Eureka. He receives royalties from Merck and Imugene.

Introduction

Most natural viruses are tropic for a specific organ. Thus, hepatitis viruses and meningitis viruses infect mainly liver and neural tissues, respectively. Oncolytic viral therapy has focused on natural or engineered viruses that selectively infect cancer. In recent years, investigators have found many promising viruses that are now under human clinical trials. Such viruses can specifically infect, replicate within, and directly kill cancer cells through lysis and apoptosis [1]. OVs can also increase local expression of checkpoint targets, increase local chemokine and cytokine production, and attract native cytotoxic T cells to tumor targets [2]. In this way, OVs turn tumors into immunologically "hot" sites. Thus, it is also not surprising that OVs synergize with checkpoint inhibitors for amplified anticancer effects [3]. One OV, T-VEC (Imlygic, Amgen, Thousand Oaks, CA, USA), has been approved by the US FDA to treat melanoma [4], and many others are under study.

Many immune cells are being studied as therapeutic agents for cancer. Two of the most promising are chimeric antigen receptor T (CAR-T) -cells and natural killer (NK) cells. CAR-T cells are genetically engineered T cells that express receptors that target surface antigens expressed on cancer cells in a major histocompatibility complex (MHC)-independent

TABLE 1 Potential interactions between oncolytic viruses (OV) and immune cell therapies (NK, CAR-T).

Negative interactions	➢ The killing of tumor cells by NK or CAR-T decreases cell factories needed for viral replication ➢ Activation of IFN-gamma by OV decreases NK or CAR-T function ➢ Direct cytotoxic effect by OV or NK or CAR-T, decreasing cancer killing
Positive effect	➢ OV changing tumor microenvironment to enhance NK or CAR-T function ➢ OV expresses transgenes at sites of tumor • to attract NK/CAR-T (chemokines) • to enhance maturation and persistence of NK/CAR-T (cytokines) • to enhance tumor infiltration by NK/CAR-T (hyaluronidase, etc.) ➢ NK/CAR-T act as carriers of OVs • To improve targeting of tumor • To evade immune reaction targeting virus

manner. Most current programs rely on T cells derived from peripheral blood mononuclear cells, transduced by retroviruses or lentiviruses to introduce the CAR transgene, and then expanded to create an autologous T cell product [5]. CAR-Ts are now approved for B-cell lymphoma or leukemia based on successful trials targeting CD-19 in these diseases [6]. Other targets under active investigation include CEA [7], TAG-72 [8], mesothelin [9], chlorotoxin [10], among others. Stem cell-derived CAR-Ts are also being investigated as an allogeneic product to reduce the cost and other logistic hurdles in this process [11].

Since the 1970s, NK cells are known as an immune cell population that can mediate cytotoxicity without prior activation [12]. NK cells account for 5%–10% of the peripheral blood leukocytes and target cancer cells in an MHC-independent fashion. Consequently, they can also target cancer stem cells as well as undifferentiated or poorly differentiated tumors. Many of the actions of NK cells are through receptor-mediated mechanisms that are enhanced by exposure to cytokines such as IL-2, IL-12, IL-15, and IL-18. NK cells cause tumor cell death via the release of lytic granules and local expression of cytokines and chemokines (IFNγ, TNF, GM-CSF, and CCL5), activating and attracting other immune cells such as T-helper cells and dendritic cells. NK cells have been used in cancer patients since the 1980s when IL-2 activated lymphokine-activated killer cells were infused. Autologous, allogeneic, as well as NK cell lines are in human clinical trials.

Combining OVs and cell therapies can have many theoretical advantages (Table 1). OVs can enhance CAR-T or NK targeting and persistence. CAR-T and NK can potentially enhance cell lysis and spread of OVs within the tumor. OVs can also be engineered to contain transgenes that enhance immune cell maturation and persistence [13]. Finally, attempts are underway to use immune cells as carriers for OVs to evade anti-OV immune mechanisms and enhance tumor targeting [14].

CAR-T cell therapy and oncolytic viruses in cancer therapy

The potential synergistic effects of OVs with CAR-T fall into two categories (Table 2). First, the general impact of OVs on innate immunity should add to that of CAR T-cell-directed cell kill. OVs clearly can produce tumor cytoreduction by immunogenic cell death and

TABLE 2 The rationale for synergies actions between OVs and CAR-T therapy.

General effects of OV on innate immunity	➢ Induction of immunogenic cell death in cancer cells ➢ Tumor cytoreduction ➢ Induction of local inflammation
Transgene-mediated effects for engineered OVs to enhance CAR-T function	➢ Delivery of cytokines supportive of professional dendritic cells ➢ Delivery of growth and maturation factors for CAR-T function and persistence ➢ Delivery of CAR T-cell attracting chemokines ➢ Expression of T-cell engagers ➢ Delivery of immune co-stimulatory molecules ➢ Delivery of molecules targeting immune checkpoints ➢ Delivery of tumor targets for CAR-T

inducing local inflammation. These effects should help in overcoming the local tumor-immunosuppressive environment.

Many investigators are also attempting to exploit the capability of OVs in delivering transgenes to further augment local inflammation that could enhance CAR-T function (Table 3). Watanabe et al. studied pancreatic cancer in a preclinical model combining mesothelin-directed CAR-T cells and cytokine-armed oncolytic adenoviruses. In concept, this group attempted to use the OV to directly kill cancer and deliver a cytokine payload to try to reverse the immunosuppressive tumor microenvironment (TME) of pancreatic cancer [9]. Mesothelin is a promising target for CAR T-cell therapy, as it is overexpressed in most pancreatic cancers, mesotheliomas, ovarian cancers, and some lung cancers. The authors hypothesized that combining *meso*-CAR-T cells with an oncolytic adenovirus expressing immunostimulatory cytokines TNF-α and IL-2 (Ad5/3-E2F-D24-TNFa-IRES-IL2, or OAd-TNFa-IL2) would improve efficacy. OAd-TNFa-IL2 enhanced the antitumor efficacy of meso-CAR-T cells in a xenograft model, and efficacy was associated with increased tumor-infiltrating lymphocytes (TILs) and prolonged T cell function. None of the mice treated with combined OAd-TNFa-IL2 and meso-CAR-T died of tumor metastasis. Furthermore, in a

TABLE 3 Preclinical studies combining oncolytic viruses with CAR-T cells.

Virus	Tumor	CAR antigen	References
Onc.Ad-EGFR BiTE	Pancreatic ductal carcinoma/colorectal carcinoma	Folate receptor alpha (FR-α)	Wing et al. [15]
Onc.Ad-TNFα/IL2	Pancreatic ductal carcinoma	Mesothelin (meso)	Watanabe et al. [9]
Onc.Ad-Rantes/IL15	Neuroblastoma	Ganglioside GD2	Nishio et al. [16]
CAdVEC-αPDL1	Prostate, squamous cell carcinoma	Human epidermal growth factor 2 (HER2)	Tanoue et al. [17]
CAdVEC-IL12p70/αPDL1	Head and neck squamous cell carcinoma	Human epidermal growth factor 2 (HER2)	Rosewell et al. [18]
CF33-CD19	Triple-negative breast cancer	CD19	Park et al. [6]

syngeneic mouse model Ad-mTNFa-mIL2 increased both CAR-T cell and host T cell infiltration to the tumor and altered host tumor immune status with M1 polarization of macrophages and increased dendritic cell maturation. These findings indicate that combining cytokine-armed oncolytic adenovirus to enhance the efficacy of CAR-T cell therapy is a promising approach to overcome the immunosuppressive TME for the treatment of pancreatic ductal adenocarcinoma.

Nishio et al. attempted to improve the efficacy of CAR-T therapy by using an OV (Ad5D24) armed with the chemokine RANTES and the cytokine IL15. They reasoned that the chemokine RANTES would increase the trafficking of CAR-T cells to the tumor sites, and the IL-15 would enhance the persistence of CAR-T cells [16]. They used a preclinical model of neuroblastoma and found that the adenovirus Ad5D24 exerted a potent, dose-dependent, cytotoxic effect on tumor cells and did not alter CAR-T cell survival. When Ad5D24 was used in combination with a CAR-T specific for the tumor antigen GD2 (GD2.CAR-T cells), the persistence of CAR-T was accompanied by increased overall survival of tumor-bearing mice. These preclinical data support the use of this innovative biologic platform of combination immunotherapy for solid tumors.

Investigators have also used OVs for local checkpoint inhibition. CAR-T cells usually produce proinflammatory cytokines that increase the expression of T-cell checkpoint signals such as PD-L1, which may inhibit their functionality against solid tumors. Tanoue et al. delivered an adenovirus armed to express a PD-L1 mini-body [17]. These investigators used human tumor xenograft models of pancreatic cancer to evaluate the proinflammatory properties of an oncolytic adenovirus that expressed PD-L1 blocking mini-antibody (mini-body; HD*PDL1*). Coadministration of the CAd-VEC*PDL1* (HD*PDL1* and oncolytic adenovirus) elicited oncolytic effects along with the production of PD-L1 mini-body at the tumor site. Combining CAd-VEC*PDL1* with HER2.CAR-T cells enhanced antitumor activity compared to treatment with either agent alone. The benefits of locally produced PD-L1 mini-body by CAd-VEC*PDL1* could not be replicated by infusion of anti-PD-L1 IgG plus HER2. These data show superiority of local production of PD-L1 mini-body by CAd-VEC*PDL1* combined with administration of prostate cancer-directed CAR-T cells.

This group then designed an adenovirus that delivered the PD-L1 blocking antibody and the immunostimulatory cytokine IL-12 [18]. They hypothesized that treating tumors with the oncolytic adenovirus (CAd12_PDL1) that promotes local oncolysis and expresses IL-12 and anti-PD-L1 would enhance the antitumor activity of HER2-specific CAR-T cells (HER2.CAR). In a head and neck squamous cell carcinoma (HNSCC) xenograft model, combined local treatment with CAd12_PDL1 and systemic HER2.CAR-T cell infusion improved survival to >100 days compared with approximately 25 days with either approach alone. This combination also controlled both primary and metastasized tumors in an orthotopic model of HNSCC.

Wing and colleagues attempted to improve CAR-T efficacy by combining it with an OV-encoded bispecific T-cell engager (BiTE) [15]. They hypothesized that an oncolytic adenovirus armed with an EGFR-targeting, bispecific T-cell engager (OAd-BiTE) would improve the efficacy of CAR-T-cell therapy. They used a CAR-T targeting the folate receptor alpha (FR-α). They demonstrated that OAd-BiTE-mediated oncolysis significantly improved CAR-T activation and proliferation, with a resultant increase in cytokine production and cytotoxicity. BiTEs secreted from infected cells redirected CAR-T cells toward EGFR in the absence of FR-α. The combinatorial approach improved antitumor efficacy and prolonged survival in murine models. These results demonstrated that combining a BiTE-expressing OV with adoptive CAR-T-cell therapy overcomes a fundamental limitation of CAR-T cells for heterogeneous tumors that do not all express the CAR-T target. These data also encourage the study of OVs expressing BiTEs as therapies in solid tumors.

An exciting approach to combining OV and CAR-T therapy was recently reported by our group. Park et al. delivered the CD19 antigen, which naturally is only found in B lymphocytes, to solid tumors using an OV [6]. Using an OV highly specific in tumor infection and highly efficient in transgene expression, we delivered the target for the only approved CAR-T therapy. Calling this the "Liquid to solid" approach, we are attempting to take a CAR-T therapy for B-cell leukemias and lymphomas and use it for any solid tumor. In this way, we are trying to overcome the lack of tumor-restricted tumor antigens in solid tumors. We engineered an OV to express a non-signaling, truncated CD19 (CD19t) protein for tumor-selective delivery, enabling targeting by CD19-CAR-T cells [6]. Infecting tumor cells with an oncolytic vaccinia virus coding for CD19t (OV19t) produced de novo CD19 at the cell surface before virus-mediated tumor lysis. Cocultured CD19-CAR-T cells secreted cytokines and exhibited potent cytolytic activity against infected tumors. Using several mouse tumor models, delivery of OV19t was shown to promote tumor control after CD19-CAR T cell administration. OV19t induced local immunity characterized by tumor infiltration of endogenous and adoptively transferred T cells. CAR-T cell-mediated tumor killing also caused release of virus from dying tumor cells, which propagated tumor expression of CD19t. Our study features a combination immunotherapy approach using OVs to promote de novo CAR-T cell targeting of solid tumors.

A recent paper by Evgin and colleagues highlights a potential obstacle for the combined use of OVs and CAR-T [19]. To test whether the inflammatory nature of OVs and their ability to remodel the TME may help recruit and potentiate the functionality of CAR-T cells, they found that infection with a vesicular stomatitis virus expressing the transgene for interferon β (VSVmIFNβ) was associated with attrition of murine EGFRvIII CAR-T cells in a B16EGFRvIII model. This is despite inducing a robust proinflammatory shift in the chemokine profile. The type I interferon (IFN) expressed following infection led to therapeutic interference. The investigators emphasized the importance of understanding physiologic feedback mechanisms meant to prevent immunopathology. They noted that IFN expression could produce upregulation of inhibitory receptors on T cells, leading to intrinsic negative feedback limiting T cell activity against tumor [19]. Thus, OVs and CAR-Ts will need to be studied in human trials to determine optimal timing and dose for best anticancer efficacy.

Thus, potentiating effects of combined OV and CAR-T therapy can result from natural viral effects on innate immunity or from engineered effects through transgene expression. A counter-regulatory effect of an overly vigorous inflammatory reaction may result in detrimental effects on CAR-T function or survival (Fig. 1). A variety of doses and timing of administration of OV and CAR-T will need to be examined in each clinical trial to determine the net effects to best harness the potential of this combination immunotherapy for humans.

NK cells and oncolytic viruses in cancer therapy

NK cells can cause tumor cell death via several mechanisms, including direct interaction with tumor cells via the release of lytic granules. NK cells can also secrete cytokines and

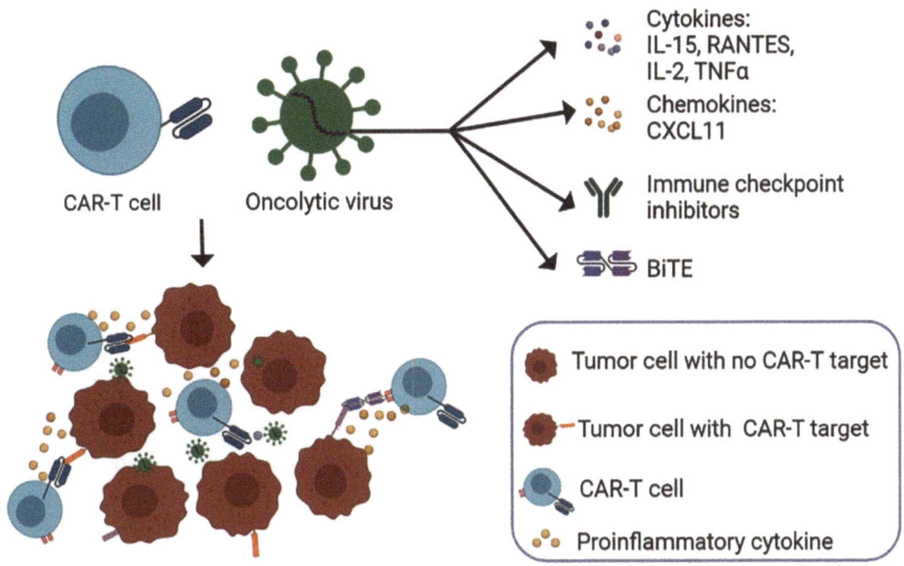

FIG. 1 Oncolytic virus interactions with CAR-T. Combination of CAR-T cells and oncolytic virus (OV) for the treatment of solid tumors. Cancer treatment with OVs and CAR-T cell therapy results in tumor debulking, immunogenic cell death, and reverted tumor immunosuppression. OVs can be genetically modified to deliver therapeutic transgenes into the tumor microenvironment to enhance and attract T-cells, improve T-cell function, and produce local checkpoint inhibition. Preclinical studies combining CAR-T cells with OVs armed with cytokines, chemokines, BiTEs, or immune checkpoint inhibitors resulted in enhanced therapeutic outcomes.

chemokines such as IFNγ, TNF, GM-CSF, and CCL5, activating and attracting other immune cells such as T-helper cells and dendritic cells. The relationship between NK cells and dendritic cells has secondary influences on T cell differentiation and activation. In addition, it is known that the presence of NK cells is integral to an optimal antitumor response from CD8+ T cells.

NK cells are involved in the efficacy of many current therapeutics. NKs are end effectors of anti-PD-L1 checkpoint inhibition [20] and are also important in effective anti-HER2 therapy for breast cancer. The presence of pretreatment tumor-infiltrating NK cells correlates with higher pathologic complete response rates and a greater disease-free survival, indicating that these NK cells predict responses to anti-HER2-based neoadjuvant therapy [21].

With the increase in understanding of the vital role of NKs as effectors for cancer therapies, investigators have tried to use activated NKs for cancer therapy since the 1980s. NK cells have been used in cancer patients as IL-2 activated lymphokine-activated killer cells, particularly for melanoma and renal cell carcinoma [22]. Despite numerous studies demonstrating the anticancer effects of NK cell-based therapies, there remain many barriers to full adoption. One challenge to the widespread use of NK cells has been the low percentage of NK cells in the peripheral blood, requiring significant ex vivo expansion to obtain clinically relevant quantities. The need for cryopreservation can also affect the cytotoxicity of NK cell preparations. Recent advances in the generation of stem-cell-derived NK cells and the improved expansion, activation, and longevity of NK cells using combinations of IL-12, IL-15, and IL-18 have renewed interest in NK cell adoptive immunotherapy.

The vast array of immune escape mechanisms employed by tumors limits the use of NK cells. The secretion of immunosuppressive cytokines by tumor cells such as PGE2, IDO, adenosine, and TGFβ can negatively modulate NK function. Modulation of the TME with the upregulation of regulatory T cells or myeloid-derived suppressor cells creates an immunosuppressive milieu to further reduce NK cell function. The downregulation of neoantigens targeted by NK cells also decreases their ability to detect cancer cells. Recently, investigators have looked to OVs as a possible potentiating agent for NK therapy. In particular, strategies have been devised to use OVs to eliminate immunosuppressive regulatory T cells and myeloid-derived suppressor cells, improve NK cell trafficking into solid tumors, and provide cytokines and growth factors to improve NK cell activation and proliferation (Fig. 2).

Of note, the host NK cells have a complex relationship with oncolytic viral therapies. Activated NK cells can adversely affect oncolytic viral therapy by limiting the extent of virus-mediated oncolysis or by preferentially killing virus-infected tumor cells (Fig. 3). Conversely, many studies have shown that OVs induce recruitment of NK cells which amplify cancer killing. Furthermore, some OVs have been shown to directly infect and mature dendritic cells, which in turn activate NK cells leading to oncosuppressive effects (Fig. 3). The general belief is that a lower number of NK cells at the time of initial oncolytic viral therapy is best to allow for viral infection of tumor cells, but a significant NK cell population is needed after the initial phase of oncolytic viral therapy to allow for the activation of tumoricidal host immune responses. Strategies using a sequenced infusion of OV first, followed by antitumor activated NK cells, seek to optimize conditions for cancer killing.

Four studies (summarized in Table 4) have evaluated the combination of NK cell therapy with oncolytic viral therapy.

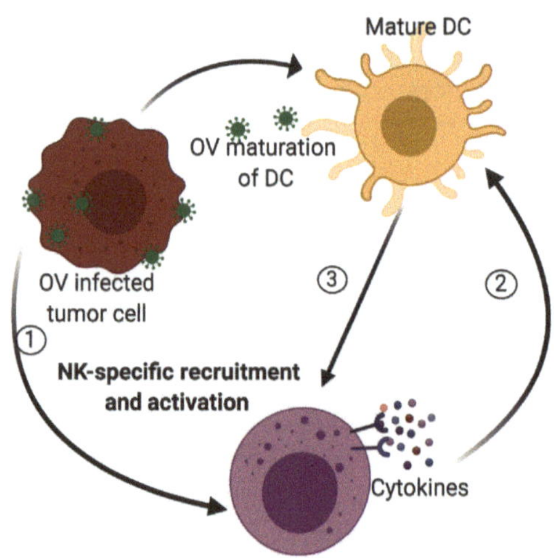

FIG. 2 Oncolytic viruses can be engineered to enhance NK function. Engineered OVs can ① secrete chemokines to attract NKs. Cytokine secretion from the activated NKs can ② lead to the maturation of dendritic cells, which can in turn ③ lead to further NK activation. Engineered OVs can secrete growth factors to enhance NK concentrations and survival. Engineered OVs can secrete immunostimulatory cytokines to change the tumor microenvironment (TME) for improved NK function.

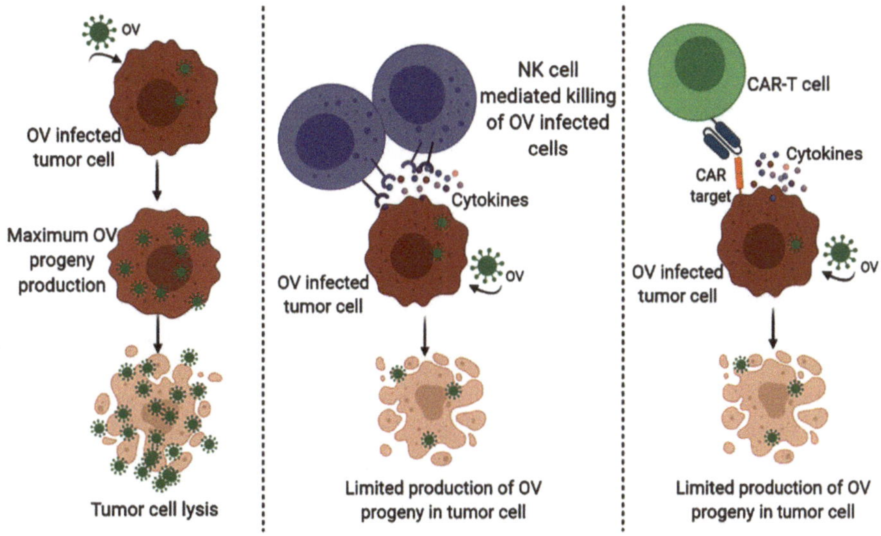

FIG. 3 NKs/CAR-Ts can negatively affect oncolytic viral therapy. Effective OV therapy relies on infection, replication of the virus, lysis of tumor targets, and propagation of infection by OV progeny. NKs and CAR-Ts may kill cancer targets before maximum production of OV progeny.

TABLE 4 Preclinical studies combining OVs and NK therapy.

Tumor	NK cell therapy	OV therapy	Results	Author/Year
MDA-MB-231 (breast cancer) brain metastases	Intratumoral EGFR-CAR NK-92 cells	Intratumoral oHSV	Improved tumor eradication and overall survival with CAR-NK + oHSV combination therapy	Chen 2016 [23]
GBM30 (glioblastoma) intracranial	Intra-tumoral NK cells	Intraperitoneal bortezomib + intratumoral oHSV	Improved survival with the addition of NK cells to mice treated with bortezomib and oHSV	Yoo 2016 [24]
Caco-2 (colon carcinoma)	Cytokine-induced killer	CRAd-CCL20-IL15	Improved tumor suppression	Ye 2017 [13]
HCT-116 (cervical cancer) subcutaneous	Intravenous NK-CCR5 cells	Intratumoral OV-ffLuc-CCL5	Improved survival and observed complete response rates with NK-CCR5 and OV-ffluc-CCL5 combination therapy	Li 2020 [25]

EGFR-CAR, *epidermal growth factor receptor—chimeric antigen receptor;* oHSV, *oncolytic herpes simplex virus;* NK-CCR5 cells, *NK cells transduced to express CCR5;* OV-ffLuc-CCL5, *oncolytic vaccinia virus engineered to express CCL5.*

Chen et al. studied NK cells engineered to express the second generation of EGFR-CAR. The combination of EGFR-CAR NK cells and a herpes simplex 1 OV was more effective at killing breast cancer. In a murine model of brain metastases from breast cancer, the combination effectively improved tumor eradication and overall survival compared to each therapy in isolation [23].

Yoo et al. studied the use of the proteasome inhibitor bortezomib and an oncolytic herpes simplex virus-1 (oHSV)-expressing GM-CSF on NK cell cancer immunotherapy [24]. Combination treatment with bortezomib- and oHSV-induced necroptotic cell death, mitochondrial ROS production, JNK phosphorylation, and significantly enhanced NK cell activation. Adjuvant NK cell therapy of mice treated with bortezomib and oHSV improved antitumor efficacy.

Ye et al. examined the use of an oncolytic adenovirus in delivering local CCL20 and IL-15 to enhance cytokine-induced killer cell therapy [13]. The CRAdCCL20-IL15 produces local expression of the chemokine CCL20 to attract NKs and other effector immune cells. The IL-15 then enhances the maturation and persistence of the NKs. The combination of cytotoxic effector cells and CRAd-CCL20-IL15 showed more significant antitumor potential than cytotoxic effector cells or CRAd-CCL20-IL15 alone. This combined treatment resulted in significant tumor regression in mouse models.

Li et al. also utilized an OV to produce local chemokines (CCR5 and CCL5) to the tumor site to increase NK trafficking to tumor targets. They hypothesized that enhancing chemokine expression may improve the therapeutic efficacy of NK cells. They found that in NK cells from patients with various solid tumors or healthy subjects, CCR5 was expressed at low levels before and after expansion in vitro. CCR5-engineered NK cells showed enhanced tumor infiltration and antitumor effects. They also constructed a CCL5-expressing oncolytic vaccinia virus. Supernatants from tumor cells infected by the CCL5-modified vaccinia virus enhanced the directional movement of CCR5-overexpressed NK cells. More importantly, NK cells were resistant to the vaccinia virus and their functions were not affected after being in contact. This is a proof of concept study [25] demonstrating the ability of virally expressed chemokines to increase NK cell homing and therapeutic effects. This study in a mouse cervical cancer model showed improved survival and, in some cases, complete responses [25].

These early studies show the potential of combination therapy with NK cell and oncolytic viral therapy. Additional modifications to OVs are underway to further enhance synergy between the two therapies (Fig. 4). In particular, OVs carrying transgenes for growth factors, cytokines, and/or chemokines to further attract and activate tumor-infiltrating NK cells have been engineered. OVs could also be used to alter the TME to decrease the population of regulatory T cells and myeloid-derived suppressor cells, resulting in the reduced immunosuppression of NK cells and potentially allow for a more durable response. Future human studies will be needed to validate the safety and efficacy of these approaches.

Immune cells as carriers of OV in cancer therapy

Many methods are also in development to provide vehicles to traffic oncolytic viral therapies to their targets. This is particularly important for viruses that are not specifically targeted to tumors. For adenoviral constructs, for example, the natural affinity in vivo is to travel to hepatocytes [26]. Thus, carrier vehicles that transport viruses to tumors offer

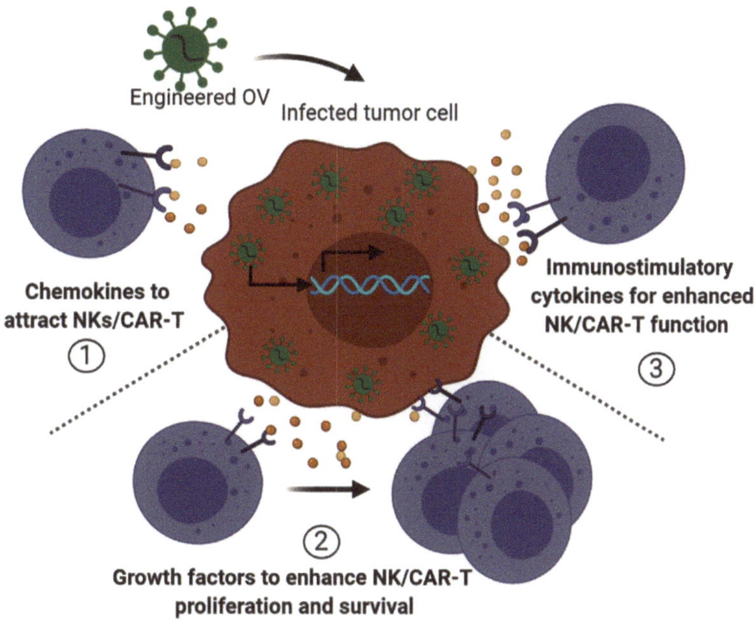

FIG. 4 Oncolytic viruses can be engineered to enhance NK and CAR-T function. ① Engineered OVs can secrete chemokines to attract NKs and CAR-T cells, ② as well as growth factors to enhance NK/CAR-T proliferation and survival. ③ Engineered OVs can secrete immunostimulatory cytokines to change TME for enhanced NK/CAR-T function.

protection from sequestration of all viruses to the liver rather than to the therapeutic target [27]. An antiviral immune response mounts as viruses are administered repeatedly. Carriers can also protect against preformed antibodies and T cells that would clear viruses before reaching tumor targets. In this regard, neural stem cells [28], blood cell ghosts [29], and nanoparticles [30] have been studied as potential carriers. In this chapter, we will discuss only T cells in such a carrier role.

CAR-Ts as carriers: Investigators have examined the use of CAR-T cells as carriers for OVs. It was hoped that loading CAR-Ts with OVs would allow delivery of two anticancer agents at tumor sites to enhance therapeutic effects. VanSeggelen and colleagues studied the use of CAR-Ts for delivery of vesicular stomatitis virus (VSV) and of Vaccinia viruses [31]. Loading of OV into human CAR-T cells did not alter cytokine production by CAR-T cells. OV-loaded CAR-T cells retained their ability to kill tumor targets specifically. Importantly, OV-loaded CAR-T cells deposit virus on tumor targets [31].

Cytokine induced killer (CIK) cells as carriers: Sampath et al. utilized a cytokine-induced killer (CIK) preloaded with an oncolytic vaccinia virus as a combined OV and immune cell therapy. The CIK cells pretreated with the vaccinia virus retained the capacity to infiltrate the tumor. The vaccinia virus-containing CIK showed a more significant tumoricidal activity than the virus or CIK delivered alone. When injected alone, viruses spread only to the neoplastic cells adjacent to the vasculature, while preinfected CIK cells can penetrate deep into the tumor mass [14].

Thorne and his colleagues have used tumor-trafficking immune cells to deliver oncolytic viral therapies to their tumor targets efficiently. The virus-loaded CIKs are meant to be both a therapeutic agent and also an OV carrier vehicle. These investigators showed that not only do the CIK help in targeting but they also help protect the virus against preexisting antiviral immunity [32]. Using a model of previously immunized mice, combination immune cell-viral therapy resulted in increased tumor infiltration of effector T-cells and a general reduction in the levels of known immune-suppressive lymphocyte populations. The CIK cells preloaded with the OV were also an efficient in situ viral factory.

Conclusion

The combination of OV and immune cell therapy has the potential to overcome the limitations of each therapy in isolation while harnessing the tumoricidal effect of both innate and adaptive immunotherapy. Early studies have shown promise for such an approach and generated data and mechanistic knowledge to encourage future human trials to combine OV and cell therapy to treat cancer.

References

[1] Chaurasiya S, Chen NG, Fong Y. Oncolytic viruses and immunity. Curr Opin Immunol 2018;51:83–90.

[2] Chaurasiya S, Chen NG, Lu J, Martin N, Shen Y, Kim SI, Warner SG, Woo Y, Fong Y. A chimeric poxvirus with J2R (thymidine kinase) deletion shows safety and anti-tumor activity in lung cancer models. Cancer Gene Ther 2020;27:125–35.

[3] Ribas A, Dummer R, Puzanov I, Vander Walde A, Andtbacka RHI, Michielin O, Olszanski AJ, Malvehy J, Cebon J, Fernandez E, Kirkwood JM, Gajewski TF, Chen L, Gorski KS, Anderson AA, Diede SJ, Lassman ME, Gansert J, Hodi FS, Long GV. Oncolytic virotherapy promotes intratumoral T cell infiltration and improves anti-PD-1 immunotherapy. Cell 2017;170(1109–1119), e10.

[4] Andtbacka RH, Kaufman HL, Collichio F, Amatruda T, Senzer N, Chesney J, Delman KA, Spitler LE, Puzanov I, Agarwala SS, Milhem M, Cranmer L, Curti B, Lewis K, Ross M, Guthrie T, Linette GP, Daniels GA, Harrington K, Middleton MR, Miller Jr WH, Zager JS, Ye Y, Yao B, Li A, Doleman S, Vander Walde A, Gansert J, Coffin RS. Talimogene laherparepvec improves durable response rate in patients with advanced melanoma. J Clin Oncol 2015;33:2780–8.

[5] Brown CE, Mackall CL. CAR T cell therapy: inroads to response and resistance. Nat Rev Immunol 2019;19:73–4.

[6] Park AK, Fong Y, Kim SI, Yang J, Murad JP, Lu J, Jeang B, Chang WC, Chen NG, Thomas SH, Forman SJ, Priceman SJ. Effective combination immunotherapy using oncolytic viruses to deliver CAR targets to solid tumors. Sci Transl Med 2020;12.

[7] Holzinger A, Abken H. CAR T cells targeting solid tumors: carcinoembryonic antigen (CEA) proves to be a safe target. Cancer Immunol Immunother 2017;66:1505–7.

[8] Shu R, Evtimov VJ, Hammett MV, Nguyen NN, Zhuang J, Hudson PJ, Howard MC, Pupovac A, Trounson AO, Boyd RL. Engineered CAR-T cells targeting TAG-72 and CD47 in ovarian cancer. Mol Ther Oncolytics 2021;20:325–41.

[9] Watanabe K, Luo Y, Da T, Guedan S, Ruella M, Scholler J, Keith B, Young RM, Engels B, Sorsa S, Siurala M, Havunen R, Tahtinen S, Hemminki A, June CH. Pancreatic cancer therapy with combined mesothelin-redirected chimeric antigen receptor T cells and cytokine-armed oncolytic adenoviruses. JCI Insight 2018;3.

[10] Wang D, Starr R, Chang WC, Aguilar B, Alizadeh D, Wright SL, Yang X, Brito A, Sarkissian A, Ostberg JR, Li L, Shi Y, Gutova M, Aboody K, Badie B, Forman SJ, Barish ME, Brown CE. Chlorotoxin-directed CAR T cells for specific and effective targeting of glioblastoma. Sci Transl Med 2020;12.

[11] DiNofia AM, Grupp SA. Will allogeneic CAR T cells for CD19(+) malignancies take autologous CAR T cells 'off the shelf'? Nat Rev Clin Oncol 2021;18:195–6.

[12] Greenberg AH, Playfair JH. Spontaneously arising cytotoxicity to the P-815-Y mastocytoma in NZB mice. Clin Exp Immunol 1974;16:99–109.
[13] Ye JF, Qi WX, Liu MY, Li Y. The combination of NK and CD8+T cells with CCL20/IL15-armed oncolytic adenoviruses enhances the growth suppression of TERT-positive tumor cells. Cell Immunol 2017;318:35–41.
[14] Sampath P, Li J, Hou W, Chen H, Bartlett DL, Thorne SH. Crosstalk between immune cell and oncolytic vaccinia therapy enhances tumor trafficking and antitumor effects. Mol Ther 2013;21:620–8.
[15] Wing A, Fajardo CA, Posey Jr AD, Shaw C, Da T, Young RM, Alemany R, June CH, Guedan S. Improving CART-cell therapy of solid tumors with oncolytic virus-driven production of a bispecific T-cell engager. Cancer Immunol Res 2018;6:605–16.
[16] Nishio N, Diaconu I, Liu H, Cerullo V, Caruana I, Hoyos V, Bouchier-Hayes L, Savoldo B, Dotti G. Armed oncolytic virus enhances immune functions of chimeric antigen receptor-modified T cells in solid tumors. Cancer Res 2014;74:5195–205.
[17] Tanoue K, Rosewell Shaw A, Watanabe N, Porter C, Rana B, Gottschalk S, Brenner M, Suzuki M. Armed oncolytic adenovirus-expressing PD-L1 mini-body enhances antitumor effects of chimeric antigen receptor T cells in solid tumors. Cancer Res 2017;77:2040–51.
[18] Rosewell Shaw A, Porter CE, Watanabe N, Tanoue K, Sikora A, Gottschalk S, Brenner MK, Suzuki M. Adenovirotherapy delivering cytokine and checkpoint inhibitor augments CAR T cells against metastatic head and neck cancer. Mol Ther 2017;25:2440–51.
[19] Evgin L, Huff AL, Wongthida P, Thompson J, Kottke T, Tonne J, Schuelke M, Ayasoufi K, Driscoll CB, Shim KG, Reynolds P, Monie DD, Johnson AJ, Coffey M, Young SL, Archer G, Sampson J, Pulido J, Perez LS, Vile R. Oncolytic virus-derived type I interferon restricts CAR T cell therapy. Nat Commun 2020;11.
[20] Bi J, Tian Z. NK cell dysfunction and checkpoint immunotherapy. Front Immunol 2019;10:1999.
[21] Peipp M, Derer S, Lohse S, Staudinger M, Klausz K, Valerius T, Gramatzki M, Kellner C. HER2-specific immunoligands engaging NKp30 or NKp80 trigger NK-cell-mediated lysis of tumor cells and enhance antibody-dependent cell-mediated cytotoxicity. Oncotarget 2015;6:32075–88.
[22] Frederiksen KS, Lundsgaard D, Freeman JA, Hughes SD, Holm TL, Skrumsager BK, Petri A, Hansen LT, McArthur GA, Davis ID, Skak K. IL-21 induces in vivo immune activation of NK cells and CD8(+) T cells in patients with metastatic melanoma and renal cell carcinoma. Cancer Immunol Immunother 2008;57:1439–49.
[23] Chen X, Han J, Chu J, Zhang L, Zhang J, Chen C, Chen L, Wang Y, Wang H, Yi L, Elder JB, Wang QE, He X, Kaur B, Chiocca EA, Yu J. A combinational therapy of EGFR-CAR NK cells and oncolytic herpes simplex virus 1 for breast cancer brain metastases. Oncotarget 2016;7:27764–77.
[24] Yoo JY, Jaime-Ramirez AC, Bolyard C, Dai H, Nallanagulagari T, Wojton J, Hurwitz BS, Relation T, Lee TJ, Lotze MT, Yu JG, Zhang J, Croce CM, Yu J, Caligiuri MA, Old M, Kaur B. Bortezomib treatment sensitizes oncolytic HSV-1-treated tumors to NK cell immunotherapy. Clin Cancer Res 2016;22:5265–76.
[25] Li F, Sheng Y, Hou W, Sampath P, Byrd D, Thorne S, Zhang Y. CCL5-armed oncolytic virus augments CCR5-engineered NK cell infiltration and antitumor efficiency. J Immunother Cancer 2020;8.
[26] Descamps D, Benihoud K. Two key challenges for effective adenovirus-mediated liver gene therapy: innate immune responses and hepatocyte-specific transduction. Curr Gene Ther 2009;9:115–27.
[27] Osali A, Zhiani M, Ghaebi M, Meymanat M, Esmaeilzadeh A. Multidirectional strategies for targeted delivery of oncolytic viruses by tumor infiltrating immune cells. Pharmacol Res 2020;161, 105094.
[28] Ahmed AU, Thaci B, Alexiades NG, Han Y, Qian S, Liu F, Balyasnikova IV, Ulasov IY, Aboody KS, Lesniak MS. Neural stem cell-based cell carriers enhance therapeutic efficacy of an oncolytic adenovirus in an orthotopic mouse model of human glioblastoma. Mol Ther 2011;19:1714–26.
[29] Kogure K, Okuda O, Itoh T, Hayashi K, Ueno M. Development of a membrane fusible drug carrier from erythrocytes by the spontaneous transfer of viral fusion protein from influenza virus-infected cells. Biol Pharm Bull 1997;20:581–3.
[30] Teng Z, Sun S, Luo X, Zhang Z, Seo H, Xu X, Huang J, Dong H, Mu S, Du P, Zhang Z, Guo H. Bi-functional gold nanocages enhance specific immunological responses of foot-and-mouth disease virus-like particles vaccine as a carrier and adjuvant. Nanomedicine 2021;33, 102358.
[31] Van Seggelen H, Tantalo DG, Afsahi A, Hammill JA, Bramson JL. Chimeric antigen receptor-engineered T cells as oncolytic virus carriers. Mol Ther Oncolytics 2015;2:15014.
[32] Thorne SH, Liang W, Sampath P, Schmidt T, Sikorski R, Beilhack A, Contag CH. Targeting localized immune suppression within the tumor through repeat cycles of immune cell-oncolytic virus combination therapy. Mol Ther 2010;18:1698–705.

CHAPTER

16

Natural killer cells in the treatment of glioblastoma: Diverse antitumor functions and potential clinical applications

Barbara Breznik[a,b], *Metka Novak*[a], *Bernarda Majc*[a,c], *Anamarija Habič*[a,c], *and Anahid Jewett*[b,d]

[a]Department of Genetic Toxicology and Cancer Biology, National Institute of Biology, Ljubljana, Slovenia [b]Division of Oral Biology and Medicine, The Jane and Jerry Weintraub Center for Reconstructive Biotechnology, University of California Los Angeles (UCLA), Los Angeles, CA, United States [c]Jozef Stefan International Postgraduate School, Ljubljana, Slovenia [d]The Jonsson Comprehensive Cancer Center, UCLA School of Dentistry and Medicine, Los Angeles, CA, United States

Abstract

Natural killer (NK) cells are cytotoxic lymphocytes of innate immunity that recognize and kill stressed, infected, and cancer cells. Moreover, NK cells orchestrate innate and adaptive immune responses and shape the tumor microenvironment. So far, adoptive transfer of NK cells has been shown to be effective and safe in patients with hematologic malignancies. In solid tumors, including the most lethal primary brain tumor glioblastoma, we are only in the early stages of exploiting the diverse antitumor potential of NK cells for treatment purposes. NK cells have the ability to target tumor-initiating and therapy-resistant cancer stem cells. In this review, we introduce the role of NK cells and the biology of glioblastoma and how this knowledge can lead to enhanced activity and in vivo persistence of NK cells that can be applied in NK cell-based immunotherapeutic platforms. Moreover, we discuss and present recent advances in NK cell-based therapies for primary brain tumors and the development of personalized glioblastoma models to test immunotherapeutic and combination strategies.

https://doi.org/10.1016/B978-0-12-822620-9.00003-3

Abbreviations

CSC	cancer stem cell
CAR	chimeric antigen receptor
CpG-ODN	CpG oligodeoxynucleotide
COX	cyclooxygenases
CTLA-4	cytotoxic T lymphocyte-associated antigen 4
ESC	embryonic stem cell
EGFR	epidermal growth factor receptor
GBM	glioblastoma
GSC	glioblastoma stem cell
GvHD	graft versus host disease
GM-CSF	granulocyte-macrophage colony-stimulating factor
HSC	hematopoietic stem cell
HDACi	histone deacetylase inhibitor
HLA	human leukocyte antigen
IDO	indoleamine 2,3-dioxygenase
iPSC	induced pluripotent stem cells
ILC3	innate lymphoid cell 3
IFN-γ	interferon gamma
IL	interleukin
KIR	killer immunoglobulin-like receptor
LAG-3	lymphocyte activation gene 3
MIC	major histocompatibility complex class I polypeptide-related sequence
MHC	major histocompatibility complex
MES	mesenchymal
mAb	monoclonal antibody
MDSC	myeloid-derived suppressor cell
NCR	natural cytotoxicity receptor
NK cell	natural killer cell
PAMP	pathogen-associated molecular pattern
PDX	patient-derived xenograft
PD-L1/2	PD-1 ligand 1/2
PFN	perforin
PBMC	peripheral blood mononuclear cell
PVRL2	poliovirus receptor-related 2
PD-1	programmed cell death protein 1
scFv	single-chain variable fragment
TIGIT	T cell immunoreceptor with immunoglobulin and ITIM domains
Treg	T regulatory cell
FDA	the food and drug administration
mTOR	mammalian target of rapamycin
TGF-β	transforming growth factor beta
TME	tumor microenvironment
TNF	tumor necrosis factor
TAM	tumor-associated macrophage
ULBP	UL16-binding protein

Conflict of interest

No potential conflicts of interest were disclosed.

Introduction

Cancer immunotherapy has emerged as a revolutionary treatment approach in recent years, showing promising clinical results in patients with various hematologic and solid tumors. The goal of immunotherapy is to enhance the patient's immune response to the tumor, overcome tumor immune resistance, and eliminate cancer cells. Several immunotherapeutic strategies, including immune checkpoint inhibitors, chimeric antigen receptor (CAR) T-cells, oncolytic viruses, and vaccines, have been used to combat cancer in recent years [1–4]. In contrast, natural killer (NK) cells have historically been overlooked as a therapeutic tool against cancer, but show great potential for treating aggressive primary brain tumors by targeting therapy-resistant cancer stem cells (CSCs) and orchestrating the functions of innate and adaptive immune system. In addition, NK cell-based therapy has been shown to be safe and effective in the treatment of advanced-stage leukemia [1,5–9].

However, there is still a large gap in understanding the biology of NK cells and their various functions related to primary brain tumors and their immunosuppressive microenvironment, as well as in vitro activation and expansion to improve in vivo functions of NK cells for clinical use. In this review, the biology of NK cells and the pathobiology of glioblastoma (GBM) are presented and how the knowledge gained may lead to the application of NK cell-based immunotherapy for the treatment of aggressive solid tumors.

Natural killer cells: Basic biology and dysfunction in cancer

Biology of natural killer cells

Natural killer (NK) cells are large granular lymphocytes of the innate immune system that directly kill target cells without prior exposure and are "ready to kill". NK cells mediate cytotoxicity and act as regulators of both innate and adaptive immune responses. They play a key role in antiviral and antitumor immunity [10]. NK cells arise from hematopoietic progenitor cells in bone marrow, spleen, liver, secondary lymphoid organs, thymus, tonsils, and intestine [11]. They comprise 5%–20% of peripheral blood mononuclear cells (PBMCs) [10] and are phenotyped as $CD3^-$ and $CD56^+$ cells. Peripheral blood NK cells are phenotypically and functionally diverse and are divided into two main groups: $CD56^{dim}$ and $CD56^{bright}$. The first subset, $CD16^{bright}/CD56^{dim}$, represents the majority (90%) of the NK cell population. These are believed to be the mature form of NK cells and are capable of performing all of their effector functions. On the other hand, the remaining NK cell population, which is less mature and does not express CD16 receptor, is only weakly cytotoxic and produces high levels of proinflammatory cytokines [5,10].

NK cell function is regulated by the balance of interactions between activating and inhibitory receptors on their surface and ligands on target cells [5]. Unlike T cells, NK cells do not carry a T cell receptor but are restricted to a peptide epitope presented by molecules of the major histocompatibility complex (MHC). The sum of stimulatory and inhibitory signals via the NK cell receptors results in either a response or tolerance to the target cells. With the help of these immune receptors, NK cells are able to recognize and spontaneously kill damaged and stressed infected or tumor cells without prior sensitization [5,9,12–14].

The most potent NK cell receptor is CD16, which is also known as the low-affinity IgG Fc region receptor III (FcγRIII). CD16 is the only receptor that enables activation of NK cells on its own and thus NK cells do not require additional activation by another receptor. NK cells are activated by cross-linking of the CD16 receptor after it has bound to the Fc regions of IgG antibodies on opsonized cells. This process is referred to as antibody-dependent cell-mediated cytotoxicity (ADCC). Several NK cell-based therapies exploit this process with monoclonal antibodies to enhance killing by NK cells via ADCC [15]. Other important activating NK cell receptors, including NKG2D, DNAM1, killer immunoglobulin-like receptors (KIRs), natural cytotoxicity receptors (NCRs), and the NKG2 family of receptors that form heterodimers with CD94, and their ligands are listed in Table 1 [1,5,9,13,15]. Inhibitory NK cell receptors provide

TABLE 1 List of receptors of NK cells and their ligands.

Receptor	Ligands	Expression on other cells
Activating receptors		
CD16/FcγRIII	Fc domain of IgG antibodies	Activated monocytes and macrophages
KIR 2DS1–2DS5 and 3DS1	HLA-C	Effector and memory T cells
NKG2C/CD159c	HLA-E	Effector and memory T cells
NKG2D/CD314	MICA/B, ULBP1–6	T cells and activated macrophages
DNAM1/CD226	CD112/PVRL2, CD155/PVR	T cells and mononcytes
NKp46/CD335	unknown bacterial components and unknown extracellular ligands	αβ and γδ T cell subsets and ILC3
NKp44/CD336	NKp44L, unknown bacterial components, viral envelope glycoproteins	αβ and γδ T cell subsets, ILC3 and plasmacytoid dendritic cells
NKp30/CD337	B7-H6	αβ and γδ T cell subsets and epithelial cells
4-1BB/CD137	4-1BBL	T cells, dendritic cells, and endothelial cells
Inhibitory receptors		
KIR 2DL1–2DL3, 2DL5 and 3DL1–3DL3	HLA-A,-B,-C	Effector and memory T cells
NKG2A/CD159a	HLA-E	Effector and memory T cells

TABLE 1 List of receptors of NK cells and their ligands—cont'd

Receptor	Ligands	Expression on other cells
TIGIT	CD112/PVRL2, CD155/PVR	T cells
Tactile/CD96	CD155/PVR	T cells
PD-1	PD-L1, PD-L2	Activated T and B cells and myeloid cells
Activating and inhibitory receptors		
KIR 2DL4	HLA-G	Effector T cells

Abbreviations: HLA, *human leukocyte antigen;* ILC3, *innate lymphoid cell 3;* KIR, *killer cell immunoglobulin-like receptor;* MIC, *major histocompatibility complex (MHC) class I polypeptide-related sequence;* PD-1, *programmed cell death protein 1;* PD-L1/2, *PD-1 ligand 1/2;* PVR, *poliovirus receptor;* PVRL2, *poliovirus receptor-related 2;* TIGIT, *T cell immunoreceptor with immunoglobulin and ITIM domains;* ULBP, *UL16-binding protein.*

immunological self-tolerance and negative feedback that balances signals from activating receptors. Several inhibitory receptors are triggered by self-MHC class I molecules (HLA-A, -B, -C, -E, and -G). For this reason, HLA-mismatched NK cells from healthy donors are preferred for adoptive cancer immunotherapy, as these NK cells do not recognize cancer cells as self and bypass inhibitory signals [1,5,16]. Other inhibitory receptors include PD-1, KIRs, NKG2A, tactile, and TIGIT (Table 1).

NK cells have the ability to control cancer progression directly via lysis of cancer cells and indirectly via regulation of innate and adaptive antitumor immune responses. The killing of target cells (cancer cells) by NK cells is mediated by granules containing membrane-disrupting proteins perforins and proteolytic enzymes granzymes. Upon interaction of NK cells with the cancer cell, an immunological synapse is formed and granules containing perforins and granzymes are exocytosed, leading to lysis of the cancer cell [17]. NK cell killing can also be mediated by tumor necrosis factor (TNF) family factors. Activated NK cells express and secrete death-inducing ligands, including FAS and TNF-related apoptosis-inducing ligand (TRAIL), which, when bound to death receptors on target cells, induce apoptosis [18].

Once NK cells are activated, they secrete cytokines and growth factors. One of the best characterized is interferon gamma (IFN-γ), although several others are also secreted, such as TNF-α, interleukins (ILs)-6, -8, and -10, chemokines C—C motif ligand 3 (CCL3), CCL5 (RANTES), and granulocyte-macrophage colony-stimulating factor (GM-CSF) that orchestrate innate and adaptive immune responses as well as shape the tumor microenvironment (TME) [6,7,15,19]. NK cells interact with macrophages, B cells, T cells, and dendritic cells and influence their phenotype and functions. For example, NK cells activate macrophages through secretion of IFN-γ, induce dendritic cell maturation, and activate cytotoxic T cells. Through cellular cross-talk with dendritic cells, NK cells exert a major influence on the adaptive immune response. NK cells shape the dendritic cell response by both directly killing immature dendritic cells and inducing their maturation [20]. Moreover, NK cells stimulate dendritic cell recruitment to the TME by secreting the chemoattractants CCL5 and XCL1, thereby promoting cancer immune control [19].

Natural killer cell dysfunction in cancer

Despite the activity of NK cells in controlling tumor growth and killing neoplastic cancer cells, NK cell functions are impaired in cancer patients under the influence of progressive tumor, its immunosuppression and aggressive treatments [21]. Defects in NK cell functions have been observed in preneoplastic stages of cancer development. These findings suggest that NK cell dysfunction contributes to malignant cancer progression, in addition to genetic, environmental, and other factors [5,6].

Circulating NK cells are recruited to the tumor under the influence of chemokines secreted by the tumor. NK cells express several cell surface chemokine receptors CXCR3, CXCR4, CX3CR1, and CCR3-CCR5, which can respond to the gradient of CCL2-CCL5, fractalkine, and CXCL12 and promote tumor infiltration of NK cells [22]. Trafficking and infiltration of NK cells into the TME are often impaired in cancer patients by various mechanisms, such as downmodulation of the secretion of NK cell chemoattractants in tumors and NK cell surface chemokine receptors, as well as the presence of stromal barriers in tumors [20].

NK cells can kill tumor cells or exert other functions, including secretion of cytokines, when infiltrated into the TME. After interaction with cancer cells, NK cells downregulate CD16 expression and lose their cytotoxic ability, but retain the potential to secrete high levels of cytokines and growth factors. These NK cells are immature and are also referred to as split anergized. They differ from anergized T cells in that they do not lose all effector functions like T cells when anergized. Split anergized NK cells are still an important source of IFN-γ, TNF-α, and other cytokines that regulate the functions of other cells in the TME [23]. NK cells eventually become completely dysfunctional in the immunosuppressive TME, as they also exhibit a loss of cytokine secretion. Similar dysfunction of NK cells has also been observed in the peripheral blood of cancer patients. Namely, NK cells isolated from the peripheral blood of patients with various types of cancer lose their cytotoxic function and their ability to secrete IFN-γ [24–26]. Importantly, activation and expansion of primary NK cells from cancer patients is limited in vitro, as NK cells from cancer patients exhibit significantly lower levels of expansion, cytotoxicity, and secretion of cytokines than primary NK cells from healthy donors [24,25]. Consistent with these findings, the use of allogeneic NK cells is a more promising approach, as NK cells from healthy individuals are not subject to immunosuppression in cancer patients and have a higher potential to exert their functions in vivo.

Use of natural killer cell-based therapeutic platforms for cancer immunotherapy

Adoptive cell transfer of NK cells has been shown to be safe and efficient in patients with hematologic cancers [27,28]. Currently, great efforts are being made to translate the successes in hematologic malignancies to solid tumors, including malignancies of the brain. Ideally, NK cell-based therapies should improve NK cell activity and function in cancer patients, but NK cells are already dysfunctional in cancer patients and exhibit poor infiltration into solid tumors [1,24,26]. Therefore, the leading approach to enhance NK cell-mediated immune responses against tumors is adoptive cell transfer of NK cells isolated from the same patients (autologous NK) or healthy donors (allogeneic NK cells), activated and expanded ex vivo, and then administered to patients. Alternative allogeneic sources of NK cells include cord blood NK cells, NK cell lines, stem cell-derived NK cells from umbilical cord blood stem cells or induced pluripotent stem cells, and CAR NK cells [1,9,13,29]. For adoptive cell transfer, HLA-mismatched NK cells from healthy donors are preferred, which do not recognize tumor cells as "self" and can bypass inhibitory signals [12].

Compared with T cells in the natural antitumor immune response or adoptive cellular therapy, NK cells possess equivalent effector functions but have several advantages. For example, NK cells have broader reactivity against tumors and lower specificity because NK cells lack T cell receptors. NK cells are efficient at eliminating early-stage tumors as well as metastatic tumor cells. Moreover, NK cells are safer as they do not carry a high risk of graft versus host disease (GvHD) and cytokine release syndrome. On the other hand, NK cells are usually found in lower numbers in tumors, have lower proliferation and recall response, and are more difficult to genetically engineer compared to T cells [5,12,15,20,30,31]. Another important advantage of NK cells is their potential for off-the-shelf therapy.

To improve the efficacy of NK cell therapies in solid tumors, several approaches have been pursued to enhance their cytotoxicity, cytokine production, and persistence in vivo. Early preclinical and clinical studies have shown that cytokines prime NK cells to promote their effector functions and thereby activate antitumor immunity. For example, IL-2, IL-15, and IFN-α are stimulatory cytokines mainly explored in clinical trials and have been shown to activate and expand NK and cytotoxic T cells in vitro and in vivo [20]. Modified recombinant IL-2 and IL-15 cytokines such as PEGylated IL-2 and IL-15 superagonist fusion complex IL-15/IL-15Rα-Sushi-Fc (ALT-803) have been shown to promote antitumor immunity in cancer patients and have been shown to be safe in clinical trials [32,33].

NK cells are activated by ADCC through contact with activating receptors, such as the highly potent CD16. Monoclonal antibodies targeting cancer cells act through two mechanisms of action. The Fab fragment of the monoclonal antibody binds a signaling receptor on the surface of the cancer cell and promotes apoptosis and growth arrest, while the Fc region recruits effector immune cells that lead to tumor eradication. The most common antigens associated with cancer cells against which monoclonal antibodies have been successfully used are EGFR, HER2, CD19, CD20, CD33, and CD133 [1]. Bispecific (BiKEs) or trispecific (TriKEs) killer cell engagers are an alternative strategy to bind the receptor CD16 and trigger ADCC-like responses. These molecules are bi- or tri-specific engager proteins consisting of multiple antibody-derived antigen-targeting domains, usually single-chain variable fragments (scFvs). One scFv domain targets an NK cell activating receptor (CD16) and another binds to a specific cancer-associated antigen or TME-associated antigen. Several strategies using BiKEs and TriKEs are currently being tested in preclinical and clinical studies, including targeting a single or two cancer antigens, targeting two TME antigens, or targeting two cancer and TME antigens [34,35].

Another approach to increase NK cell activity in vivo is to target immune checkpoints that inhibit NK cell activity and promote their exhaustion. NK cells express a variety of inhibitory receptors such as KIRs, NKG2A, CD96, PD-1, cytotoxic T lymphocyte-associated antigen 4 (CTLA-4), lymphocyte activation gene 3 (LAG-3), and T cell immunoglobulin and mucin domain-containing molecules 3 (TIM-3) [1]. In general, the relationships between inhibitory signaling pathways and the clinical utility of targeting these inhibitory signaling pathways in NK cells in cancer patients remain to be explored, and currently several agents targeting immune checkpoint inhibitors on NK cells are in clinical trials. For example, a humanized anti-NKG2A blocking antibody, alone or in combination with other therapeutics, is in multiple clinical trials for various cancers and has been shown to promote antitumor immunity by increasing the activity of cytotoxic T and NK cells [36,37]. PD-1 is expressed on the surface of NK cells in tumors [9]. Monoclonal antibodies against PD-1 and PD-L1 have shown clinical success in the treatment of various cancers, and NK cells have been identified as key mediators of antibody PD-L1 induced cytotoxicity against cancer cells via the ADCC mechanism [9,38].

NK cells expressing chimeric antigen receptors (CARs) directed against tumor surface antigens are another approach to improve NK cell activity and show encouraging clinical responses in hematologic cancers [27]. CARs enable recognition of the target antigen and subsequent immune cell activation without antigen presentation by the MHC. These properties are mediated by the specific CAR structure consisting of an extracellular antigen-binding domain fused with an intracellular signaling domain. The latter has been refined through several CAR generations. In contrast to CAR T-cells, CAR NK cells can be generated from allogeneic donors [39,40]. When it comes to antigens expressed on tumor cells but also in normal tissues, CAR NK cell therapy might represent a safer therapeutic approach compared to CAR T-cells due to the limited expansion potential and lifespan of NK cells [12]. Irradiation of CAR NK cells prior to therapeutic application has been shown to prevent their proliferation and permanent engraftment while maintaining high and specific cytotoxicity to target cells [41]. Unlike CAR T cells, which can cause severe toxicities in patients such as cytokine release syndrome, GvHD, and others, treatment with CAR NK has not been associated with adverse immunological events [12,27,28]. Recently, umbilical cord blood-derived HLA mismatched anti-CD19 CAR NK cells proved safe and showed a high response rate in patients with lymphoid tumors [28]. Research is now focused on further enhancing the efficacy of CAR NK cells by overcoming mechanisms of functional exhaustion and prolonging their in vivo persistence [27,42].

NK cell-based adoptive transfer can be combined with standard treatment approaches such as radio- and chemotherapy and other immunotherapeutic strategies such as oncolytic viruses, monoclonal antibodies, immune checkpoint inhibitors, and CAR T-cells, with the goal of inhibiting tumor growth and progression [5]. Another advantage of NK cell-based adoptive cell transfer is the recognition and elimination of tumor-initiating and therapy-resistant CSCs by NK cells [15,43], which will be discussed in more detail in the following chapters.

Glioblastoma remains an incurable primary brain tumor

Therapeutic resistance

GBMs, isocitrate dehydrogenase (IDH)1 wild-type grade IV gliomas, are the most common form of malignant primary brain tumors in adults [44,45]. GBMs respond poorly to standard cancer therapy, including maximal surgical resection, temozolomide chemotherapy, and radiation therapy, and have one of the shortest survival rates of any cancer [46]. Despite novel treatment modalities based on the 2005 Stupp protocol targeting novel mechanistic vulnerabilities, 5-year patient survival is less than 5% [47,48]. Therapeutic resistance and recurrence of GBM are due to rapid tumor growth, invasion, inter- and intratumor heterogeneity and the presence of GBM stem cells (GSCs) [49–52]. Moreover, current immunotherapeutic strategies against GBM are not efficient due to the high level of immunosuppression in patients and the heterogeneity and plasticity of GBM, which lacks specific and persistently expressed tumor-specific antigens.

The molecular subtypes of GBM molecularly categorized by The Cancer Genome Atlas (TCGA) are defined by three subtypes—proneural (PN), characterized by abnormalities in *platelet-derived growth factor alpha receptor (PDGFRA)* gene, classical (CL), characterized by abnormalities in *epidermal growth factor receptor (EGFR)* gene, and mesenchymal (MES), characterized by deletions of *neurofibromin 1 (NF1)* gene. GBM subtypes differ in terms of treatment resistance and patient survival, with MES having the shortest overall survival and PN the longest [53]. GBM subtypes are differentially expressed in individual cells within a tumor, forming a dynamic intratumoral heterogeneity that influences GBM biology, prognosis, and especially treatment resistance [54].

GSCs, also known as tumor-propagating cells or tumor-initiating cells, are maintained in hypoxic and perivascular niches [55,56] and are responsible for GBM recurrence and therapy resistance, mainly due to their DNA repair and multidrug resistance mechanisms [51,57]. GSCs have the ability to self-renew and express multiple developmental- and stemness-related signaling pathways [55,57–60]. The cell surface marker CD133 (Prominin-1) was the first marker used for isolation of GSCs [60–62]. Since then, other markers for detection and isolation of GSCs have been confirmed such as CD15 (SSEA-1), CD44, α6 integrin, CD9, SOX2, nestin, and A2B5 [59,63–67]. Galdieri et al. have shown that GSCs highly expressing CD15, CD44, CD133, and α6 integrin have the greatest self-renewal capacity, WNT inhibitor sensitivity, and in vivo tumorigenicity, but it is still unknown how these individual surface markers define GSC populations and whether expression of different combinations of these markers is associated with different functional capacities of GSCs [64]. GBM TME plays a key role in shaping cellular states and GSC plasticity, but it is still unclear whether it plays a role in selecting for survival of specific GSCs or whether GSCs simply adapt within the new TME [58]. GSCs also evade chemotherapy and radiotherapy through dormancy and upregulation of cell surface transporters. GSCs also evade antitumor immune responses by downregulating MHC class I molecules and specific antigens and increasing the expression of PD-L1 on their cell surface. GSCs are known to secrete immunosuppressive factors into the TME, thereby inhibiting the antitumor activity of immune cells [3,61]. NK cells do not recognize the MHC I molecules on GSC cells and therefore kill GSCs [5,7,13,43].

Complex immunosuppressive microenvironment

To sustain an effective antitumor response, immune cells must be present in sufficient numbers and activated to exert their killing functions. These processes, and thus the success of adoptive cell therapies, are hampered by specific immune evasive properties of cancer cells and by the immunosuppressive TME [68]. GBM exhibit a specific immune landscape that differs from cancers in other parts of the body. This is due to its specific location in the brain and the blood–brain barrier that control immune cell trafficking and the entry of drugs into the central nervous system [69]. The brain is no longer considered an immune-privileged organ, as there is an active interaction between the brain and the immune system [70,71]. On closer inspection, the TME is not only composed of cancer cells but also noncancerous cell types such as astrocytes, neurons, endothelial cells, pericytes, fibroblasts, mesenchymal stem cells, and immune cells that interact with cancer cells to promote tumor progression and resistance to therapy [72–74].

The most abundant immune cells in GBM are tumor-associated macrophages (TAMs), which divide into monocyte-derived macrophages and microglia and can account for up to 50% of cells in GBM TME [75–79]. Together with other immune cells such as T cells, NK cells [77,80,81], dendritic cells [73,82], and myeloid-derived suppressor cells (MDSCs) [83], they have an impact on the TME and thus on the response to cancer therapy. Immune cells support tumor growth and invasion by secreting cytokines and extracellular vesicles and by connecting nanotubes [84]. Four different immune-based subtypes have been discovered in GBM [85]. Immune subtype 1 is enriched in microglia, macrophages, and lymphocytes, while immune subtype 2 has higher levels of macrophages and lower numbers of lymphocytes. Immune subtype 3 is similar to immune subtype 2 but has a higher neuronal count. The most pronounced immune subtype is subtype 4 with the lowest numbers of all immune cell types. When combining immune subtypes with GBM subtypes, the most aggressive and refractory MES subtype was enriched with both immune subtypes 1 and 2 [85].

NK cells are a component of GBM TME, but occur in relatively small amounts and account for approximately 2% of all tumor-infiltrating immune cells [86]. Nevertheless, a strong correlation has been found between activated NK cells and survival of GBM patients [87]. In orthotopic GBM xenograft models, NK cells have also been shown to prevent systemic tumor metastasis [81]. In GBM TME, two populations of NK cells were found. One is immature $CD56^{int/bright}$ $CD16^-$ NK cells and the other is highly cytotoxic populations of $CD56^{int}CD16^+$ NK cells. Immature NK cells are more common in GBM, whereas cytotoxic NK cells are more common in low-grade gliomas and brain metastases [88].

Immunosuppressive GBM TME is the major challenge to overcome for neuro immunotherapy. Immunosuppressive transformed stromal and immune cells, including MDSCs, TAMs, T regulatory cells (Tregs), tumor-associated fibroblasts, and mesenchymal stem cells, as well as secreted immunosuppressive factors impair NK and T cell activity [13,89,90,124,168]. T and NK cells found in GBM TME are inactive and exhausted [91]. In addition to lymphopenia, decreased CD4+ T cells and increased Tregs, the number of NK cells in the peripheral blood of GBM patients is already decreased after temozolomide and radiation therapy [21].

GBMs secrete several immunosuppressive factors such as transforming growth factor-beta (TGF-ß), IL-10, adenosine, indoleamine 2,3-dioxygenase (IDO), cyclooxygenases (COX), and prostaglandin E2, which impair the functions of NK and T cells [19,107]. GBMs can also avoid the immune system by expressing the enzyme indoleamine 2,3-dioxygenase 1 (IDO1), which is involved in tryptophan metabolism. Tryptophan metabolites induce anergy of effector T cells while promoting the maturation and activation of Tregs [148]. TGF-ß is one of the major cytokines that decreases the activity of NK and CD8+ T cells by downregulating the activating receptors NKp30, NKG2D, and DNAM1 [99,102,109,170,174]. These receptors have been found to be significantly decreased on tumor-derived NK cells compared to NK cells from peripheral blood of the same patient [68]. Although detectable in patient-derived cancer cell lines in vitro [108], surface expression of MICA and ULBP2 on cancer cells is downregulated by TGF-β, whereas these ligands can also be cleaved from the tumor cell surface by metalloproteinases in the TME [105,109]. TGF-ß can also inhibit NK cell function by targeting the serine and threonine kinase mammalian target of rapamycin (mTOR), a crucial signal integrator for proinflammatory and antiinflammatory cytokines, such as IL-15 [5]. For NK cells to survive and proliferate at sufficient levels, the growth factor IL-15, which is absent in TME, must be present [9].

Recently, cystatin F, a member of the cystatin type II family and inhibitor of the cysteine peptidases cathepsins L, C, and H, which activate precursor forms of granzymes and perforins in cytotoxic lymphocytes, has been linked to GBM immunosuppression. Cystatin F is upregulated in GBM TME and GSCs, and decreased NK cell-mediated cytotoxicity was observed following cystatin F uptake by GBM cells [158].

Very important in large solid tumors is hypoxia with nonvascularized areas of low oxygen concentration where NK activity is inhibited. Low oxygen and nutrient deficiency in the TME limit NK cell cytotoxicity and cytokine secretion. The main fuel for NK cells is glucose, which is restricted in the TME where high lactate concentrations, amino acid deficiency and low pH are present, altering NK cell metabolism [5,164].

Natural killer cell-based immunotherapy for primary brain tumors

NK-based immunotherapy has recently attracted attention as one of the potential approaches to treat GBM. However, combined treatments may be required to sufficiently enhance NK activity and prevent immune evasion of tumor cells [94]. Several preclinical and clinical studies are currently ongoing in primary brain tumors, including high-grade gliomas and GBM (Tables 2 and 3). The results of the published studies are briefly described in the following paragraphs.

One possible approach in NK cell therapy is to stimulate the natural NK antitumor response, for example, by immunoinducing adjuvants such as CpG oligodeoxynucleotides (CpG-ODNs). Like natural pathogen-associated molecular patterns (PAMPs), CpG-ODNs bind to Toll-like receptor 9, leading to immune cell activation, increased antigen presentation, and inflammatory cytokine production [131]. In a mouse glioma model, an increase in NK cells and their cytotoxicity was observed after multiple intratumoral injections of CpG-ODNs at low doses. Remarkably, the surviving animals showed long-term tumor remission and were protected from tumor rechallenge [92]. Single administration of CpG-ODN has been previously studied in clinical trials, where it showed tolerable toxicity and partial response in some patients with recurrent gliomas [97,98], but did not prolong survival of patients with newly diagnosed GBM [167].

Ishikawa et al. reported of the first clinical study investigating the effects of autologous NK cell therapy in patients with recurrent malignant gliomas. Peripheral blood mononuclear cells (PBMCs) were isolated from the patients' blood and NK cell-rich effector cells were expanded ex vivo. Injection of these cells followed by a low dose of IFN-β was found to be safe but only partially effective [121]. Recently, Khatua et al. published the results of a phase I clinical trial (NCT02271711) of intraventricular infusions of autologous ex vivo expanded NK cells in children with recurrent medulloblastoma and ependymoma and reported no dose-limiting toxicities [127]. Another phase I clinical trial has been initiated (NCT04254419) to evaluate the

TABLE 2 An overview of in vivo preclinical studies of NK cell-based immunotherapeutic approaches for GBM.

Therapy type	Treatment	NK cell source	Model	Results summary	Reference (s)
Adoptive NK cell therapy	Intravenous or intratumoral transplantation of NK cells	Primary human NK cells from healthy donors	Orthotopic xenograft model in BALB/c-nude or NSG mice	Administration of NK cells induced apoptosis of GBM cells and prevented systemic metastasis	[81]
	Intracranial and systemic administration of NK cells pretreated with IL-2 and heat shock protein 70 (HSP70)	Primary rat NK cells	Orthotopic rat C6 glioma model in Wistar rats	Efficient homing of NK cells at the tumor site, amelioration of the tumor	[159]
Counteracting the immunosuppressive TME, sensitizing GBM cells and promoting natural NK cell responses	Intratumoral injection of IL-21	/	Orthotopic GL261 GBM model in C57BL/6 mice	GBM rejection in 77% of mice	[103]
	Vaccination with a recombinant adeno-associated virus vector encoding IL-12	/	Orthotopic DBTRG GBM xenograft model in nude mice	Increase of activated NK cells, decreased tumor growth rate and increased presumed-survival	[100]
	Intratumoral injections of CpG oligodeoxynucleotides	/	Orthotopic GL261 GBM model in C57BL/6 mice	Tumor eradication in 70% of mice, surviving animals were protected from intracranial rechallenge	[92]
	Treatment of GBM cells with trichostatin A	/	Subcutaneous LNT-229 xenograft GBM model in NMRI nude mice	Upregulation of NKG2D ligands MICA and ULBP2 on GBM cells, delayed tumor growth	[118]

TABLE 2 An overview of in vivo preclinical studies of NK cell-based immunotherapeutic approaches for GBM—cont'd

Therapy type	Treatment	NK cell source	Model	Results summary	Reference (s)
Combined approaches	Adoptive NK cell therapy plus mAb9.2.27 against chondroitin sulfate proteoglycan 4	Primary rat NK cells	Orthotopic GBM xenograft model in nude rats	Diminished tumor growth, prolonged animal survival	[149]
	viriotherapy plus bortezomib treatment and allogeneic NK cell adjuvant therapy	Primary human NK cells from healthy donors	Orthotopic patient-derived GBM xenograft model in nude mice	Enhanced NK cell activation, improved animal survival	[173]
	Bortezomib and allogeneic NK cell adoptive transfer	Primary human NK cells from healthy donors	Subcutaneous or intracranial U87 xenograft GBM model in NSG mice	Upregulation of NKG2D-activating stress ligands MICA and MICB, delayed tumor growth	[137]
	Bortezomib and autologous NK cell adoptive transfer	Primary human NK cells from the patient whose cells were used for the xenograft model	Orthotopic patient-derived GBM xenograft model in NOD/SCID mice	Upregulation of stress-ligands and TNF-related apoptosis-inducing ligand receptor (TRAIL-R2), augmented IFN-γ release, prolonged animal survival	[145]
	Intracranial administration of *TGFBR2*-knock-out NK cells in combination with inhibitors of integrin or TGF-β signaling	Primary human NK cells from healthy donors	Orthotopic patient-derived xenograft model in NSG mice	Prevention of GSC-induced NK cell dysfunction, effective control of the tumor	[42]
CAR-NK cell therapy	EGFRvIII-directed CAR NK cells expressing chemokine receptor CXCR4	Lentivirally transduced cell line YTS	Subcutaneous U87-MG^{EGFRvIII} or U87-MG$^{EGFRvIII/SDF-1\alpha}$ GBM xenograft model in NMRI nude mice	Complete tumor remission in a number of mice, increased survival	[143]

Continued

TABLE 2 An overview of in vivo preclinical studies of NK cell-based immunotherapeutic approaches for GBM—cont'd

Therapy type	Treatment	NK cell source	Model	Results summary	Reference (s)
	EGFRvIII-directed CAR NK cells	Lentivirally transduced cell line KHYG-1	Subcutaneous U87-MG^{EGFRvIII} GBM xenograft model in NOG mice	No effect on tumor volume, pseudo-palisading pattern	[144]
	CAR NK cells recognizing EGFR and EGFRvIII	Lentivirally transduced cell line NK-92	Orthotopic U251 and patient-derived GBM xenograft model in NSG mice	Suppressed tumor growth, prolonged animal survival	[115]
	CAR NK cells recognizing EGFR, EGFRvIII or both	Lentivirally transduced cell line NK-92	Orthotopic GBM xenograft model in NSG mice, produced by intracranial injection of LNT-229^{EGFR}, LNT-229^{EGFRvII}I or a 1:1 mixture of both cell lines	Dual-specific CAR NK cells were superior to monospecific CAR NK cells, prolonged animal survival and prevented rapid immune escape	[112]
	ErbB2-directed CAR NK cells	Lentivirally transduced cell line NK-92	Subcutaneous and orthotopic LN-319 GBM xenograft model in NSG mice and subcutaneous syngeneic GL261/ErbB2 GBM model in C57BL/6 mice	Complete tumor regression in a majority of cases, long-term protection against tumor rechallenge	[175]
	Dual-specific CAR NK cells (directed against disialoganglioside (GD2) and NKG2D), enabling local release of an antibody fragment that inhibits the activity of CD73 in the TME	Lentivirally transduced primary NK cells from healthy donors	Subcutaneous and orthotopic patient-derived GBM xenograft model in NSG mice	Delayed or arrested tumor growth	[169]

TABLE 3 Clinical trials of adoptive NK cell-based therapies for primary brain tumors.

Treatment	NK cell source	Route of administration	Disease	Study phase	Study status	Clinical trial identifier and/or reference
Autologous NK cells therapy plus adjuvant low-dose IFN-β	Primary NK cells from patient peripheral blood	Focal and intravenous injection or intravenous injection alone	Recurrent malignant glioma	Phase I	Completed	[121]
Autologous NK cells therapy	Primary NK cells from patient peripheral blood	Intraventricular infusions	Recurrent medulloblastoma and ependymoma	Phase I	Completed	NCT02271711 [127]
Autologous NK cells therapy	Primary NK cells from patient peripheral blood	Intratumoral application	Recurrent high grade glioma	Phase I	Not yet recruiting	NCT04254419
Hematopoietic stem cell-derived NK cells therapy	Hematopoietic stem cell-derived NK cells	Intravenous or intratumoral administration	Recurrent GBM	Phase I	Active, not recruiting	NCT04489420
Adoptive cell therapy with engineered NK cells containing deleted TGF-βR2 and NR3C1	Cord blood-derived NK cells	Intratumoral application	Recurrent GBM	Phase I	Not yet recruiting	NCT04991870
ErbB2-specific CAR NK cell therapy	NK-92	Application into the resection cavity	Recurrent or refractory ErbB2-positive GBM	Phase I	Recruiting	NCT03383978

intratumoral applications of autologous ex vivo expanded NK cells in children with recurrent high-grade glioma. However, to bypass the MHC class I-mediated NK cell inhibition, the use of allogeneic NK cells may be preferred over the use of autologous NK cells [111,134]. The Food and Drug Administration (FDA) has recently granted CYNK-001, an off-the-shelf nongenetically modified allogeneic placental hematopoietic stem cell-derived NK cell therapy, for the treatment of adult patients with recurrent GBM. The therapy is currently being evaluated in a phase I clinical trial (NCT04489420) evaluating both intravenous and intratumoral administration of CYNK-001 cells. Alternatively, MHC class I-mediated inhibition of NK cells could be avoided by blocking inhibitory receptor-ligand interactions. An anti-KIR-specific monoclonal antibody (lirilumab) has been shown to stimulate NK cell activity against tumor cells [161], and similar effects have been observed with the humanized anti-NKG2A monoclonal antibody (monalizumab), which prevents the interaction between

NKG2A on NK cells and HLA-E on tumor cells [36]. However, none of these antibodies have been tested in GBM to date.

To achieve better antitumor efficacy, it may also be beneficial to upregulate activating molecules on the surface of NK cells. This can be achieved by direct cytokine stimulation of NK cells or indirectly by modulating the immune landscape within the TME. For example, intratumoral injection of IL-21 enhanced the cytotoxic response of NK cells in a murine model of GBM and significantly inhibited tumor growth [103]. Enhanced activation of NK cells was also observed after vaccination with a recombinant adeno-associated viral vector encoding the proinflammatory cytokine IL-12 [100]. Pretreatment of NK cells with IL-2 and heat shock protein 70 (HSP70) significantly improved their functionality in vitro and in vivo. Activated NK cells were able to cross the blood–brain barrier and infiltrate the tumor [159]. Conversely, several studies targeted one of the major immunosuppressive cytokines in the GBM TME, TGF-β [109,170]. Its silencing prevented downregulation of the activating receptor NKG2D on NK cells and upregulated the expression of MICA in GBM cells. NK cells isolated from mice implanted with TGF-β-deficient GBM cells showed an activated phenotype [109]. In another study, NK cells from umbilical cord blood were modified to express the dominant-negative TGF-β receptor II (DNRII). These cells expressed the activating receptors NKG2D and DNAM1 and retained their cytolytic activity against GBM even in the presence of TGF-β, holding great potential for clinical applications [174]. In vitro, their functionality was also demonstrated in the presence of TGF-β-rich medulloblastoma-conditioned medium [150]. Recently, a phase I clinical trial (NCT04991870) was initiated to evaluate the safety, tolerability, and response to genetically engineered NK cells with deleted TGF-βR2 and NR3C1 in patients with recurrent GBM.

To prevent immune evasion of cancer cells, NK cell therapy can be combined with drugs that make cancer cells (more) susceptible to lysis by NK cells. One such example is trichostatin A, a histone deacetylase inhibitor (HDACi), which by itself enhances apoptosis of GBM cells [160]. However, the compound also induces upregulation of NKG2D ligands MICA and ULBP2 in cancer cells, immunizing them against NK cell-mediated lysis. Treatment with this drug delayed tumor growth in xenotransplanted nude mice [118]. It is also possible that the increased levels of activating MICA after treatment with trichostatin A are related to the decreased expression of MMPs, which are known to cleave MICA from the cell surface [172]. The proteasome inhibitor bortezomib has been tested in several studies and found to upregulate the expression of DNAM1 and NKG2D-activating stress ligands [137,145] and TNF-related apoptosis-inducing ligand receptor (TRAIL-R2) on GBM cells [145]. In vivo, combination therapy of bortezomib and NK transfer delayed tumor growth [137] and prolonged animal survival [145]. In another study, proteasome blockade was tested in combination with viral therapy using an oncolytic herpes simplex virus (oHSV) strain and NK cell therapy. This approach also resulted in increased activation of NK cells, which secreted higher levels of TNF-α and IFN-γ, and prolonged murine survival [173].

NK cells not only exert killing functions but also modulate the TME via the secretion of cytokines, thus influencing other immune cell types, e.g., macrophages and microglia. Therefore, administration of activated NK cells may represent a strategy to stimulate endogenous immune responses and enhance the efficacy of monoclonal antibody (mAb)-based immunotherapies [128,149]. The combination of NK cells and mAb9.2.27 against chondroitin sulfate proteoglycan 4 (CSPG4 or NG2) inhibited tumor growth and prolonged animal survival compared to vehicle and monotherapy controls in preclinical animal models of GBM. Combination treatment increased secretion of the proinflammatory cytokines IFN-γ and TNF-α and recruitment of proinflammatory macrophages, which then presumably promoted or differentiated into cytotoxic antigen-presenting $ED1^+$ $ED2^{low}$ MHC II^+ microglia [149].

Targeting glioblastoma stem cells

One of the most important properties of NK cells is their ability to select and kill GSCs. Due to their resistance to conventional therapies, GSCs are increasingly recognized as a crucial target in the development of new therapeutic approaches for GBM. Several in vitro studies have shown that GSCs are more susceptible to NK cell-mediated lysis compared to their differentiated counterparts [43,80,93,166]. This is crucial for the elimination of GSCs, as these cells generally do not express or downregulated surface-specific antigens under specific TME triggers or therapies. GSCs are characterized by their plasticity and are therefore difficult to target by specific CAR T-cells and other immunotherapeutic approaches [42,43,59,113,166]. Performing an NK cytotoxicity assay on a mixed cell population with GSCs and normal neural progenitor cells Close et al. demonstrated that IL-15-activated NK cells preferentially attack GSCs [68]. To date, the mechanisms underlying the sensitivity of GSCs to NK cells are poorly understood.

Several immunomodulatory cell surface molecules are expressed on GSCs. These include ligands recognized by activating NK cell receptors, e.g., the DNAM1 ligands PVR, Nectin-2 and CD112 [43,80], NKG2D ligands [43,165], and low levels of ligands for the NKp30/NCR3, NKp46/NCR1, and NKp44/NCR2 receptors [43,171]. However, GSCs also express low amounts of MHC class I molecules [80,93,165,171]. Avril et al. have shown by antibody-based blockade of MHC class I that these molecules can protect GSCs from lysis by resting NK cells [93], although this was ruled out in another study using a similar approach but a different blocking antibody [80]. In both studies, GSCs were effectively destroyed by preactivated NK cells. Preactivation induced the expression of additional activating receptors and was achieved either by specific cytokines (IL-2 or IL-15) [80] or by lectins [93]. Nevertheless, even resting NK cells were able to kill GSCs via ADCC, as demonstrated by the use of the therapeutic anti-EGFR antibody Cetuximab [93,166].

Several studies show that GSCs upregulate the expression of MHC class I [165,171] and the expression of ligands for activating NK receptors NKp30, NKp44, NKp46, and NKG2D after treatment with IFN-γ [171]. In rodent GBM models, intratumoral *IFN-γ* gene transfer resulted in slower rate of the tumor growth [104] or even complete tumor regression [153]. The enhanced immune response has been mainly attributed to T cells [153], but could also involve NK cells, which have been shown to effectively lyse GSCs sensitized by IFN-γ pretreatment in vitro [171].

In addition to directly killing GSC, NK cells also shape the phenotype of GSC by secreting cytokines. We have shown that split anergized NK cells affect CSCs differentiation. It has been shown that NK cell phenotype and function can change in response to stem cell signals, transforming NK cells from cytotoxic to cytokine-producing cells. The so-called split anergy was induced in NK cells co-cultured with oral squamous cell carcinoma CSCs or alternatively in NK cells treated in vitro with IL-2 and anti-CD16 mAb, resulting in decreased expression of

CD16 on the surface of NK cells and markedly increased secretion of IFN-γ [23]. Ultimately, split anergized NK cells promote differentiation of surviving CSCs and upregulation of CD54, PD-L1, and MHC class I expression, thus promoting CSC resistance to NK cell killing [23]. Resistance to NK cell-mediated cytotoxicity was also observed in GSCs in vitro when cultured in supernatants of split anergized NK cells [166]. Differentiation of GSCs resulted in higher expression of CD54 and MHC class I and increased production of IL-6 and IL-8, which was accompanied by decreased secretion of IFN-γ [7]. Although NK cells may not completely eradicate GSCs due to split anergy, differentiation of GSCs to a more differentiated state may significantly increase the efficacy of other therapeutic approaches such as chemotherapy and T cell-based immunotherapy [130,152]. Nevertheless, the ability of NK cells to differentiate GSCs may be limited due to inadequate production of IFN-γ, possibly leading to reversion of differentiated cells back to a stem-like state [7,23]. Combined approaches will probably be needed to completely eliminate GSCs.

CAR NK cells

A number of (pre)clinical studies on CAR NK cells are ongoing, including some studies in patients with GBM (Tables 2 and 3) [12,40]. Several antigens are being explored as CAR NK cell targets in GBM, including epidermal growth factor receptor (EGFR) or its mutant variant III (EGFRvIII) and ErbB2 (HER2), further described below.

EGFR is one of the most frequently mutated genes in GBM. Besides gene amplifications, deletion of exons 2–7 is the most common genetic alteration of the gene, resulting in a truncated and constitutively active protein, EGFRvIII. Since EGFR and EGFRvIII are overexpressed in a large proportion of GBMs, they are the target of a number of therapeutic approaches [106]. Müller et al. transduced the NK cell line YTS with an EGFRvIII-targeted CAR. CAR NK cells exhibited specific cytotoxicity against EGFRvIII-positive tumor cells in vitro and delayed tumor growth in a mouse GBM model. The efficacy of NK cell therapy was further enhanced when CAR-equipped YTS cells were modified to concomitantly express the chemokine receptor CXCR4. This modification promoted NK cell chemotaxis toward the CXCR4 ligand CXCL12 (SDF-1α) produced by the tumor [143]. Another NK cell line engineered to express a different variant of the EGFRvIII-specific CAR was the human NK cell line KHYG-1 (EvCAR-KHYG-1), which also performed well when tested in GBM cell lines in vitro [8]. However, these CAR NK cells failed to inhibit subcutaneous tumor growth in in vivo xenografts, but increased the necrotic area and induced a progression-like phenotype. Nevertheless, further experiments in orthotopic GBM models are needed to provide information on the antitumor effect of EvCAR-KHYG-1 [144]. NK cells were also engineered to express CARs that recognize both wild-type and mutant EGFRvIII. NK cell lines NK-92 and NKL modified in this manner were significantly more cytotoxic against EGFR-positive GSCs compared to unmodified controls and produced higher levels of IFN-γ when cultured with $EGFR^+$ GSCs. Intratumoral injection of NK-92-EGFR-CAR cells significantly prolonged the survival of mice bearing either wtEGFR or EGFRvIII-expressing tumors [115]. Another study compared the efficacy of CAR NK-92 cells expressing a second-generation CAR targeting either EGFR (NK-92/R1), EGFRvIII (NK-92/MR1-1), or an epitope common to both antigens (NK-92/225). In GBM xenografted mice, all of these CAR NK cells delayed tumor growth and prolonged animal survival, with dual-specific effectors performing better than monospecific CAR NKs. Furthermore, dual-specific NK cells in the mixed xenografts consisting of EGFR and EGFRvIII GBM cells prevented rapid immune escape that occurred when treated with monospecific effectors [112]. Targeting both forms of EGFR may be more applicable in the clinic, as EGFR is not always mutated and/or is not mutated in all tumor cells.

ErbB2 (HER2) is another member of the EGFR-related receptor tyrosine kinases. ErbB2 is expressed in a large proportion of GBM tumors but not in normal brain tissue. The ErbB2-specific NK-92 CAR NK cell line was generated by Schönfeld et al. [41], which expressed a second-generation CAR consisting of an ErbB2-specific antibody FRP5 and the signaling domains CD28 and CD3ζ (CAR 5.28.z). This cell line showed selective cytotoxicity against various ErbB2-expressing tumor cells, including GBM cells [175]. Local application of modified CAR NK cells resulted in prolonged symptom-free survival in GBM xenograft mouse models and induced tumor regression in the majority of immunocompetent mice. The latter were protected from tumor rechallenge at day 126, indicating the induction of long-lasting immune protection [175]. Promising preclinical data lead the way toward the evaluation of ErbB2-specific NK-92/5.28.z CAR NK cells for GBM in the clinic. Currently, the phase I CAR2BRAIN clinical trial (NCT03383978) is ongoing in patients with recurrent or refractory ErbB2-positive GBM to evaluate the safety and tolerability of ErbB2-specific NK-92/5.28.z CAR NK cells and to determine the maximum tolerated dose or the maximum feasible dose. The preirradiated CAR NK cells are locally applied into the resection cavity during a relapse surgery. In addition, an expansion cohort is planned to study repeated applications of NK-92/5.28.z CAR NK cells administered weekly through an implanted Rickham reservoir and catheter [12].

Supercharged natural killer cells

We have recently developed a platform that enables the activation and expansion of very potent cytotoxic cells in vitro. Osteoclasts as a feeder layer in a combination of sonicated probiotic bacteria were used to generate large numbers of NK cells from primary peripheral blood with high cytotoxic activity, termed supercharged NK cells. Moreover, these supercharged NK cells secrete high levels of cytokines and differentiate CSCs [24–26]. We have also shown that osteoclast-expanded NK cells have greater potential to directly kill tumor cells, while they also express lower levels of CD16 receptors compared to freshly isolated primary NK cells [24]. Human osteoclasts secrete IL-15, IL-12, IL-18, and IFN-α, but not IFN-γ, and express lower levels of MHC class I and II, CD14, CD11b, and CD54, and minimally upregulate surface expression of MHC class I when treated with either the combination of TNF-α and IFN-γ or activated NK cell supernatants known to increase MHC class I expression. Low expression of MHC class I together with increased release of IL-15, IL-12, IL-18, and IFN-α may represent some of the underlying mechanisms by which osteoclasts are able to expand highly activated NK cells. Osteoclasts also exhibit higher expression of NKG2D ligands [5,24–26].

The effects of supercharged NK cells were tested in vivo in tumor-bearing humanized BLT mice. Supercharged NK cells inhibited pancreatic tumor growth, lysed CSCs, and also induced CSCs differentiation of the remaining tumor cells that are more susceptible to chemotherapy (increase in cell surface expression of PD-L1, CD54, and MHC class I), as well as increased infiltration of immune cells into the TME in humanized mice [26]. We are now in the process of fully characterizing the supercharged NK cells and testing their antitumor function in 3D models of GBM (manuscript submitted).

Preclinical tumor models and the translation of natural killer cell therapy into the clinic

Many promising preclinical immunotherapies, such as immune checkpoint inhibitors and vaccines, have failed in phase III clinical trials [2]. While CAR technology has shown remarkable success in patients with hematologic malignancies ("liquid cancers"), the application of the same approach in solid tumors such as GBM has encountered several challenges, such as the lack of tumor-specific target antigens [139], tumor heterogeneity and plasticity, and immunosuppressive TME [110]. Recent advances in the development of new immunotherapy approaches therefore require the creation of appropriate personalized preclinical models that reliably reflect the complexity of the tumor as well as the precise molecular and cellular heterogeneity and its microenvironment. Currently, numerous GBM tumor models are available for testing immunotherapies, including established cell lines, patient-derived tumor cells, heterotypic spheroids, organoids, syngeneic mouse models, xenografts, and humanized mice.

Three-dimensional glioblastoma stem cell models

Tumorospheres obtained from GSCs represent a simple yet relevant model, typically obtained from single-cell suspensions that self-assemble or are forced to aggregate. It accounts for the cellular heterogeneity of the tumor and allows for better prediction of response to therapy [156]. Tumorospheres provide a more tumor-like environment characterized by an outer proliferating zone, an intermediate quiescent zone, and an inner necrotic core that mimics the conditions of GBM tumors in vivo [136], features not seen in conventional 2D cultures. Cell–cell interactions within the GSC tumorospheres exhibit profiles of signaling pathways that influence proliferation and survival [116], as well as gene expression patterns that are nearly identical to those of the parental tumor. In addition, the 3D organization of the cells provides a physical barrier that prevents drug penetration and thus influences response to therapy [114]. The major limitation of these models is the absence of adjacent nonmalignant cells that form the TME, i.e., stromal cells, including astrocytes, neurons, endothelial cells, brain resident microglia and infiltrated peripheral immune cells, limiting the study of cancer-stromal cell interactions. Other TME cell types, such as immune cells, can also be integrated into so-called heterotypic spheroids to study their killing capabilities and the effects of potential immunotherapies. For example, in the study by Morimoto et al. [142], the authors investigated the potential of NK cell-based immunotherapy using spheroids derived from T98G, U87MG, LN-18, and U251MG cell lines with upregulated stem cell properties.

Organoids

Organoids are 3D multicellular tissue constructs that recapitulate the original organ-like architecture and its functionality in vitro [162]. These preclinical models are essential for more efficient translation of basic cancer research into new treatments for cancer patients. There are several approaches to create organoids, from pluripotent embryonic stem cells and induced pluripotent stem cells [101] to adult stem cells and resected tumor tissue [155], which have the ability to self-organize and self-renew. The first GBM organoids to study heterogeneity and hypoxic gradient were established from finely minced tumor samples using a Matrigel-based 3D method [119]. Next, GBM organoids have been generated by genetic engineering of cerebral organoids to induce tumor formation [95,146], or by co-culturing patient-derived GSCs with human ESC and iPSC-derived cerebral organoids [135]. In addition, organoids can be grown with high efficiency from patient-derived tumor tissue, potentially enabling patient-specific drug testing and the development of individualized treatment regimens, including immunotherapy. In the study by Jacob et al. [122,123], the authors established GBM organoids that recapitulated the histological features, cellular diversity, gene expression, and mutational profiles of the corresponding parental tumors. Because this model preserves the immune microenvironment, such as macrophages/microglia and T cells, and other cells of the TME, including oligodendrocyte precursor cells, astrocytes, oligodendrocytes, neurons, and CD31+ cells of the vasculature, it can be used as an effective preclinical model for precision medicine of GBM. In addition, they describe a method to study response to immunotherapies using GBM organoids with CAR T-cells in co-cultures. As an alternative to CAR T-cells, CAR NK cells may be more effective in treating solid tumors due to the presence of inhibitory receptors that restrict their killing of normal cells. Tumor antigen-specific cytotoxicity was investigated in the study by Schanlzger et al. [157] using CAR-NK-92 cells targeting colorectal cancer organoids expressing EGFRvIII. This was the first study to use the CAR-NK platform to evaluate efficacy and tumor specificity in an organoid model. The establishment of immunocompetent cancer organoids is an active area of research and an urgent need. Such novel models fill a gap in preclinical research and enable functional and translational studies for immunotherapies.

Animal models

A syngeneic mouse tumor model is a preclinical in vivo model that plays a critical role in testing and understanding the immune response to novel therapies before they enter clinical trials. These models are produced by transplanting in vitro expanded genetic or chemical/viral induced mouse tumor cell lines into the same immunocompetent inbred mouse strain [147]. The advantages of syngeneic mouse models include their rapid tumor growth, reproducibility, low cost, and ability to be genetically manipulated. The most commonly used syngeneic GBM tumor models in preclinical research for immunotherapy are K-Ras mutant/p53 mutant GL261 mice [92,103,175]. Bulk RNA sequencing analysis revealed a strong enrichment of differentially expressed genes for various immune pathways compared to naïve control mice, a higher frequency of activated microglia and T cells, and a lower frequency of Tregs and antigen-presenting cells [126]. The fact that the GL261 tumor model is highly immunogenic leads to results that do not correlate well with clinical findings [138]. At the same time, there are important differences between the mouse and human immune systems, including differences in signaling pathways and receptor expression in immune cells [133]. Another major concern is that the complexity and heterogeneity of human GBM may not be fully replicated in such models [132]. Therefore, further studies are needed to evaluate the suitability of these models for testing immunotherapies in GBM.

Patient-derived xenografts (PDXs) are even more relevant than classical cancer cell line xenografts because they preserve heterogeneity within the tumor, faithfully recapitulate patient response to drugs, and capture the genetic diversity of GBM [125]. These models are based on subcutaneous or intracranial transplantation of tumor cells (cell lines or patient-derived cells), organoids, or tissues from patients into nonobese/diabetic mice (NOD), severe combined immunodeficient mice (SCID), or the combination NOD/SCID and NOD/SCID/IL-2 receptor γ null (NSG) mice. The athymic nude mouse has a depleted population of T cells [120], relatively intact populations of B cells, dendritic cells, and granulocytes and increased activity of NK cells and macrophages [151]. SCID mice lack mature B and T cells [96], while NSG mice have decreased NK cell function in addition to loss of B and T cells to reduce the risk of GvHD. However, in the context of testing immunotherapies, these models are unable to accurately predict response to therapy due to the use of immunodeficient host strains and the lack of functional elements of the immune system [154]. Recent studies show that the use of humanized mice can overcome these challenges [117].

Fully autologous humanized mouse tumor models are generated by transplanting human tumor cell lines, CSCs, or human PDX tumors into immunodeficient mice, including NOD and NSG mice, with an HLA-matched human immune system. This is done by transplanting human PBMCs isolated from adult human blood or CD34+ hematopoietic stem cells (HSCs) and transplanting autologous human thymus grafts into immunodeficient mice. Transplanted CD34+ HSCs in immunocompromised mice differentiate into human helper T cells, cytotoxic T cells, B cells, monocytes, NK cells, dendritic cells [176] and also into human microglia and macrophages [140]. They are used to test new CAR technologies in vivo for safety, efficacy, immune response, design, combination therapies, and mechanism of action before entering a clinical trial [141]. The long-term revolution in optimizing the humanized mouse model in onco-immunology would provide an unprecedented platform for cancer immunotherapy.

Selection of the source and expansion of natural killer cells for clinical application

NK cells have been used as adoptive immunotherapy in several clinical trials. It is not straightforward to provide purified NK cells with preserved cytotoxic activity and in sufficient quantity for use in the clinic. Moreover, their antitumor properties need to be preactivated and modified. For example, NK cells from different sources can be stimulated with different cytokines. In addition to cytokines, NK cells are cultured with autologous cells or together with different types of growth-inactivated or allogeneic feeder cells to further enhance their expansion ex vivo [9]. There are several sources of therapeutic NK cells. Umbilical cord blood NK cells are one of the many potential sources of therapeutic NK cells, but their therapeutic efficacy is still under investigation. Clonal NK cell lines such as NK-92, KHYG-1, and YT are an alternative source of allogeneic NK cells, and the NK-92 cell line has been extensively tested in clinical trials. Irradiated NK-92 cells have been observed to kill tumor cells in cancer patients, although irradiation limits the in vivo persistence of these cells to a maximum of 48h, which is a potential barrier to sustained clinical efficacy [163]. To overcome these obstacles, an approach to generate NK cells from CD34+ HPCs was developed. This approach allows for the production of an unlimited amount of homogeneous NK cells and is much easier to genetically manipulate than primary NK cells [129]. Following the clinical successes achieved with CAR T cell therapies, it is now possible to genetically modify NK cells ex vivo to further enhance their antitumor activity, for example, by integrating CARs against specific tumor antigens, as described above. Although CAR NK cells are a promising future "off-the-shelf" drug, there are NK cell-specific challenges, such as potential loss of viability and activity due to the freeze–thaw process and lack of in vivo persistence.

Conclusions and future directions

NK cell-based therapeutic platforms are a promising and safe approach, and we are only beginning to understand the diverse potential of NK cells in cancer therapy. Initial clinical trials in patients with hematologic cancers show high response rates and safety of NK cell-based approaches. However, large patient cohorts are still needed to validate these results. NK cells are a promising approach to treat aggressive GBM because they (1) target tumor-initiating and refractory CSCs without the need for approaches that generate immunogenic antigens and enable cell priming with appropriate co-stimulatory signals as required for T- or dendritic cell-based approaches, (2) have multiple functions that, in addition to directly killing cancer cells, orchestrate innate and adoptive immunity, shape the TME, and have the ability to differentiate CSCs, (3) have broader reactivity and lower specificity than T cells to target heterogeneous populations of GBM cells lacking specific antigens, (4) are safe and allogeneic NK cells can be derived from cells that have not been previously exposed to cancer-induced immunosuppression as seen in autologous approaches, and (5) can be combined with other strategies such as oncolytic viruses, immune checkpoint inhibitors, radio- and chemotherapy, and CAR T-cells (Fig. 1).

To translate NK cell immunotherapy to solid tumors and GBM in particular, we need to better understand NK cell biology and how immunosuppressive TME and the absence of growth factors and oxygen suppress NK cell antitumor functions. In addition, future combined treatments need to provide growth factors and activating signals to increase NK cell efficacy and in vivo persistence. As GBM are highly heterogeneous, there is also an urgent need to develop clinically relevant personalized tumor models to test NK cell-based approaches in preclinical studies before applying them to patients.

Overall, the field of NK cell-based immunotherapy has enormous potential for application in cancer therapy. As the number of preclinical and clinical studies in this field grows, so does the understanding of NK cell biology and the specific roles of NK cells in antitumor immunity. All this is crucial to design and develop new efficient therapies and to exploit the full potential of NK cells.

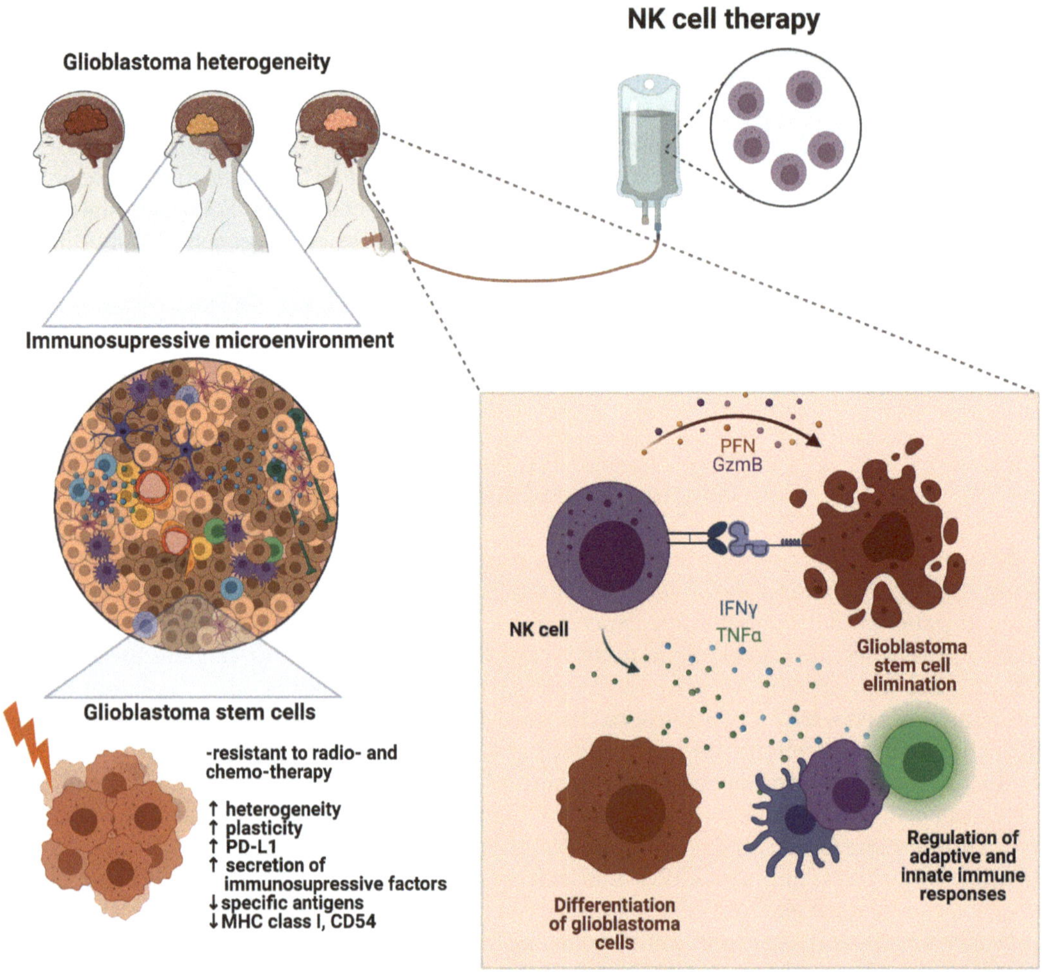

FIG. 1 Natural killer cell-based platforms are a promising approach for the treatment of glioblastoma due to their ability to select and eliminate resistant glioblastoma stem cells, broader reactivity to target heterogeneous cancer cells, ability to boost innate and adoptive antitumor immunity, and differentiate glioblastoma stem cells. Abbreviations: *IFNγ*: interferon γ, *MHC* I: major histocompatibility complex class I, *NK cell*: natural killer cell, *PD-L1*: programmed death-ligand 1, *PFN*: perforin, *TNFα*: tumor necrosis factor α. *Image was created by BioRender.*

Acknowledgments

This work was supported by the Slovenian Research Agency (Grant program P1-0245, Postdoctoral project Z3-1870, Grant project J3-2526, BI-US/19-21-021). We acknowledge prof. Tamara Lah Turnšek for carefully reading and commenting the manuscript.

Author contributions

Conceptualization: B.B. and A.J.; writing—original draft preparation: B.B., M.N., B.M., and A.H.; writing—review and editing: B.B., M.N., B.M., A.H., and A.J.; visualization: B.B.; all authors have read and agreed to the published version of the manuscript.

References

[1] Guillerey C, Huntington ND, Smyth MJ. Targeting natural killer cells in cancer immunotherapy. Nat Immunol 2016;17:1025–36. https://doi.org/10.1038/ni.3518.

[2] Lim M, Xia Y, Bettegowda C, Weller M. Current state of immunotherapy for glioblastoma. Nat Rev Clin Oncol 2018;15:422–42. https://doi.org/10.1038/s41571-018-0003-5.

[3] Majc B, Novak M, Jerala NK, Jewett A, Breznik B. Immunotherapy of glioblastoma: current strategies and challenges in tumor model development. Cells 2021;10:1–22. https://doi.org/10.3390/cells10020265.

[4] Schalper KA, Rodriguez-Ruiz ME, Diez-Valle R, López-Janeiro A, Porciuncula A, Idoate MA, et al. Neoadjuvant nivolumab modifies the tumor immune microenvironment in resectable glioblastoma. Nat Med 2019;25:470–6. https://doi.org/10.1038/s41591-018-0339-5.

[5] Jewett A, Kos J, Kaur K, Turnsek TL, Breznik B, Senjor E, et al. Multiple defects of natural killer cells in cancer patients: anarchy, dysregulated systemic immunity, and immunosuppression in metastatic cancer. Crit Rev Immunol 2020;40:93–133. https://doi.org/10.1615/CritRevImmunol.2020033391.

[6] Kaur K, Chang HH, Topchyan P, Cook JM, Barkhordarian A, Eibl G, et al. Deficiencies in natural killer cell numbers, expansion, and function at the pre-neoplastic stage of pancreatic cancer by KRAS mutation in the pancreas of obese mice. Front Immunol 2018;9:1–12. https://doi.org/10.3389/fimmu.2018.01229.

[7] Kozlowska AK, Tseng HC, Kaur K, Topchyan P, Inagaki A, Bui VT, et al. Resistance to cytotoxicity and sustained release of interleukin-6 and interleukin-8 in the presence of decreased interferon-γ after differentiation of glioblastoma by human natural killer cells. Cancer Immunol Immunother 2016;65:1085–97. https://doi.org/10.1007/s00262-016-1866-x.

[8] Murakami T, Nakazawa T, Natsume A, Nishimura F, Nakamura M, Matsuda R, et al. Novel human NK cell line carrying CAR targeting EGFRvIII induces antitumor effects in glioblastoma cells. Anticancer Res 2018;38:5049–56. https://doi.org/10.21873/ANTICANRES.12824.

[9] Myers JA, Miller JS. Exploring the NK cell platform for cancer immunotherapy. Nat Rev Clin Oncol 2021;18:85–100. https://doi.org/10.1038/s41571-020-0426-7.

[10] Dogra P, Rancan C, Ma W, Toth M, Senda T, Carpenter DJ, et al. Tissue determinants of human NK cell development, function, and residence. Cell 2020;180:749–763.e13. https://doi.org/10.1016/j.cell.2020.01.022.

[11] Scoville SD, Freud AG, Caligiuri MA. Modeling human natural killer cell development in the era of innate lymphoid cells. Front Immunol 2017;8:4–11. https://doi.org/10.3389/fimmu.2017.00360.

[12] Burger MC, Zhang C, Harter PN, Romanski A, Strassheimer F, Senft C, et al. CAR-engineered NK cells for the treatment of glioblastoma: turning innate effectors into precision tools for cancer immunotherapy. Front Immunol 2019;10:2683. https://doi.org/10.3389/FIMMU.2019.02683.

[13] Golán I, de la Fuente LR, Costoya JA. NK cell-based glioblastoma immunotherapy. Cancers (Basel) 2018;10:522. https://doi.org/10.3390/CANCERS10120522.

[14] Morvan MG, Lanier LL. NK cells and cancer: you can teach innate cells new tricks. Nat Rev Cancer 2016;16:7–19. https://doi.org/10.1038/nrc.2015.5.

[15] Kaur K, Safaie T, Ko M-W, Wang Y, Jewett A. ADCC against MICA/B is mediated against differentiated oral and pancreatic and not stem-like/poorly differentiated tumors by the NK cells; loss in cancer patients due to down-modulation of CD16 receptor. Cancers (Basel) 2021;13:239. https://doi.org/10.3390/CANCERS13020239.

[16] Ljunggren HG, Kärre K. In search of the 'missing self': MHC molecules and NK cell recognition. Immunol Today 1990;11:237–44. https://doi.org/10.1016/0167-5699(90)90097-S.

[17] Voskoboinik I, Whisstock JC, Trapani JA. Perforin and granzymes: function, dysfunction and human pathology. Nat Rev Immunol 2015;15:388–400. https://doi.org/10.1038/nri3839.

[18] Cretney E, Takeda K, Yagita H, Glaccum M, Peschon JJ, Smyth MJ. Increased susceptibility to tumor initiation and metastasis in TNF-related apoptosis-inducing ligand-deficient mice. J Immunol 2002;168:1356–61. https://doi.org/10.4049/jimmunol.168.3.1356.

[19] Böttcher JP, Bonavita E, Chakravarty P, Blees H, Cabeza-Cabrerizo M, Sammicheli S, et al. NK cells stimulate recruitment of cDC1 into the tumor microenvironment promoting cancer immune control. Cell 2018;172:1022–1037.e14. https://doi.org/10.1016/j.cell.2018.01.004.

[20] Bald T, Krummel MF, Smyth MJ, Barry KC. The NK cell–cancer cycle: advances and new challenges in NK cell–based immunotherapies. Nat Immunol 2020;21:835–47. https://doi.org/10.1038/s41590-020-0728-z.

[21] Fadul CE, Fisher JL, Gui J, Hampton TH, Côté AL, Ernstoff MS. Immune modulation effects of concomitant temozolomide and radiation therapy on peripheral blood mononuclear cells in patients with glioblastoma multiforme. Neuro Oncol 2011;13:393–400. https://doi.org/10.1093/NEUONC/NOQ204.

[22] Bernardini G, Antonangeli F, Bonanni V, Santoni A. Dysregulation of chemokine/chemokine receptor axes and NK cell tissue localization during diseases. Front Immunol 2016;7:1–9. https://doi.org/10.3389/fimmu.2016.00402.

[23] Tseng H-C, Bui V, Man Y-G, Cacalano N, Jewett A. Induction of split anergy conditions natural killer cells to promote differentiation of stem cells through cell–cell contact and secreted factors. Front Immunol 2014;5:269. https://doi.org/10.3389/FIMMU.2014.00269.

[24] Kaur K, Cook J, Park SH, Topchyan P, Kozlowska A, Ohanian N, et al. Novel strategy to expand super-charged NK cells with significant potential to lyse and differentiate cancer stem cells: differences in NK expansion and function between healthy and cancer patients. Front Immunol 2017;8:297. https://doi.org/10.3389/FIMMU.2017.00297/FULL.

[25] Kaur K, Ko MW, Ohanian N, Cook J, Jewett A. Osteoclast-expanded super-charged NK-cells preferentially select and expand CD8+ T cells. Sci Rep 2020;10:1–17. https://doi.org/10.1038/s41598-020-76702-1.

[26] Kaur K, Kozlowska AK, Topchyan P, Ko MW, Ohanian N, Chiang J, et al. Probiotic-treated super-charged NK cells efficiently clear poorly differentiated pancreatic tumors in Hu-BLT mice. Cancers (Basel) 2020;12:63. https://doi.org/10.3390/cancers12010063.

[27] Biederstädt A, Rezvani K. Engineering the next generation of CAR-NK immunotherapies. Int J Hematol 2021;114:554–71. https://doi.org/10.1007/s12185-021-03209-4.

[28] Liu E, Marin D, Banerjee P, Macapinlac HA, Thompson P, Basar R, et al. Use of CAR-transduced natural killer cells in CD19-positive lymphoid tumors. N Engl J Med 2020;382:545–53. https://doi.org/10.1056/nejmoa1910607.

[29] Barry KC, Hsu J, Broz ML, Cueto FJ, Binnewies M, Combes AJ, et al. A natural killer–dendritic cell axis defines checkpoint therapy–responsive tumor microenvironments. Nat Med 2018;24:1178–91. https://doi.org/10.1038/s41591-018-0085-8.

[30] Deguine J, Breart B, Lemaître F, Di Santo JP, Bousso P. Intravital imaging reveals distinct dynamics for natural killer and CD8+ T cells during tumor regression. Immunity 2010;33:632–44. https://doi.org/10.1016/J.IMMUNI.2010.09.016.

[31] Rosenberg J, Huang J. CD8+ T cells and NK cells: parallel and complementary soldiers of immunotherapy. Curr Opin Chem Eng 2018;19:9–20. https://doi.org/10.1016/j.coche.2017.11.006.

[32] Bentebibel S, Hurwitz ME, Bernatchez C, Haymaker C, Hudgens CW, Kluger HM, et al. A first-in-human study and biomarker analysis of NKTR-214, a novel IL2Rβγ-biased cytokine, in patients with advanced or metastatic solid tumors. Cancer Discov 2019;9:711–21. https://doi.org/10.1158/2159-8290.CD-18-1495.

[33] Romee R, Cooley S, Berrien-elliott MM, Westervelt P, Verneris MR, Wagner JE, et al. First-in-human phase 1 clinical study of the IL-15 superagonist complex ALT-803 to treat relapse after transplantation. 2018;131:2515–27. https://doi.org/10.1182/blood-2017-12-823757.

[34] Gauthier L, Morel A, Anceriz N, Rossi B, Blanchard-Alvarez A, Grondin G, et al. Multifunctional natural killer cell engagers targeting NKp46 trigger protective tumor immunity. Cell 2019;177:1701–1713.e16. https://doi.org/10.1016/j.cell.2019.04.041.

[35] Sarhan D, Brandt L, Felices M, Guldevall K, Lenvik T, Hinderlie P, et al. 161533 TriKE stimulates NK-cell function to overcome myeloid-derived suppressor cells in MDS. Blood Adv 2018;2:1459–69. https://doi.org/10.1182/BLOODADVANCES.2017012369.

[36] André P, Denis C, Soulas C, Bourbon-Caillet C, Lopez J, Arnoux T, et al. Anti-NKG2A mAb is a checkpoint inhibitor that promotes anti-tumor immunity by unleashing both T and NK cells. Cell 2018;175, 1731–43.e13. https://doi.org/10.1016/j.cell.2018.10.014.

[37] Haanen JB, Cerundolo V. NKG2A, a new kid on the immune checkpoint block. Cell 2018;175:1720–2. https://doi.org/10.1016/j.cell.2018.11.048.

[38] Juliá EP, Amante A, Pampena MB, Mordoh J, Levy EM. Avelumab, an IgG1 anti-PD-L1 immune checkpoint inhibitor, triggers NK cell-mediated cytotoxicity and cytokine production against triple negative breast cancer cells. Front Immunol 2018;9:1–12. https://doi.org/10.3389/fimmu.2018.02140.

[39] Elahi R, Heidary AH, Hadiloo K, Esmaeilzadeh A. Chimeric antigen receptor-engineered natural killer (CAR NK) cells in cancer treatment; recent advances and future prospects. Stem Cell Rev Rep 2021;17:2081–106. https://doi.org/10.1007/s12015-021-10246-3.

[40] Marofi F, Al-Awad AS, Rahman HS, Markov A, Abdelbasset WK, Enina YI, et al. CAR-NK cell: a new paradigm in tumor immunotherapy. Front Oncol 2021;11:673276. https://doi.org/10.3389/FONC.2021.673276.
[41] Schönfeld K, Sahm C, Zhang C, Naundorf S, Brendel C, Odendahl M, et al. Selective inhibition of tumor growth by clonal NK cells expressing an ErbB2/HER2-specific chimeric antigen receptor. Mol Ther 2015;23:330–8. https://doi.org/10.1038/MT.2014.219.
[42] Shaim H, Shanley M, Basar R, Daher M, Gumin J, Zamler DB, et al. Targeting the αv integrin/TGF-β axis improves natural killer cell function against glioblastoma stem cells. J Clin Invest 2021;131, e142116. https://doi.org/10.1172/JCI142116.
[43] Haspels HN, Rahman MA, Joseph JV, Navarro AG, Chekenya M. Glioblastoma stem-like cells are more susceptible than differentiated cells to natural killer cell lysis mediated through killer immunoglobulin-like receptors-human leukocyte antigen ligand mismatch and activation receptor-ligand interactions. Front Immunol 2018;9:1–17. https://doi.org/10.3389/fimmu.2018.01345.
[44] Weller M, van den Bent M, Preusser M, Le Rhun E, Tonn JC, Minniti G, et al. EANO guidelines on the diagnosis and treatment of diffuse gliomas of adulthood. Nat Rev Clin Oncol 2020;18:170–86. https://doi.org/10.1038/s41571-020-00447-z.
[45] Wen PY, Weller M, Lee EQ, Alexander BM, Barnholtz-Sloan JS, Barthel FP, et al. Glioblastoma in adults: a Society for Neuro-Oncology (SNO) and European Society of Neuro-Oncology (EANO) consensus review on current management and future directions. Neuro Oncol 2020;22:1073–113. https://doi.org/10.1093/NEUONC/NOAA106.
[46] Louis DN, Perry A, Reifenberger G, von Deimling A, Figarella-Branger D, Cavenee WK, et al. The 2016 World Health Organization classification of tumors of the central nervous system: a summary. Acta Neuropathol 2016;131:803–20. https://doi.org/10.1007/s00401-016-1545-1.
[47] Stupp R, Mason WP, van den Bent MJ, Weller M, Fisher B, Taphoorn MJB, et al. Radiotherapy plus concomitant and adjuvant temozolomide for glioblastoma. N Engl J Med 2005;352:987–96. https://doi.org/10.1056/NEJMoa043330.
[48] Thomas AA, Brennan CW, DeAngelis LM, Omuro AM. Emerging therapies for glioblastoma. JAMA Neurol 2014;71:1437–44. https://doi.org/10.1001/jamaneurol.2014.1701.
[49] Chen J, Li Y, Yu TS, McKay RM, Burns DK, Kernie SG, et al. A restricted cell population propagates glioblastoma growth after chemotherapy. Nature 2012;488:522–6. https://doi.org/10.1038/nature11287.
[50] Galli R, Binda E, Orfanelli U, Cipelletti B, Gritti A, De Vitis S, et al. Isolation and characterization of tumorigenic, stem-like neural precursors from human glioblastoma. Cancer Res 2004;64:7011–21. https://doi.org/10.1158/0008-5472.CAN-04-1364.
[51] Lathia J, Mack S. Cancer stem cells in glioblastoma. Genes Dev 2015;29:1203–17. https://doi.org/10.1101/gad.261982.115.tumors.
[52] Osuka S, Van Meir EG. Overcoming therapeutic resistance in glioblastoma: the way forward. J Clin Invest 2017;127:415–26. https://doi.org/10.1172/JCI89587.
[53] Verhaak RGW, Hoadley KA, Purdom E, Wang V, Qi Y, Wilkerson MD, et al. Integrated genomic analysis identifies clinically relevant subtypes of glioblastoma characterized by abnormalities in PDGFRA, IDH1, EGFR, and NF1. Cancer Cell 2010;17:98–110. https://doi.org/10.1016/j.ccr.2009.12.020.
[54] Patel AP, Tirosh I, Trombetta JJ, Shalek AK, Gillespie SM, Wakimoto H, et al. Single-cell RNA-seq highlights intratumoral heterogeneity in primary glioblastoma. Science 2014;344:1396–401. https://doi.org/10.1126/science.1254257.
[55] Hira VVV, Aderetti DA, van Noorden CJF. Glioma stem cell niches in human glioblastoma are periarteriolar. J Histochem Cytochem 2018;66:349–58. https://doi.org/10.1369/0022155417752676.
[56] Hira VVV, Breznik B, Vittori M, Loncq de Jong A, Mlakar J, Oostra R-J, et al. Similarities between stem cell niches in glioblastoma and bone marrow: rays of hope for novel treatment strategies. J Histochem Cytochem 2019;68:33–57. https://doi.org/10.1369/0022155419878416.
[57] Bao S, Wu Q, McLendon RE, Hao Y, Shi Q, Hjelmeland AB, et al. Glioma stem cells promote radioresistance by preferential activation of the DNA damage response. Nature 2006;444:756–60. https://doi.org/10.1038/nature05236.
[58] Dirkse A, Golebiewska A, Buder T, Nazarov PV, Muller A, Poovathingal S, et al. Stem cell-associated heterogeneity in Glioblastoma results from intrinsic tumor plasticity shaped by the microenvironment. Nat Commun 2019;10:1–16. https://doi.org/10.1038/s41467-019-09853-z.

[59] Podergajs N, Motaln H, Rajčević U, Verbovšek U, Koršič M, Obad N, et al. Transmembrane protein CD9 is glioblastoma biomarker, relevant for maintenance of glioblastoma stem cells. Oncotarget 2015;7:593–609. https://doi.org/10.18632/oncotarget.5477.
[60] Zeppernick F, Ahmadi R, Campos B, Dictus C, Helmke BM, Becker N, et al. Stem cell marker CD133 affects clinical outcome in glioma patients. Clin Cancer Res 2008;14:123–9. https://doi.org/10.1158/1078-0432.CCR-07-0932.
[61] Vora P, Venugopal C, Salim SK, Tatari N, Bakhshinyan D, Singh M, et al. The rational development of CD133-targeting immunotherapies for glioblastoma. Cell Stem Cell 2020;26:832–844.e6. https://doi.org/10.1016/J.STEM.2020.04.008.
[62] Yuan X, Curtin J, Xiong Y, Liu G, Waschsmann-Hogiu S, Farkas DL, et al. Isolation of cancer stem cells from adult glioblastoma multiforme. Oncogene 2004;23:9392–400. https://doi.org/10.1038/sj.onc.1208311.
[63] Brennan CW, Verhaak RGW, McKenna A, Campos B, Noushmehr H, Salama SR, et al. The somatic genomic landscape of glioblastoma. Cell 2013;155:462–77. https://doi.org/10.1016/j.cell.2013.09.034.
[64] Galdieri L, Jash A, Malkova O, Mao DD, DeSouza P, Chu YE, et al. Defining phenotypic and functional heterogeneity of glioblastoma stem cells by mass cytometry. JCI Insight 2021;6, e128456. https://doi.org/10.1172/JCI.INSIGHT.128456.
[65] Son MJ, Woolard K, Nam D-H, Lee J, Fine HA. SSEA-1 is an enrichment marker for tumor-initiating cells in human glioblastoma. Cell Stem Cell 2009;4:440–52. https://doi.org/10.1016/J.STEM.2009.03.003.
[66] Tchoghandjian A, Baeza N, Colin C, Cayre M, Metellus P, Beclin C, et al. A2B5 cells from human glioblastoma have cancer stem cell properties. Brain Pathol 2010;20:211–21. https://doi.org/10.1111/J.1750-3639.2009.00269.X.
[67] Venere M, Fine HA, Dirks PB, Rich JN. Cancer stem cells in gliomas: identifying and understanding the apex cell in cancer's hierarchy. Glia 2011;59:1148–54. https://doi.org/10.1002/GLIA.21185.
[68] Close HJ, Stead LF, Nsengimana J, Reilly KA, Droop A, Wurdak H, et al. Expression profiling of single cells and patient cohorts identifies multiple immunosuppressive pathways and an altered NK cell phenotype in glioblastoma. Clin Exp Immunol 2020;200:33–44. https://doi.org/10.1111/cei.13403.
[69] Engelhardt B, Ransohoff RM. Capture, crawl, cross: the T cell code to breach the blood–brain barriers. Trends Immunol 2012;33:579–89. https://doi.org/10.1016/J.IT.2012.07.004.
[70] Dantzer R. Neuroimmune interactions: from the brain to the immune system and vice versa. Physiol Rev 2018;98:477–504. https://doi.org/10.1152/PHYSREV.00039.2016.
[71] Heimberger AB, Sampson JH. Immunotherapy coming of age: what will it take to make it standard of care for glioblastoma? Neuro Oncol 2011;13:3–13. https://doi.org/10.1093/NEUONC/NOQ169.
[72] Broekman ML, Maas SLN, Abels ER, Mempel TR, Krichevsky AM, Breakefield XO. Multidimensional communication in the microenvirons of glioblastoma. Nat Rev Neurol 2018;14:482–95. https://doi.org/10.1038/s41582-018-0025-8.
[73] Quail DF, Joyce JA. The microenvironmental landscape of brain tumors. Cancer Cell 2017;31:326–41. https://doi.org/10.1016/J.CCELL.2017.02.009.
[74] Teng J, Da Hora CC, Kantar RS, Nakano I, Wakimoto H, Batchelor TT, et al. Dissecting inherent intratumor heterogeneity in patient-derived glioblastoma culture models. Neuro Oncol 2017;19:820–32. https://doi.org/10.1093/neuonc/now253.
[75] Darmanis S, Sloan SA, Croote D, Mignardi M, Chernikova S, Samghababi P, et al. Single-cell RNA-Seq analysis of infiltrating neoplastic cells at the migrating front of human glioblastoma. Cell Rep 2017;21:1399–410. https://doi.org/10.1016/J.CELREP.2017.10.030.
[76] Gieryng A, Pszczolkowska D, Bocian K, Dabrowski M, Rajan WD, Kloss M, et al. Immune microenvironment of experimental rat C6 gliomas resembles human glioblastomas. Sci Rep 2017;7:1–14. https://doi.org/10.1038/s41598-017-17752-w.
[77] Klemm F, Maas RR, Bowman RL, Kornete M, Soukup K, Nassiri S, et al. Interrogation of the microenvironmental landscape in brain tumors reveals disease-specific alterations of immune cells. Cell 2020;181:1643–1660.e17. https://doi.org/10.1016/J.CELL.2020.05.007.
[78] Ott M, Prins RM, Heimberger AB. The immune landscape of common CNS malignancies: implications for immunotherapy. Nat Rev Clin Oncol 2021;18:729–44. https://doi.org/10.1038/s41571-021-00518-9.
[79] Phillips JP, Eremin O, Anderson JR. Lymphoreticular cells in human brain tumours and in normal brain. Br J Cancer 1982;45:61–9. https://doi.org/10.1038/bjc.1982.8.

[80] Castriconi R, Daga A, Dondero A, Zona G, Poliani PL, Melotti A, et al. NK cells recognize and kill human glioblastoma cells with stem cell-like properties. J Immunol 2009;182:3530–9. https://doi.org/10.4049/JIMMUNOL.0802845.
[81] Lee SJ, Kang WY, Yoon Y, Jin JY, Song HJ, Her JH, et al. Natural killer (NK) cells inhibit systemic metastasis of glioblastoma cells and have therapeutic effects against glioblastomas in the brain. BMC Cancer 2015;15:1–13. https://doi.org/10.1186/s12885-015-2034-y.
[82] Linnerbauer M, Wheeler MA, Quintana FJ. Astrocyte crosstalk in CNS inflammation. Neuron 2020;108:608–22. https://doi.org/10.1016/J.NEURON.2020.08.012.
[83] Mi Y, Guo N, Luan J, Cheng J, Hu Z, Jiang P, et al. The emerging role of myeloid-derived suppressor cells in the glioma immune suppressive microenvironment. Front Immunol 2020;11:737. https://doi.org/10.3389/FIMMU.2020.00737.
[84] Matias D, Balça-Silva J, da Graça GC, Wanjiru CM, Macharia LW, Nascimento CP, et al. Microglia/astrocytes–glioblastoma crosstalk: crucial molecular mechanisms and microenvironmental factors. Front Cell Neurosci 2018;12:235. https://doi.org/10.3389/fncel.2018.00235.
[85] Wang LB, Karpova A, Gritsenko MA, Kyle JE, Cao S, Li Y, et al. Proteogenomic and metabolomic characterization of human glioblastoma. Cancer Cell 2021;39:509–528.e20. https://doi.org/10.1016/J.CCELL.2021.01.006.
[86] Kmiecik J, Poli A, Brons NHC, Waha A, Eide GE, Enger PØ, et al. Elevated CD3+ and CD8+ tumor-infiltrating immune cells correlate with prolonged survival in glioblastoma patients despite integrated immunosuppressive mechanisms in the tumor microenvironment and at the systemic level. J Neuroimmunol 2013;264:71–83. https://doi.org/10.1016/j.jneuroim.2013.08.013.
[87] Mostafa H, Pala A, Högel J, Hlavac M, Dietrich E, Westhoff MA, et al. Immune phenotypes predict survival in patients with glioblastoma multiforme. J Hematol Oncol 2016;9:77. https://doi.org/10.1186/S13045-016-0272-3.
[88] Friebel E, Kapolou K, Unger S, Núñez NG, Utz S, Rushing EJ, et al. Single-cell mapping of human brain cancer reveals tumor-specific instruction of tissue-invading leukocytes. Cell 2020;181:1626–1642.e20. https://doi.org/10.1016/j.cell.2020.04.055.
[89] Alban TJ, Bayik D, Otvos B, Rabljenovic A, Leng L, Jia-Shiun L, et al. Glioblastoma myeloid-derived suppressor cell subsets express differential macrophage migration inhibitory factor receptor profiles that can be targeted to reduce immune suppression. Front Immunol 2020;11:1191. https://doi.org/10.3389/FIMMU.2020.01191.
[90] El Andaloussi A, Lesniak MS. CD4+CD25+FoxP3+ T-cell infiltration and heme oxygenase-1 expression correlate with tumor grade in human gliomas. J Neurooncol 2007;83:145–52. https://doi.org/10.1007/s11060-006-9314-y.
[91] Adhikaree J, Moreno-Vicente J, Kaur AP, Jackson AM, Patel PM. Resistance mechanisms and barriers to successful immunotherapy for treating glioblastoma. Cells 2020;9:263. https://doi.org/10.3390/cells9020263.
[92] Alizadeh D, Zhang L, Brown CE, Farrukh O, Jensen MC, Badie B. Induction of anti-glioma natural killer cell response following multiple low-dose intracerebral CpG therapy. Clin Cancer Res 2010;16:3399–408. https://doi.org/10.1158/1078-0432.CCR-09-3087.
[93] Avril T, Vauleon E, Hamlat A, Saikali S, Etcheverry A, Delmas C, et al. Human glioblastoma stem-like cells are more sensitive to allogeneic NK and T cell-mediated killing compared with serum-cultured glioblastoma cells. Brain Pathol 2012;22:159–74. https://doi.org/10.1111/j.1750-3639.2011.00515.x.
[94] Barrow AD, Colonna M. Tailoring natural killer cell immunotherapy to the tumour microenvironment. Semin Immunol 2017;31:30–6. https://doi.org/10.1016/J.SMIM.2017.09.001.
[95] Bian S, Repic M, Guo Z, Kavirayani A, Burkard T, Bagley JA, et al. Genetically engineered cerebral organoids model brain tumor formation. Nat Methods 2018;15:631–9. https://doi.org/10.1038/s41592-018-0070-7.
[96] Bosma GC, Custer RP, Bosma MJ. A severe combined immunodeficiency mutation in the mouse. Nature 1983;301:527–30. https://doi.org/10.1038/301527a0.
[97] Carpentier A, Laigle-Donadey F, Zohar S, Capelle L, Behin A, Tibi A, et al. Phase 1 trial of a CpG oligodeoxynucleotide for patients with recurrent glioblastoma. Neuro Oncol 2006;8:60–6. https://doi.org/10.1215/S1522851705000475.
[98] Carpentier A, Metellus P, Ursu R, Zohar S, Lafitte F, Barrié M, et al. Intracerebral administration of CpG oligonucleotide for patients with recurrent glioblastoma: a phase II study. Neuro Oncol 2010;12:401–8. https://doi.org/10.1093/NEUONC/NOP047.

[99] Castriconi R, Cantoni C, Della Chiesa M, Vitale M, Marcenaro E, Conte R, et al. Transforming growth factor beta 1 inhibits expression of NKp30 and NKG2D receptors: consequences for the NK-mediated killing of dendritic cells. Proc Natl Acad Sci U S A 2003;100:4120–5. https://doi.org/10.1073/PNAS.0730640100.
[100] Chiu T-L, Lin S-Z, Hsieh W-H, Peng C-W. AAV2-mediated interleukin-12 in the treatment of malignant brain tumors through activation of NK cells. Int J Oncol 2009;35:1361–7. https://doi.org/10.3892/IJO_00000454.
[101] Clevers H. Modeling development and disease with organoids. Cell 2016;165:1586–97. https://doi.org/10.1016/j.cell.2016.05.082.
[102] Crane CA, Han SJ, Barry JJ, Ahn BJ, Lanier LL, Parsa AT. TGF-β downregulates the activating receptor NKG2D on NK cells and CD8+ T cells in glioma patients. Neuro Oncol 2010;12:7–13. https://doi.org/10.1093/neuonc/nop009.
[103] Daga A, Orengo AM, Gangemi RMR, Marubbi D, Perera M, Comes A, et al. Glioma immunotherapy by IL-21 gene-modified cells or by recombinant IL-21 involves antibody responses. Int J Cancer 2007;121:1756–63. https://doi.org/10.1002/IJC.22901.
[104] Ehtesham M, Samoto K, Kabos P, Acosta FL, Gutierrez MA, Black KL, et al. Treatment of intracranial glioma with in situ interferon-gamma and tumor necrosis factor-alpha gene transfer. Cancer Gene Ther 2002;9:925–34. https://doi.org/10.1038/sj.cgt.7700516.
[105] Eisele G, Wischhusen J, Mittelbronn M, Meyermann R, Waldhauer I, Steinle A, et al. TGF-β and metalloproteinases differentially suppress NKG2D ligand surface expression on malignant glioma cells. Brain 2006;129:2416–25. https://doi.org/10.1093/BRAIN/AWL205.
[106] Eskilsson E, Røsland GV, Solecki G, Wang Q, Harter PN, Graziani G, et al. EGFR heterogeneity and implications for therapeutic intervention in glioblastoma. Neuro Oncol 2018;20:743–52. https://doi.org/10.1093/NEUONC/NOX191.
[107] Fares J, Fares MY, Fares Y. Natural killer cells in the brain tumor microenvironment: defining a new era in neuro-oncology. Surg Neurol Int 2019;10:1–4. https://doi.org/10.25259/SNI-97-2019.
[108] Friese MA, Lutz SZ, Naumann U, Aulwurm S, Bischof F, Bühring H-J, et al. MICA/NKG2D-mediated immunogene therapy of experimental gliomas. Cancer Res 2003;63:8996–9006.
[109] Friese MA, Wischhusen J, Wick W, Weiler M, Eisele G, Steinle A, et al. RNA interference targeting transforming growth factor-beta enhances NKG2D-mediated antiglioma immune response, inhibits glioma cell migration and invasiveness, and abrogates tumorigenicity in vivo. Cancer Res 2004;64:7596–603. https://doi.org/10.1158/0008-5472.CAN-04-1627.
[110] Gajewski TF, Woo SR, Zha Y, Spaapen R, Zheng Y, Corrales L, et al. Cancer immunotherapy strategies based on overcoming barriers within the tumor microenvironment. Curr Opin Immunol 2013;25:268–76. https://doi.org/10.1016/J.COI.2013.02.009.
[111] Geller M, Miller J. Use of allogeneic NK cells for cancer immunotherapy. Immunotherapy 2011;3:1445–59. https://doi.org/10.2217/IMT.11.131.
[112] Genßler S, Burger MC, Zhang C, Oelsner S, Mildenberger I, Wagner M, et al. Dual targeting of glioblastoma with chimeric antigen receptor-engineered natural killer cells overcomes heterogeneity of target antigen expression and enhances antitumor activity and survival. Onco Targets Ther 2015;5, e1119354. https://doi.org/10.1080/2162402X.2015.1119354.
[113] Gimple RC, Bhargava S, Dixit D, Rich JN. Glioblastoma stem cells: lessons from the tumor hierarchy in a lethal cancer. Genes Dev 2019;33:591–609. https://doi.org/10.1101/gad.324301.119.
[114] Grantab R, Sivananthan S, Tannock IF. The penetration of anticancer drugs through tumor tissue as a function of cellular adhesion and packing density of tumor cells. Cancer Res 2006;66:1033–9. https://doi.org/10.1158/0008-5472.CAN-05-3077.
[115] Han J, Chu J, Chan WK, Zhang J, Wang Y, Cohen JB, et al. CAR-engineered NK cells targeting wild-type EGFR and EGFRvIII enhance killing of glioblastoma and patient-derived glioblastoma stem cells. Sci Rep 2015;5:11483. https://doi.org/10.1038/SREP11483.
[116] Hanahan D, Coussens LM. Accessories to the crime: functions of cells recruited to the tumor microenvironment. Cancer Cell 2012;21:309–22. https://doi.org/10.1016/J.CCR.2012.02.022.
[117] da Hora CC, Schweiger MW, Wurdinger T, Tannous BA. Patient-derived glioma models: from patients to dish to animals. Cells 2019;8:1177. https://doi.org/10.3390/CELLS8101177.
[118] Höring E, Podlech O, Silkenstedt B, Rota IA, Adamopoulou E, Naumann U. The histone deacetylase inhibitor trichostatin a promotes apoptosis and antitumor immunity in glioblastoma cells. Anticancer Res 2013;33:1351–60.

[119] Hubert CG, Rivera M, Spangler LC, Wu Q, Mack SC, Prager BC, et al. A three-dimensional organoid culture system derived from human glioblastomas recapitulates the hypoxic gradients and cancer stem cell heterogeneity of tumors found in vivo. Cancer Res 2016;76:2465–77. https://doi.org/10.1158/0008-5472.CAN-15-2402.
[120] Huszthy PC, Daphu I, Niclou SP, Stieber D, Nigro JM, Sakariassen PØ, et al. In vivo models of primary brain tumors: pitfalls and perspectives. Neuro Oncol 2012;14:979–93. https://doi.org/10.1093/NEUONC/NOS135.
[121] Ishikawa E, Tsuboi K, Saijo K, Harada H, Takano S, Nose T, et al. Autologous natural killer cell therapy for human recurrent malignant glioma. Anticancer Res 2004;24:1861–71.
[122] Jacob F, Salinas RD, Zhang DY, Nguyen PTT, Schnoll JG, Wong SZH, et al. A patient-derived glioblastoma organoid model and biobank recapitulates inter- and intra-tumoral heterogeneity. Cell 2019;180:188–204.e22. https://doi.org/10.1016/j.cell.2019.11.036.
[123] Jacob FP, Ming G, Song H. Generation and biobanking of patient-derived glioblastoma organoids and their application in CAR-T cell testing. Nat Protoc 2020;15:4000–33. https://doi.org/10.1038/s41596-020-0402-9.
[124] Jacobs JFM, Idema AJ, Bol KF, Grotenhuis JA, de Vries IJM, Wesseling P, et al. Prognostic significance and mechanism of Treg infiltration in human brain tumors. J Neuroimmunol 2010;225:195–9. https://doi.org/10.1016/j.jneuroim.2010.05.020.
[125] Joo KM, Kim J, Jin J, Kim M, Seol HJ, Muradov J, et al. Patient-specific orthotopic glioblastoma xenograft models recapitulate the histopathology and biology of human glioblastomas in situ. Cell Rep 2013;3:260–73. https://doi.org/10.1016/J.CELREP.2012.12.013.
[126] Khalsa JK, Shah K. Immune profiling of syngeneic murine and patient GBMs for effective translation of immunotherapies. Cells 2021;10:491. https://doi.org/10.3390/cells10030491.
[127] Khatua S, Cooper LJN, Sandberg DI, Ketonen L, Johnson JM, Rytting ME, et al. Phase I study of intraventricular infusions of autologous ex vivo expanded NK cells in children with recurrent medulloblastoma and ependymoma. Neuro Oncol 2020;22:1214–25. https://doi.org/10.1093/neuonc/noaa047.
[128] Kmiecik J, Navarro AG, Poli A, Planagumà JP, Zimmer J, Chekenya M. Combining NK cells and mAb9.2.27 to combat NG2-dependent and anti-inflammatory signals in glioblastoma. Oncoimmunology 2014;3, e27185. https://doi.org/10.4161/ONCI.27185.
[129] Knorr DA, Ni Z, Hermanson D, Hexum MK, Bendzick L, Cooper LJN, et al. Clinical-scale derivation of natural killer cells from human pluripotent stem cells for cancer therapy. Stem Cells Transl Med 2013;2:274–83. https://doi.org/10.5966/SCTM.2012-0084.
[130] Kozlowska A, Topchyan P, Kaur K, Tseng HC, Teruel A, Hiraga T, et al. Differentiation by NK cells is a prerequisite for effective targeting of cancer stem cells/poorly differentiated tumors by chemopreventive and chemotherapeutic drugs. J Cancer 2017;8:537–54. https://doi.org/10.7150/jca.15989.
[131] Lai C-Y, Yu G-Y, Luo Y, Xiang R, Chuang T-H. Immunostimulatory activities of CpG-oligodeoxynucleotides in teleosts: toll-like receptors 9 and 21. Front Immunol 2019;10:179. https://doi.org/10.3389/FIMMU.2019.00179.
[132] Lenting K, Verhaak R, ter Laan M, Wesseling P, Leenders W. Glioma: experimental models and reality. Acta Neuropathol 2017;133:263–82. https://doi.org/10.1007/s00401-017-1671-4.
[133] Li E, Lin L, Chen C-W, Ou D-L. Mouse models for immunotherapy in hepatocellular carcinoma. Cancers (Basel) 2019;11:1800. https://doi.org/10.3390/cancers11111800.
[134] Lim O, Jung MY, Hwang YK, Shin E-C. Present and future of allogeneic natural killer cell therapy. Front Immunol 2015;6:286. https://doi.org/10.3389/FIMMU.2015.00286.
[135] Linkous A, Balamatsias D, Snuderl M, Edwards L, Miyaguchi K, Milner T, et al. Modeling patient-derived glioblastoma with cerebral organoids. Cell Rep 2019;26:3203–3211.e5. https://doi.org/10.1016/j.celrep.2019.02.063.
[136] Louis DN. Molecular pathology of malignant gliomas. Annu Rev Pathol Mech Dis 2006;1:97–117. https://doi.org/10.1146/ANNUREV.PATHOL.1.110304.100043.
[137] Luna JI, Grossenbacher SK, Sturgill IR, Ames E, Judge SJ, Bouzid LA, et al. Bortezomib augments natural killer cell targeting of stem-like tumor cells. Cancers (Basel) 2019;11:1–20. https://doi.org/10.3390/cancers11010085.
[138] Maes W, Van Gool SW. Experimental immunotherapy for malignant glioma: lessons from two decades of research in the GL261 model. Cancer Immunol Immunother 2010;60:153–60. https://doi.org/10.1007/S00262-010-0946-6.
[139] Majzner RG, Mackall CL. Tumor antigen escape from CAR T-cell therapy. Cancer Discov 2018;8:1219–26. https://doi.org/10.1158/2159-8290.CD-18-0442.

[140] Mathews S, Branch Woods A, Katano I, Makarov E, Thomas MB, Gendelman HE, et al. Human Interleukin-34 facilitates microglia-like cell differentiation and persistent HIV-1 infection in humanized mice. Mol Neurodegener 2019;14:12. https://doi.org/10.1186/s13024-019-0311-y.
[141] Mhaidly R, Verhoeyen E. Humanized mice are precious tools for preclinical evaluation of CAR T and CAR NK cell therapies. Cancers (Basel) 2020;12:1–22. https://doi.org/10.3390/CANCERS12071915.
[142] Morimoto T, Nakazawa T, Matsuda R, Nishimura F, Nakamura M, Yamada S, et al. Evaluation of comprehensive gene expression and NK cell-mediated killing in glioblastoma cell line-derived spheroids. Cancers (Basel) 2021;13:4896. https://doi.org/10.3390/CANCERS13194896.
[143] Müller N, Michen S, Tietze S, Töpfer K, Schulte A, Lamszus K, et al. Engineering NK cells modified with an EGFRvIII-specific chimeric antigen receptor to overexpress CXCR4 improves immunotherapy of CXCL12/SDF-1α-secreting glioblastoma. J Immunother 2015;38:197–210. https://doi.org/10.1097/CJI.0000000000000082.
[144] Nakazawa T, Murakami T, Natsume A, Nishimura F, Morimoto T, Matsuda R, et al. KHYG-1 cells with EGFRvIII-specific CAR induced a pseudoprogression-like feature in subcutaneous tumours derived from glioblastoma-like cells. Anticancer Res 2020;40:3231–7. https://doi.org/10.21873/ANTICANRES.14304.
[145] Navarro AG, Espedal H, Joseph JV, Trachsel-Moncho L, Bahador M, Gjertsen BT, et al. Pretreatment of glioblastoma with bortezomib potentiates natural killer cell cytotoxicity through TRAIL/DR5 mediated apoptosis and prolongs animal survival. Cancers (Basel) 2019;11:1–25. https://doi.org/10.3390/cancers11070996.
[146] Ogawa J, Pao GM, Shokhirev MN, Verma IM. Glioblastoma Model Using Human Cerebral Organoids. Cell Rep 2018;23:1220–9. https://doi.org/10.1016/j.celrep.2018.03.105.
[147] Oh T, Fakurnejad S, Sayegh ET, Clark AJ, Ivan ME, Sun MZ, et al. Immunocompetent murine models for the study of glioblastoma immunotherapy. J Transl Med 2014. https://doi.org/10.1186/1479-5876-12-107.
[148] Ozawa Y, Yamamuro S, Sano E, Tatsuoka J, Hanashima Y, Yoshimura S, et al. Indoleamine 2,3-dioxygenase 1 is highly expressed in glioma stem cells. Biochem Biophys Res Commun 2020;524:723–9. https://doi.org/10.1016/J.BBRC.2020.01.148.
[149] Poli A, Wang J, Domingues O, Planagumà J, Yan T, Rygh CB, et al. Targeting glioblastoma with NK cells and mAb against NG2/CSPG4 prolongs animal survival. Oncotarget 2013;4:1527–46. https://doi.org/10.18632/ONCOTARGET.1291.
[150] Powell AB, Yadavilli S, Saunders D, Van Pelt S, Chorvinsky E, Burga RA, et al. Medulloblastoma rendered susceptible to NK-cell attack by TGFβ neutralization. J Transl Med 2019;17:1–10. https://doi.org/10.1186/S12967-019-2055-4.
[151] Richmond A, Su Y. Mouse xenograft models vs GEM models for human cancer therapeutics. Dis Model Mech 2008;1:78–82. https://doi.org/10.1242/DMM.000976.
[152] Ruscetti M, Leibold J, Bott MJ, Fennell M, Kulick A, Salgado NR, et al. NK cell-mediated cytotoxicity contributes to tumor control by a cytostatic drug combination. Science 2018;362:1416–22. https://doi.org/10.1126/SCIENCE.AAS9090.
[153] Saleh M, Jonas NK, Wiegmans A, Stylli SS. The treatment of established intracranial tumors by in situ retroviral IFN-γ transfer. Gene Ther 2000;7:1715–24. https://doi.org/10.1038/sj.gt.3301273.
[154] Salinas RD, Durgin JS, O'Rourke DM. Potential of glioblastoma-targeted chimeric antigen receptor (CAR) T-cell therapy. CNS Drugs 2020;34:127–45. https://doi.org/10.1007/s40263-019-00687-3.
[155] Sato T, Stange DE, Ferrante M, Vries RGJ, Van Es JH, Van Den Brink S, et al. Long-term expansion of epithelial organoids from human colon, adenoma, adenocarcinoma, and Barrett's epithelium. Gastroenterology 2011;141:1762–72. https://doi.org/10.1053/J.GASTRO.2011.07.050.
[156] Saygin C, Matei D, Majeti R, Reizes O, Lathia JD. Targeting cancer stemness in the clinic: from hype to Hope. Cell Stem Cell 2019;24:25–40. https://doi.org/10.1016/J.STEM.2018.11.017.
[157] Schnalzger TE, de Groot MH, Zhang C, Mosa MH, Michels BE, Röder J, et al. 3D model for CAR-mediated cytotoxicity using patient-derived colorectal cancer organoids. EMBO J 2019;38, e100928. https://doi.org/10.15252/EMBJ.2018100928.
[158] Senjor E, Perišić Nanut M, Breznik B, Mitrović A, Mlakar J, Rotter A, et al. Cystatin F acts as a mediator of immune suppression in glioblastoma. Cell Oncol 2021;44:1051–63. https://doi.org/10.1007/s13402-021-00618-9.
[159] Sharifzad F, Soura M, Saeid M, Fakharian E, Taghikhani A, Sharifzad A, et al. HSP70/IL-2 treated NK cells effectively cross the blood brain barrier and target tumor cells in a rat model of induced glioblastoma multiforme (GBM). Int J Mol Sci 2020;21:2263. https://doi.org/10.3390/ijms21072263.
[160] Sharma V, Koul N, Joseph C, Dixit D, Ghosh S, Sen E. HDAC inhibitor, scriptaid, induces glioma cell apoptosis through JNK activation and inhibits telomerase activity. J Cell Mol Med 2010;14:2151–61. https://doi.org/10.1111/J.1582-4934.2009.00844.X.

[161] Sola C, Chanuc F, Thielens A, Fuseri N, Morel Y, Bléry M, et al. Anti-tumoral efficacy of therapeutic human anti-KIR antibody (Lirilumab/BMS-986015/IPH2102) in a preclinical xenograft tumor model. J Immunother Cancer 2013;1:P40. https://doi.org/10.1186/2051-1426-1-S1-P40.
[162] de Souza N. Organoids. Nat Methods 2018;15:23. https://doi.org/10.1038/nmeth.4576.
[163] Suck G, Odendahl M, Nowakowska P, Seidl C, Wels WS, Klingemann HG, et al. NK-92: an 'off-the-shelf therapeutic' for adoptive natural killer cell-based cancer immunotherapy. Cancer Immunol Immunother 2015;65:485–92. https://doi.org/10.1007/S00262-015-1761-X.
[164] Terrén I, Orrantia A, Vitallé J, Zenarruzabeitia O, Borrego F. NK cell metabolism and tumor microenvironment. Front Immunol 2019;10:2278. https://doi.org/10.3389/FIMMU.2019.02278.
[165] Di Tomaso T, Mazzoleni S, Wang E, Sovena G, Clavenna D, Franzin A, et al. Immunobiological characterization of cancer stem cells isolated from glioblastoma patients. Clin Cancer Res 2010;16:800–13. https://doi.org/10.1158/1078-0432.CCR-09-2730.
[166] Tseng HC, Inagaki A, Bui VT, Cacalano N, Kasahara N, Man YG, et al. Differential targeting of stem cells and differentiated glioblastomas by NK cells. J Cancer 2015;6:866–76. https://doi.org/10.7150/jca.11527.
[167] Ursu R, Carpentier A, Metellus P, Lubrano V, Laigle-Donadey F, Capelle L, et al. Intracerebral injection of CpG oligonucleotide for patients with de novo glioblastoma—a phase II multicentric, randomised study. Eur J Cancer 2017;73:30–7. https://doi.org/10.1016/J.EJCA.2016.12.003.
[168] Vitale M, Cantoni C, Pietra G, Mingari MC, Moretta L. Effect of tumor cells and tumor microenvironment on NK-cell function. Eur J Immunol 2014;44:1582–92. https://doi.org/10.1002/EJI.201344272.
[169] Wang J, Toregrosa-Allen S, Elzey BD, Utturkar S, Lanman NA, Bernal-Crespo V, et al. Multispecific targeting of glioblastoma with tumor microenvironment-responsive multifunctional engineered NK cells. Proc Natl Acad Sci U S A 2021;118, e2107507118. https://doi.org/10.1073/pnas.2107507118.
[170] Wilson EB, El-Jawhari JJ, Neilson AL, Hall GD, Melcher AA, Meade JL, et al. Human tumour immune evasion via TGF-β blocks NK cell activation but not survival allowing therapeutic restoration of anti-tumour activity. PLoS One 2011;6, e22842. https://doi.org/10.1371/JOURNAL.PONE.0022842.
[171] Wu A, Wiesner S, Xiao J, Ericson K, Chen W, Hall WA, et al. Erratum: expression of MHC I and NK ligands on human CD133 + glioma cells: possible targets of immunotherapy. J Neurooncol 2007;83:133. https://doi.org/10.1007/s11060-006-9307-x.
[172] Yamanegi K, Yamane J, Kobayashi K, Ohyama H, Nakasho K, Yamada N, et al. Downregulation of matrix metalloproteinase-9 mRNA by valproic acid plays a role in inhibiting the shedding of MHC class I-related molecules A and B on the surface of human osteosarcoma cells. Oncol Rep 2012;28:1585–90. https://doi.org/10.3892/OR.2012.1981.
[173] Yoo JY, Jaime-Ramirez AC, Bolyard C, Dai H, Nallanagulagari T, Wojton J, et al. Bortezomib treatment sensitizes oncolytic HSV-1 treated tumors to NK cell immunotherapy. Clin Cancer Res 2016;22:5265–76. https://doi.org/10.1158/1078-0432.CCR-16-1003.
[174] Yvon ES, Burga R, Powell A, Cruz CR, Fernandes R, Barese C, et al. Cord blood natural killer cells expressing a dominant negative TGF-β receptor: implications for adoptive immunotherapy for glioblastoma. Cytotherapy 2017;19:408–18. https://doi.org/10.1016/J.JCYT.2016.12.005.
[175] Zhang C, Burger MC, Jennewein L, Genßler S, Schönfeld K, Zeiner P, et al. ErbB2/HER2-specific NK cells for targeted therapy of glioblastoma. J Natl Cancer Inst 2016;108:1–12. https://doi.org/10.1093/jnci/djv375.
[176] Zhao Y, Shuen TWH, Toh TB, Chan XY, Liu M, Tan SY, et al. Development of a new patient-derived xenograft humanised mouse model to study human-specific tumour microenvironment and immunotherapy. Gut 2018;67:1845–54. https://doi.org/10.1136/gutjnl-2017-315201.

CHAPTER

17

Immunotherapy using CAR T: What we have learned from trials and where we are heading

Amy Zhu[a], Rebecca Bellis[a], Jasmeen Saini[a], Christina Fong[b], and Prasad S. Adusumilli[a,c]

[a]Thoracic Service, Department of Surgery, Memorial Sloan Kettering Cancer Center, New York, NY, United States [b]Department of Pediatrics, Memorial Sloan Kettering Cancer Center, New York, NY, United States [c]Center for Cell Engineering, Memorial Sloan Kettering Cancer Center, New York, NY, United States

Abstract

Chimeric antigen receptor (CAR) T-cell therapy is successful in the treatment of liquid cancers and is being investigated in solid tumors. While immunotherapy by way of immune checkpoint inhibitors is now standard of care therapy for some solid tumors, there are substantial differences in the design, conduct, and interpretation of immune checkpoint inhibitor clinical trials compared to CAR T-cell clinical trials. In this chapter, we describe the key comparisons between cell therapy and immune checkpoint inhibitor clinical trials and highlight the lessons learned from conducting CAR T-cell trials in a high-volume cancer center. Utilizing this learning experience, we provide guidelines for optimizing the clinical flow of CAR T-cell trials. We believe that a structured approach for CAR T-cell trials in solid tumors as described here can be adopted by other cell therapy trials and provide insights to advancing immunotherapy.

Abbreviations

CAR	chimeric antigen receptor
CLL	chronic lymphocytic leukemia
CRS	cytokine release syndrome
ICI	immune checkpoint inhibitor
irAE	immune-related adverse events
PBMC	peripheral blood mononuclear cells
PD-1	programmed cell death 1
PD-L1	programmed cell death 1 ligand
scFV	single-chain variable fragment

https://doi.org/10.1016/B978-0-12-822620-9.00012-4

Conflicts of interest

P.S.A. has received research funding from ATARA Biotherapeutics, has served on the Scientific Advisory Board or as consultant to ATARA Biotherapeutics, Bayer, Carisma Therapeutics, Imugene, ImmPACT Bio, and Takeda Therapeutics, and has patents, royalties, and intellectual property on mesothelin-targeted CARs and T-cell therapies, method for detection of cancer cells using virus, and pending patent applications on T-cell therapies. All the other authors declare no disclosures.

Better understanding the influence of the immune microenvironment on cancer evolution and response to therapy has resulted in the development of therapeutic agents that can modulate immune responses and facilitate killing of tumor cells [1–3]. While immune checkpoint inhibitor (ICI) monoclonal antibodies, which can reverse tumor-mediated inhibition of cytotoxic T cells, have played a major role in solid tumor immunotherapy [4], advances in genetic engineering have resulted in the generation of synthetic receptors—chimeric antigen receptors (CARs)—that are able to prolong survival in patients with chemotherapy-refractory liquid tumors [5–7]. The lessons learned from studying CAR T-cell therapy of liquid tumors, combined with the better understanding of the solid tumor immune microenvironment, can lead to advances in solid tumor immunotherapy in the near future. Herein, we discuss what has been learned from immunotherapy trials to date, with a focus on the conduct of and logistical issues specific to trials of CAR T-cell therapy. As we only briefly describe the advances in CAR design, readers are encouraged to refer to the excellent available review articles for more in-depth knowledge [8–10].

Landscape of CAR T-cell clinical trials

Cell therapies—in particular, CAR T-cells—account for the majority of therapies being developed in immuno-oncology. As of March 2020, of 1483 cell therapy agents currently in development, more than half are CAR T-cell agents [11]. A review of https://clinicaltrials.gov using the search terms "CAR" and "cancer" indicates there are 809 registered interventional CAR T-cell trials, of which 74% are for hematological malignancies and 26% are for solid tumors [12]. The majority of CAR T-cell clinical trials are in phase I or phase I/II (86%) (Fig. 1 and Table 1). While these trials are ongoing to provide evidence of safety and antitumor efficacy, investigators are taking advantage of emerging knowledge from other immunotherapy trials to incorporate advanced features into the design of CARs to further enhance antitumor immunity.

New developments in CAR design

CARs are synthetic receptors that allow CAR-expressing immune cells to recognize and destroy cells displaying the target antigen [10]. The candidate target antigen is expressed mainly on the target of interest—the cancer cell—with no or low-level expression on normal cells, to minimize "on-target/off-tumor" toxicity [13]. The target antigen's cell surface density and epitope location are critical in designing and selecting the single-chain variable fragment (scFv), an important component of the CAR [10] that affects its target-binding capacity. The key structures of a CAR are (1) the antigen-binding domain, (2) a hinge attached to a transmembrane domain, and (3) the intracellular signaling domain—all of which can be modified to optimize the CAR's efficacy. The antigen-binding domain mediates antigen recognition and is

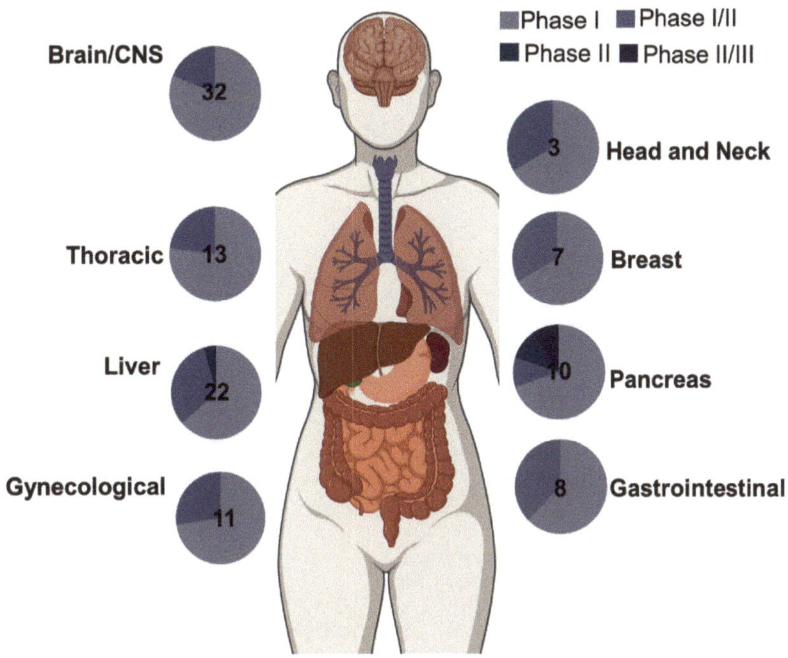

FIG. 1 A schematic overview of the most common solid tumors targeted by CAR T-cell therapy, by clinical trial phase(s). The number represents the number of clinical trials targeting the organ or organ system. Data were extracted from clinicaltrials.gov. CNS, central nervous system (figure created with BioRender.com).

TABLE 1 Summary of CAR T-cell clinical trials registered in clinicaltrials.gov by phase ($N=808$).

Disease	Phase I	Phase I/II	Phase II	Phase II/III	Phase III	N/A
Hematologic cancers						
Leukemia	161	104	22	2	2	16
Lymphoma	175	103	28	0	3	15
Myeloma	69	33	15	0	3	6
Solid tumors						
Bladder	0	1	0	0	0	0
Brain/CNS	26	6	0	0	0	0
Breast	5	2	0	0	0	0
Esophageal	0	1	0	0	0	0
GI	6	2	0	0	0	0
Gynecological	8	3	0	0	0	0
Head and neck	0	2	1	0	0	0
Liver	14	7	1	0	0	1
Melanoma	3	0	0	0	0	0

Continued

TABLE 1 Summary of CAR T-cell clinical trials registered in clinicaltrials.gov by phase ($N=808$)—cont'd

Disease	Phase I	Phase I/II	Phase II	Phase II/III	Phase III	N/A
Multiple tumor types[a]	30	12	0	0	0	0
Neuroblastoma	9	2	0	0	0	1
Pancreas	7	1	1	1	0	1
Peritoneal	1	0	0	0	0	0
Prostate	8	0	0	0	0	0
Renal	2	1	0	0	0	0
Sarcoma	1	2	0	0	0	0
Thoracic	10	3	0	0	0	0
Thyroid	1	0	0	0	0	0
Unspecified[b]	10	13	0	0	0	2

[a] *Clinical trials including two or more disease types.*
[b] *Clinical trials not specific to a disease or organ; targeting an antigen.*
CNS, central nervous system; GI, gastrointestinal; N/A, not applicable.

composed of a scFV linked to a monoclonal antibody's variable heavy and light chains [9,10]. Whereas high scFV specificity and affinity are essential for antigen recognition and lymphocyte activation, a disproportionately high scFV affinity can cause activation-induced cell death of the CAR T-cell itself. The scFV is connected to a hinge by a linker region, typically composed of glycine and serine peptides, which affords flexibility and solubility for enhanced scFV-target binding. The length and type of the hinge serve to optimize the synapse of the CAR and the target [14]. This hinge is anchored to a transmembrane domain, which determines the stability of the CAR and connects the antigen-binding domain to the intracellular domain. The intracellular signaling domain consists of an activation domain, often the CD3ζ chain of the T-cell receptor, and at least one costimulatory domain, such as CD28 or 4-1BB. These costimulatory domains enhance the CAR T-cell's function, expansion, persistence, and memory. CAR constructs can be further engineered to express "armor," usually a cytokine or cell surface protein that functions to stimulate the host's intrinsic immune response or dampen a tumor's immunosuppressive microenvironment [9]. In addition to the modification of these various structures, gene editing techniques such as CRISPR-Cas9 have allowed the design of CARs that can overcome known causes of lymphocyte exhaustion. For example, CRISPR-Cas9 has been used to integrate CAR constructs specifically into the T-cell receptor alpha chain locus, resulting in controlled CAR expression and decreased ligand-independent tonic signaling [15]. In addition, CRISPR-Cas9 has been used to silence the gene encoding programmed cell death 1 (PD-1), effectively addressing exhaustion caused by immune checkpoint blockade [16].

CAR T-cell trials vs ICI trials

Thanks to the widespread clinical use of ICIs, many facilities are familiar with immunotherapy and the clinical monitoring of it. ICIs, including anticytotoxic T lymphocyte antigen

4, anti-PD-1, and antiprogrammed cell death 1 ligand (PD-L1) antibodies, are currently approved by the FDA for multiple cancers, including melanoma, nonsmall cell lung cancer, hepatocellular cancer, and head and neck cancer [4]. Increasingly, ICI agents are being combined with either chemotherapeutic agents or targeted therapies, resulting in improved survival [17]. For example, incidence-based mortality from nonsmall cell lung cancer decreased 6.3% annually from 2013 through 2016, correlated with advances with therapies [18].

Achieving widespread implementation of CAR T-cell therapy, in line with that of ICI agents, faces multiple challenges. The key differences between ICI agent therapy and CAR T-cell therapy clinical trials are summarized in Table 2. Patients on ICI trials do not require cellular apheresis and are able to start treatment immediately. Most CAR T-cell trials include autologous therapies, requiring patients to undergo apheresis for the collection of peripheral blood mononuclear cells (PBMCs). Whereas ICI trials do not require preconditioning, CAR T-cell trials require preconditioning lymphodepletion before T-cell infusion to enhance the ability of CAR T-cells to engraft and proliferate and to transiently reduce the number of regulatory T cells, which may otherwise suppress the function of CAR T-cells. The most common lymphodepletion regimen used is cyclophosphamide and fludarabine, followed by single-agent cyclophosphamide or fludarabine [19]. Steroids commonly used alongside these lymphodepleting agents for the treatment of severe nausea or other effects are typically avoided in T-cell therapy trials as they may compromise ensuing CAR T-cell function.

Depending on the agent, ICIs are administered on a recurring schedule every 3–4 weeks until sustained therapy response, undue toxicity, or disease progression is observed. CAR T-cells are usually administered as a single dose or divided over a few days. Whereas ICIs are exclusively administered intravenously, CAR T-cells can be administered through various routes, including intravenous or regional [20]. Common regional injection sites for solid tumor trials include intracerebral, intraperitoneal, and intratumoral [19]. ICIs are administered as outpatient treatments, whereas CAR T-cells may be administered in an inpatient or outpatient setting.

The most common toxicities associated with ICIs are immune-related adverse events (irAEs) affecting multiple organs [21,22]. Although irAEs may vary with the type of ICI agent

TABLE 2 Summary of key clinical differences between immune checkpoint inhibitor and CAR T-cell clinical trials.

Component	Immune checkpoint inhibitor trials	CAR T-cell trials
Source	Off the shelf	Autologous—apheresis Allogenic—off-the-shelf
Preconditioning regimen	No	Yes
Dose regimen	Recurring	Single
Delivery	Intravenous	Intravenous Regional
Treatment setting	Outpatient	Outpatient Inpatient
Onset of adverse events	1–3 months	1 week
Treatment of adverse events	Steroids	Steroids, IL6 blockade

used, common irAEs include skin rash, pruritus, diarrhea, colitis, hypothyroidism, and pneumonitis. Toxicities may occur 2–12 weeks after the receipt of ICI agents, depending on the agent, frequency, and number of doses. Corticosteroids are typically used to treat mild irAEs, although treatment may be discontinued altogether if toxicity continues. Cytokine release syndrome (CRS) and neurotoxicity are the most common adverse events associated with CAR T-cells [23]. CRS results from the release of cytokines from activated CAR T-cells, and its hallmark sign is fever. The onset of CRS usually occurs within the first week after infusion, and symptoms typically resolve within 1 week. Treatments for CRS include the IL-6 receptor antibody tocilizumab and corticosteroids. Symptoms of neurotoxicity include aphasia, tremor, and lethargy, which can occur within 1–3 weeks after CAR T-cell infusion and are usually resolved within 1 week with treatment. Neurotoxicity is managed with supportive care and corticosteroids. Owing to the high number of CAR T-cell treatments that have been approved by the FDA, clinicians at many centers are increasingly confident in treating these adverse events successfully without long-term impairment.

Lessons learned in optimizing the clinical flow of CAR T-cell trials

The enrollment of patients in CAR T-cell clinical trials requires multiple steps; hence, coordination among multiple parties is required for successful conduct of the trial. The key factors are presented in Fig. 2.

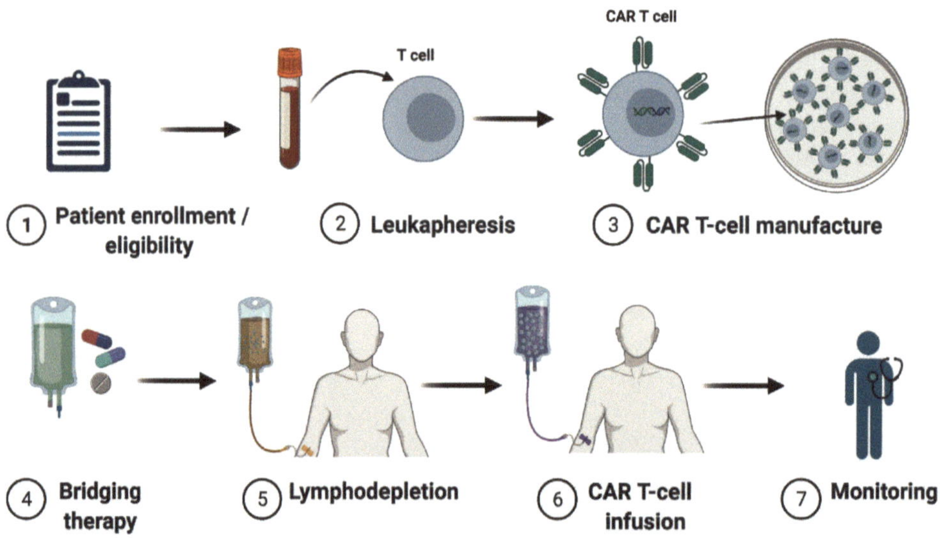

FIG. 2 A schematic overview of the clinical flow of a patient enrolling in a CAR T-cell clinical trial. (1) After undergoing pretreatment eligibility assessments, patients undergo (2) leukapheresis in which PBMCs are harvested and (3) sent to a manufacturing facility for production. (4) Patients may receive bridging therapy while the cell product is being manufactured. (5) Patients undergo lymphodepleting chemotherapy. (6) The CAR T-cells are administered to the patient following lymphodepletion. (7) The patient is then closely monitored with a long-term follow-up plan (figure created with BioRender.com).

Patient screening: In addition to an assessment of overall health and adequate organ function, CAR T-cell clinical trials include screening for additional factors. Since CAR T-cells target a specific antigen, patients must undergo screening for antigen expression in tissue and/or serum. Tumor heterogeneity, common in solid tumors, poses a screening challenge [24]. The use of target antigen expression intensity and distribution, reflected as the "H" score, as eligibility criteria may limit accrual, especially for trials that require expression of two or more target antigens.

In some trials, patients may receive a repeat dose of CAR T-cells. The administration of a second dose of CD19-targeted CAR T-cells in patients with B-cell malignancies was shown to be feasible and safe and induced responses in 39% of patients [25]. Although planned repeat dosing within a short period can proceed without repeat antigen testing, patient screening and evaluation of tumor antigen expression must be repeated when a prolonged time has elapsed or when the disease has relapsed. By means of "antigen escape," tumor cells can downregulate the target antigen or express alternative forms of the target antigen [26]. The presence of these escape mechanisms may complicate the identification of patients who are likely to benefit from a repeat dose of CAR T-cells. Another challenge during screening for repeat dosing in patients with solid tumors is the identification of the specific site to biopsy. In addition to tumor antigen expression heterogeneity across multiple organs, immunotherapy may result in confusing patterns of response and what appears to be progression on imaging due to inflammation [27]. In patients with solid tumors with multiple sites of metastases, it may be difficult to determine which sites are true progression and which sites are inflammation. In general, biopsy of solid organ metastases is preferred over biopsy of lymph nodes, as the lymph nodes frequently show granulomatous inflammation.

Leukapheresis: After undergoing pretreatment eligibility assessments, patients undergo leukapheresis in which PBMCs are harvested. Delays in leukapheresis may occur if patients require central venous catheter access and need to undergo an additional procedure for catheter placement. Since most CAR T-cell trials treat patients with refractory or relapsed cancers, it is unlikely that enrolled patients are treatment naïve. Disease burden, duration of prior therapy (such as chemotherapies and ICI agents), and time since the last exposure to therapy may affect T-cell quality. Protocols require a washout period between the last dose of treatment and leukapheresis. Further study may be warranted to identify optimal treatment washout and timing for leukapheresis. The composition of the apheresis product can affect the final manufactured product. For example, the depletion of monocytes from the leukapheresis product has been shown to increase CAR transduction efficiency, improve consistency, and increase the level of CAR expression [28]. However, the highly debated aspect of specific CD4:CD8 ratio or phenotype has yet to translate to a treatment benefit, and there is no consensus protocol, to date, defining the cellular composition of the apheresis or manufacturing product [29].

Manufacturing and quality assurance: The T cells are isolated and activated before being transduced to express the CAR. The transduced CAR T-cells are expanded to reach the dose size before cryopreservation, which is followed by quality and assurance testing. Production times may vary depending on the construct, protocol, and manufacturing facility. Prolonged or delayed production times can be clinically problematic if the patient's disease is progressing rapidly. Another challenge that may potentially delay a patient's treatment is cell viability. Different protocols vary in viability criteria and treating a patient with a CAR T-cell product below the threshold of viability requires permission from the sponsor, institutional review board, and FDA. Some protocols allow for remanufacturing or releukapheresis if the product does not meet viability requirements. Further study is warranted to determine which factors affect cell viability and whether viability contributes to safety and efficacy outcomes. Such efficacy outcomes can be measured using cytotoxicity assays. Assays, including chromium-51 release assays, bioluminescence assays, and impedance assays, are used to detect the cancer cell-killing ability of an effector cell, which can lead to reduction in tumor burden [30]. Assay requirements for investigational cell therapy products vary.

The complex manufacturing procedures of CAR T-cells limit the number of manufacturing facilities that have the expertise and infrastructure in accordance to Good Manufacturing Practice guidelines. Current manufacturing procedures of isolation, activation, transduction, and expansion involve multiple hands-on steps that require skilled personnel that increases the risk of human error, cell contamination, and cell loss [31,32]. To address these issues, the CliniMACS Prodigy (Miltenyi Biotech), which uses a closed single-use tubing set, has been adapted to simplify procedures and automate cell production on a clinical scale [31,33]. It has been shown that the expansion kinetics, transduction efficiency, and cell phenotypes of manufactured CD19 CAR T-cells were comparable to those of current manufacturing processes [31]. A Phase I study for patients with relapsed or refractory non-Hodgkin lymphoma showed that the production CAR T-cells targeting two antigens, CD19 and CD20, was feasible with CliniMACS Prodigy [34]. Benefits of CliniMACS Prodigy include decreased risk of contamination, shortened production times, and expanded access to institutions.

Bridging therapy: Patients may benefit from receiving bridging therapy, such as standard of care treatment, for disease control while CAR T-cells are being manufactured. In addition to disease control, bridging therapy has been shown to enhance CAR T-cell efficacy. The effect of bridging therapy was evaluated in a cohort of patients with chronic lymphocytic leukemia (CLL) who received ibrutinib before administration of CD19-directed CAR T-cells. Patients who received $\geq$5cycles of ibrutinib had better expansion of CD19-directed CAR T-cells (associated with decreased expression of the immunosuppressive molecule PD-1) on T cells and of CD200 on B-CLL cells [35]. Concurrent administration of ibrutinib did not impair CAR T-cell function in vitro but did improve CAR T-cell engraftment, tumor clearance, and survival in mouse models of resistant acute lymphocytic leukemia and CLL. However, patients may develop toxicities from bridging therapy that preclude participation in the trial.

Development of institutional infrastructure to conduct successful CAR T-cell trials

With an increasing number of trials across the spectrum of diseases, institutions are developing infrastructure to adopt uniform approaches and best practices for the successful development and conduct of CAR T-cell trials. The main components of such infrastructure are represented in Fig. 3.

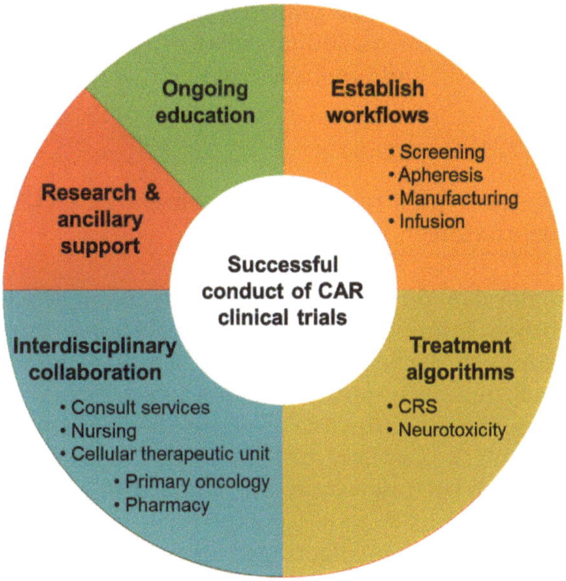

FIG. 3 An institutional framework for approaching CAR T-cell clinical trials. For the successful conduct of CAR T-cell clinical trials, a cellular therapeutic program is recommended to provide standardization in workflows, treatment algorithms, collaboration, research support, and ongoing education to maintain a high quality of patient care, treatment, and research outcomes. CRS, cytokine release syndrome.

I. Standardized workflows across the institution that are flexible to different disease models and CAR constructs

Screening workflow: An efficient screening workflow is needed to ensure that patients can access and enroll in CAR T-cell trials. Referral calls should be directed to a centralized group of experienced coordinators and nurse navigators, who can provide education on trials and initiate screening. During the consultation, patients need to be educated on the risks of the therapy and on the process and duration of CAR T-cell manufacturing. A tentative treatment plan for leukapheresis and infusion should account for washout of chemotherapy or ICI agents and should be provided to patients and the patients' primary oncologist.

Leukapheresis and manufacturing workflow: When scheduling leukapheresis, it is important to consider treatment washout periods and the availability of leukapheresis and manufacturing facilities to collect and process isolated PBMCs. To avoid potential delays in the treatment of patients who need central venous catheter placement, it is recommended to include vein assessment as part of the trial screening eligibility assessments. It is essential to maintain communication with the manufacturing facility for updates on the CAR product, potential delays, and cell viability.

Infusion workflow: Planning for the infusion of CAR T-cells involves coordination among multiple parties. It is important to confirm with the manufacturing facility the expected date of product completion, cell viability, and the expected date of product shipment to the institution. Confirming cell viability in advance will ensure that there is enough time to request permission to begin treatment from the institutional review board and FDA, if necessary. If the patient has been receiving bridging therapy, the clinical trial team should communicate with the patient's primary oncologist that the treatment washout period has been followed according to the protocol. The CAR T-cell product should be prepared before the patient receives lymphodepleting chemotherapy. If the infusion is administered in the inpatient setting, it is essential to coordinate with the admitting and inpatient units.

II. Standardized grading and treatment algorithms for the management of CRS and neurotoxicity

For a time, there was no consensus grading system for CRS or neurotoxicity. Because the CRS grading criteria, as defined by the Common Terminology Criteria for Adverse Events [36], are associated with the infusion of agents such as chemotherapy and ICIs and are not applicable to cellular therapies, several revised grading systems were developed, including Lee, Penn, and CARTOX (Table 3) [37–39]. This made it difficult to compare toxicities across different studies, constructs, and diseases. The American Society for Transplantation and Cellular Therapy consensus grading system has since been developed to standardize grading toxicities associated with CAR T-cells (Table 4) [40].

In addition to standardized grading of toxicities, treatment algorithms are necessary to ensure consistent patient care. Although treatment algorithms may differ among institutions, the consensus first-line treatment for grade $\geq$2 CRS is tocilizumab. For fever, infectious disease workup and empiric antibiotics are recommended. If CRS has not improved with initial administration of tocilizumab, an additional dose of tocilizumab should be given or corticosteroids should be considered.

TABLE 3 Summary of previously published grading systems for cytokine release syndrome.

Grading system	Grade 1	Grade 2	Grade 3	Grade 4
CTCAE version 4.03	Mild reaction; infusion interruption not indicated; intervention not indicated	Therapy or infusion interruption indicated but responds promptly to symptomatic treatment (antihistamines, NSAIDs, narcotics, iv fluids); prophylactic medications indicated for <24h	Prolonged (e.g., not rapidly responsive to symptomatic medication and/or brief interruption of infusion); hospitalization indicated for clinical sequelae (e.g., renal impairment, pulmonary infiltrates)	Life-threatening consequences; pressor or ventilatory support indicated
CTCAE version 5.0	Fever with or without constitutional symptoms	Hypotension responding to fluids; hypoxia responding to <40% O_2	Hypotension managed with one pressor; hypoxia requiring $\geq$40% O_2	Life-threatening consequences; urgent intervention indicated
Lee criteria	Symptoms are not life threatening and require symptomatic treatment only (fever, nausea,	Symptoms require and respond to moderate intervention: • Oxygen requirement <40% FiO_2 OR	Symptoms require and respond to aggressive intervention: • Oxygen requirement 40% FiO2 OR	Life-threatening symptoms: • Requirement for ventilator support OR

TABLE 3 Summary of previously published grading systems for cytokine release syndrome—cont'd

Grading system	Grade 1	Grade 2	Grade 3	Grade 4
	fatigue, headache, myalgias, and malaise)	• Hypotension responsive to iv fluids or low dose of one vasopressor OR • Grade 2 organ toxicity[a]	• Hypotension requiring high-dose or multiple vasopressors OR • Grade 3 organ toxicity[a] or grade 4 transaminitis	• Grade 4 organ toxicity[a] (excluding transaminitis)
Penn criteria	Mild reaction: Treated with supportive care, such as antipyretics, antiemetics	Moderate reaction: Some signs of organ dysfunction (grade 2 creatinine or grade 3 LFTs) related to CRS and not attributable to any other condition Hospitalization for management of CRS-related symptoms, including neutropenic fever and need for iv therapies (not including fluid resuscitation for hypotension)	More severe reaction: Hospitalization required for management of symptoms related to organ dysfunction, including grade 4 LFTs or grade 3 creatinine, related to CRS and not attributable to any other condition Hypotension treated with multiple fluid boluses or low-dose vasopressors Coagulopathy requiring fresh frozen plasma, cryoprecipitate, or fibrinogen concentrate Hypoxia requiring supplemental oxygen (nasal cannula oxygen, high-flow oxygen, CPAP, or BiPAP)	Life-threatening complications such as hypotension requiring high-dose vasopressors Hypoxia requiring mechanical ventilation
CARTOX criteria	Temperature 38°C Grade 1 organ toxicity[b]	Hypotension responds to iv fluids or low-dose vasopressor Hypoxia requiring FiO_2 <40% Grade 2 organ toxicity[b]	Hypotension needing high-dose or multiple vasopressors Hypoxia requiring FiO_2 40% Grade 3 organ toxicity[b] or grade 4 transaminitis	Life-threatening hypotension Needing ventilator support Grade 4 organ toxicity[b] except grade 4 transaminitis

[a] *As per CTCAE version 4.03.*
[b] *Cardiac (tachycardia, arrhythmias, heart block, low ejection fraction), respiratory (tachypnea, pleural effusion, pulmonary edema), gastrointestinal (nausea, vomiting, diarrhea), hepatic (increased serum alanine aminotransferase, aspartate aminotransferase, bilirubin level), renal (acute kidney injury, increased serum creatinine, decreased urine output), dermatologic (rash), or coagulopathy (disseminated intravascular coagulation).*
BiPAP, bilevel-positive airway pressure; CPAP, continuous positive airway pressure; CRS, cytokine release syndrome; CTCAE, Common Terminology Criteria for Adverse Events; LFT, liver function test.

TABLE 4 American Society for Transplantation and Cellular Therapy cytokine release syndrome grading system.

CRS Parameter	Grade 1	Grade 2	Grade 3	Grade 4
Fever[a]	Temperature ≥38°C	Temperature ≥38°C	Temperature ≥38°C	Temperature ≥38°C
With:				
Hypotension	None	Not requiring vasopressors	Requiring one vasopressor with or without vasopressin	Requiring multiple vasopressors (excluding vasopressin)
And/or:[b]				
Hypoxia	None	Requiring low-flow nasal cannula[c] by blow-by	Requiring high-flow nasal cannula[c], facemask, non-rebreather mask, or Venturi mask	Requiring positive pressure (e.g., CPAP, BiPAP, intubation, and mechanical ventilation)

[a] *Fever is defined as temperature ≥38°C not attributable to any other cause. In patients who have CRS then receive antipyretics or anticytokine therapy such as tocilizumab or steroids, fever is no longer required to grade subsequent CRS severity. In this case, CRS grading is driven by hypotension and/or hypoxia.*
[b] *CRS grade is determined by the more severe event: hypotension or hypoxia not attributable to any other cause. For example, a patient with a temperature of 39.5°C, hypotension requiring one vasopressor, and hypoxia requiring low-flow nasal cannula is classified as having grade 3 CRS.*
[c] *Low-flow nasal cannula is defined as oxygen delivered at ≤6 L/min. Low-flow also includes blow-by oxygen delivery, sometimes used in pediatrics. High-flow nasal cannula is defined as oxygen delivered at >6 L/min.*
Organ toxicities associated with cytokine release syndrome (CRS) may be graded according to Common Terminology Criteria for Adverse Events version 5.0, but they do not influence CRS grading. BiPAP, bilevel positive airway pressure; CPAP, continuous positive airway pressure.

III. Interdisciplinary collaboration

Consult services: Consult services, including critical care, neurology, and infectious diseases, should be included in the care continuum to detect and manage toxicities specific to CAR T-cell therapy. Other services, such as interventional radiology, may also be critically involved, particularly in trials that involve regional injection of CAR T-cells.

Nursing service for CAR T-cell therapy: Nurses play a critical role in the intake, treatment, and monitoring of patients on CAR T-cell trials. Nurses involved in CAR T-cell trials are not only well versed in CAR constructs and protocol specifies but they also have knowledge of multiple other diseases. Nurses conduct appropriate patient selection, provide patient and family education, and identify and manage toxicities. They also serve as the primary point of communication between the patient and their families, investigators, and the research team.

Dedicated cellular therapy unit: Whereas some institutions administer CAR T-cells through the transplant program, other institutions have developed a separate clinical unit, such as a cellular therapy program, to manage patients receiving cellular therapy agents. A cellular therapy program helps ensure that protocols, procedures, quality management, and clinical outcomes are consistent. Regularly scheduled meetings to discuss patient toxicities and outcomes with a designated team of cellular therapy physicians, investigators, and nurses maintains cohesiveness and high standards of care throughout the service. Most importantly, such a program can reduce repetitive, inefficient methods and prevent delays in delivering cellular products and treatment of related adverse events.

Role of the primary oncology team: Collaboration with the primary oncology team is essential for the seamless handoff of patient care. If the patient is receiving therapy before leukapheresis or is receiving bridging therapy before the infusion, the primary oncologist needs to be aware of the washout periods to withhold treatment doses. The primary oncologist should also communicate to the CAR T-cell team of any adverse events that occur during therapy that may affect the patient's eligibility in the trial.

Pharmacy: Collaboration with pharmacies is critical to ensure the rapid availability of tocilizumab and suicide-switch agents to manage complications associated with CAR T-cell therapy. Satellite pharmacies are encouraged at treatment sites and on inpatient floors.

IV. Research and ancillary support

Ancillary staff are essential for the management of CAR T-cell trials. A dedicated research team is needed to ensure data integrity, maintain regulatory documents, and manage communication with the protocol sponsor, manufacturing facility, institutional review board, and FDA. A dedicated team of coordinators is also needed to manage the day-to-day of office triage and referral calls and clinic management.

V. Ongoing education

As CAR T-cell constructs and trials continue to develop, ongoing education is critical. It is important to educate the staff in emergency rooms, urgent care centers, intensive care units, and inpatient units on the management of patients receiving CAR T-cell therapy. Additionally, as CAR T-cell therapy is increasingly investigated in patients with solid tumors, the education of and learning processes among cell therapy unit personnel and solid tumor treatment staff will cross-connect and become mutual.

Where we are heading

We can expect the number of CAR T-cell agents in preclinical and clinical development to continue to rapidly increase. At present, most available cellular therapies that have been developed are autologous. As structural engineering and genome editing techniques become more sophisticated, the therapeutic potential of CAR T-cell therapy will further expand to include not only individualized treatments but also universally available, off-the-shelf CARs [41]. Autologous CAR T-cells, several of which have been approved by the FDA, have the benefit of bypassing the risk of graft vs host disease, a condition in which a donor's cells attack the recipient's cells. Challenges to the use of these autologous CARs include product variability, treatment delays, and individual production. The quantity and quality of cells collected from patients may be insufficient for the treatment of those with treatment-related myelosuppression, advanced disease, or progressive infections. The lengthy manufacturing time needed to produce suitable individualized CAR products may not always meet the short timeframe mandated by clinically urgent situations. In contrast, allogeneic, off-the-shelf CARs carry the inherent risks of graft vs host disease and rejection of the donor by the recipient's cells; however, they also have the potential to address issues of product inconsistency, treatment delays, manufacturing costs, and limited access. Since 2019, there has been a 74% increase in preclinical and a 91% increase in phase I development of allogenic agents [11]. Most current allogeneic products are produced from cells collected from the peripheral blood of healthy donors. Induced pluripotent stem cells, which have the capacity to self-renew and to generate T cells, have recently been identified as a novel allogeneic source [42]. Induced pluripotent stem cell therapies are still being investigated in preclinical studies but have the potential to become a readily available, banked product—a characteristic that could substantially mitigate costs and expand patient access to CAR immunotherapies.

Another future direction of investigation involves CAR T-cells used in combination with other agents, such as ICIs [43,44]. A combination strategy using mesothelin-targeted CAR T- cells has been investigated in mouse models of mesothelioma. Tumor cells upregulate PD-L1 in response to T cell-secreted cytokines, and CAR T-cells upregulate PD-1. Upregulated PD-L1 on tumor cells binds and inhibits T cells. The addition of anti-PD-1 agents after infusion of CAR T-cells resulted in enhanced antitumor efficacy and functional persistence of mesothelin CAR T-cells in mouse models [45]. These preclinical results are currently being studied in NCT02414269, in which patients are receiving the ICI agent pembrolizumab after CAR T-cell infusion.

As our understanding of immunotherapy continues to improve, the number of trials involving CAR T-cell therapy, ICIs, or combinations of immunotherapy continues to increase. Continued research into CAR design, mechanisms of exhaustion, and interactions with the immune microenvironment is critical. As we have shown, CAR T-cell trials are unique from trials of ICI agents and chemotherapies. To keep step with the rapid pace of development and not squander the work that has already been done, we must strive to learn from the challenges faced by ongoing trials and apply this knowledge to ensure the successful development and conduct of future trials of CAR T-cell therapy.

Funding support

P.S.A.'s laboratory work is supported by grants from the National Institutes of Health (P30 CA008748, R01 CA236615-01, and R01 CA235667), the U.S. Department of Defense (BC132124, LC160212, CA170630, and CA180889), the Comedy vs Cancer Foundation, the Dalle Pezze Foundation, the Derfner Foundation, the Geoffrey Beene Foundation, the Memorial Sloan Kettering Technology Development Fund, the Miner Fund for Mesothelioma Research, the Mr. William H. Goodwin and Alice Goodwin, the Commonwealth Foundation for Cancer Research, and the Experimental Therapeutics Center of Memorial Sloan Kettering Cancer Center.

Disclosures

Memorial Sloan Kettering Cancer Center has licensed intellectual property related to mesothelin-targeted CARs and T-cell therapies to ATARA Biotherapeutics and has associated financial interests.

References

[1] Suzuki K, Kadota K, Sima CS, et al. Clinical impact of immune microenvironment in stage I lung adenocarcinoma: tumor interleukin-12 receptor beta2 (IL-12Rbeta2), IL-7R, and stromal FoxP3/CD3 ratio are independent predictors of recurrence. J Clin Oncol 2013;31(4):490–8.

[2] Bograd AJ, Suzuki K, Vertes E, et al. Immune responses and immunotherapeutic interventions in malignant pleural mesothelioma. Cancer Immunol Immunother 2011;60(11):1509–27.

[3] Ujiie H, Kadota K, Nitadori JI, et al. The tumoral and stromal immune microenvironment in malignant pleural mesothelioma: A comprehensive analysis reveals prognostic immune markers. Onco Targets Ther 2015;4(6), e1009285.

[4] Vaddepally RK, Kharel P, Pandey R, Garje R, Chandra AB. Review of indications of FDA-approved immune checkpoint inhibitors per NCCN guidelines with the level of evidence. Cancers (Basel) 2020;12(3).

[5] Park JH, Riviere I, Gonen M, et al. Long-term follow-up of CD19 CAR therapy in acute lymphoblastic leukemia. N Engl J Med 2018;378(5):449–59.

[6] Schuster SJ, Svoboda J, Chong EA, et al. Chimeric antigen receptor T cells in refractory B-Cell lymphomas. N Engl J Med 2017;377(26):2545–54.

[7] Maude SL, Laetsch TW, Buechner J, et al. Tisagenlecleucel in children and young adults with B-cell lymphoblastic leukemia. N Engl J Med 2018;378(5):439–48.

[8] Bailey SR, Maus MV. Gene editing for immune cell therapies. Nat Biotechnol 2019;37(12):1425–34.

[9] Larson RC, Maus MV. Recent advances and discoveries in the mechanisms and functions of CAR T cells. Nat Rev Cancer 2021;21(3):145–61.

[10] Rafiq S, Hackett CS, Brentjens RJ. Engineering strategies to overcome the current roadblocks in CAR T cell therapy. Nat Rev Clin Oncol 2020;17(3):147–67.

[11] Yu JX, Upadhaya S, Tatake R, Barkalow F, Hubbard-Lucey VM. Cancer cell therapies: the clinical trial landscape. Nat Rev Drug Discov 2020;19(9):583–4.

[12] US National Library of Medicine. Clinicaltrials.gov. Accessed March 10, 2021.

[13] Morello A, Sadelain M, Adusumilli PS. Mesothelin-targeted CARs: driving T cells to solid tumors. Cancer Discov 2016;6(2):133–46.

[14] Jensen MC, Riddell SR. Designing chimeric antigen receptors to effectively and safely target tumors. Curr Opin Immunol 2015;33:9–15.

[15] Eyquem J, Mansilla-Soto J, Giavridis T, et al. Targeting a CAR to the TRAC locus with CRISPR/Cas9 enhances tumour rejection. Nature 2017;543(7643):113–7.

[16] Rupp LJ, Schumann K, Roybal KT, et al. CRISPR/Cas9-mediated PD-1 disruption enhances anti-tumor efficacy of human chimeric antigen receptor T cells. Sci Rep 2017;7(1):737.

[17] Zappasodi R, Merghoub T, Wolchok JD. Emerging concepts for immune checkpoint blockade-based combination therapies. Cancer Cell 2018;33(4):581–98.

[18] Howlader N, Forjaz G, Mooradian MJ, et al. The effect of advances in lung-cancer treatment on population mortality. N Engl J Med 2020;383(7):640–9.

[19] Schaft N. The landscape of CAR-T cell clinical trials against solid tumors-a comprehensive overview. Cancers (Basel) 2020;12(9).

[20] Adusumilli PS, Cherkassky L, Villena-Vargas J, et al. Regional delivery of mesothelin-targeted CAR T cell therapy generates potent and long-lasting CD4-dependent tumor immunity. Sci Transl Med 2014;6(261).

[21] Marin-Acevedo JA, Chirila RM, Dronca RS. Immune checkpoint inhibitor toxicities. Mayo Clin Proc 2019;94 (7):1321–9.

[22] Marrone KA, Ying W, Naidoo J. Immune-related adverse events From immune checkpoint inhibitors. Clin Pharmacol Ther 2016;100(3):242–51.

[23] Neelapu SS. Managing the toxicities of CAR T-cell therapy. Hematol Oncol 2019;37(Suppl 1):48–52.

[24] Kachala SS, Bograd AJ, Villena-Vargas J, et al. Mesothelin overexpression is a marker of tumor aggressiveness and is associated with reduced recurrence-free and overall survival in early-stage lung adenocarcinoma. Clin Cancer Res 2014;20(4):1020–8.

[25] Gauthier J, Bezerra ED, Hirayama AV, et al. Factors associated with outcomes after a second CD19-targeted CAR T-cell infusion for refractory B-cell malignancies. Blood 2021;137(3):323–35.

[26] Majzner RG, Mackall CL. Tumor antigen escape from CAR T-cell therapy. Cancer Discov 2018;8(10):1219–26.

[27] Dromain C, Beigelman C, Pozzessere C, Duran R, Digklia A. Imaging of tumour response to immunotherapy. Eur Radiol Exp 2020;4(1):2.
[28] Noaks E, Petticone C, Kotsopoulou E, Bracewell DG. Enriching leukapheresis improves T cell activation and transduction efficiency during CAR T processing. Mol Ther Meth Clin Dev 2021;20:675–87.
[29] Turtle CJ, Hanafi LA, Berger C, et al. CD19 CAR-T cells of defined CD4+:CD8+ composition in adult B cell ALL patients. J Clin Invest 2016;126(6):2123–38.
[30] Kiesgen S, Messinger JC, Chintala NK, Tano Z, Adusumilli PS. Comparative analysis of assays to measure CAR T-cell-mediated cytotoxicity. Nat Protoc 2021.
[31] Mock U, Nickolay L, Philip B, et al. Automated manufacturing of chimeric antigen receptor T cells for adoptive immunotherapy using CliniMACS prodigy. Cytotherapy 2016;18(8):1002–11.
[32] Zhu F, Shah N, Xu H, et al. Closed-system manufacturing of CD19 and dual-targeted CD20/19 chimeric antigen receptor T cells using the CliniMACS Prodigy device at an academic medical center. Cytotherapy 2018;20 (3):394–406.
[33] Apel M, Brüning M, Granzin M, et al. Integrated Clinical Scale Manufacturing System for Cellular Products Derived by Magnetic Cell Separation, Centrifugation and Cell Culture. Chem Ing Tech 2013;85(1-2):103–10.
[34] Shah NN, Zhu F, Taylor C, et al. A Phase 1 study with point-of-care manufacturing of dual targeted, tandem anti-CD19, anti-CD20 chimeric antigen receptor modified T (CAR-T) cells for relapsed, refractory, non-hodgkin lymphoma. Blood 2018;132(Supplement 1):4193.
[35] Fraietta JA, Beckwith KA, Patel PR, et al. Ibrutinib enhances chimeric antigen receptor T-cell engraftment and efficacy in leukemia. Blood 2016;127(9):1117–27.
[36] Common Terminology Criteria for Adverse Events (CTCAE) version 4.03. http://evs.nci.nih.gov/ftp1/CTCAE/CTCAE_4.03_2010-06-14_QuickReference_8.5x11.pdf. Published 2010. Accessed.
[37] Lee DW, Gardner R, Porter DL, et al. Current concepts in the diagnosis and management of cytokine release syndrome. Blood 2014;124(2):188–95.
[38] Porter D, Frey N, Wood PA, Weng Y, Grupp SA. Grading of cytokine release syndrome associated with the CAR T cell therapy tisagenlecleucel. J Hematol Oncol 2018;11(1):35.
[39] Neelapu SS, Tummala S, Kebriaei P, et al. Chimeric antigen receptor T-cell therapy - assessment and management of toxicities. Nat Rev Clin Oncol 2018;15(1):47–62.
[40] Lee DW, Santomasso BD, Locke FL, et al. ASTCT consensus grading for cytokine release syndrome and neurologic toxicity associated with immune effector cells. Biol Blood Marrow Transplant 2019;25(4):625–38.
[41] Ruella M, Kenderian SS. Next-generation chimeric antigen receptor T-cell therapy: going off the shelf. BioDrugs 2017;31(6):473–81.
[42] Themeli M, Kloss CC, Ciriello G, et al. Generation of tumor-targeted human T lymphocytes from induced pluripotent stem cells for cancer therapy. Nat Biotechnol 2013;31(10):928–33.
[43] Grosser R, Cherkassky L, Chintala N, Adusumilli PS. Combination immunotherapy with CAR T cells and checkpoint mlockade for the treatment of solid yumors. Cancer Cell 2019;36(5):471–82.
[44] Ramello MC, Haura EB, Abate-Daga D. CAR-T cells and combination therapies: what's next in the immunotherapy revolution? Pharmacol Res 2018;129:194–203.
[45] Cherkassky L, Morello A, Villena-Vargas J, et al. Human CAR T cells with cell-intrinsic PD-1 checkpoint blockade resist tumor-mediated inhibition. J Clin Invest 2016;126(8):3130–44.

CHAPTER

18

NKT cell: Success and promises in transplantation and immunotherapy

Samuel Zeng[a,*], *Zhe Li*[a,*], *Derek Lee*[a], *and Lili Yang*[a,b,c,d]

[a]Department of Microbiology, Immunology, and Molecular Genetics, University of California Los Angeles (UCLA), Los Angeles, CA, United States [b]Eli and Edythe Broad Center of Regenerative Medicine and Stem Cell Research, University of California Los Angeles (UCLA), Los Angeles, CA, United States [c]Jonsson Comprehensive Cancer Center, David Geffen School of Medicine, University of California Los Angeles (UCLA), Los Angeles, CA, United States [d]Molecular Biology Institute, University of California Los Angeles (UCLA), Los Angeles, CA, United States

Abstract

Natural killer T (NKT) cells comprise a small population of αβ T lymphocytes. They bridge the innate and adaptive immune systems and exert potent immunomodulatory functions against many diseases, including autoimmune/alloimmune reactions and cancer. This has made NKT cells attractive agents for reducing graft-vs-host disease in the context of allogeneic cell transplantation, as well as for treating cancer. Various approaches have been applied in clinical trials to target NKT cells, and the recent advances in chimeric antigen receptor have also shown great potential in increasing the efficacy of NKT cells. In this chapter, we first introduce the biology and importance of NKT cells in both immune regulation and antitumor immunity. Then, we summarize the preclinical data and clinical trials utilizing NKT cell-based immunotherapies. Finally, we discuss the challenges and future work that could be done to unleash the full potential of NKT cells for transplantation and cancer immunotherapy.

Abbreviations

allo-HCT	allogeneic hematopoietic cell transplantation
APCs	antigen-presenting cells
ATG	antithymocyte globulin
CAR	chimeric antigen receptor
CRS	cytokine release syndrome

*S.Z. and Z.L. contributed equally to this manuscript.

https://doi.org/10.1016/B978-0-12-822620-9.00016-1

DCs	dendritic cells
DN	CD4-CD8-
FasL	Fas ligand
GvHD	graft vs host disease
GvT	graft-vs-tumor
HSC	hematopoietic stem cell
IFN-γ	interferon γ
IL	interleukin
iNKT	invariant NKT
iPSC	induced pluripotent stem cell
KIR	killer cell immunoglobulin-like receptor
MDSCs	myeloid-derived suppressor cells
MHC	major histocompatibility complex
MIC	MHC class I chain-related molecule
MOAs	mechanisms of actions
NKR	NK receptor
NKT	natural killer T
TAMS	tumor-associated macrophages
TCR	T-cell receptor
Th	T helper
TLI	total lymphoid irradiation
TME	tumor microenvironment
TNFα	tumor necrosis factor α
TRAIL	TNFα-related apoptosis-inducing ligand
ULBP	unique long-binding protein
α-GalCer	α-galactosylceramide

Conflict of interest

L.Y. is a founder, board member, and stockholder of Appia Bio and a scientific advisory board member of Amberstone Biosciences. All other authors declare no competing interests. None of the declared companies contributed to or directed any of the research reported in this article.

Biology of NKT cells

Natural killer T (NKT) cells are a unique subset of αβ T lymphocytes characterized by their expression of both an αβ T-cell receptor (TCR) and NK surface markers. This small population of T cells (~0.1%–1% in mouse blood and ~0.01%–1% in human blood) arise from the same lymphoid precursor pool as conventional T cells and mature in the thymus [1,2]. However, unlike conventional T cells that recognize peptide antigens, NKT cells respond to lipid antigens presented by the nonpolymorphic major histocompatibility complex (MHC) class I-like molecule CD1d. During an immune response, NKT cells are some of the first cells to be activated and can rapidly produce copious amounts of cytokines and chemokines, thereby functioning as a "bridge" linking the innate and adaptive immune responses [3].

Subtypes of NKT cells

Two major categories of NKT cells have been identified based on their TCR expression. Type I or invariant NKT (iNKT) cells express a restricted TCR recombinant comprising a semi-invariant TCR α chain (Vα14-Jα18 in the mouse, Vα24-Jα18 in humans) paired to a limited repertoire of Vβ chains (Vβ 2, 7, or 8.2 in mice and Vβ 11 in humans) and are the most prevalent type of NKT cells [2,3]. iNKT cells can be further divided based on their CD4 and CD8 expressions. Human have $CD4^+CD8^-$ ($CD4^+$), $CD4^-CD8^+$ ($CD8^+$), and $CD4^-CD8^-$ (DN) iNKT populations, whereas mice only have $CD4^+$ and DN populations [2]. For both mice and humans, iNKT cells are also functionally defined by their ability to recognize the prototypic antigen α-galactosylceramide (α-GalCer) presented on CD1d [4,5]. α-GalCer is a strong activator of iNKT cells and has been widely used to study iNKT cell biology. Type II NKT cells, in contrast, express more diverse TCRs and are therefore also called variant NKT cells [6–8]. Notably, Type II NKT cells do not respond to α-GalCer. Given this and the lack of specific markers, study of this subpopulation has been more challenging. Currently, no experimental tools are available to identify and analyze the entire population of Type II NKT cells.

From studies of animal models and humans, both subsets of NKT cells have been shown to play diverse roles in the immune system, sometimes synergistically but also antagonistically. However, because Type I NKT cells are better characterized and their functions better studied, the remaining chapter will focus on discussing Type I NKT cells (iNKT cells).

Activation of iNKT cells

There are three main mechanisms to activating iNKT cells: (1) via TCR/CD1d/lipid stimulation, (2) via cytokine stimulation, and (3) via activating NK receptor stimulation [9–12]. iNKT cells can be directly activated in a TCR-dependent manner and without the need for costimulation or cytokine receptor engagement through exposure to potent glycolipid antigens (e.g., α-GalCer) presented by CD1d [3]. However, in a more physiological context, iNKT cells are stimulated by microbial or self-lipid antigens that are weakly immunogenic and thus requires a second activation signal from pro-inflammatory cytokines [13]. Sometimes, cytokines alone (e.g., IL-12, IL-18, IL-23, and IL-25) are sufficient to activate iNKT cells even in the absence of TCR engagement. Several studies have found that iNKT cells release IFN-γ when stimulated by IL-12 during viral infections [14,15]. The heightened capacity of iNKT cells to respond to cytokines is due to their elevated expression of functional receptors at baseline. Because iNKT cells also express activating and inhibitory NK receptors (NKR) and killer cell immunoglobulin-like receptors (KIR), they can become activated when the summation of stimulatory signals overcomes the inhibitory signals. Activating NKRs recognize a variety of MHC-like molecules and cellular targets frequently referred to as "stress proteins." For example, DNAM-1 recognizes the poliovirus receptor and Nectin-2; the NKG2D receptor recognizes MHC class I-like molecules (MIC) A and B and unique long-binding proteins (ULBPs). Inhibitory NKRs and KIRs, in contrast, often bind to HLA molecules [12].

Effector functions of iNKT cells

In contrast to conventional T cells that emerge from the thymus "naïve," recently emigrated iNKT cells are able to perform their effector functions immediately without priming [2,16,17]. Activated NKT cells are known to secrete copious amounts of cytokines, including T helper (Th)1-like (IFN-γ), Th2-like (IL-4, IL-13), Th17-like (IL-17, IL-22), and regulatory (IL-10) cytokines [2,18]. The mechanisms of cell activation, the location, and the iNKT cell subsets all affect what cytokines are produced. For example, the activation of iNKT cells with the potent agonist α-GalCer leads to the production of both Th1- and Th2-like cytokines, whereas activation involving both recognition of the endogenous ligand/CD1d complexes and cytokine costimulation leads to the polarized production of Th1-like IFN-γ, but not Th2-like cytokines [13,19]. Additionally, both human $CD8^+$ and DN iNKT cells have been found to predominately express Th1 cytokines, whereas human $CD4^+$ iNKT cells predominately express Th2 cytokines, although this distinction is less apparent with mouse iNKT cells [20,21]. Interestingly, subsets expressing different cytokines, transcription factors, and surface markers appear to be acquired during thymic development rather than as a result of peripheral experience [2,22].

In addition to cytokines, iNKT cells also produce cytolytic proteins such as perforin and granzyme B, and surface molecules involved in cytotoxicity such as Fas ligand (FasL) and tumor necrosis factor α (TNFα)-related apoptosis-inducing ligand (TRAIL) [23,24]. Owing to their ability to produce an array of effector molecules, iNKT cells can profoundly influence many other cell types, including dendritic cells (DCs), macrophages, neutrophils, NK cells, and T and B cells and orchestrate immune responses during infection, autoimmune disease, allergy and cancer [2]. Because they affect many different immunological processes, iNKT cells have great clinical potentials, most notably their ability to reduce graft vs host disease (GvHD) without loss of graft-vs-tumor (GvT) in allogeneic transplantation and their capability to enhance antitumor immunity [2,3,25,26].

Mechanisms and relevant studies of iNKT cells suppressing GvHD in allogeneic transplantation

Allogeneic hematopoietic cell transplantation (allo-HCT) remains the only curative treatment for many patients with hematological malignancies. However, development of GvHD continues to be the leading cause of morbidity and mortality, making identification of mechanism to prevent GvHD without loss of GvT of great clinical significance [27,28]. Multiple studies have identified iNKT cells to be capable of fulfilling this unique niche [29].

Murine studies

Preclinical studies have demonstrated iNKT cells can simultaneously prevent GvHD while retaining GvT effect following allo-HCT. Selective depletion of host conventional T cells through fractionated total lymphoid irradiation (TLI) and antithymocyte globulin (ATG), and thereby selectively enriching host iNKT cells, prior to allo-HCT, attenuated GvHD through host iNKT cell elevation of IL-4 secretion and polarization of donor T cells toward a Th2 cytokine pattern. This, in turn, inhibited donor T cell early expansion and infiltration of GvHD target organs [30]. Administration of synthetic iNKT TCR ligand α-GalCer on the day of transplantation was also found to significantly reduced mortality and morbidity from GvHD [31–34]. Likewise, the addition of donor or third-party $CD4^+$ iNKT cells to the allograft was found to prevent GvHD by inhibiting T-cell proliferation, promoting Th2-biased cytokine response, and expanding donor MDSCs [35,36]. Notably, enrichment of the iNKT cells, whether host, donor or third party, did not abrogate donor T-cell GvT [30,35,36]. In all these studies, it was found that iNKT cells promoted donor T cell polarization toward a Th2 cytokine pattern, expanded donor Treg populations, and significantly reduced production of IFN-γ and tumor necrosis factor (TNF)-alpha [30,34,36,37]. Myeloid-derived suppressor cells (MDSCs) or $CD8^+$ (DC were also found to play important roles in the interplay between iNKT cells and Tregs [36,38].

Human studies

The protective role of human iNKT cells against GvHD has also been highlighted by several clinical studies. Nonmyeloablative conditioning with TLI/ATG prior to allo-HCT was associated with a higher iNKT/T cell ratio, increased IL-4 production, decreased incidence of GvHD, and retained GvT effect [27,39]. Patients with acute GvHD were found to have reduced numbers of total iNKT cells [40], whereas enhanced iNKT cell reconstitution following allo-HCT positively correlated with reduction in GvHD without loss of GvT effect [41]. Separately, low $CD4^-$ iNKT cell numbers in donor graft was associated with clinically significant GvHD in patients receiving HLA-identical allo-HCT [42]. Thus, increasing the iNKT cell numbers of allograft that contain few iNKT cells may provide an attractive strategy for suppressing GvHD while preventing leukemia relapse.

Human $CD4^+$ and $CD4^-$ iNKT cells likely mediate distinct effects that collectively results in a beneficial effector response for transplant recipients. $CD4^-$ iNKT cells are Th1 biased-secreting more IFN-γ than IL-4 and preferentially expressing perforin—and is thought to both promote GvT and suppress GvHD. Supporting this notion are in vitro studies demonstrating that $CD4^-$ iNKT cells display direct cytotoxicity against CD1d-expressing mature myeloid DCs [42]. Human iNKT cells also express KIRs, including KIRDL4, KIR3DL2 and KIR2DL1 [43]. In an allogeneic setting, donor $CD4^-$ iNKT cells may kill host APCs in a TCR-CD1d and KIR-dependent manners to downregulate GvHD. $CD4^+$ iNKT cells, on the other hand, can ameliorate GvHD through production of IL-4 [20], polarizing pathogenic donor T cells toward an anti-inflammatory Th2 response and promoting the expansion of regulatory T cells [32,44].

Phases 1 and 2A clinical trials aimed at providing α-GalCer to recipients at the time of transplantation to reduce incidence of GvHD have revealed the approach to be safe, although further research is needed to determine the most appropriate transplant setting and optimal dosing strategy [45].

Mechanisms of iNKT cells in antitumor immunity

Recent progress in understanding the activation and effector functions of iNKT cells has led to an increased appreciation of their antitumor potentials [10,17,26,46]. There are compelling evidences suggesting the significant role of iNKT cells in tumor surveillance. In humans, the frequency of iNKT cells is decreased in patients with solid tumors (including melanoma, colon, lung, breast, and head and neck cancers) and hematologic malignancies (including leukemia, multiple myeloma, and myelodysplastic syndromes), and increased iNKT cell numbers are associated with a better prognosis [12,47,48]. In addition to correlative human data, the role of iNKT cells in exerting antitumor activity has been well characterized using murine models of tumors [47,49]. Direct tumor killing, cytokine-mediated regulation of effector cells, and the modulation of immunosuppressive tumor microenvironment (TME) are currently considered the three major mechanisms of action (MOAs) of iNKT cells in combating cancer [26].

Direct cytotoxicity against tumor cells

iNKT cells are known to exert a direct tumor killing effect against $CD1d^+$ tumors, mainly through the production of cytolytic molecules such as perforin and granzyme B [23,50,51], and the interaction of death-inducing receptors including Fas and TRAIL receptors [24,52,53]. The CD1d expression level in tumor cells has been demonstrated to be an important determinant for iNKT cell-mediated cytotoxicity. The correlation between reduced CD1d expression and enhanced tumor progression has been reported in a variety of tumor models [54–60], indicating that CD1d downregulation may be an important cause of tumor evasion from iNKT cell-mediated immunosurveillance. More than half of all iNKT cells also express NK activating receptors, such as NKG2D, enabling direct cytotoxicity against tumors expressing corresponding ligands [20,61].

Regulation of antitumor effector cells

In addition to their capacity to directly kill malignant cells, iNKT cells are known to enhance the immunogenic activities of a variety of immune cell subsets [62]. As mentioned above, iNKT cells can rapidly produce copious amounts of Th1 and Th2 cytokines upon activation, leading to reciprocal activation and modulation of both innate and adaptive immune cells [2,63–66]. One of the major contributions of iNKT cells to immune surveillance is linked to DC maturation. Most DCs found in the TME are immature, and maturation of DCs is essential to initiating a sufficient T cell mediated response [67]. Indeed, activation of iNKT cells causes the upregulation of the IL-12 receptor and CD40L and induces the maturation of DCs via a CD40-CD40L interaction. Mature DCs then express higher costimulatory molecules CD40, CD80, CD86, CD70 and release more IL-12, which in turn further activate iNKT cells and amplify IFN-γ responses, leading to a positive feedback loop for Th1 immunity [68–70]. Furthermore, maturated DCs can induce the transactivation of NK cells and contributes to the priming of CD8 T cells [71,72], allowing the establishment of both innate and adaptive immune responses to eliminate MHC-negative and -positive tumor cells, respectively.

Modulation of immunosuppressive TME

iNKT cells can also augment antitumor immunity by counteracting cells contributing to an immunosuppressive TME. The immunosuppressive microenvironment not only promotes tumor growth and migration but also helps the tumor cells evade the surveillance of the host immunity and resists immunotherapy [73,74]. Tumor-associated macrophages (TAMS) constitute a significant portion of cell populations in the TME and serve as major tumor-promoting immune cells [75]. It has been shown that iNKT cells can kill CD1d-expressing TAMs in primary human neuroblastoma samples, in part by relieving the immunosuppressive TME and limiting metastases [76]. In addition to TAMs, iNKT cells can also alter the numbers and effects of myeloid-derived suppressor cells (MDSCs), also known to be a population of myeloid cells with immunosuppressive properties [77]. In a model of influenza A virus infection, adoptive transfer of iNKT cells reduced the immunosuppressive activity of MDSCs [78]. In another study using murine tumor models, iNKT cells facilitated the conversion of immunosuppressive

MDSCs into immunogenic antigen-presenting cells (APCs), eliciting successful antitumor immunity and providing the basis for alternative cell-based vaccines [79].

iNKT cell-based immunotherapy for treating cancer

Preclinical studies have shown iNKT cells can induce strong antitumor immune responses. Activated iNKT cells increase IFN-γ production and cytokine secretion, which further enhances activation of other immune cells such as NK cells, T cells, and B cells. Direct administration of α-GalCer, activation of iNKT cells via α-GalCer-loaded DCs, autologous transfer of expanded iNKT cells, CAR-iNKT cells, hematopoietic stem cell (HSC) derived iNKT cells, and induced pluripotent stem cell (iPSC)-derived iNKT cells have all been studied as potential cancer treatments.

α-GalCer-loaded DCs therapy

Multiple clinical trials based on α-GalCer-DC transfer have demonstrated the clinically relevant antitumor effects and the safety of this approach. Several preclinical and clinical studies have reported using ex vivo stimulation and loading of autologous DCs with α-GalCer prior to administration of DC vaccines [80–91]. A detailed analysis showed that injecting these vaccines into mice can augment the frequency of iNKT cells and circulating IFN-γ producing cells in a solid tumor preclinical model [83]. In humans, α-GalCer-loaded DCs can increase the infiltration of lymphocytes into the tumor microenvironment and promote iNKT cell-induced immune memory upon secondary administration [86]. In a Phase I/II trial, Chang et al. intravenously injected monocyte-derived mature DCs loaded with α-GalCer into five patients with advanced myeloma [88]. In all patients, the frequency of iNKT cells increased over a 100-fold and the number of iNKT cells stayed above baseline for 3 months. Elevated IFN-γ levels were also detected. These data demonstrate the feasibility of using α-GalCer-loaded DCs to improve iNKT-based therapies, including iNKT cells, in clinical settings.

Autologous adoptive transfer therapy

The autologous adoptive transfer approach seeks to increase iNKT cell numbers in cancer patients by harvesting patients' own PBMCs, expanding the PBMC iNKT cells by stimulation with α-GalCer and/or cytokines, and then infusing the enriched iNKT cell population back into the patient. This approach has been shown to be more effective in expanding iNKT cells than i.v. administration of α-GalCer [92–95]. In a preclinical study, adoptive transfer of IL-12 activated iNKT cells prevented hepatic metastasis of melanoma in mice [83]. This study also suggested the involvement of direct cytotoxic mechanisms rather than cytokine-mediated immune responses at the effector phase of the iNKT cell-mediated antitumor activity.

Motohashi et al. performed a Phase I study with autologous activated iNKT cell therapy for nonsmall cell lung cancer patients [92]. No severe side effects were observed during this study and the number of IFN-γ producing cells in the blood increased in some patients after the administration of activated iNKT cells. This study demonstrated the safety and feasibility of this approach, although no patient developed a partial or complete response. Subsequent adoptive iNKT cell-based clinical trials began treating patients with α-GalCer loaded DCs in addition to iNKT cells [93–95]. In these studies, the treatments did not cause major adverse effects: only one patient had a grade 3 toxicity in a phase II trial of 10 patients [94]. Increased numbers of iNKT cells were seen after several infusions, but the effects were transient in the same phase II study. At the end of this trial, some patients had no evidence of disease, some had partial responses, and some died of disease [94]. These clinical trials demonstrated that the adoptive transfer of autologous iNKT cells was feasible, well tolerated, and had some beneficial clinical effects. However, further studies are clearly needed to improve clinical outcomes.

A variation on this approach, the Yang laboratory sought to increase circulating iNKT cells through infusion of TCR-engineered HSCs [96,97]. Smith et al. demonstrated the feasibility of dramatically increasing iNKT cells in mice through TCR gene engineering of autologous HSCs [96]; Zhu et al. utilized a BLT (human bone marrow-liver-thymus engrafted NOD/SCID/$\gamma c-/-$ mice) humanized mouse model to support the engraftment of engineered human HSCs and give rise to human iNKT cells [97]. iNKT TCR-engineered HSCs could generate a clonal population of iNKT cells. These HSC-iNKT cells closely resembled endogenous iNKT cells, could deploy multiple mechanisms to attack tumor cells, and effectively suppressed tumor growth in vivo in multiple tumor mouse models [97]. Preclinical safety studies showed no toxicity or tumorigenicity of the HSC-iNKT cell therapy [97].

CAR-iNKT therapy

T cells engineered with chimeric antigen receptors (CARs) represent a novel class of immunotherapeutics that have shown promising clinical results and have obtained FDA approval for treating blood cancers such as acute lymphoblastic leukemia and B cell lymphoma [98]. CAR T-cells can respond quickly to CAR-specific tumor antigenic stimulation and rapidly produce cytokines to enable an effective antitumor response. However, there are several major limitations to the current CAR T-cell therapy, including serious side effects such as the cytokine release syndrome (CRS), lethal neurotoxicity, and potential antigen escape. Additionally, the therapy is also less effective against solid tumors than against blood cancers [99]. Furthermore, allogeneic application of CAR-T therapy is restricted by concerns of GvHD [100,101].

iNKT cells as a platform for CAR immunotherapy represents a novel approach, as the addition of a CAR increases their tumor targeting specificity, in vivo persistence, and potentially tumor infiltrating capability [102]. Importantly, owing to antitumor capabilities inherent to iNKT cells, CAR-iNKT cells can theoretically attack tumor cells through multiple mechanisms, thereby reducing the chance for antigen escape. In addition to the specificity provided by the CAR, their endogenous TCR would allow iNKT cells to target $CD1d^+$ tumors, using either perforin, granzyme B, Fas ligand, or TNF-α-mediated cytotoxic pathways. iNKT cells also have indirect antitumoral effects by targeting $CD1d^+$ TAMS and tumor-promoting myeloid cells [60,76,103]. In addition, iNKT cells express NK receptors such as NKG2D, enabling direct cytotoxicity against tumors expressing NKG2D ligands [20,61].

iNKT cells also have several other advantages as CAR carriers, including their ability to target nonlymphoid tumors: their interaction with tissue chemokines CCL2 and CCL20 allows them to migrate into nonlymphoid tissues [104,105]. It has also been shown that iNKT cells can promote the maturation of DCs through CD40-CD40L and CD1d/lipid antigen-TCR interactions [106]. In the tumor microenvironment, where most DCs are immature, maturation of DCs in turn can lead to DC activation of NK cells and iNKT cells by expressing costimulatory molecules (CD40, CD80, and CD86) and cytokines, further enhancing the antitumor effects of these cells [107]. Additionally, it is well established that iNKT cells have limited capacity to cause GvHD; instead, iNKT cell presence is associated with reduced GvHD in clinical trials [40,41,108,109]. The presence of iNKT cells after hematopoietic stem cell transfer is predictive for survival with a reduction in GvHD [40,41,108,109]. Furthermore, iNKT cells interact with the monomorphic CD1d [110–114] and are therefore not MHC-restricted, making them highly suitable for "off-the-shelf" therapy [4,115,116].

The Metelitsa group provides the first evidence for the feasibility of engineering iNKT cells with CAR and expanding them to a clinical scale. GD2-CARs and CD19-CARs were inserted into human iNKT cells and tested against neuroblastoma and $CD19^+$ lymphoma in vitro and in preclinical murine models [117,118]. CAR-iNKT cells were capable of killing target tumor cells without causing GvHD, whereas conventional CAR-T cells cleared the tumor, but recipients also succumbed to GvHD. Furthermore, the endogenous TCR of CAR-iNKT cells remained functional, such that they were also able to kill $CD1d^+$ tumor cells [117]. A subsequent study published by the same group coexpressed IL-15 with GD2-CAR in iNKT cells [119]. Coexpression of IL-15 reduced expression of exhaustion markers in CAR-iNKT cells and enhanced in vivo persistence and localization to tumor sites. Owing to their results, the first-in-human CAR-iNKT cell clinical trial is underway [119]. In their interim report consisting of three patients, the group observed no dose-limiting toxicities, expansion of CAR-iNKT cells in vivo, and localized to tumors [120]. One patient also demonstrated an objective response with regression of bone metastatic lesions. These initial results suggest that CAR-iNKT cells can be expanded to clinical scale and safely applied to treat patients with cancer. Another study reported by the Karadimitris group further demonstrated that CD19-CAR iNKT cells can be activated by CD1d and CD19-dependent stimulation, which provided a dual targeting strategy for CD19-CAR-iNKT cells against CD1d expressing lymphomas in vitro and in vivo [121].

Stem cell-derived iNKT therapy

As previously noted, adoptive CAR T-cell therapy is currently limited to autologous administration. Having to make individualized CAR T-cell therapy is expensive, time consuming, and labor intensive [115]. In addition, many cancer patients are nonideal candidates owing to the chemotherapy or irradiation they received beforehand that could negatively impact the quality of their T cells. The use of donor-derived allogeneic CAR therapy provides an opportunity to treat cancer patients with "off-the-shelf" cell products. However, the risk of GvHD remains a major concern for allogeneic CAR T-cell therapy due to HLA mismatches between the donor and the patient [115,122]. Advances in gene editing have allowed scientists to remove the endogenous TCR from allogeneic CAR T-cells, lowering the chances of developing GvHD. However, 100% endogenous TCR removal is not guaranteed. Because having less than 1% of TCR-expressing T cells is sufficient to cause GvHD, patients receiving these products remain at risk for developing GvHD [115,122]. Although NK cells, iNKT cells, and γδ T cells may serve as better candidates for allogeneic adoptive transfer therapy, cell products derived from donors still have residual conventional αβ T cells.

To overcome these issues, allogeneic cell therapy generated from self-renewal stem cells, such as iPSCs, could be an alternative approach. iPSCs can be stored as master cell banks, continuously differentiating and generating cells for treatments, which avoids using primary lymphoid cells from patients or healthy donors as the cell source [123]. This approach standardizes manufacturing procedures and provides a solution to donor-to-donor variations, cell number, and dosage limitations of the cell products. iPSCs can also be genetically modified by gene editing technology prior to the differentiation step. Genetically engineered cell clones can be isolated and expanded as stable cell sources in which all clones contain 100% purity of desirable characteristics [123].

It has been demonstrated that T cells and NK cells can be successfully differentiated from iPSCs in vitro [123–130]. Recently, the Kaneko group reported the generation of human iNKT cells from iPSCs [131]. They reprogrammed iNKT cells to pluripotency and then re-differentiated the cells into iNKT cells in vitro. The iPSC-derived iNKT cells showed TCR-dependent proliferation and IFN-γ production after α-GalCer stimulation. In vitro cytotoxicity against the tumor cell lines K562 (leukemia) and U937 (lymphoma) was also observed [131]. In a separate study, the Fujii group also generated functional iPSC-iNKT cells and then tested them in vitro and in vivo [132]. These cells possessed significant antitumor activity in a K562 xenograft mouse model and activated human NK cells by adjuvant effect [132]. The data from this preclinical study also led to the first clinical trial using iPSC-derived iNKT cells to treat patients with head and neck cancer. This Phase I trial is expected to last for 2 years and involves 4–18 patients.

Challenges and perspectives

Recent progress in our understanding of iNKT cells has highlighted the clinical potential of iNKT cells for both allogeneic transplantation and iNKT cell-based cancer therapies. However, promising findings in preclinical studies have not yet convincingly translated to similar outcomes in human clinical trials. On one hand, application of iNKT cells for the prevention of GvHD in allo-HCT has remained largely limited to correlative studies, with a significant barrier being the relative inaccessibility of iNKT cells; comprising 0.1%–1% of human PBMCs, it remains an enormous effort to extract a sufficient number of donor iNKT cells and would also significantly complicate the already strenuous task of obtaining donor hematopoietic stem cells. Notably, most preclinical studies of donor iNKT cells obtain the necessary quantity through isolation from donor spleen and liver, which understandably, are inaccessible clinically. Expansion of host iNKT cells via α-GalCer stimulation underwent Phases I and II clinical trials showing limited side effects, but also limited efficacy [45]. And unfortunately, third-party iNKT cells that can be used to supplement the allograft remain under development.

Nevertheless, the inaccessibility of iNKT cells is likely to be resolved in the near future given the enormous interest in iNKT cells, particularly for their potential in cancer immunotherapy.

Many challenges, however, also limit the overall performance of iNKT cell-based immunotherapies, including insufficient antitumor activity, limited in vivo persistence and immunosuppressive TME. Various approaches are currently being explored to address these concerns.

The addition of CARs to iNKT cells to more potently direct iNKT cell-mediated cytotoxicity against refractory tumors holds promise to revolutionize cancer treatment. This approach has been demonstrated to enhance antitumor efficacy against neuroblastoma in preclinical studies and is under further investigation in clinical trials. Compared to conventional CAR-T therapy, CAR-iNKT cells are less likely to cause GvHD. However, iNKT are also highly inflammatory and may induce the cytokine release/storm syndrome (CRS) as a side-effect to rapid tumor lysis. Indeed, CRS is a common concern, especially for all CAR-T therapies. So far, the fast-evolving CAR-T cell therapy has accumulated valuable clinical experiences on managing CRS (e.g., anti-IL-6 antibody treatment) that can be adapted for guiding the iNKT cell-based therapies.

Expansion and persistence of iNKT cells following infusion appear to be another determinant of clinical response. Importantly, both mouse and human studies have highlighted the central role of IL-15 in iNKT cell homeostasis [133,134]. So far, transgenic expression of IL-15 in adoptively transferred iNKT cells has improved iNKT in vivo persistence without causing significant toxicity and is therefore a potential approach for general application to maximize NKT persistence and efficacy [119].

Although iNKT cells are known to modulate immunosuppressive cells in the TME, they are frequently suppressed in cancer patients due to nutrient deprivation, hypoxia, acidity, and accumulation of toxic by-products of catabolism [74]. One strategy that can partially reverse the dysfunction of iNKT cells is through blocking inhibitory pathways such as PD-1 and CTLA-4. While these "immune checkpoints" are predominantly defined in the context of $CD8^+$ T cells, iNKT cells also showed upregulated PD-1 expression following stimulation with α-GalCer [135–137]. Indeed, blockade of PD-1 at the time of α-GalCer injection prevents the anergy of iNKT cells [135–137]. Considering the broad use of checkpoint inhibitors in the clinic, future combination treatments may synergize the antitumor efficacies of iNKT cells. Another strategy involves the blockade of inhibitory cytokines present in the TME, such as TGF-β. The use of dominant negative receptors on T cells is also being actively explored and has shown some promises [138], suggesting a potential role in enhancing iNKT cell-mediated antitumor immunity.

Furthermore, the possibility of third-party "off-the-shelf" iNKT products derived from standardized, allogeneic sources can improve the practicality of iNKT cell therapy. With increasing focus on improving the persistence and functions of iNKT cells, it is likely that iNKT cells will move to the forefront of immunotherapy over the next few years.

Acknowledgments

L.Y. receives research funding support from The University of California, Los Angeles (BSCRC Innovation Awards, BSCRC/JCCC Ablon Scholars Award), National Institutes of Health (DP2 CA196335, P50 CA092131, P01 CA132681), California Institute for Regenerative Medicine (RB5-07089, TRAN1-08533, DISC2-11157), Stop Cancer Foundation (Research Career Development Award), and Prostate Cancer Foundation (GTSN Challenge Award for Lethal

Prostate Cancer). S.Z is a predoctoral fellow supported by the University of California, Los Angeles Medical Scientist Training Program Grant (T32-GM008042). Z.L. is a postdoctoral fellow supported by the University of California, Los Angeles Tumor Immunology Training Grant (USHHS Ruth L. Kirschstein Institutional National Research Service Award #T32 CA 009120). D.L. is a predoctoral fellow supported by the University of California, Los Angeles Microbial Pathogenesis Training Grant (Ruth L. Kirschstein National Research Service Award AI007323).

References

[1] Godfrey DI, Stankovic S, Baxter AG. Raising the NKT cell family. Nat Immunol 2010;11(3):197–206.
[2] Brennan PJ, Brigl M, Brenner MB. Invariant natural killer T cells: an innate activation scheme linked to diverse effector functions. Nat Rev Immunol 2013;13(2):101–17.
[3] Bendelac A, Savage PB, Teyton L. The biology of NKT cells. Annu Rev Immunol 2007;25:297–336.
[4] Kawano T, Cui J, Koezuka Y, Toura I, Kaneko Y, Motoki K, et al. CD1d-restricted and TCR-mediated activation of Vα14 NKT cells by glycosylceramides. Science 1997;278(5343):1626–9.
[5] Brossay L, Chioda M, Burdin N, Koezuka Y, Casorati G, Dellabona P, et al. CD1d-mediated recognition of an α-galactosylceramide by natural killer T cells is highly conserved through mammalian evolution. J Exp Med 1998;188(8):1521–8.
[6] Dasgupta S, Kumar V. Type II NKT cells: a distinct CD1d-restricted immune regulatory NKT cell subset. Immunogenetics 2016;68(8):665–76.
[7] Marrero I, Ware R, Kumar V. Type II NKT cells in inflammation, autoimmunity, microbial immunity, and cancer. Front Immunol 2015;6:316.
[8] Dhodapkar MV, Kumar V. Type II NKT cells and their emerging role in health and disease. J Immunol 2017;198 (3):1015–21.
[9] Kumar A, Suryadevara N, Hill TM, Bezbradica JS, Van Kaer L, Joyce S. Natural killer T cells: an ecological evolutionary developmental biology perspective. Front Immunol 2017;8:1858.
[10] Bedard M, Salio M, Cerundolo V. Harnessing the power of invariant natural killer T cells in cancer immunotherapy. Front Immunol 2017;8:1829.
[11] Reilly EC, Wands JR, Brossay L. Cytokine dependent and independent iNKT cell activation. Cytokine 2010;51 (3):227–31.
[12] Krijgsman D, Hokland M, Kuppen PJ. The role of natural killer T cells in cancer—a phenotypical and functional approach. Front Immunol 2018;9:367.
[13] Brigl M, Bry L, Kent SC, Gumperz JE, Brenner MB. Mechanism of CD1d-restricted natural killer T cell activation during microbial infection. Nat Immunol 2003;4(12):1230–7.
[14] Wesley JD, Tessmer MS, Chaukos D, Brossay L. NK cell–like behavior of Vα14i NK T cells during MCMV infection. PLoS Pathog 2008;4(7), e1000106.
[15] Tyznik AJ, Tupin E, Nagarajan NA, Her MJ, Benedict CA, Kronenberg M. Cutting edge: the mechanism of invariant NKT cell responses to viral danger signals. J Immunol 2008;181(7):4452–6.
[16] Rossjohn J, Pellicci DG, Patel O, Gapin L, Godfrey DI. Recognition of CD1d-restricted antigens by natural killer T cells. Nat Rev Immunol 2012;12(12):845–57.
[17] Wolf BJ, Choi JE, Exley MA. Novel approaches to exploiting invariant NKT cells in cancer immunotherapy. Front Immunol 2018;9:384.
[18] Stetson DB, Mohrs M, Reinhardt RL, Baron JL, Wang Z-E, Gapin L, et al. Constitutive cytokine mRNAs mark natural killer (NK) and NK T cells poised for rapid effector function. J Exp Med 2003;198(7):1069–76.
[19] Nagarajan NA, Kronenberg M. Invariant NKT cells amplify the innate immune response to lipopolysaccharide. J Immunol 2007;178(5):2706–13.
[20] Gumperz JE, Miyake S, Yamamura T, Brenner MB. Functionally distinct subsets of CD1d-restricted natural killer T cells revealed by CD1d tetramer staining. J Exp Med 2002;195(5):625–36.
[21] Lee PT, Benlagha K, Teyton L, Bendelac A. Distinct functional lineages of human Vα24 natural killer T cells. J Exp Med 2002;195(5):637–41.
[22] Watarai H, Sekine-Kondo E, Shigeura T, Motomura Y, Yasuda T, Satoh R, et al. Development and function of invariant natural killer T cells producing TH 2- and TH 17-cytokines. PLoS Biol 2012;10(2), e1001255.
[23] Kawano T, Cui J, Koezuka Y, Toura I, Kaneko Y, Sato H, et al. Natural killer-like nonspecific tumor cell lysis mediated by specific ligand-activated Vα14 NKT cells. Proc Natl Acad Sci 1998;95(10):5690–3.

[24] Wingender G, Krebs P, Beutler B, Kronenberg M. Antigen-specific cytotoxicity by invariant NKT cells in vivo is CD95/CD178-dependent and is correlated with antigenic potency. J Immunol 2010;185(5):2721–9.

[25] Terabe M, Berzofsky JA. The immunoregulatory role of type I and type II NKT cells in cancer and other diseases. Cancer Immunol Immunother 2014;63(3):199–213.

[26] Nair S, Dhodapkar MV. Natural killer T cells in cancer immunotherapy. Front Immunol 2017;8:1178.

[27] Lowsky R, Takahashi T, Liu YP, Dejbakhsh-Jones S, Grumet FC, Shizuru JA, et al. Protective conditioning for acute graft-versus-host disease. N Engl J Med 2005;353(13):1321–31.

[28] Magenau J, Reddy P. Next generation treatment of acute graft-versus-host disease. Leukemia 2014;28(12):2283–91.

[29] Mavers M, Maas-Bauer K, Negrin RS. Invariant natural killer T cells as suppressors of graft-versus-host disease in allogeneic hematopoietic stem cell transplantation. Front Immunol 2017;8:900.

[30] Lan F, Zeng D, Higuchi M, Higgins JP, Strober S. Host conditioning with total lymphoid irradiation and antithymocyte globulin prevents graft-versus-host disease: the role of CD1-reactive natural killer T cells. Biol Blood Marrow Transplant 2003;9(6):355–63.

[31] Morecki S, Panigrahi S, Pizov G, Yacovlev E, Gelfand Y, Eizik O, et al. Effect of KRN7000 on induced graft-vs-host disease. Exp Hematol 2004;32(7):630–7.

[32] Hashimoto D, Asakura S, Miyake S, Yamamura T, Van Kaer L, Liu C, et al. Stimulation of host NKT cells by synthetic glycolipid regulates acute graft-versus-host disease by inducing Th2 polarization of donor T cells. J Immunol 2005;174(1):551–6.

[33] Duramad O, Laysang A, Li J, Ishii Y, Namikawa R. Pharmacologic expansion of donor-derived, naturally occurring CD4+ Foxp3+ regulatory T cells reduces acute graft-versus-host disease lethality without abrogating the graft-versus-leukemia effect in murine models. Biol Blood Marrow Transplant 2011;17(8):1154–68.

[34] Hirai T, Ishii Y, Ikemiyagi M, Fukuda E, Omoto K, Namiki M, et al. A novel approach inducing transplant tolerance by activated invariant natural killer T cells with costimulatory blockade. Am J Transplant 2014;14(3):554–67.

[35] Schneidawind D, Pierini A, Alvarez M, Pan Y, Baker J, Buechele C, et al. CD4+ invariant natural killer T cells protect from murine GVHD lethality through expansion of donor CD4+ CD25+ FoxP3+ regulatory T cells. Blood J Am Soc Hematol 2014;124(22):3320–8.

[36] Schneidawind D, Baker J, Pierini A, Buechele C, Luong RH, Meyer EH, et al. Third-party CD4+ invariant natural killer T cells protect from murine GVHD lethality. Blood 2015;125(22):3491–500.

[37] Leveson-Gower DB, Olson JA, Sega EI, Luong RH, Baker J, Zeiser R, et al. Low doses of natural killer T cells provide protection from acute graft-versus-host disease via an IL-4–dependent mechanism. Blood J Am Soc Hematol 2011;117(11):3220–9.

[38] Weber M, Rudolph B, Stein P, Yogev N, Bosmann M, Schild H, et al. Host-derived CD8+ dendritic cells protect against acute graft-versus-host disease after experimental allogeneic bone marrow transplantation. Biol Blood Marrow Transplant 2014;20(11):1696–704.

[39] Kohrt HE, Turnbull BB, Heydari K, Shizuru JA, Laport GG, Miklos DB, et al. TLI and ATG conditioning with low risk of graft-versus-host disease retains antitumor reactions after allogeneic hematopoietic cell transplantation from related and unrelated donors. Blood J Am Soc Hematol 2009;114(5):1099–109.

[40] Haraguchi K, Takahashi T, Hiruma K, Kanda Y, Tanaka Y, Ogawa S, et al. Recovery of V α 24+ NKT cells after hematopoietic stem cell transplantation. Bone Marrow Transplant 2004;34(7):595–602.

[41] Rubio M-T, Moreira-Teixeira L, Bachy E, Bouillié M, Milpied P, Coman T, et al. Early posttransplantation donor-derived invariant natural killer T-cell recovery predicts the occurrence of acute graft-versus-host disease and overall survival. Blood 2012;120(10):2144–54.

[42] Chaidos A, Patterson S, Szydlo R, Chaudhry MS, Dazzi F, Kanfer E, et al. Graft invariant natural killer T-cell dose predicts risk of acute graft-versus-host disease in allogeneic hematopoietic stem cell transplantation. Blood J Am Soc Hematol 2012;119(21):5030–6.

[43] Patterson S, Chaidos A, Neville DC, Poggi A, Butters TD, Roberts IA, et al. Human invariant NKT cells display alloreactivity instructed by invariant TCR-CD1d interaction and killer Ig receptors. J Immunol 2008;181(5):3268–76.

[44] Pillai AB, George TI, Dutt S, Strober S. Host natural killer T cells induce an interleukin-4–dependent expansion of donor CD4+ CD25+ Foxp3+ T regulatory cells that protects against graft-versus-host disease. Blood J Am Soc Hematol 2009;113(18):4458–67.

[45] Chen Y-B, Efebera YA, Johnston L, Ball ED, Avigan D, Lekakis LJ, et al. Increased Foxp3+ Helios+ regulatory T cells and decreased acute graft-versus-host disease after allogeneic bone marrow transplantation in patients receiving sirolimus and RGI-2001, an activator of invariant natural killer T cells. Biol Blood Marrow Transplant 2017;23(4):625–34.

[46] Bae E-A, Seo H, Kim I-K, Jeon I, Kang C-Y. Roles of NKT cells in cancer immunotherapy. Arch Pharm Res 2019;1-6.
[47] Berzins SP, Smyth MJ, Baxter AG. Presumed guilty: natural killer T cell defects and human disease. Nat Rev Immunol 2011;11(2):131–42.
[48] Lam PY, Nissen MD, Mattarollo SR. Invariant natural killer T cells in immune regulation of blood cancers: harnessing their potential in immunotherapies. Front Immunol 2017;8:1355.
[49] Vivier E, Ugolini S, Blaise D, Chabannon C, Brossay L. Targeting natural killer cells and natural killer T cells in cancer. Nat Rev Immunol 2012;12(4):239–52.
[50] Bassiri H, Das R, Guan P, Barrett DM, Brennan PJ, Banerjee PP, et al. iNKT cell cytotoxic responses control T-lymphoma growth in vitro and in vivo. Cancer Immunol Res 2014;2(1):59–69.
[51] Dao T, Mehal WZ, Crispe IN. IL-18 augments perforin-dependent cytotoxicity of liver NK-T cells. J Immunol 1998;161(5):2217–22.
[52] Leite-de-Moraes MC, Hameg A, Arnould A, Machavoine F, Koezuka Y, Schneider E, et al. A distinct IL-18-induced pathway to fully activate NK T lymphocytes independently from TCR engagement. J Immunol 1999;163(11):5871–6.
[53] Hagihara M, Gansuvd B, Ueda Y, Tsuchiya T, Masui A, Tazume K, et al. Killing activity of human umbilical cord blood-derived TCRValpha24+ NKT cells against normal and malignant hematological cells in vitro: a comparative study with NK cells or OKT3 activated T lymphocytes or with adult peripheral blood NKT cells. Cancer Immunol Immunother 2002;51(1):1–8.
[54] Spanoudakis E, Hu M, Naresh K, Terpos E, Melo V, Reid A, et al. Regulation of multiple myeloma survival and progression by CD1d. Blood J Am Soc Hematol 2009;113(11):2498–507.
[55] Tiper IV, Webb TJ. Histone deacetylase inhibitors enhance CD1d-dependent NKT cell responses to lymphoma. Cancer Immunol Immunother 2016;65(11):1411–21.
[56] Nowak M, Arredouani MS, Tun-Kyi A, Schmidt-Wolf I, Sanda MG, Balk SP, et al. Defective NKT cell activation by CD1d+ TRAMP prostate tumor cells is corrected by interleukin-12 with alpha-galactosylceramide. PLoS One 2010;5(6), e11311.
[57] Miura S, Kawana K, Schust DJ, Fujii T, Yokoyama T, Iwasawa Y, et al. CD1d, a sentinel molecule bridging innate and adaptive immunity, is downregulated by the human papillomavirus (HPV) E5 protein: a possible mechanism for immune evasion by HPV. J Virol 2010;84(22):11614–23.
[58] Hix LM, Shi YH, Brutkiewicz RR, Stein PL, Wang C-R, Zhang M. CD1d-expressing breast cancer cells modulate NKT cell-mediated antitumor immunity in a murine model of breast cancer metastasis. PLoS One 2011;6(6), e20702.
[59] Yang P-M, Lin P-J, Chen C-C. CD1d induction in solid tumor cells by histone deacetylase inhibitors through inhibition of HDAC1/2 and activation of Sp1. Epigenetics 2012;7(4):390–9.
[60] Metelitsa LS. Anti-tumor potential of type-I NKT cells against CD1d-positive and CD1d-negative tumors in humans. Clin Immunol 2011;140(2):119–29.
[61] Kuylenstierna C, Björkström NK, Andersson SK, Sahlström P, Bosnjak L, Paquin-Proulx D, et al. NKG2D performs two functions in invariant NKT cells: direct TCR-independent activation of NK-like cytolysis and co-stimulation of activation by CD1d. Eur J Immunol 2011;41(7):1913–23.
[62] Parekh VV, Lalani S, Kaer LV. The in vivo response of invariant natural killer T cells to glycolipid antigens. Int Rev Immunol 2007;26(1–2):31–48.
[63] Hermans IF, Silk JD, Gileadi U, Salio M, Mathew B, Ritter G, et al. NKT cells enhance CD4+ and CD8+ T cell responses to soluble antigen in vivo through direct interaction with dendritic cells. J Immunol 2003;171 (10):5140–7.
[64] Fujii S-i, Shimizu K, Smith C, Bonifaz L, Steinman RM. Activation of natural killer T cells by α-galactosylceramide rapidly induces the full maturation of dendritic cells in vivo and thereby acts as an adjuvant for combined CD4 and CD8 T cell immunity to a coadministered protein. J Exp Med 2003;198(2):267–79.
[65] Coquet JM, Kyparissoudis K, Pellicci DG, Besra G, Berzins SP, Smyth MJ, et al. IL-21 is produced by NKT cells and modulates NKT cell activation and cytokine production. J Immunol 2007;178(5):2827–34.
[66] Chung Y, Kim B-S, Kim Y-J, Ko H-J, Ko S-Y, Kim D-H, et al. CD1d-restricted T cells license B cells to generate long-lasting cytotoxic antitumor immunity in vivo. Cancer Res 2006;66(13):6843–50.
[67] de Jong EC, Smits HH, Kapsenberg ML. Dendritic cell-mediated T cell polarization. In: Springer Seminars in Immunopathology. Springer; 2005. p. 289–307.
[68] Kitamura H, Iwakabe K, Yahata T, Nishimura S-i, Ohta A, Ohmi Y, et al. The natural killer T (NKT) cell ligand α-galactosylceramide demonstrates its immunopotentiating effect by inducing interleukin (IL)-12 production by dendritic cells and IL-12 receptor expression on NKT cells. J Exp Med 1999;189(7):1121–8.

[69] Fujii SI, Shimizu K, Hemmi H, Steinman RM. Innate Vα14+ natural killer T cells mature dendritic cells, leading to strong adaptive immunity. Immunol Rev 2007;220(1):183–98.
[70] Cui J, Shin T, Kawano T, Sato H, Kondo E, Toura I, et al. Requirement for Vα14 NKT cells in IL-12-mediated rejection of tumors. Science 1997;278(5343):1623–6.
[71] Taraban VY, Martin S, Attfield KE, Glennie MJ, Elliott T, Elewaut D, et al. Invariant NKT cells promote CD8+ cytotoxic T cell responses by inducing CD70 expression on dendritic cells. J Immunol 2008;180(7):4615–20.
[72] Semmling V, Lukacs-Kornek V, Thaiss CA, Quast T, Hochheiser K, Panzer U, et al. Alternative cross-priming through CCL17-CCR4-mediated attraction of CTLs toward NKT cell-licensed DCs. Nat Immunol 2010;11 (4):313–20.
[73] Binnewies M, Roberts EW, Kersten K, Chan V, Fearon DF, Merad M, et al. Understanding the tumor immune microenvironment (TIME) for effective therapy. Nat Med 2018;24(5):541–50.
[74] Joyce JA, Fearon DT. T cell exclusion, immune privilege, and the tumor microenvironment. Science 2015;348 (6230):74–80.
[75] Noy R, Pollard JW. Tumor-associated macrophages: from mechanisms to therapy. Immunity 2014;41(1):49–61.
[76] Song L, Asgharzadeh S, Salo J, Engell K, Wu H-w, Sposto R, et al. Vα24-invariant NKT cells mediate antitumor activity via killing of tumor-associated macrophages. J Clin Invest 2009;119(6):1524–36.
[77] Gabrilovich DI, Ostrand-Rosenberg S, Bronte V. Coordinated regulation of myeloid cells by tumours. Nat Rev Immunol 2012;12(4):253–68.
[78] De Santo C, Salio M, Masri SH, Lee LY-H, Dong T, Speak AO, et al. Invariant NKT cells reduce the immunosuppressive activity of influenza A virus-induced myeloid-derived suppressor cells in mice and humans. J Clin Invest 2008;118(12):4036–48.
[79] Ko H-J, Lee J-M, Kim Y-J, Kim Y-S, Lee K-A, Kang C-Y. Immunosuppressive myeloid-derived suppressor cells can be converted into immunogenic APCs with the help of activated NKT cells: an alternative cell-based antitumor vaccine. J Immunol 2009;182(4):1818–28.
[80] Kawano T, Nakayama T, Kamada N, Kaneko Y, Harada M, Ogura N, et al. Antitumor cytotoxicity mediated by ligand-activated human Vα24 NKT cells. Cancer Res 1999;59(20):5102–5.
[81] Toura I, Kawano T, Akutsu Y, Nakayama T, Ochiai T, Taniguchi M. Cutting edge: inhibition of experimental tumor metastasis by dendritic cells pulsed with α-galactosylceramide. J Immunol 1999;163(5):2387–91.
[82] Nishimura T, Kitamura H, Iwakabe K, Yahata T, Ohta A, Sato M, et al. The interface between innate and acquired immunity: glycolipid antigen presentation by CD1d-expressing dendritic cells to NKT cells induces the differentiation of antigen-specific cytotoxic T lymphocytes. Int Immunol 2000;12(7):987–94.
[83] Shin T, Nakayama T, Akutsu Y, Motohashi S, Shibata Y, Harada M, et al. Inhibition of tumor metastasis by adoptive transfer of IL-12-activated Vα14 NKT cells. Int J Cancer 2001;91(4):523–8.
[84] Smyth MJ, Crowe NY, Pellicci DG, Kyparissoudis K, Kelly JM, Takeda K, et al. Sequential production of interferon-γ by NK1. 1+ T cells and natural killer cells is essential for the antimetastatic effect of α-galactosylceramide. Blood J Am Soc Hematol 2002;99(4):1259–66.
[85] Giaccone G, Punt CJ, Ando Y, Ruijter R, Nishi N, Peters M, et al. A phase I study of the natural killer T-cell ligand alpha-galactosylceramide (KRN7000) in patients with solid tumors. Clin Cancer Res 2002;8(12):3702–9.
[86] Nieda M, Okai M, Tazbirkova A, Lin H, Yamaura A, Ide K, et al. Therapeutic activation of Vα24+ Vβ11+ NKT cells in human subjects results in highly coordinated secondary activation of acquired and innate immunity. Blood 2004;103(2):383–9.
[87] Ishikawa A, Motohashi S, Ishikawa E, Fuchida H, Higashino K, Otsuji M, et al. A phase I study of α-galactosylceramide (KRN7000)-pulsed dendritic cells in patients with advanced and recurrent non-small cell lung cancer. Clin Cancer Res 2005;11(5):1910–7.
[88] Chang DH, Osman K, Connolly J, Kukreja A, Krasovsky J, Pack M, et al. Sustained expansion of NKT cells and antigen-specific T cells after injection of α-galactosyl-ceramide loaded mature dendritic cells in cancer patients. J Exp Med 2005;201(9):1503–17.
[89] Uchida T, Horiguchi S, Tanaka Y, Yamamoto H, Kunii N, Motohashi S, et al. Phase I study of α-galactosylceramide-pulsed antigen presenting cells administration to the nasal submucosa in unresectable or recurrent head and neck cancer. Cancer Immunol Immunother 2008;57(3):337–45.
[90] Motohashi S, Nagato K, Kunii N, Yamamoto H, Yamasaki K, Okita K, et al. A phase I-II study of α-galactosylceramide-pulsed IL-2/GM-CSF-cultured peripheral blood mononuclear cells in patients with advanced and recurrent non-small cell lung cancer. J Immunol 2009;182(4):2492–501.

[91] Nagato K, Motohashi S, Ishibashi F, Okita K, Yamasaki K, Moriya Y, et al. Accumulation of activated invariant natural killer T cells in the tumor microenvironment after α-galactosylceramide-pulsed antigen presenting cells. J Clin Immunol 2012;32(5):1071–81.
[92] Motohashi S, Ishikawa A, Ishikawa E, Otsuji M, Iizasa T, Hanaoka H, et al. A phase I study of in vitro expanded natural killer T cells in patients with advanced and recurrent non-small cell lung cancer. Clin Cancer Res 2006;12(20):6079–86.
[93] Kunii N, Horiguchi S, Motohashi S, Yamamoto H, Ueno N, Yamamoto S, et al. Combination therapy of in vitro-expanded natural killer T cells and α-galactosylceramide-pulsed antigen-presenting cells in patients with recurrent head and neck carcinoma. Cancer Sci 2009;100(6):1092–8.
[94] Yamasaki K, Horiguchi S, Kurosaki M, Kunii N, Nagato K, Hanaoka H, et al. Induction of NKT cell-specific immune responses in cancer tissues after NKT cell-targeted adoptive immunotherapy. Clin Immunol 2011;138(3):255–65.
[95] Exley MA, Friedlander P, Alatrakchi N, Vriend L, Yue S, Sasada T, et al. Adoptive transfer of invariant NKT cells as immunotherapy for advanced melanoma: a phase I clinical trial. Clin Cancer Res 2017;23(14):3510–9.
[96] Smith DJ, Liu S, Ji S, Li B, McLaughlin J, Cheng D, et al. Genetic engineering of hematopoietic stem cells to generate invariant natural killer T cells. Proc Natl Acad Sci 2015;112(5):1523–8.
[97] Zhu Y, Smith DJ, Zhou Y, Li Y-R, Yu J, Lee D, et al. Development of hematopoietic stem cell-engineered invariant natural killer T cell therapy for cancer. Cell Stem Cell 2019;25(4):542–557.e9.
[98] June CH, Sadelain M. Chimeric antigen receptor therapy. N Engl J Med 2018;379(1):64–73.
[99] Ma S, Li X, Wang X, Cheng L, Li Z, Zhang C, et al. Current progress in CAR-T cell therapy for solid tumors. Int J Biol Sci 2019;15(12):2548.
[100] Neelapu SS, Tummala S, Kebriaei P, Wierda W, Gutierrez C, Locke FL, et al. Chimeric antigen receptor T-cell therapy—assessment and management of toxicities. Nat Rev Clin Oncol 2018;15(1):47.
[101] Bedoya F, Frigault MJ, Maus MV. The flipside of the power of engineered T cells: observed and potential toxicities of genetically modified T cells as therapy. Mol Ther 2017;25(2):314–20.
[102] Martinez M, Moon EK. CAR T cells for solid tumors: new strategies for finding, infiltrating, and surviving in the tumor microenvironment. Front Immunol 2019;10:128.
[103] Robertson FC, Berzofsky JA, Terabe M. NKT cell networks in the regulation of tumor immunity. Front Immunol 2014;5:543.
[104] Song L, Ara T, Wu H-W, Woo C-W, Reynolds CP, Seeger RC, et al. Oncogene MYCN regulates localization of NKT cells to the site of disease in neuroblastoma. J Clin Invest 2007;117(9):2702–12.
[105] Liu D, Song L, Wei J, Courtney AN, Gao X, Marinova E, et al. IL-15 protects NKT cells from inhibition by tumor-associated macrophages and enhances antimetastatic activity. J Clin Invest 2012;122(6):2221–33.
[106] Cerundolo V, Silk JD, Masri SH, Salio M. Harnessing invariant NKT cells in vaccination strategies. Nat Rev Immunol 2009;9(1):28–38.
[107] Tyznik AJ, Verma S, Wang Q, Kronenberg M, Benedict CA. Distinct requirements for activation of NKT and NK cells during viral infection. J Immunol 2014;192(8):3676–85.
[108] Pillai AB, George TI, Dutt S, Teo P, Strober S. Host NKT cells can prevent graft-versus-host disease and permit graft antitumor activity after bone marrow transplantation. J Immunol 2007;178(10):6242–51.
[109] De Lalla C, Rinaldi A, Montagna D, Azzimonti L, Bernardo ME, Sangalli LM, et al. Invariant NKT cell reconstitution in pediatric leukemia patients given HLA-haploidentical stem cell transplantation defines distinct CD4+ and CD4− subset dynamics and correlates with remission state. J Immunol 2011;186(7):4490–9.
[110] Brennan PJ, Tatituri RV, Brigl M, Kim EY, Tuli A, Sanderson JP, et al. Invariant natural killer T cells recognize lipid self antigen induced by microbial danger signals. Nat Immunol 2011;12(12):1202.
[111] Cox D, Fox L, Tian R, Bardet W, Skaley M, Mojsilovic D, et al. Determination of cellular lipids bound to human CD1d molecules. PLoS One 2009;4(5), e5325.
[112] Fox LM, Cox DG, Lockridge JL, Wang X, Chen X, Scharf L, et al. Recognition of lyso-phospholipids by human natural killer T lymphocytes. PLoS Biol 2009;7(10), e1000228.
[113] Muindi K, Cernadas M, Watts GF, Royle L, Neville DC, Dwek RA, et al. Activation state and intracellular trafficking contribute to the repertoire of endogenous glycosphingolipids presented by CD1d. Proc Natl Acad Sci 2010;107(7):3052–7.
[114] Yuan W, Kang S-J, Evans JE, Cresswell P. Natural lipid ligands associated with human CD1d targeted to different subcellular compartments. J Immunol 2009;182(8):4784–91.
[115] Torikai H, Cooper LJ. Translational implications for off-the-shelf immune cells expressing chimeric antigen receptors. Mol Ther 2016;24(7):1178–86.

[116] Shissler SC, Lee MS, Webb TJ. Mixed signals: co-stimulation in invariant natural killer T cell-mediated cancer immunotherapy. Front Immunol 2017;8:1447.
[117] Heczey A, Liu D, Tian G, Courtney AN, Wei J, Marinova E, et al. Invariant NKT cells with chimeric antigen receptor provide a novel platform for safe and effective cancer immunotherapy. Blood 2014;124(18):2824–33.
[118] Tian G, Courtney AN, Jena B, Heczey A, Liu D, Marinova E, et al. CD62L+ NKT cells have prolonged persistence and antitumor activity in vivo. J Clin Invest 2016;126(6):2341–55.
[119] Xu X, Huang W, Heczey A, Liu D, Guo L, Wood M, et al. NKT cells coexpressing a GD2-specific chimeric antigen receptor and IL15 show enhanced in vivo persistence and antitumor activity against neuroblastoma. Clin Cancer Res 2019;25(23):7126–38.
[120] Heczey A, Courtney AN, Montalbano A, Robinson S, Liu K, Li M, et al. Anti-GD2 CAR-NKT cells in patients with relapsed or refractory neuroblastoma: an interim analysis. Nat Med 2020;26(11):1686–90.
[121] Rotolo A, Caputo VS, Holubova M, Baxan N, Dubois O, Chaudhry MS, et al. Enhanced anti-lymphoma activity of CAR19-iNKT cells underpinned by dual CD19 and CD1d targeting. Cancer Cell 2018;34(4):596–610.e11.
[122] Qasim W, Zhan H, Samarasinghe S, Adams S, Amrolia P, Stafford S, et al. Molecular remission of infant B-ALL after infusion of universal TALEN gene-edited CAR T cells. Sci Transl Med 2017;9(374).
[123] Iriguchi S, Kaneko S. Toward the development of true "off-the-shelf" synthetic T-cell immunotherapy. Cancer Sci 2019;110(1):16–22.
[124] Themeli M, Kloss CC, Ciriello G, Fedorov VD, Perna F, Gonen M, et al. Generation of tumor-targeted human T lymphocytes from induced pluripotent stem cells for cancer therapy. Nat Biotechnol 2013;31(10):928–33.
[125] Montel-Hagen A, Seet CS, Li S, Chick B, Zhu Y, Chang P, et al. Organoid-induced differentiation of conventional T cells from human pluripotent stem cells. Cell Stem Cell 2019;24(3):376–389.e8.
[126] Nagano S, Maeda T, Ichise H, Kashima S, Ohtaka M, Nakanishi M, et al. High frequency production of T cell-derived iPSC clones capable of generating potent cytotoxic T cells. Mol Ther Methods Clin Dev 2020;16:126–35.
[127] Li Y, Hermanson DL, Moriarity BS, Kaufman DS. Human iPSC-derived natural killer cells engineered with chimeric antigen receptors enhance anti-tumor activity. Cell Stem Cell 2018;23(2):181–192.e5.
[128] Zhu H, Kaufman DS. An improved method to produce clinical-scale natural killer cells from human pluripotent stem cells. In: In Vitro Differentiation of T-Cells. Springer; 2019. p. 107–19.
[129] Bernareggi D, Pouyanfard S, Kaufman DS. Development of innate immune cells from human pluripotent stem cells. Exp Hematol 2019;71:13–23.
[130] Zhu H, Blum RH, Bjordahl R, Gaidarova S, Rogers P, Lee TT, et al. Pluripotent stem cell-derived NK cells with high-affinity noncleavable CD16a mediate improved antitumor activity. Blood J Am Soc Hematol 2020;135 (6):399–410.
[131] Kitayama S, Zhang R, Liu T-Y, Ueda N, Iriguchi S, Yasui Y, et al. Cellular adjuvant properties, direct cytotoxicity of re-differentiated Vα24 invariant NKT-like cells from human induced pluripotent stem cells. Stem Cell Rep 2016;6(2):213–27.
[132] Yamada D, Iyoda T, Vizcardo R, Shimizu K, Sato Y, Endo TA, et al. Efficient regeneration of human Vα24+ invariant natural killer T cells and their anti-tumor activity in vivo. Stem Cells 2016;34(12):2852–60.
[133] Matsuda JL, Gapin L, Sidobre S, Kieper WC, Tan JT, Ceredig R, et al. Homeostasis of V α 14 iNKT cells. Nat Immunol 2002;3(10):966–74.
[134] Baev DV, Peng X-h, Song L, Barnhart JR, Crooks GM, Weinberg KI, et al. Distinct homeostatic requirements of CD4+ and CD4-subsets of Vα24-invariant natural killer T cells in humans. Blood 2004;104(13):4150–6.
[135] Durgan K, Ali M, Warner P, Latchman YE. Targeting NKT cells and PD-L1 pathway results in augmented anti-tumor responses in a melanoma model. Cancer Immunol Immunother 2011;60(4):547–58.
[136] Parekh VV, Lalani S, Kim S, Halder R, Azuma M, Yagita H, et al. PD-1/PD-L blockade prevents anergy induction and enhances the anti-tumor activities of glycolipid-activated invariant NKT cells. J Immunol 2009;182 (5):2816–26.
[137] Chang W-S, Kim J-Y, Kim Y-J, Kim Y-S, Lee J-M, Azuma M, et al. Cutting edge: programmed death-1/ programmed death ligand 1 interaction regulates the induction and maintenance of invariant NKT cell anergy. J Immunol 2008;181(10):6707–10.
[138] Kloss CC, Lee J, Zhang A, Chen F, Melenhorst JJ, Lacey SF, et al. Dominant-negative TGF-β receptor enhances PSMA-targeted human CAR T cell proliferation and augments prostate cancer eradication. Mol Ther 2018;26 (7):1855–66.

CHAPTER

19

Tumor-infiltrating lymphocytes: Prognostic considerations and current trials as adoptive cell therapy

Timothy D. Folsom[a,b,c,d], *Branden S. Moriarity*[a,b,c,d], *Timothy K. Starr*[b,e], *Emil Lou*[b,f], and *Beau R. Webber*[a,b,c,d]

[a]Department of Pediatrics, University of Minnesota, Minneapolis, MN, United States [b]Masonic Cancer Center, University of Minnesota, Minneapolis, MN, United States [c]Center for Genome Engineering, University of Minnesota, Minneapolis, MN, United States [d]Stem Cell Institute, University of Minnesota, Minneapolis, MN, United States [e]Department of Obstetrics and Gynecology and Women's Health, Minneapolis, MN, United States [f]Department of Medicine, Division of Hematology, Oncology, and Transplantation, Minneapolis, MN, United States

Abstract

Cancer remains one of the leading causes of mortality globally. In recent years, new treatments have been developed that capitalize on the patient's own immune system to fight cancer including antibody-based immune checkpoint inhibition and adoptive cell therapy (ACT) using various cellular components of the immune system. Over the past few decades, it has been demonstrated that ACT using tumor-infiltrating lymphocytes (TIL) can successfully treat melanomas in patients for whom first and second line chemotherapies were not effective. Treatment of other solid tumors with TIL therapy has achieved durable cures in patients with advanced, metastatic disease; however, these outcomes are the exception and significant challenges remain. New advances in genetic engineering, particularly the use of CRISPR/Cas9 are currently being used to develop new TIL products to expand TIL-based ACT beyond melanoma treatment. In this review, we describe the history of ACT using TILs, TILs as a prognostic factor for cancer survival, new methods of improving TILs for ACT, and the current state of clinical research.

Abbreviations

ACT	adoptive cell therapy
APC	antigen-presenting cell
CADPS2	calcium-dependent secretion activator 2

https://doi.org/10.1016/B978-0-12-822620-9.00005-7

CAR chimeric antigen receptor
CAS9 CRISPR-associated protein 9
CCRT cisplatin-based concurrent chemotherapy
CDKN2A cyclin-dependent kinase inhibitor 2a
CEA carcinoembryonic antigen
CRC colorectal cancer
CRISPR clustered regularly interspaced short palindromic repeats
CRS cytokine release syndrome
CTSB cathepsin B
CXCL1 CXC-motif chemokine ligand 1
CXCL8 CXC-motif chemokine ligand 8
CXCR2 CXC chemokine receptor 2
DC dendritic cell
DFS disease-free survival
DSS disease-specific survival
EOC epithelial ovarian cancer
ERBB2IP Erb-B2 receptor tyrosine kinase 2 interacting protein
FDA Food and Drug Administration
FOXP3 forkhead box P3
GAPDH glyceraldehyde-3-phosphate dehydrogenase
GAS7 growth arrest-specific gene 7
GM-CSF granulocyte macrophage colony-stimulating factor
Gp100 glycoprotein 100
HER2 human epidermal growth factor 2
HGSOC high-grade serous ovarian cancer
HLA human leukocyte antigen
HPV human papillomavirus
ICI immune checkpoint inhibitors
IL-1 interleukin 1
IL-10 interleukin 10
IL-13 interleukin 13
IL-15 interleukin 15
IL-2 interleukin 2
IL-5 interleukin 5
IL-6 interleukin 6
IL-7 interleukin 7
INF-γ interferon gamma
IT intratumoral compartment
KIAA0368 proteasome-associated protein ECM29 homolog
KRAS Kirsten rat sarcoma viral oncogene homolog
LAK lymphocyte-activated killer cell
MAGE-A4 MAGE family member A4
MART-1 melanoma antigen recognized by T-cells
MS mass spectrometry
NK natural killer cell
NSCLC nonsmall cell lung cancer
NY-ESO-1 New York esophageal cancer 1
OR objective response
ORR overall response rate
OS overall survival
PBMC peripheral blood mononuclear cell
pCR pathological complete response
PD-1 programmed cell death 1
PD-L1 programmed-death ligand 1
PFS progression-free survival

PPP1R3B	protein phosphatase 1 regulatory subunit 3B
SLC32A2	solute carrier family 32
TCR	T-cell receptor
TIL	tumor-infiltrating lymphocyte
TME	tumor microenvironment
TNBC	triple negative breast cancer
TNF-α	tumor necrosis factor alpha
TRAIL	TNF-related apoptosis-inducing ligand
Treg	T regulatory cells
TRM	tissue resident memory
WES	whole exome sequencing

Conflict of interest

B.S.M., B.R.W., and T.K.S. have received research funds from Intima Biosciences related to the use and development of CRISPR/Cas9 edited TIL for the treatment of cancer.

Introduction

Cancers develop in a heterogeneous cellular environment involving dynamic interactions between tumor, stroma, and multiple immune cell types. While effective immune system activation can lead to tumor clearance, tumors implement multiple strategies to evade recognition and continue uncontrolled growth [1–7]. Over the past several decades, researchers have investigated how to harness the potency of the immune system to develop new treatments for cancer.

Adoptive cell therapy (ACT) is emerging as an effective therapeutic modality for the treatment of cancer. ACT is a cell-based intervention that transfers either a patient's own (autologous transfer) or a donor's (allogeneic transfer) immune cells to boost the patient's immune response. Immune cells, including dendritic cells (DCs), natural killer (NK), and T-lymphocytes can be modified and/or expanded *ex vivo* before being infused as therapeutic agents to mediate antitumor activity. ACT using T cells can be broadly classified into three different types: expanded natural TILs; T-cell receptor (TCR)-engineered T-cells and chimeric antigen receptor (CAR)-engineered T-cells [8]. While we will include discussion of both CARs and TCRs, the focus of the current chapter is on TILs. Treatments involving all three types are currently being tested and refined clinically to produce more effective cancer treatments. With ACT using TIL, upwards of 10^{11} antitumor lymphocytes can be grown ex vivo, which releases them from the inhibitory factors present in vivo within the tumor microenvironment (TME), and selected/modified for enhanced tumor recognition and effector functions to improve antitumor activity [9].

Following expansion of bulk, selected, and/or genetically engineered TIL, there are three basic components of TIL-based ACT: lymphodepletion, infusion of TIL, and subsequent dosing of interleukin (IL)-2 [9–13] (Fig. 1). Lymphodepletion commonly involves administration of high-dose cyclophosphamide and fludarabine that promotes survival of the infused TIL product [14]. Moreover, lymphodepletion induces the production of IL-7 and IL-15, which promote the expansion of infused TIL in the absence of endogenous T-cells and may also eliminate competition for homeostatic cytokine, i.e., elimination of cytokine sinks [15,16]. IL-2 has been used as a therapeutic agent on its own for metastatic melanoma, non-Hodgkin lymphoma, and kidney cancer, possibly due to its activation of T-cells [17,18]. The infusion of

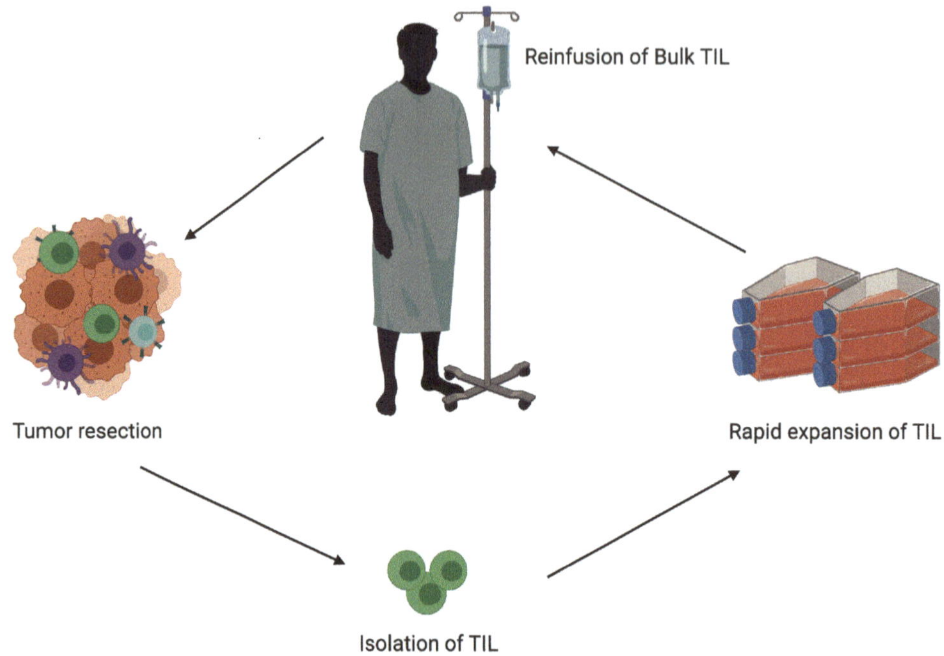

FIG. 1 Outline of TIL-based ACT. Tumors are resected from the patient and TIL are isolated and undergo a rapid expansion protocol ex vivo. When enough TIL are grown (approximately 10^9–10^{11}), the patient undergoes lymphodepletion followed by TIL Infusion. After the infusion of TIL, the patient receives an IL-2 infusion to promote T-cell persistence.

IL-2 with TIL has been shown to increase T-cell persistence and is now commonly used in conjunction with TIL therapy [19,20]. In the following section, the history of discoveries that led to the development of this three-part procedure for TIL-based ACT will be discussed.

History of ACT

The history of ACT began with the discovery that treatment with chemotherapy or radiation along with bone marrow or infant spleen cells could prolong survival in mice injected with leukemic cells [21]. Subsequent experiments demonstrated the efficacy of immunocompetent cells derived from spleen, bone marrow, and lymph nodes in prolonging survival in both murine models and human subjects [22–27]. Concomitant with these discoveries was the finding of the key role played by lymphocytes, including T-cells, in mediating these antitumor responses in rodent [28–30] and human [27,31] studies. Subsequent use of chemotherapy in preclinical experiments involving transfer of T-cells [32,33] demonstrated the importance of host inhibitory factors and set the stage for the inclusion of lymphodepletion regimens in ACT [9].

A key step in enabling further development of ACT was the identification of IL-2 as a factor that promoted T-cell growth [34]. This discovery allowed for the expansion of T-cells in culture to provide large numbers of cells for transfer. A subsequent series of experiments by

Rosenberg and colleagues demonstrated that T-cells grown in the presence of IL-2 could lyse fresh syngeneic or autologous tumors [35–39]. These findings led to the first demonstration that the administration of what Rosenberg et al. [40] described as autologous lymphokine-activated killer cells (LAKs) and recombinant lymphokine IL-2 resulted in objective regression of cancer in 11 out of 25 patients with metastatic cancer.

The Rosenberg group subsequently found that lymphocytes isolated from within the tumor stroma (i.e., TILs) and expanded in IL-2 were 50-100X better able to recognize and destroy liver tumors in mice when compared with LAKs [41]. Rosenberg's group found that human TIL could be expanded using IL-2 by up to 95,652-fold and maintain their antitumor cytotoxic function as demonstrated by potency against fresh autologous melanoma tumors [42]. The first demonstration of ACT using TILs from autologous tumor cells that were expanded in vitro and administered with IL-2 was published in 1988 [43]. Objective regression of cancer was observed in 60% of patients with metastatic melanoma who had not previously been treated with IL-2 and in 40% of patients for whom IL-2 therapy had previously failed [43]. A further study of 86 patients with metastatic melanoma found a 34% objective response (OR) rate following treatment with autologous TILs plus high-dose IL-2 infusions [44]. The frequency of treatment response was significantly higher in patients that received TILs from younger cultures (fewer than 33 days in culture), TILs with shorter doubling times and TILs that exhibited greater lysis of autologous tumors [44]. Moreover, patients who received TILs from subcutaneous tumor deposits had a 49% OR rate vs only 17% of patients who received TILs derived from lymph nodes [44].

The addition of lymphodepletion prior to the infusion of TIL and IL-2 was the next step in the evolution of TIL therapy [45,46]. An initial phase I clinical trial used a chemotherapy regimen of cyclophosphamide for 2 days followed by fludarabine for 5 days with subsequent infusion of TILs and either low or high doses of IL-2 [45]. While no patients displayed an OR to treatment, several displayed transient reductions of tumor size [45]. Furthermore, the researchers found that high doses of IL-2 were better tolerated when accompanied by preconditioning chemotherapy [45]. However, a larger study involving 93 patients and the same regimen of cyclophosphamide for 2 days followed by fludarabine for 5 days and infusion of TIL and IL-2 had better results with 22% of patients showing a complete tumor regression and OR was 49% for patients with chemotherapy alone, 52% with 2 Gy irradiation, and 78% for patients with 12 Gy irradiation [46]. Factors that contributed to enhanced response to lymphodepletion include persistence of infused cells in the circulation at 1 month and the number of CD8+ cells infused [46]. Lymphodepletion had previously been shown to increase CD8+ cell survival [45,47] as well as induce IL-7 and IL-15 cytokine expression, both of which promote T-cell expansion [15,16].

TILs as a prognostic factor for survival in selected cancers

As previously mentioned, Rosenberg and colleagues found that TIL were more potent than LAK in fighting tumors [41–43]. Since these findings, studies have shown that the presence of TIL and/or specific subtypes of TIL are associated with increased survival in various cancer types (Table 1).

TABLE 1 Tumor infiltrating lymphocytes and prognosis of solid tumors.

Cancer	Cells	N	Results	Reference
Melanoma	Brisk vs nonbrisk TIL	264	OR of 11.3 for brisk vs 3.5 for nonbrisk	[48]
Melanoma	Brisk vs nonbrisk TIL	285	↑Survival for patients with high levels of TIL	[50]
Melanoma	Brisk vs nonbrisk TIL	2845	↑Survival for brisk and nonbrisk vs no TIL Death from melanoma 50% reduced with brisk and 30% reduced with nonbrisk	[51]
Melanoma	TIL	259	↑Survival for patients with TIL	[52]
Melanoma	Brisk vs nonbrisk TIL	1241	↑OS and RFS in patients with brisk TIL	[53]
Melanoma	Brisk vs nonbrisk TIL	7633	Brisk TIL correlated with better OS and DSS vs nonbrisk	[54]
NSCLC	CD4+, CD8+	109	↑Survival with Invasion by both CD4+, CD8+ TIL	[57]
NSCLC	CD3+, CD8+, CD57+, CD68+, CD79a+, S100+	710	↑POS with IT invasion of high levels of CD3+ and S100+ TIL	[58]
Lung	TIL	1290	↑Survival assoc. with high TIL in squamous cell carcinoma	[59]
NSCLC	CD4+, CD8+, Ki-67/CD8+	178	↓Survival assoc. with CD8+, Ki-67/CD8+ TIL in cancer nests; ↑survival assoc. with CD4+ in stroma	[60]
NSCLC	FOXP3+ Treg	87	Treg associated with poorer prognosis; ↑ risk of relapse	[61]
NSCLC	FOXP3+	953	↑FOXP3/CD3 ratio assoc. with ↑ risk of relapse	[62]
NSCLC	TIL	189	High TIL assoc. with ↑ ORR, OS, PFS	[63]
NSCLC	TIL from stroma & tumor	144	Stromal TIL a better marker for OS and PFS than tumoral TIL	[64]
Breast	CD8+, FOXP3+	12,439	Presence of CD8+ TIL ↓ relative risk of death in TNBC and HER2+ subtypes	[69]
Breast	TIL	3771	↑ TIL assoc. with survival of TMBC and HER2+ subtypes and predicted response to neoadjuvant chemotherapy in all subtypes.	[70]
Breast	TIL, CD3+	368	High TIL assoc. with better response to anthracycline-based chemotherapy; High CD3+ intraepithelial TIL assoc. with improved DFS	[71]
TNBC	TIL, CD4+, CD8+, FOXP3+	10,258	High TIL assoc. with ↑ DFS, OS, pCR; ↑ CD8+ assoc. with ↑ DFS; ↑ CD4+ assoc. with ↑ DFS and OS; FOXP3+ assoc. with ↑ DFS	[72]
TNBC	TIL	1152	↑ TIL concentration assoc. with improved DFS, OS	[73]
Gastric	CD3+, CD8+, CD20+, CD68+, GranzymeB+, FOXP3+	52	↑ FOXP3+ Treg in tumor stroma increased survival; ↓stromal CD68+/FoxP3+ assoc. with improved survival	[76]

TABLE 1 Tumor infiltrating lymphocytes and prognosis of solid tumors—cont'd

Cancer	Cells	N	Results	Reference
MSI-hGC	CD8+, FOXP3+	99	↑ OS with high concentration of CD8+ or FOXP3+ vs low concentration	[77]
Gastric	CD3+, CD8+, CD20+, CD45RO+	220	↑ Survival with high concentration CD3+, CD8+, CD45RO+ TIL	[78]
Gastric	FOXP3+ Tregs	133	↑ FOXP3 Tregs in IT assoc. with shorter OS and lymph node metastasis	[79]
Gastric	CD3+, CD8+, CD20+, CD57+, CD45RO+, FOXP3+	4185	↑ Survival with CD3+ TIL, and FOXP3+ TIL are present in IT; ↓survival with FOXP3+ TIL is present in extra-tumoral compartment	[80]
Gastric	TIL	2835	High TIL assoc. with ↑OS, ↓depth of invasion, lymph node involvement, and early metastasis stage	[81]
CRC	TIL	447	Five-year survival of 92%, 65%, and 36% for pronounced, moderate, and little lymphocytic invasion	[83]
CRC	TIL	2369	Tumors with TIL/HPF of two or greater had higher CRC-specific survival and OS	[85]
CRC	CD3+, CD8+, FOXP3+	21,015	↑OS with high TIL; ↑OS with CD8+ and FOXP3+ TIL at tumor center; ↑DFS for CD3+ TIL at tumor center and invasive margin.	[87]
HGSOC	CD3+, CD4+, CD8+, CD103+	6004	Multivariate: ↑OS and PFS assoc. with intraepithelial CD8+ TIL; stromal CD8+ assoc. with ↑OS Univariate: ↑OS assoc. with intraepithelial CD3+, CD4+, CD8+, CD103+ TIL and with stromal CD4+ and CD8+ TIL	[88]
Cervical	CD3+, CD4+, CD8+, CD20+, CD206+, FOXP3+	55	↑OS and PFS associated with infiltration of CD3+, CD4+, CD8+, CD206+, FOXP3+ TIL	[89]

↓, decrease;↑, increase; Assoc. associated; CRC, colorectal cancer; DFS, disease-free survival; DSS, disease-specific survival; HPF, high powered field; IT, intra-tumoral compartment; MSI-hGC, microsatellite instability high gastric cancer; OR, odds ratio; ORR, overall response rate; OS, overall survival; PFS, progression free survival; POS, post-operative survival; RFS, recurrence free survival; TC, tumor center.

TILs from melanomas have been extensively studied and treatment of melanomas with ACT has provided the earliest and greatest successes. Extent of TIL infiltration into tumors in melanomas have been classified as brisk (i.e., are present throughout the tumor during its vertical growth phase), nonbrisk (i.e., TIL present in one or more foci during the vertical growth phase), or absent (i.e., entirely not present or present but not infiltrating the tumor) [48,49]. The odds ratio for survival with brisk TIL was found to be 11.3 vs 3.5 for nonbrisk infiltrate [48]. This scoring system has subsequently been validated in studies that included 5000 patients [50–53], reporting that TIL was a significant prognostic factor for survival. A more recent meta-analysis involving 13 observational studies enrolling 7633 patients found that in univariate analysis brisk TIL grade was correlated with better 5-year overall survival (OS), 5-year disease-specific survival (DSS) and endpoint DSS when compared with nonbrisk or absent TIL [54]. Moreover, in multifactor analysis, brisk TIL was associated with better DSS, while nonbrisk and absent TIL were associated with poor DSS [54].

Globally, lung cancer is the most common cause of cancer-related deaths [55]. Nonsmall cell lung cancer (NSCLC) is the most common form of lung cancer, representing 85% of lung cancer cases [56]. Notably, high CD3+, CD4+, and CD8+ infiltrations have been associated with a more favorable prognosis in lung cancer cases [57–60], while regulatory T cells (Tregs) were associated with poorer prognosis [61]. Additionally, a higher ratio of forkhead box P3 (FOXP3) to CD3+ cells was associated with a higher risk of relapse [62]. Two recent studies involving immune checkpoint inhibitors (ICI) (i.e., nivolumab and pembrolizumab) have similarly identified TIL as a prognostic factor for survival [63,64]. In a cohort of 189 patients with NSCLC, high TIL was associated with OS, progression-free survival (PFS), and overall response rate (ORR) [63]. A second study that included a subcohort of 144 patients who were examined for stromal and tumor TILs found stromal TILs were a better marker for PFS and OS than tumor TILs, tumor programmed death-ligand 1 (PD-L1), or stromal PD-L1 [64].

Breast cancer is a leading cause of death in women and remains largely incurable in advanced stages [65]. Human epidermal growth factor receptor 2 (HER2)+ and luminal-like breast cancer are considered poorly immunogenic with low levels of TIL and a low burden of neoantigens [66–68]. However, the triple-negative breast cancer (TNBC) subtype is more immunogenic, and a strong correlation between TILs and a better prognosis has been established [68–71]. Recent meta-analyses have provided further evidence of a relationship between TILs and survival from TNBC [72,73]. A meta-analysis involving seven studies and 1152 patients found that high levels of TIL were associated with long-term OS and disease-free survival (DFS) [73]. A second meta-analysis involving 37 studies found that patients with high TIL had greater pathological complete response (pCR), OS, and DFS [72]. Regarding subtypes, higher CD8+ levels were associated with greater DFS, while higher CD4+ levels were associated with greater OS and DFS and higher FOXP3+ levels were associated with DFS but not OS [72].

Gastric cancers are the fifth most diagnosed form of carcinomas, ranking third among global cancer deaths [74]. The high rate of mortality in gastric cancers is likely tied to the fact that upward of 60% of patients present with late-stage cancer at the time of diagnosis [75]. In gastric cancer, there have been conflicting results regarding TILs as prognostic factors for survival [76–79]. For example, CD8+ TIL have either been associated with better OS [77,78] or having no correlation with OS [76]. Similarly high density of FOXP3+ TILs has been associated with poorer OS [79], while Kim et al. [77] found that high-density FOXP3+ TIL were associated with high OS. To better determine the prognostic value of TILs in gastric cancer, a meta-analysis of 31 observational studies including 4185 patients was conducted [80]. The authors found that the amount of CD3+ TIL in the intratumoral compartment (IT) was the most significant prognostic marker for survival, while CD4+ density was not significantly correlated with survival [80]. With respect to FOXP3+ TIL, there were bidirectional prognostic roles with high FOXP3+ TIL in the IT showing a positive correlation while there was a negative correlation in the extratumoral compartment [80]. A more recent meta-analysis that looked at TILs as a whole rather than specific subtypes found that high TILs were associated with superior OS when compared with low TILs [81]. High TILs were also associated with lower depth of invasion, less lymph node involvement, and earlier tumor node metastasis [81].

Advances in screening and treatment of colorectal cancer (CRC) have improved patient survival rates; however, relapse rates of 20%–30% occur in patients with localized cancer [82]. TILs have long been shown to have a prognostic value in patients with CRC [83–85]. A meta-analysis of 22 studies involving 5108 patients with CRC examined the types of TIL and the location of TIL as prognostic factors for cancer survival [86]. Zhao et al. [86] found that CD8+ and FOXP3+ TILs correlated with improved OS and DFS while CD3+ TIL did not. In terms of location, high-density FOXP3+ TILs in the tumor stroma correlated positively with OS, while high-density CD3+ TILs in the invasive margin were associated with improved DFS, and high-density CD8+ TILs in the tumor center were associated with better DFS [86]. A larger meta-analysis that included 43 studies describing 21,015 patients found that high TIL levels were associated with greater OS when compared against low TIL numbers [87]. CD3+ TIL at the tumor center and at the invasive margin were associated with greater DFS [87]. Additionally, both CD8+ and FOXP3+ TIL located at the tumor center were associated with cancer-specific survival and OS [87].

Gynecological cancers, particularly ovarian cancer and cervical cancer, remain the leading cause of cancer deaths in women. Recently, TIL data from both cancer types have been the subject of meta-analyses to determine its prognostic role on survival. In a systematic review of 19 studies comprising 6004 patients with high-grade serous ovarian cancer (HGSOC) comparing TIL-positive vs TIL-negative patients, multivariate analysis found that intraepithelial CD8+ TILs were positively correlated with OS and PFS, while stromal CD8+ TILs positively correlated with OS [88]. In univariate analysis, intraepithelial CD3+, CD4+, CD8+, and CD103+ TILs were positively correlated with OS, as were stromal CD4+ and CD8+ TILs [88]. In a retrospective study in patients with advanced cervical cancer who were undergoing cisplatin-based concurrent chemotherapy (CCRT), Ohno et al. [89] found that infiltration by CD8+ TILs was associated with pelvic lymph node metastases while abundant infiltration by CD3+, CD4+, CD8+, CD206+, and FOXP3+ TIL were indicators of better PFS and OS.

In comparing TIL across cancers, a few trends emerge. First, high TIL levels are associated with better OS, DFS, and ORR. In terms of cell subtypes, increased levels of CD3+, CD4+, and CD8+ were additional factors associated with a better prognosis and survival. FOXP3+ TIL levels had varying effects from being associated with an increased risk of relapse [62] and poorer OS [77,80] to improved survival [86,89]. The location of TIL also impacted survival rates. This is particularly true for FOXP3+ TIL [80,86,87].

Preparation of TIL for ACT

Historically, lymphocytes used for ACT were either TIL or isolated from peripheral blood mononuclear cells (PBMCs), yielding a largely unmanipulated product for infusion [14]. Only around 30% of TILs are, in fact, tumor reactive [10]. Hence, unselected TIL products are heterogeneous mixtures of reactive and nonreactive cells, which can limit antitumor efficacy, particularly in cancers with lower mutational burden. To address this limitation, significant effort has been made to develop TIL products enriched with T-cells that recognize tumor-specific antigens, which has been shown to increase antitumor efficacy compared to bulk TIL [90–92]. Moreover, selective expansion of TIL expressing the activation marker CD137/4-1BB or CD8+PD-1+CD103+ tissue-resident memory (TRM) phenotype has also been shown to enhance clinical outcomes [93–96].

One of the most advanced strategies in TIL development is to select for and expand TILs that recognize unique tumor antigens created by somatic mutation—known as neoantigens [97,98]. Since neoantigens are expressed only by tumors, TIL directed against neoantigens are more likely to act specifically against cancer cells. Several neoantigen-specific TILs have shown a critical role in producing an effective response, including TIL that recognize a mutated cyclin-dependent kinase inhibitor 2A (CDKN2A) in melanoma patients [99], TIL that recognized mutated growth arrest-specific gene 7 (GAS7) and glyceraldehyde-3-phosphate dehydrogenase (GAPDH) in a patient with metastatic lesions in lung and soft tissue [100], and mutated protein phosphatase 1 regulatory subunit 3B (PPP1R3B) in a melanoma patient [101]. Rosenberg's group has published several scientific case reports that have demonstrated clinical efficacy of ACT using neoantigen-selected TIL, including TIL that recognized a mutation in Erb-B2 receptor tyrosine kinase 2 interacting protein (ERBB2IP) in a patient with metastatic cholangiocarcinoma [91] TIL that recognized the Kirsten rat sarcoma viral oncogene homolog (KRAS) G12D mutation in a patient with metastatic CRC [102], and TILs reactive against mutated solute carrier family 32 (SLC32A2), proteasome-associated protein ECM29 homolog (KIAA0368), calcium-dependent secretion activator 2 (CADPS2), and cathepsin B (CTSB) in a patient with metastatic breast cancer [103].

Initial approaches to identify neoantigen-specific T-cells utilized tumor cell cDNA libraries, but were time and labor intensive [99,100]. Newer strategies have been developed to reduce the workload and increase efficiency. One strategy combines whole-exome sequencing (WES) to identify tumor-specific mutations and select for high confidence mutations via RNA-seq variant frequency with mass spectrometry (MS) to isolate and identify HLA-1 correlated proteins [104–106]. Similarly, the Rosenberg laboratory has employed WES/RNA-Seq followed by tandem minigene/peptide pool screening with autologous antigen-presenting cells (APCs) to identify neoantigen-specific T-cells [103,107–111] (Fig. 2). A second strategy has been the creation of an off-the-shelf peptide library by screening high-frequency mutations in nine types of solid tumor cancers to identify genes with frequencies of greater than 10% [112]. A third strategy involves identification of T-cell antigens through trogocytosis, in which target cells capture cell surface membranes from T-cells, which has the advantage of identifying the neoantigen and TCR at the same time [113].

Gene editing approaches including the use of transposons and clustered regularly interspaced short palindromic repeats (CRISPR)/CRISPR-associated protein 9 (Cas9) have been applied to TIL to produce TIL that are more cytotoxic, have enhanced homing ability, and reduce cell exhaustion [114] (Fig. 3). Enhanced cytotoxicity has been achieved by creating TIL that secrete tumor necrosis factor alpha (TNF-α) [115]. In another experiment, transfection of TNF-related apoptosis-inducing ligand (TRAIL) and IL-2 enhanced cytotoxicity of TIL against renal cell carcinoma tumors in vitro [116]. Engineering T-cells to express receptors for chemokines secreted by tumor cells such as CXC chemokine receptor 2 (CXCR2), which binds several chemokines including CXC motif chemokine ligand 1 (CXCL1), and CXC motif

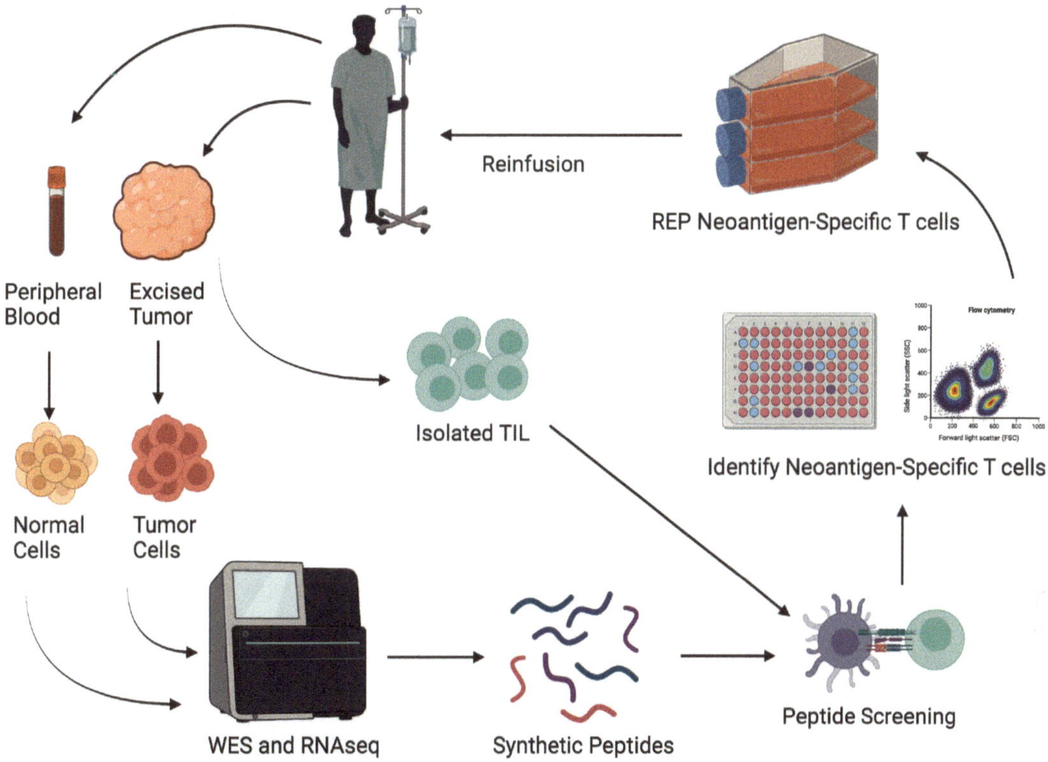

FIG. 2 Identification of neoantigen-specific T-cells. First, peripheral blood and tumor samples are obtained from the patient. TIL are isolated from the tumor and are either used as bulk TIL or undergo gene editing to enhance their cytotoxic activity. Normal cells are compared against tumor cells using WES and RNAseq to identify tumor-specific neoantigens. Synthetic peptides of these neoantigens are then generated. APCs bearing the peptides are co-cultured with TIL to identify neoantigen-specific T-cells. These T-cells are then expanded and reinfused in the patient to selectively target tumors.

chemokine ligand 8 (CXCL8) resulted in improved migration to tumors and enhanced killing activity in vitro [117] and in rodent models [118,119]. A sustained T-cell response is required for successful antitumor activity. The interaction of the programmed cell death-1 (PD-1) receptor on T-cells with its ligand on tumor cells inhibits T-cell function. T-cells that were engineered to remove PD-1 showed improved tumor removal and increased expression of TNF-α, granulocyte macrophage colony-stimulating factor (GM-CSF), and INF-γ in vitro [120].

ACT therapy

As described earlier, ACT using TILs was originally developed against melanoma and continues to be optimized. Combination of TIL ACT with various drugs including dacarbazine,

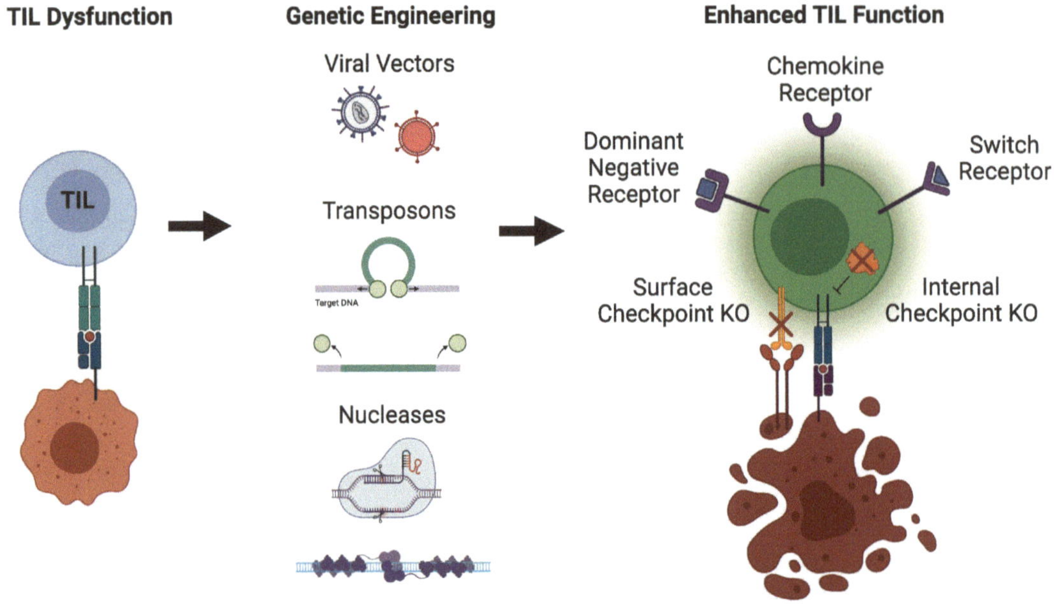

FIG. 3 Gene editing of T-cells to enhance anti-tumor function. Gene editing techniques including the use of transposons or CRISPR Cas9 are combined with delivery technologies including the use of viral vectors such as lentiviruses and retroviruses, electroporation, or the use of nanoparticles. Several modifications to the T-cells are possible including: (1) enhanced release of TNF-alpha or IFN-gamma to attack tumor cells; (2) improved homing of T-cells to tumors by increased expression of chemokine receptors such as CXCR2; (3) improved persistence through the deletion of inhibitory receptors such as PD1; (4) expression of dominant negative receptors to block the effects of suppressive cytokines such as TGF-beta; (5) expression of switch receptors which bind negative molecules released by cancer cells but further stimulate T-cells; or (6) internal checkpoint inhibition.

interferon-a, IL-2, and ipilimumab has shown clinical efficacy in melanoma [8,15,46,121,122]. There are less data available for the use of TIL either on its own or in combination with drug treatment for other solid tumors. However, reactive TIL have been isolated from primary renal cell carcinoma [123], breast [124], NSCLC [125], gastrointestinal [126,127], pancreatic [128], and head and neck squamous cell carcinoma [129] tumors and have demonstrated efficacy against tumors in vitro.

ACTs using TILs have shown some efficacy in the treatment of solid tumors in gynecological cancers and recently in NSCLC. ACTs using TILs were first used as a treatment for epithelial ovarian cancer (EOC), the most lethal gynecologic malignancy [31], in 1991 [130]. TILs were administered by i.v. with or without cisplatin-based chemotherapy resulted in seven cases with complete regression (with no recurrence for more than 15 months) and two cases of partial regression [130]. Another early study involving patients with EOC found that infusion of TIL resulted in significant OS and DFS when compared with a control group [131]. Three recent studies have demonstrated the potential for TILs to treat ovarian and cervical cancers [132–134]. A study of six patients with platinum-resistant OC were treated with TIL following lymphodepletion and followed by a subsequent IL-2 infusion, found that treatment was well-tolerated and that four patients had stable disease for 3 months and two achieved stability for 5 months [133]. A second study that combined TIL ACT in combination with checkpoint inhibitors in six patients with late-stage HGSOC, resulted in one patient experiencing a partial response, and five achieving disease stability for up to 12 months, suggesting a promising new treatment [132]. A phase II trial tested TIL therapy for human papilloma virus (HPV)-associated carcinomas which included one cervical cancer cohort vs a noncervical cancer cohort with both groups receiving a lymphodepletion regimen accompanied by a high dose of IL-2 [134]. In the cervical cancer cohort 28% of patients displayed an objective tumor response vs 18% in the noncervical cancer cohort [134]. Clinical response was correlated with the HPV reactivity of T-cells [134]. A recent Phase I trial demonstrated that TIL-based ACT may be effective for NSCLC [135]. Twenty patients whose cancer had progressed after nivolumab monotherapy were infused with autologous TIL. Two patients achieved a complete response that lasted more than 1.5 years and eleven patients had a reduction in tumor burden, suggesting that TIL-based ACT may represent a new treatment for patients with NSCLC [135].

Clinical trials have demonstrated prolonged tumor regression in patients with TCRs directed against melanoma-associated antigen recognized by T cells 1 (MART-1 ;melanoma) [136], glycoprotein 100 (gp100; melanoma) [137], NY-ESO-1 (New York esophageal cancer 1; melanoma, synovial sarcoma, multiple myeloma) [138,139], MAGE family member A4 (MAGE-A4; esophageal cancer) [140], and carcinoembryonic antigen (CEA; CRC carcinoma) [141]. Further studies are ongoing to capitalize on these findings (Table 2).

New TIL products are currently in the pipeline. Data from a phase II study on the efficacy of the TIL product Lifileucel (Iovance Biotherapeutics) was recently presented at the 2020 American Society of Clinical Oncology Annual Meeting found an overall response rate of 36.4% following a single infusion when accompanied by lymphodepletion and IL-2 infusion in patients with higher tumor burden [142]. Another product by Iovance, LN-145, demonstrated a 44% OR rate in patients with cervical cancer who had undergone at least one previous line of chemotherapy [143].

TABLE 2 TCRs that have demonstrated clinical efficacy.

Target	Cancer	Outcome	Reference
CEA	CRC carcinoma	One patient had objective regression of lung and liver metastases	Parkhurst et al. [141]
Gp100	Melanoma	Objective cancer regression observed in 30% of patients receiving human TCR and 19% in patients receiving murine TCR	Johnson et al. [137]
MAGE-A4	Esophageal	Three out of 10 patients survived more than 27 months	Kageyama et al. [140]
MART-1	Melanoma	Two out of 15 patients displayed objective regression of metastatic lesions	Morgan et al. [136]
NY-ESO-1	Myeloma	Clinical responses were observed in 80% of patients with median PFS of 19.1 months	Rapoport et al. [138]
NY-ESO-1	Synovial cell carcinoma and melanoma	Four out of six patients with synovial cell carcinoma and five of 11 patients with melanoma showed an objective clinical response. Two melanoma patients displayed a complete regression	Robbins et al. [139]

Safety and tolerability of ACT

Toxicities associated with ACT fall into three general categories: toxicity associated with the lymphodepletion regimen; toxicity associated with the TIL product and accompanying IL-2 infusion; and toxicity associated with cytokines [12]. Moreover, there are varying toxicity profiles depending on whether TCR or CAR are used [144]. Additionally, a minority of patients receiving ACT may experience viral infections including *Pneumocystis carinii*, herpes zoster, and respiratory syncytial virus [122].

Most patients who undergo a nonmyeloablative lymphodepletion chemotherapy regimen experience hematological side effects including in pancytopenia, febrile neutropenia, lymphopenia, and prolonged depression of CD4 T-cells [15,121,122,145–148]. These symptoms can be managed using good clinical practices [149,150] including the transfusion of blood products and treatment with granulocyte-colony stimulating factor (G-CSF) [15,16,121,122,145–148].

Nonhematological side effects of lymphodepletion include diarrhea, hyperbilirubinemia, and fludarabine-induced neurotoxicity [121,122,146].

It is often difficult to distinguish toxicities associated with TIL products themselves from the accompanying infusion of IL-2 [15,43,122,151]. Autoimmune reactions, largely due to destruction of melanocytes, have also been documented in response to TIL infusion with vitiligo occurring in 35% of cases [122] and uveitis occurring in 15% of cases [152]. The most common side effects associated with IL-2 infusion include chills, high fever, hypotension, and edema, resulting from systemic inflammation, all of which can be treated with supportive measures [153]. IL-2 toxicity can negatively affect multiple organ systems including the heart, lungs, kidneys, and central nervous system [10]. However, IL-2-related toxicities have been minimized with the adoption of lymphodepleting regimens as a patient's own lymphocytes are the main source of cytokines that produce interleukin-related side effects [45].

Cytokine release syndrome (CRS) is a potentially life-threatening side effect of immune-based therapies and TIL transfer [154]. Levels of inflammatory cytokines including IL-1, IL-5, IL-6, IL-10, IL-13, INFγ, and TNFα at a range of 5–69-fold over baseline have been reported with IL-6 acting as a central mediator of toxicity [154]. This release of cytokines is the result of the activation of TIL and is more prevalent in ACT involving CAR T-cells [155–158]. Symptomatology and severity of CRS varies widely and may include fever, fatigue, nausea, headache, skin rashes, and life-threatening complications including cardiac dysfunction (tachycardia, diminished cardiac output), adult respiratory distress syndrome, neurologic toxicity, renal and/or hepatic failure [154]. The severity of CRS has been correlated with tumor burden [159,160]. Life-threatening or severe cases of CRS have been effectively treated with tocilizumab, a monoclonal antibody directed against the IL-6 receptor or siltuximab a monoclonal antibody directed against IL-6 [154,160,161]. Corticosteroid therapy constitutes a second-line treatment option for CRS [154].

Future directions

Currently, there are several challenges to the development of TIL-based ACT. First, despite promising studies, this strategy has not yet been approved by the FDA to treat solid tumors.

Moreover, it is a costly procedure and essentially only available at large academic medical centers where there is expertise to produce TIL products and manage adverse events including potential CRS [162]. One way to reduce costs is to outsource TIL production to manufacturing centers that are dedicated to this purpose [163]. Indeed, current efforts to obtain FDA approval have focused on streamlining and optimizing manufacturing processes [164]. Genetic engineering of TIL to make them more effective at homing in on targets, enhance their cytotoxicity, and reduce T-cell exhaustion represents an important aspect of current and future research to produce better TIL products [114]. As manufacturing processes are streamlined and TIL products become more effective, the potential for TIL-based ACT will produce an important new therapeutic cancer treatment.

References

[1] Balkwill F, Mantovani A. Inflammation and cancer: back to virchow? Lancet 2001;357:539–45.
[2] Dean RM, Bishop MR. Allogeneic hematopoietic stem cell transplantation for lymphoma (Internet). Clinical Lymphoma 2004;4:238–49. Available from https://doi.org/10.3816/clm.2004.n.004.
[3] Gonzalez H, Hagerling C, Werb Z. Roles of the immune system in cancer: from tumor initiation to metastatic progression. Genes Dev 2018;32:1267–84.
[4] Kaplan DH, Shankaran V, Dighe AS, Stockert E, Aguet M, Old LJ, Schreiber RD. Demonstration of an interferon gamma-dependent tumor surveillance system in immunocompetent mice. Proc Natl Acad Sci U S A 1998;95:7556–61.
[5] Schreiber RD, Old LJ, Smyth MJ. Cancer immunoediting: integrating immunity's roles in cancer suppression and promotion. Science 2011;331:1565–70.
[6] Shankaran V, Ikeda H, Bruce AT, White JM, Swanson PE, Old LJ, Schreiber RD. IFNg and lymphocytes prevent primary tumour development and shape tumour immunogenicity. Nature 2001;410:1107–11.
[7] Vesely MD, Kershaw MH, Schreiber RD, Smyth MJ. Natural innate and adaptive immunity to cancer. Annu Rev Immunol 2011;29:235–71.
[8] Andersen R, Donia M, Ellebaek E, Borch TH, Kongsted P, Iversen TZ, Hölmich LR, Hendel HW, Met Ö, Andersen MH, Thor Straten P, Svane IM. Long-lasting complete responses in patients with metastatic melanoma after adoptive cell therapy with tumor-infiltrating lymphocytes and an attenuated IL2 regimen. Clin Cancer Res 2016;22:3734–45.
[9] Rosenberg SA, Restifo NP. Adoptive cell transfer as personalized immunotherapy for human cancer. Science 2015;348:62–8.
[10] Kumar A, Watkins R, Vilgelm AE. Cell therapy with TILs: training and taming T cells to fight cancer. Front Immunol 2021;12, 690499.
[11] Met Ö, Jensen KM, Chamberlain CA, Donia M, Svane IM. Principles of adoptive T cell therapy in cancer. Semin Immunopathol 2019;41:49–58.
[12] Rohaan MW, van den Berg JH, Kvistborg P, Haanen JBAG. Adoptive transfer of tumor-infiltrating lymphocytes in melanoma: a viable treatment option. J Immunother Cancer 2018;6:102.
[13] Sarivalasis A, Morotti M, Mulvey A, Imbimbo M, Coukos G. Cell therapies in ovarian cancer. Ther Adv Med Oncol 2021;13, 17588359211008399.
[14] Tsimberidou A-M, Van Morris K, Vo HH, Eck S, Lin Y-F, Rivas JM, Andersson BS. T-cell receptor-based therapy: an innovative therapeutic approach for solid tumors. J Hematol Oncol 2021;14:102.
[15] Dudley ME, Yang JC, Sherry R, Hughes MS, Royal R, Kammula U, Robbins PF, Huang J, Citrin DE, Leitman SF, Wunderlich J, Restifo NP, Thomasian A, Downey SG, Smith FO, Klapper J, Morton K, Laurencot C, White DE, Rosenberg SA. Adoptive cell therapy for patients with metastatic melanoma: evaluation of intensive myeloablative chemoradiation preparative regimen. J Clin Oncol 2008;26:5233–9.
[16] Gattinoni L, Finkelstein SE, Klebanoff CA, Antony PA, Palmer DC, Spiess PJ, Hwang LN, Yu Z, Wrzesinski C, Heimann DM, Surh CD, Rosenberg SA, Restifo NP. Removal of homeostatic cytokine sinks by lymphodepletion enhances the efficacy of adoptively transferred tumor-specific CD8+ T cells. J Exp Med 2005;202:907–12.

[17] Atkins MB, Kunkel L, Sznol M, Rosenberg SA. High-dose recombinant interleukin-2 therapy in patients with metastatic melanoma: long-term survival update. Cancer J Sci Am 2000;6(Suppl 1):S11–4.
[18] Rosenberg SA. IL-2: the first effective immunotherapy for human cancer. J Immunol 2014;192:5451–8.
[19] Chapuis AG, Ragnarsson GB, Nguyen HN, Chaney CN, Pufnock JS, Schmitt TM, Duerkopp N, Roberts IM, Pogosov GL, Ho WY, Ochsenreither S, Wölfl M, Bar M, Radich JP, Yee C, Greenberg PD. Transferred WT1-reactive CD8+ T cells can mediate antileukemic activity and persist in post-transplant patients. Sci Transl Med 2013;5, 174ra27.
[20] Yee C, Thompson JA, Byrd D, Riddell SR, Roche P, Celis E, Greenberg PD. Adoptive T cell therapy using antigen-specific CD8 T cell clones for the treatment of patients with metastatic melanoma: in vivo persistence, migration, and antitumor effect of transferred T cells (Internet). Proc Natl Acad Sci U S A 2002;99:16168–73. Available from: https://doi.org/10.1073/pnas.242600099.
[21] Barnes DW, Corp MJ, Loutit JF, Neal FE. Treatment of murine leukemia with X rays and homologous bone marrow; preliminary communication. Br Med J 1956;2:626–7.
[22] Boranić M. Transplantability of leukaemia from leukaemic mice after irradiation and injection of allogeneic spleen cells. Rev Eur Etud Clin Biol 1970;15:104–9.
[23] Bortin MM, Rimm AA, Saltzstein EC, Rodey GE. Graft vs leukemia. 3. Graft vs leukemia. 3. Apparent independent anthost and antileukemia activity of transplanted immunocompetent cells. Transplantation 1973;16:182–8.
[24] Bortin MM, Rimm AA, Rose WC, Saltzstein EC, Graft-vs-leukemia V. Absence of antileukemic effect using allogeneic h-2-identical immunocompetent cells. Transplantation 1974;18:280–3.
[25] de Vries MJ, Vos O. Treatment of mouse lymphosarcoma by total-body X-irradiation and by injection of bone marrow and lymph-node cells. J Natl Cancer Inst 1958;21:1117–29.
[26] Thomas ED, Buckner CD, Banaji M, Clift RA, Fefer A, Flournoy N, Goodell BW, Hickman RO, Lerner KG, Neiman PE, Sale GE, Sanders JE, Singer J, Stevens M, Storb R, Weiden PL. One hundred patients with acute leukemia treated by chemotherapy, total body irradiation, and allogeneic marrowtransplantation. Blood 1977;49:511–33.
[27] Weiden PL, Flournoy N, Donnall Thomas E, Prentice R, Fefer A, Dean Buckner C, Storb R. Antileukemic effect of graft-vs-host disease in human recipients of allogeneic-marrow grafts (Internet). N Engl J Med 1979;300:1068–73. Available from https://doi.org/10.1056/nejm197905103001902.
[28] Bortin MM, Truitt RL, Rimm AA, Bach FH. Graft-vs-leukaemia reactivity induced by alloimmunisation without augmentation of graft-vs-host reactivity. Nature 1979;281:490–1.
[29] Delorme EJ, Alexander P. Treatment OF primary fibrosarcoma in the rat with immune lymphocytes. Lancet 1964;2:117–20.
[30] Korngold R, Sprent J. Lethal graft-vs-host disease after bone marrow transplantation across minor histocompatibility barriers in mice. prevention by removing mature T cells from marrow. J Exp Med 1978;148:1687–98.
[31] Southam CM, Brunschwig A, Levin AG, Dizon QS. Effect of leukocytes on transplantability of human cancer. Cancer 1966;19:1743–53.
[32] Prendergast GC, Jaffee EM. Cancer immunotherapy: immune suppression and tumor growth. Elsevier; 2011. p. 428.
[33] Fernandez-Cruz E, Woda BA, Feldman JD. Elimination of syngeneic sarcomas in rats by a subset of T lymphocytes. J Exp Med 1980;152:823–41.
[34] Morgan D, Ruscetti F, Gallo R. Selective in vitro growth of T lymphocytes from normal Human bone marrows. Science 1976;193:1007–8.
[35] Ettinghausen SE, Lipford 3rd EH, Mulé JJ, Rosenberg SA. Recombinant interleukin 2 stimulates in vivo proliferation of adoptively transferred lymphokine-activated killer (LAK) cells. J Immunol 1985;135:3623–35.
[36] Grimm EA, Mazumder A, Zhang HZ, Rosenberg SA. Lymphokine-activated killer cell phenomenon. lysis of natural killer-resistant fresh solid tumor cells by interleukin 2-activated autologous human peripheral blood Lymphocytes. J Exp Med 1982;155:1823–41.
[37] Lotze MT, Grimm EA, Mazumder A, Strausser JL, Rosenberg SA. Lysis of fresh and cultured autologous tumor by human lymphocytes cultured in T-cell growth factor. Cancer Res 1981;41:4420–5.
[38] Rosenstein M, Yron I, Kaufmann Y, Rosenberg SA. Lymphokine-activated killer cells: lysis of fresh syngeneic natural killer-resistant murine tumor cells by lymphocytes cultured in interleukin 2. Cancer Res 1984;44:1946–53.
[39] Yron I, Wood Jr TA, Spiess PJ, Rosenberg SA. In vitro growth of murine t cells. v. the isolation and growth of lymphoid cells infiltrating syngeneic solid tumors. J Immunol 1980;125:238–45.

[40] Rosenberg SA, Lotze MT, Muul LM, Leitman S, Chang AE, Ettinghausen SE, Matory YL, Skibber JM, Shiloni E, Vetto JT. Observations on the systemic administration of autologous lymphokine-activated killer cells and recombinant interleukin-2 to patients with metastatic cancer. N Engl J Med 1985;313:1485–92.
[41] Rosenberg SA, Spiess P, Lafreniere R. A new approach to the adoptive immunotherapy of cancer with tumor-infiltrating lymphocytes. Science 1986;233:1318–21.
[42] Muul LM, Spiess PJ, Director EP, Rosenberg SA. Identification of specific cytolytic immune responses against autologous tumor in humans bearing malignant melanoma. J Immunol 1987;138:989–95.
[43] Rosenberg SA, Packard BS, Aebersold PM, Solomon D, Topalian SL, Toy ST, Simon P, Lotze MT, Yang JC, Seipp CA. Use of tumor-infiltrating lymphocytes and interleukin-2 in the immunotherapy of patients with metastatic melanoma. A preliminary report. N Engl J Med 1988;319:1676–80.
[44] Rosenberg SA, Yannelli JR, Yang JC, Topalian SL, Schwartzentruber DJ, Weber JS, Parkinson DR, Seipp CA, Einhorn JH, White DE. Treatment of patients with metastatic melanoma with autologous tumor-infiltrating lymphocytes and interleukin 2. J Natl Cancer Inst 1994;86:1159–66.
[45] Dudley ME, Wunderlich JR, Yang JC, Hwu P, Schwartzentruber DJ, Topalian SL, Sherry RM, Marincola FM, Leitman SF, Seipp CA, Rogers-Freezer L, Morton KE, Nahvi A, Mavroukakis SA, White DE, Rosenberg SA. A phase i study of nonmyeloablative chemotherapy and adoptive transfer of autologous tumor antigen-specific T lymphocytes in patients with metastatic melanoma. J Immunother 2002;25:243–51.
[46] Rosenberg SA, Yang JC, Sherry RM, Kammula US, Hughes MS, Phan GQ, Citrin DE, Restifo NP, Robbins PF, Wunderlich JR, Morton KE, Laurencot CM, Steinberg SM, White DE, Dudley ME. Durable complete responses in heavily pretreated patients with metastatic melanoma using T-cell transfer immunotherapy. Clin Cancer Res 2011;17:4550–7.
[47] Rosenberg SA, Dudley ME. Adoptive cell therapy for the treatment of patients with metastatic melanoma. Curr Opin Immunol 2009;21:233–40.
[48] Clark Jr WH, Elder DE, Guerry 4th D, Braitman LE, Trock BJ, Schultz D, Synnestvedt M, Halpern AC. Model Predicting survival in stage I melanoma based on tumor progression. J Natl Cancer Inst 1989;81:1893–904.
[49] Elder DE, Guerry 4th D, VanHorn M, Hurwitz S, Zehngebot L, Goldman LI, LaRossa D, Hamilton R, Bondi EE, Clark Jr WH. The role of lymph node dissection for clinical stage I malignant melanoma of intermediate thickness (1.51-3.99 mm). Cancer 1985;56:413–8.
[50] Clemente CG, Mihm Jr MC, Bufalino R, Zurrida S, Collini P, Cascinelli N. Prognostic value of tumor infiltrating lymphocytes in the vertical growth phase of primary cutaneous melanoma. Cancer 1996;77:1303–10.
[51] Thomas NE, Busam KJ, From L, Kricker A, Armstrong BK, Anton-Culver H, Gruber SB, Gallagher RP, Zanetti R, Rosso S, Dwyer T, Venn A, Kanetsky PA, Groben PA, Hao H, Orlow I, Reiner AS, Luo L, Paine S, Ollila DW, Wilcox H, Begg CB, Berwick M. Tumor-infiltrating lymphocyte grade in primary melanomas is independently associated with melanoma-specific survival in the population-based genes, environment and melanoma study. J Clin Oncol 2013;31:4252–9.
[52] Tuthill RJ, Unger JM, Liu PY, Flaherty LE, Sondak VK. Southwest oncology group. risk assessment in localized primary cutaneous melanoma: a southwest oncology group study evaluating nine factors and a test of the clark logistic regression prediction model. Am J Clin Pathol 2002;118:504–11.
[53] Weiss SA, Han SW, Lui K, Tchack J, Shapiro R, Berman R, Zhong J, Krogsgaard M, Osman I, Darvishian F. Immunologic heterogeneity of tumor-infiltrating lymphocyte composition in primary melanoma. Hum Pathol 2016;57:116–25.
[54] Sun Q, Sun H, Wu N, Cong L, Cong X. Prognostic significance of tumor-infiltrating lymphocyte grade in melanoma: a meta-analysis. Dermatology 2020;236:481–92.
[55] Jemal A, Bray F, Center MM, Ferlay J, Ward E, Forman D. Global cancer statistics (Internet). CA Cancer J Clin 2011;61:69–90. Available from https://doi.org/10.3322/caac.20107.
[56] Goldstraw P, Ball D, Jett JR, Le Chevalier T, Lim E, Nicholson AG, Shepherd FA. Non-small-cell lung cancer. Lancet 2011;378:1727–40.
[57] Hiraoka K, Miyamoto M, Cho Y, Suzuoki M, Oshikiri T, Nakakubo Y, Itoh T, Ohbuchi T, Kondo S, Katoh H. Concurrent infiltration by CD8+ T cells and CD4+ T cells is a favourable prognostic factor in non-small-cell lung carcinoma. Br J Cancer 2006;94:275–80.
[58] Johnson SK, Kerr KM, Chapman AD, Kennedy MM, King G, Cockburn JS, Jeffrey RR. Immune cell infiltrates and prognosis in primary carcinoma of the lung. Lung Cancer 2000;27:27–35.
[59] Ruffini E, Asioli S, Filosso PL, Lyberis P, Bruna MC, Macrì L, Daniele L, Oliaro A. Clinical significance of tumor-infiltrating lymphocytes in lung neoplasms. Ann Thorac Surg 2009;87:365–71. discussion 371–372.

[60] Wakabayashi O, Yamazaki K, Oizumi S, Hommura F, Kinoshita I, Ogura S, Dosaka-Akita H, Nishimura M. CD4+ T cells in cancer stroma, not CD8+ T cells in cancer cell nests, are associated with favorable prognosis in human non-small cell lung cancers. Cancer Sci 2003;94:1003–9.
[61] Tao H, Mimura Y, Aoe K, Kobayashi S, Yamamoto H, Matsuda E, Okabe K, Matsumoto T, Sugi K, Ueoka H. Prognostic potential of FOXP3 expression in non-small cell lung cancer cells combined with tumor-infiltrating regulatory T cells. Lung Cancer 2012;75:95–101.
[62] Suzuki K, Kadota K, Sima CS, Nitadori J-I, Rusch VW, Travis WD, Sadelain M, Adusumilli PS. Clinical impact of immune microenvironment in stage I lung adenocarcinoma: tumor Interleukin-12 Receptor β2 (IL-12Rβ2), IL-7R, and stromal FoxP3/CD3 ratio are independent predictors of recurrence. J Clin Oncol 2013;31:490–8.
[63] Gataa I, Mezquita L, Rossoni C, Auclin E, Kossai M, Aboubakar F, Le Moulec S, Massé J, Masson M, Radosevic-Robin N, Alemany P, Rouanne M, Bluthgen V, Hendriks L, Caramella C, Gazzah A, Planchard D, Pignon J-P, Besse B, Adam J. Tumour-infiltrating lymphocyte density is associated with favourable outcome in patients with advanced non-small cell lung cancer treated with immunotherapy. Eur J Cancer 2021;145:221–9.
[64] Hashemi S, Fransen MF, Niemeijer A, Ben Taleb N, Houda I, Veltman J, Becker-Commissaris A, Daniels H, Crombag L, Radonic T, Jongeneel G, Tarasevych S, Looysen E, van Laren M, Tiemessen M, van Diepen V, Maassen-van den Brink K, Thunnissen E, Bahce I. Surprising impact of stromal TIL's on immunotherapy efficacy in a real-world lung Cancer study. Lung Cancer 2021;153:81–9.
[65] Lord SJ, Kiely BE, Pearson S-A, Daniels B, O'Connell DL, Beith J, Bulsara MK, Houssami N. Metastatic breast cancer incidence, site and survival in Australia, 2001–2016: a population-based health record linkage study protocol (Internet). BMJ Open 2019;9, e026414. Available from: https://doi.org/10.1136/bmjopen-2018-026414.
[66] Luen SJ, Salgado R, Fox S, Savas P, Eng-Wong J, Clark E, Kiermaier A, Swain SM, Baselga J, Michiels S, Loi S. Tumour-infiltrating lymphocytes in advanced HER2-positive breast cancer treated with pertuzumab or placebo in addition to trastuzumab and docetaxel: a retrospective analysis of the CLEOPATRA study. Lancet Oncol 2017;18:52–62.
[67] Miyoshi Y, Shien T, Ogiya A, Ishida N, Yamazaki K, Horii R, Horimoto Y, Masuda N, Yasojima H, Inao T, Osako T, Takahashi M, Tomioka N, Wanifuchi-Endo Y, Hosoda M, Doihara H, Yamashita H. Associations in tumor infiltrating lymphocytes between clinicopathological factors and clinical outcomes in estrogen receptor-positive/human epidermal growth factor receptor type 2 negative breast cancer. Oncol Lett 2019;17:2177–86.
[68] Nagarajan D, McArdle S. Immune landscape of breast cancers (Internet). Biomedicines 2018;6:20. Available from: https://doi.org/10.3390/biomedicines6010020.
[69] Ali HR, Provenzano E, Dawson S-J, Blows FM, Liu B, Shah M, Earl HM, Poole CJ, Hiller L, Dunn JA, Bowden SJ, Twelves C, Bartlett JMS, Mahmoud SMA, Rakha E, Ellis IO, Liu S, Gao D, Nielsen TO, Pharoah PDP, Caldas C. Association between CD8+ T-cell infiltration and breast cancer survival in 12,439 patients. Ann Oncol 2014;25:1536–43.
[70] Denkert C, von Minckwitz G, Darb-Esfahani S, Lederer B, Heppner BI, Weber KE, Budczies J, Huober J, Klauschen F, Furlanetto J, Schmitt WD, Blohmer J-U, Karn T, Pfitzner BM, Kümmel S, Engels K, Schneeweiss A, Hartmann A, Noske A, Fasching PA, Jackisch C, van Mackelenbergh M, Sinn P, Schem C, Hanusch C, Untch M, Loibl S. Tumour-infiltrating lymphocytes and prognosis in different subtypes of breast cancer: a pooled analysis of 3771 patients treated with neoadjuvant therapy. Lancet Oncol 2018;19:40–50.
[71] West NR, Milne K, Truong PT, Macpherson N, Nelson BH, Watson PH. Tumor-infiltrating lymphocytes predict response to anthracycline-based chemotherapy in estrogen receptor-negative breast cancer. Breast Cancer Res 2011;13:R126.
[72] Gao G, Wang Z, Qu X, Zhang Z. Prognostic value of tumor-infiltrating lymphocytes in patients with triple-negative breast cancer: a systematic review and meta-analysis. BMC Cancer 2020;20:179.
[73] Lotfinejad P, Asghari Jafarabadi M, Abdoli Shadbad M, Kazemi T, Pashazadeh F, Sandoghchian Shotorbani S, Jadidi Niaragh F, Baghbanzadeh A, Vahed N, Silvestris N, Baradaran B. Prognostic role and clinical significance of tumor-infiltrating lymphocyte (TIL) and programmed death ligand 1 (PD-L1) expression in triple-negative breast cancer (TNBC): a systematic review and meta-analysis study. Diagnostics (Basel) (Internet) 2020;10(9). Available from https://doi.org/10.3390/diagnostics10090704.
[74] Bray F, Ferlay J, Soerjomataram I, Siegel RL, Torre LA, Jemal A. Global cancer statistics 2018: GLOBOCAN estimates of incidence and mortality worldwide for 36 cancers in 185 countries. CA Cancer J Clin 2018;68:394–424.
[75] Oh S-Y, Lee J-H, Lee H-J, Kim TH, Huh Y-J, Ahn H-S, Suh Y-S, Kong S-H, Kim GH, Ahn SJ, Kim SH, Choi Y, Yang H-K. Natural history of gastric cancer: observational study of gastric cancer patients not treated during follow-up. Ann Surg Oncol 2019;26:2905–11.

[76] Haas M, Dimmler A, Hohenberger W, Grabenbauer GG, Niedobitek G, Distel LV. Stromal regulatory T-cells are associated with a favourable prognosis in gastric cancer of the cardia. BMC Gastroenterol 2009;9:65.
[77] Kim K-J, Lee KS, Cho HJ, Kim YH, Yang HK, Kim WH, Kang GH. Prognostic implications of tumor-infiltrating FoxP3+ regulatory T cells and CD8+ cytotoxic t cells in microsatellite-unstable gastric cancers. Hum Pathol 2014;45:285–93.
[78] Lee HE, Chae SW, Lee YJ, Kim MA, Lee HS, Lee BL, Kim WH. Prognostic implications of type and density of tumour-infiltrating lymphocytes in gastric cancer. Br J Cancer 2008;99:1704–11.
[79] Zhou S, Shen Z, Wang Y, Ma H, Xu S, Qin J, Chen L, Tao H, Zhen Z, Chen G, Zhang Z, Li R, Xiao H, Zhong C, Yang Y, Liang C. CCR7 expression and intratumoral FOXP3+ regulatory T cells are correlated with overall survival and lymph node metastasis in gastric cancer. PLoS One 2013;8, e74430.
[80] Zheng X, Song X, Shao Y, Xu B, Chen L, Zhou Q, Hu W, Zhang D, Wu C, Tao M, Zhu Y, Jiang J. Prognostic role of tumor-infiltrating lymphocytes in gastric cancer: a meta-analysis (Internet). Oncotarget 2017;8:57386–98. Available from: https://doi.org/10.18632/oncotarget.18065.
[81] Tian C, Jing H, Wang C, Wang W, Cui Y, Chen J, Sha D. Prognostic role of tumour-infiltrating lymphocytes assessed by H&E-stained section in gastric cancer: a systematic review and meta-analysis. BMJ Open 2021;11, e044163.
[82] Benson AB, Venook AP, Al-Hawary MM, Cederquist L, Chen Y-J, Ciombor KK, Cohen S, Cooper HS, Deming D, Engstrom PF, Garrido-Laguna I, Grem JL, Grothey A, Hochster HS, Hoffe S, Hunt S, Kamel A, Kirilcuk N, Krishnamurthi S, Messersmith WA, Meyerhardt J, Miller ED, Mulcahy MF, Murphy JD, Nurkin S, Saltz L, Sharma S, Shibata D, Skibber JM, Sofocleous CT, Stoffel EM, Stotsky-Himelfarb E, Willett CG, Wuthrick E, Gregory KM, Freedman-Cass DA. NCCN guidelines insights: colon cancer, version 2.2018. J Natl Compr Canc Netw 2018;16:359–69.
[83] Jass JR. Lymphocytic infiltration and survival in rectal cancer. J Clin Pathol 1986;39:585–9.
[84] Joost P, Bendahl P-O, Halvarsson B, Rambech E, Nilbert M, editors. Efficient and reproducible identification of mismatch repair deficient colon cancer: validation of the mmr index and comparison with other predictive models (Internet). BMC Clin Pathol 2013;13. Available from https://doi.org/10.1186/1472-6890-13-33.
[85] Rozek LS, Schmit SL, Greenson JK, Tomsho LP, Rennert HS, Rennert G, Gruber SB. Tumor-infiltrating lymphocytes, crohn's-like lymphoid reaction, and survival from colorectal cancer (Internet). J Natl Cancer Inst 2016;108. Available from: https://doi.org/10.1093/jnci/djw027.
[86] Zhao L, Cao YJ. Engineered T cell therapy for cancer in the clinic. Front Immunol 2019;10:2250.
[87] Idos GE, Kwok J, Bonthala N, Kysh L, Gruber SB, Qu C. The prognostic implications of tumor infiltrating lymphocytes in colorectal cancer: a systematic review and meta-analysis. Sci Rep 2020;10:1–14.
[88] Hao J, Yu H, Zhang T, An R, Xue Y. Prognostic impact of tumor-infiltrating lymphocytes in high grade serous ovarian cancer: a systematic review and meta-analysis. Ther Adv Med Oncol 2020;12, 1758835920967241.
[89] Ohno A, Iwata T, Katoh Y, Taniguchi S, Tanaka K, Nishio H, Nakamura M, Morisada T, Chen G, Saito M, Yaguchi T, Kawakami Y, Aoki D. Tumor-infiltrating lymphocytes predict survival outcomes in patients with cervical cancer treated with Concurrent chemoradiotherapy. Gynecol Oncol 2020;159:329–34.
[90] Muranski P, Borman ZA, Kerkar SP, Klebanoff CA, Ji Y, Sanchez-Perez L, Sukumar M, Reger RN, Yu Z, Kern SJ, Roychoudhuri R, Ferreyra GA, Shen W, Durum SK, Feigenbaum L, Palmer DC, Antony PA, Chan C-C, Laurence A, Danner RL, Gattinoni L, Restifo NP. Th17 cells are long lived and retain a stem cell-like Molecular signature. Immunity 2011;35:972–85.
[91] Tran E, Turcotte S, Gros A, Robbins PF, Lu Y-C, Dudley ME, Wunderlich JR, Somerville RP, Hogan K, Hinrichs CS, Parkhurst MR, Yang JC, Rosenberg SA. Cancer immunotherapy based on mutation-specific cd4+t cells in a patient with epithelial cancer. Science 2014;344:641–5.
[92] Veatch JR, Lee SM, Fitzgibbon M, Chow I-T, Jesernig B, Schmitt T, Kong YY, Kargl J, Houghton AM, Thompson JA, McIntosh M, Kwok WW, Riddell SR. Tumor-infiltrating BRAFV600E-specific CD4+ T cells correlated with complete clinical response in melanoma. J Clin Invest 2018;128:1563–8.
[93] Duhen T, Duhen R, Montler R, Moses J, Moudgil T, de Miranda NF, Goodall CP, Blair TC, Fox BA, McDermott JE, Chang S-C, Grunkemeier G, Leidner R, Bell RB, Weinberg AD. Co-expression of CD39 and CD103 identifies tumor-reactive CD8 T cells in human solid tumors. Nat Commun 2018;9:1–13.
[94] Malik BT, Byrne KT, Vella JL, Zhang P, Shabaneh TB, Steinberg SM, Molodtsov AK, Bowers JS, Angeles CV, Paulos CM, Huang YH, Turk MJ. Resident memory T cells in the skin mediate durable immunity to melanoma. Sci Immunol (Internet) 2017;2(10). https://doi.org/10.1126/sciimmunol.aam6346. Available from.

[95] Webb JR, Milne K, Watson P, Deleeuw RJ, Nelson BH. Tumor-infiltrating lymphocytes expressing the tissue resident memory marker CD103 are associated with increased survival in high-grade serous ovarian cancer. Clin Cancer Res 2014;20:434–44.
[96] Ye Q, Song D-G, Poussin M, Yamamoto T, Best A, Li C, Coukos G, Powell Jr DJ. CD137 accurately identifies and enriches for naturally occurring tumor-reactive T cells in tumor. Clin Cancer Res 2014;20:44–55.
[97] Schumacher TN, Schreiber RD. Neoantigens in cancer immunotherapy. Science 2015;348:69–74.
[98] Schumacher TN, Scheper W, Kvistborg P. Cancer neoantigens. Annu Rev Immunol 2019;37:173–200.
[99] Huang J, El-Gamil M, Dudley ME, Li YF, Rosenberg SA, Robbins PF. T cells associated with tumor regression recognize frameshifted products of the CDKN2A tumor suppressor gene locus and a mutated HLA class I gene product. J Immunol 2004;172:6057–64.
[100] Zhou J, Dudley ME, Rosenberg SA, Robbins PF. Persistence of multiple tumor-specific t-cell clones is associated with complete tumor regression in a melanoma patient receiving adoptive cell transfer therapy. J Immunother 2005;28:53–62.
[101] Lu Y-C, Yao X, Li YF, El-Gamil M, Dudley ME, Yang JC, Almeida JR, Douek DC, Samuels Y, Rosenberg SA, Robbins PF. Mutated PPP1R3B is recognized by T cells used to treat a melanoma patient who experienced a durable complete tumor regression. J Immunol 2013;190:6034–42.
[102] Tran E, Robbins PF, Lu Y-C, Prickett TD, Gartner JJ, Jia L, Pasetto A, Zheng Z, Ray S, Groh EM, Kriley IR, Rosenberg SA. T-cell transfer therapy targeting mutant KRAS in cancer. N Engl J Med 2016;375:2255–62.
[103] Zacharakis N, Chinnasamy H, Black M, Xu H, Lu Y-C, Zheng Z, Pasetto A, Langhan M, Shelton T, Prickett T, Gartner J, Jia L, Trebska-McGowan K, Somerville RP, Robbins PF, Rosenberg SA, Goff SL, Feldman SA. Immune recognition of somatic mutations leading to complete durable regression in metastatic breast cancer. Nat Med 2018;24:724–30.
[104] Bassani-Sternberg M, Bräunlein E, Klar R, Engleitner T, Sinitcyn P, Audehm S, Straub M, Weber J, Slotta-Huspenina J, Specht K, Martignoni ME, Werner A, Hein R, H. Busch D, Peschel C, Rad R, Cox J, Mann M, Krackhardt AM. Direct identification of clinically relevant neoepitopes presented on native human melanoma tissue by mass spectrometry. Nat Commun 2016;7:1–16.
[105] Gros A, Tran E, Parkhurst MR, Ilyas S, Pasetto A, Groh EM, Robbins PF, Yossef R, Garcia-Garijo A, Fajardo CA, Prickett TD, Jia L, Gartner JJ, Ray S, Ngo L, Wunderllich JR, Yang JC, Rosenberg SA. Recognition of human gastrointestinal cancer neoantigens by circulating PD-1+ lymphocytes. J Clin Invest 2019;129:4992–5004.
[106] Yadav M, Jhunjhunwala S, Phung QT, Lupardus P, Tanguay J, Bumbaca S, Franci C, Cheung TK, Fritsche J, Weinschenk T, Modrusan Z, Mellman I, Lill JR, Delamarre L. Predicting immunogenic tumour mutations by combining mass spectrometry and exome sequencing. Nature 2014;515:572–6.
[107] Cafri G, Yossef R, Pasetto A, Deniger DC, Lu Y-C, Parkhurst M, Gartner JJ, Jia L, Ray S, Ngo LT, Jafferji M, Sachs A, Prickett T, Robbins PF, Rosenberg SA. Memory T cells targeting oncogenic mutations detected in peripheral blood of epithelial cancer patients. Nat Commun 2019;10:1–9.
[108] Leko V, Cafri G, Yossef R, Paria B, Hill V, Gurusamy D, Zheng Z, Gartner JJ, Prickett TD, Goff SL, Robbins P, Lu Y-C, Rosenberg SA. Identification of neoantigen-reactive T lymphocytes in the peripheral blood of a patient with glioblastoma. J Immunother Cancer (Internet) 2021;9(7). https://doi.org/10.1136/jitc-2021-002882. Available from.
[109] Lu Y-C, Zheng Z, Robbins PF, Tran E, Prickett TD, Gartner JJ, Li YF, Ray S, Franco Z, Bliskovsky V, Fitzgerald PC, Rosenberg SA. An efficient single-cell RNA-seq approach to identify neoantigen-specific T cell receptors. Mol Ther 2018;26:379–89.
[110] Parkhurst MR, Robbins PF, Tran E, Prickett TD, Gartner JJ, Jia L, Ivey G, Li YF, El-Gamil M, Lalani A, Crystal JS, Sachs A, Groh E, Ray S, Ngo LT, Kivitz S, Pasetto A, Yossef R, Lowery FJ, Goff SL, Lo W, Cafri G, Deniger DC, Malekzadeh P, Ahmadzadeh M, Wunderlich JR, Somerville RPT, Rosenberg SA. Unique neoantigens arise from somatic mutations in patients with gastrointestinal cancers. Cancer Discov 2019;9:1022–35.
[111] Tran E, Ahmadzadeh M, Lu Y-C, Gros A, Turcotte S, Robbins PF, Gartner JJ, Zheng Z, Li YF, Ray S, Wunderlich JR, Somerville RP, Rosenberg SA. Immunogenicity of somatic mutations in human gastrointestinal cancers. Science 2015;350(6266):1387–90.
[112] Chen F, Zou Z, Du J, Su S, Shao J, Meng F, Yang J, Xu Q, Ding N, Yang Y, Liu Q, Wang Q, Sun Z, Zhou S, Du S, Wei J, Liu B. Neoantigen identification strategies enable personalized immunotherapy in refractory solid tumors. J Clin Invest 2019;129:2056–70.
[113] Li G, Bethune MT, Wong S, Joglekar AV, Leonard MT, Wang JK, Kim JT, Cheng D, Peng S, Zaretsky JM, Su Y, Luo Y, Heath JR, Ribas A, Witte ON, Baltimore D. T cell antigen discovery via trogocytosis. Nat Methods 2019;16:183–90.

[114] Jiménez-Reinoso A, Nehme-Álvarez D, Domínguez-Alonso C, Álvarez-Vallina L. Synthetic TILs: engineered tumor-infiltrating lymphocytes with improved therapeutic potential. Front Oncol 2020;10, 593848.

[115] Treisman J, Hwu P, Yannelli JR, Shafer GE, Cowherd R, Samid D, Rosenberg SA. Upregulation of tumor necrosis factor-alpha production by retrovirally transduced human tumor-infiltrating lymphocytes using trans-retinoic acid. Cell Immunol 1994;156:448–57.

[116] Tian J-Q, Wang Z-P, Rodriguez R, Fu J-S, Lu J-Z, Ma B-L. In vitro enhanced cytotoxicity of tumor-infiltrating lymphocytes transfected with tumor necrosis factor-related apoptosis-inducing ligand and/or interleukin-2 gene in human renal cell carcinoma (Internet). Urology 2006;67:1093–8. Available from: https://doi.org/10.1016/j.urology.2005.11.030.

[117] Kershaw MH, Wang G, Westwood JA, Pachynski RK, Tiffany HL, Marincola FM, Wang E, Young HA, Murphy PM, Hwu P. Redirecting migration of T cells to chemokine secreted from tumors by genetic modification with CXCR2. Hum Gene Ther 2002;13:1971–80.

[118] Idorn M, Skadborg SK, Kellermann L, Halldórsdóttir HR, Holmen Olofsson G, Met Ö, Thor SP. Chemokine receptor engineering of T cells with CXCR2 improves homing towards subcutaneous human melanomas in xenograft mouse model. Oncoimmunology 2018;7, e1450715.

[119] Peng W, Ye Y, Rabinovich BA, Liu C, Lou Y, Zhang M, Whittington M, Yang Y, Overwijk WW, Lizée G, Hwu P. Transduction of tumor-specific T cells with CXCR2 chemokine receptor improves migration to tumor and antitumor immune responses. Clin Cancer Res 2010;16:5458–68.

[120] Beane JD, Lee G, Zheng Z, Mendel M, Abate-Daga D, Bharathan M, Black M, Gandhi N, Yu Z, Chandran S, Giedlin M, Ando D, Miller J, Paschon D, Guschin D, Rebar EJ, Reik A, Holmes MC, Gregory PD, Restifo NP, Rosenberg SA, Morgan RA, Feldman SA. Clinical scale zinc finger nuclease-mediated gene editing of PD-1 in tumor infiltrating lymphocytes for the treatment of metastatic melanoma. Mol Ther 2015;23:1380–90.

[121] Besser MJ, Shapira-Frommer R, Itzhaki O, Treves AJ, Zippel DB, Levy D, Kubi A, Shoshani N, Zikich D, Ohayon Y, Ohayon D, Shalmon B, Markel G, Yerushalmi R, Apter S, Ben-Nun A, Ben-Ami E, Shimoni A, Nagler A, Schachter J. Adoptive transfer of tumor-infiltrating lymphocytes in patients with metastatic melanoma: intent-to-treat analysis and efficacy after failure to prior immunotherapies. Clin Cancer Res 2013;19(17):4792–800.

[122] Dudley ME, Wunderlich JR, Yang JC, Sherry RM, Topalian SL, Restifo NP, Royal RE, Kammula U, White DE, Mavroukakis SA, Rogers LJ, Gracia GJ, Jones SA, Mangiameli DP, Pelletier MM, Gea-Banacloche J, Robinson MR, Berman DM, Filie AC, Abati A, Rosenberg SA. Adoptive cell transfer therapy following non-myeloablative but lymphodepleting chemotherapy for the treatment of patients with refractory metastatic melanoma. J Clin Oncol 2005;23:2346–57.

[123] Andersen R, Westergaard MCW, Kjeldsen JW, Müller A, Pedersen NW, Hadrup SR, Met Ö, Seliger B, Kromann-Andersen B, Hasselager T, Donia M, Svane IM. T-cell responses in the microenvironment of primary renal cell carcinoma-implications for adoptive cell therapy. Cancer Immunol Res 2018;6:222–35.

[124] Lee HJ, Kim Y-A, Sim CK, Heo S-H, Song IH, Park HS, Park SY, Bang WS, Park IA, Lee M, Lee JH, Cho YS, Chang S, Jung J, Kim J, Lee SB, Kim SY, Lee MS, Gong G. Expansion of tumor-infiltrating lymphocytes and their potential for application as adoptive cell transfer therapy in human breast cancer. Oncotarget 2017;8:113345–59.

[125] Ben-Avi R, Farhi R, Ben-Nun A, Gorodner M, Greenberg E, Markel G, Schachter J, Itzhaki O, Besser MJ. Establishment of adoptive cell therapy with tumor infiltrating lymphocytes for non-small cell lung cancer patients. Cancer Immunol Immunother 2018;67:1221–30.

[126] Turcotte S, Gros A, Hogan K, Tran E, Hinrichs CS, Wunderlich JR, Dudley ME, Rosenberg SA. Phenotype and function of T cells infiltrating visceral metastases from gastrointestinal cancers and melanoma: implications for adoptive cell transfer therapy. J Immunol 2013;191:2217–25.

[127] Turcotte S, Gros A, Tran E, Lee C-CR, Wunderlich JR, Robbins PF, Rosenberg SA. Tumor-reactive CD8+ T cells in metastatic gastrointestinal cancer refractory to chemotherapy. Clin Cancer Res 2014;20:331–43.

[128] Hall M, Liu H, Malafa M, Centeno B, Hodul PJ, Pimiento J, Pilon-Thomas S, Sarnaik AA. Expansion of tumor-infiltrating lymphocytes (TIL) from human pancreatic tumors. J Immunother Cancer 2016;4:61.

[129] Junker N, Andersen MH, Wenandy L, Dombernowsky SL, Kiss K, Sørensen CH, Therkildsen MH, Von Buchwald C, Andersen E, Straten PT, Svane IM. Bimodal ex vivo expansion of T cells from patients with head and neck squamous cell carcinoma: a prerequisite for adoptive cell transfer. Cytotherapy 2011;13:822–34.

[130] Aoki Y, Takakuwa K, Kodama S, Tanaka K, Takahashi M, Tokunaga A, Takahashi T. Use of adoptive transfer of tumor-infiltrating lymphocytes alone or in combination with cisplatin-containing chemotherapy in patients with epithelial ovarian cancer. Cancer Res 1991;51:1934–9.

[131] Fujita K, Ikarashi H, Takakuwa K, Kodama S, Tokunaga A, Takahashi T, Tanaka K. Prolonged disease-free period in patients with advanced epithelial ovarian cancer after adoptive transfer of tumor-infiltrating lymphocytes. Clin Cancer Res 1995;1:501–7.
[132] Kverneland AH, Pedersen M, Westergaard MCW, Nielsen M, Borch TH, Olsen LR, Aasbjerg G, Santegoets SJ, van der Burg SH, Milne K, Nelson BH, Met Ö, Donia M, Svane IM. Adoptive cell therapy in combination with checkpoint inhibitors in ovarian cancer. Oncotarget 2020;11:2092–105.
[133] Pedersen M, Westergaard MCW, Milne K, Nielsen M, Borch TH, Poulsen LG, Hendel HW, Kennedy M, Briggs G, Ledoux S, Nøttrup TJ, Andersen P, Hasselager T, Met Ö, Nelson BH, Donia M, Svane IM. Adoptive cell therapy with tumor-infiltrating lymphocytes in patients with metastatic ovarian cancer: a pilot study. Oncoimmunology 2018;7, e1502905.
[134] Stevanović S, Helman SR, Wunderlich JR, Langhan MM, Doran SL, Kwong MLM, Somerville RPT, Klebanoff CA, Kammula US, Sherry RM, Yang JC, Rosenberg SA, Hinrichs CS. A phase II study of tumor-infiltrating lymphocyte therapy for human papillomavirus-associated epithelial cancers. Clin Cancer Res 2019;25:1486–93.
[135] Creelan BC, Wang C, Teer JK, Toloza EM, Yao J, Kim S, Landin AM, Mullinax JE, Saller JJ, Saltos AN, Noyes DR, Montoya LB, Curry W, Pilon-Thomas SA, Chiappori AA, Tanvetyanon T, Kaye FJ, Thompson ZJ, Yoder SJ, Fang B, Koomen JM, Sarnaik AA, Chen D-T, Conejo-Garcia JR, Haura EB, Antonia SJ. Tumor-infiltrating lymphocyte treatment for anti-PD-1-resistant metastatic lung cancer: a phase 1 trial. Nat Med 2021;27:1410–8.
[136] Morgan RA, Dudley ME, Wunderlich JR, Hughes MS, Yang JC, Sherry RM, Royal RE, Topalian SL, Kammula US, Restifo NP, Zheng Z, Nahvi A, de Vries CR, Rogers-Freezer LJ, Mavroukakis SA, Rosenberg SA. Cancer regression in patients after transfer of genetically engineered lymphocytes. Science 2006;314:126–9.
[137] Johnson LA, Morgan RA, Dudley ME, Cassard L, Yang JC, Hughes MS, Kammula US, Royal RE, Sherry RM, Wunderlich JR, Lee C-CR, Restifo NP, Schwarz SL, Cogdill AP, Bishop RJ, Kim H, Brewer CC, Rudy SF, VanWaes C, Davis JL, Mathur A, Ripley RT, Nathan DA, Laurencot CM, Rosenberg SA. Gene therapy with human and mouse T-CELL receptors mediates cancer regression and targets normal tissues expressing cognate antigen. Blood 2009;114:535–46.
[138] Rapoport AP, Stadtmauer EA, Binder-Scholl GK, Goloubeva O, Vogl DT, Lacey SF, Badros AZ, Garfall A, Weiss B, Finklestein J, Kulikovskaya I, Sinha SK, Kronsberg S, Gupta M, Bond S, Melchiori L, Brewer JE, Bennett AD, Gerry AB, Pumphrey NJ, Williams D, Tayton-Martin HK, Ribeiro L, Holdich T, Yanovich S, Hardy N, Yared J, Kerr N, Philip S, Westphal S, Siegel DL, Levine BL, Jakobsen BK, Kalos M, June CH. NY-ESO-1- specific TCR-engineered T cells mediate sustained antigen-specific antitumor effects in myeloma. Nat Med 2015;21:914–21.
[139] Robbins PF, Morgan RA, Feldman SA, Yang JC, Sherry RM, Dudley ME, Wunderlich JR, Nahvi AV, Helman LJ, Mackall CL, Kammula US, Hughes MS, Restifo NP, Raffeld M, Lee C-CR, Levy CL, Li YF, El-Gamil M, Schwarz SL, Laurencot C, Rosenberg SA. Tumor regression in patients with metastatic synovial cell sarcoma and melanoma using genetically engineered lymphocytes reactive with NY-ESO-1. J Clin Oncol 2011;29:917–24.
[140] Kageyama S, Ikeda H, Miyahara Y, Imai N, Ishihara M, Saito K, Sugino S, Ueda S, Ishikawa T, Kokura S, Naota H, Ohishi K, Shiraishi T, Inoue N, Tanabe M, Kidokoro T, Yoshioka H, Tomura D, Nukaya I, Mineno J, Takesako K, Katayama N, Shiku H. Adoptive transfer of MAGE-A4 T-cell receptor gene-transduced lymphocytes in patients with recurrent esophageal cancer. Clin Cancer Res 2015;21:2268–77.
[141] Parkhurst MR, Yang JC, Langan RC, Dudley ME, Nathan D-AN, Feldman SA, Davis JL, Morgan RA, Merino MJ, Sherry RM, Hughes MS, Kammula US, Phan GQ, Lim RM, Wank SA, Restifo NP, Robbins PF, Laurencot CM, Rosenberg SA. T cells targeting carcinoembryonic antigen can mediate regression of metastatic colorectal cancer but induce severe transient colitis. Mol Ther 2011;19:620–6.
[142] Sarnaik A, Khushalani NI, Chesney JA, Lewis KD, Medina TM, Kluger HM, Thomas SS, Domingo Musibay E, Pavlick AC, Whitman ED, Martin-Algarra S, Corrie PG, Lutzky J, Hamid O, Wu R, Shi W, Fardis M, Weber JS, Larkin JMG, Kirkwood JM. Long-term follow up of lifileucel (ln-144) cryopreserved autologous tumor infiltrating lymphocyte therapy in patients with advanced melanoma progressed on multiple prior therapies. J Clin Orthod 2020;38:10006.
[143] Jazaeri AA, Zsiros E, Amaria RN, Artz AS, Edwards RP, Wenham RM, Slomovitz BM, Walther A, Thomas SS, Chesney JA, Morris R, Matsuo K, Gaillard S, Rose PG, Donas JG, Tromp JM, Tavakkoli F, Li H, Fardis M, Monk BJ. Safety and efficacy of adoptive cell transfer using autologous tumor infiltrating lymphocytes (LN-145) for treatment of recurrent, metastatic, or persistent cervical carcinoma. J Clin Orthod 2019;37:2538.
[144] Harris DT, Kranz DM. adoptive T cell therapies: a comparison of t cell receptors and chimeric antigen receptors. Trends Pharmacol Sci 2016;37:220–30.

[145] Dudley ME, Gross CA, Langhan MM, Garcia MR, Sherry RM, Yang JC, Phan GQ, Kammula US, Hughes MS, Citrin DE, Restifo NP, Wunderlich JR, Prieto PA, Hong JJ, Langan RC, Zlott DA, Morton KE, White DE, Laurencot CM, Rosenberg SA. CD8+ enriched "young" tumor infiltrating lymphocytes can mediate regression of metastatic melanoma. Clin Cancer Res 2010;16:6122–31.

[146] Dudley ME, Gross CA, Somerville RPT, Hong Y, Schaub NP, Rosati SF, White DE, Nathan D, Restifo NP, Steinberg SM, Wunderlich JR, Kammula US, Sherry RM, Yang JC, Phan GQ, Hughes MS, Laurencot CM, Rosenberg SA. Randomized selection design trial evaluating CD8+-enriched vs unselected tumor-infiltrating lymphocytes for adoptive cell therapy for patients with melanoma. J Clin Oncol 2013;31:2152–9.

[147] Ellebaek E, Iversen TZ, Junker N, Donia M, Engell-Noerregaard L, Met Ö, Hölmich LR, Andersen RS, Hadrup SR, Andersen MH, thor Straten P, Svane IM. AAdoptive cell therapy with autologous tumor infiltrating lymphocytes and low-dose interleukin-2 in metastatic melanoma patients. J Transl Med 2012;10:169.

[148] Hong JJ, Rosenberg SA, Dudley ME, Yang JC, White DE, Butman JA, Sherry RM. Successful treatment of melanoma brain metastases with adoptive cell therapy. Clin Cancer Res 2010;16:4892–8.

[149] Haanen JBAG, Carbonnel F, Robert C, Kerr KM, Peters S, Larkin J, Jordan K. ESMO guidelines committee. management of toxicities from immunotherapy: ESMO clinical practice guidelines for diagnosis, treatment and follow-up. Ann Oncol 2018;29:iv264–6.

[150] Wolf B, Zimmermann S, Arber C, Irving M, Trueb L, Coukos G. Safety and tolerability of adoptive cell therapy in cancer (Internet). Drug Safety 2019;42:315–34. Available from https://doi.org/10.1007/s40264-018-0779-3.

[151] Topalian SL, Solomon D, Avis FP, Chang AE, Freerksen DL, Linehan WM, Lotze MT, Robertson CN, Seipp CA, Simon P. Immunotherapy of patients with advanced cancer using tumor-infiltrating lymphocytes and recombinant interleukin-2: a pilot study. J Clin Oncol 1988;6:839–53.

[152] Yeh S, Karne NK, Kerkar SP, Heller CK, Palmer DC, Johnson LA, Li Z, Bishop RJ, Wong WT, Sherry RM, Yang JC, Dudley ME, Restifo NP, Rosenberg SA, Nussenblatt RB. Ocular and systemic autoimmunity after successful tumor-infiltrating lymphocyte immunotherapy for recurrent. Metastatic Melanoma Ophthalmology 2009;116:981–989.e1.

[153] Marabondo S, Kaufman HL. High-dose interleukin-2 (IL- 2) for the treatment of melanoma: safety considerations and future directions. Expert Opin Drug Saf 2017;16:1347–57.

[154] Lee DW, Gardner R, Porter DL, Louis CU, Ahmed N, Jensen M, Grupp SA, Mackall CL. Current concepts in the diagnosis and management of cytokine release syndrome. Blood 2014;124:188–95.

[155] Brentjens R, Yeh R, Bernal Y, Riviere I, Sadelain M. Treatment of chronic lymphocytic leukemia with genetically targeted autologous T cells: case report of an unforeseen adverse event in a phase I clinical trial. Mol Ther 2010;18:666–8.

[156] Davila ML, Riviere I, Wang X, Bartido S, Park J, Curran K, Chung SS, Stefanski J, Borquez-Ojeda O, Olszewska M, Qu J, Wasielewska T, He Q, Fink M, Shinglot H, Youssif M, Satter M, Wang Y, Hosey J, Quintanilla H, Halton E, Bernal Y, Bouhassira DCG, Arcila ME, Gonen M, Roboz GJ, Maslak P, Douer D, Frattini MG, Giralt S, Sadelain M, Brentjens R. Efficacy and toxicity management of 19-28z CAR T cell therapy in B cell acute lymphoblastic leukemia. Sci Transl Med 2014;6, 224ra25.

[157] Grupp SA, Kalos M, Barrett D, Aplenc R, Porter DL, Rheingold SR, Teachey DT, Chew A, Hauck B, Wright JF, Milone MC, Levine BL, June CH. Chimeric antigen receptor-modified T cells for acute lymphoid leukemia. N Engl J Med 2013;368:1509–18.

[158] Morgan RA, Yang JC, Kitano M, Dudley ME, Laurencot CM, Rosenberg SA. Case report of a serious adverse event following the administration of T cells transduced with a chimeric antigen receptor recognizing ERBB2. Mol Ther 2010;18:843–51.

[159] Brentjens RJ, Davila ML, Riviere I, Park J, Wang X, Cowell LG, Bartido S, Stefanski J, Taylor C, Olszewska M, Borquez-Ojeda O, Qu J, Wasielewska T, He Q, Bernal Y, Rijo IV, Hedvat C, Kobos R, Curran K, Steinherz P, Jurcic J, Rosenblat T, Maslak P, Frattini M, Sadelain M. CD19-targeted T cells rapidly induce molecular remissions in adults with chemotherapy-refractory acute lymphoblastic leukemia. Sci Transl Med 2013;5, 177ra38.

[160] Maude SL, Frey N, Shaw PA, Aplenc R, Barrett DM, Bunin NJ, Chew A, Gonzalez VE, Zheng Z, Lacey SF, Mahnke YD, Melenhorst JJ, Rheingold SR, Shen A, Teachey DT, Levine BL, June CH, Porter DL, Grupp SA. Chimeric antigen receptor T cells for sustained remissions in leukemia. N Engl J Med 2014;371:1507–17.

[161] Le RQ, Li L, Yuan W, Shord SS, Nie L, Habtemariam BA, Przepiorka D, Farrell AT, Pazdur R. FDA approval summary: tocilizumab for treatment of chimeric antigen receptor T cell-induced severe or life-threatening cytokine release syndrome. Oncologist 2018;23:943–7.

[162] Lindenberg MA, Retèl VP, van den Berg JH, Geukes Foppen MH, Haanen JB, van Harten WH. Treatment with tumor-infiltrating lymphocytes in advanced melanoma: evaluation of early clinical implementation of an advanced therapy medicinal product. J Immunother 2018;41:413–25.
[163] Lindenberg M, Retèl V, Rohaan M, van den Berg J, Haanen J, van Harten W. Evaluating different adoption scenarios for TIL-therapy and the influence on its (early) cost-effectiveness. BMC Cancer 2020;20:712.
[164] Hopewell EL, Cox C, Pilon-Thomas S, Kelley LL. Tumor-infiltrating lymphocytes: streamlining a complex manufacturing process. Cytotherapy 2019;21:307–14.

CHAPTER

20

Molecular remission using personalized low-dose immunotherapy with minimal toxicities for poor prognosis hematological and solid tumor cancers

Dipnarine Maharaj[a], *Vineet Polineni*[a], *Maria M. Abreu*[b], *and Jacqueline Gouvea*[a]

[a]The Maharaj Institute of Immune Regenerative Medicine, Boynton Beach, FL, United States
[b]Institute for NeuroImmune Medicine, Nova Southeastern University, Ft. Lauderdale, FL, United States

Abstract

The prognosis for patients with advanced hematological and solid tumor cancers is poor. In this review, we describe a novel personalized N-of-1 treatment of advanced malignancies using low-dose immunotherapy combined with reduced dose targeted therapy when available, guided by liquid biopsy of plasma-derived cell-free circulating tumor DNA with mutational profiling, to simultaneously improve the patient's immune response to their disease and directly reduce the tumor burden while minimizing toxicities.

Abbreviations

AFP	alpha-fetoprotein
AIHA	autoimmune hemolytic anemia
ALL	acute lymphoblastic leukemia
Allo-HCT	allogeneic hematopoietic cell transplantation
B-ALL	precursor B-cell acute lymphoblastic leukemia
BCC	basal cell carcinoma
BEP	bleomycin, etoposide, cisplatin combination chemotherapy regimen
BID	twice per day
CAR	chimeric antigen receptor

https://doi.org/10.1016/B978-0-12-822620-9.00002-1

CCC	clear cell carcinoma
CLL	chronic lymphocytic leukemia
CT	computerized tomography
ctDNA	cell-free circulating tumor DNA
DLI	donor lymphocyte infusion
EC	endometrial carcinoma
EGCT	extragonadal germ cell tumor
ER	estrogen receptor
FCR	fludarabine, cyclophosphamide, rituximab combination chemotherapy regimen
FEC-DH	fluorouracil, epirubicin, cyclophosphamide, docetaxel, and Herceptin combination chemotherapy regimen
FISH	fluorescence in situ hybridization
G-CSF	granulocyte-colony stimulating factor
GCT	germ cell tumor
GNAS	guanine nucleotide-binding α-subunit
GVHD	graft-vs-host disease
HCC	hepatocellular carcinoma
HER2	human epidermal growth factor receptor 2
HRAS	Harvey rate sarcoma viral oncogene homolog
IFN-γ	interferon-gamma
IHC	immunohistochemistry
IL-2	interleukin-2
IL-6	interleukin-6
InO	inotuzumab ozogamicin
ITMTX	intrathecal methotrexate
laBCC	locally advanced basal cell carcinoma
mBCC	metastatic basal cell carcinoma
MRD	minimal residual disease
MRI	magnetic resonance imaging
NGS	next-generation sequencing
NK	natural Killer
OS	overall survival
PARP	poly ADP-ribose polymerase
PBMC	peripheral blood mononuclear cell
PET	positron emission tomography
PFS	progression-free survival
Ph +	Philadelphia positive
PR	progresterone receptor
rIL-2	recombinant interleukin-2
RT-PCR	reverse transcription polymerase chain reaction
SEER	surveillance, epidemiology, and end results
TBI	total body irradiation
TKI	tyrosine kinase inhibitor
TNF-R1	tumor necrosis factor receptor 1
U-CLL	unmutated heavy chain chronic lymphocytic leukemia
uMRD	undetectable minimal residual disease
WBC	white blood cell

Conflict of interest

No potential conflicts of interest were disclosed.

Introduction

The prognosis for patients with advanced hematological and solid tumor cancers is poor despite recent developments in standard care treatments, including systemic chemotherapy, surgical interventions, radiation therapy, immunotherapy, and targeted therapies. The current standard of care treatments can be highly toxic, with deleterious side effects including reduced immune system functionality. Patients who are refractory to these treatments are often reluctant to undergo the additional toxicities of different standard of care modalities. Therefore, there is an urgent need for effective, personalized treatments with low toxicity that preserve the quality of life. In recent years, the use of targeted therapies or immunotherapies guided by diagnostic mutational profiling has become prevalent to match the demand for a more personalized, less toxic approach to cancer treatment.

In this review, we describe a novel personalized approach to the treatment of advanced malignancies using low-dose immunotherapy combined with reduced dose targeted therapy when available, guided by liquid biopsy of plasma-derived cell-free circulating tumor DNA (ctDNA) with mutational profiling, to simultaneously improve the patient's immune response to their disease and directly reduce the tumor burden while minimizing toxicities. We have previously shown the efficacy of this approach in treating chronic lymphocytic leukemia (CLL) and acute lymphoblastic leukemia (ALL) [1,2]. We provide updates on these cases as well as our experience with a series of additional case reports of malignancies, including breast cancer, hepatocellular carcinoma, extragonadal germ cell tumors, basal cell carcinoma, endometrial adenocarcinoma, and ovarian cancer.

Liquid biopsy and next-generation sequencing (NGS) is a recent advancement in mutational analysis with a simultaneous mapping of multiple biomarkers of genomic alterations, rather than one biomarker at a time. It has shown potential for improved cancer diagnosis and therapeutic decision making, owing to its enhanced sensitivity in DNA mutation detection and fast turnaround of samples compared to the current gold standard methods. NGS of ctDNA is indicated in advanced cancer patients to prevent repeated invasive biopsies when the initial biopsy is inadequate, unobtainable for genomic testing, or uninformative, or when the patient's cancer progressed despite treatment. Its clinical utility is derived from reducing the costs, complications, and delays associated with invasive tissue biopsies for genomic testing. Beyond this technology's benefit as a diagnostic tool, studies have shown the potential for NGS and liquid biopsy to match or qualify patients for targeted therapies, as well as predict a patient's response to treatment [3,4].

Currently, there are many developments in the field of natural killer (NK) cell immunotherapy [5]. Primary successes in NK cell therapy-based treatments for hematological malignancies have led to efforts to replicate such success in solid cancers, which has shown promise in an adoptive transfer therapy setting [5]. NK cell-based therapies typically have better safety profiles than the standard of care treatments for hematological and solid tumor malignancies, often significantly reducing autoimmune responses and sparing graft-vs-host disease (GVHD) reactions [5,6]. NK cells are highly cytotoxic lymphocytes of the innate immune system that exhibit high levels of antitumor activity and are implicated in the immune response in many hematological and solid tumor cancers [1,2,5]. NK cells can be divided into subpopulations depending upon CD16 and CD56 expression, with $CD56^{bright}CD16^{dim/-}$ cells specialized for cytokine production and exhibiting low levels of perforin, and $CD56^{dim}CD16^{+}$

cells have high levels of cytotoxic ability and high levels of perforin [1,6]. In our personalized approach, we utilize the patient's peripheral blood immune evaluation including B cells, T cells, and NK cells and subsets. NK cell cytotoxicity being an important component of the personalized low-dose interleukin-2 (IL-2) immunotherapy. Studies have shown the potential to improve NK cell function in the treatment of malignancies through the use of cytokines such as IL-2 [7].

IL-2 is a cytokine with pleiotropic effects and functions as an activator of NK and cytotoxic T cells [1,2]. IL-2-activated NK cells increase in count, have enhanced cytotoxic capabilities, and exhibit increased lytic potential [1,2,8]. Therapeutic uses of IL-2 have utilized two different strategies by varying the dose of IL-2; ultralow doses to reduce autoimmune responses through regulatory T cell stimulation and low doses to augment NK cell-mediated antitumor immune responses [2,7]. This dualistic anti- and proimmune function of IL-2 based on dosing and frequency can be utilized in a personalized therapeutic approach in immune-oncology which can augment NK cell antitumor responses [1,2]. A novel combined, personalized approach of low-dose IL-2 immunotherapy and low-dose targeted therapy can improve innate NK cytotoxic antitumor response while directly reducing tumor burden with significantly reduced toxicities to the patient compared to the standard of care regimens in several types of malignancies. We discuss in this review that the administration of such immunotherapies that can augment the immune system in such a way that is not toxic to the patient and helps to increase innate and adaptive immunity while the immune system remains strong, unlike the use of highly toxic standard of care treatments.

Case reports

Poor prognosis unmutated heavy chain chronic lymphocytic leukemia (U-CLL)

CLL is the most common leukemia diagnosed in adults. The standard of care for CLL, with variation based upon disease progression, age, and comorbidities, is a combination chemotherapy approach "FCR" including fludarabine, cyclophosphamide, and rituximab [9]. Although initially efficacious in treating systemic disease, these treatments often adversely impact the host's immune function resulting in vulnerability to chronic infections, CLL relapse, and the development of secondary cancer [1]. Additionally, patients can experience significant toxic side effects and may need to delay or discontinue treatment [10]. Patients who are refractory or unable to tolerate chemotherapy experience significantly reduced progression-free survival (PFS) [10]. The introduction of targeted therapies such as Ibrutinib, a Bruton's tyrosine kinase inhibitor, and Venetoclax, a BCL-2 inhibitor, have changed the landscape of CLL treatments by increasing PFS and overall survival outcomes. CLL is frequently associated with innate and adaptive immune response defects, with overall decreases in normal B cells, T cells, and natural killer cells [1]. In addition, patients with CLL typically have deficient NK cell cytotoxicity, and studies have demonstrated the responsiveness of NK cell function to cytokine stimulation [11]. In this case report, we describe the therapeutic benefit of maintaining the NK cell function with our novel approach.

A 56-year-old woman first attended our clinic in October 2015 with a history of B-CLL. In March 2009, blood counts revealed leukocytosis. In May 2012, the patient developed enlarged neck lymph nodes, and subsequent blood counts revealed a white blood cell count of 19 K/μL with 8.31 K/μL abnormal lymphocytes measured by flow cytometry with a monoclonal lambda B cell population co-expressing CD5 and CD23 consistent with CLL. Other positive markers included CD19, CD20, CD11c, and CD45. Fluorescence in situ hybridization (FISH) analysis in July 2012 showed a heterozygous deletion of the D13S319 locus (13q14.3) in 33.7% of cells (normal <3.1%). Computerized tomography (CT) scans showed extensive pathologic anterior and posterior cervical adenopathy. The treatment plan was to observe and monitor blood counts and to perform a clinical examination every 3 months. In August 2014, the patient developed severe fatigue, dizziness, and anemia, requiring a transfusion of 2 units of packed red blood cells. Subsequently, the patient started treatment with Imbruvica (ibrutinib), 420 mg once daily.

In October 2015, the patient attended our clinic for evaluation of her disease. She was in clinical remission. In January 2016, ibrutinib was discontinued due to a severe skin rash on the arms and scalp tenderness. In late December 2016 through early January 2017, the patient had a recurrence of enlarged neck lymph nodes. Blood counts showed an elevated white blood cell (WBC) count of 27.1 K/μL and a low platelet count of 121 K/μL, with flow cytometry showing abnormal CLL cells at 9.27 K/μL. FISH showed a heterozygous deletion of the D13S319 locus (13q14.3). Molecular pathology showed unmutated IGHV, indicative of aggressive disease.

Treatment options discussed included responses and side effects of fludarabine, cyclophosphamide, rituximab, bendamustine, CD20 monoclonal antibodies, targeted therapy including venetoclax, and idelalisib, bone marrow transplantation, CAR-T cell therapy, no therapy, or experimental personalized low-dose immunotherapy. The patient refused chemotherapy or treatments with potentially severe side effects. She gave informed consent for experimental personalized low-dose immunotherapy, including publication of results. Peripheral blood immune evaluation showed an increase in absolute monocyte count of 1.355 K/μL (normal range: 0.55–1.00 K/μL). While NK cells were increased at 628 cells/μL (normal range: 98–294 cells/μL), there was decreased functional NK cell cytotoxicity to 4.56% (normal range: 16%–39.9%). CD8+ T cells were 364 cells/μL (normal range: 219–731 cells/μL), CD4+ T cells levels were high at 1210 cells/μL (normal range: 599–1199 cells/μL), and CD4:CD8 ratio of 3.32. The median percentage of regulatory T cells (CD4+CD25+) was low at 0.18% (normal range: 4.93%–7.29%), and NKT cells numbers at 1014.0 cells/μL (normal range: 593–1375 cells/μL). Plasma cytokines showed undetectable levels of interferon-gamma (IFN-γ) and elevated levels of IL-10 at 11.25 pg/mL.

In February 2017, the patient began low-dose immunotherapy. The cycles consisted of daily low-dose subcutaneous recombinant IL-2 (rIL-2) injections of 10–20,000 IU/kg for 5 days per week in addition to oral low-dose lenalidomide 10 mg per day. Variation of the duration, dosing, and frequency of administration of cycles of rIL-2 was based on peripheral blood immune panel results. Disease progression was monitored through peripheral blood counts, flow cytometry, FISH, and molecular analysis.

In October 2017, the patient had received 4 cycles of rIL-2 injections, and peripheral blood analysis showed significant improvements with an undetectable level of the abnormal CLL heterozygous deletion of the D13S319 locus (13q14.3). Molecular pathology showed a mutated IGHV. However, there was persistence of SF3B1 mutation, which is associated with a poor prognosis. Further treatment with low-dose rIL-2 was recommended, but the patient did not comply.

In March 2018, flow cytometry showed that the absolute number of abnormal CLL cells was 0.08 K/μL and FISH showed recurrence of heterozygous deletion of the D13S319 locus

(13q14.3). Molecular pathology showed progression to an unmutated IGHV and persistence of the SF3B1 mutation. Peripheral blood immune evaluation showed that NK cell function was 6.74% and this improved to 11.45% in June 2018, in response to 2 cycles of low-dose rIL-2. In June 2018, molecular pathology showed persistence of unmutated IGHV and the SF3B1 mutation. Further low-dose rIL-2 was recommended, but the patient decided not to continue.

In May 2019, the patient had a recurrence of fatigue and enlarged neck lymph nodes. Blood counts revealed an elevated WBC of 56.7K/μL and low platelets of 123K/μL. Flow cytometry showed a major increase of 49.6K/μL abnormal CLL cells, and FISH showed a heterozygous deletion of the D13S319 locus (13q14.3). Molecular pathology revealed a persistent SF3B1 mutation and unmutated IGHV. Immune panel results showed that her NK cell function (6.57%) had also decreased during the period when the patient was off treatment with low-dose rIL-2.

In June of 2019, the patient had extreme fatigue caused by Coombs positive autoimmune hemolytic anemia (AIHA) secondary to CLL. The immune hemolysis resolved in response to prednisone, and she significantly improved with hemoglobin levels rising from 5.2g/dL during the peak of hemolysis to 10.7g/dL. As a precaution, further treatment with low-dose lenalidomide was discontinued because of the risk of hemolysis reported with lenalidomide at standard doses. The patient agreed to standard dose monotherapy with venetoclax with a weekly dosage increase (20, 50, 100, 200, and 300mg). At the 300mg dose of venetoclax, the patient experienced side effects of severe neutropenia (0.1K/μL) requiring granulocyte-colony stimulating factor (G-CSF) administration, as well as thrombocytopenia (100K/μL). Venetoclax was discontinued. After her blood counts had improved, the patient requested restarting immunotherapy with low-dose rIL-2. A low dose of venetoclax at 50mg per day was added to minimize its toxicity.

In October 2019, the patient was asymptomatic. She had received 4 cycles of low-dose rIL-2 and daily venetoclax 50mg with minimal toxicity. Blood counts revealed a WBC of 15.4K/μL and normal platelets of 160K/μL. Flow cytometry showed a major response to treatment with minimal residual disease (MRD) of 0.33K/μL absolute abnormal CLL cells and FISH showed a heterozygous deletion of the D13S319 locus (13q14.3). Molecular pathology showed mutated IGHV. However, there was a persistent SF3B1 and ATM mutation.

In December 2019, after two further cycles of low-dose rIL-2 and low-dose venetoclax, the patient experienced minor side effects. Blood counts revealed a normal white blood count of 4.72K/μL, hemoglobin count of 13.60g/dL with a low platelet count of 137K/μL. Flow cytometry analysis showed undetectable MRD (uMRD) with no abnormal CLL cells. FISH analyses showed no evidence of the previously observed deletions, and molecular pathology was negative for previously seen mutations. Peripheral blood immune evaluation showed an absolute monocyte count of 0.36K/μL and a normal NK cell count at 139cells/μL, but NK cell cytotoxicity reduced at 5.89%. However, the patients NK cell subsets (CD56high, CD16−) showed increased levels at 14.21%, (CD56high, CD16+) at 14.36%, (CD56low, CD16+) low at 4.31% and (CD56low, CD16−) increased at 12.86%. Other cell populations, including CD4+ T cells were normal at 625cells/μL, CD8+ T cells decreased to 129cells/μL. Plasma cytokines showed an increase in IFN-γ at 33.09pg/mL and a decrease in IL-10 at 11.02pg/mL.

Further analyses were performed on the patient's peripheral blood in October 2019 and compared to similar analyses performed in December 2019. The percentages of all different subsets of immune effectors improved and resembled that of the healthy controls. From October 2019 to December 2019, the most significant increase detected was in CD3+ T cells, which went from 17% to 51%, B cells (CD19+) significantly improved from 70.2% to within the normal range of 13.76%. Significant improvement in the IFN-γ spots and IFN-γ release from both peripheral blood mononuclear cells (PBMCs) and NK cells were also seen. These results suggested significant improvements in the quantitative and qualitative defects within the innate and adaptive immune response system when the patient was treated with multiple sequential cycles of low-dose rIL-2.

Nine months after treatment, the patient remained in remission and IGVH gene was found to be mutated. However, in February 5, 2021, after 5 months without any treatment, molecular testing done showed recurrent SF3B1k700E mutation, at a level of 7.3%, and also ATM splice site c.6198 + 1G > A mutation, at a level of 6.1%. FISH analysis revealed a recurrence of 13q14 deletion at a level of 18.5% (normal <3.1%). Flow cytometry showed 29.9% monoclonal B-cell population with an immunophenotype suggestive of chronic lymphocytic leukemia. The patient was restarted on cycles of low-dose rIL-2 and oral low-dose venetoclax, 50 mg per day.

In March 2021, after 2 cycles of treatment immune panel showed CD3+ T cells increased to 893 cells/μL, NK cells increased to 179 cells/μL, and cytotoxic T cells normalized to 225 cells/μL from an abnormal low. Plasma cytokines showed IFN-γ normalized to 7.21 pg/mL, and IL-17 increased from 0 to 22.80 pg/mL Flow cytometry showed a reduction in monoclonal B cells from 29.9% to 17%. FISH showed a large reduction in 13q14 deletion from 18.5% to 5.7%, molecular testing pathology showed SF3B1 mutation decreased from 7.3% to 4.9%, and the ATM mutation decreased to 8.6% from 14.0%. IGVH was not detected. The patient remains clinically well with no symptoms related to CLL.

Precursor B-cell acute lymphoblastic leukemia

ALL is a hematological malignancy stemming from lymphoid progenitor cells, and it is often associated with a poor prognosis in adults [2,12]. Precursor B-cell ALL (B-ALL) has been shown to comprise approximately 75% of total adult ALL cases [12]. Long-term survival rates of adults with B-cell ALL remain low at 40% with current standard of care. For patients with Philadelphia-positive (Ph+) B-ALL, allogeneic hematopoietic cell transplantation (allo-HCT) has become standard of care due to poor outcomes in OS with chemotherapy alone [2,12]. After allo-HCT, there is a significant risk of graft-vs-host disease and relapse results in a very poor OS of less than 1 year [2]. Targeted therapies such as tyrosine kinase inhibitors (TKIs) and immunotherapies such as chimeric antigen receptor (CAR) T cells have shown promise in B-ALL treatment, but there are still significant, unique toxicities [2,12].

In patients with precursor B-ALL, studies have shown a deficiency in NK cytotoxicity [13]. Therefore, B-ALL patients who are refractory to the standard of care may benefit from therapies aimed at increasing NK cell cytotoxicity. In this case report, we present a patient with (Ph+) B-ALL who remains in molecular remission as a result of our novel treatment approach over 6 years following diagnosis.

A 39-year-old man with a history of Ph+ precursor B-cell ALL attended our clinic in March 2017. In June 2015, he attended an emergency department with a 2-week history of fatigue, lethargy, backache, leg, and rib pain refractory to opioids. Blood counts revealed leukocytosis with white cell count of 67×10^9 cells/L with 34% lymphoblasts measured by flow cytometry. Bone marrow biopsy showed precursor B-cell ALL, Ph+ (*t* 9; 22). The patient was enrolled in the UKALL14, version 6 Protocol (NCT01085617: ClinicalTrials.gov) including a five-drug regimen in adults with de novo ALL between 25 and 65 years. Induction phase I included PEG-asparaginase (1000 IU/m^2) on days 4 and 18, daunorubicin 30 mg/m^2 and vincristine 1.4 mg/m^2 on days 1, 8, 15, and 22, dexamethasone 10 mg/m^2 days 1, 4, 8, 11, 15, and 18, and intrathecal methotrexate (ITMTX) 12.5 mg on day 14. The patient received continuous imatinib 400 mg escalating to 600 mg daily throughout the treatment. Cyclophosphamide 1000 mg/m^2 was given on days 1 and 15, Ara-C 75 mg/m^2 on days 2, 5, 9, 12, 16, 19, 23, and 26, and mercaptopurine 60 mg/m^2 throughout whole treatment and intrathecal methotrexate on days 1, 8, 15, and 22. After phase II induction, the patient achieved complete molecular remission with negative BCR-ABL1 p190 transcripts by reverse transcription polymerase chain reaction (RT-PCR). Complications included constipation, febrile neutropenia, and pneumonia. Remission was consolidated with myeloablative conditioning, including cyclophosphamide/total body irradiation (TBI), followed by a matched (brother) allogeneic stem cell transplant. He developed posttransplant hyponatremia, total body irradiation somnolence, and grade IV acute cutaneous GVHD, which was managed with steroids. Posttransplant maintenance included imatinib, infection prophylaxis, and GVHD prophylaxis with methotrexate and cyclosporine. Disease relapse occurred at 6 months posttransplant with *BCR-ABL1* p190 transcript expression at a level of 18.23%, confirmed on bone marrow biopsy at 9 months post allo-HCT. Imatinib was replaced by the TKI dasatinib 140 mg daily and followed by donor lymphocyte infusion (DLI) (3×10^6 CD3 cells/kg), resulting in second remission with negative MRD by flow cytometry. Three months later, disease progression occurred with expression for *BCR-ABL1* p190 in 163,275 copies of control gene. He was given an overall life expectancy of less than 1 year.

The patient attended our clinic for evaluation and treatment in March 2017. Options discussed included responses and side effects of CAR-T cell therapy, antibody-drug conjugates, such as inotuzumab ozogamicin (InO) or blinatumomab, chemotherapy, a second HCT, and/or DLI, no therapy and palliative care or experimental personalized low-dose immunotherapy. The patient gave informed consent for the experimental immunotherapy including publication of results. The treatment consisted of daily low-dose subcutaneous rIL-2 with a variation of dosing and frequency of administration based upon an extensive peripheral blood immune panel, which includes NK cells, NK cell cytotoxicity, B cells, T cells, and regulatory T (Treg) cells to selectively stimulate a graft-vs-leukemia response while minimizing GVHD. Simultaneously, cytokine levels including IFN-γ in plasma were measured. The patient received 4 cycles (4–7 weeks) of rIL-2 injections, 10–20,000 IU/kg, 5 days per week. The daily dose and duration of every cycle were determined by the results of the peripheral blood immune panels. Cycles 1 and 2 were 6 weeks, cycle 3 was 7 weeks, and cycle 4 was 4 weeks. Dasatinib was continued at 140 mg daily. Expression of BCR-ABL1 p190 transcript was monitored using RT-PCR throughout his rIL-2 treatment. Results showed NK cell activity improvement from no activity detected before cycle 1 of rIL-2 to 5.08% after cycle 3 and to 9.68% after cycle 4. The $CD56^{bright}$CD3-NK cells were high before starting rIL-2 and remained within the upper to normal range throughout the treatment. IFN-γ increased from 0 pg/mL before cycle 1 to a peak of 6.6 pg/mL at the end of cycle 1 and remained elevated through cycles 2, 3, and 4 with a level of 1.9 pg/mL at the end of the cycle 4. $CD2^+CD26^+$ increased from 7.1% before initiation of cycle 1 of rIL-2 to 63.4% after cycle 4. $CD4^+CD25^+$ Tregs which were at the upper end of the normal range showed a progressive decrease to the lower end.

After cycle two bone marrow showed no abnormal lymphoid cells by flow cytometry, normal cytogenetics, and no detectable levels of BCR-ABL, p190. Following cycle three peripheral blood showed no detectable levels of BCR-ABL, p190 or p210 transcripts consistent with a complete molecular and cytogenetic remission with recombinant IL2 and a TKI. Immunotherapy with rIL-2 and TKI was tolerated by the patient with the only side effect being erythema (Grade 1) at the injection site. Dasatinib was continued at 140 mg daily. The patient continued to have negative test results for MRD testing in London, England. Fifty-six months after starting rIL-2, the patient is well, asymptomatic, and he possesses a normal quality of life.

Basal cell carcinoma

Basal cell carcinoma (BCC) is a malignant keratinocyte carcinoma and the most common cancer diagnosed in the United States [14]. Fortunately, long-term survival in BCC patients is fairly high, with 5-year survival rates of 87.1%; surgical excision is the primary intervention [15]. However, locally advanced basal cell carcinoma (laBCC) and metastatic basal cell carcinomas (mBCC) in particular have proven to be particularly challenging to treat, and median long-term survival was found to be 87 months in patients with regional metastases and significantly lower at 24 months for patients with distant and metastatic BCC [15]. Surgical and radiotherapy interventions have proved ineffective in these cases, but recent advances have highlighted Hedgehog/smoothened inhibitors such as vismodegib and sonidegib as therapeutic avenues to increase survival in laBCC and mBCC, although they are also associated with significant toxicities [15]. Patients with mBCC who responded to vismodegib treatment experienced significant improvement with a median survival of 33.4 four months in a phase 2 clinical trial [15]. Immunotherapies have also begun to emerge as a potential treatment for advanced BCC patients who are refractory to Hedgehog inhibitors, although there is currently a paucity of research in this field [15]. In patients with BCC, the role of NK cells in coordinating the immune response has been well established [16]. There exists a current need for research into the level of NK cell dysfunction in BCCs. However, the therapeutic benefit of NK cell therapies in improving the immune response to solid tumor malignancies has been well established [16]. Here, we describe a patient who was refractory to standard of care treatments and experienced significant progression of his disease. The patient achieved remission in response to our novel approach of low-dose immunotherapy in combination with a targeted therapy.

A 70-year-old man first attended our clinic in April 2021 with a history of mBCC. The patient was diagnosed with BCC in 2013 and treated with vismodegib (Erivedge)—a hedgehog inhibitor from 4/2013 to 12/2015, to which he responded well, and magnetic resonance imaging (MRI) showed that his tumor decreased in size from 11 × 22 mm to 5 × 15 mm. In December 2015, he stopped Erivedge because of side effects and was monitored until August 2017 when, with progression, he restarted Erivedge and continued on it until October 2018. In November 2018, he participated in the phase 2 REGN2810 trial studying the monoclonal antibody for programmed cell death at Stanford Hospital, and he exited the trial in November 2019 when it was ineffective in treating his condition. In March 2020, he started 200 mg of sonidegib, 5 days of the week, and 100 mg per day of itraconazole. He was reevaluated in March of 2021 and was told that his disease was progressing, and he needed to start chemotherapy.

In April 2021, the patient attended our clinic for evaluation and treatment. Options discussed included responses and side effects of radiation therapy, chemotherapy, no therapy and palliative care, and experimental personalized low-dose immunotherapy. Peripheral blood immune evaluation showed that the CD5+ CD19+ count was low at 2 cells/μL (normal: 15–178 cells/μL). $CD4^+CD25^+$ regulatory T cell count was 1.22% (normal: 4.93%–7.29%). The NK cell count was normal at 138 cells/μL (normal: 98–294 cells/μL), but NK cell cytotoxicity was very low at 5.11% (normal: 16.3%–39.9%). The cytokine analysis showed elevated levels of proinflammatory cytokines with interleukin-6 (IL-6) at 12.11 pg/mL (normal: $\leq$7.1 pg/mL) and tumor necrosis factor receptor 1 (TNF-R1) at 732.96 pg/mL (normal: 235.5–602.8 pg/mL). Immune evaluation showed immune dysregulation with abnormalities in his innate and adaptive immune system, with a severe decrease in his natural killer cell function. The treatment plan was to correct, or improve, his immune function to stop disease progression. The treatment plan consisted of cycles of low-dose rIL-2 immunotherapy, where the patient received daily low-dose subcutaneous rIL-2 injections of 10–20,000 IU/kg for 5 days per week. In addition, the patient was maintained on Sonidegib (Odomzo) 200 mg and then was changed to a low dose of Erivedge, at 150 mg/day, because of side effects. Variation of the duration, dosing, and frequency of administration of cycles of rIL-2 was based on peripheral blood immune panel results. In addition, liquid biopsy for NGS of ctDNA showed 0.2% amplification of SMO S387N (highly associated with basal cell carcinoma), 0.8% amplification of FBXW7 H139R (highly associated with colon adenocarcinoma) and 0.2% amplification of ATM H2554R (highly associated with lung adenocarcinoma). Although some of these mutations are associated with different types of cancers, the cancer cells secreting the DNA being sequenced were most likely coming from the basal cell carcinoma.

After the patient's specialized treatment of low-dose rIL-2 immunotherapy, he had a repeat of the peripheral blood immune evaluation and it revealed that CD5+CD19+ count was still low but had increased to 8 cells/μL from 2 cells/μL. The C4+CD25+ increased to 13.95% and NK cell count increased to 253 cells/μL. The NK cell cytotoxicity remained below the normal level but showed a 100% increase to 10.92% from the prior level of 5.11%. Similarly, NK subsets all increased with CD56 Bright CD16− from 6.59% to 24.48%, CD56 Bright CD16+ from 6.62% to 26.63%, CD56 Dim CD16+ from 5.22% to 12.32%, and CD56 Dim CD16− increased from 6.59% to 26.03%. Liquid biopsy for NGS of ctDNA showed undetectable amplification of SMO S387 and ATM H2554R mutations, as well as a decrease in FBXW7 H139R mutation from 0.8% amplification to 0.5% amplification. The patient remains well clinically with no symptoms related to his BCC.

High-grade clear cell carcinoma of the endometrium

Endometrial cancers (ECs) comprise the majority of gynecological cancers diagnosed in the United States, and their incidence is rising [17]. Endometrial cancers have been divided into two broad categories, type 1 and type 2 [17,18]. The majority of ECs are type 1, estrogen-dependent, while type 2 are typically estrogen-independent, poorly differentiated, and more aggressive [17,18]. Clear cell carcinoma (CCC) is a rare type 2 EC comprising less than 5% of all ECs [17]. CCC, however, is commonly associated with aggressive disease progression and poor clinical outcomes, with 5-year overall survival at 62.5% compared to 83.2% for patients with all stages of EC [18,19]. The standard of care for patients with CCC of the endometrium is total abdominal hysterectomy and bilateral salpingo-oophorectomy, with comprehensive surgical staging including peritoneal washing, omentectomy, and pelvic and para-aortic lymphadenectomy. Currently, there are no approved targeted therapies available, given the rarity of the disease and lack of data on the disease pathology [19]. For patients with EC that refuse or are refractory to standard of care treatments including chemotherapy or radiation, survival has proven to be significantly low. The NK cell profile of patients with EC typically shows loss of cytotoxic function along with reduced cytolytic effector production [20]. Therefore, improving the NK cell function has been shown to be a promising avenue of treatment for these patients. Here, we present a patient with EC whom we were able to treat with our novel approach to improve and maintain her NK cell function.

In July 2019, a 70-year-old woman had a routine Pap smear and the pathology showed malignant adenocarcinoma, most consistent with endometrial origin. In August 2019, a pelvic ultrasound revealed a rounded area of increased echogenicity that appeared to be associated with the cervix. There was a heterogeneous appearance of the endometrium, with an area near the fundus of the uterus which was suspicious for EC. On September 9, 2019, the patient attended our clinic for evaluation and treatment. Options discussed included responses and side effects of an abdominal hysterectomy and bilateral salpingo-oophorectomy with comprehensive surgical staging, lymphadenectomy, chemotherapy, radiation therapy, no therapy and palliative care, or experimental personalized low-dose immunotherapy. The patient had a liquid biopsy, which showed 0.3% BRCA2 N802fs mutation, which is a driver mutation in ovarian and/or uterine cancer and 0.2% EGFR Y978Y mutation (synonymous alteration). On October 22, 2019, a tissue biopsy of the endometrial lesion showed a high-grade clear cell carcinoma. The patient had a total abdominal radical hysterectomy, bilateral salpingo-oophorectomy, sentinel lymph node dissection, omentectomy, and appendectomy. The patient refused chemotherapy or treatments with potentially severe side effects. She gave informed consent for experimental personalized low-dose immunotherapy, including publication of results. Peripheral blood immune evaluation test showed NK cell cytotoxicity was markedly decreased at 1.69% (normal: 16.3%–39.9%), CD2+CD26+ (T cell + NK cells expressing dipeptidyl peptidase) cell count was low at 563 cells/μL (normal: 593–1375 cells/μL), CD4+CD25+ Treg count was 2.319% (normal range: 4.93%–7.29%), and plasma cytokines levels for IFN-γ were 14.76 pg/mL (normal: $\leq$14.8 pg/mL). On February 10, 2020, the patient received daily low-dose subcutaneous rIL-2 injections of 10–20,000 IU/kg for 5 days per week. Variation of the duration, dosing, and frequency of administration of cycles of rIL-2 was based on peripheral blood immune panel results. On April 23, 2020, a repeat liquid biopsy showed the BRCA2 N802fs mutation was still present at a level of 0.3% and the EGFR Y978Y mutation was undetectable. However, a new mutation, TERT R72H (variant of uncertain significance) was 0.2%. The patient continued additional cycles of low-dose rIL-2 immunotherapy. On January 28, 2021, peripheral blood immune panel test results showed an NK cell count of 1039 cells/μL. NK cell cytotoxicity increased to 13.31%; CD2+CD26+ count increased to 1540 cells/μL. The CD4+CD25+ Treg count was 14.04% and IFN-γ was 9.03 pg/mL. Liquid biopsy showed undetectable levels of the BRCA2 N802fs, TERT R72H, and EGR Y978Y mutations. The patient remains clinically well and with no evidence of CCC.

Ovarian cancer

Ovarian cancer is a gynecological cancer with one of the highest mortality rates and can affect women of all ages [21]. However, over 90% occur in postmenopausal women [21]. Less than 50% of women diagnosed with ovarian cancer survive for 5 years or more [21]. The standard of care for ovarian cancers is dependent upon the severity of the disease at diagnosis; however, patients typically receive debulking surgery, with adjuvant platinum-taxane chemotherapy and other chemotherapy regimens based upon their mutation status [22]. TKIs and antiangiogenic agents are potential targeted therapies; however, the data on their efficacy is still scarce [22]. Additionally, patients with stage 2 ovarian cancer have a 30% chance of recurrent disease. Although the exact nature of NK cell deficiencies in ovarian cancer has yet to be definitely established, studies have shown significant promise in NK cell-mediated immunotherapies as adjunct treatments to improve survival in ovarian cancer patients, particularly those with recurrent disease [23,24]. Here, we present a patient with significant immune dysfunction that we treated with our novel approach to improve and maintain her NK cell function.

A 72-year-old woman with Stage II Ovarian Cancer, diagnosed in June 2009, underwent surgical resection of the ovary, followed by 4 cycles of Carboplatin combination chemotherapy. The patient was disease free until 2012 when she experienced abdominal discomfort and an MRI showed recurrence of the ovarian cancer. She had surgical resection with negative margins of the recurrent cancer at the site of the original disease in the left lower abdominal quadrant with negative margins. Surgical resection was followed by Gemcitabine, Carboplatin, and Topotecan combination chemotherapy. In 2014, she had a recurrence of ovarian cancer with extension into the urinary bladder and colon. She had surgical resection of the recurrent lesions with partial colon and urinary bladder resections. She was told that she needed chemotherapy with Gemcitabine, Carboplatin, and Topotecan. She chose not to receive this chemotherapy because of the toxicities.

On May 5, 2015, the patient attended our clinic for evaluation and treatment. Options discussed included responses and side effects of additional surgeries, chemotherapy, no therapy, palliative care, BMSCTI 08001 clinical trial for solid tumor cancers or experimental personalized low-dose immunotherapy. An MRI of the pelvis on July 22, 2015, showed a significant increase in the pelvic mass with an increase in size from 4.1×4 cm to 5.3×7 cm and invasion of the bladder. A second lesion also increased in size from $2.8 \times 4 \times 5.3$ cm to $3.1 \times 0.8 \times 6.2$ cm. There was no pelvic adenopathy or free fluid. Analysis of the surgical tissue biopsy of the tumor showed mutations of the BRCA1 and BRCA2 genes. The patient refused chemotherapy or treatments with potentially severe side effects. She gave informed consent for experimental personalized low-dose immunotherapy with targeted therapy, including publication of results.

In July 2015, the patient received daily low-dose subcutaneous rIL-2 injections of 10–20,000 IU/kg for 5 days per week. Variation of the duration, dosing, and frequency of administration of cycles of rIL-2 was based on peripheral blood immune panel results. Following IL-2 treatment, the patient's immune panel results showed that NK cell cytotoxicity was normal at 17.65% (normal: 16.3%–39.9%). In September 2015, the patient was started on Olaparib 100 mg twice per day (BID). Olaparib is an oral inhibitor of poly (ADP-ribose) polymerase (PARP) proteins that play a key role in DNA repair and genomic stability. By blocking these vital PARP-mediated tumor cell DNA repair mechanisms, Olaparib leads to "synthetic lethality" and selective tumor cell death [25]. After 3 months, Olaparib dosage was reduced to 50 mg BID to minimize the toxicities. In February 2016, a pelvic MRI showed significantly decreased size of the pelvic masses with a residual 2.6 cm soft tissue mass at the left vaginal cuff with likely associated vesicovaginal fistula. Abdominal MRI showed no evidence of metastatic disease in the abdomen. Serial Ca-125 measurements showed a decrease from 552 U/mL to normal levels by March 2016 (Fig. 1). MRIs of the abdomen and pelvis in September 2016, June 2017, and August 2019 showed no findings of recurrent disease with stable and negative examinations. The patient remains clinically well with no evidence of ovarian cancer for over 6 years since her recurrence in 2015, following the initiation of personalized treatment with low-dose immunotherapy and oral-targeted therapy.

Extragonadal germ cell tumor

Extragonadal germ cell tumors (EGCTs) are rare subsets comprising less than 5% of adult germ cell tumors (GCTs) [26]. EGCTs are typically more aggressive and frequently found in the anterior mediastinal and retroperitoneal regions [26]. Furthermore, they exhibit a poor response to conventional chemotherapy regimens administered for GCTs [27]. EGCTs refractoriness to conventional chemotherapy regimens, along with the yet-unclear pathogenesis of the disease present a unique clinical and therapeutic challenge. EGCTs can be further divided into pure seminomas and nonseminoma tumors [27]. Nonseminoma tumors are typically more aggressive and can present with yolk sac tumors and teratomas [26,27]. The standard of care for EGCTs is orchiectomy followed by chemotherapy regimens including bleomycin, etoposide, and cisplatin (BEP) [27]. Patients with early disease typically respond well to the BEP regimen and 5-year PFS for intermediate-risk stage IIC or stage III seminoma patients was 83% in a trial of 84 patients [27]. However, patients in a separate trial with advanced

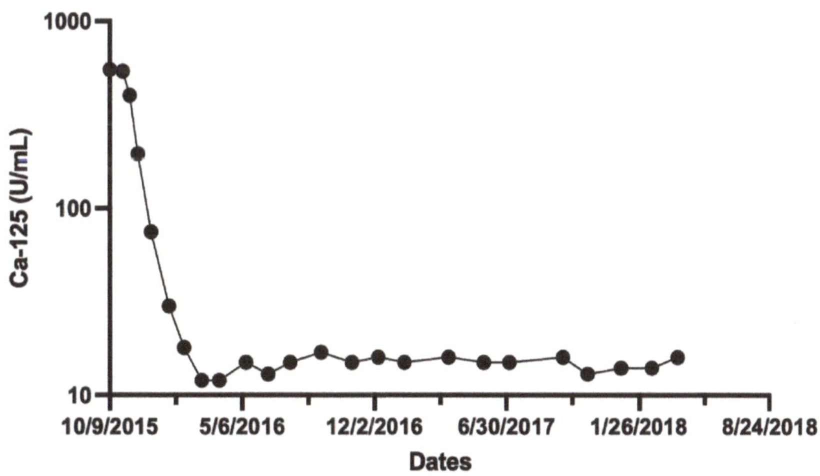

FIG. 1 Normalization of serial Ca-125 tumor marker measurements in case 5 with a diagnosis of ovarian cancer after treatment with low-dose immunotherapy and low-dose targeted therapy.

nonseminomatous germ cell tumors or pure seminomas requiring salvage therapy had a 3-year overall survival of 23% [28]. Advanced GCTs have been characterized through immunohistochemistry and gene expression profiling to exhibit decreased T cell and NK cell signatures [29]. Here, we describe a patient who achieved remission in response to our novel approach of low-dose immunotherapy in combination with a targeted therapy.

In August 2006, a 21-year-old male was diagnosed with extra gonadal germ cell tumor with involvement of retroperitoneal lymph nodes, mediastinal lymph nodes, and a superior vena cava syndrome. Serum alpha-fetoprotein (AFP) was 3429 ng/mL. He received bleomycin, etoposide, and cisplatin (BEP) chemotherapy from August 2006 to December 2006 with side effects including ear ringing, nausea, anorexia, and hair loss. The patient achieved complete remission following chemotherapy. In March 2008, CT scans demonstrated disease recurrence, and the patient had a retroperitoneal lymph node dissection that revealed a mature teratoma with yolk sac component and involvement of seven lymph nodes. In February 2010, his AFP had increased to 48 ng/mL and in September 2010 he had neck pain caused by a large supraclavicular mass. Biopsy revealed mature cartilaginous tissue consistent with a mature teratoma. In May 2012, neck dissection was performed via a sternotomy and the tumor consisted of mature teratoma (70%) and yolk sac tumor (30%). Between 2015 and 2016, repeat CT scans of the chest and biopsies showed continued growth of the tumor and the patient had surgical resection with CT monitoring for disease recurrence. In May 2018, the patient had a relapse of the EGCT with a biopsy confirming seminoma in the groin. A positron emission tomography (PET) scan also showed increased activity of tumors in the chest.

In April 2019, the patient attended our clinic for evaluation and treatment. Options discussed included responses and side effects of chemotherapy, no therapy and palliative care, or experimental personalized low-dose immunotherapy. The patient refused chemotherapy or treatments with potentially severe side effects. He gave informed consent for experimental personalized low-dose immunotherapy and targeted therapy, including publication of results. A peripheral blood immune evaluation showed CD2+CD26+ (T cells + NK cells expressing dipeptidyl peptidase) count was low at 201 cells/μL (normal: 593–1375 cells/μL). Total B cell count was low at 57 cells/μL (normal: 128–400 cells/μL). Total T cell (CD3+) count was low at 380 cells/μL (normal: 989–1899 cells/μL), NK cell enumeration was low at 62 cells/μL (normal: 98–294 cells/μL), and NK cell cytotoxicity was low at 2.49% (normal: 16%–39.9%). A liquid biopsy showed medium amplification (below the 50th percentile of amplifications) of KRAS mutation, which is altered in 15.95% of all cancers. The patient received consecutive cycles of daily low-dose subcutaneous rIL-2 injections of 10–20,000 IU/kg for 5 days per week. Variation of the duration, dosing, and frequency of administration of cycles of rIL-2 was based on his peripheral blood immune panel results. After the immunotherapy cycles, the NK cell count increased to 212 cells/μL and NK cell cytotoxicity increased to 10.97%. In July 2019, after two additional cycles of immunotherapy, the NK cell count increased to 263 cells/μL and NK cell cytotoxicity was 7.77%. Plasma IFN-γ increased to 29.96 pg/mL from 5.73 pg/mL. In December 2019, at 6 months post treatment, liquid biopsy showed that the KRAS mutation amplification was not detectable. The patient's peripheral blood immune panel showed that the CD2+CD26+ count increased to 737 cells/μL and CD4+CD25+ Treg cell count increased to 10.43%. Total B cell count was 55 cells/μL and total T cell (CD3+) count increased to 980 cells/μL, NK cell count increased to 215 cells/μL, and NK cell cytotoxicity increased to 8.51%; plasma IFN-γ was detected at 10.26 pg/mL. In May 2020, liquid biopsy showed that the KRAS mutation remained undetected. However, EGFR mutation amplification was found and measured to be low (below the 50th percentile of EGFR amplifications). The patient received additional treatment with low-dose rIL2 immunotherapy treatment and in July 2020, a peripheral blood immune panel revealed that his CD2+CD26+ count remained normal at 665 cells/μL and his CD4+CD25+ Treg cell count increased to 15.32%. Total B cell count was 36 cells/μL, total T cell (CD3+) count was increased at 862 cells/μL, NK cell count was increased at 152 cells/μL, and NK cell cytotoxicity was 4.93%; plasma IFN-γ level was detected at 10.39 pg/mL. Liquid biopsy showed that the KRAS amplification remained undetected and the EGFR (exon 19 deletions or exon 21 L858R mutations) amplification remained low (below the 50th percentile of EGFR mutations) with a slight decrease in value. For the EGFR mutation, the patient was started on low-dose Afatinib 30 mg per day which he tolerated well with minimal toxicities. In addition, in January 2021, the patient received proton beam radiation therapy to the residual right pelvis mass and, following this therapy, CT images of the patient's abdomen and pelvis showed significant interval decrease in size of masses. CT images of the chest showed that in the mass there appeared to have developed increased necrosis and the mass appeared to be nearly completely cystic. The patient remains clinically well.

Breast cancer

Breast cancer is the most common cancer found in women, and almost 30% of patients with early-stage breast cancer have recurrent disease [30]. The standard of care systemic treatment of breast cancer includes cytotoxic, hormonal, and immunotherapeutic agents [31]. Systemic agents are active at the beginning of therapy in 90% of primary breast cancers and 50% of metastases. However, after a variable period, progression occurs. At that point, resistance to therapy is not only common but expected [30]. Overall 5-year survival has risen to 84% in the United States due to advances in treatments and diagnostic markers [32]. However, for patients with invasive or metastatic breast cancers who develop resistance to therapy, 5-year survival rates are as low as 28% with a large variance based upon individual patient conditions [33].

NK cell deficiencies are typically seen in breast cancer patients [34]. Most breast tumors show low levels of NK cell infiltration, and studies have shown the capability of IL-2 enhanced natural killer cell cytotoxicity in lysing breast cancer cells [34,35]. Here we present three patients with a diagnosis of breast cancer at various stages, who were treated with our novel approach of low-dose immunotherapy resulting in halting of disease progression.

Case 1

A 66-year-old woman first attended our clinic in August 2016 with a history of multifocal left breast cancer T1N0M0 diagnosed in March 2016. In February 2016, she had a routine screening mammogram which did not show any abnormalities. However, with her family history of breast cancer (sister and maternal aunt), additional diagnostic testing was done with bilateral breast ultrasound and MRI with contrast. In February 2016, bilateral breast ultrasound showed a 6 mm speculated oval solid mass with associated calcifications present in the left upper outer hemisphere, and on MRI, a 6 mm suspicious enhancing mass in the left retroareolar region and 6 mm enhancing mass and adjacent 4 mm mass in the middle left upper hemisphere. Also, a 7 mm enhancing mass and a 3 mm focus of enhancement in the left upper quadrant, 5 mm focus of enhancement in the posterior right upper outer quadrant and a 3 mm focus of enhancement in the right retroareolar region.

In March 2016, core needle biopsies of the left breast, MRI enhancing masses showed invasive ductal carcinoma, estrogen receptor (ER) >90% positive, progesterone receptor (PR) >80% positive, human epidermal growth factor receptor 2 (HER2)/neu 2+ by immunohistochemistry (IHC), FISH amplified with a ratio of 2.0 in 2. MRI-guided biopsy of the right breast showed fibrocystic changes, including ductal hyperplasia, microcysts, columnar cell change, and stromal fibrosis and associated calcifications.

Genetic testing with an Ashkenazi panel for BRCA mutations was negative. The patient had surgical oncology, plastic surgery, and medical oncology consultations to evaluate her options. She was recommended to have left needle localization partial mastectomy, sentinel lymph node biopsy, possible axillary lymph node dissection, and possible lymphovenous anastomosis, and oncoplastic reduction. Chemotherapy after surgery was to include Taxol and Herceptin. In June 2016, the patient chose to have complete bilateral mastectomies with reconstruction. The patient refused chemotherapy and radiation because of the multiple toxicities including damage to her immune system and decided to explore immunotherapy options.

On August 30, 2016, the patient attended our clinic for evaluation and treatment. Options discussed included responses and side effects of chemotherapy, radiation therapy, no therapy, or experimental personalized low-dose immunotherapy. The patient refused chemotherapy, radiation, or treatments with potentially severe side effects. She gave informed consent for experimental personalized low-dose immunotherapy, including publication of results. The patient had an immune panel evaluated in August 2016, which revealed normal numbers of B cells, T cells, NK cells, and natural killer (NK) cell activity at 26.13% (normal range: 16%–39.9%). After 6 months of follow-up with no therapy, immune testing in February 2017 revealed decreased natural killer (NK) cell activity at 9.75%. The patient received daily low-dose subcutaneous rIL-2 injections of 10–20,000 IU/kg for 5 days per week. Variation of the duration, dosing, and frequency of administration of cycles of rIL-2 was based on the peripheral blood immune panel results. Following 2 cycles of IL-2 treatment in 2017, the patient had repeat immune panel evaluation in March 2018, which showed NK cell activity at 15.28% and later in July 2018 at 26.9%.

In August 2019, the patient had repeat immune panel evaluation which revealed normal natural killer (NK) cell function at 24.19%.

In January 2020, the patient complained of a small pea-sized lump on her left breast near the surgical scar, along with an itching sensation at the site of the lump. Evaluations including PET/CT Skull-Base-Mid Thigh showed no evidence of metastatic or recurrent disease. MRI of the breasts showed minimal background parenchymal enhancement, with no significant adenopathy seen bilaterally and no "cysts" or skin abnormalities seen in left breast. In February 2020, the patient underwent left breast skin excisional biopsy of the palpable region near the surgical scar and sentinel lymph node biopsy. The pathology of the excised lesion revealed small foci of residual carcinoma within a walled-off area of fibrous tissue with clear margins and negative sentinel node. The patient had a repeat immune panel evaluation which revealed NK cell function at 12.53%. This improved to 24.19% after the patient received three additional cycles of treatment. Liquid biopsy testing in March 2020 revealed no tumor-related somatic alterations in the patient's circulating blood. In March 2021, repeat testing showed that the patient's NK function had decreased to 5.19% and increased to normal NK cell activity at 28.73% after an additional low-dose IL-2cycle. Repeat liquid biopsy detected no tumor-related somatic alterations consistent with disease remission.

Case 2

In June 2017, a 49-year-old woman was diagnosed with multifocal estrogen receptor (ER) negative, progesterone receptor (PR) negative, human epidermal growth factor receptor 2 (HER2) positive breast cancer. The patient was treated with neoadjuvant chemotherapy consisting of 5-fluorouracil, epirubicin, cyclophosphamide, docetaxel, and herceptin (FEC-DH). The FEC chemotherapy was completed in February 2018, and it was associated with a side effect of severe bone pain. The patient discontinued Herceptin because of increased liver enzymes and extreme bone pain. Right mastectomy and sentinel node biopsy performed on March 2018 showed residual ypT1a N0 disease, with 2.5mm of invasive ductal carcinoma, grade II, lymphovascular invasion negative, and two lymph nodes negative. The patient had a prophylactic left mastectomy with bilateral immediate reconstruction. Adjuvant radiation was not administered.

In April 2019, the patient attended our clinic for evaluation and treatment. Options discussed included responses and side effects of chemotherapy, radiation, no therapy and palliative care, or experimental personalized low-dose immunotherapy. The patient refused chemotherapy, radiation, or treatments with potentially severe side effects. She gave informed consent for experimental personalized low-dose immunotherapy, including publication of results. A peripheral blood immune test revealed a low total B cell count at 101 cells/μL (normal: 128–400cells/μL), total T cell (CD3+) count was low at 819cells/μL (normal: 989–1899cells/μL). NK cell count was normal at 189cells/μL (normal: 98–294cells/μL), but NK cell cytotoxicity was low at 9.61% (normal: 16%–39.9%). The patient also had a liquid biopsy test which showed 0.1% ctDNA of guanine nucleotide-binding α-subunit (GNAS) R201H mutation, which is associated with breast cancer. In May 2019, the patient received daily low-dose subcutaneous rIL-2 injections of 10–20,000IU/kg for 5 days per week. Variation of the duration, dosing and frequency of administration of cycles of rIL-2 was based on her peripheral blood immune panel results. After the low-dose rIL-2cycles the patient's NK cell count increased to 212cells/μL and NK cell cytotoxicity increased to 15.1%. In September 2019, following additional low-dose rIL-2cycles, the patient's NK cell count increased to 286cells/μL and NK cell cytotoxicity increased to 28.36%. In October 2020, the patient's NK cell count decreased to 189cells/μL and NK cell cytotoxicity to 6.67%. Following an additional cycle of immunotherapy NK cell count was 161 cells/μL, and NK cell cytotoxicity increased to 9.79%. Following the patient's immunotherapy, a liquid biopsy test showed no tumor-related somatic alterations. The patient breast cancer remains in complete remission.

Case 3

In July of 2018, a 68-year-old woman was diagnosed with stage III infiltrating ductal carcinoma of the left breast. The patient first noticed a lump in her left breast in July 2018, and a biopsy revealed ductal carcinoma. Staging investigations, including a PET scan, showed a mass in the left lingular area, and a biopsy showed a follicular lymphoma. The patient had surgical resection of the lymphoma, a double mastectomy. CT scans of the chest and abdomen were negative. The patient declined chemotherapy and radiation. In February 2019, the patient had breast implant placement. In September 2019, repeat CT scans of her chest and abdomen were negative.

On December 18, 2019, the patient attended our clinic for evaluation and treatment. Options discussed included responses and side effects of chemotherapy, no therapy and palliative care, or experimental personalized low-dose immunotherapy. The patient refused chemotherapy, radiation, or treatments with potentially severe side effects. She gave informed consent for experimental personalized low-dose immunotherapy, including publication of results.

In January 2020, a peripheral blood immune test showed that the patient's NK cell count was normal at 219 cells/μL (normal: 98–294 cells/μL), but NK cell function was low at 4.90% (normal: 16%–39.9%). A liquid biopsy showed alterations of two biomarkers, NRAS G12D at 0.08% ctDNA amplification and TP53 V216M at 0.2% ctDNA amplification. The TP53 V216M and NRAS G12D mutations are associated with breast carcinoma. The patient also had flow cytometry analysis, FISH, and NGS, which identified the BCL2 translocation characteristic of follicular lymphoma and associated with decreased survival [36]. The patient received daily low-dose subcutaneous rIL-2 injections of 10–20,000 IU/kg for 5 days per week. Variation of the duration, dosing, and frequency of administration of cycles of rIL-2 was based on the peripheral blood immune panel results. Following 2 cycles of low-dose rIL-2 treatment, the patient had repeat immune panel evaluation which showed the NK cell function had improved significantly to 17.58%.

The patient was followed at 6-month intervals with repeat immune testing and to correct abnormalities in NK cell function. In January 2021, the patient was treated with low-dose rIL-2 immunotherapy and the NK cell function improved from 6.55% to 15.36%. A liquid biopsy in May 2021, which showed that the TP53 V216M mutation remained at 0.2% ctDNA and the NRAS G12D mutation decreased to 0.07% ctDNA. Repeat FISH and NGS did not detect any clinically significant variants, including the BCL2 translocation previously identified. Here, we describe a patient who achieved molecular remission of her non-Hodgkin's lymphoma and no progression of her breast cancer mutations in response to our novel approach of low-dose immunotherapy. We continue to personalize the patient's low-dose immunotherapy treatment based on immune function and NK cell cytotoxicity.

Hepatocellular carcinoma

Hepatic malignancies are some of the most common and deadliest cancers found worldwide, with hepatocellular carcinoma (HCC) being the most prevalent primary liver cancer [37]. Harvey rat sarcoma viral oncogene homolog (HRAS) mutations have been shown to be one of the main cancer genes involved in liver tumorigenesis [38]. A study recently found

HRAS to be associated with a high risk for metastatic recurrence of HCC as compared to a low-risk group [39]. HCC has historically been one of the deadlier cancer prognoses to date due to a dearth of effective systemic therapies [38]. The standard of care for patients with advanced HCC is systemic therapies such as sorafenib or Lenvatinib, with a recent meta-analysis of phase 3 trials showing overall survival rates in advanced HCC patients treated with sorafenib to be 10 months [40,41]. Future studies for advanced HCC treatment should aim to explore targeted therapies and immunotherapies with low toxicity profiles that can be personalized with the use of mutational analysis such as next-generation sequencing.

Studies have previously demonstrated the dysfunction and dysregulation of NK cells that are implicit in the development, progression and prognosis and of HCC [42,43]. Indeed, it has even been suggested that reduced NK cell cytotoxicity and cytokine production may serve as predictors for HCC incidence [42]. Here, we describe a patient who achieved significant correction of NK cell cytotoxicity and remission in response to our novel approach of low-dose immunotherapy.

A 43-year-old female received an MRI of the abdomen in November 2017, which revealed multiple small liver cysts. A large left liver lobe mass lesion was found, which did not follow typical hemangioma appearance on pulse sequences, and even postcontrast imaging was not characteristic. Therefore, with the presence of a central scar-like area, focal nodular hyperplasia was considered as a differential diagnosis. The lesion was recommended to be investigated with a nuclear scan to rule out atypical hemangioma vs focal nodular hyperplasia. A biopsy would also be valuable to rule out certain types of HCC (fibrolamellar) that can present similarly.

An MRI of the liver on January 8, 2018, revealed an unchanged lobulated left hepatic arterial enhancing lesion measuring 8.8 × 6.0 cm with a central nonenhancing scar and minimal associated restricted diffusion, which becomes isointense to liver parenchyma on delayed imaging with retention of hepatocellular-specific contrast agent, consistent with focal nodular hyperplasia. Unchanged, additional few scattered hepatic cysts and segment 6 probable hemangioma were found. No suspicious lesions were noted on this exam.

The patient attended our clinic in October 2018, for evaluation and treatment. Options discussed included responses and side effects of chemotherapy, radiation therapy, no therapy and palliative care, or experimental personalized low-dose immunotherapy. The patient refused chemotherapy, radiation, or treatments with potentially severe side effects. She gave informed consent for experimental personalized low-dose immunotherapy, including publication of results.

Peripheral blood immune testing showed decreased levels of total B cells at 99 cells/μL (normal range: 128–400 cells/μL) and decreased NK cell cytotoxicity at 8.24% (normal range: 16.3%–39.9%). A liquid biopsy showed 0.1% of HRAS A59T mutation in the ctDNA consistent with diagnosis of hepatocellular carcinoma. Based upon the deficiencies in her immune system, the patient received experimental rIL-2 immune modulation therapy in order to correct her immune function.

The patient received 2 cycles of rIL-2 treatment and the natural killer function significantly increased to 16.28%. The patient further received rIL-2 treatment to decrease the level of mutation found in ctDNA. The second treatment proceeded from mid-October to early November 2019. Natural killer cell function was 10.36%. Repeat liquid biopsy with mutational testing in ctDNA revealed no mutations detected. In September 2021, the patient's natural killer function significantly increased to 16.28% and repeat liquid biopsy again, revealed no mutations detected in ctDNA. We continue to personalize the patient's low-dose immunotherapy treatment based on immune function and NK cell cytotoxicity. The patient remains clinically well.

Discussion

We report complete clinical and molecular remissions with personalized low-dose immunotherapy treatment using rIL-2 to increase NK cell function and NK cell levels in 10 patients with poor prognosis hematological and solid tumor malignancies. These patients with advanced cancers had disease progression and significant toxicities after receiving standard of care treatments for their disease. Fifty percent of cases received low-dose immunotherapy combined with low-dose targeted therapy to which a specific mutation was identified. The other cases responded as well to low-dose immunotherapy alone indicating that the improvement of the immune dysfunction and natural killer cell function was effective in inducing molecular and clinical remissions. All patients had minimal toxicities and improved quality of life in response to our novel treatment (Table 1).

We have previously shown the efficacy of low-dose immunotherapy in combination with low-dose targeted therapy in treatment of hematological malignancies with cases 1 and 2, in patients with CLL and ALL, respectively [1,2]. In these cases, we showed that individualized immunotherapy in an N-of-1 approach focused on the objective determination of the optimal therapy for a single individual, could improve cancer therapeutic responses and outcomes by stratifying care based on each patient's very specific tumor genetic and immunological profiles. In case 1, a poor prognosis U-CLL the patient received a combination of low-dose rIL-2 immunotherapy with low-dose venetoclax and achieved molecular remission after six uninterrupted cycles of treatment. In case 2, a patient with Philadelphia chromosome (Ph+) B-cell ALL relapsed after matched donor allo-HCT, a donor lymphocyte infusion, and two TKIs. The patient achieved complete molecular and cytogenetic remission after completing 4 cycles of low-dose recombinant IL-2 in combination with a low-dose TKI targeted therapy with only minimal adverse events. In this report, we provide additional outcome data showing that cases 1 and 2 have been able to maintain remission and a good quality of life.

Based on treatment results in hematological malignancies, we have used the same personalized approach to treat a variety of solid tumor malignancies including BCC, CCC, EGCT, HCC, ovarian, and breast cancer. The patients received varying cycles of low-dose rIL-2 in combination with low-dose-targeted therapy to minimize related side effects while increasing the activity of immune effector cells and directly destroying the tumor cells as indicated by decreases in or molecular remission of cancer driver mutations detected by liquid biopsies.

The consistent finding in our cases is the altered phenotype and diminished functioning of NK cell cytotoxicity and cytokine production, which results in dysregulation in T cells and B of the adaptive immune system. All the tumor types in this report have been shown to demonstrate low levels of NK cell function and the capability of IL-2 enhanced natural killer cell cytotoxicity in lysing cancer cells and primary tumors, including recurrent and invasive carcinomas has been documented [16,20,34,35,42,44–47].

TABLE 1 Patient demographics, performance status, previous treatment, targeted therapy and low-dose immunotherapy characteristics, range of NK cell cytotoxicity and NKT cell count, mutation and molecular remission after treatment, survival, and toxicity grading.

Case report	Diagnosis	Age[a]	Gender	Performance status at initiation of low-dose immunotherapy (ECOG)	Previous treatment	Number of cycles[b]	Lowest natural killer cell cytotoxicity (normal range: 16.3%–39.9%)	Highest natural killer cell cytotoxicity (normal range: 16.3%–39.9%)	Lowest NKT cell count (normal range: 593–1375 cells/μL)	Highest NKT cell count (normal range: 593–1375 cells/μL)	Mutation	Molecular remission after IL-2 treatment	Targeted therapy	Time since molecular remission (months)	Survival since diagnosis (months)	Toxicity of low-dose targeted therapy and immuno-therapy (Grade 0–4)
1	Poor prognosis unmutated heavy chain chronic lymphocytic leukemia	56	Female	1	Ibrutinib	6	0.97%	24.68%	151	1014	Unmutated IGHV, SF3B1, ATM	SF3B1 decreased, ATM and IGVH mutations undetected	Low-dose Venetoclax	24	115	1
2	Precursor B-cell acute lymphoblastic leukemia	39	Male	1	UKALL14 Clinical Trial: NCT01085617	4	No activity	9.68%	213	835	BCR-ABL1, p190	BCR-ABL1, p190 undetected	Low-dose Dasatinib	56	78	1
3	Basal cell carcinoma	70	Male	1	Vismodegib, Cemiplimab, Sonidegib	5	5.11%	26.52%	532	1364	SMO S387N, FBXW7 H139R, ATM H2554R	SMO and ATM mutations undetected decrease in FBXW7 H139R	Low-dose Sonidegib, low-dose Erivedge	8	104	2
4	High-grade clear cell carcinoma of the endometrium	70	Female	0	Total abdominal radical hysterectomy, bilateral salpingo-oophorectomy, sentinel lymph node dissection, omentectomy, and appendectomy	9	1.69%	27.71%	444	1540	BRCA2 N802fs EGFR Y97Y TERT R72H	BRCA2, EGFR, and TERT mutations undetected	N/A	12	29	1
5	Ovarian cancer	72	Female	1	Oophorectomy Carboplatin, Gemcitabine, Topotecan chemotherapy 2 Surgical resections of recurrence with partial colon and urinary bladder resection	1	13.23%	17.65%	143	503	BRCA1 BRCA2	Post-treatment ctDNA analysis not performed	Low-dose Olaparib	72	150	1

Continued

TABLE 1 Patient demographics, performance status, previous treatment, targeted therapy and low-dose immunotherapy characteristics, range of NK cell cytotoxicity and NKT cell count, mutation and molecular remission after treatment, survival, and toxicity grading—cont'd

Case report	Diagnosis	Age[a]	Gender	Performance status at initiation of low-dose immunotherapy (ECOG)	Previous treatment	Number of cycles[b]	Lowest natural killer cell cytotoxicity (normal range: 16.3%–39.9%)	Highest natural killer cell cytotoxicity (normal range: 16.3%–39.9%)	Lowest NKT cell count (normal range: 593–1375 cells/μL)	Highest NKT cell count (normal range: 593–1375 cells/μL)	Mutation	Molecular remission after IL-2 treatment	Targeted therapy	Time since molecular remission (months)	Survival since diagnosis (months)	Toxicity of low-dose targeted therapy and immunotherapy (Grade 0–4)
6	Extragonadal germ cell tumor	21	Male	1	BEP combination chemotherapy Retroperitoneal lymph node dissection	5	2.07%	10.97%	201	737	KRAS EGFR	KRAS mutation undetected decrease in EGFR mutation	Low-dose Afatinib	24	183	1
7	Breast cancer (Case 1)	66	Female	1	Complete bilateral mastectomy with reconstruction	6	5.19%	28.73%	140	2002	ER (>90% positive) HER2/neu 2+ PR (>80% positive)	ER, HER2, and PR mutations undetected	N/A	37	69	1
8	Breast cancer (Case 2)	49	Female	1	FEC-DH combination chemotherapy Prophylactic left mastectomy	3	4.76%	28.36%	556	937	HER2 GNAS R201H	HER2, GNAS mutations undetected	N/A	10	54	1
9	Breast cancer (Case 3), non-Hodgkin's lymphoma	68	Female	1	Surgical resection of lymphoma, bilateral mastectomy	3	4.90%	17.58%	1210	2095	NRAS G12D TP53 V216M BCL2 mutation	NRAS decreased, BCL2 mutation undetected	N/A	7	41	1
10	Hepatocellular carcinoma	44	Female	0	None	2	6.59%	16.28%	930	3467	HRAS A59T amplification	HRAS mutation undetected	N/A	25	49	1

[a] *Age at which the patient presented to the clinic.*
[b] *Number of continuous cycles to reach molecular remission.*

In case 3 with advanced BCC, the patient presented with significantly reduced NK cell cytotoxicity after a poor response to Vismodegib, a phase 2 monoclonal antibody trial, Sonidegib and Itraconazole. After 4 cycles of low-dose immunotherapy, the patient's NK cell cytotoxicity increased by over 100% from 5.11% to 10.92%. Liquid biopsy NGS showed undetectable SMO S387 and ATM H2554R mutations.

Case 4 with high-grade CCC of the endometrium, the patient had a significantly diminished NK cell cytotoxicity at 1.69%, after abdominal radical hysterectomy, bilateral salpingo-oophorectomy, sentinel lymph node dissection, omentectomy, and appendectomy. After 8 cycles of low-dose rIL-2 following which the patient's NK cell function significantly increased to 8.12% endometrial cancer driver mutations BRCA1, BRCA2, and EGFR were undetectable.

Case 5, with a metastatic ovarian cancer and progression of disease following multiple rounds of surgery and chemotherapy with Gemcitabine, Carboplatin, and Topotecan, the patient presented with a low NK cell function, which recovered to normal at 17.65% after receiving low-dose IL-2. The tumor showed BRCA 1, 2 mutations and the addition of targeted treatment with Olaparib resulted in complete remission, which has been maintained for over 6 years since starting low-dose immunotherapy.

Case 6, with recurrent EGCT after multiple surgeries and BEP chemotherapy NK cell cytotoxicity, was significantly decreased at 2.49% and increased to a high of 10.97% (200% increase) resulting in KRAS mutation decreasing to undetectable levels after low-dose immunotherapy. With the addition of the targeted therapy, Afatinib to the low-dose immunotherapy the EGFR mutation continued to decrease and was maintained at low levels.

Cases 7, 8, and 9, with breast cancer of varying stages including triple-positive T1N0M0, HER2+ T1aN0, and stage IIIB ductal carcinoma, respectively, all had decreased NK cell. After treatment with varying cycles of low-dose rIL-2, the natural killer cell function of each case increased and there were significant reductions in cancer driver mutations to the lowest or undetectable levels. Case 9 with an additional diagnosis of non-Hodgkin's lymphoma had a complete remission of her BCL2 mutation which is maintained.

In case 10, with HCC and HRAS mutation, the patient had significantly diminished NK cell function. After receiving cycles of low-dose rIL-2, NK cell function increased to 18% and HRAS mutation levels were undetectable.

Fifty percent of our patients were treated with the combination of varying cycles of low-dose rIL-2 with targeted therapy for the specific detected mutation. It could be argued that the targeted inhibition of driver cancer mutation proliferation played the major role in clearance of ctDNA, which allowed these patients to achieve undetectable levels of mutations associated with disease remission. In the other half of our cases, low-dose rIL-2 immunotherapy alone was an effective treatment in maintaining antitumor effect by stimulating and maintaining natural killer function to achieve undetectable levels of mutations associated with disease remission (Fig. 2).

In all cases, we showed that an individualized or personalized N-of-1 treatment approach focused on the objective determination of the optimal therapy either rIL-2 alone or rIL-2 and a targeted therapy for a single individual, could improve responses and outcomes by basing the numbers and duration of treatment cycles on each patient very specific tumor mutation and natural killer cell function. When the responses of our individualized treatments are compared with the best standard of care in cases such as relapsed refractory Ph + ALL with dismal

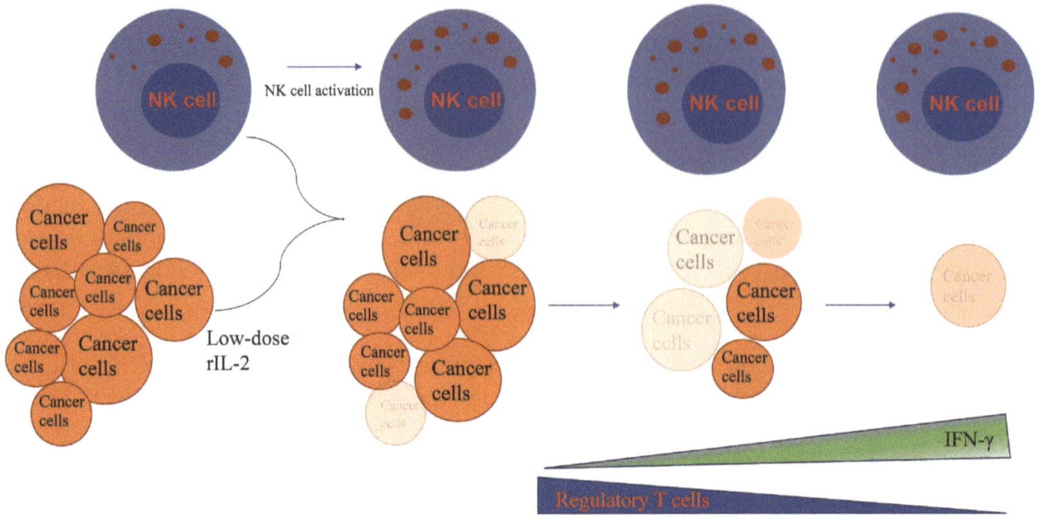

FIG. 2 Decrease in cancer cells with activation of natural killer cell function after low-dose immunotherapy.

outcomes of a median survival of less than 1 year and stage 4 ovarian cancer with a 5-year disease-free survival of 17%, our novel treatment approach provides a significant improvement in outcomes with disease-free survivals of over 6 years for these individual cases [2].

This report describes a novel approach for treating a multiple relapsed or poor prognosis patients with a variety of hematological and solid tumors by using a personalized treatment of measuring functional NK cytotoxicity while tailoring the dose and frequency of rIL-2 in response to these effector cells. Our report serves as a basis for low toxicity immunotherapeutic options in patients with poor prognosis disease.

Conclusions

Standard of care for advanced hematological and solid tumor malignancies typically involves systemic chemotherapies with significant toxicities and side effect profiles. We have previously presented a novel approach utilizing personalized immunotherapy with rIL-2, in combination with low-dose targeted therapy, which resulted in sustained remission in hematological malignancies including poor prognosis U-CLL and Ph+B-cell ALL. We now show that the same approach can be utilized in solid tumor malignancies to achieve lasting molecular remissions with minimal toxicities.

Future perspectives

Current estimates utilizing data from the National Cancer Institute's Surveillance, Epidemiology, and End Results (SEER) Program project a 49% increase in annual cancer cases from 2015 to 2050, with the highest increase in cases among adults over 75 years of age in the United

States [48]. As older adults represent the largest risk group and comprise a larger segment of the US demographic makeup year after year, strategies for risk reduction, early screening, and improved treatment options become increasingly important.

The advent of NGS has shown potential for significant improvements in cancer diagnostics, allowing for identification of over 50 cancer-related genes with a single blood test [49,50]. Early, noninvasive screening for ctDNA in healthy individuals is also being explored to reduce progression to severe disease and cancer mortality [51]. Furthermore, research shows that the immune system begins to decline as early as age 40. Measuring the immune system to identify and correct dysfunction will be an important risk reduction strategy as cancer incidence and mortality continues to rise. Stem cell banking can also be utilized for future restoration of immune function as the immune system declines with aging and risk of cancer incidence rises.

This case series shows the efficacy of a two-pronged approach using personalized immunotherapy with rIL-2, in combination with low-dose targeted therapies, in achieving cytogenetic and molecular remission with minimal side effects in patients with poor prognosis solid tumor and hematological malignancies. Along with the need for improved risk reduction and diagnostic screening, the results achieved by this unique approach support the need for future exploration of these combined therapeutic options to improve outcomes and minimize toxicities in poor prognosis cancer patients.

References

[1] Maharaj D, Srinivasan G, Abreu MM, Ko MW, Jewett A, Gouvea J. Molecular remission using low-dose immunotherapy with minimal toxicities for poor prognosis IGHV—unmutated chronic lymphocytic leukemia. Cell 2020;10(1):10. https://doi.org/10.3390/cells10010010.

[2] Maharaj D, Vianna P, DeCarvalho G, Pourkalbassi D, Hickey C, Gouvea J. Molecular remission using low-dose immunotherapy for relapsed refractory Philadelphia chromosome-positive precursor B-cell acute lymphoblastic leukemia post-allogeneic stem cell transplant. Future Sci OA 2019;5(5), FSO380. https://doi.org/10.2144/fsoa-2019-0009.

[3] Gagan J, Van Allen EM. Next-generation sequencing to guide cancer therapy. Genome Med 2015;7 (1):80. https://doi.org/10.1186/s13073-015-0203-x.

[4] Thompson JC, Carpenter EL, Silva BA, et al. Serial monitoring of circulating tumor DNA by next-generation gene sequencing as a biomarker of response and survival in patients with advanced NSCLC receiving Pembrolizumab-based therapy. JCO Precis Oncol 2021;5:510–24. https://doi.org/10.1200/PO.20.00321.

[5] Bald T, Krummel MF, Smyth MJ, Barry KC. The NK cell-cancer cycle: advances and new challenges in NK cell-based immunotherapies. Nat Immunol 2020;21(8):835–47. https://doi.org/10.1038/s41590-020-0728-z.

[6] Liu S, Galat V, Galat4 Y, Lee YKA, Wainwright D, Wu J. NK cell-based cancer immunotherapy: from basic biology to clinical development. J Hematol Oncol 2021;14(1):7. https://doi.org/10.1186/s13045-020-01014-w.

[7] Fehniger TA, Bluman EM, Porter MM, et al. Potential mechanisms of human natural killer cell expansion in vivo during low-dose IL-2 therapy. J Clin Invest 2000;106(1):117–24. https://doi.org/10.1172/JCI6218.

[8] Yu TK, Caudell EG, Smid C, Grimm EA. IL-2 activation of NK cells: involvement of MKK1/2/ERK but not p38 kinase pathway. J Immunol 2000;164(12):6244–51. https://doi.org/10.4049/jimmunol.164.12.6244.

[9] Jain N, O'Brien S. Initial treatment of CLL: integrating biology and functional status. Blood 2015;126 (4):463–70. https://doi.org/10.1182/blood-2015-04-585067.

[10] Badoux XC, Keating MJ, Wang X, et al. Fludarabine, cyclophosphamide, and rituximab chemoimmunotherapy is highly effective treatment for relapsed patients with CLL. Blood 2011;117(11):3016–24. https://doi.org/10.1182/blood-2010-08-304683.

[11] Sportoletti P, De Falco F, Del Papa B, et al. NK cells in chronic lymphocytic leukemia and their therapeutic implications. Int J Mol Sci 2021;22(13):6665. https://doi.org/10.3390/ijms22136665.

[12] Hefazi M, Litzow MR. Recent advances in the biology and treatment of B-cell acute lymphoblastic leukemia. Blood Lymphat Cancer 2018;8:47–61. https://doi.org/10.2147/BLCTT.S170351.
[13] Pastorczak A, Domka K, Fidyt K, Poprzeczko M, Firczuk M. Mechanisms of immune evasion in acute lymphoblastic leukemia. Cancer 2021;13(7):1536. https://doi.org/10.3390/cancers13071536.
[14] Smith V, Walton S. Treatment of facial basal cell carcinoma: a review. J Skin Cancer 2011;2011:1–7. https://doi.org/10.1155/2011/380371.
[15] Peris K, Fargnoli MC, Garbe C, et al. Diagnosis and treatment of basal cell carcinoma: European consensus-based interdisciplinary guidelines. Eur J Cancer 2019;118:10–34. https://doi.org/10.1016/j.ejca.2019.06.003.
[16] Shaver KA, Croom-Perez TJ, Copik AJ. Natural killer cells: the linchpin for successful cancer immunotherapy. Front Immunol 2021;12, 679117. https://doi.org/10.3389/fimmu.2021.679117.
[17] Brooks RA, Fleming GF, Lastra RR, et al. Current recommendations and recent progress in endometrial cancer. CA Cancer J Clin 2019. https://doi.org/10.3322/caac.21561 [Published online May 10, 2019:caac.21561].
[18] Abdulfatah E, Sakr S, Thomas S, et al. Clear cell carcinoma of the endometrium: evaluation of prognostic parameters in a multi-institutional cohort of 165 cases. Int J Gynecol Cancer 2017;27(8):1714–21. https://doi.org/10.1097/IGC.0000000000001050.
[19] Gadducci A, Cosio S, Spirito N, Cionini L. Clear cell carcinoma of the endometrium: a biological and clinical enigma. Anticancer Res 2010;30(4):1327–34.
[20] Degos C, Heinemann M, Barrou J, et al. Endometrial tumor microenvironment alters human NK cell recruitment, and resident NK cell phenotype and function. Front Immunol 2019;10:877. https://doi.org/10.3389/fimmu.2019.00877.
[21] Doubeni CA. Diagnosis and management of ovarian cancer. Am Fam Physician 2016;93(11):8.
[22] Della Pepa C, Tonini G, Pisano C, et al. Ovarian cancer standard of care: are there real alternatives? Chin J Cancer 2015;34(1):17–27. https://doi.org/10.5732/cjc.014.10274.
[23] Nersesian S, Glazebrook H, Toulany J, Grantham SR, Boudreau JE. Naturally killing the silent killer: NK cell-based immunotherapy for ovarian cancer. Front Immunol 2019;10:1782. https://doi.org/10.3389/fimmu.2019.01782.
[24] Hoogstad-van Evert JS, Bekkers R, Ottevanger N, Jansen JH, Massuger L, Dolstra H. Harnessing natural killer cells for the treatment of ovarian cancer. Gynecol Oncol 2020;157(3):810–6. https://doi.org/10.1016/j.ygyno.2020.03.020.
[25] Munroe M, Kolesar J. Olaparib for the treatment of BRCA-mutated advanced ovarian cancer. Am J Health Syst Pharm 2016;73(14):1037–41. https://doi.org/10.2146/ajhp150550.
[26] Chahoud J, Kohli M. Managing extragonadal germ cell tumors in male adults. AME Med J 2020;5:8. https://doi.org/10.21037/amj.2020.01.01.
[27] Siddiqui BA, Zhang M, Pisters LL, Tu SM. Systemic therapy for primary and extragonadal germ cell tumors: prognosis and nuances of treatment. Transl Androl Urol 2020;9(Suppl 1):S56–65. https://doi.org/10.21037/tau.2019.09.11.
[28] Rodney AJ, Tannir NM, Siefker-Radtke AO, et al. Survival outcomes for men with mediastinal germ-cell tumors: the University of Texas M.D. Anderson Cancer center experience. Urol Oncol 2012;30(6):879–85. https://doi.org/10.1016/j.urolonc.2010.08.005.
[29] Chahoud J, Zhang M, Shah A, Lin SH, Pisters LL, Tu SM. Managing seminomatous and nonseminomatous germ cell tumors. Curr Opin Oncol 2018;30(3):181–8. https://doi.org/10.1097/CCO.0000000000000446.
[30] Gonzalez-Angulo AM, Morales-Vasquez F, Hortobagyi GN. Overview of resistance to systemic therapy in patients with breast cancer. Adv Exp Med Biol 2007;608:1–22. https://doi.org/10.1007/978-0-387-74039-3_1.
[31] Liu M, Wang CB, Xie F, Peng Y, Wang S, Chinese Society of Breast Surgery. Clinical practice guidelines for diagnosis and treatment of invasive breast cancer: Chinese Society of Breast Surgery (CSBrS) practice guidelines 2021. Chin Med J 2021;134(9):1009–13. https://doi.org/10.1097/CM9.0000000000001498.
[32] Allemani C, Sant M, Weir HK, et al. Breast cancer survival in the US and Europe: a CONCORD high-resolution study. Int J Cancer 2013;132(5):1170–81. https://doi.org/10.1002/ijc.27725.
[33] Xu AJ, DeSelm CJ, Ho AY, et al. Overall survival of breast cancer patients with locoregional failures involving internal mammary nodes. Adv Radiat Oncol 2019;4(3):447–52. https://doi.org/10.1016/j.adro.2019.02.004.
[34] Garcia-Chagollan M, Carranza-Torres IE, Carranza-Rosales P, et al. Expression of NK cell surface receptors in breast cancer tissue as predictors of resistance to antineoplastic treatment. Technol Cancer Res Treat 2018;17, 1533033818764499. https://doi.org/10.1177/1533033818764499.

[35] Shenouda MM, Gillgrass A, Nham T, et al. Ex vivo expanded natural killer cells from breast cancer patients and healthy donors are highly cytotoxic against breast cancer cell lines and patient-derived tumours. Breast Cancer Res 2017;19(1):76. https://doi.org/10.1186/s13058-017-0867-9.
[36] Correia C, Schneider PA, Dai H, et al. BCL2 mutations are associated with increased risk of transformation and shortened survival in follicular lymphoma. Blood 2015;125(4):658–67. https://doi.org/10.1182/blood-2014-04-571786.
[37] Ananthakrishnan A, Gogineni V, Saeian K. Epidemiology of primary and secondary liver cancers. Semin Interv Radiol 2006;23(1):47–63. https://doi.org/10.1055/s-2006-939841.
[38] Zhang Y, Qiu Z, Wei L, et al. Integrated analysis of mutation data from various sources identifies key genes and signaling pathways in hepatocellular carcinoma. PLoS One 2014;9(7), e100854. https://doi.org/10.1371/journal.pone.0100854 [Wong N, ed.].
[39] Sui G, Ma X, Liu S, Niu H, Dong Q. Study of the correlation between H-ras mutation and primary hepatocellular carcinoma. Oncol Lett 2012;4(4):779–82. https://doi.org/10.3892/ol.2012.832.
[40] Likhitsup A, Razumilava N, Parikh ND. Treatment for advanced hepatocellular carcinoma: current standard and the future. Clin Liver Dis 2019;13(1):13–9. https://doi.org/10.1002/cld.782.
[41] Cabibbo G, Cucchetti A, Cammà C, et al. Outcomes of hepatocellular carcinoma patients treated with sorafenib: a meta-analysis of phase III trials. Future Oncol 2019;15(29):3411–22. https://doi.org/10.2217/fon-2019-0287.
[42] Sun C, Sun HY, Xiao WH, Zhang C, Tian ZG. Natural killer cell dysfunction in hepatocellular carcinoma and NK cell-based immunotherapy. Acta Pharmacol Sin 2015;36(10):1191–9. https://doi.org/10.1038/aps.2015.41.
[43] Juengpanich S, Shi L, Iranmanesh Y, et al. The role of natural killer cells in hepatocellular carcinoma development and treatment: a narrative review. Transl Oncol 2019;12(8):1092–107. https://doi.org/10.1016/j.tranon.2019.04.021.
[44] Brasseur K, Gévry N, Asselin E. Chemoresistance and targeted therapies in ovarian and endometrial cancers. Oncotarget 2017;8(3):4008–42. https://doi.org/10.18632/oncotarget.14021.
[45] Baci D, Bosi A, Gallazzi M, et al. The ovarian cancer tumor immune microenvironment (TIME) as target for therapy: a focus on innate immunity cells as therapeutic effectors. Int J Mol Sci 2020;21(9):E3125. https://doi.org/10.3390/ijms21093125.
[46] Chovanec M, De Giorgi U, Mego M. Immune-related concepts in biology and treatment of germ-cell tumors. Adv Urol 2018;2018:3718165. https://doi.org/10.1155/2018/3718165.
[47] Muntasell A, Cabo M, Servitja S, et al. Interplay between natural killer cells and anti-HER2 antibodies: perspectives for breast cancer immunotherapy. Front Immunol 2017;8:1544. https://doi.org/10.3389/fimmu.2017.01544.
[48] Weir HK, Thompson TD, Stewart SL, White MC. Cancer incidence projections in the United States between 2015 and 2050. Prev Chronic Dis 2021;18, 210006. https://doi.org/10.5888/pcd18.210006.
[49] Lanman RB, Mortimer SA, Zill OA, et al. Analytical and clinical validation of a digital sequencing panel for quantitative, highly accurate evaluation of cell-free circulating tumor DNA. PLoS One 2015;10(10), e0140712. https://doi.org/10.1371/journal.pone.0140712.
[50] Nakagaki T, Tamura M, Kobashi K, et al. Targeted next-generation sequencing of 50 cancer-related genes in Japanese patients with oral squamous cell carcinoma. Tumor Biol 2018;40(9), 101042831880018. https://doi.org/10.1177/1010428318800180.
[51] Chen X, Gole J, Gore A, et al. Non-invasive early detection of cancer four years before conventional diagnosis using a blood test. Nat Commun 2020;11(1):3475. https://doi.org/10.1038/s41467-020-17316-z.

Index

Note: Page numbers followed by *f* indicate figures and *t* indicate tables.

D

E

F

O

T

U

V

W

X

Z

Kolibri 360 GmbH
Ferdinand-Jühlke-Straße 7
99095 Erfurt, Deutschland
produktsicherheit@kolibri360.de